Handbook of Research on Biomimetics and Biomedical Robotics

Maki Habib
The American University in Cairo, Egypt

A volume in the Advances in Computational Intelligence and Robotics (ACIR) Book Series

IGI Global
DISSEMINATOR OF KNOWLEDGE

Published in the United States of America by
IGI Global
Engineering Science Reference (an imprint of IGI Global)
701 E. Chocolate Avenue
Hershey PA, USA 17033
Tel: 717-533-8845
Fax: 717-533-8661
E-mail: cust@igi-global.com
Web site: http://www.igi-global.com

Library of Congress Cataloging-in-Publication Data

Names: Habib, Maki K., 1955- editor.
Title: Handbook of research on biomimetics and biomedical robotics / Maki Habib, editor.
Description: Hershey, PA : Engineering Science Reference, [2018] | Includes bibliographical references.
Identifiers: LCCN 2017013545| ISBN 9781522529934 (hardcover) | ISBN 9781522529941 (ebook)
Subjects: LCSH: Biomimetics--Research. | Biomedical engineering.
Classification: LCC QP517.B56 H36 2018 | DDC 610.28--dc23 LC record available at https://lccn.loc.gov/2017013545

This book is published in the IGI Global book series Advances in Computational Intelligence and Robotics (ACIR) (ISSN: 2327-0411; eISSN: 2327-042X)

British Cataloguing in Publication Data
A Cataloguing in Publication record for this book is available from the British Library.

All work contributed to this book is new, previously-unpublished material. The views expressed in this book are those of the authors, but not necessarily of the publisher.

For electronic access to this publication, please contact: eresources@igi-global.com.

Advances in Computational Intelligence and Robotics (ACIR) Book Series

Ivan Giannoccaro
University of Salento, Italy

ISSN:2327-0411
EISSN:2327-042X

Mission

While intelligence is traditionally a term applied to humans and human cognition, technology has progressed in such a way to allow for the development of intelligent systems able to simulate many human traits. With this new era of simulated and artificial intelligence, much research is needed in order to continue to advance the field and also to evaluate the ethical and societal concerns of the existence of artificial life and machine learning.

The **Advances in Computational Intelligence and Robotics (ACIR) Book Series** encourages scholarly discourse on all topics pertaining to evolutionary computing, artificial life, computational intelligence, machine learning, and robotics. ACIR presents the latest research being conducted on diverse topics in intelligence technologies with the goal of advancing knowledge and applications in this rapidly evolving field.

Coverage

- Synthetic Emotions
- Natural Language Processing
- Artificial Intelligence
- Evolutionary computing
- Brain Simulation
- Computational Logic
- Fuzzy systems
- Agent technologies
- Robotics
- Computational Intelligence

IGI Global is currently accepting manuscripts for publication within this series. To submit a proposal for a volume in this series, please contact our Acquisition Editors at Acquisitions@igi-global.com or visit: http://www.igi-global.com/publish/.

The Advances in Computational Intelligence and Robotics (ACIR) Book Series (ISSN 2327-0411) is published by IGI Global, 701 E. Chocolate Avenue, Hershey, PA 17033-1240, USA, www.igi-global.com. This series is composed of titles available for purchase individually; each title is edited to be contextually exclusive from any other title within the series. For pricing and ordering information please visit http://www.igi-global.com/book-series/advances-computational-intelligence-robotics/73674. Postmaster: Send all address changes to above address.

Titles in this Series

For a list of additional titles in this series, please visit: www.igi-global.com/book-series

Handbook of Research on Biomimetics and Biomedical Robotics
Maki Habib (The American University in Cairo, Egypt)
Engineering Science Reference • ©2018 • 400pp • H/C (ISBN: 9781522529934) • US $325.00 (our price)

Androids, Cyborgs, and Robots in Contemporary Culture and Society
Steven John Thompson (University of Maryland University College, USA)
Engineering Science Reference • ©2018 • 286pp • H/C (ISBN: 9781522529736) • US $205.00 (our price)

Developments and Trends in Intelligent Technologies and Smart Systems
Vijayan Sugumaran (Oakland University, USA)
Engineering Science Reference • ©2018 • 351pp • H/C (ISBN: 9781522536864) • US $215.00 (our price)

Handbook of Research on Modeling, Analysis, and Application of Nature-Inspired Metaheuristic Algorithms
Sujata Dash (North Orissa University, India) B.K. Tripathy (VIT University, India) and Atta ur Rahman (University of Dammam, Saudi Arabia)
Engineering Science Reference • ©2018 • 538pp • H/C (ISBN: 9781522528579) • US $265.00 (our price)

Concept Parsing Algorithms (CPA) for Textual Analysis and Discovery Emerging Research and Opportunities
Uri Shafrir (University of Toronto, Canada) and Masha Etkind (Ryerson University, Canada)
Information Science Reference • ©2018 • 139pp • H/C (ISBN: 9781522521761) • US $130.00 (our price)

Handbook of Research on Applied Cybernetics and Systems Science
Snehanshu Saha (PESIT South Campus, India) Abhyuday Mandal (University of Georgia, USA) Anand Narasimhamurthy (BITS Hyderabad, India) Sarasvathi V (PESIT- Bangalore South Campus, India) and Shivappa Sangam (UGC, India)
Information Science Reference • ©2017 • 463pp • H/C (ISBN: 9781522524984) • US $245.00 (our price)

Handbook of Research on Machine Learning Innovations and Trends
Aboul Ella Hassanien (Cairo University, Egypt) and Tarek Gaber (Suez Canal University, Egypt)
Information Science Reference • ©2017 • 1093pp • H/C (ISBN: 9781522522294) • US $465.00 (our price)

Handbook of Research on Soft Computing and Nature-Inspired Algorithms
Shishir K. Shandilya (Bansal Institute of Research and Technology, India) Smita Shandilya (Sagar Institute of Research Technology and Science, India) Kusum Deep (Indian Institute of Technology Roorkee, India) and Atulya K. Nagar (Liverpool Hope University, UK)
Information Science Reference • ©2017 • 627pp • H/C (ISBN: 9781522521280) • US $280.00 (our price)

IGI Global
DISSEMINATOR OF KNOWLEDGE

701 East Chocolate Avenue, Hershey, PA 17033, USA
Tel: 717-533-8845 x100 • Fax: 717-533-8661
E-Mail: cust@igi-global.com • www.igi-global.com

List of Contributors

Table of Contents

Detailed Table of Contents

Biologically inspired systems, known as "biomimetics" or the "mimicry of nature," is an interdisciplinary scientific research field inspired by nature and featured by the technology outcome (hardware and software) and lies at the interface of biology, physics, chemistry, information, and engineering sciences. Biomimetics is initiated by making nature a model of inspiration that would immensely help conscious abstraction of new innovative principles and creative design ideas and concepts that help developing new techniques and functionalities, seeking new paradigms and methods, designing new materials, and developing new streams of intelligent machines, robots, systems, devices, algorithms, etc. Biologically inspired approaches create a new reality with great development and application potential with the goal of identifying specific desirable qualities and attributes in biological systems and using them in the design of new products and systems. This chapter provides the importance of biomimetic as an interdisciplinary field and its evolution, advances, challenges, and constraints along with the associated enabling technologies supporting its growth. In addition, it introduces scientific ideas and directions of research activities in the field. The chapter also presents key developments in the field of biomimetic robots and underlines the challenges facing it.

In this chapter, the lower limb exoskeleton is studied. The roles of the exoskeleton both as a measurement device for studying human locomotion and as an assistive device that restores the human ability to walk are discussed. Particular attention is given to the investigation of the role of the pressure sensors and other devices that allow us to measure normal reactions at the contact points with the supporting surface and also detect these contacts. The way the geometry of the supporting surface affects the sensors system of the robot is considered, and new designs for feet sensor system are proposed. These include elastic foot, a foot with actuated sensors, and a foot with spring-damper systems.

Active orthosis (exoskeleton) is an assistive device with a wearable structure, corresponding to the natural motions of the human. This chapter focuses on developing an active/assistive orthosis system (AOS) enhancing movement. The AOS design is inspired by the biological musculoskeletal system of human upper and lower limbs and mimics the muscle-tendon-ligament structure. The exoskeleton structure includes left and right upper limb, left and right lower limb, and central exoskeleton structure for human torso and waist and provides support, balance, and control of different segments of the body. The device was fabricated with light materials and powered by pneumatic artificial muscles that provide more than fifteen degrees of freedom for the different joints. The active orthotic systems (AOS) can operate in three modes: motion tracking system with data exchange with virtual reality; haptic and rehabilitation device; and assistive mode with active orthosis in cases of impaired muscles.

Hybrid exoskeletons are a recent development, combining electrically controlled actuation with functional electrical stimulation, which potentially offers great benefits for muscular rehabilitation. This chapter presents a review on the state of the art of upper-limb hybrid exoskeletons with a particular focus on stroke rehabilitation. The current needs of the stroke rehabilitation field are discussed and the ability of hybrid exoskeletons to provide a solution to some of the gaps in this field is explored. Due to the early stage of development which most hybrid exoskeletons are in, little research has yet been done in control methods used for them. In particular, more investigation is needed with regards to the potential benefit of hybrid exoskeletons as a patient-monitoring and rehabilitation assist-as-need tool.

This chapter deals with robotics for assisting patients in their recovery after stroke. The first section outlines the key physiotherapist movements that are needed in post-stroke treatment. Sets of experimental data are discussed for defining the main specifications for post-stroke rehabilitation devices. Then, a detailed overview is provided on the existing key achievements and solutions concerning rehabilitation robotics for post-stroke treatment. This section is followed by a detailed critical analysis aiming at the definition of new concepts for cost-oriented user-friendly solutions.

The role of retrograde intra-renal surgery (RIRS) with flexible ureterorenoscopy (FURS) in the management of kidney stone formation has garnered a prevalent use among urologists thanks to new improvements in medical equipment; however, it still remains a challenging technique. The purpose of this chapter is to overview robotic assistance for RIRS and mainly focus on ELMED's robotic device by thoroughly describing it and investigating its effects on surgical performance based on the IDEAL (idea, development, evaluation, assessment, long-term study) framework. The robotic device, Avicenna Roboflex, is a teleoperated robot consisting of a surgeon's console (master control console – MCC) and manipulator arm (MA) to house and couple with the FURS. It provides a suitable and safe platform for robotic-assisted (RA) RIRS with significant improvements in ergonomics and potentially prolongs equipment durability which may lead to its general acceptance and preference of use by many surgeons in various countries including Turkey, Germany, France, the UAE, and Romania.

In the modern scenario of neurological rehabilitation, which requires affordable solutions oriented toward promoting home training, the Institute of Industrial Technologies and Automation (ITIA) of the Italian National Research Council (CNR) developed a line of prototypal devices for the rehabilitation of the upper limb, called "Arm." Arm devices were conceived to promote rehabilitation at affordable prices by capturing all the main features of the state-of-the-art devices. In fact, Arm devices focus on the main features requested by a robot therapist: mechanical adaptation to the patient, ranging from passive motion to high transparency, assist-as-needed and resistive modalities; proper use of sensors for performance monitoring; easy-to-use, modular, and adaptable design. These desirable features are combined with low-cost, additive manufacturing procedures, with the purpose of meeting the requirements coming from research on neuro-motor rehabilitation and motor control and coupling them with the recent breakthrough innovations in design and manufacturing.

This chapter presents a physical simulator for needle steering in brachytherapy. As the user inserts the needle in a phantom tissue, images of the needle and prostate shape reconstructed from 2D transverse ultrasound images are displayed online in a semi-transparent mirror. During insertion, the user sees the images as if they were floating inside the phantom accounting for scale and orientation. The ultrasound images of the needle are combined with a needle-tissue interaction model that predicts the needle deflection further along the insertion process. The necessary manoeuvres that bring the needle towards its intended target location are displayed to the user along with the actual needle location. This platform allows the user to test different manual and robotic-assisted needle steering techniques. Reported experimental results confirm the accuracy of the system in reconstructing and overlaying images onto the phantom.

In making neural networks learn nonlinear relations effectively, it is desired to have appropriate training sets. In the proposed method, after a certain number of iterations, input-output pairs having worse errors are extracted from the original training set and form a new temporary set. From the following iteration, the temporary set is applied to the neural networks instead of the original set. In this case, only pairs with worse errors are used for updating the weights until the mean value of errors decreases to a desired level. Once the learning is conducted using the temporary set, the original set is applied again instead of the temporary set. The effectiveness of the proposed approach is demonstrated through simulations using kinematic models of a leg module with a serial link structure and an industrial robot.

This chapter explains how the MEMS microrobot system could perform the walking behavior of ants. MEMS microrobot system consists of micro-mechanical systems and micro-electro systems. The micro-mechanical systems mimic the alternating tripod gait of an ant by the shape memory alloy-type rotary actuator and the link mechanism. The micro-electro systems mimic the electrical activity of biological neural networks using the artificial neural networks IC. The artificial neural networks IC generates the driving pulses of shape memory alloy-type rotary actuator without using software programs. The micro-mechanical systems and micro-electro systems are integrated as a robot system. As a result, the authors show that the MEMS microrobot system could perform the ant-like walking behavior with a speed of 20 mm/min. The MEMS microrobot system was 0.079 g in weight, 4 mm width, 4 mm length, and 5 mm height in size. The robot system needs only the electrical power source as an external device.

dynamics of the system. This chapter combines the comparison of the use of the predictive control based on dynamical study and the decoupling of the dynamical model, with auxiliary parallelograms, for locating the center of mass of the mechanism using springs in order to achieve the balancing of each leg. Virtual model was implemented and its kinematic and dynamic motion analyzed through simulation of an exoskeleton, aimed at lower limbs, for training and rehabilitation of the human gait, in which the dynamic model of anthropomorphic mechanism and predictive control architecture with robust control is already developed.

Since a nonholonomic system such as a robot with two independent driving wheels includes complicated nonlinear terms generally, it is hard to realize a stable and tractable controller design. However, about a dynamic control method for the motion planning, it is guaranteed that a nonholonomic-controlled object can always be converged to an arbitrary point using a control method based on an invariant manifold. Based on it, the method called "kinodynamic motion planning" was proposed to converge the states of the two-wheeled mobile robot to the arbitrary target position while avoiding obstacles by combining the control based on the invariant manifold and the HPF. In this chapter, how to combine the invariant manifold control and the concept of the HPF is explained in detail, and the usefulness of the proposed approach is verified through some simulations.

This chapter describes a novel dual-arm motion discrimination method that combines posterior probabilities estimated independently for left and right arm movements, and its application to control a robotic manipulator. The proposed method estimates the posterior probability of each single-arm motion through learning using recurrent probabilistic neural networks. The posterior probabilities output from the networks are then combined based on motion dependency between arms, making it possible to calculate a joint posterior probability of dual-arm motions. With this method, all the dual-arm motions consisting of each single-arm motion can be discriminated through leaning of single-arm motions only. In the experiments performed, the proposed method was applied to the discrimination of up to 50 dual-arm motions. The results showed that the method enables relatively high discrimination performance. In addition, the possibility of applying the proposed method for a human-robot interface was confirmed through operation experiments for the robotic manipulator using dual-arm motions.

Wheeled mobile robots (WMRs) are very interesting regarding different applications from in-house activities in assisting elderly people and patients to space exploration. While the design concept and the application of the WMRs determine specifications of the robot, the positional errors occur during the WMR motion. The positional errors are inevitable, as they are caused by imperfections in design to fabrication; therefore, there is a need to rectify them using calibration techniques such as odometry, camera-based error detection, or using gyroscope and compasses. This chapter focuses on the use of odometry as it provides improved short-term accuracy with high sampling rates while it is more economical and requires fewer landmarks to localize the WMR. The context provides an overview of WMRs mechanisms, differential and omnidirectional drive, and then introduces an odometry-based method to correct the motion of both types of WMRs. Experimental results on four robots exhibited that positional error was significantly improved. Using analysis of variance (ANOVA) test, the authors could not detect any change in error improvement when the robot changed.

Given the special working environments and application functions of the amphibious robot, an improved RGB-D visual tracking algorithm with dual trackers is proposed and implemented in this chapter. Compressive tracking (CT) was selected as the basis of the proposed algorithm to process colour images from a RGB-D camera, and a Kalman filter with a second-order motion model was added to the CT tracker to predict the state of the target, select candidate patches or samples, and reinforce the tracker's robustness to high-speed moving targets. In addition, a variance ratio features shift (VR-V) tracker with a Kalman prediction mechanism was adopted to process depth images from a RGB-D camera. A visible and infrared fusion mechanism or feedback strategy is introduced in the proposed algorithm to enhance its adaptability and robustness. To evaluate the effectiveness of the algorithm, Microsoft Kinect, which is a combination of colour and depth cameras, was adopted for use in a prototype of the robotic tracking system.

With the development of micro-electro-mechanical systems (MEMS), wireless communication technology, and embedded systems, wireless sensor network (WSN) has been a focus for research among various fields. Wireless sensor technology inspires many innovations for industrial applications and medical robots, and breaks many limitations and inconvenience for such sensing devices. A self-powered height sensing system with ZigBee technology is presented. It specifically targets to replace an original wired system with an integrated wireless sensor that is comprised of all necessary parts: sensing module, processing module, RF transceiver, and power supply. First, the authors present the system framework design including the layout of a wireless sensor node based on ZigBee. Second, with the vibration environment, a self-powered generator was developed through the comparison between piezoelectric and electromagnetic generation. Then several experiments are conducted to test and analyze the feasibility of the whole system. Finally, a future upgraded design is proposed to improve system performance.

This chapter reports the authors' experimental results on analyzing the human goal-finding process in continuous learning. The objective of this research is to make clear the mechanism of continuous learning. To fill in the missing piece of reinforcement learning framework for the learning robot, the authors focus on two human mental learning processes, awareness as pre-learning process and reflection as post-learning process. To observe mental learning processes of a human, the authors propose a new method for visualizing them by the reflection subtask for human to be aware of the goal-finding process in continuous learning with invisible mazes. The two-layered task is introduced. The first layer is the main task of continuous learning designing the environmental mastery task to accomplish the goal for any environment. The second layer is the reflection subtask to make clear the goal-finding process in continuous learning. The reflection cost is evaluated to analyze it.

This chapter presents the way to design a learning support system toward acquiring a creative skill on learning. There are two research goals. One is to establish designing the creative learning task. The other is to make clear the human sense of creativity. As the background of this research, the jobs with high creativity or social skills will remain in the future. However, acquiring human's creativity is too difficult for computers. To solve this problem, the authors focus on the way to utilize higher creativity of human than that of computers. The main method is the visualization of learning traces to support awareness for creativity on the learning. The authors conducted the preliminary learning experiment with three human subjects. After that, the questionnaire and the hearing investigation were conducted. As the future work, the authors are planning to conduct an updated version of the experiment.

Preface

As a result of years and through effective natural selection, nature tested every field of science and engineering leading to inventions with solutions for everything that can work well, adapt and last long. The evolution of nature led to resolve of many research and development challenges. Biomimetics as an emerging area of research development that studies and examines nature, its models, elements, processes, systems, structures, mechanisms, etc. to take inspiration from, or emulate, nature's best biological ideas and products in order to solve modern science and engineering problems. Biomimetics incorporates materials, behaviors, structures, concept, sensing and techniques drawn from naturally made substances, and resembles the aspiration of biological systems in structure, mechanism and/or function. Biologically inspired approaches will create new reality with great potential. In addition, biomedical robotics research and development area is evolving rapidly with an increasing number of robotic systems with different capabilities that perform wide variety of tasks, such as, surgical robots, rehabilitative and prosthetic devices, neurosurgery, preparation of bone surfaces, biomedical image analysis, image-guided prenatal cardiac interventions and telesurgery, assisting living. In addition, robots have also been developed to improve the accuracy of radiation treatment, organizing and dispensing food and drugs, as well as other materials within the hospital or health care. Finally, the use of robotics provides a remote presence of a physician.

This handbook includes 20 chapters that contribute with the state-of-art and up-to-date knowledge on research advancement in the field of biomimetics and biomedical robotics. The chapters provide relevant theoretical knowledge, practices, technological evolution and new findings. Furthermore, the handbook helps to prepare engineers and scientists who are looking to develop innovative, challenging, intelligent, bioinspired systems and value added ideas for autonomous and smart interdisciplinary products and systems to meet today's and future most pressing challenges.

Chapter 1 is titled "Biomimetics and the Evolution of Robotics and Intelligent Systems." This chapter presents biomimetics as an evolving interdisciplinary area of research development that studies and examines nature, its models, elements, processes, systems, structures, mechanisms, etc. to take inspiration from, or emulate, nature's best biological ideas in order to solve scientific and engineering problems. It also highlights some of the challenges and considerations for future development in the field of biomimetics.

Chapter 2 is titled "Investigation of Human Locomotion With a Powered Lower Limb Exoskeleton." This chapter studied lower limb exoskeleton both as a measurement device for studying human locomotion and as an assistive device that restores human ability to walk. In addition, the chapter considered the way the geometry of the supporting surface affects the sensory system of the robot through different development of new foot design structures.

Chapter 3 is titled "Active Assistive Orthotic System: (Exoskeleton) Enhancing Movement." Active orthosis (exoskeleton) is assistive device with a wearable structure to help generating the natural human motion. The chapter focuses on developing an active/assistive orthosis system (AOS) with enhanced movement. The design is bioinspired by musculoskeletal system of human upper and lower limbs, and mimics the muscle-tendon-ligament structure. It was fabricated using light materials and powered by pneumatic artificial muscles with more than fifteen degrees of freedom. AOS can operate in three modes.

Chapter 4 is titled "Hybrid Exoskeletons for Upper Limb Stroke Rehabilitation." This chapter presents a review on the state of the art of upper-limb hybrid exoskeletons with a particular focus on current need of stroke rehabilitation and how the hybrid exoskeletons can provide a solution to difficulties in this field. The chapter shows in particular that more investigation is needed with regards to the potential benefit of hybrid exoskeletons as a patient-monitoring and rehabilitation assist-as-need tool.

Chapter 5 is titled "Modeling a Predictive Control of Human Locomotion Based on the Dynamic Behavior." This work presents the development of a lower limb orthosis based on the continuous dynamic behavior and on the events presented on the human locomotion. Computational model was developed to approach the different functioning models related to the bipedal anthropomorphic gait. Virtual model was implemented and its kinematic and dynamic motion analyzed through simulation of an exoskeleton, aimed at lower limbs, for training and rehabilitation of the human gait, in which it is already developed the dynamic model of anthropomorphic mechanism and predictive control architecture with robust control.

Chapter 6 is titled "Towards Cost-Oriented User-Friendly Robotic Systems for Post-Stroke Rehabilitation." This chapter deals with robotics for assisting patients in their recovery after stroke with focus on key physiotherapist movements that are needed in post-stroke treatment. In addition, it discusses sets of experimental data for defining the main specifications for post-stroke rehabilitation devices. Also the chapter presents key achievements in the field, critical analysis and solutions concerning rehabilitation robotics for post-stroke treatment aiming at the definition of new concepts for developing cost-oriented user-friendly robotic system.

Chapter 7 is titled "The 'Arm' Line of Devices: For Neurological Rehabilitation." This chapter introduces the development effort of a line of prototypal devices for the rehabilitation of the upper-limb, called "Arm". Arm devices focus on the main features requested by a robot therapist: mechanical adaptation to the patient, ranging from passive motion to high transparency, assist-as-needed and resistive modalities; proper use of sensors for performance monitoring; easy-to-use, modular and adaptable design. These desirable features are combined with low-cost, additive manufacturing procedures, with the purpose of meeting the requirements coming from research on neuro-motor rehabilitation and motor control and coupling them with the recent breakthrough innovations in design and manufacturing.

Chapter 8 is titled "Lithotripsy of Renal Stones With Avicenna Roboflex Robotic-Assisted Retrograde Intra-Renal Surgery (RA-RIRS)." The work in this chapter introduces an overview on robotic assistance for RIRS with focus to introduce ELMED's robotic device with its effects on surgical performance based on the IDEAL (idea, development, evaluation, assessment, long-term study) framework. Also, the chapter presents the robotic device, Avicenna Roboflex, is a teleoperated robot consisting of a surgeon's console (Master Control Console – MCC) and manipulator arm (MA) to house and couple with the FURS. It provides a suitable and safe platform for robotic assisted (RA) RIRS with significant improvements in ergonomics and potentially prolongs equipment durability.

Chapter 9 is titled "Brachytherapy Needle Steering Guidance Using Image Overlay." This chapter presents a physical simulator for needle steering in brachytherapy. As the user inserts the needle in a phantom tissue, images of the needle and prostate shape reconstructed from 2D transverse ultrasound images are displayed online in a semi-transparent mirror. As the user inserts the needle in a phantom tissue, images can be seen as the user is floating inside the phantom accounting for scale and orientation. The ultrasound images of the needle are combined with a needle-tissue interaction model that predicts the needle deflection further along the insertion process. This platform allows the user to test different manual and robotically assisted needle steering techniques. Reported experimental results confirm the accuracy of the system in reconstructing and overlaying images onto the phantom.

Chapter 10 is titled "Neural Networks to Solve Nonlinear Inverse Kinematic Problems." This chapter proposes a new neural based learning approach to solve nonlinear inverse kinematic problems. The effectiveness of the proposed approach is demonstrated through simulations using kinematic models of a leg module with serial link structure and an industrial robot.

Chapter 11 is titled "Ant-Like Walking Behavior of MEMS Microrobot With Artificial Neural Networks." This chapter explains the development of MEMS microrobot system and how to mimic the alternating tripod gait of ant walking behavior using shape memory alloy-type rotary actuator and the link mechanism. The micro-electro systems mimic the electrical activity of biological neural networks using the artificial neural networks IC. The results show MEMS microrobot system could perform the ant-like walking behavior by a speed of 20 mm/min.

Chapter 12 is titled "Bio-Inspired Snake Robots: Design, Modelling, and Control." This chapter goal is to introduce the complexity and the fundamental design, modelling and control approaches for efficient snake robot locomotion in cluttered environments. As a contribution, the chapter presents shape-based modelling approaches and a number of available control schemes for operation in unknown environments are presented, which aims to motivate scholars to start working on snake robots. Some ideas about future research plans are also proposed to help operating in real world environment.

Chapter 13 is titled "Review of Friction and Surface Properties of Snakeskin." Natural systems tend to conserve energy; as such it is necessary to mitigate the effects of friction during motion. This chapter attempts to review the major findings about the tribology of the legless reptiles obtained through the last two decades.

Chapter 14 is titled "Kinodynamic Motion Planning for a Two-Wheel-Drive Mobile Robot." This chapter presents the development of kinodynamic motion planning to support convergence of the states of the two-wheeled mobile robot to an arbitrary target position while avoiding obstacles by combining the control based on the invariant manifold and the HPF.

Chapter 15 is titled "Discrimination of Dual-Arm Motions Using a Joint Posterior Probability Neural Network for Human-Robot Interfaces." This chapter describes a novel dual-arm motion discrimination method that combines posterior probabilities estimated independently for left and right arm movements, and its application to control a robotic manipulator. With this method, all the dual-arm motions consisting of each single-arm motion can be discriminated through leaning of single-arm motions only. The method was applied to the discrimination of up to 50 dual-arm motions and the results showed that the method enables relatively high discrimination performance.

Chapter 16 is titled "Off-Line Calibration of Autonomous Wheeled Mobile Robots: From Theory to Experiment." This chapter provides an overview of wheeled mobile robots (WMRs) mechanisms, differential drive and omnidirectional. In addition, it introduces an odometry-based method to correct the motion of both types of WMRs. Experimental results on four robots exhibited that positional error was significantly improved.

Chapter 17 is titled "A Real-Time Compressive Tracking System for Amphibious Spherical Robots." An improved RGB-D visual tracking algorithm with dual trackers is proposed and implemented in this chapter. Compressive tracking (CT) was selected as the basis of the proposed algorithm to process color images from a RGB-D camera, and a Kalman filter with a second-order motion model used to predict the state of the target, select candidate patches or samples and reinforce the tracker's robustness to high-speed moving targets. A visible and infrared fusion mechanism or feedback strategy is introduced in the proposed algorithm to enhance its adaptability and robustness. Microsoft Kinect, which is a combination of color and depth cameras, was adopted for use in a prototype of the robotic tracking system.

Chapter 18 is titled "Self-Powered Height Sensor With ZigBee Networks for Intelligent Systems." A self-powered height sensing system with ZigBee technology is presented in this chapter. It specifically targets to replace an original wired system with an integrated wireless sensor which is comprised of all necessary parts: sensing module, processing module, RF transceiver and power supply. Experiments several were conducted to test and analyze the feasibility of the developed system.

Chapter 19 is titled "Analyzing the Goal-Finding Process of Human Learning With the Reflection Subtask." This chapter presents experimental results on analyzing the human's goal finding process in continuous learning. The objective of this research is to make clear the mechanism of continuous learning. To fill in the missing piece of reinforcement learning framework for the learning robot, the authors focus on two human mental learning processes, awareness as pre-learning process and reflection as post-learning process. To observe mental learning processes of a human, the authors propose a new method for visualizing them by the reflection subtask for human to be aware of the goal finding process in continuous learning with invisible mazes. The two-layered task is introduced. First layer is the main task of continuous learning designing the environmental mastery task to accomplish the goal for any environment. Second layer is the reflection subtask to make clear the goal finding process in continuous learning. The reflection cost is evaluated to analyze it.

Chapter 20 is titled "Designing the Learning Goal Space for Human Toward Acquiring a Creative Learning Skill." This chapter introduces the way to design the learning support system toward acquiring a creative skill on learning. There are two research goals. One is to establish designing the creative learning task and the other is to make clear the human sense of creativity.

Maki K. Habib
The American University in Cairo, Egypt

Chapter 1
Biomimetics and the Evolution of Robotics and Intelligent Systems

Maki K. Habib
The American University in Cairo, Egypt

Fusaomi Nagata
Tokyo University of Science, Japan

ABSTRACT

Biologically inspired systems, known as "biomimetics" or the "mimicry of nature," is an interdisciplinary scientific research field inspired by nature and featured by the technology outcome (hardware and software) and lies at the interface of biology, physics, chemistry, information, and engineering sciences. Biomimetics is initiated by making nature a model of inspiration that would immensely help conscious abstraction of new innovative principles and creative design ideas and concepts that help developing new techniques and functionalities, seeking new paradigms and methods, designing new materials, and developing new streams of intelligent machines, robots, systems, devices, algorithms, etc. Biologically inspired approaches create a new reality with great development and application potential with the goal of identifying specific desirable qualities and attributes in biological systems and using them in the design of new products and systems. This chapter provides the importance of biomimetic as an interdisciplinary field and its evolution, advances, challenges, and constraints along with the associated enabling technologies supporting its growth. In addition, it introduces scientific ideas and directions of research activities in the field. The chapter also presents key developments in the field of biomimetic robots and underlines the challenges facing it.

INTRODUCTION

Nature is featured by huge resources and long experience to preserve biological strategies with successful evolution and discard fail ones. Biomimicry focuses on the study of nature models and species for aspiration to develop bioinspired technologies that are capable solve effectively human ongoing and

DOI: 10.4018/978-1-5225-2993-4.ch001

challenging problems. Nature found answers for many engineering problems and it is always inspires human seeking innovative and effective solutions. Insect and flying birds lead to innovative aeronautics development. Collision avoidance sensors mimic the whiskers of rodents help to develop collision avoidance techniques (Bar-Cohen, 2011). Nature tested every field of science and engineering leading to inventions that work well, can adapt and long lasting. In addition, living-creatures exhibit high agility, flexibility, robustness, efficiency, and the ability to learn and adapt to improve performance and survive in the real world, and hence the evolution of nature led to the introduction of highly efficient biological mechanisms and open the door for new development possibilities in materials, mechanisms, algorithms, fabrication processes, design, education, etc. (Bar-Cohen, 2011; Habib 2011; Neves & Francke, 2012; Yurtkuran et al., 2013).

Nature has always served as a model for mimicking or inspiration to humans in their efforts to improve life quality. Adopting mechanisms and capabilities from nature and the use of innovative technological solutions, led to effective materials, structures, tools, mechanisms, machines, processes, algorithms and techniques, control, systems, intelligence, and many other benefits. Humans throughout history have always sought to mimic or bioinspired by the appearance, mobility, functionality, intelligent operation, and thinking process of biological creatures (Bar-Cohen, 2003; Habib et al., 2007a; Habib et al., 2007b; Habib, 2011). Matured conventional technologies are associated with constraints and inadequate performance, and this fosters the demand for new solutions to maximize functionality while minimizing costs in energy and materials. The need to seek for new solutions is driving science to consider nature as a biologically inspired model. The driving force behind attempting to merge biological principles and physics applications stems from the recognition that there are a number of areas where biological methods are more efficient, environmentally friendly, and overall superior to current technology. Hence, understanding biological systems presents unique opportunities for wide range development of potential ideas, paradigms, concepts and methods for engineering solutions, and to create new generations of smart materials, novel advanced structures, intelligent devices and technologies. Engineers are increasingly turning to biologists to understand and learn how living organisms function and solve problems. This enables the fusion of the best solutions from nature with artificially engineered components to develop systems that are better in function and efficiency than existing conventional approaches. The new approach of using ideas from nature towards innovative development in science and technology has been given a number of names. At early stage the term bionics was used to mean the study of basic principles in nature and emerge with applications of principles and processes to the needs of mankind (Papanek, 1971). Bionics comes from biology and has recently been used in the field of medicine where its principles have been used to assist with the production of replacement organs and other body parts. Then, the term biomimetics was coined from *bios* to mean life, and *mimesis* to mean imitate and it has been used mainly in science and engineering. In addition, the term biomimicry was defined as innovation inspired by nature (Bewnyus, 1997). Benyus suggests looking to nature as "a model, measure, and mentor" to promote the concept of learning from nature as a possible methodology, emphasizing sustainability and sustainable design. Bionics, biomimetics, biognosis and biomimicry all follow nature for design ideas (Knight, 2009).

Researchers in engineering sciences diverge in precisely how they define biomimetics (Paulson, 2004). However, the use of inspiration, instead of mimics, represents accurate reflection of the biomimetics field since mimicry and strict adherence to biological designs is neither possible nor always desirable. Biomimetics incorporates building novel materials at nature's scale, developing concepts and techniques drawn from naturally made substances, and resembles biological systems in structure, mechanism and/

or function. In the field of sensing and actuations, biologically inspired devices can provide efficient ways of converting mechanical energy into electrical or chemical forms and vice versa. Scientists hope this blending may one day lead to stronger, cost- and energy-efficient, and intelligent products that are attractive ecologically. Biomimetics uses ideas of nature's evolved way of doing things to offer potential improvements of our life and the tools we use. The field of biomimetics has spawned a number of innovations far superior to what the human mind alone could have devised. The extreme competition of natural selection describes that loss and efficiency are not tolerated in natural systems, unlike many of the technologies devised by humans. In automotive industries, gas-powered cars are only about 20 percent efficient, that is, only 20 percent of the thermal-energy content of the gasoline is converted into mechanical work.

This chapter provides the importance of Biomimetic as an interdisciplinary field and the associated enabling technologies supporting its growth with highlight on robotics. In addition, it introduces scientific evolution of the field, development, ideas and directions of bioinspired research activities to solve human problems. The Chapter also presents key developments in the field of Biomimetic at different levels and underlines the challenges facing it.

BACKGROUND

Today, researchers are looking for any insights they can be inspired by biological concepts and find out how biological species do things that the current class of robots cannot and can solve with aim to transfer such inspiration into a new innovative technologies and developing a new class of biologically inspired robots that exhibit greater robustness in performance in dynamics and unstructured environments than today's robots. Such new generation of robots expected to be substantially compliant and stable than current robots, and will take advantage of new developments in materials, fabrication technologies, sensors, actuators, reasoning, etc (Habib et al, 2007a; Habib et al, 2007b; Habib, 2011). Applications include autonomous or semi-autonomous tasks such as reconnaissance and demining for small, insect-like robots and human interaction tasks at a larger scale. Technologies that allow developing biologically inspired system are increasingly emerging, such as artificial muscles, AI and computational intelligence, and computer vision as well as biomimetic capabilities in materials science, mechanics, electronics, computing science, information technology and many others. Making biologically inspired intelligent robots requires understanding the biological models as well as advancements in analytical modeling, graphic simulation and the physical implementation of the related technology. However, engineering such biomimetic intelligent creatures as realistic robots was hampered by physical and technological constraints and shortcomings and it is still a challenge. Making creatures, such as robots and toys that look and behave like the biological model are greatly inspired by science fiction (books, movics, toys, animatronics, etc.). These models have created perceptions and expectations that are far beyond the reach of current engineering capabilities, which are constrained by laws of physics and current technology. However, engineering such biomimetic intelligent creatures as realistic robots are still involves many challenges. Innovative robot design should not be restricted by an animal's model design; rather the fusion of the best solutions from nature with artificially engineered components.

Researchers in the field of robotics can apply the available research outcomes and lessons that already learnt from biology. How birds fly, how fish swim, how dolphins locate objects, and how humans walk, etc. might best be discovered and understood by trying to understand and abstract the key knowledge from relevant bio-systems, which can reproduce these activities and use it in developing devices that maximize efficiency and performance. Advances in low-cost, and power-efficient computing systems allow researchers to develop new robots and techniques that function effectively outside laboratory environment, aiming to support potential applications dealing with the complex dynamics of real world problems.

A wide range of biologically inspired robots has been developed. Some of the developed robot designs are inspired by the biological shape and motion mechanisms selected from the bio-systems family. Such robots may perform combinations of locomotion techniques including legged (two legs (Park & Kim, 1998), four legs (Raibert, 1986; Buehler, 2000), six legs (Buehler, 2000; Cham, et al., 2004; Simmons & Krotkov, 1991), eight legs (Klaassen et al., 2003), etc.), hopping (Raibert, 1986; Schwind and Koditchek, 1995; Cherouvim and Papadopoulos, 2006), walking and jumping (Aschenbeck et al., 2006), running (Raibert, 1986), climbing (Luk et al., 1996), walking and climbing (Hirose et al., 1991), rolling (Bhattacharya & Agrawal, 2000; Michaud & Caron, 2002), snake like (Hirose & Fukushima, 2002), crawling (Galvez et al., 2001), swimming (underwater, shallow water) (Videler, 1993; Anderson et al., 1998; Ayers et al., 2000a; Ayers et al., 2000b; Neuhaus et al., 2004), flying (Yan, et al., 2001; Michelson & Naqvi, 2003; Madangopal et al., 2005), etc. Other development was inspired by behaviors of bio-systems, such as, sensing, navigation, formation, tracking, etc. (Viollet & Franceschini, 2001; Iida, 2003; Shen et al., 2004; Lewinger et al., 2005; Lewinger et al., 2006).

Current robots lack the robustness and performance of even the simplest insect or animal when operating in dynamic, unstructured and complex environments. The engineering of biomimetic intelligent creatures as realistic robots was hampered by physical limitations and technological constraints, and it is still a challenge. Even if the design of biomimetic robots is inspired from biology, however, their realizations were, in most cases, compromised by their complexity and fragility that results from the use of traditional engineering materials and manufacturing methods (Cham et al, 2004).

BIOMIMETICS AND ENABLING TECHNOLOGIES

The growing effect of interdisciplinary technology is changing the world across all dimension of life: social, health, education, manufacturing, industry, economic, political, personal, etc. The revolution of information technology, information availability and utilities continue to influence the world in all life dimensions. Biotechnology is revolutionizing our life by enabling us to identify, understand, manipulate, improve, and control every aspect of living organisms. Smart materials, agile manufacturing, and nanotechnology are inspiring, enabling, and offering the promise of a new dimension in innovation of new devices with unforeseen capabilities that we have not seen. Biotechnology relies heavily on laboratory equipment providing lab-on-a-chip analysis as well as progress in bioinformatics. Many of the advances in lowering energy intensity have come from developments in the materials and chemical sciences, such as new magnetic materials; high strength, lightweight alloys and composites; novel electronic materials; and new catalysts, with a host of energy technology applications. Materials sciences are helping in the development of energy generation, conversion, transmission, and use, while chemical sciences provide the necessary understanding of the interactions of atoms, molecules, and ions with photons and electrons; the making and breaking of chemical bonds in gas phase, in solutions, at interfaces, and on surfaces; and the energy transfer processes within and between molecules.

Human body can be viewed as complex assembles of molecular-level machines (Goodsell, 1996). Biolmolecular motors are individual protein complexes that are ultimately responsible for all active biological motion involving internal material transport. They perform tasks vital to life of the organism, such as muscle contraction, damage repair, cell division, intracellular transport and genomic transcription (Videler, 1993). The success of future molecule-driven actuators most likely lies in the development of artificial molecular motors because of their ability to provide large forces from low voltage inputs while featuring bistable actuation characteristics and molecular design flexibility. In order to develop mechanical parts in the nanometer range, and thereby create nano-electromechanical systems (NEMS) (Graighead, 2000) that may be used beyond those satisfied by current MEMS devices, different fabrication methods have been initiated (Goodsel, 1996; Sauvage, J.-P. & Dietrich-Buchecker, 1999; Huang et al., 2005). Biomimetic synthetics are being used to encourage bone regeneration and interaction with human implants and orthopaedic implants. The smaller the particle size of the material used, the better the result as far as tissue growth and regeneration. Recent developments are using nanotechnology to produce molecular level changes to materials in order to mimic natural found materials

Nano- and smart materials technologies are changing the way structures and electronics are made, and promise the development of useful components, products and systems for desired characteristics that are highly stronger, smaller (composed of just a few atoms and molecules), smarter, high resolution, lighter weight, faster, accurate, reliable, structural functionality and flexible, and self-repairing, multi-functional, user and environment friendly, efficient energy conversion and storage, more survivable, customizable, etc. The revolution in genomics research has the potential to provide entirely new ways of producing forms of energy, sequestering carbon, and generating materials that require less energy to produce. It includes research to investigate the underlying biological processes of plants and microorganisms, potentially leading to new processes and products for energy applications.

Different materials with sensing and actuation capabilities are increasingly used to combine these capabilities in response to various needs, such as, new smart and specialized equipment and systems; monitoring and diagnosis for health, industry, structures and environment; security and tracking; realizing new clothes that can respond to weather and health or comfort, interface with information systems, monitor vital signs, deliver medicines, and protect wounds; personal identification; supporting new trends in building intelligent transportation systems, etc. Rapid prototyping, and flexible manufacturing together with embedded sensors and systems, has provided a means for accelerated and affordable design and development of complex components and systems. Hardware advances for exponentially smaller, faster, and cheaper semiconductors are supporting the growth of information technology. This trend in technology increases the availability of low-cost computing, high memory devices, and enable the development of ubiquitous embedded sensors and computational systems in consumer products, intelligent system, appliances, and environments. The evolution of these technologies is leading to the development of new trend of instrumentation and measurement technologies along with chemical, fluidic, optical, mechanical, and biological integrated micro- and nano-scale components, which are integrated with computational logic in chip designs. Increases in materials performance for power sources, sensing, and actuation could also enable new and more sophisticated classes of robots and remotely guided vehicles based on biological models. A potential long-term solution to overcome possible limitation in computational power is to shift the basis of computation to devices that take advantage of various quantum effects. In addition, another approach known as molecular electronics would use chemically assembled logic switches organized in large numbers to form sophisticated devices and computers. These concepts are attractive because of the huge number of parallel, low-power devices that can be developed.

The biomimetics field is highly interdisciplinary and all of the mentioned technologies that contribute to its progress are highly inter-related and heavily synergized, which makes the progress of each area leads to accelerate progress in each of the other areas. In addition, biomimetics is demanding to increase the importance of continued interdisciplinary education, training and research. The progress of the mentioned fields is highly contributing to improve life quality by having better disease control, new medicines and treatments, gene therapy, age mitigation and reversal, prosthetics, bionic implants, animal transplants, and many other advances. The accelerating pace of the advancements in the field of biomimetics seems to make evident that the emergence of machines as our peers is imminent. Further advancement and the actual realization of these possibilities depends on a number of factors, including local acceptance of technological change, levels of technology and infrastructure investments, market drivers and limitations, and technology breakthroughs and advancements.

BIOINSPIRED MUSCLES

Muscles are key to the mobility and manipulation capabilities of biological creatures, and when developing biologically inspired robots, it is important to have actuators that allow the emulation of the behavior and performance of real muscles. The potential to make such actuators is becoming increasingly feasible with the emergence of new development, such as the electroactive polymers (EAP), which are also known as artificial muscles (Bar-Cohen, 2001). The emergences of EAP have demonstrated promise for a variety of applications. The potential of this technology is immense as electroactive polymers that perform muscular activities by expanding and contracting silently based on variable voltage input levels.

EAP can be divided into two major categories based on their activation mechanism electronic and ionic. The electronic EAP, such as electrostrictive, electrostatic, piezoelectric, and ferroelectric, are driven by Coulomb forces. These types of EAP materials can be made to hold the induced displacement while activated under a DC voltage, making them suitable for robotic applications. These materials have a greater mechanical energy density and they can be operated in air with no major constraints. However, the electronic EAP require a high activation fields (>100-V/μm) that may be close to the breakdown level. In contrast to the electronic EAP, ionic EAP are materials that involve mobility or diffusion of ions and they consist of two electrodes and electrolyte. The activation of the ionic EAP show large deformation in the presence of low applied voltage as low as 1-2 Volts, and exhibit low impedance while mostly a bending displacement is induced. They operate best in a humid environment and can be made as self-contained, encapsulated actuators to operate in dry environments as well. Examples of ionic EAP include gels, polymer-metal composites, conductive polymers, and carbon nano-tubes. Their resistance increases with decreasing temperature, a property that is opposite to all metallic conductors. They have been modeled as both capacitive and resistive element actuators that behave like biological muscles and provide an attractive means of actuation as artificial muscles for biomechanics and biomimetics applications. Their disadvantages are the need to maintain wetness and they pose difficulties to sustain constant displacement under activation of a DC voltage (except for conductive polymers) (Bar-Cohen, 2001).

The basic EAP, used as artificial muscle (AM) and called EPAM, is made up of a proprietary dielectric elastomer that is coated on both sides with another expandable film of a conducting electrode. When voltage is applied to the two electrodes a Maxwell pressure is created upon the dielectric layer. The elastic dielectric polymer acts as an incompressible fluid expanding in the planar directions. Electrical force is converted to mechanical actuation and motion. These same polymers have produced electric power in response to large deformations with high coupling efficiency and high energy density. EAP artificial muscle gets significant displacement from this electrostatic pressure as compared to other technologies, such as conventional electromagnetic (EM) actuators, but from other emerging technologies like piezo electric (PZT) and shape memory alloys (SMA).

The overall displacement is a function of the area of EPAM, and the force exerted is a function of the number of layers of EPAM. These layers can be constructed in multiple planar configurations. Furthermore, the electrode layer of the EPAM can be patterned to achieve specific envelopes of motion. A significant advantage EPAM has over electromagnetic actuators is energy density, that is, more energy created per unit mass of the actuator itself. These materials have functional similarities to biological muscles, including resilience, damage tolerance, and large actuation strains.

As polymers, EAP materials can be easily formed in various shapes, their properties can be engineered and they can potentially be integrated with micro-electro-mechanical-system (MEMS) sensors to produce smart actuators. Their most attractive feature is their ability to emulate the operation of biological muscles with high fracture toughness, large actuation strain and inherent vibration damping. The good electromechanical response of these materials, as well as other characteristics, such as good environmental tolerance and long-term durability, suggest a wide range of possible applications. Unfortunately, the EAP materials that have been developed so far are still exhibiting low conversion efficiency, are not robust, and there is a need to develop an adequate understanding of EAP materials' behavior, as well as processing and characterization techniques (Bar-Cohen, 2002). In order to develop better actuators, a discipline of visco-elastic engineering and control strategies will need to be developed to supplant the traditional engineering of rigid structures. Figure 1 shows JPL robotic hand has been developed to demonstrate the use of EAP actuators

Figure 1. JPL robotic hand as a platform to demonstrate the use of EAP actuators
Source: Bar-Cohen, 2001

BIOINSPIRATION AND INNOVATION

Nature continues to inspire researchers to initiate new studies, and create new ideas, paradigms and concept for engineering solutions. Biological organisms often exhibit properties that would be desirable in many systems. They function with high degrees of autonomy, while all biological organisms have mechanisms for self-repair and all multi-cellular organisms grow from an initial state that is much less phenotypically complex than their final states.

There are completely different levels on which biology can be considered as a source of ideas. The interest lies not just in the abstraction of useful ideas from the living world but also in the process by which this is done. Significant progress has been made while many aspects of biology are still beyond our understanding and capability,

Some of the biological inspired and innovative research efforts that are under consideration are,

1. Sophisticated inspirations to develop aerodynamic structures and systems came through the shape of seeds due their capability to travel the wind currents. The design of a mobility method for operating a robot on Mars using wind rather than a power consuming mechanism was inspired from tumbleweed plant (see Figure 2-a). Since there is significant wind on Mars, NASA was inspired by tumbleweed to design a spherical robot that is able to use the natural winds on Mars to propel itself around, rolling like tumbleweed over boulders instead of sidestepping them like conventional rovers. The robot would be lightweight, large and inflatable, and could carry instrument payload in its interior, such as cameras, water-seeking instruments, etc. It can traverse great distances with a minimal use of power, and when it reached a spot of particular interest, the robot could partially deflate to stop, and then re-inflate once it wrapped up its close-up observations. Such rovers offer

Figure 2. Bio-inspiration from plants a. The tumbleweed (left) offered an inspiration for NASA to develop a Mars Tumbleweed Rover (right) Source: Lorenz et al, 2003; b. Winged seed of the tropical Alsomitra Macrocarpa Source: Armstrong, 2007.

(a) The tumbleweed (left) offered an inspiration for NASA to develop a Mars Tumbleweed Rover (right) (Lorenz et al, 2003)

an effective and simple means of gathering data over large spatial extents of Earth, Mars, and other solar system bodies (Jack, 2001; Lorenz et al., 2003). Tests of a 1.5-meter ball in Greenland were done, where it completed a more than 100km autonomous traverse across an ice sheet. It was earlier tested in desert and the researchers determined that a sphere four times that size could climb over sizeable rocks and roll up slopes with an incline of more than 25 degrees, using only the thin, but robust, winds of Mars. The rolling Mars robot could reach speeds of 35 km/h (22 mph) and come to a halt by releasing gas.

The stable wing platform designed by Ignaz and Igo Etrich in 1904, was reportedly inspired from the large (about 15 cm span) winged seed of Alsomitra macrocarpa, which grows on islands in tropical pacific. Figure 2-b shows winged seed of Alsomitra macrocarpa (Liana seeds). The seed with its outgrowths functions is capable of gliding for significant distances through the air of the rain forest in wide circles. In addition, other plants generate compounds that fight off infection from fungi, insects, and others.

2. The Lobster's acute sense of smell in the turbulent ocean environment inspired researchers to develop a chemical-tracking, underwater robot based on lobster biology, the Robolobster. Such development constitutes two directions, one focuses on developing the sensory-chemical tracking system, and the other building the biologically inspired locomotion. In addition, in order to enable the Robolobster to operate under water autonomously for reasonable amount of time, researchers made it as close to autonomous as possible. The developers reported that Robolobster II is computationally and energetically autonomous for missions up to five hours (See Figure 3). Parallel to this development, another team of researchers is working to apply the concept of central pattern generators to help generating rhythmic gaits and walking trajectories closer to that of the biological lobster (Ayers et al., 2000). RoboLobster is small, has delicate, wiry artificial muscles that move its legs. These wiry artificial muscles are made of a nickel-titanium alloy called nitinol that contracts when electricity is applied. It has the advantage of dealing with real-world environments.

Figure 3. Robolobster
Source: Ayers et al., 2000

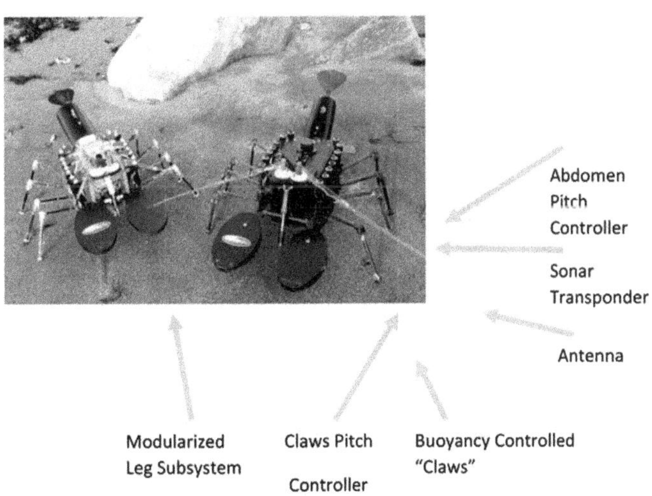

3. Robotics has proven to be an excellent instrument for performing safe and successful surgeries with the added benefit of minimally invasive surgery. The most common pathologies that are currently solved through robotic surgery are fibroid tumors, uterine cancer and cervical cancer. While conventional instruments are rigid and only achieve straight-line trajectories, the biomimetic design will enable to develop small size, smart and flexible devices that can achieve curved paths connecting any entry point to any target within sensitive organs such as brain to be followed autonomously. Researchers inspired by nature and supported by the evolution of new technology, such as nanotechnology, are working to develop small and flexible bio-Mechatronic robotic systems with the ability to do complex maneuvers and to make surgery less invasive.

Furthermore, inspired by elephant trunks, mammalian tongues, tentacles, octopus arms, and appendages which possess remarkable dexterous manipulation abilities without a typical skeletal structure, continuum robots seek to mimic the remarkable abilities of these organisms through designs that lack rigid links composing traditional robots (Jones et al., 2006). Figure 4 shows a flexible robot arm inspired by elephant trunk.

4. Cockroaches are effective at maneuvering, and they can go anywhere. The mechanical simplicity of their bodies and the ease at which they can maneuver over obstacles inspired researchers to model robots after the cockroach. Cockroaches typically employ an alternating tripod gait and they rely heavily on passive mechanical properties to achieve dynamic stability. The sprawled posture with large forces in the horizontal plane, and the compliance and damping in the limbs and joints, serve as "preflexes" (Kubow & Full, 1999).

One emphasis was to look at insect locomotion, posture and bouncing for ideas about how to build flexible and efficient walking robots. The Sprawl family of robots (Sprawlita (Clark et al., 2001),

Figure 4. Trunkbot each segment can be manipulated by a controller giving it similar capability to an elephant's trunk
Source: MIT

AquaSprawl and iSprawl (Kim et al., 2004)) was developed at the Center for Design Research at Stanford University. These six-legged robots draw their inspiration from the physical construction and mechanical design principles that are responsible for the robustness of the cockroach. The iSprawl (Yan et al., 2001) as shown in figure 5 is the first fully autonomous member of the Sprawl family. It is small, about 11 centimetres in length, light weight (0.3 kg), and can run at 15 body-lengths/second. Examination of the trajectory of the center of mass and the ground reaction forces for iSprawl show that it is stable and with bouncing locomotion closer to that in the biological insect.

5. Crickets use sound to track potential mates. Researchers are investigating this phenomenon so that robots that could be used in search and rescue efforts. The idea scenario is to deploy small robots to safely locate voices calling for help, or detect breathing in a constricted environment such as ruined buildings following an earthquake or explosion. The design places sensing microphones close together like cricket ears so the robot can use the Doppler effects to locate sound. A Cricket-inspired robot is shown in Figure 6 (Paulson, 2004). The robot is approximately three inches long,

Figure 5. iSprawl
Source: Yan et al., 2001

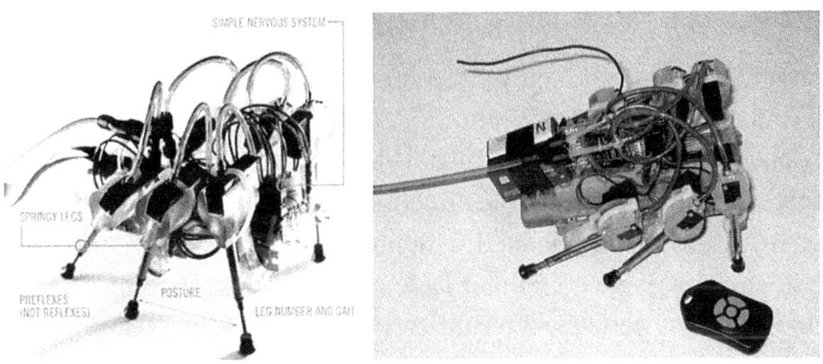

Figure 6. The Cricket robot
Source: Paulson, 2004

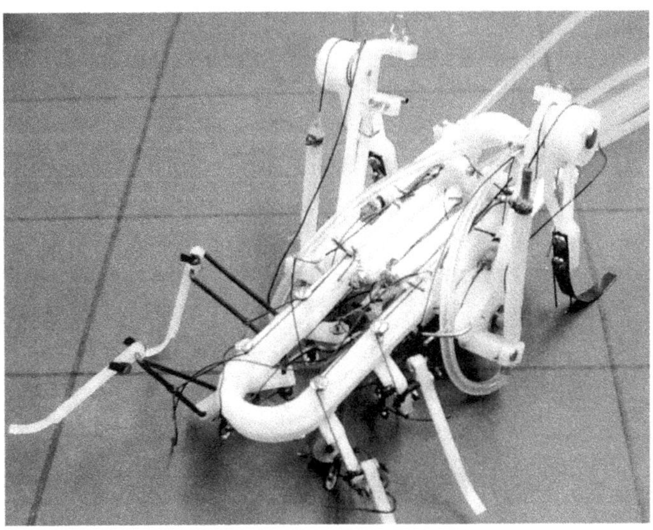

designed for both walking and jumping to navigate terrain with features much larger than itself. This three-inch version of the platform can run up to 10 body lengths per second and comes in a jumping version as well. The next step is to eliminate radio control in favor of autonomous operation.

6. The tiny hairs at the ends of a gecko's toes allow the lizard to grab onto smooth surfaces, scale walls and hang from ceilings. It can even run upside down on a ceiling of polished glass. This is due to the way tiny chemical forces are exploited by tiny hairs called setae (Sarfati, 2003) that enables it to stick and unstuck on surfaces. Yet to understand how their toes and the hairs fall off and self-clean, and how come they do not mat, or stick together. Inspiration is going on to build gecko like substance for adhesion that support wall climbing robot, and researchers at CMU have managed to duplicate the gecko foot surface structure to some extent using a very fine nanoprobe and an electron microscope to make a tiny wax mould for a resin, which sets to form artificial hairs (Graham-Rowe, 2003; Unver et al., 2006). However, the artificial hairs are not yet strong enough to support heavy weight. Another group of researchers from UK and Russia have prepared a self-cleaning adhesive tape modeled on the gecko's foot. Their tape (The tape had fibres 2 microns long, with a diameter of around 0.5 microns and spaced 1.6 microns apart, on a film of polyimide 5 microns thick), with a contact area of only 0.5 cm^2 on glass can bear a load of more than 100 grams (Geim et al, 2003). However, the tape is not durable enough to attach and detach more than a few times, unlike the real gecko as it is necessary to be cleaned regularly.

A multi university project was initiated to develop a possible robot inspired by Gecko at Stanford University and it is called Sickybot (See Figure 7) (Kim et al, 2007). The robot is a glass-climbing quadruped robot equipped with Gecko Tape. The developed platform is an embodiment of principles of adhesion-hierarchical compliance, controllable directional adhesion and distribution of forces. Stickybot inherits close characteristics to that associated with natural Gecko species, and it comprises 12-active actuators, 8 DOF serial and 4 passive compliant DOFs in leg and 16 segmented toes controlled via two stage differential cable-driven system. Gecko runs at 24cm/s and climbs 0.8m/min on glass, acrylic and similar smooth surfaces.

7. Hopping can cover large distances and thus provide significant mobility. For the purpose of exploring a different mobility paradigm which may present advantages over conventional wheel and leg locomotion, hopping was considered as an efficient way to achieve mobility in low-gravity environments and perform science and imaging via rolling. A small hopping robot with frog-like abilities that moves by a combination of rolls and hops to its desired destination could be the answer

Figure 7. Gecko and Sickybot
Source: Kim et al., 2007

someday (NASA-JP, 2000). Hence, NASA has develop a robot called 'Frogbot' or 'Hopbot'. It uses a single motor to hop, right itself and swivel about on a round foot before hopping again (See Figure 8). This helps the robot to go over something several times its body size. It is equipped with a camera, solar panels, sensors and onboard computer to execute commands autonomously, making the robot ideally suitable for exploration of distant planets, comets and asteroids. The robot has the ability to point in any direction, compressing itself down with a geared, six-bar linkage (which gives the robot its distinct frog-like appearance) before launching at a 45-degree angle. No matter how it lands the robot can quickly right itself by using a two-phase system to regain its footing, flipping first onto its back and then raising itself up with the help of a flap that doubles as a solar array. Weighing less than 1.3 kilograms. It has a spring between its knees that allows it bend its legs and hop. When the spring releases, the frogbot takes a 1.8-meter hop on Earth, which could become a 6 meter leap under low-gravity conditions on planets like Mars, depending on terrain. However, the control of the hopping direction was not satisfactory due to friction in the pointing mechanism and too much dependence on the surface type. Other generations of Frogbot are under consideration aiming for better design performance.

8. While wheel locomotion mechanisms allow reaching great distances and speeds, wheeled vehicles are subjected to great limitations with regards to traversing complex terrain with large obstacles and objects. Hence, it is necessary to look for alternatives by studying the navigational systems and locomotive strategies of animals and insects to design the next generation of autonomous robots and vehicles. Understanding high-mobility, robustness, learning, adaptation, etc. of biological systems helps to develop robots that can make difficult missions possible, such as search-and-rescue, detection, sampling, and removal of biohazards and mines.

Researchers have studied, in extensive detail, rodent behavior while performing a range of spatial navigation tasks and how they represent their location within the world. Flying insects display unparal-

Figure 8. The Frogbot hopping robot prototype
Source: NASA-JPL, 2000

leled efficiency and effectiveness when navigating through environments. These miniature organisms represent incredibly robust flight control systems and navigation techniques when faced with uncertainties. A bee when navigating sees the velocity on each side matches, and hence it flies the middle path. In nature a bee will always fly through the center of a window, never to one side, because it equalizes its velocity on both sides as it passes obstacles.

The collective robotics approach was inspired from the social characteristics of insects and animals, mainly ants, bees and birds. Research on ant societies, their structure and dynamics have revealed a great deal of knowledge about how such simple insects living in such complex social organization are able to perform a plethora of tasks with high level of efficiency.

9. Biological snakes occupy a wide variety of ecological niches, ranging from arid desert to tropical jungle as well as swimming in rivers and oceans. Abandoning limbs and developing elongated spines has proved an effective survival strategy, allowing snakes to hunt underground in confined tunnels, above ground in grassy fields and up in the tree-tops, even falling in a controlled glide from one tree to the next. A bioinspired design with long but thin body with small, waterproof, modular and self-contained (power, actuation, sensing and computation) segments makes them well suited to a large number of applications. Some of these applications may include, search and rescue in urban and natural environments, planetary surface exploration, minimally invasive surgery / examination, pipe inspection /cable routing, etc. The modular segments of snake robots can help to adjust its length by adding or removing segments, as well as to replace defective segments. Shigeo Hirose introduced the idea of the artificial articulated body that could perform several snake-like functions. Hirose's group built several mechanisms to demonstrate the feasibility of this idea, most notably the Active Cord Mechanisms (ACM) (Hirose,1993). ACM III incorporates 20 joints, and each joint consists of servomechanisms that can bend to the left and right (See Figure 9). In order to facilitate the locomotion of the device, each segment rests on a pair of wheels and in Klaassen et al., 2003 the ACM III can be seen weaving its way through several tight passageways. The research group of Hirose developed several generations of snake like robots including ACM IV-VII, ACM R1 and R2, and the Slim Slime Robot (SSR) (Ohno and Hirose, 2000). The SSR composed of six modules, pneumatically operated, with three internal bellows capable of generating pitch, yaw and axial extension/contraction of the module. The maximum deviation of those parameters is ±30 degree for pitch and yaw. SSR has a total length from 1120 mm to 730 mm, a total mass of 12kg and its maximum speed is about 60mm/sec. This design represents a transition from HDOF devices to fundamentally continuous devices. The SSR robot is shown in Figure 10. The introduc-

Figure 9. ACM III snake is approximately 2 m long and weighs 28 kg
Source: Hirose, 1993

tion of modules capable of 3-dimensional motions opened up new possibilities for locomotive gait patterns. Recently, the group developed a unique design of a snake robot ACM-R5 that been demonstrated to swim and crawl on ground (See Figure 11). ACM-R5 has 18 DOFs. The robot size is 1750 mm (length) x 80 mm (diameter), weight 7.5 Kg, and it is equipped with paddles and passive wheels around the body (Yamada et al., 2005). Each joint unit has CPU, battery, motors, so they can operate independently. Through communication lines each unit exchanges signals and automatically recognizes its number from the head, and how many units join the system. However a large number of problems are remained for realization of practical snake-like robots, both in software and hardware.

10. The assembly, inspection, and maintenance requirements of permanent installations in space demand robots that would potentially posses mobility in complex terrains. Conducting sample acquisition and analysis attributed to legged animals including grasping and object manipulation, and provide a high level of operational flexibility relative to mass and volume. Such demands point to robots that are dexterous, have significant processing and sensing capabilities, and can be easily reconfigured, both physically and algorithmically (such as force control for mobility and manipulation, and adaptive visual feedback). A series of LEMUR (Limbed Excursion Mobile Utility Robot) has been developed at JPL to address these challenges (See Figure 12) (NASA-JPL, 2006). The physical layout of the LEMUR I consists of six 4 DOFs limbs arranged axi-symmetrically about a hexagonal body platform. The current Lemur IIa platform represents the jumping-off point toward two more-advanced robotic platforms equipped with multi-functional tools and multiple cameras. This

Figure 10. The SSR robot
Source: Ohno & Hirose, 2000

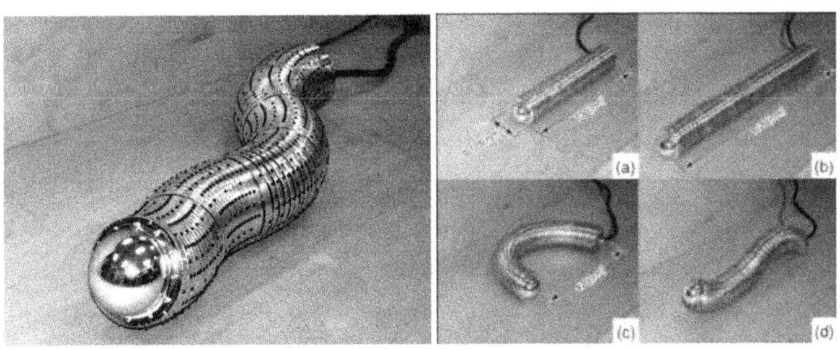

Figure 11. Illustrates crawling and swimming ACM-R5 snake robot
Source: Yamada et al., 2005

Figure 12. LEMUR rovers
Source: NASA-JPL, 2006

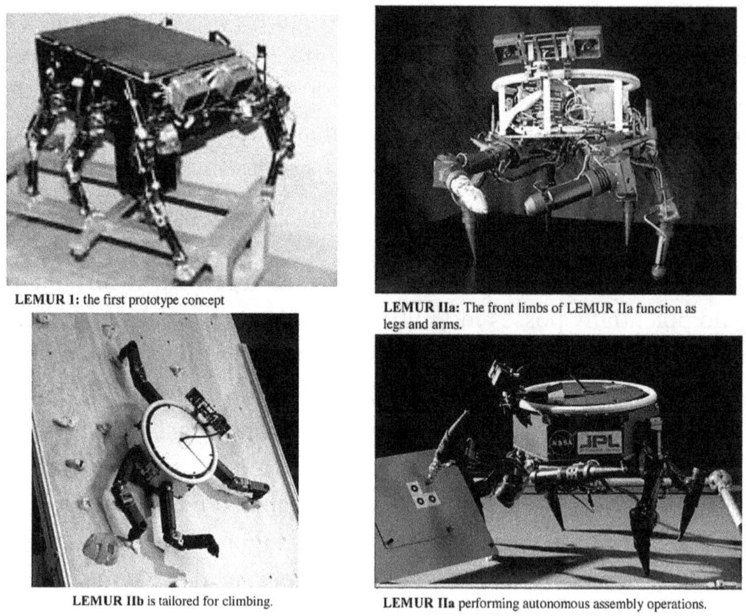

LEMUR 1: the first prototype concept

LEMUR IIa: The front limbs of LEMUR IIa function as legs and arms.

LEMUR IIb is tailored for climbing.

LEMUR IIa performing autonomous assembly operations.

spider looking robot also has 6 legs, each of which has interchangeable end-effectors to perform the required mission. The axis-symmetric layout is a lot like a starfish or octopus, and it has a panning camera system that allows omni-directional movement and manipulation operations. In addition, the front limbs of Lemur IIa function as legs and arms. Lemur IIb, a 4-limbed robotic system has been developed for the purpose to investigate several aspects of climbing-system design.

11. Fixed wing aerial Mars rovers would have to fly at over 250 MPH just to stay aloft in the rarefied Mars atmosphere. This makes landing on the rocky surface almost impossible, thereby precluding sample inspection/gathering. Also, the high-speed flight means that dwell time on any particular area will be difficult-- a negative feature that is compounded by the fact that turns in the thin atmosphere will require enormous radii (Michelson & Reece, 1998). An electromechanical multimode (could land, take samples/recharge/or communicate, and then take off to continue the survey mission) insect has been developed at Georgia Tech Research Institute (Michelson and Reece, 1998; Cassanova, 2002). It even has the potential of returning to its launch point for refueling, downloading of data, or transferring of samples. This mechanical flying insect is known as Entomopter. The Entomopter concept uses the same lift generating means that insects do on Earth to generate lift within the Mars environment. Unlike aircraft or birds, insects generate lift by the continuous formation and shedding of vortices on their wings. This vortex formation and shedding produces very high wing lift coefficients on the order of 5 compared to maximum lift coefficients of 1 to 1.2 for conventional airfoils. This very high lift generating capability is what allows insects to fly, hover and maneuver as they do (Cassanova, 2002). Entomopter is based around a new development called Reciprocating Chemical Muscle (RCM) which is capable of generating autonomic wing beating from a chemical energy source. Through direct conversion, the RCM also provides small amounts of electricity for onboard systems and further provides differential lift enhancement on the wings to achieve roll

and hence, steered flight. An Entomopter can achieve high lift with rapidly flapping wings, and therefore allows the fuselage to move slowly in relation to the ground. The Entomopter will extend the rover's eyes and will allow the rover to choose its path ahead more intelligently. The rover will be able to move more rapidly with less risk.

For the Entomopter to fly, it will need to flap its wings at a specified rate, thereby producing and shedding the vortices that will generate the lift. The motion of the wings is a fairly power intensive process so a given flight mission for the Entomopter will be short (on the order of 10 to 15 minutes). These short flight times are due to the amount of fuel it is estimated the Entomopter can carry. Because of these short flight times, the Entomopter would need to be operated as part of a system. It is envisioned that this system would consist of one or more Entomopter vehicles that operate in conjunction with a base vehicle such as a lander or rover (see Figure 13). This base vehicle would provide refueling capability to the Entomopters as well as act as a data relay for the science data and samples the Entomopters collect.

12. Researchers have found that the flipper of the humpback whale is a more efficient wing design than the current model used by the aeronautics industry on airplanes. Engineers are working to apply the aerodynamic findings to future airplane and automotive design. Similarly, engineers at Airbus have used the rough skin of the shark as inspiration in developing a striated foil coating for the wings of aircraft, a design which has resulted in six percent less friction and improved fuel efficiency.

Figure 13. Animation and implementation model of Entomopter
Source: Cassanova, 2002

Animation of Entomopter mission.

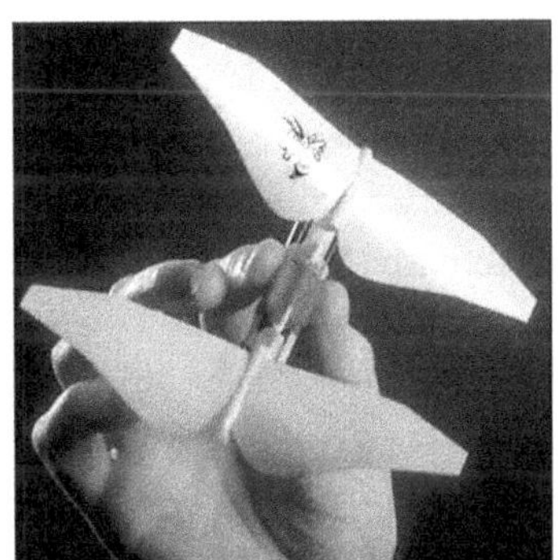

Stereolithographic kinematically correct model of the terrestrial Entomopter

Animation of sample collection.

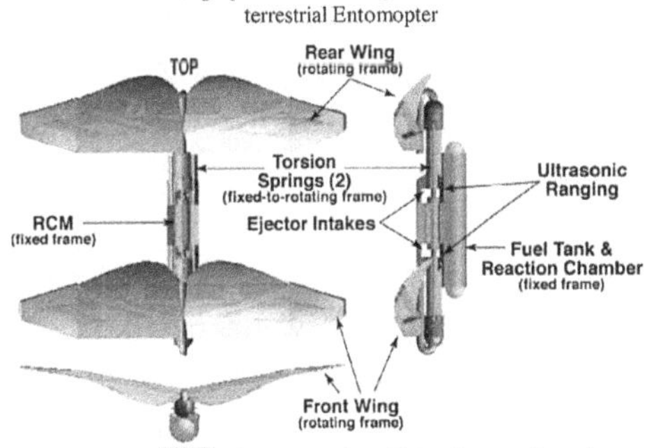

Entomopter-based Mars flyer configuration

13. Hair-cells are used for many biological functions such as, invertebrate hearing and equilibrium, insect flow sensing, fish lateral line flow sensing, and insect priopreceptors. Fish can hear and see, and they rely on a series of specialized flow sensors strung along the sides of their bodies, called the lateral line system. A fish natural lateral-line system consists of about 100 sense cells that run from the gills to the tail on each side of the fish with aim to provide guidance for synchronized swimming, predator and obstacle avoidance, and prey detection and tracking. The cells within the lateral-line detect subtle water movements, and from the different times that the individual cells are stimulated by these movements the fish's brain is able to reconstruct an image of what caused them and then help to navigate fish. A research team at the University of Illinois at Urbana-Champaign has built an artificial lateral line that could be integrated with a submarine or underwater robot aiming to extend their ability to detect and track moving underwater targets, and avoid collisions with moving or stationary objects. The artificial lateral line consists of an integrated linear array of micro fabricated flow sensors, with the sizes of individual sensors and spacing between them matching those of their biological counterpart.
14. Most of the developed underwater vehicles exclusively use propellers or jets to move in water and are powered by electromechanical devices. At the same time, most of marine habitants use radically different means of locomotion. Biomimetic studies and observations from fish and cetaceans have provided a wealth of information on how these animals employ their flapping tails and several fins to produce propulsive and maneuvering forces. Recent work on live animals has provided important information on the resulting flow structures (Fish & Hui, 1991; Videler, 1993, Drucker & Lauder, 1999; Mao et al, 2004). Enhanced propulsion for engineered systems may be possible by bio-inspired mechanisms involved with oscillating propellers, flexible wings, boundary layer stabilization, laminar flow maintenance, and active vortex control. In addition, natural propulsive systems can be self-stabilizing and self-correcting (Fish, 2006).
15. Modern vision sensors generate huge amount of data and therefore they need heavy computational power to process the information. Looking to have sensors that demand less processing power and are lighter, the vibrating eyes of the tropical jumping spiders have inspired a new breed of vision sensors that could give the next generation of Mars rover sharper eyesight. The spider vibrates its thin and long retina back and forth to see more clearly than would otherwise be possible. Researchers at California Institute of Technology in Pasadena built an array comprising 32 rows of 32 pixels each on a silicon image-sensor chip and clamped a metal frame onto it, and they mounted a lens inside the frame at a fixed distance to the surface of the chip, attached by four springs, while the lens focuses an image onto the chip. When the unit is shaken, the lens starts vibrating at the springs' resonant frequency, about 300 hertz, moving the image over the pixels. The effect of the vibration looks as if each pixel is moving elliptically, sampling a larger area of the image than if it were stationary. Each pixel generates a train of digital pulses from which a signal processor extracts features of the image using knowledge of the position of the springs. Experimental results show that, although the developed sensor is smaller than a matchbox, the vibrating sensor's images are almost as sharp as those of an immobile sensor with 256 by 256 pixels.
16. Spider silk is an unusually strong, resilient, and elastic fiber protein that is only surpassed by synthetic high-performance fibers. Silk fibers are nearly as strong as the manmade materials, and have an unbeatable capacity for absorbing energy, also called resilience, which can be quantified by the area under the stress-strain curve measured in a tensile test. This ability to store energy, and the fact that most of the energy is dissipated as the fiber deforms, enable spiders to intercept and

catch their prey, absorbing their kinetic energy. This property makes silk fibers attractive for many applications in which energy absorption is the design parameter. Another interesting feature of silk fibers is the way they are produced: spiders spin silk in an aqueous medium, at room temperature, and from common materials. In addition, silk is recyclable and biodegradable.

17. Many manmade drills are rotary. They typically require two motors and a lot of force or thrust while using them. But small animals like insects do not have very much weight, and hence they cannot exert a lot of thrust onto their drills. In addition, the wood wasp uses a long tube called an ovipositor to drill into the bark of a tree so that it can lay its eggs. The wasp's ovipositor mechanism has been converted into a novel type of drill. It is basically a percussive drill, rather than a rotary drill, and the initial tests yield higher efficiency than man-made drills. The developed drill may have applications in space missions.
18. Mountain goats have the ability to scale up steep and rocky terrain. What is going on in the goat's foot is very interesting and researchers are trying to answer how it gets traction on very steep surfaces, and consequently how a robot might achieve similar poise.
19. Teeth are complex in design and have a lot of structural features, such as geometry, contact surface, orientation of dentin, depth of enamel, etc. All these features are strongly connected with mechanical properties of teeth to reach optimal results. Fully understanding and analysis of these characteristics can aid in further designing restorative materials and composite materials.
20. Nature helps to explore an impact-resistant armor with higher strength structure than contemporary armors found on tanks and shields. The new armor is modeled after the ancient design of the abalone shell. Researchers discovered the intricate micro-architecture that makes up the shell's mother-of-pearl appearance.
21. Bell Labs scientists reported that a deep-sea sponge known as euplectella produces optical fibers similar to those found in telecommunications networks. The natural fibers aren't transparent enough for real-world applications, but they are far more durable than commercial optical fiber. Researchers at MIT have also found inspiration in abalone. They teased out the biochemical process the mollusk uses to build its shell. Supported with that knowledge, they genetically engineered a virus to coat itself in a shell of semiconducting nanoparticles. Burning away the virus's "scaffold" causes the nanoparticles to fall together and form a nanowire (Mao et al., 2004).
22. Water harvesting system to recycle steam from cooling towers and allowing buildings to collect their own water supplies from the air was inspired by the way the Namib Desert Beetle of Namibia harvests water from desert fogs. The beetle lives in a driest location that receives a mere half an inch of rain a year yet can harvest water from fogs that blow in gales across the land several mornings each month (Binns, 2006). The ability of Stenocara beetle to pull water from fog is due to bumps on its wing scales that have high tolerance of heat, water-loving tips and water-shedding sides. As morning fog sweeps across the desert floor, the water sticks to the peaks bumps, eventually forming droplets. The bumps are surrounded by waxy water-repelling channels. When a bump collects enough water to form a big droplet, that is thinner than a human hair and is heavy enough, it rolls down from the top of the peaks and is channeled to a spot on the beetle's back that leads straight to its mouth. The Namib Desert beetle has dueling surfaces on its back. The smooth parts repel water, while the top bumpy parts collect water. University of Oxford and the UK defense research firm QinetiQ (UK) have developed plastic water-harvesting sheets that mimic the beetle's bumps by dipping glass into solutions of charged polymers, imitating the porous and smooth coating of

> the beetle. The material is then coated with a Teflon-like substance, making it water-repellent. The developed material is useful for capturing water in cooling towers and industrial condensers, arid agricultural systems, and buildings in fog-rich areas. The results based on the developed prototypes of such surfaces have shown that these surfaces are several times more effective than "net harvesting" methods used to collect water from fog in remote mountain towns of Peru and Chile. Net harvesting is done by using fine polypropylene meshes stretched between two poles. Fog passes through the mesh, forming large drops of water which run down into a gutter leading to a reservoir. Net harvesting is a fairly crude and inefficient way of collecting water.

In addition, researchers studying the infrared reflectors on the beetle's shell that may help it survive in the Desert by reflecting heat. Such reflectors could possibly lead to the production of heat-resistant materials useful for many applications such as rocket design. Furthermore, scientists at M.I.T. are looking to add new chemical functionalities to spray coatings have turned to the beetle for inspiration (Cohen, 2008). They have created a material that can capture and control tiny amounts of water.

CONCLUSION

Biomimetics's approach as new multidisciplinary field proves through research, development and applications that represent an important source of ideas, solutions and technological advancement in many fields including the development of innovative robotics. The inspiration of nature is expected to continue leading to technological improvements and impact is expected to be felt in every aspect of human life. By studying the biological system, one may be able to understand and derive relevant principles and use the strength of these principles to help solve non-biological problems.

The biggest challenge facing the development of Biomimetics robots and systems is the currently available paradigm of technology, materials and methods of fabrication as they are still in their infancy compared to nature's evolution. When energy, economy, and greater locomotor performance are desired in engineered systems, innovative and bioinspired solutions from nature may serve as good alternative for new technologies. Biomaterials are expected to become the dominant focus of materials research, as it would lead to down-sizing of engineered smart components and the up-scaling incorporation of biomimetic concepts and processes.

For the purpose to achieve efficient and innovative design and development inspired by nature, it is important to have proper understanding of nature evolution, challenges, effectiveness and constraints of biological solutions and understand different evolutionary strategies with the possibility to integrate multiple source of information and conclude successful design and development approaches and solve current and future problems facing the world. Finally, Biomimetics inspired research and development required collaboration of experts of different field such as, biologists, engineers, chemists, and physicists to collaborate in a scientific context aiming to develop new and innovative systems design, technology, smart materials, etc.

REFERENCES

Anderson, A. M., Streitlien, K., Barrett, D. S., & Triantafyllou, M. S. (1998). Oscillating Foils of High Propulsive Efficiency. *Journal of Fluid Mechanics*, *360*, 41–72. doi:10.1017/S0022112097008392

Armstrong, W. P. (2007). Do Pearls Come From Coconuts? *The Drifting Seed*, *13*(1), 13–17.

Aschenbeck, K. S., Kern, N. I., Bachmann, R. J., & Quinn, R. D. (2006). Design of a Quadruped Robot Driven by Air Muscles. *The IEEE/RAS-EMBS International Conference on Biomedical Robotics and Biomechatronics (BioRob'06)*, 875-880. doi:10.1109/BIOROB.2006.1639201

Ayers, J., Wilbur, C., & Olcott, C. (2000b). Lamprey Robots. In *The International Symposium on Aqua Biomechanisms*. Tokai University.

Ayers, J., Witting, J., McGruer, N., Olcott, C., & Massa, D. (2000a). Lobster Robots. In *The International Symposium on Aqua Biomechanisms*. Tokai University.

Bar-Cohen, Y. (Ed.). (2001). Electroactive Polymer (EAP) Actuators as Artificial Muscles: Reality, Potential and Challenges. SPIE Press.

Bar-Cohen, Y. (2002, January). Biologically Inspired Robots as Artificial Inspectors. *e-Journal of NDT*, *7*(1).

Bar-Cohen, Y. (2003). Biologically Inspired Intelligent Robotics. *The SPIE Smart Structures Conference*.

Bar-Cohen, Y. (2011). *Biomimetics: Nature-Based Innovation*. CRC Press.

Benyus, J. (1997). *Biomimicry: Innovation Inspired by Nature*. New York: William Morrow & Company, Inc.

Bhattacharya, S., & Agrawal, S. K. (2000). Spherical rolling robot: A design and motion planning studies. *IEEE Transactions on Robotics and Automation*, *16*(16), 835–839. doi:10.1109/70.897794

Binns, C. (2006). New Material Collects Water From Thin Air. *LiveScience*. Retrieved from http://www.livescience.com/animals/060616_beetle_material.html

Buehler, M. (2000). Dynamic Locomotion with One, Four and Six-Legged Robots. *Journal of the Robotics Society of Japan*, *20*(3), 15–20.

Cassanova, R. A. (2002). *Planetary Exploration using Biomimetics an Entomopter for Flight on Mars*. Phase II Final Report, NASA Institute for Advanced Concepts Project NAS5-98051, October 2002.

Cham, J. G., Karpick, J. K., & Cutkosky, M. R. (2004). Stride period adaptation for a biomimetic running hexapod. *The International Journal of Robotics Research*, *23*(2), 141–153. doi:10.1177/0278364904041323

Cherouvim, N., & Papadopoulos, E. (2006). Energy Saving Passive-Dynamic Gait for a One-Legged Hopping Robot. *Robotica*, *24*(04), 491–498. doi:10.1017/S026357470500250X

Clark, J. E., Cham, J. G., Bailey, S. A., Froehlich, E. M., Nahata, P. K., Full, R. J., & Cutkosky, M. R. (2001). Biomimetic Design and Fabrication of a Hexapedal Running Robot. *The IEEE International Conference on Robotics and Automation (ICRA'2001)*, 4, 3643–3649. doi:10.1109/ROBOT.2001.933183

Cohen, R. (2008). *Beetle-Based Water Harvesting*. Retrieved from http://www.rexresearch.com/qinetiq/qinetiq.htm

Drucker, E. G., & Lauder, G. V. (1999). Locomotor Forces on a Swimming Fish: Three-Dimensional Vortex Wake Dynamics Quantified using DPIV. *The Journal of Experimental Biology*, *202*, 2393–2412. PMID:10460729

Fish, F. E. (2006). Limits of Nature and Advances of Technology: What does Biomimetics have to Offer to Aquatic Robots? *ABBl*, *3*(1), 49–60.

Fish, F. E., & Hui, C. A. (1991). Dolphin Swimming: A Review. *Mammal Review*, *21*(4), 181–195. doi:10.1111/j.1365-2907.1991.tb00292.x

Galvez, J. A., Gonzalez de Santos, P., & Pfeiffer, F. (2001). Intrinsic Tactile Sensing for the Optimization of Force Distribution in a Pipe Crawling Robot. *IEEE/ASME Transactions on Mechatronics*, *6*(1), 26–35. doi:10.1109/3516.914388

Geim, A. K., Dubonos, S. V., Grigorieva, I. V., Novoselov, K. S., Zhukov, A. A., & Shapoval, S. Y. (2003). Microfabricated Adhesive Mimicking Gecko Foot-Hair. *Nature Materials*, *2*(7), 461–463. doi:10.1038/nmat917 PMID:12776092

Goodsell, D. S. (1996). *Our molecular Nature: The body's Motors*. Copernicus, NY: Machines and Messages. doi:10.1007/978-1-4612-2336-8

Graham-Rowe, D. (2003, May). Fancy A Walk on the Ceiling? *New Scientist*, *178*(2395), 15.

Graighead, H. G. (2000). Nanoelectromechanical Systems. *Science*, *290*(5496), 1532–1535. doi:10.1126/science.290.5496.1532 PMID:11090343

Habib, M. K. (2011). Biomimetics: Innovations and Robotics. *Int. J. Mechatronics and Manufacturing Systems*, *4*(2), 113–134. doi:10.1504/IJMMS.2011.039263

Habib, M. K., & Watanabe, K., & Izumi, K. (2007a). Biomimetics Intelligent Robots and Biological Inspiration. *International Conference on Control, Instrumentation and Mechatronics 'CIM'2007'*, 824-832.

Habib, M. K., Watanabe, K., & Izumi, K. (2007b). Biomimetics Robots: From Bio-inspiration to Implementation. *The 33rd Annual Conference of the IEEE Industrial Electronics Society IECON'2007*, 143-148.

Hirose, S. (1993). *Biologically inspired robots (snake-like locomotor and manipulator)*. Oxford University Press.

Hirose, S., & Fukushima, E. F. (2002). Snakes and Strings: New Robotic Components for Rescue Operations. *The 41st Annual SICE Conference*, 338-343.

Hirose, S., Nagakubo, A., & Toyama, R. (1991). Machine that can Walk and Climb on Floors, Walls and Ceilings. *The 5th International Conference on Advanced Robotics*, 753-758.

Huang, T. J., Liu, Y., Brough, B., Flood, A. H., Bonvallet, P., Tseng, H.-R., ... Ho, C.-M. (2005). *A Nano-Chemo-Mechanical Actuator Based on Artificial Molecular Machines*. Miami, FL: MEMS.

Iida, F. (2003). Biologically Inspired Visual Odometer for Navigation of a Flying Robot. *Journal of Robotics and Autonomous Systems*, *44*(3-4), 201–208. doi:10.1016/S0921-8890(03)00070-8

Jack, A. J. (2001). Inflatable Robotics for Planetary Applications. *The 6th International Symposium on Artificial Intelligence, Robotics, and Automation in Space*. I-SAIRAS, Canadian Space Agency.

Jones, B. A., McMahan, W., & Walker, I. D. (2006). Practical Kinematics for Real-Time Implementation of Continuum Robots. *The 2006 IEEE International Conference on Robotics and Automation (ICRA'2006),* 1840-1847.

Kim, S., Clark, J. E., & Cutkosky, M. R. (2004). iSprawl: Autonomy, and the Effects of Power Transmission. *The 7th International Conference on Climbing and Walking Robots CLAWAR'2004*.

Kim, S., Spenko, M., Trujillo, S., Heyneman, B., Mattoli, V., & Cutkosky, M. R. (2007). Whole body adhesion: Hierarchical, Directional and Distributed control of Adhesive forces for a climbing robot. *The IEEE International Conference on Robotics and Automation (ICRA'2007)*, 1268-1273. doi:10.1109/ROBOT.2007.363159

Klaassen, B., Linnemann, R., Spenneberg, D., & Kirchner, F. (2003). Biologically Inspired Robot Design and Modeling. *The International Conference on Advanced Robotics*, 576-581.

Knight, A. (2009). *Hidden Histories: The Story of Sustainable Design*. Retrieved from http://www.csa.com/discoveryguides/design/review.pdf

Kubow, T. M., & Full, R. J. (1999). The Role of the Mechanical System in Control: A Hypothesis of Self-Stabilization in Hexapedal Runners. *Philosophical Transactions of the Royal Society of London. Series B, Biological Sciences*, *354*(1385), 849–861. doi:10.1098/rstb.1999.0437

Lewinger, W. A., Harley, C. M., Ritzmann, R. E., Branicky, M. S., & Quinn, R. D., R. D (2005). Insect-like Antennal Sensing for Climbing and Tunneling Behavior in a Biologically-Inspired Mobile Robot. *The IEEE International Conference on Robotics and Automation (ICRA'05)*, 4176-4181. doi:10.1109/ROBOT.2005.1570761

Lewinger, W. A., Watson, M. S., & Quinn, R. D. (2006). Obstacle Avoidance Behavior for a Biologically-Inspired Mobile Robot using Binaural Ultrasonic Sensors. *The IEEE/RSJ International Conference on Intelligent Robots and Systems (IROS'06)*, 6. doi:10.1109/IROS.2006.282461

Lorenz, R. D., Jones, J. A., & Wu, J. J. (2003). Mars Magnetometry from a Tumbleweed Rover. *IEEE Aerospace Conference*, 597-602.

Luk, B. L., Collie, A. A., Piefort, V., & Virk, G. S. (1996). Robug III: A Tele-Operated Climbing and Walking Robot. *UKACC International Conference on Control*, 1, 347-352. doi:10.1049/cp:19960577

Madangopal, R., Khan, Z. A., & Agrawal, S. K. (2005). Biologically Inspired Design of Small Flapping Wing Air Vehicles using Four-Bar Mechanisms and Quasi-Steady Aerodynamics. *Journal of Mechanical Design*, *127*, 809–816.

Mao, C., Solis, D. J., Reiss, B. D., Kottmann, S. T., Sweeney, R. Y., Hayhurst, A., ... Belcher, A. M. (2004). Virus-Based Toolkit for the Directed Synthesis of Magnetic and Semiconducting Nanowires. *Science, 303*(5655), 213–217. doi:10.1126/science.1092740 PMID:14716009

Michaud, F., & Caron, S. (2002). Roball, the Rolling Robot. *Journal of Autonomous Robots, 12*(2), 211–222.

Michelson, R. C., & Naqvi, M. A. (2003). Beyond Biologically-Inspired Insect Flight. von Karman Institute for Fluid Dynamics RTO/AVT Lecture Series on Low Reynolds Number Aerodynamics on Aircraft Including Applications in Emerging UAV Technology.

Michelson, R. C., & Reece, S. (1998). Update on Flapping Wing Micro Air Vehicle Research-Ongoing Work to Develop a Flapping Wing, Crawling Entomopter. *13th Bristol International RPV/UAV Systems Conference*, 30.1-30.12.

NASA-JPL. (2000). *Leaping into the Future: One Hop at a Time*. Retrieved from http://www.jpl.nasa.gov/releases/2000/frog.html

NASA-JPL. (2006). *Limber Robot Might Hitchhike to Space*. Retrieved from http://www.jpl.nasa.gov/news/features.cfm?feature=1101

Neuhaus, P. D., O'Sullivan, M., Eaton, D., Carff, J., & Pratt, J. E. (2004). Concept Designs for Underwater Swimming Exoskeletons. *The IEEE International Conference on Robotics and Automation (ICRA'2004)*, 4893-4898. doi:10.1109/ROBOT.2004.1302493

Neves, J. C. L., & Francke, I. C. M. (2012). Creative Product Design Using Biomimetics. *WIT Transactions on Ecology and the Environment, 160*, 149–155. doi:10.2495/DN120141

Ohno, H., & Hirose, S. (2000). Study on Slim Slime Robot. *Proceedings of RSJ/IEEE International Conference on Intelligent Robots and Systems (IROS'2000)*. 2218-2223.

Papanek, V. (1971). *Design for the real world: human ecology and social change*. London: Thames and Hudson.

Park, J. H., & Kim, K. D. (1998). Biped Robot Walking Using Gravity-Compensated Inverted Pendulum Mode and Computed Torque Control. *The IEEE International Conference on Robotics and Automation (ICRA'1998)*, 3528–3533. doi:10.1109/ROBOT.1998.680985

Paulson, L. D. (2004). *Biomimetic Robots*. IEEE Computer.

Raibert, M. H. (1986). *Legged Robots that Balance*. Cambridge, MA: MIT Press.

Sarfati, J. (2003). Gecko Foot Design—Could it Lead to a Real Spiderman'? *Creation, 26*(1), 54–55.

Sauvage, J.-P., & Dietrich-Buchecker, C. (1999). Molecular Catenanes Rotaxanes and Knots. Wiely-VCH. doi:10.1002/9783527613724

Schwind, W. J., & Koditchek, D. E. (1995). Control of Forward Velocity for a Simplified Planar Hopping Robot. *The IEEE International Conference on Robotics and Automation (ICRA'1995)*, 691–696. doi:10.1109/ROBOT.1995.525364

Shen, W. M., Will, P., Galstyan, A., & Chuong, C. M. (2004). Hormone-Inspired Self-Organization and Distributed Control of Robotic Swarms. *Autonomous Robots*, *17*(1), 93–105. doi:10.1023/B:AURO.0000032940.08116.f1

Simmons, R., & Krotkov, E. (1991). An Integrated Walking System for the Ambler Planetary Rover. *The IEEE International Conference on Robotics & Automation (ICRA'1991)*, 2086–2091. doi:10.1109/ROBOT.1991.131935

Triantafyllou, M. S., Techet, A. H., & Hover, F. S. (2004). Review of Experimental Work in Biomimetic Foils. *IEEE Journal of Oceanic Engineering*, *29*(3), 585–594. doi:10.1109/JOE.2004.833216

Unver, O., Uneri, A., Aydemir, A., & Sitti, M. (2006). Geckobot: A Gecko Inspired Climbing Robot using Dry Adhesives. *The IEEE Robotics and Automation Conference (ICRA'2006)*, 2329-2335.

Videler, J. (1993). *Fish Swimming*. London, UK: Chapman and Hall. doi:10.1007/978-94-011-1580-3

Viollet, S., & Franceschini, N. (2001). *Aerial Minirobot that Stabilizes and Tracks with a Bio-Inspired Visual Scanning Sensor. In* B. Webb & T. Consi (Eds.), *Biorobotics* (pp. 67–73). Cambridge, MA: MIT Press.

Yamada, H., Chigisaki, S., Mori, M., Takita, K., Ogami, K., & Hirose, S. (2005). Development of Amphibious Snake-like Robot ACM-R5. In *Proceedings of 36th International Symposium on Robotics (ISR'2005).* TH3C4.

Yan, J., Wood, R. J., Avadhanula, S., Sitti, M., & Fearing, R. S. (2001). Towards Flapping Wing Control of a Micromechanical Flying Insect. *The IEEE Robotics and Automation Conference (ICRA'2001)*, 3901-3908. doi:10.1109/ROBOT.2001.933225

Yurtkuran, S., Kirh, G., & Taneli, T. (2013). Learning from Nature: Biomimetic Design in Architectural Education. *Procedia: Social and Behavioral Sciences*, *89*, 633–639. doi:10.1016/j.sbspro.2013.08.907

Chapter 2
Investigation of Human Locomotion With a Powered Lower Limb Exoskeleton

Sergey Fedorovich Jatsun
Southwest State University, Russia

Andrey Yatsun
Southwest State University, Russia

Sergei Savin
Southwest State University, Russia

ABSTRACT

In this chapter, the lower limb exoskeleton is studied. The roles of the exoskeleton both as a measurement device for studying human locomotion and as an assistive device that restores the human ability to walk are discussed. Particular attention is given to the investigation of the role of the pressure sensors and other devices that allow us to measure normal reactions at the contact points with the supporting surface and also detect these contacts. The way the geometry of the supporting surface affects the sensors system of the robot is considered, and new designs for feet sensor system are proposed. These include elastic foot, a foot with actuated sensors, and a foot with spring-damper systems.

INTRODUCTION

Nowadays there are tens of millions of people who live with the injuries of the musculoskeletal system. It is often, but not always, possible to restore the function of musculoskeletal system with special training devices (Bogue, 2009; Mikołajewska & Mikołajewski, 2011; Tung 2013). One of the ways to restore walking ability to the patient is based on the use of lower limb exoskeleton, which allow the user to perform complicated motions, such as verticalization, walking and others (Bogue, 2015; Ferrati et al., 2013; Pons, 2008; Carpino et al., 2013).

DOI: 10.4018/978-1-5225-2993-4.ch002

The development of devices that allow the user who is suffering from a dysfunction of musculoskeletal system to move is being done in USA, Japan, South Korea, Germany, Russia and other countries. While the attempts to create exoskeletons had been made for a long time only relatively recently there had been a progress in creating vertically stable prototypes. One of the main design issues is ensuring the vertical balance of the system (Barbareschi et al., 2015; Wang et al., 2013; Jatsun et al., 2016a, 2016b; Panovko et al., 2016). Because of this issue the development of lower limb exoskeletons needs to be based on theoretical studies of the vertical balancing problem and on walking simulation results.

At the moment there are two different types of exoskeletons that gain prominence. First type represents combined motion scheme, where the vertical balance is achieved using crutches help by the exoskeleton's user. Examples of such devices are Ekso Bionics, ReWalk, and ExoAtlet (Esquenazi et al., 2012; Gwynne, 2013), developed and produced by laboratories and commercial companies.

This motion type requires the user to be well trained, having developed shoulder and arm muscles, thus limiting the number of patients who are capable of using it. It also impedes the walking comfort.

The second type of exoskeletons is different as there the exoskeleton itself allows the system to remain vertically balanced, without requirement for the user to play any active role in the motion. Therefore this type of exoskeleton does not require the use of crutches. Such devices have been developed by company Rex Bionics, shown in Figure 1 on the left (Viteckova et al., 2013; Contreras-Vidal & Grossman, 2013), and at the Robotics and Mechatronics lab at Southwest State University, where ExoLite lower limb exoskeleton had been developed, shown in figure 1 on the right (Jatsun et al., 2015, 2016c, 2016d). These two devices allow the user to walk comfortably without crutches.

These exoskeletons can be used to perform verticalization, walking straight forward and backwards, walking upstairs and performing other types of motions while leaving the hands free (Jatsun et al., 2016e, 2016f, 2016g). Development of this type of exoskeletons is a difficult task as their development is based on the study of human walking biomechanics.

Figure 1. General view of Rex and ExoLite exoskeletons

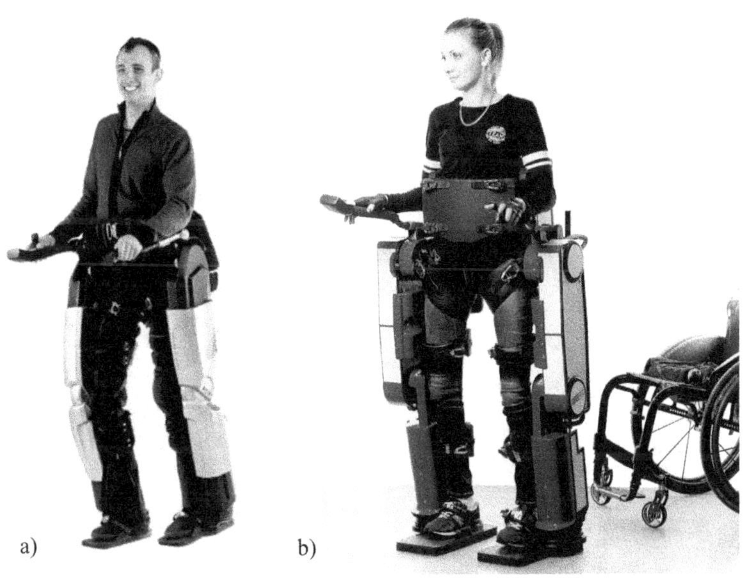

a) b)

HUMAN WALKING BIOMECHANICS

There are different methods that allow to study human walking motion. Walking kinematics is studied using contact-based and contactless sensors that measure joint angles. This can be done using micro-electromechanical gyroscopes, accelerometers and inertial measurement units. These devices allow to determine orientations of the parts of the human body. An important method for studying kinematics of human walking motion is marker technology (Jatsun et al., 2016h), that allows to track the positions of visual markers using cameras and combine this data with input from various sensors placed on the body. It is also possible to use specially constructed exoskeletons designed to measure and record motions of the user. An example of such exoskeletons is ExoMeasurer device, produced in Southwest State University (see Figure 2).

Additionally to inertial measuring units this device has encoders at each joint and pressure sensors at its feet.

Dynamical characteristics of walking are studied with so-called force platform (Herzog et al., 1989; Jung et al., 2014; de Castro et al., 2014). When feet are resting on the force platform, it can, depending on its design, register vertical and horizontal components of reaction force. To register forces acting at different points on the foot, pressure sensors mounted in shoe soles are used (de Castro et al., 2014).

When walking humans put their weight on one leg and then the other, cyclically. The leg that currently supports human weight is called supporting leg. The other leg that is moving forward is called swing leg. The period when both feet are in contact with the surface is called a double support phase. Single support phase and double support phase together form a step cycle. The period when each leg went through swing phase is called a double step cycle.

Figure 2. General view of ExoMeasurer device in use

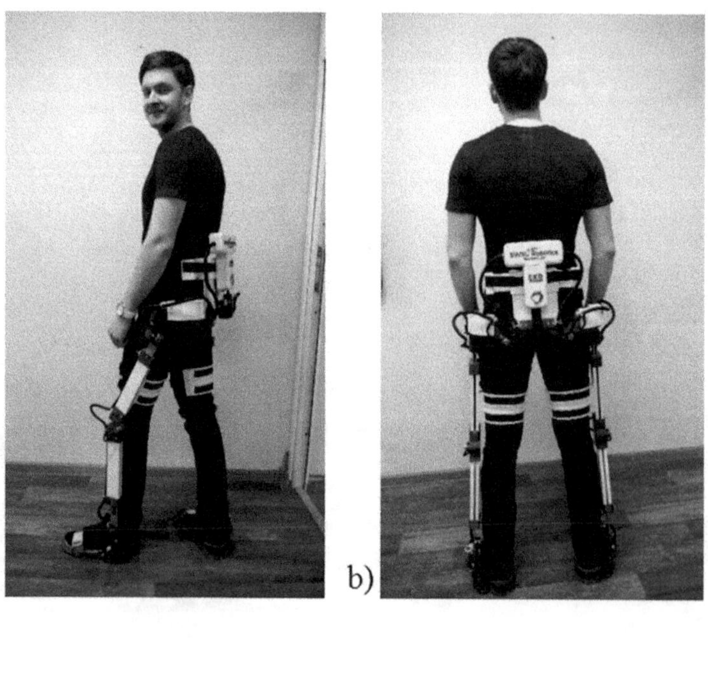

a) b)

Normal dynamic walking is characterized by 60% time spent in double support phase (both feet touch the ground) and 40% spent in swing phase. In so-called static walk the time spent in double support phase goes up to 90%.

Let us consider body movements in sagittal plane during step cycle. The beginning of the cycle corresponds to the heel landing on the ground. In the case when the right foot starts and ends the step cycle, it is then called a double step. The time it takes is determined by rhythmicity (as defined in the following paragraph). The double step includes two support phases and two swing phases.

REACTION FORCE AND PARAMETERS GOVERNING THE HUMAN GAIT

There are a number of works on biomechanics that study human locomotion (Inman et al., 1981; Rose & Gamble, 1994; Whittle, 2014; Winter, 2009). Below we give a few definitions.

Reaction force (RF) — is the force acting on the body from the direction of the support. If when standing the reaction force equals the weight of the body, then inertial forces are added to this force. The reaction of the support is a force distributed over a certain area, but it is approximately possible to consider this force as concentrated at one point. The point of application of the vector of the reduced reaction force at the foot is called the pressure center.

Reaction forces can be measured when walking using special sensors. Comparison of the reaction force and step kinematics makes it possible to evaluate the magnitude of the torque at a particular joint of the lower extremities using mathematical modeling methods.

The dynamic characteristics of gait are evaluated by studying reaction forces that reflect the interaction of gravitational and inertial forces involved in the structure of the locomotor act. The minimum value of the reaction force at different speeds as a percentage of body weight is as follows: at a slow speed - about 100%, at an arbitrary speed - 70% and at a rapid speed - 40%.

The walking process is characterized by the movement of the center of mass from the heel to the toe, while contact between the heel and the supporting surface is more point like, so the pressure force registered by the sensor in the heel is greater than the readings of the other two sensors. As the roll passes through the heel, most of the load moves to the medial part of the heel, then the point of application of the reduced normal reaction is shifted to the second metatarsal bone, which is typical for the support phase on the entire foot. The load moves to the first metatarsal bone and the big toe in the support phase at the front section. Folding the first finger and pushing away from the support completes the support step phase.

The pattern of support reaction is largely determined by the functional state of the lower limb muscles and the innervations of the gait (Dubrovkiy & Fedorova, 2008). It is known that the parameters determining walking are: the line (trajectory) of the displacement of the center of mass of the body, step length, turning angle of the foot, support base and the speed of movement and the rhythm of walking (Herzog et al., 1989).

The support base is the distance between two parallel lines drawn through the centers of the heel support which are perpendicular to the direction of movement. The support base determines the stability of the human body.

The foot turn is the angle formed by the line of movement and the line passing through the the center of the heel and the point between the first and second finger. The greater the foot turn, the greater the support base, but the less efficient is the walking (and vice versa).

The step length is the distance between the center of the heel of one leg and the center of the heel of the opposite leg.

The rhythmicity is the ratio of the duration of the swing phase of one leg to the length of the swing phase of the other leg. It determines the number of steps per minute. For an adult it is around 113 steps per minute (Dubrovkiy & Fedorova, 2008).

The walking speed is the number of big steps per unit time measured in steps per minute or kilometers per hour.

Figure 3 shows a diagram of a two-coordinate anthropomorphic foot that allows rotation of the human foot relative to the shin with respect to two axes, whose point of intersection coincides with the center of the ankle joint. The position of the foot of the exoskeleton 1 in space is determined by the coordinate system *Oxyz* relative to the fixed coordinate system $O_0x_0y_0z_0$. One of the axes is directed along the axis of rotation of the talus joint of the foot and enables supination and pronation, while the second is located in the transverse plane to enable rotation of the foot. This design is biologically inspired and allows the patient's foot to make movements close to the normal human movements.

It is important to determine the position of the center of mass of a person relative to the foot as a supporting element. Figure 4 (a) illustrates the displacement of the center of mass of the man-machine system when walking. The main difference of this gait is that the exoskeleton's foot practically does not rotate, but performs translational motion (Figure 4(b)). In this case, contact with the surface occurs immediately with the entire foot, and the transfer phase consists in rotating the leg of the support leg relative to the foot. This circumstance limits the step length and the speed of movement in the exoskeleton, since the angle of rotation of the shin is limited by the possibilities of the ankle joint. The supporting phase of such a walk considerably exceeds the duration of the transfer phase.

The pictures above show so-called flat feet walking, where the foot is parallel to the supporting surface prior to landing. Figure 5 illustrates this type of walking showing experimental data for two experiments: walking on horizontal surface and walking on inclined surface.

Figure 3. Two-coordinate anthropomorphic foot (left): 1 - stop, 2 - ankle retainer bracket, 3 - ankle retainer, 4 - shin, γ - angle of subtalar joint

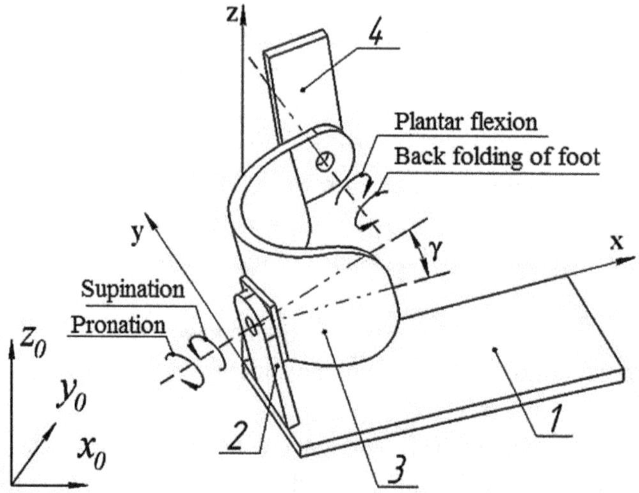

Figure 4. a) Displacement of the center of mass of the human-machine system while walking; b) he positions of the human foot during a step in the exoskeleton

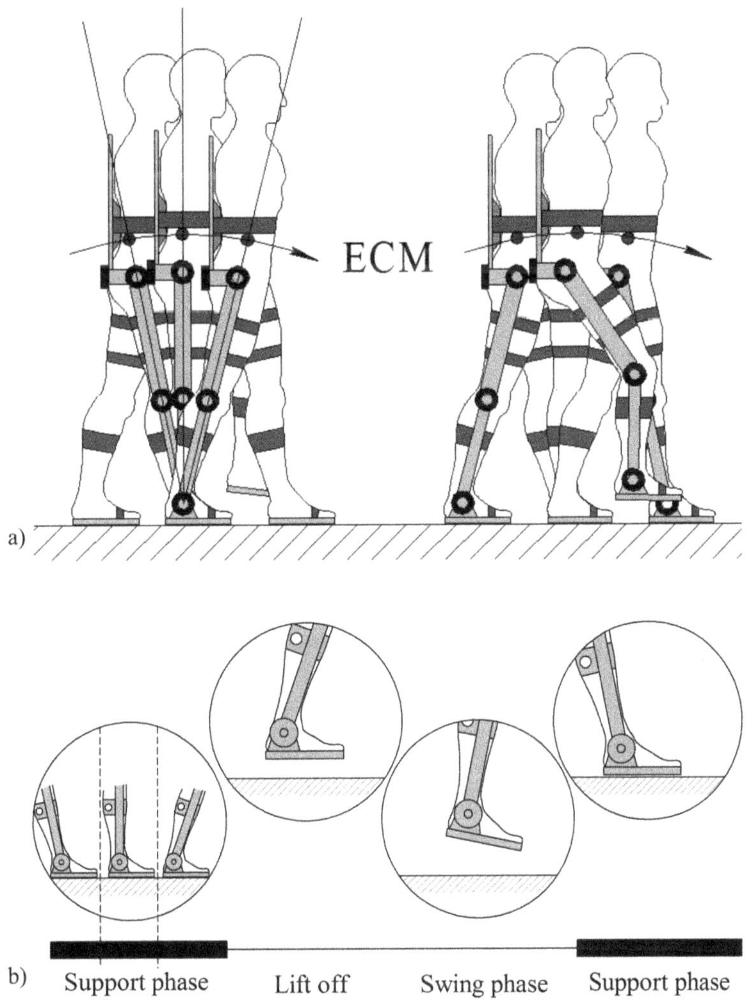

We can observe rapid changes in foot angles before it rests on the supporting surface. These changes are corresponding to the transition between heel landing and whole foot connecting to the ground. On the inclined surface this dynamics is absent, as the foot comes in contact with the supporting surface being parallel to it. These effects are varying between individuals and depend on the form of the rigid shoe.

DEFINITION OF THE CONCEPTS OF STABILITY AND EQUILIBRIUM

In the general case, equilibrium or balance means a state of a system in which the impacts on it are compensating each other. Equilibrium can be stable or unstable. Stability is the ability of the system to maintain its current state under the influence of external perturbation.

Figure 5. Time function for angle defining foot orientation in sagittal plane; a) human walking on horizontal surface, b) human walking on inclined surface

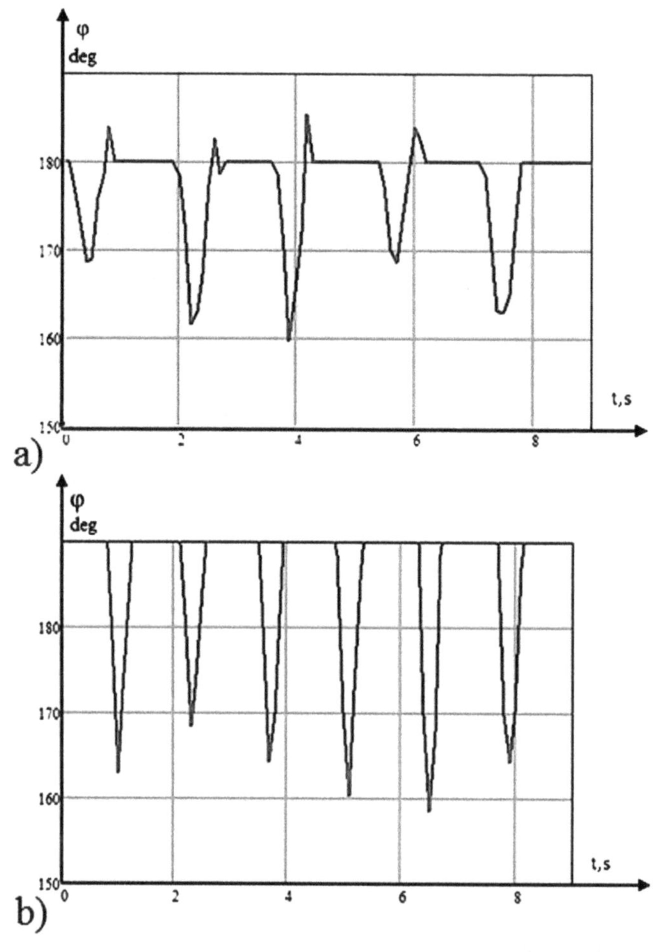

The motion of controlled bioelectromechanical systems, such as exoskeletons, depends on the acting forces and the initial conditions, i.e. on the positions and speeds of points of the system at the moment of the beginning of the motion. Knowing these forces and the initial conditions makes it possible to calculate how the system will move. That calculated motion is called unperturbed.

However, since all measurements are made with some degree of accuracy, the true values of the initial conditions will usually be slightly different in practice from the calculated ones. The deviations of the initial conditions from their calculated values that arise due to various reasons are called initial perturbations. In addition, the forces from the external environment caused by factors unaccounted for in the model also act on the system during motion. The motion performed by the system in the presence of these perturbations is called the disturbed motion.

The issues of ensuring the patient's stability during movement with the help of an exoskeleton are discussed in detail in a number of works (Panovko et al., 2016; Jatsun et al., 2015). It is important to clearly define the concept of "stability of the vertical position of the exoskeleton". For this purpose, we distinguish two types of movements: slow and fast. It is possible to determine these motions by introducing a dynamics coefficient, which we define by the formula:

$$v = \frac{\Phi}{Mg}, \quad (1)$$

where v is the dynamic coefficient, Φ is the modulus of the reduced inertia force and Mg is the weight of the exoskeleton.

Let us consider the case of slow motions, in which the inertia forces caused by the accelerated motion of the links of the exoskeleton are small as compared to the gravitational forces, i.e. the dynamic coefficient v is close to zero. For simplicity, let us consider the movement of the exoskeleton in the sagittal plane. The center of mass position (CMP) of the exoskeleton for slow motions is shown in Figure 6.

The condition for maintaining the vertical stability of the exoskeleton is:

$$x_2 \geq x_c \geq x_1, \quad (2)$$

where x_1, x_2 are the coordinates of the edge points of the support surface along the x-axis. That is, the projection of CMP on the horizontal plane must be inside the support surface.

The concept of the "stability of the vertical position of the exoskeleton" can be related to the position of the center of mass relative to the support surface. If the projection of the center of mass is inside the support surface, then the exoskeleton is stable and if not – unstable.

The nature of the behavior of the exoskeleton changes if motion which causes acceleration occurs (as shown in figure 7). In the general case, the forces of inertia are distributed, and are estimated by reducing these forces to the center of mass to obtain the known principal inertial forces vector given by the formula:

$$\vec{\Phi} = -M\frac{d^2}{dt^2}\vec{r}_c \quad (3)$$

Figure 6. The position of the center of mass of the exoskeleton with insignificant accelerations (static balance)

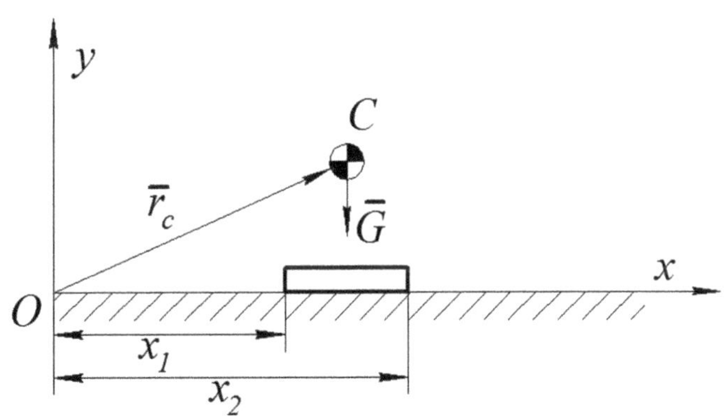

Figure 7. The position of the center of mass of the exoskeleton under significant accelerations (dynamic balance)

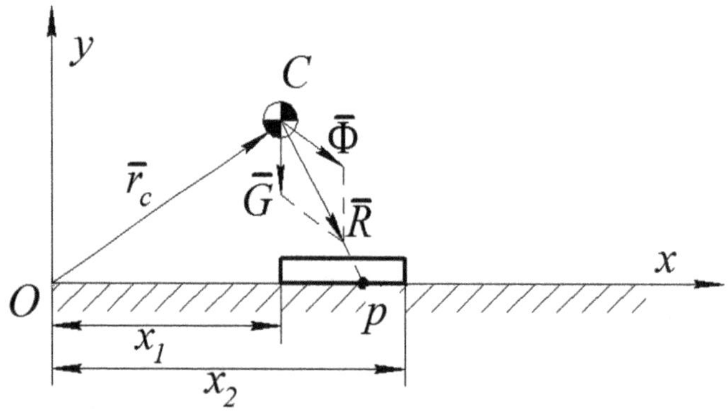

where $\vec{\Phi}$ — is the inertial forces vector expressed in N, M — total mass of the exoskeleton in kg, $\frac{d^2}{dt^2}\vec{r}_c$ is the acceleration of the center of mass in m/s^2, $\vec{G}$ - force of gravity acting on the mechanism in N, $\vec{\Phi} = \begin{bmatrix} \Phi_x \\ \Phi_y \end{bmatrix}$, $\vec{G} = \sum m_i \vec{g}$.

The value of inertia forces during walking can be significant and comparable to the gravitational forces. The direction of the main vector and the principal moment of the inertia forces vary with time and depend on the type of movement performed by the exoskeleton. To assess the stability of the exoskeleton in the vertical position, taking into account the inertial forces, Vukobratovich et al. (2004, 2012) proposed the concept of the ZMP (zero momentum point) determined by the formula:

$$p = x_c - \frac{\Phi_x}{G + \Phi_y} \cdot z_c, \tag{4}$$

$$x_2 \geq p \geq x_1,$$

where p — zero momentum point, expressed in m, $\vec{r}_c = \begin{bmatrix} x_c & z_c \end{bmatrix}^T$ — is the radius vector defining the position of the center of mass of the mechanism,

Both the CMP and ZMP are virtual points whose positions in space can only be determined by computational methods based on the kinematic and dynamic analysis of the CMP and ZMP respectively. The position of these points inside the support surface allows estimating the stability margin (SM) of the exoskeleton. The location of CMP or ZMP in the center of the supporting surface provides maximum SM. The closer the CMP or ZMP are to the boundary of the reference surface, the lower the level of external disturbances necessary for the loss of the exoskeleton's stability. It is obvious that on the boundary of the region small perturbations will lead the system to an unstable position.

The stability margin in the vertical plane is estimated with the aid of the SM criterion given by the formula:

$$SM = a_0 + a_1 x + a_2 x^2, \tag{5}$$

$$a_0 = -\frac{4x_1 x_2}{(x_1 - x_2)^2},\ a_1 = \frac{4x_1 + 4x_2}{(x_1 - x_2)^2},\ a_2 = -\frac{4}{(x_1 - x_2)^2}. \tag{6}$$

We consider, as an example of the application of the SM criterion, a reference surface located along one coordinate x, whose boundaries are defined by conditions: $x_1 = 0$, $x_2 = 0.4$.

A graphic depiction of the dependence of the stability margin on the position of the CMP projection on the support surface is shown in Figure 8.

It can be clearly seen that the maximum margin of stability is achieved under the condition that $x =$ 0.2. When the position of the CMP projection is shifted to the boundaries of the region of the support surface, a sharp decrease in the stability margin occurs. Therefore, it is important to limit the region of the support surface in which the value is larger than a critical value, SM > SM_0 (region 1 in Figure 8, SM0 = 0.4). Thus, the area of the support surface available for safe operation is much smaller than the original.

In three-dimensional space, the surface has a more complicated shape in the form of a shell shown in Figure 9, and depends on the type of the boundary of the support surface.

Figure 8. Dependence of the stability margin on the position of the CMP projection on the support surface

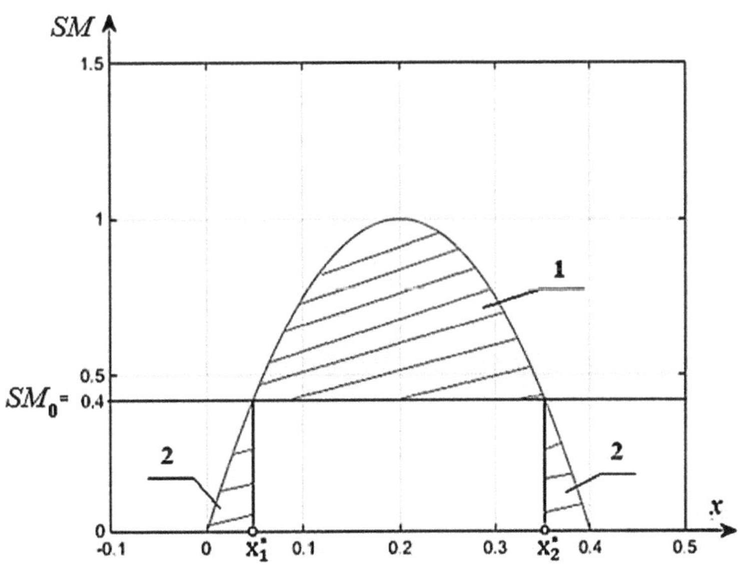

Figure 9. Graph of SM in space in the case of a surface with a rectangular boundary (a - principal view, b - spatial image, c - top view)

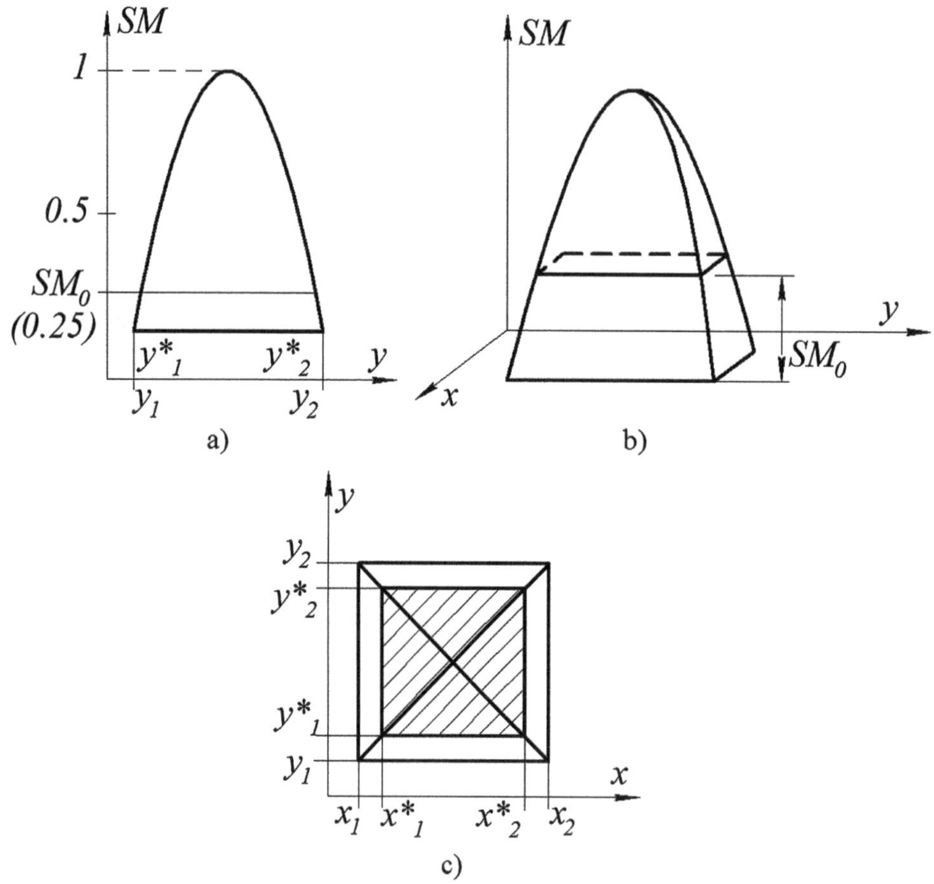

DETERMINATION OF REACTION FORCES

The movement of CMP and ZMP along the reference surface alters the shape of the diagram of normal and tangential reactions (Figure 10). If the acceleration during motion is small, then the frictional forces are insignificant, but when moving with accelerations it is necessary to take into account not only normal reactions, but frictional forces also.

These forces can be measured using special sensors and thereby assess the stability margin. Therefore, if the exoskeleton is equipped with such meters, it is possible to control stability during movement by determining the normal reactions acting on the exoskeleton foot when performing the given exercises.

To estimate the distribution field of the normal reaction on the reference surface a system of sensors forming a tactile matrix, based on which it is possible to determine whether CMP and ZMP are in the support plane or not is used (Figure 11).

The greatest problems with the determination of support reactions arise when the exoskeleton moves along an uneven surface. Figure 12 shows the scheme of normal reactions for the case when foot 1 touches the support surface at two points, while at least one of the pressure sensors is not in contact with

Figure 10. Reaction of the support surface a) under minor accelerations (static balance) b) in the presence of acceleration of the center of mass in the horizontal plane (dynamic balance)

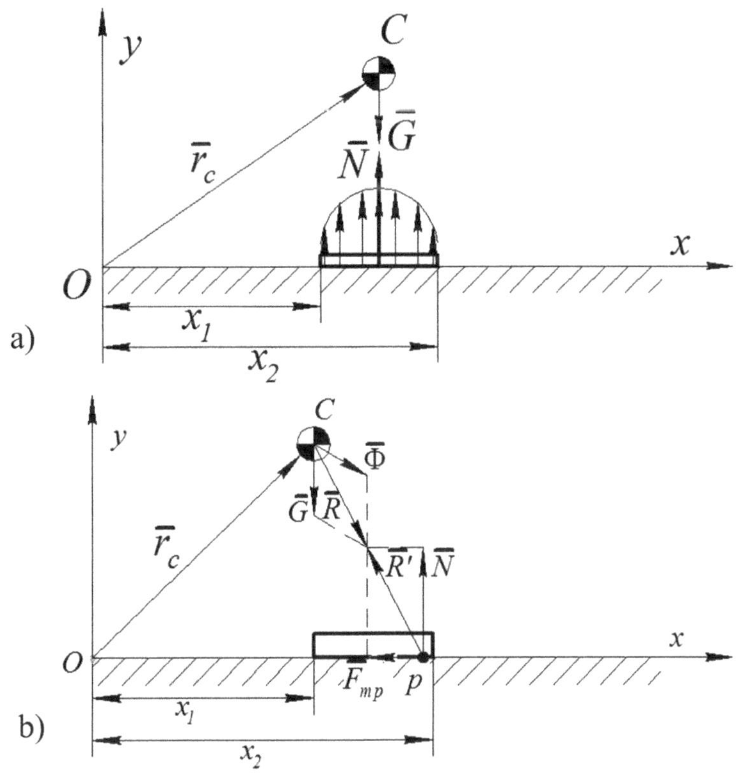

Figure 11. Arrangement of pressure sensors on the feet and illustration of normal reactions

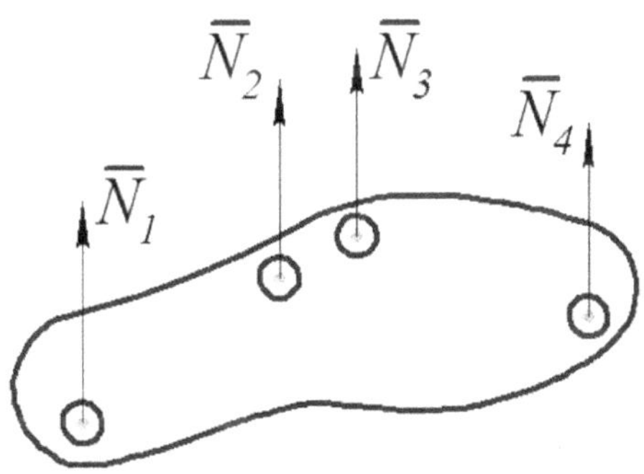

the support surface (SS) and is not involved in the measurements, so the information on the reactions is significantly distorted.

Note that pressure sensors cannot perform measurements because they do not touch the support surface. The presence of contact between sensors and the support surface depends on its shape, the location of the sensors and the orientation and position of the foot. In the case when the support surface is a plane (or is piecewise flat) it is sufficient to orient the foot parallel to the plane of the SS. A modified foot structure can be used when the support surface is not a plane to prevent transition to the modes shown in Figure 12.

Figure 12. Scheme of interaction of an inflexible foot and a supporting surface for cases when the foot touches the SS: a) at two points, b) at one point: 1 - foot, 2 - pressure sensors

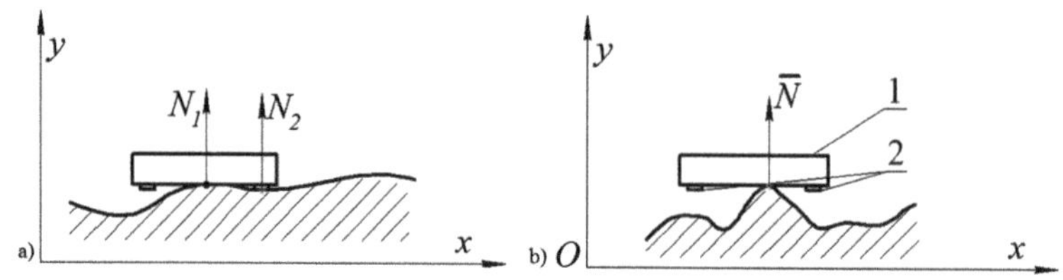

To ensure stability in these cases, it is necessary to use a special design of the foot suspension and the so-called flexible or deformable foot, see Figure 13. In some cases, a foot with elastic sensor sole that adapts to the road surface profile is also suitable. The version of the “flexible foot” includes a multi-layer composite structure of the sole that allows the use of foot muscles, thus making the walk more natural.

Figure 13 shows a flexible foot, whose shape can be deformed by the forces applied to it.

As shown in Figure 13, modes, under which only one of the sensors measures are possible even when a flexible foot is used. Figure 13 (b) shows situation where a reaction force N is applied to the foot directly, avoiding pressure sensors, which may lead to an incorrect measurement.

Figure 14 shows a foot with an elastic sensory sole that adapts to the shape of the supporting surface without changing its orientation.

There are three ways of implementing this kind of elastic sensor sole. The first is to have a passive spring-damper-like system that would push the sensors towards the ground. It is shown in Figure 15 (a). The second way is to use an actuated system where motors are used to push the sensors towards the ground (see Figure 15 (b). The last possibility is to combine the two mentioned setups, see Figure 15 (c).

Figure 13. The interaction of the flexible foot and the support surface for cases when the foot touches the support surface: a) at two points with sensors, b) one point with sensors and one point without sensors: 1 - foot, 2 - pressure sensors, 3 - support surface

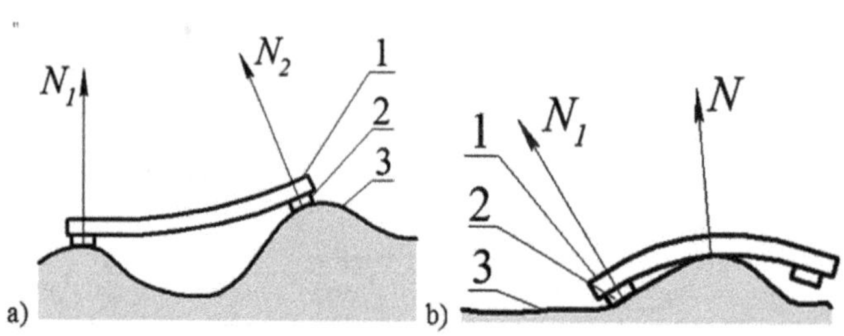

Figure 14. A diagram of elastic sensor sole interacting with the supporting surface: 1 – rigid sole, 2 – force sensors mounted on deformable supports, 3 – supporting surface

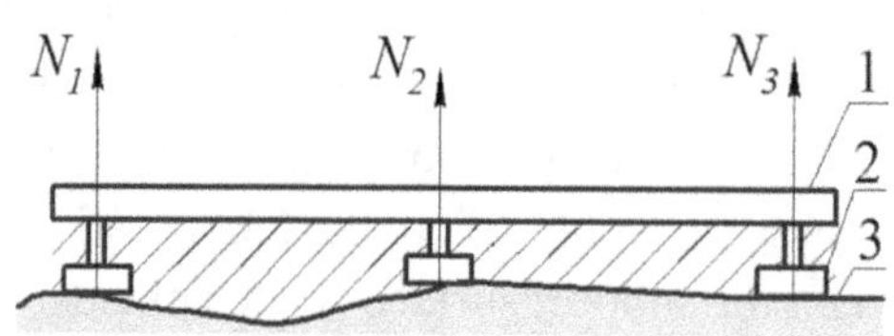

Figure 15. Different possible designs of elastic sensor sole: 1 – rigid sole, 2 – force sensors mounted on deformable supports, 3 – supporting surface

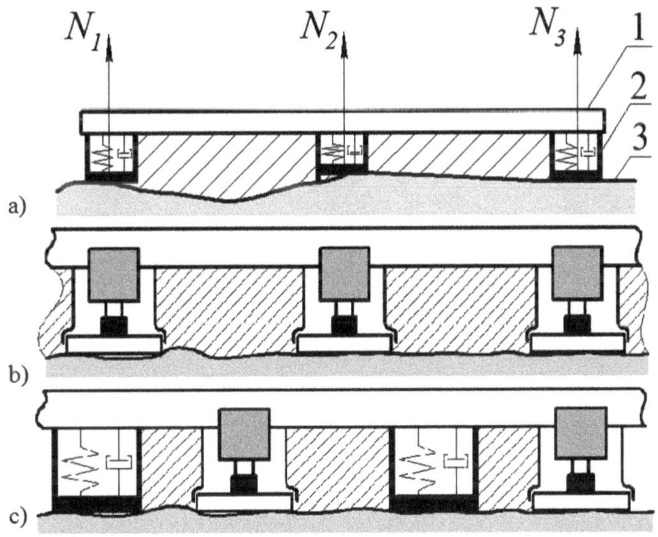

As is well known, the patient's gait in the exoskeleton should begin with lifting the patient from the sitting position and then transfer to the vertical position. The patient then takes the first step with the help of the exoskeleton and this movement must take place in two planes: the frontal plane to transfer the center of gravity to one foot, and then in the sagittal plane movement to move the released leg forward. The control system should monitor the position of the center of mass of the man-exoskeleton system and exclude the exit of this point from the support plane, otherwise the person falls.

Let us consider the positions of the CMP that ensure a stable position of the exoskeleton for various modes of motion (Figure 16). If the CMP lies inside the support surface (the shaded area), then the exoskeleton is stable.

Figure 17 shows the position of CMP in static and dynamic walking. During dynamic gait the CMP can leave the support surface, for stability the ZMP must be inside the support surface. The dynamic walking ensures CMP movement along a smooth trajectory, which under certain conditions tends to a straight line. At the same time, static walking assumes that the CMP moves along some piecewise linear trajectory. In this case, planning such a trajectory requires the use of complex algorithms based on the planning for each step.

INFORMATION SYSTEM OF THE EXOSKELETON

Next, we consider the structure of the information system of the exoskeleton with an active foot as one of the main elements. The foot should be equipped with local navigation tools for accurately determining the distance from the foot to the support surface in both the vertical and horizontal planes. For this purpose, two rangefinders are mounted on the foot, as shown in Figure 18.

Figure 19 shows the exoskeleton information system with video camera 1, angle sensors 2, three pressure sensors 3, an inertial navigation system implemented on the IMU module 4 and rangefinders 5.

Figure 16. The position of the center of mass a) when walking on the spot; 1-support surface, 2 - position of the CMP; b) during verticalization and forward gait (one step); 1- support surface, 2 - position of CMP

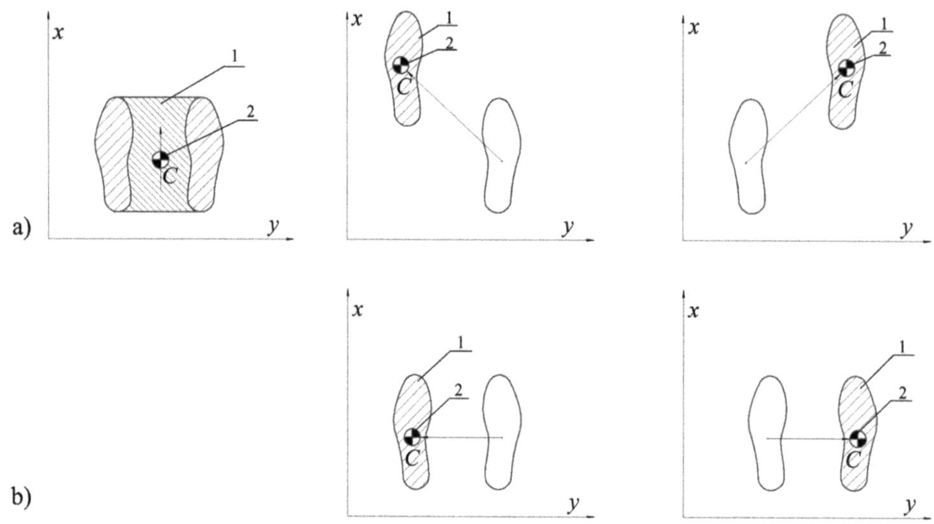

Figure 17. Trajectory of motion of the center of mass when walking forward: 1 – during dynamic gait; 2 - during static gait

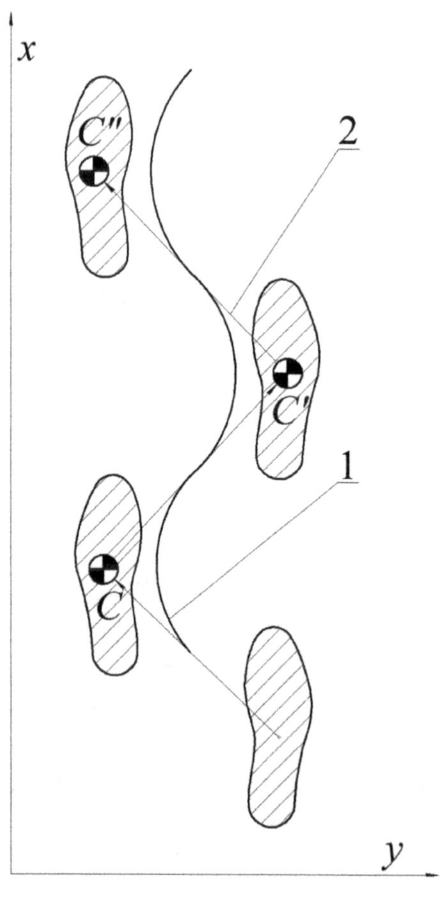

Figure 18. Mounting of a) the rangefinder b) the horizontal rangefinder

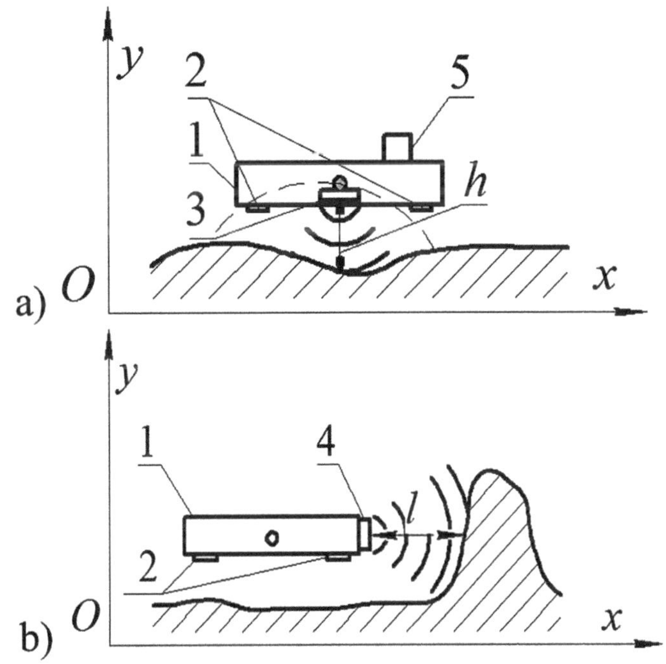

Figure 19. Exoskeleton information system: 1-video camera, 2-angle sensors, 3-pressure sensors, 4-IMU-module, 5-rangefinder, PA-power amplifier

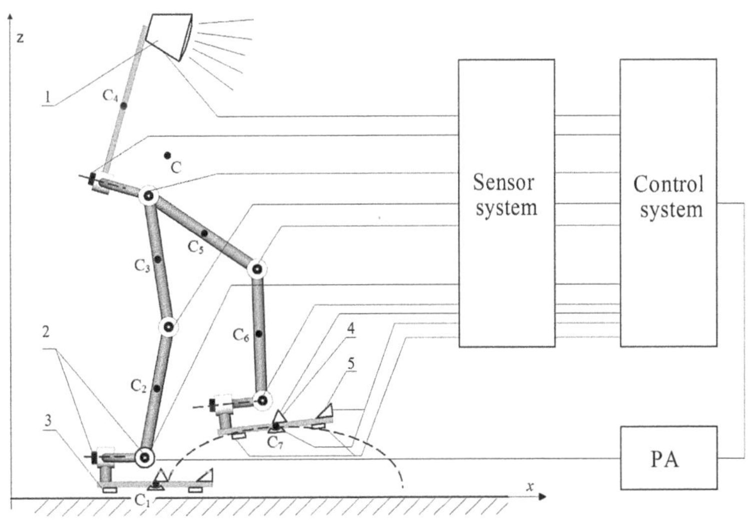

The movement of the patient in the exoskeleton is considered both in the frontal and sagittal planes. To this end, based on the assumptions made, various design schemes of the device have been developed, generalized coordinates have been determined and both global and local coordinate systems have been identified. Exoskeletons are characterized by a large number of degrees of freedom, spatial motion of the links of the mechanism and a network of electric drives and sensory devices. Therefore, to design an exoskeleton a well developed theoretical foundation is needed. At the same time, special attention should be given to the problems of motion control algorithms, taking into account the person-exoskeleton interaction.

The sensory system of the exoskeleton includes units for data processing and analysis, tailored for the on-board sensors. The use of this processing system in particular enables the exoskeleton to interact with the surrounding environment with or without the patient's participation, plan the trajectory of motion, taking into consideration the information about obstacles and other factors influencing the stability of patient's movement.

SCHEME OF AN ACTIVE TWO-COORDINATE FOOT

A key role in ensuring the stability of the exoskeleton in an upright position is played by the structural features of the active foot. The difficulty of designing the foot is that it is necessary to provide at least two degrees of freedom in rotational joints of the foot relative to the shin. The magnitude of the torques providing such a rotation must be significant (of the range of 60-100 Nm). Additionally, the weight and dimensions of the drives should be minimal. It is also important to determine the location of the rotation axes in the hinges of the foot that enable supination and pronation. It is equally important to choose the transmission of electric foot drives that produce controlled rotation of the foot relative to the hinge axes.

A toothed belt based mechanism of the Rolamite type is used to create a prototype of the active foot as a converter of the motor shaft's rotation to the angular rotation of the foot (Jatsun, 2016g). A diagram of the active foot is shown in Figure 20.

Figure 20. General view of the two-coordinate foot with two active hinges: 1, 2 - ankle joint drives, 3 - exoskeleton shin, 4 - heel support, 5 - foot, 6 - hinge body, 7 - pressure sensors, 8 - obstacle sensors

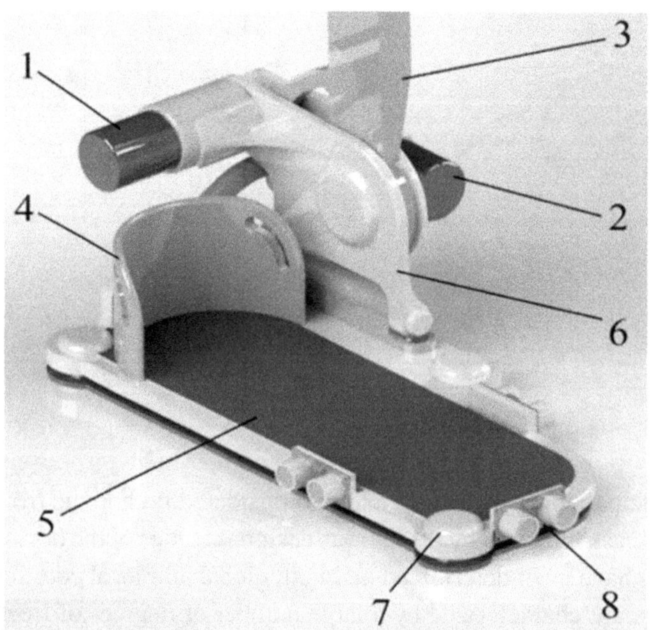

Foot 5, equipped with local navigation system installed on shin 3 and comprises actuators 1 and 2 located on hinges 6. The normal reactions are measured with the pressure sensors 7 at three points. The distance to obstacles is determined using the obstacle sensor 8.

Human gait is largely determined by the interaction of the foot with the supporting surface and as previously shown, contact with the support surface begins with the heel, then in the second stage, the entire foot contacts, and in the third stage, the foot contacts the support surface with the fingers. In fact, the foot "rolls" along the reference surface and in certain modes slipping is possible.

When walking, there is a significant deformation of the foot and a significant change in the area of the supporting surface, affects the stability of movement. However, humans easily carry out gait in both dynamic and quasi-static modes.

Exoskeleton gait considered here is significantly different from the ordinary human gait, primarily due to the fact that the exoskeleton foot must establish contact with the supporting surface immediately over its entire area in order to ensure a stable gait. The system and algorithms for controlling human-exoskeleton gait are imperfect, so the movement is rather slow and the gait is close to the quasistatic.

The analysis of the biomechanical principles of human movement makes it possible to formulate the requirements of an exoskeleton that enables a person with damage to the musculoskeletal system to walk upright and also to formulate quantitative criteria that allow us to assess the quality of an exoskeleton design. Such criteria include safety, comfort, controllability, autonomy, reliability and minimum mass-dimensional parameters.

Safety B_1 must be understood in a broad sense. This includes ensuring the stability of the exoskeleton in an upright position, electrical and environmental safety, minimum noise pollution, the possibility to adjust and plan the trajectory of the patient-exoskeleton movement, avoiding collision with obstacles and the ability to move over rough terrain while maintaining a stable vertical position.

Autonomy B_2 is determined by the presence of on-board energy sources, which make it possible for the exoskeleton to move in space for a sufficiently long time. The energy source must be equipped with convenient charging means and an alarm system that informs the patient of the battery level.

Comfort B_3 is the convenient posture of the patient in the exoskeleton with the help of fixing elements that guarantee normal blood circulation during prolonged use of the exoskeleton. It also takes into account the accelerations arising during motion. The trajectories of motion of the exoskeleton links should be as smooth as possible and exclude sharp changes in the positions of the links.

Controllability B_4 is determined by the accuracy of executing the given movements and is ensured by the high rigidity of the exoskeleton links and hinged joints, high-precision sensors and effective control algorithms with the optimal controllers. Measuring circuits implemented using specific protocols focus on controllability. Signal processing time should be minimal to reduce delay in the feedback channels.

Also, controllability depends on the type of user interface used in the exoskeleton to formulate control actions. This interface should be engineered taking into account the individual human features. The interface can be controlled by hand, allowing the patient to select the mode and basic parameters of movement. The use of a voice-based control system is also possible. In the future, as knowledge on neurobiological processes of muscle management accumulates, systems based on myograms, encephalograms and other signals generated by human systems and recorded by devices will be widely used.

It is important to ensure the reliability B_5 of the exoskeleton, which is a complex electromechanical system that includes numerous electric drives, mechanical gearboxes, hinges, sensors, meters of cables, electronic components, on-board computers, patient fixation devices and software. A malfunction in one of the elements of this system can jeopardize the functionality of the exoskeleton, while auxiliary backup systems in some cases increase the reliability of the system as a whole. To ensure the reliability the service centre specialists should perform the periodic maintenance of the exoskeleton.

Mass-dimensional parameters B_6 of exoskeleton relate to the market prospects of the device, the possibility of its widespread implementation and ease of use. To obtain the minimum dimensions of the device, it is necessary to use the up-to-date materials. These include composite materials, carbon plastics, aluminium, titanium and many others. It is necessary to search for new technical solutions in the field of modern electric drives that generate high torques at a relatively small angular speed of rotation of the links. It is advisable to use various kinematic schemes, including those based on parallel kinematics and that ensure high rigidity of the structure with minimum dimensions and mass of the device.

To evaluate the quality of the exoskeleton we will use the complex criterion, K that makes it possible to estimate the consumer properties of the exoskeleton and is given by the formula:

$$K = \sum_{i=1}^{6} \alpha_i B_i, \tag{7}$$

where: α_i — weight coefficients determined by the expert evaluations method and B_i — quality indicators. The parameters α_i and B_i vary in the interval 0 to 1. Then the maximum value of the criterion K corresponding to the best exoskeleton is 6, while the minimum value corresponding to the worst is 0.

CONCLUSION

In this chapter, an analysis of active exoskeletons was carried out, which showed that the patient's equilibrium in the exoskeleton is achieved in two fundamentally different ways. In the first case, additional external devices such as crutches are used, and in the second – controlled movement of the body and legs of the exoskeleton, which ensures stability in the vertical position by using special control algorithms. The analysis of the biomechanics of motion showed that the main element that guarantees stable upright human gait is the foot equipped with at least a two-coordinate drive and a hinge with two degrees of freedom relative to the shin.

An essential role in ensuring a stable human movement in the exoskeleton in the vertical plane is played by control over the reactions acting on the feet that interact with the support surface. The requirements for the exoskeleton in order to achieve upright human gait were formulated and a complex criterion that gives a quantitative assessment of the quality of the exoskeleton was proposed. A classification of gaits according to the level of inertia forces arising during walking and into quasi-static and dynamic gait was used. It was shown that with a quasi-static gait it is sufficient to control the position of the center of mass of the exoskeleton relative to the support surface. For dynamic walking, we apply the criterion of the zero moment point (ZMP), which must be inside the support surface. Planning the trajectory of the motion of the center of mass taking into account the position of the support surface for a quasi-static gait allows the use of piecewise linear trajectories passing through the centers of support surfaces. For dynamic gait, nonlinear trajectories of motion of the ZMP are applied. To assess the stability of the exoskeleton in the vertical position, we proposed a criterion called the stability margin, defined by a smooth function with a maximum at the center of the reference surface and a zero level at the boundary.

The structure of the information system of the exoskeleton comprising a vision system, an information collection and processing system and a decision-making system and local navigation system were described. The analysis of the interaction of the foot with an uneven surface showed that the sensors located discretely on the foot cannot always control the supporting reactions. For more precise control of these parameters, it is possible to use a flexible foot.

REFERENCES

Barbareschi, G., Richards, R., Thornton, M., Carlson, T., & Holloway, C. (2015). Statically vs dynamically balanced gait: Analysis of a robotic exoskeleton compared with a human. In *Engineering in Medicine and Biology Society (EMBC), 2015 37th Annual International Conference of the IEEE* (pp. 6728-6731). IEEE.

Bogue, R. (2009). Exoskeletons and robotic prosthetics: A review of recent developments. *Industrial Robot: An International Journal*, *36*(5), 421–427. doi:10.1108/01439910910980141

Bogue, R. (2015). Robotic exoskeletons: A review of recent progress. *Industrial Robot: An International Journal*, *42*(1), 5–10. doi:10.1108/IR-08-2014-0379

Carpino, G. I., Accoto, D. I., Tagliamonte, N. L., Ghilardi, G. I., & Guglielmelli, E. U. (2013). Lower limb wearable robots for physiological gait restoration: State of the art and motivations. *Medic*, *21*(2), 72–80.

Contreras-Vidal, J. L., & Grossman, R. G. (2013). NeuroRex: A clinical neural interface roadmap for EEG-based brain machine interfaces to a lower body robotic exoskeleton. In *Engineering in medicine and biology society (EMBC), 2013 35th annual international conference of the IEEE* (pp. 1579-1582). IEEE.

de Castro, M. P., Abreu, S. C., Sousa, H., Machado, L., Santos, R., & Vilas-Boas, J. P. (2014). In-shoe plantar pressures and ground reaction forces during overweight adults' overground walking. *Research Quarterly for Exercise and Sport*, *85*(2), 188–197. doi:10.1080/02701367.2014.893055 PMID:25098014

Dubrovkiy, V., & Fedorova, V. (2008). *Biomechanics*. Moscow: Vlados-press.

Esquenazi, A., Talaty, M., Packel, A., & Saulino, M. (2012). The ReWalk powered exoskeleton to restore ambulatory function to individuals with thoracic-level motor-complete spinal cord injury. *American Journal of Physical Medicine & Rehabilitation*, *91*(11), 911–921. doi:10.1097/PHM.0b013e318269d9a3 PMID:23085703

Ferrati, F., Bortoletto, R., Menegatti, E., & Pagello, E. (2013), November. Socio-economic impact of medical lower-limb exoskeletons. In *Advanced Robotics and its Social Impacts (ARSO), 2013 IEEE Workshop on* (pp. 19-26). IEEE.

Gwynne, P. (2013). Technology: Mobility machines. *Nature*, *503*(7475), 16–17. doi:10.1038/503S16a PMID:24227002

Herzog, W. A., Nigg, B. M., Read, L. J., & Olsson, E. W. (1989, February 1). Asymmetries in ground reaction force patterns in normal human gait. *Medicine and Science in Sports and Exercise*, *21*(1), 110–114. doi:10.1249/00005768-198902000-00020 PMID:2927295

Inman, V. T., Ralston, H. J., & Todd, F. (1981). *Human walking*. Williams & Wilkins.

Jatsun, S., Savin, S., Lushnikov, B., & Yatsun, A. (2016). Algorithm for motion control of an exoskeleton during verticalization. In *ITM Web of Conferences* (Vol. 6, p. 1-6). EDP Sciences.

Jatsun, S., Savin, S., Lushnikov, B., & Yatsun, A. (2016). System analysis of sagittal plane human motion wearing an exoskeleton using marker technology. In *ITM Web of Conferences* (Vol. 6). EDP Sciences.

Jatsun, S., Savin, S., & Yatsun, A. (2016). Improvement of energy consumption for a lower limb exoskeleton through verticalization time optimization. In *Control and Automation (MED), 2016 24th Mediterranean Conference on* (pp. 322-326). IEEE.

Jatsun, S., Savin, S., & Yatsun, A. (2016). Motion Control Algorithm for a Lower Limb Exoskeleton Based on Iterative LQR and ZMP method for trajectory generation. In *International Workshop on Medical and Service Robots* (pp. 305-317). Springer.

Jatsun, S., Savin, S., & Yatsun, A. (2016). Study of controlled motion of an exoskeleton performing obstacle avoidance during a single support walking phase. In *System Theory, Control and Computing (ICSTCC), 2016 20th International Conference on* (pp. 113-118). IEEE.

Jatsun, S., Savin, S., & Yatsun, A. (2016). Comparative analysis of global optimization-based controller tuning methods for an exoskeleton performing push recovery. In *System Theory, Control and Computing (ICSTCC), 2016 20th International Conference on* (pp. 107-112). IEEE.

Jatsun, S., Savin, S., Yatsun, A., & Malchikov, A. (2016). Study of controlled motion of exoskeleton moving from sitting to standing position. In *Advances in Robot Design and Intelligent Control* (pp. 165–172). Springer International Publishing. doi:10.1007/978-3-319-21290-6_17

Jatsun, S., Savin, S., Yatsun, A., & Turlapov, R. (2015). Adaptive control system for exoskeleton performing sit-to-stand motion. In *Mechatronics and its Applications (ISMA), 2015 10th International Symposium on* (pp. 1-6). IEEE.

Jatsun, S., Yatsun, A., Savin, S., & Postolnyi, A. (2016). Approach to motion control of an exoskeleton in "verticalization-to-walking" regime utilizing pressure sensors. In *Cyber Technology in Automation, Control, and Intelligent Systems (CYBER), 2016 IEEE International Conference on* (pp. 452-456). IEEE.

Jung, Y., Jung, M., Lee, K., & Koo, S. (2014). Ground reaction force estimation using an insole-type pressure mat and joint kinematics during walking. *Journal of Biomechanics*, *47*(11), 2693–2699. doi:10.1016/j.jbiomech.2014.05.007 PMID:24917473

Mikołajewska, E., & Mikołajewski, D. (2011). Exoskeletons in neurological diseases-current and potential future applications. *Advances in Clinical and Experimental Medicine*, *20*(2), 227–233.

Panovko, G. Y., Savin, S. I., Yatsun, S. F., & Yatsun, A. S. (2016). Simulation of exoskeleton sit-to-stand movement. *Journal of Machinery Manufacture and Reliability*, *45*(3), 206–210. doi:10.3103/S1052618816030110

Pons, J. L. (2008). *Wearable robots: biomechatronic exoskeletons*. John Wiley & Sons. doi:10.1002/9780470987667

Rose, J., & Gamble, J. G. (1994). *Human Walking. 2nd.* Baltimore, MD: Williams and Williams.

Tung, W. Y. (2013). *Design and Operation of Minimally Actuated Medical Exoskeletons for Individuals with Paralysis* (Doctoral dissertation). University of California, Berkeley, CA.

Viteckova, S., Kutilek, P., & Jirina, M. (2013). Wearable lower limb robotics: A review. *Biocybernetics and Biomedical Engineering*, *33*(2), 96–105. doi:10.1016/j.bbe.2013.03.005

Vucobratovic, M., & Borovac, B. (2004). Zero-moment point—thirty five years of its life. *International Journal of Humanoid Robotics*, *1*(01), 157–173. doi:10.1142/S0219843604000083

Vukobratovic, M., Borovac, B., Surla, D., & Stokic, D. (2012). *Biped locomotion: dynamics, stability, control and application* (Vol. 7). Springer Science & Business Media.

Wang, L., Wang, S., van Asseldonk, E. H., & van der Kooij, H. (2013). Actively controlled lateral gait assistance in a lower limb exoskeleton. In *Intelligent Robots and Systems (IROS), 2013 IEEE/RSJ International Conference on* (pp. 965-970). IEEE.

Whittle, M. W. (2014). *Gait analysis: An introduction*. Butterworth-Heinemann.

Winter, D. A. (2009). *Biomechanics and motor control of human movement*. John Wiley & Sons. doi:10.1002/9780470549148

Chapter 3
Active Assistive Orthotic System:
(Exoskeleton) Enhancing Movement

Ivanka Petkova Veneva
Bulgarian Academy of Sciences, Bulgaria

Dimitar Chakarov
Bulgarian Academy of Sciences, Bulgaria

Mihail Tsveov
Bulgarian Academy of Sciences, Bulgaria

Dimitar Stefanov Trifonov
Bulgarian Academy of Sciences, Bulgaria

Evgeni Zlatanov
Bulgarian Academy of Sciences, Bulgaria

Pavel Venev
Bulgarian Academy of Sciences, Bulgaria

ABSTRACT

Active orthosis (exoskeleton) is an assistive device with a wearable structure, corresponding to the natural motions of the human. This chapter focuses on developing an active/assistive orthosis system (AOS) enhancing movement. The AOS design is inspired by the biological musculoskeletal system of human upper and lower limbs and mimics the muscle-tendon-ligament structure. The exoskeleton structure includes left and right upper limb, left and right lower limb, and central exoskeleton structure for human torso and waist and provides support, balance, and control of different segments of the body. The device was fabricated with light materials and powered by pneumatic artificial muscles that provide more than fifteen degrees of freedom for the different joints. The active orthotic systems (AOS) can operate in three modes: motion tracking system with data exchange with virtual reality; haptic and rehabilitation device; and assistive mode with active orthosis in cases of impaired muscles.

DOI: 10.4018/978-1-5225-2993-4.ch003

INTRODUCTION

An exoskeleton is wearable robotic device with joints and limbs corresponding to those in the human body. The exoskeleton transmits torques to human joints by actuators and has four basic functions according to the control algorithms (Perry J., Rosen J, Burns S., 2007):

1. **Rehabilitation**: The exoskeleton fits closely to the body and fulfils tasks of physical therapy in an active or passive working mode (Tsagarakis and Caldwell, 2003, Gupta and O'Malley, 2006, Perry J. et al., 2007, Rahman M.H. et al., 2010);
2. **Master Device**: For *„master-slave"* tele-communication robot control. The interaction forces are applied to the slave exoskeleton by moving the master (Nakai et al., 1998);
3. **Auxiliary Device**: Human body amplifier that supports body weight so as to minimize the loading on the joint. The operator feels the loads accepted by the exoskeleton lighter (Lee,S. et al., 1998)
4. **Haptic Device**: The subject physically interacts with virtual objects. The interaction forces are applied by the exoskeleton actuators. It should be noted that the haptic exoskeletons found application in rehabilitation as well.

Many upper limb exoskeletons with different mechanical structure and actuation, with or without force feedback haptic devices, have been presented in the literature. The first exoskeleton arm for replication of sensations of contacts and collisions was designed by Perceptual Robotics Laboratory (Bergamasco M. et al., 1994). This is a seven degree of freedom ungrounded device, attached to operator's shoulder and torso. The operator holds onto the device with his/her palm. Hence, the device can only exert forces at the palm of the user. It uses DC motors with a cable transmissions. This haptic device was designed to the purpose of a more fully immersive simulation. The authors from PERCRO have developed another arm exoskeletons for haptic interaction with virtual environments L-Exos (Frisoli A. et al., 2005; Frisoli A. et al., 2007; Frisoli A. et al., 2009). This is a five-DoF exoskeleton with a wearable structure and anthropomorphic workspace. Another active exoskeleton as kinesthetic force/moment feedback device with 6 DoF is known as Freflex (Wright-Patterson) exoskeleton (Williams et al., 1998). A7 DoF arm type haptic interface for teleoperation and virtual reality systems was designed by university of Tokyo (Nakai et al., 1998). The newly redesigned exoskeleton presented of (Sledd Alan et al., 2006) is capable of providing kinesthetic feedback to the joints of the lower arm and wrist of the operator, it is used for robot-assisted rehabilitation and training.

A 9-DoF under-actuated exoskeleton arm using *pneumatic actuators* is developed by (Lee,S. et al., 1998). Their device allows full reproduction of the human arm's workspace by the exoskeleton. An alternate arm exoskeleton developed at the Korea Institute of Science and Technology addresses the limited wearability issues of previous designs by using parallel mechanisms and pneumatic actuators (Jeong Y. et al., 2001). However, this solution have the disadvantage of requiring an auxiliary system, limiting the possibilities of displacement of the exoskeleton. Interface-envelope MGA, of the University of Georgetown, is able to provide several superiors human capabilities (Carignan et al., 2005) indicated to the shoulder and elbow, while the MAHI of the University of Rice is oriented to the rehabilitation of the forearm (Gupta and O'Malley, 2006). Other examples of dedicated exoskeletons primarily a rehabilitation are Armin from the University of Zurich (Nef et al., 2007) and the LDCs the University of Salford which uses pneumatic actuators (Tsagarakis & Caldwell, 2003). The wearable Salford arm addresses some of the issues and limitations of earlier designs (Tsagarakis, N.G., Tsachouridis, V., &

Caldwell Dg. 2003). For example, nearly ninety percent of the human arm's workspace can be replicated with their device. Soft arm–exoskeleton was used in physiotherapy and training.

Several *human power amplifier systems*, related to exoskeleton haptic devices, have been presented in the literature (Li P. Y., 2004; Kiguchi & Fukuda, 2004). Human amplifier systems provide force feedback to the operator through a direct coupling between the amplification device and the operator. An alternative exoskeleton design that uses compliant continuum mechanisms is proposed (Kai Xu et al., 2013). Its intrinsic flexibility adapts to different human anatomy automatically.

Lower-limb exoskeletons are also currently being developed in addition to the usual military and medical applications, to assist in rehabilitation, to support factory workers while performing manual work, to increase the mobility of the elderly and even for purely entertainment purposes. The fundamental purpose of this type of exoskeleton is to compensate the lack of force in the joint (or joints) and support the user's body weight so as to minimize the loading on the joint (Low K.H., 2011) or gait assistance to a user by supplying assistive torque at the knee joint during dynamic ADLs. (Farris R. J. et al., 2011 ; Chandrapa M. et al., 2013 ; Karavas N. C. et al., 2012).

The exoskeletons described above can be combined with sub-elements and subsystems (e.g. partial exoskeletons, exoskeleton for lower body part/feet and exoskeleton for upper body part) to more complex total systems as full body exoskeletons (Toyama S et al., 2006, Kobayashi H. et al., 2008, Mineo I. et al., 2005). Each of them uses different actuators, e.g. ultrasonic motors, pneumatic rubber artificial muscles, and air bag actuators. For example Japanese exoskeleton "Robot Suit HAL" (Kawamoto H. et al., 2003; Lee S. et al., 2003) is specialized on detecting very weak corporal signals on the surface of the skin which are generated when a person attempts to move.

Nowadays there are tens of millions of people with the injuries of the musculoskeletal system. It is often, but not always, possible to restore the function of musculoskeletal system with special training devices. One of the ways to restore walking ability to the patient is based on the use of exoskeleton together with the avatar in virtual reality, which allow the user to perform complicated motions, such as verticalization, walking and others.

The known exoskeletons working as haptic devices in virtual reality are commonly for one or both upper limbs. In most cases, these devices are grounded and they provide a limited range of interaction in virtual reality. Although worn by the user, the device itself may be grounded, in which case it restricts user mobility.

The main task of our work was to create an exoskeleton for the whole body with a wearable structure and anthropomorphic workspace including: exoskeleton for upper limbs as a haptic device providing force feedback of the limbs during the interaction in a virtual reality; exoskeleton for the torso and lower limbs that minimize the load on the lower limb joints by providing assistive torques and control the virtual avatar movements.

INTEGRATION OF THE ACTIVE ORTHOSIS WITH VIRTUAL REALITY

Virtual Embodiment Concept

A participant with movement problems is embodied in an immersive virtual environment, formed by a synchronous combination of a visual virtual body seen in a virtual mirror and an active orthosis. Through real-time motion capture the virtual body moves synchronously with the real body with feedback from wearable active orthotic devices. The orthosis provides support giving the impression of a younger body with greater body balance and free movement. The visual body reinforces this. The overall illusion is ownership over a transformed body, one with improved functionality compared to the real body.

The active orthosis (exoskeleton) works as a haptic device to display the sensory information from a virtual reality to the user. To achieve lightness and inherent safety of the device pneumatic artificial muscles (PAM) with a new design and a new control scheme has been used, including positional and force feedback for each muscle. So it will fully exploit the advantages of PAM, unlike prior pneumatic driven exoskeletons.

Haptic or force-reflecting interfaces are robotic devices used to display touch or force-related sensory information from a virtual or remote environment to the user. It has long been recognized that the dynamics of the haptic device interfere with the realism of haptic feedback, and the drive to achieve low dynamics or transparency has always been regarded as a desirable feature in haptic controller design (Lawrence, 1992). By employing new materials and drive components, the mass and friction in these devices have been reduced dramatically making the current generation of haptic devices adequate for many applications such as the Phantom (Massie and Salisbury, 1994). *Force feedback control* is implemented for improving the quality of the haptic feedback in virtual reality applications (Craig R. et al. 2000). An example of a simple linear haptic device with force sensing is a six degree of freedom Cartesian manipulator developed by (Yokoi et al. 1994)

The proposed design, unlike the known solutions provides monitoring of the movements and position of the whole body, a force feedback to the units of the upper limbs and assistive moments in the joints of the lower limbs. The device consist of a full lower and upper limb orthosis equipped with specially adapted actuators, sensors and control system. The orthosis is foreseen to allow motion of the legs and arm in all anatomically plausible degrees of freedom and add power in the flexion/extension directions of the shoulder, elbow, hips, knees and ankle. One of the main achievements is the multifunctionality of the AOS that combines the motion tracking system, haptic and rehabilitation device in one wearable structure with force feed-back and anthropomorphic workspace covering the full range of human motions. The proposed system can be used for application where both motion tracking and force feedback are required, such as human interaction with virtual environments or rehabilitation tasks.

Further in this chapter the different stages of the design and development of the AOS will be described as follows:

- **An Active Orthosis for Upper Limbs**: Haptic device with a wearable structure, force-feedback and anthropomorphic workspace that can cover the full range of human motion and display sensory information from a virtual reality to the user. The orthosis consists of a novel construction produced by carbon, actuated by pneumatic actuators and controllers with force-reflecting interface;
- **An Active Orthosis for Lower Limbs**: Assistive locomotion device with a wearable structure, corresponding to the natural motion of the human lower limb from the hip to the foot. The orthosis is actuated by specially designed pneumatic actuators, sensors and a control system where the joint torque can be adjusted according to the needs of the operations;
- **A Body Balancing and Supporting Exoskeleton:** The structure has the ability to perform bending and strengthening of the body, to connect and support the upper and lower exoskeletons assuring balance of the system by inertial sensors;
- **A Control System for Active Orthosis:** With distributed control and control algorithm for joint actuation, positioning and force feed-back from VR. Joints actuation is achieved by producing

appropriate antagonistic torques through antagonistic action of pneumatic actuators applying impedance control;

- **Program "ExoInterface":** For exoskeleton calibration, communication and interaction with virtual reality. The orthotic device can be used to drive an embodied avatar and interact in virtual reality displaying sensory information.

DESIGN AND DEVELOPMENT OF WHOLE BODY EXOSKELETON

This work focuses on developing an active/assistive orthosis system (AOS) enhancing movement.

The AOS design is inspired by the biological musculoskeletal system of human upper and lower limbs, and mimics the muscle-tendon-ligament structure. The device was fabricated with light materials and powered by pneumatic artificial muscles that provide more than fifteen degrees of freedom for the different joints. The whole exoskeleton structure includes left and right upper limb, left and right lower limb and central exoskeleton structure for human torso and waist, supporting upper and lower limbs.

The total number of degrees of freedom of the whole exoskeleton structure is: h=5+5+6+6+7=29, and the number of actuated joints is: h=4+4+3+3=14. Figure 1 shows the active joints on the left side of exoskeleton and the corresponding parameters of motion - actuated joint angles, q_n and applied force at the end-effector, *F(x,y,z)*.

Mechanical System

The exoskeleton mechanical structure as a wearable device has to fulfil the following design requirements: low mass/inertia; safety; comfort; anthropomorphic extensive range of motion, etc. A model of the whole exoskeleton was designed in Solid Work (Figure 2). Based on the structure scheme in Fig1, the device was build-up like a branched serial kinematics structure consisting of rotational joints and functioning as a powered exoskeleton.

Figure 1. Whole body exoskeleton: a) kinematic structure; b) active joints; c) real prototype

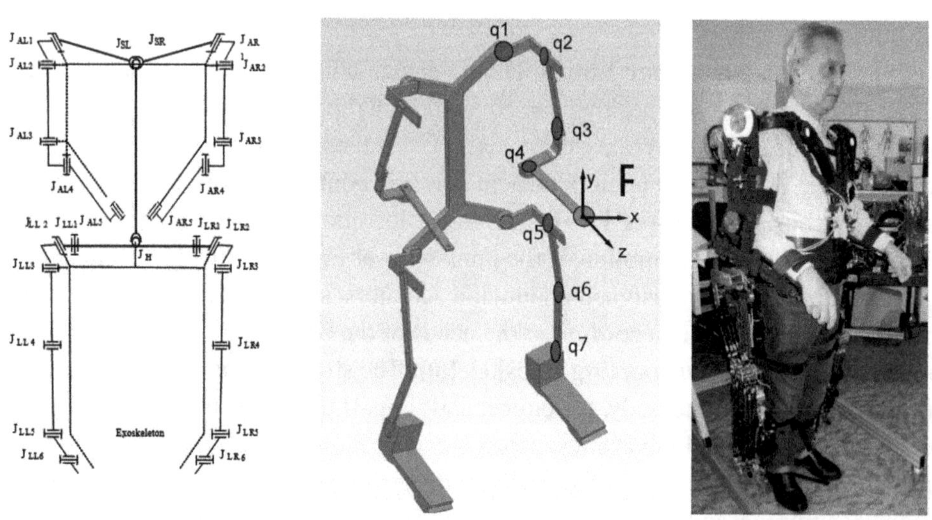

Figure 2. CAD design of the whole body exoskeleton

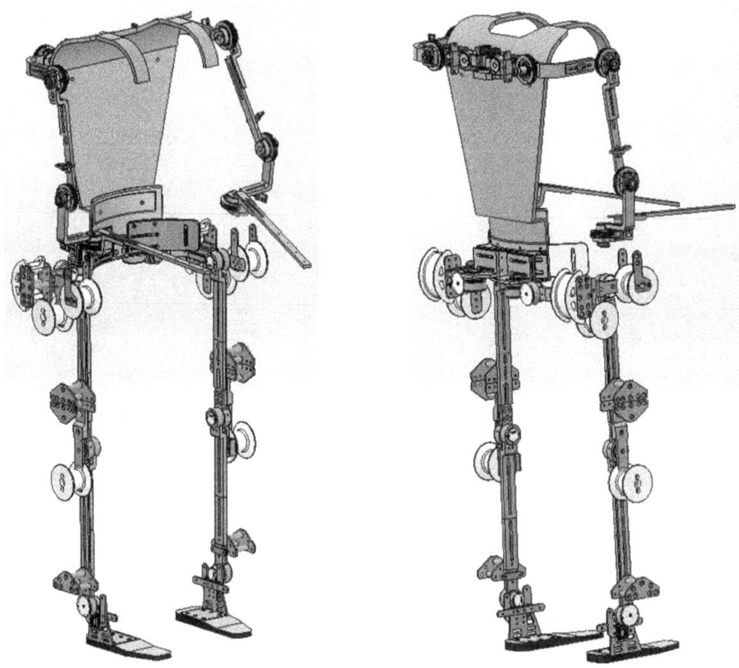

Design of an Active Orthosis for Lower Limbs (Exoskeleton Locomotion Device)

CAD design of the lower limb exoskeleton has been carried out (Figure 3b), according to the structure scheme (Figure 3a) including left and right lower limb (2), foot (3) and central exoskeleton structure corresponding to human torso and waist (1). Antagonistic acting pneumatic muscles are used for moving the limb units.

The mechanical structure of each lower limb has 6 d.o.f. corresponding to the natural motion of the human lower limb from the hip to the foot. The device is designed by a serial kinematics consisting of six rotational joints. The first three rotational joints $\mathbf{J}_{L1}$, $\mathbf{J}_{L2}$, $\mathbf{J}_{L3}$ are incident and mutually orthogonal in order to emulate the kinematics of a universal joint with the same centre of rotation of the human hip, which is fixed in the body brace behind the operator's back. The target workspace for the joint was assumed to be a spherical sector. The third and the fourth joints $\mathbf{J}_{L4}$, $\mathbf{J}_{L5}$ are selected to be two rotational joints incident and mutually parallel in order to emulate the kinematics of two rotation joints with the same centre of rotation of the human knee flexion and the human ankle flexion. The effective movement of the joints is equivalent of the basic human leg movement.

The sixth joint $\mathbf{J}_{L6}$ was assumed to be coincident with the centre of rotation of the human metacarpophalangeal joint in order to allow extension. First two rotational joints $\mathbf{J}_{L1}$, $\mathbf{J}_{L2}$ and sixth joint $\mathbf{J}_{L6}$ are not-actuated. The exoskeleton lower limb structure has **6 d.o.f.,** whith **3** actuated and **3** not-actuated joints. In this structure identical type rotation joints are used and thus three axes joints are avoided.

Figure 3. Lower limb exoskeleton. a) kinematic structure; b) solid works model; c) real prototype

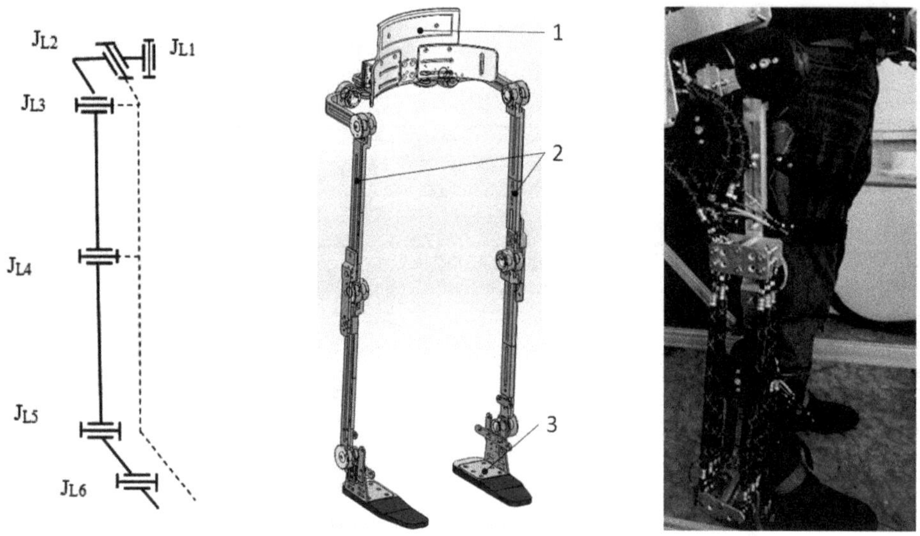

Design of an Active Orthosis for Upper Limbs (Exoskeleton for Arm Motion Device)

The mechanical arm structure has 5 d.o.f. corresponding to the natural motion of the human arm from the shoulder to the elbow but excluding the wrist and the hand.

The device is built by a serial kinematics consisting of five rotational joints (Figure 4). The effective movement of the joints are equivalent to the human arm movement. The first two rotational joints $\mathbf{J}_{a1}$, $\mathbf{J}_{a2}$ are incident and mutually orthogonal in order to emulate the kinematics of a universal joint with the same centre of rotation of the human shoulder, which is fixed in the body brace behind the operator's back and before one and target workspace for the shoulder joint was assumed to be a spherical sector. The third and the fourth joints $\mathbf{J}_{a3}$, $\mathbf{J}_{a4}$ are orthogonal in order to emulate the kinematics of the human elbow flexion and human shoulder rotation. The third joint was assumed to be coincident with the elbow joint.

Figure 4. Upper limb exoskeleton. a) kinematic structure; b) aluminium exoskeleton 3d cad model; c) plastic exoskeleton 3d cad model

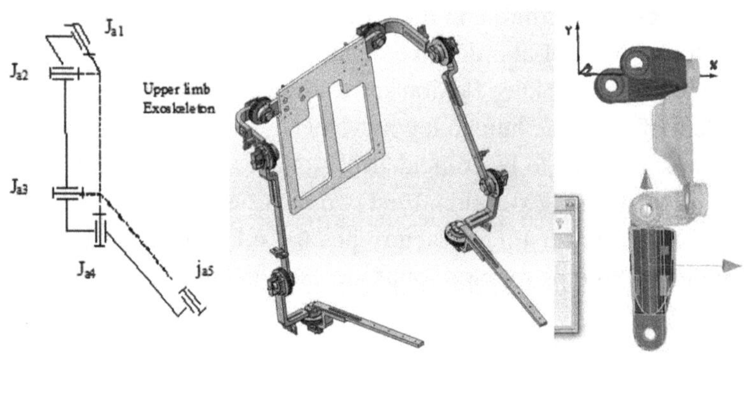

The workspace of the elbow coincides with the angle of rotation of the third joint. The fifth joint $\mathbf{j}_{a5}$ was assumed to be coincident with the forearm, in order to allow the prone- supination of the wrist. This joint is non-actuated. The exoskeleton upper limb structure has **5 d.o.f.** - 4 actuated and 1 not-actuated joints.

The first prototype was designed from aluminium (Figure 5a) and actuated by pneumatic muscles working in antagonistic manner. The second and third prototype is constructed from plastic and carbon to fulfil the design requirement as lightness and structure with low mass/inertia characteristics (Figure 5b, c). The main details are manufactured using a 3D printer from PLA plastic. The most loaded parts like axes and bearings are manufactured from steel. The exoskeleton arm segments are designed so that it can be easily used by the 'typical adult' with only minor changes to the set-up.

Design of a Body Balancing and Supporting Exoskeleton

The mechanical structure according to Figure 1, has 7 d.o.f. corresponding to the natural motion of the human pelvis - J_h and some of the human spine motions - $\mathbf{J}_{SL}$ $\mathbf{J}_{SR}$. The central exoskeleton structure consists of a central supporting body (1) with one spherical joint – $\mathbf{J}_h$. and two universal joints $\mathbf{J}_{SL}$, $\mathbf{J}_{SR}$. Joint $\mathbf{J}_h$ has 3 d.o.f. in order to emulate the kinematics of the human hip motions. Each of the two joints $\mathbf{J}_{SL}$, $\mathbf{J}_{SR}$ has 2 degrees of freedom in order to emulate the kinematics of some of the human spine motions. All three joints in the central exoskeleton structure are non-actuated, but they have compliance and damping to perform bending and support of the upper and lower exoskeletons. Supporting brackets are attached to the joints with the ability to rotate and support the joints of the upper and lower exoskeletons (Figure 7).

The body balancing and supporting exoskeleton system is provided also with two insoles with tactile switches and a set of accelerometers.

Actuation System

The designed active orthosis transmits the torques to all human joints. Joint motion/torque on the exoskeleton is achieved by producing appropriate antagonistic torques through antagonistic action of pneumatic actuators designed and developed in IM-BAS. A group of two or more pneumatic muscles are used for actuation of each joint. Two muscle groups are used to drive one single degree of freedom for the rotational joint.

Figure 5. Upper limb exoskeleton a) aluminium prototype b); c) 3D printed PLA prototype

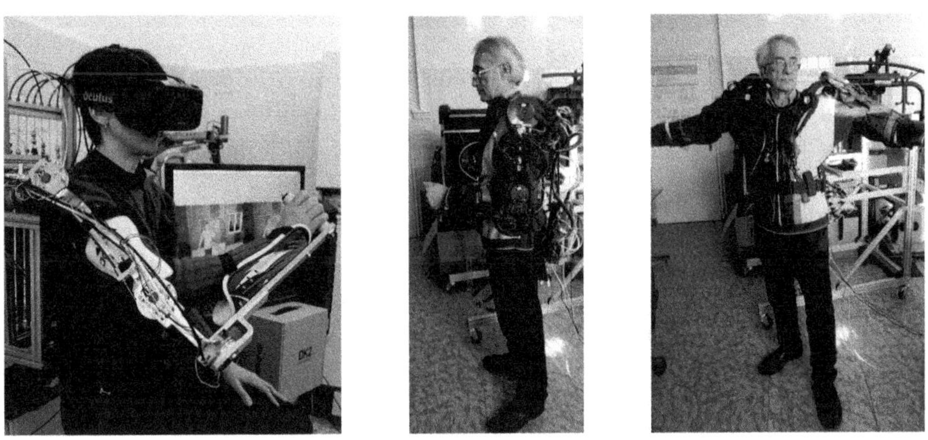

Figure 6. Body balancing and supporting exoskeleton –a) front view; b) side view; c) rear view

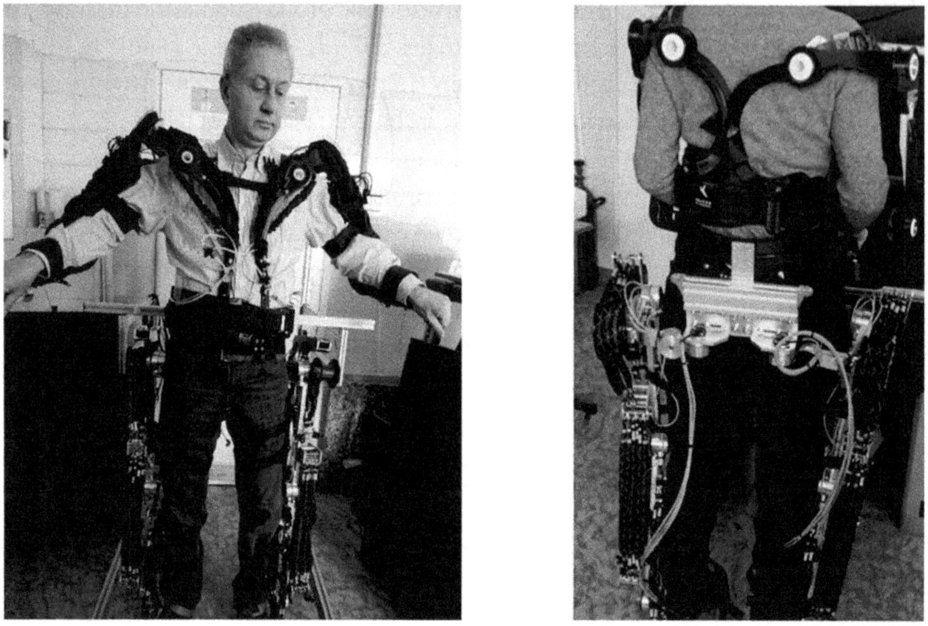

Figure 7. Hip joint with 3 d.o.f.: a) front view; b) rear view; 1 - profiled-plate for fastening of the lower limbs to the human body (made by 3D printing); 2 lever of hip joint

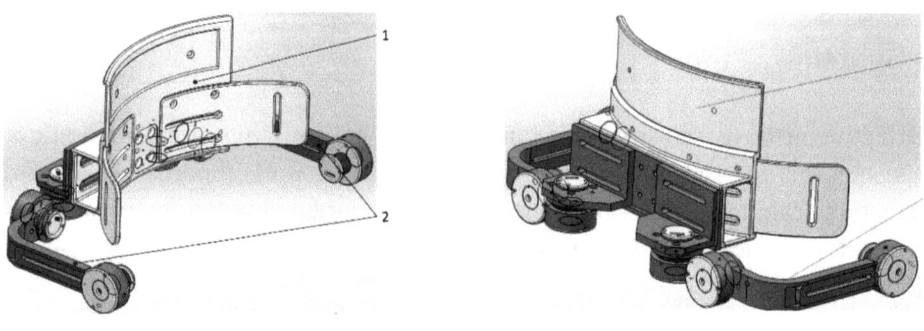

Actuation of the Upper Limbs

The upper limb exoskeleton is a tactile haptic device used to display force-related sensory information from a virtual reality. When both muscles work antagonistically in a single joint the system ensures easy joint control, bidirectional actuation without backlash and possibility for stiffness joint variation.

The aluminium prototype is actuated by pneumatic muscles working in antagonistic manner at *linear actuator influence* (Figure 8a). Two pneumatic muscles, A1 and A2, are used to drive one single degree of freedom rotation joint. Two acting elements A1 and A2 work together simulating a biceps-triceps system to provide the bidirectional motion. Flexible steel cable transmissions C1 and C2 are used for the coupling between the muscles and the pulley P. The actuators are mounted on the central body frame, behind the operator's back.

Figure 8. Actuation unit with pneumatic muscles antagonist. a) linear PMA; b) nonlinear PMA

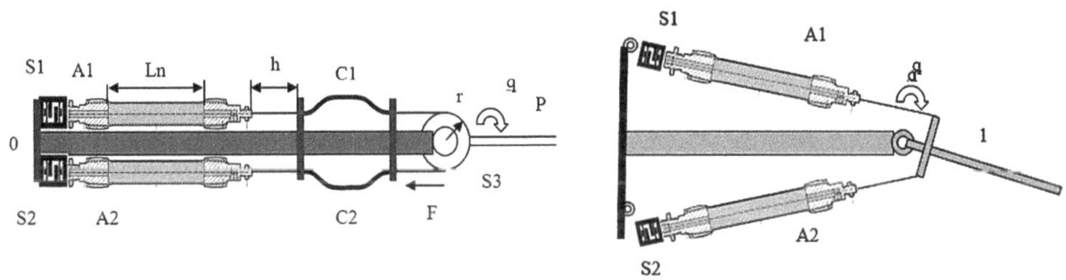

The plastic prototype is actuated by pneumatic muscles working in antagonistic manner at *nonlinear actuator influence* because there are no a pulley and steel cable transmissions. All pneumatic muscles are attached on exoskeleton links (Figure 8b). The actuators A1 and A2 form a linkage with driving limb 1.

Actuation of the Lower Limbs

Antagonistic acting pneumatic muscles are used at the same manner to move the lower limb units. The design task here was to create a lower limb with active joints, where the torque can be adjusted within wide limits according to the needs of the operations. The design problem was solved with BG Patent Reg. No 112139/06.11.2015, “Lower limb with active joints”. General view of the exoskeleton including left and right lower limbs is shown in Figure 9. The actuation structure includes pneumatic artificial muscles and tensioning pulley to reduce variable moment in the leg joints.

To go from a sitting position to a standing position, the joint torque is reduced mechanically in the following way. In the “seated” position, artificial pneumatic muscles are activated by the pressurized air and tensioned from the additional pulleys that generates maximum torque of the hip and knee joint. It leads to the upright position of the body. At standing position, the position of the thigh and the shank is changed and the torque in the joints is reduced, which corresponds to the reduced gravitational load.

An advantage of the lower limb with active joints is that the torque in the joints may be adjusted within a wide range according to the needs of the operations that allows the use of lighter pneumatic actuators. The actuators consist of parallel acting pneumatic artificial muscles, attached directly to the

Figure 9. Lower limb exoskeleton with active hip knee and ankle joints–a) 3D CAD Model; b) real prototype

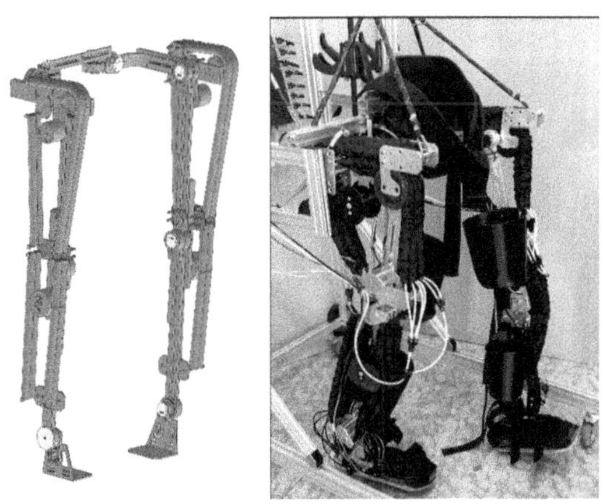

units of the limb, which creates safety in case of damage or breakage of one muscle. Another advantage is that the lower limb with active joints has a lightweight design, which creates a natural safety for the person interacting with the exoskeleton.

Calibration of the Pneumatic Muscles

A muscle tester program is used to measure and present how the force produced by the pneumatic muscle depends on the pressure applied to the muscles. Calibration of the length of the muscle and Force/Contraction characteristics are made during the test.

Experiments with different pressure (Figure 10a) and experiments with different number of muscles (Figure 10b.) are made. Experimentally obtained force/contraction diagrams for the used bundles consisting of 1 to 6 muscles at constant supply pressure $\boldsymbol{p}$ = 400 kPa and pressure 0 kPa are shown in Figure 11. Muscle contraction $\boldsymbol{C}$ is represented as a percentage of the nominal muscle length $\boldsymbol{L_n}$ determined by the outer braided layer

$$c = \frac{C}{L_n} 100 \tag{1}$$

Figure 10. Force/Contraction characteristics. a) Experiments with different pressure, b) Experiments with different number of muscles

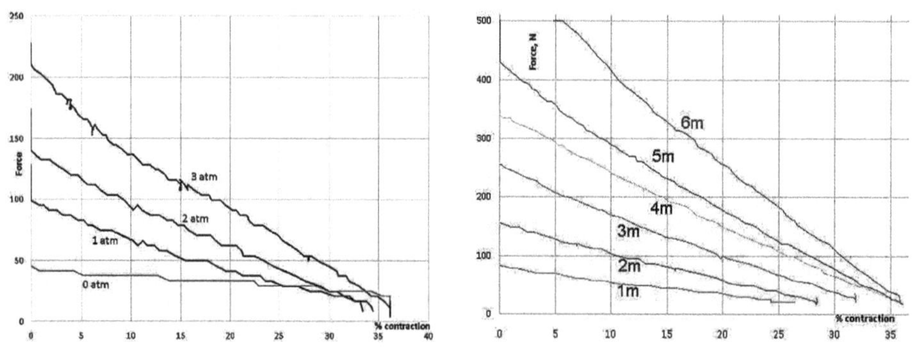

Figure 11. Force/contraction diagrams for the bundle of 1, 2, ..., 6 muscles at constant supply pressure: a) 400 kPa; b) 0 kPa

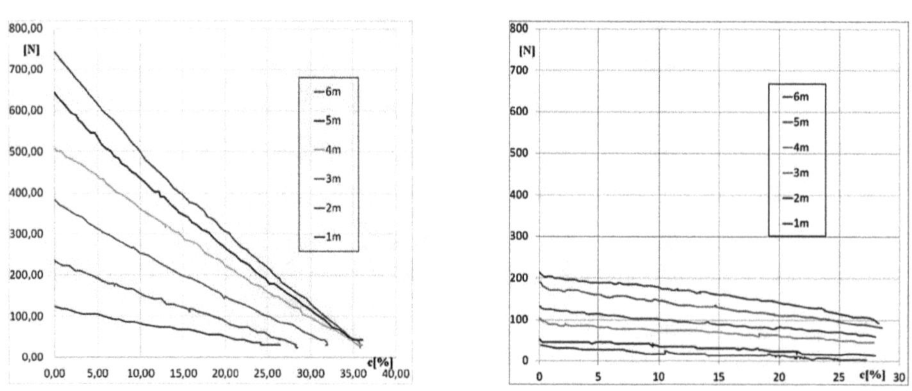

As it can be seen from the graphs, an increase in the muscle number increases the rigidity of the muscle bundle. At zero pressure the muscles show elastic properties determined of inner rubber liner.

Graphs in Figure 11 can be approximated by a quadratic functions. For replication of PAM as a nonlinear quadratic spring can be used according to the following equation

$$P = \frac{6\pi D_0^2 p}{4b^2} L(L - L_{\min}) \tag{2}$$

where D_0 is the maximum theoretical muscle diameter at a braid interweave angle of 90°, $\boldsymbol{P}$ is the pooling force of the muscles, p is the operating pressure, $\boldsymbol{b}$ is the length of one thread of the braided structure, $\boldsymbol{L}$ and $\boldsymbol{L_{min}}$ are the current muscle length and the length of the muscle when it is fully contracted. At a constant supply pressure of the muscle $p = const.$ the above equation can be written as

$$P = KL(L - L_{\min}) \tag{3}$$

where

$$K = \frac{6\pi D_0^2 p}{4b^2} \tag{4}$$

It is a ratio of permanent members.

Equation (4) represents the muscle as a nonlinear quadratic spring. According (1), (2) and (3) it can be transformed

$$P = \frac{6\pi D_0^2 p}{4b^2}(L_n - C)(C_{\max} - C)$$

and can be presented as a function of relative contraction

$$P = K\frac{L_n}{100}(100 - c)(c_{\max} - c) \tag{5}$$

or

$$P = K(L_n - \frac{L_n}{100}c)(c_{\max} - c)$$

The muscle force equation in the form (5) is used as initial function for approximation of the measured characteristics of the two type of pneumatic muscles.

Control of the AOS Platform

Modelling and Control of the Exoskeleton Arm as Haptic Device for Human Embodiment in Virtual Environment

When the system operates in force feedback mode, it can generate a virtual kinematic constraint to the user movements corresponding to the contact with a surface. Let $F = -F_r$ denote the force that the human exerts on the exoskeleton end effector (which is actually the force felt by the user). $\boldsymbol{F_r}$ is the vector of contact forces that the arm generates by the exoskeleton end-effector. During operations in virtual reality, both static and dynamic performance requirements must be guaranteed. *Impedance control* scheme is selected as the most suitable for force reflection between the user and the virtual environment. Impedance controlled systems detect the motion commanded by the operator and control the force applied by the haptic device.

An *open-loop impedance controller with model-feedforward* is selected. This type of force control is called open-loop (Carignan, C., Cleary, K.) because there is no force feedback from the device to the controller to regulate the force output of the master as it is illustrated by the control block diagram in Figure 12.

The goal of the model feedforward, is to cancel out the corresponding terms in the dynamics of the haptic device, $\boldsymbol{Z_e}$

$$[Z_e + Z_d^j - Z_e^f]\Delta q = -Q \tag{6}$$

Above $\boldsymbol{Z_e}^f$ represents the dynamic model of the exoskeleton in joint space, Q is the negative of the force applied by the human operator in joint space and represents a physical, not control, input to the haptic device:

$$Q_d = J^T F_d = J^T Z_d [\Delta X_d - \Delta X] \tag{7}$$

$\boldsymbol{Q_d}$ and $\boldsymbol{F_d}$ are the desired torque and force command in joint space, calculated in accordance with desired virtual environment impedance $\boldsymbol{Z_d}$,

$\Delta \boldsymbol{X_d}$, $\Delta \boldsymbol{X}$ are end effector motion parameters, $\boldsymbol{J}$ is the Jacobian matrix

In above equation theoretically the natural dynamics of the exoskeleton could be accurately modelled, including all components of the mechanical impedance: inertia, damping and stiffness.

Figure 12. Block scheme of open-loop impedance controller with model-feedforward

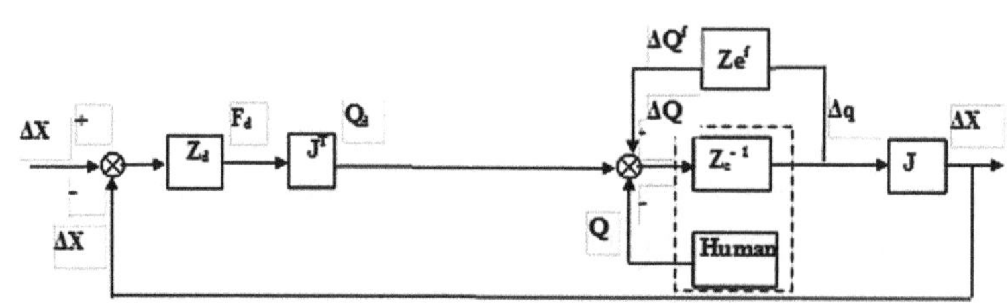

The values of the physical damping and stiffness of the actuators and transfer elements of the exoskeleton are evaluated experimentally.

In applications characterized in low speed the modelling of inertial characteristics may be omitted, and be taken into account the influence of the gravitational effects. The gravity moments are configuration-dependent because the centres of mass vectors are dependent on join positions.

Actuation System: Joint Control

Joint motion/torque on the exoskeleton is achieved by producing appropriate antagonistic torques through antagonistic action of pneumatic actuators designed and developed in IM-BAS.

Establishing a maximum torque in the joint of antagonists muscle actuators, it is assumed, that one muscle group is active in one direction, when the other is passive and vice versa at the sign change of the torque. This is achieved by switching the filling valve of one muscle group and the vending valve of other to achieve the desired joint moment. However, the muscle bundles with zero pressure always participate with a force in the joint antagonistic equilibrium, as they are elastic. If we denote by ***Pa*** and ***Pb*** forces in passive muscle groups (at zero pressure), the desired force of an active muscle bundles is calculated by equations:

$$dPa=Pb+|dP| \quad (8)$$

$$dPb=Pa+|dP|, \quad (9)$$

where $|\boldsymbol{dP}|$ is the module of the useful force, set by the desired torque in the joint $\mathbf{dP} = \mathbf{\Delta Q} / \boldsymbol{r}$, r is the radius of the pulley.

In the aluminium prototype, the control scheme of the joint torque caused by the antagonistic interaction is shown in Figure13. Forces in the muscle bundles ***Pa*** and ***Pb*** are monitored by force cells mounted to each group. In the 3D printed prototype pressure sensors have been used instead of force sensors for each muscle. The range of each muscle depends on the operating pressure. The torque regulation in the joint position are analysed by means of pressure variation $p_a=f_a(Q,q)$, $p_b=f_b(Q,q)$ which are used to control the moments in the joints.

Figure 13. Control scheme of the joint torques

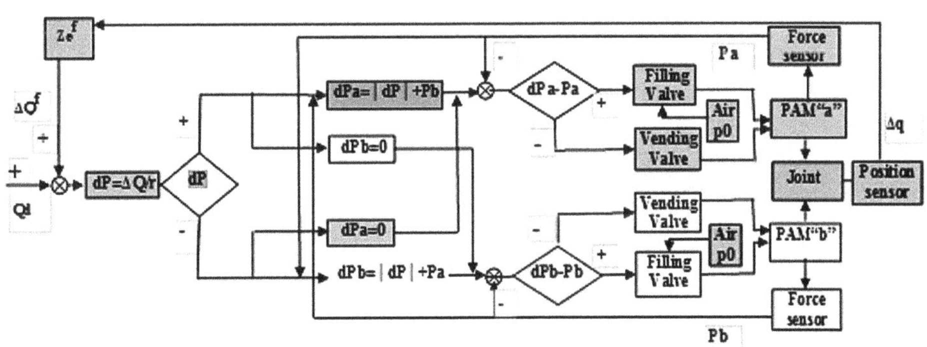

Control of Lower Limbs Exoskeleton for Motion Tracking and Assistive Walking

A Control Algorithm for lower limb exoskeleton was created, based on the following basic functions shown in the block diagram in Figure 14.

1. **Estimation of Human Intention:** Can estimate his/her walking intentions from the command panel - **g** and state of the walking phases according to foot contact sensor information **-f.**
2. **Reference Patterns of the Human Motion:** Extracted from a healthy person's walk. They are applied as reference values of human joint angles and velocities for the corresponding joint motions of a wearer.
3. **Exoskeleton Joint Motion Control:** Control of antagonistic actuation of pneumatic artificial muscles according to the block diagram in Figure 14a.
4. **Lower Limb Exoskeleton:** Attached to the human legs, including three active controllable joints - hip joint, knee joint and ankle joint for each leg.
5. **Foot Contact Sensors:** Providing information about landing phase and supporting phase of walking motion.
6. **Joint Position Sensors**: Magnetic encoders located in each leg joints to measure the joint angles.
7. **Command Panel:** Including display and buttons, to gives person walking intention.

Distributed Control of the AOS Platform

The control system of the exoskeleton contains multiple micro-controller units (MCUs) to support the following tasks for each joint: actuation, sensing, signal processing and control. The prototype sensor system provides information for joints rotation in real-time measuring angles with magnetic encoders. The Master controller determines the position of the limb and exchanges the data with the PC and virtual environment.

The exoskeleton end-effector acts as a haptic device that displays force-related sensory information from the virtual reality (VR). The *Local control system* interprets the applied forces at the end effector of the avatar and sends the commands to drive the corresponding segments of the exoskeleton. As a result the *Joint torque control* is implemented in each joint of the upper limb.

Control algorithm and program application are build up according to control schemes in Fig 15 and Fig 16, using C and C++ program code. The *Control algorithm* performs the following steps:

Figure 14. Block diagram of lower limbs exoskeleton control

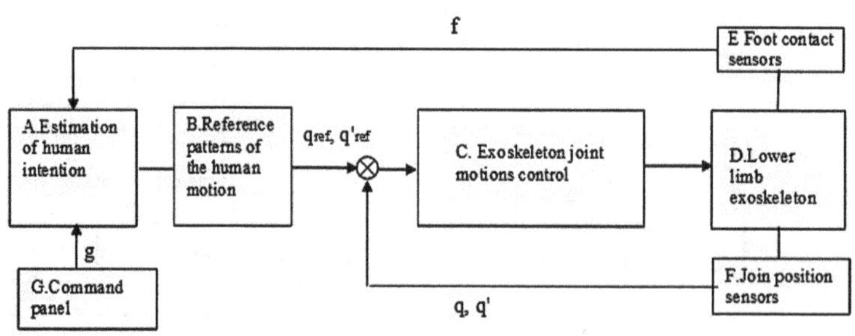

Figure 15. Distributed control

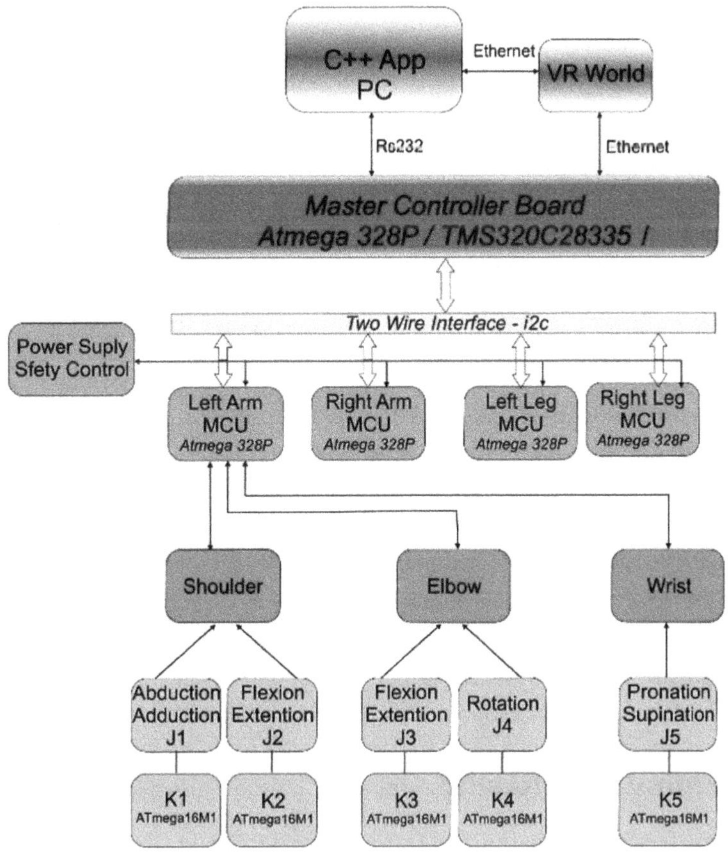

Figure 16. Joint control block scheme

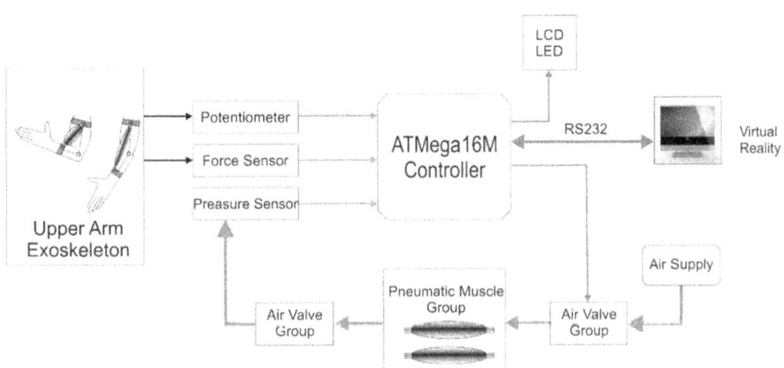

- The position sensors of the prototype joints provide information for the joint angles rotation in real-time and sends it to Exoskeleton C++program. This program transfers the data through the internet by UDP protocol in order to map the Avatar in the virtual reality;
- The Avatar in the virtual reality may perform an action or movement and thereby activate the corresponding limb sending new information about forces applied to the end effector;

- The Control algorithm analyses the motion and received information and decides on appropriate action-corrections of the joint position or new orientation of the body;
- The tactile and force sensors detect the actuators forces (joint torque) and send the appropriate signals to the control system realising joint torque control implemented on each joint.

The MCU determine the joint position by computing the torque and carry out commands to the pneumatic muscles by signalling the solenoid valves to open or close.

Emergency stop and user activated alarms are fitted to the system hardware and linked to the overall control system to ensure that the user will not be injured by the system.

Sensing System

The complete control system consists of control unit, sensing system, data acquisition and communication. The acquisition unit gathers and digitizes the information from the sensors for each active orthosis during the human movements. Control signals from the sensors measuring position of the AOS joints are received by the microcontrollers in real time. The microcontrollers receive the diagnostic information about the system and generate the commands to the drivers in order to assure automatic adaptation of the joints torque. The proposed control technology is based on feedback information from different sensor: visual, tactile, sensors for detection of position etc. Position sensors (magnetic encoders) and force sensor (stain gage) are mounted in each joint of the exoskeleton. A 3D Accelerometer sensor is attached to the subject to detect the vibrations in the x-, y-, and z-directions and displays the value in G's.

Data Exchange With Virtual Reality

For realizing the communication between *ExoInterface program* and virtual reality a *UB data exchange server* was used. Data transfer between Exoskeleton and Virtual avatar has been performed in 3D Unity virtual Engine. HMD *Oculus Rift is* used for visualization in the virtual system. All required operations: read data from controllers, send data to virtual embodiment system, receive data from virtual system and calculate data are performed at 170Hz. Actual tests have been performed between IM-BAS in Sofia and UB in Barcelona with very low latency between the Exoskeleton and the Virtual avatar. The avatar and the exoskeleton were moving in the same way and speed. The following figure shows the configuration used in the exoskeleton integration trials. A *first optical tracker* (HMD *Oculus Rift*) is used to track the head position and orientation, and a *second (Exoskeleton system) to track the rigid frame* of the upper-limb exoskeleton. The two reference frame are fused in a single reference system and together *are used to track the position and orientation of all human body.*

The information from exoskeleton sensors is obtained in real time for the right and the left hand.

Results From the Experiments

Experiments With the Aluminum Exoskeleton

To prove that the algorithm works correctly the following test were performed - any deviation of the exoskeleton end-effector from the initial starting position of the avatar in the virtual reality generates an applied force directly proportional to its deviation but with reverse direction. The force is generated as virtual spring deformation.

Figure 17. Data transfer between exoskeleton and virtual avatar in 3D Unity virtual engine

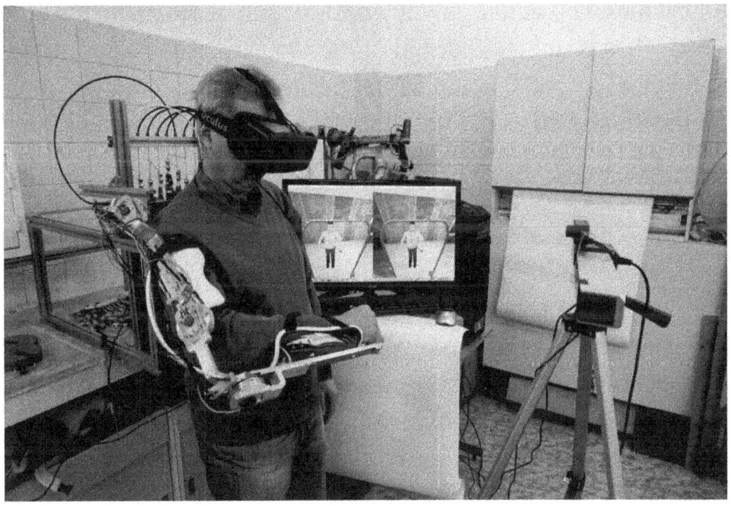

Figure 18. Impedance control demonstration with force generated from virtual reality

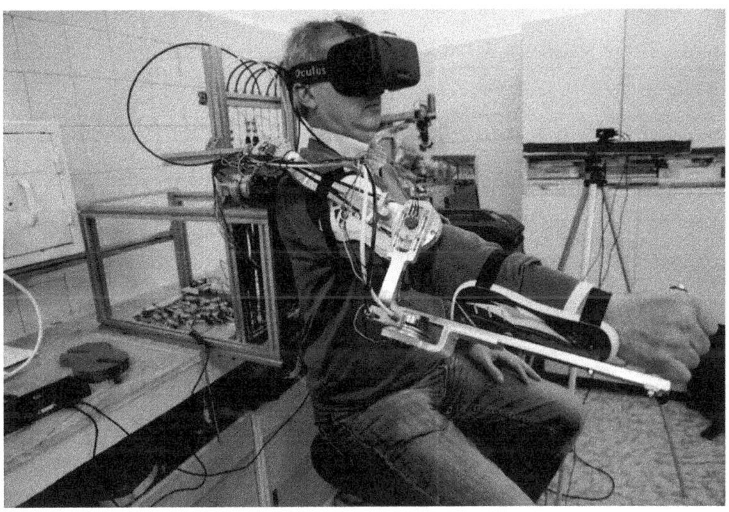

Here two local MCU for both upper limbs obtain the position and force signals from the joints and exchange the data with the PC and virtual environment. The control system interprets the applied forces at the end effector of the avatar and sends the commands to drive the corresponding segments of the exoskeleton. As a result the position of the upper limb automatically is identified and *Joint torque control* is implemented on each joint. The exoskeleton end-effector returns to the initial starting position acting like a spring.

Gravitational components of joint moments were rated according to the model for gravity compensation of the exoskeleton arm. This evaluation is made for the entire range of motion of each joint of the exoskeleton. The program calculates gravitational moments, taking into account the masses of four mobile units as well as the mass m = 1.5 kg of an extra body, located in the end-effector. This body is added to simulate the influence of human hand or an additional load applied to the end effector.

The variations of the forces for the muscles antagonists are calculated according to the control scheme of the joint moments. One muscle is active only in one direction of the moment in the joint, and the other is passive and vice versa when changing the sign of the moment. The radius of the pulley for all joints of the aluminium exoskeleton is $r = 0.0315$ m. Graphs of the muscle force variation for each of the arm joints are shown in Figure 19 as well as the relevant law of motion for each of the joint (q_1, q_2, q_3, q_4).

Experiments With Printed PLA Exoskeleton

To fulfil the design requirement as lightness and low mass/inertia characteristics a third prototype have been constructed from plastic (3D printed with PLA) and actuated by pneumatic muscles working in antagonistic manner (Figure 20). For this prototype we use pressure sensors instead of force sensors for each muscle. To prove that the algorithm works correctly the following force/ contraction diagrams are obtained for two antagonistic acting muscles. The diagrams shown in Figure 21 illustrate the operating range of each muscle depending on the operating pressure. The torque regulation possibilities in the joint or in the joint position are analysed by means of pressure variation.

Figure 19. Muscle force variations in corresponding joint movements: a) shoulder abduction; b) shoulder flexion / extension; c) elbow flexion; d) lateral / int. rotation

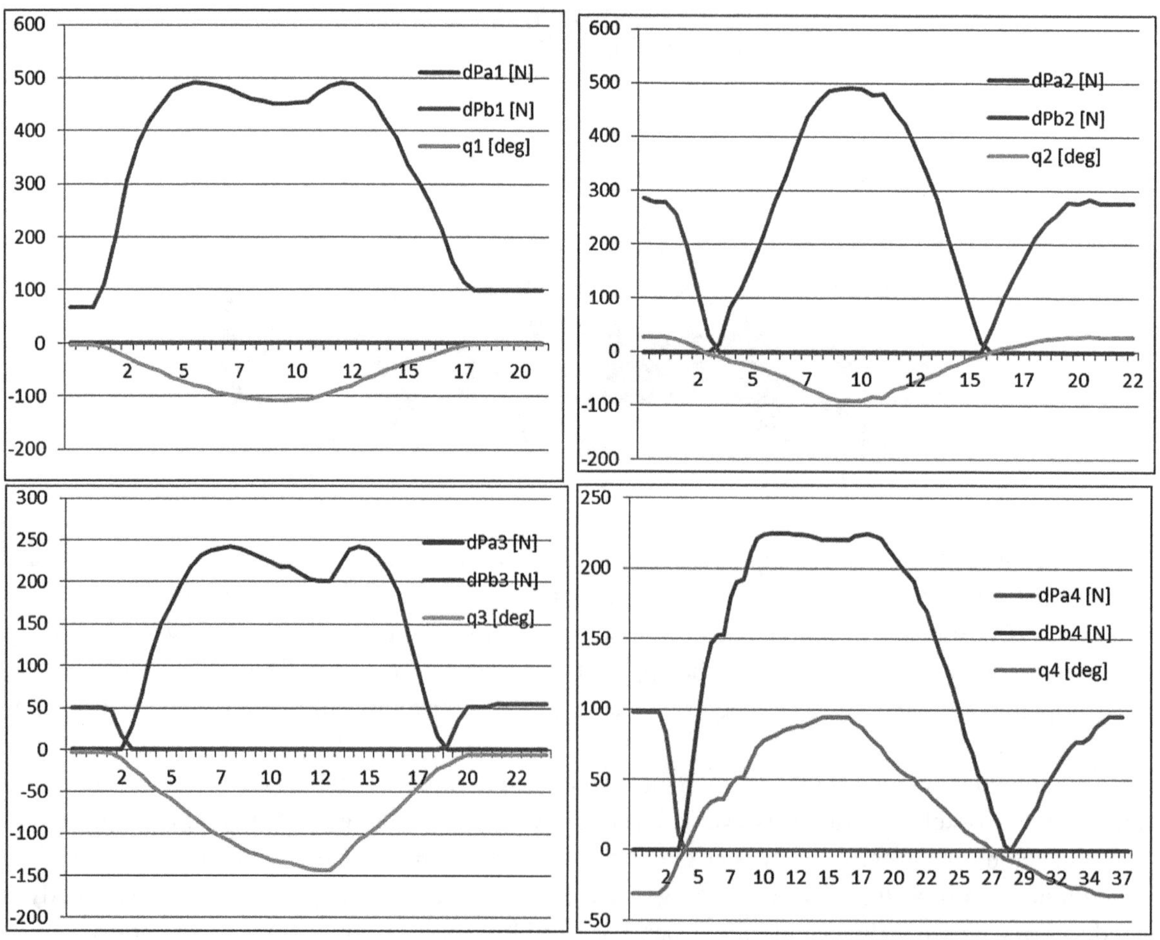

Figure 20. Active exoskeleton for upper and lower limb actuated by pneumatic muscles: a) side view; b) rear view; c) front view

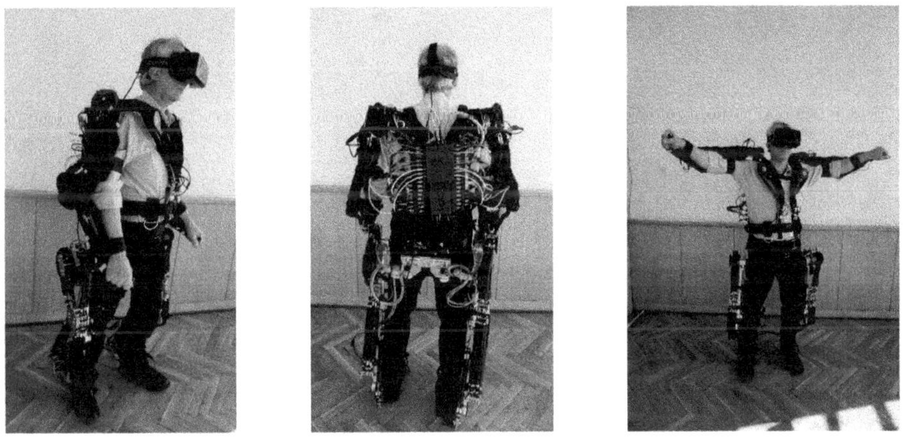

The size of angular movements into the joint q^* and the pulley radius r determine the maximum contraction of the muscle bundles:

$$C^* = rq^* \tag{10}$$

In this design, a muscle group with nominal length $Ln = 0.39$ m are used. Maximal contractions of muscles do not exceed 20% of nominal length. The maximum contractions and graphics force/ contraction (Figure 21 a, b) allow to choose muscle bundles antagonists (a, b) for each joint, so the desired moments in the joints to be maintained.

Muscle bundles with different number of muscles m, maximal muscle bundles contractions c, and stiffness coefficients k, are shown in Table 1. At this choice, the influence of antagonistic force of passive muscle bundle is taken into account.

Graphs *force/ contraction* can be approximated by quadratic functions. In the part c = 0 - 20%, they can be replicated as an approximately *linear function*. For their replication as linear functions, can be used equality:

Table 1. Maximal muscle bundles contractions, number of muscles and stiffness coefficients

		q^* [deg]	C* [m]	c%	m	k [N/m]
Shoulder abduction	a	110	0.06	15.5	6	890-5800
	b	110	0.06	15.5	2	350-1700
Shoulder flexion	a	120	0.066	16.9	6	890-5800
	b	120	0.066	16.9	5	800-4760
Elbow flection	a	142	0.078	20.0	5	800-4760
	b	142	0.078	20.0	2	350-1700
Lateral/int. rotation	a	131	0.072	18.5	3	400-3000
	b	131	0.072	18.5	4	600-3660

Figure 21. Force / contraction of muscle bundles of 1,2, ..., 6 muscles and their linear replicas at supply pressure: a) 400 kPa; b) 0 kPa

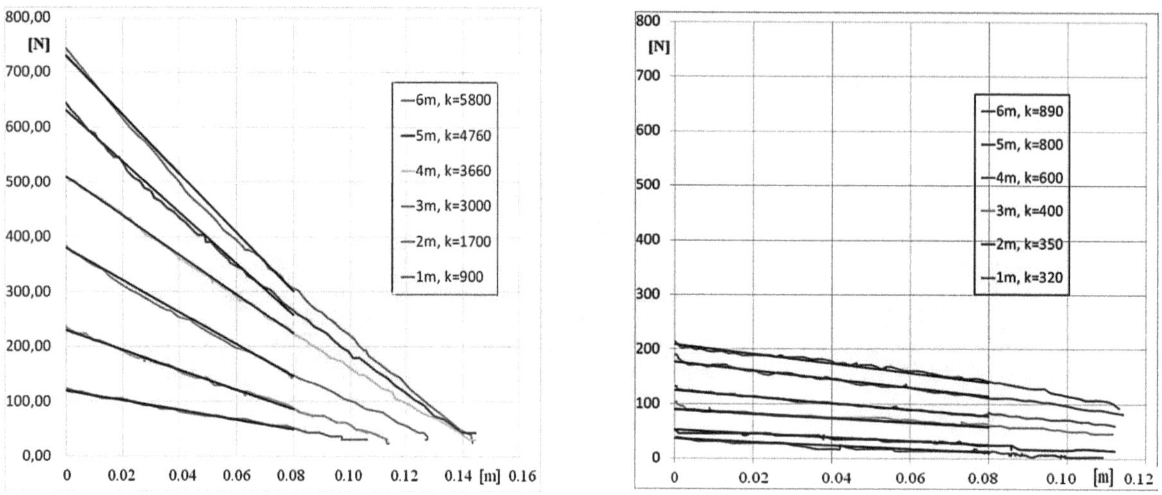

$$P = P_o - kC \tag{11}$$

where $\boldsymbol{P_0}$ is the constant force for each muscle bundle at a nominal length and the coefficient $\boldsymbol{k}$ represents the muscle bundle stiffness at a certain pressure, derived from the ratio:

$$k = \frac{\partial P}{\partial C} \tag{12}$$

Graphs *force/contraction* of muscle bundles with 1, 2, ..., 6 muscles together with their linear replicas at a supply pressure of 400 kPa are shown in Figure 21a, and in Figure 21b at a pressure of 0 kPa).

The type of selected muscle bundles and their respective stiffness coefficients under certain pressure, specified by equation (12), allow to define the joint stiffness:

$$K = r^2[k_a + k_b] \tag{13}$$

These joint stiffnesses can determine the stiffness coefficients for each joint of the *Feedforward Impedance* model.

Three-dimensional characteristics *joint torque – muscle contraction - pressure* are used for torque control of the joints by monitoring the muscles pressure and positions in the actuated joints (Figure 22). Here are obtained approximated characteristics of every joint: torque **Q**/ *muscle contraction %*/ pressure $\mathbf{p_a}$ in the muscles of one group and approximated characteristics: torque **Q**/ *muscle contraction %* / pressure $\mathbf{p_b}$ in the muscles of the second group.

Figure 22. Characteristics in 3D - joint torque/ joint position /pressure, for: a) a bundle of muscles antagonists; b) bundle of the other muscles agonists

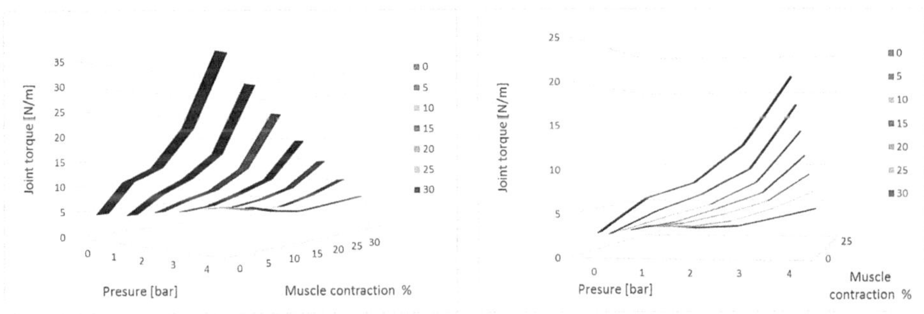

Geometry of Lower Limbs Exoskeleton With Pneumatic Actuation

The actuation structure of exoskeleton lower limbs includes pneumatic artificial muscles and tensioning pulley to reduce variable torque in the leg joints. CAD simulations have been made to determine the coordinates of the attachment points of muscles that define the moves and contractions of each muscle. Adopted is the range of movement in the active joints shown in Table2: hip flexion 90/20, knee flexion – 90/0. The lower limb geometrical data provides maximum and minimum lengths of each muscle in limit end positions of the leg. Received contraction c% of each muscle is determined by the equality

$$C = (Lmax-Lmin)*100/Lmax \tag{14}$$

These parameters are shown for all muscles in Table 2. The resulting contractions are within allowable contractions of pneumatic artificial muscles: 20 -35%.

Experiments With the Active Exoskeleton With Volunteers

Real experiment with the exoskeleton with 30 volunteers have been performed. Main objectives of the experiments are:

- To evaluate the use of Virtual Reality in middle-aged male participants and to validate effects of physical training within a virtual environment as an alternative to outdoor exercise;
- To validate the functionality of the exoskeleton;

Table 2. Muscle parameters

Muscle No	Lmax	Lmin	c%
1	0.720	0.540	25%
2	0.660	0.493	25%
3	0.400	0.270	32%
4	0.509	0.400	21%

- To validate the VR systems' (VR) from a technical point of view and usability for the intervention in the exercise of healthy people;
- To evaluate the functional effect in the realization of physical exercise within a VR environment and to compare it with outdoor exercise;
- To determine if illusory visual feedback during embodiment in the virtual character can improve physical exercises efficiency.
  Virtual Weight Lifting

A GIM virtual environment suitable for the realization of the physical exercise from a first person perspective has been created and the following experiment methodology has been developed:

- **Experiment 1: Visual Feedback Manipulation***:* The experiment consist of a set of 4 sessions where in the needed conditions the body shape and/or the weight is increasing from session to session. The real force being applied by the exoskeleton remains constant in all the conditions.
- **Experiment 2: Force Feedback Manipulation:** The experiment consists of a set of 4 sessions where in the needed conditions the force to be applied to lift the same virtual weight is increasing from session to session.

- **Exercise Design:** Due to the orthotic system design, the best way to simulate a resistance/force is the following exercise with arms in a standing pose shown in Figure 23.

Measures Taken. In order to evaluate the progression during the sessions and to compare with the control groups the following set of variables is recorded:

- Forces applied and retrieved from the exoskeleton
- Tracking data for later playback and evaluation.
- Embodiment questionnaires

Participants have been recruited with similar body proportions and sizes since the exoskeleton is not fully adaptable to all different body shapes and sizes. The participants are male and middle aged (20-50). *Exclusion criteria* Participants with physical problems that impede the correct realization of experiment exercises, participants with epilepsy or severe visual problems.

Virtual Weight Lifting Results. The effect of embodiment in different types of bodies on personal strength has been examined. An experiment where participants (n = 30) were either embodied in a weak or strong body has been completed in VR. The measure was how long they could hold up a weight under these two conditions. Results suggest that taking account of the *Body Mass Index* the stronger body is associated with longer duration time.

Figure 23. Muscle Experiments with the active exoskeleton: a) Exercise design; b) Virtual Gymnasium –Pilates Exercise; c) Virtual Weight Lifting;

FUTURE PLANS AND RESEARCH DIRECTIONS

In order to be able to drive different body segments the control system must be able to 'understand' the goal of the action, to identify the segment that need to be manipulated and to actuate the corresponding joint. The proposed method for control of the joint actuation is connected with the following: the user has a visual, tactile and/or force information from sensors. Being in the virtual reality he decided to perform an action or movement and activates the corresponding limb. The sensors detect the movement and send the appropriate signals to the control system. The control system 'understands' the intentions of the user and sends the commands to drive the corresponding segments of the exoskeleton.

Impedance control scheme is selected as the most suitable for force reflection between the user and the virtual environment. In the feedforward impedance, the natural dynamics of the exoskeleton is accurately modelled, including the components: damping, stiffness and gravitation.

Force control is implemented in the joints of the exoskeleton, which guarantees the determined forces of reflection during permanent contact with human. The exoskeleton compliance doesn't affect achieving these forces, but being independent of servo-responses, provides a natural safety in the interaction with human.

For the future, the optimization of the exoskeleton design is planned for the upper and lower limbs, body balancing and supporting exoskeleton as well as the control algorithm of the AOS platform. A new scenario for the embodiment and control of the exoskeleton through the virtual reality is foreseen to be created.

CONCLUSION

The design and development of an active/assistive orthotic system has been carried out. The orthotic system consists of a novel lightweight wearable structure for the whole body with anthropomorphic workspace and includes: exoskeleton for upper limbs as a haptic device providing force feedback of the limbs during the interaction in a virtual reality; exoskeleton for the torso and lower limbs that minimize the load on the lower limb joints by providing assistive torques and control the virtual avatar movements.

The device consist of a full lower and upper limb orthosis equipped with specially adapted pneumatic actuators, sensors and control system. The basic features of the orthosis segments are the lightweight shaped bodies with the ability to rotate and support the joints. The orthosis allows motion of the legs and arm in all anatomically plausible degrees of freedom and add power in the shoulder, elbow, hips, knees and ankle motions. The participant intention is sensed by a combination of joint angle sensors, accelerometers and force sensors that gives a measure of the force applied by the different muscles on the lower and upper limbs.

Developed orthotic system is human friendly and injury-free. The increased safety of the orthotic system is achieved by the passive compliance approach. Compliant antagonistic actuation, close to that of human muscular system with spring like properties has been used. Pneumatic muscle actuators used as power source for the system provide a clean, low cost actuation source with a high power/weight ratio and natural safety due to the inherent compliance.

Program "ExoInterface" for exoskeleton calibration, communication and interaction with virtual reality has been developed. The orthotic device can be used to drive an embodied avatar and interact in virtual reality displaying sensory information.

One of the main achievements is the multifunctionality of the AOS that combines the motion tracking system, haptic and rehabilitation device in one wearable structure with force feed-back and anthropomorphic workspace covering the full range of human motions.

The proposed system can be used for application where both motion tracking and force feedback are required, such as human interaction with virtual environments or rehabilitation tasks and assistive movements. The system would be of great importance to people with limited mobility as well.

The system would be of great importance to people with limited mobility for assistive and rehabilitation tasks

ACKNOWLEDGMENT

This research was supported by the European Commission through FP7 [grant number 257695]; Bulgarian Science Found [grant number DN 07/9].

REFERENCES

Baumann, R., & Clavel, R. (1998). Haptic interface for virtual reality based minimally invasive surgery simulation. *Proc. IEEE Int. Conf. on Robotics and Automation*, 381-386. doi:10.1109/ROBOT.1998.676435

Bergamasco, M., Allotta, B., Bosio, L., Ferretti, L., Perrini, G., Prisco, G. M., . . . And, G. Sartini, (1994). An Arm Exoskeleton System for Teleoperation and Virtual Environment Applications. *IEEE Int'l Conf. Robot. Automat., 2*, 1449–1454.

Carignan, C. R., & Cleary, K. R. (2000, February 23). Closed- Loop Force Control for Haptic Simulation of Virtual Environments. *Haptics-e, 1*(2), 1–14. http://www.haptics-e.org

Chakarov. (2015). *Torque measuring device*. Patent Application (BG, N°111934/20.02.2015).

Chandrapa, M., Xiao, Ch., & Wenhui, W. (2013). Preliminary Evaluation of Intelligent Intention Estimation Algorithms for an Actuated Lower-Limb Exoskeleton. *International Journal of Advanced Robotic Systems, 10*(2), 147. doi:10.5772/56063

Farris, R. J., Quintero, H. A., & Goldfarb, M. (2011). Preliminary evaluation of a powered lower limb orthosis to aid walking in paraplegic individuals. *IEEE Journal on Neural Systems and Rehabilitation Engineering, 19*(6), 652–659. doi:10.1109/TNSRE.2011.2163083 PMID:21968791

Frisoli, Borelli, Montagner, Marcheschi, Procopio, Salsedo, … Rossit. (2007). Arm rehabilitation with a robotic exoskeleleton in Virtual Reality. *Proc. of the 2007 IEEE 10th International Conference on Rehabilitation Robotics.*

Frisoli, A., Rocchi, F., Marcheschi, S., Dettori, A., Salsedo, F., & Bergamasco, M. (2005). A new force-feedback arm exoskeleton for haptic interaction in virtual environments. *WHC 2005. First Joint Eurohaptics Conf. and Symposium on Haptic Interfaces for Virtual Environment and Teleoperator Systems*, 195–201. doi:10.1109/WHC.2005.15

Frisoli, A., Salsedo, F., Bergamasco, M., Rossi, B., & Carboncini, M. C. (2009, June). Fabio Salsedo, Massimo Bergamasco, Bruno Rossi And Maria C. Carboncini, (2009). A force-feedback exoskeleton for upper-limb rehabilitation in virtual reality. *Applied Bionics and Biomechanics, 6*(2), 115–126. doi:10.1155/2009/378254

Gupta, A., & O'Malley, M. (2006, June). Design of a haptic arm exoskeleton for training and rehabilitation. *IEEE/ASME Transactions on Mechatronics, 11*(3), 280–289. doi:10.1109/TMECH.2006.875558

Hiroshi, K., Hasegawa, S., & Nozaki, H. (2008). Development and Application of a Muscle Force Enhancement Wear: Muscle Suit. *Proceedings of the 11th Symposium on Construction Robotics in Japan*, 93-100.

Ishii, M., Yamamoto, K., & Hyodo, K. (2005). Stand-Alone Wearable Power Assist Suit –Development and Availability-. *Journal of Robotics and Mechatronics*, *17*(5), 575–583. doi:10.20965/jrm.2005.p0575

Jeong, Y., Lee, Y., Kim, K., Hong, Y., & And, J. Park, (2001). A 7 DOF Wearable Robotic Arm using Pneumatic Actuators. *Proceedings of the 32nd ISR International Symposium on Robotics*, 388-393.

Kawamoto, , & Lee, , Kanbe, & Sankai. (2003). Power assist method for hal-3 using emg-based feedback controller. *IEEE International Conference, Man and Cybernetics,* 2, 1648–1653.

Kiguchi, K., & Fukuda, T. (2004). A 3 DOF Exoskeleton for Upper Limb Motion Assist: Consideration of the Effect of Bi-Articular Muscles. *IEEE Int'l Conf. Robot. Automat., 3*, 2424–2429. doi:10.1109/ROBOT.2004.1307424

Lawrence, D. (1992). Stability and transparency in bilateral teleoperation. *Proc. IEEE Conf. Decision and Control.*

Lee, S., & Sankai, Y. (2003). The natural frequency-based power assist control for lower body with hal-3. *IEEE International Conference on Systems, Man and Cybernetics,* 2, 1642–1647. doi:10.1109/ICSMC.2003.1244648

Li, P. Y. (2004). Design and Control of a Hydraulic Human Power Amplifier. *ASME International Mechanical Engineering Conference and Exposition.* doi:10.1115/IMECE2004-60868

Low, K. H. (2011). Robot-assisted gait rehabilitation: From exoskeletons to gait systems. *Defense Science Research Conference and Expo (DSR)*, 1–10. doi:10.1109/DSR.2011.6026886

Massie, T., & Salisbury, J. K. (1994). The PHANTOM haptic interface: A device for probing virtual objects. *Proc. ASME Winter Annual Meeting: Symposium on Haptic Interfaces for Virtual Environment and Teleoperator Systems.*

Nakai, A., Ohashi, T., & Hashimoto, H. (1998). 7 dof arm type haptic interface for teleoperation and virtual reality systems. *Proc. IEEE/RSJ International Conference on Intelligent Robots and Systems*, 2, 1266–1271. doi:10.1109/IROS.1998.727473

Nef, Mihelji, & Riener. (2007). Armin: A robot for patient-cooperative arm therapy. *Medical & Biological Engineering & Computing, 45*, 887–900.

Perry, J., Rosen, J., & Burns, S. (2007, August). Upper-limb powered exoskeleton design. *IEEE/ASME Transactions on Mechatronics*, *12*(4), 408–417. doi:10.1109/TMECH.2007.901934

Rahman, Ouimet, Saad, & Kenne. (2010). *Development and Control of a Wearable Robot for Rehabilitation of Elbow and Shoulder Joint Movements*. IEEE.

Ranchev. (2015). *A fluid drive device*. Patent Application (BG, N°112191/28.12.2015).

Shigeki, T., & Yonetake, J. (2006). UltraSonicMotor Powered Assisted Suit System. *Society of Biomechanism Japan, 30*(4), 189–193. doi:10.3951/sobim.30.189

Sledd, A., & O'Malley, M. K. (2006). Performance Enhancement of a Haptic Arm Exoskeleton. *Symposium on Haptic Interfaces for Virtual Environment and Teleoperator Systems.*

Tsagarakis, N. G., & Caldwell, D. (2003). Development and control of a "soft-actuated" exoskeleton for use in physiotherapy and training. *Autonomous Robots*, *15*(1), 21–33. doi:10.1023/A:1024484615192

Tsagarakis, N. G., Tsachouridis, V., & Caldwell, D. G. (2003). Modeling and control of a pneumatic muscle actuated joint using on/off solenoid valves. *Proceedings of the IEEE International Conference on Advanced Robotics*, 929–934.

Veneva. (2015). *Lower limb with active joints*. Patent Application (BG, N°112139/06.11.2015).

Williams, R., Murphy, M., North, D., Berlin, J., & Krier, M. (1998). Kinesthetic force/moment feedback via active exoskeleton. *Proc. 1998 Image Conference*.

Yokoi, H., Yamashita, J., Fukui, Y., & Shimojo, M. (1994). Development of the virtual shape manipulating system. *Proc. of the 4th Int. Conf. on Artifcial Reality and Tele-Existence (ICAT'94)*, 43-48.

Chapter 4
Hybrid Exoskeletons for Upper Limb Stroke Rehabilitation

Ashley M. Stewart
University of Canterbury, New Zealand

Christopher G. Pretty
University of Canterbury, New Zealand

Mark Adams
Burwood Hospital, New Zealand

XiaoQi Chen
University of Canterbury, New Zealand

ABSTRACT

Hybrid exoskeletons are a recent development, combining electrically controlled actuation with functional electrical stimulation, which potentially offers great benefits for muscular rehabilitation. This chapter presents a review on the state of the art of upper-limb hybrid exoskeletons with a particular focus on stroke rehabilitation. The current needs of the stroke rehabilitation field are discussed and the ability of hybrid exoskeletons to provide a solution to some of the gaps in this field is explored. Due to the early stage of development which most hybrid exoskeletons are in, little research has yet been done in control methods used for them. In particular, more investigation is needed with regards to the potential benefit of hybrid exoskeletons as a patient-monitoring and rehabilitation assist-as-need tool.

INTRODUCTION

Stroke is the second largest cause of disability worldwide after dementia (World Heart Federation, 2015). Each year the costs of stroke on society are considerable, both financially and with regards to the wellbeing of the affected individual, their families, and their friends. Approximately 1 in 6 people will suffer a stroke during their lives and these numbers are only likely to increase as the number of elderly increase (World Stroke Campaign, 2015). Effective rehabilitation after a stroke is important if the costs of stroke on society are to be minimised.

DOI: 10.4018/978-1-5225-2993-4.ch004

A stroke can cause many impairments which interfere with a person's ability to perform tasks with their affected limbs. These will be described in the next few paragraphs. Temporary hemiparesis is common among stroke survivors. Regaining strength and movement in the affected side takes time and can be improved with the use of rehabilitation therapy involving repetitive and function-specific tasks (Senelick, 2010). Patients who undergo stroke rehabilitation generally have a few sessions a week with a physiotherapist where they undertake various exercises. Many patients, particularly the external patients, will also be assigned exercises to complete outside of their regular sessions with the physiotherapist.

The abilities of stroke patients can vary quite considerably. While it is more common for a patient to regain strength in their shoulder before the elbow and fingers respectively there are also many exceptions. The rate at which patients improve and the joints with which they struggle with can vary significantly from patient to patient. Muscle atrophy is one of the most common issues that occurs after a stroke. Muscle atrophy is the wasting away of the muscles which leads to a loss in the size and strength of the muscle. For each day a patient is in hospital lying in bed with minimal activity approximately 13% of muscular strength is lost (Ellis, Jackson, Liu, Molloy, & Paterson, 2013). Regaining muscle strength and motor control are very important if a patient is to regain independence in their day to day life. Muscle atrophy can only be prevented by physically working the muscles either through the patient's own volition or the use of Functional Electrical Stimulation (FES).

Problems that patients have after a stroke are not just limited to lack of strength in the affected limbs. Some patients can be very capable of movement but lack sensation. This makes it difficult for a patient to know how much force to apply when manipulating objects such as holding a glass of water. It can also make it difficult for a patient to perform movements with their limb without observing where the limb is. This increases the mental energy required to undertake everyday tasks and can even pose a danger to the patient's limb as they are not always aware of where it is. Other problems include muscle spasticity. Pharmacological and non-pharmacological treatments can combat spasticity.

A lack of control or strength in one joint may affect the ability of a patient to perform movements with another joint. For example, lack of control or strength at the wrist may make performing exercises with the hand difficult as the forearm may not be in the ideal positioned to allow a strong grip. Some patients may adopt compensatory patterns of movement due to their impairments. Over a longer period, these compensatory patterns can lead to long term issues such as pain. Some patients experience involuntary movements where the limb will move unpredictably without the patient intending it to. This can cause damage to the limb and may be embarrassing for some people. In addition, patients can often also suffer from reduced mental cognition and increased mental fatigue.

There are many different types of training strategies and devices which can be employed as tools to help combat the problems which stroke patients face. Despite all the interventions available however, there remain several gaps in the field of stroke rehabilitation. The current tools and rehabilitation methods used, as well as the gaps in the stroke rehabilitation field, will be described in the background section of this chapter. Following on from this the potential of hybrid exoskeletons to help fill some of these gaps will be discussed, an evaluation of the state of research with regards to upper limb hybrid exoskeletons will be undertaken, and future directions suggested.

BACKGROUND

Modern Stroke Technologies and Rehabilitation Methods

Devices used to aid stroke patients may be divided into passive devices and active devices. Passive devices, sometimes involving lockable and unlockable brakes or springs, can be used to limit a patient's movement either by locking a joint in place or by preventing a joint from moving too far in a direction. This can prevent a patient from hyper-extending or hyper-flexing or alternatively may be used to allow a patient to focus on a specific movement in isolation. Sometimes simple strapping tape may be used to remind a patient not to hyper-extend or hyper-flex. Passive devices can also be used to manage tremor or provide an elastic force for resistive training (Rocon et al., 2007).Active devices are devices which can move a patient's limbs on their own without help from the patient. Active devices are used to assist patients with movements that they may not be able to perform themselves or to provide resistive forces. Active devices may involve the use of functional electrical stimulation (FES), or an actuator, such as an electric motor, or pneumatic artificial muscle. Active devices are often heavier than passive devices due to the weight of the actuators however, they tend to be more useful for patients in the early stages of rehabilitation who have limited movement.

The training which a stroke patient will undergo can be either resistive or assistive. Resistive training involves the use of passive or active devices to provide a force against the movement of the patient. Coloured elastic bands are a common way of performing resistive training where one end will be tied to a static point and the patient will perform movements that stretch the rubber band requiring the patient to work harder to perform the movement. Alternatively resistive devices such as robots or exoskeletons may provide either a passive elastic force or an active (motorised/pneumatic/hydraulic) force in the opposite direction to which the patient wants to move (Carpi, Mannini, & De-Rossi, 2008). Resistive training is a good way for a patient with some movement already to build more muscular strength.

Assistive training involves the use of primarily active devices to provide an aiding force in the same direction as the movement of the patient. Assistive training can also include passive gravity balancing devices which reduce opposing forces for movement of limbs away from Earth (Gijbels et al., 2011). Assistive devices may be purely for rehabilitative training, for help with activities of daily living (ADL) and specific tasks, or a combination of the two. Assistive training may involve a physiotherapist providing the assistive forces, the use of functional electrical stimulation (FES), or an actuated robotic device.

Robotic devices have the potential to offer significant benefit to stroke rehabilitation when used in conjunction with physicians (Loureiro, Harwin, Nagai, & Johnson, 2011). They are capable of providing a higher frequency of training and allow for a safer way for a patient to practice movement without having to be directly supervised by a health professional (Loureiro et al., 2011; Maciejasz, Eschweiler, Gerlach-Hahn, Jansen-Troy, & Leonhardt, 2014). Robotic devices may include end-effector devices or exoskeletons. End-effector devices are located distally to the patient's limb. They are often simpler in design however as the movement of end effector devices increases in variety and range, the control becomes much more complex if injuries and redundant configurations are to be avoided. The general design of end-effector devices lends itself to be much less susceptible to portability than that of exoskeleton devices. Therefore, end-effectors are predominately used for rehabilitation purposes as opposed to assistance with ADL (Vaca Benitez et al., 2013). Exoskeletons are fitted more closely to the patient's body and closely mimic natural human movement. They are more complicated in structure but allow for independent control of different joints and much more precise control over a patient's limb (Maciejasz et al., 2014). However, many exoskeleton devices are also often limited in portability due to energy requirements and actuator weight.

The use of electromyography (EMG) to control and/or initiate movement can be used in conjunction with either active robotic devices or Functional electrical stimulation (FES). For a healthy person, and for many stroke patients, an attempt to move their limb will result in the production of muscle action potentials (MUAPs) at the site of the related muscle. These action potentials, measured in volts, and often in the range of 0-10mV peak, can be read using electrodes and signal processing and offer an indication of the subject's intended movement (De Luca, 2002). Despite the presence of the signals, the patient often lacks the strength to perform the movement. The EMG signals appear shortly before movement actually takes place thus it is possible to use EMG to initiate and control a device which can help the individual perform their intended movement in real-time (Chandrapal, 2012). This can allow for a larger variety of different movements to be produced by assistive devices rather than pre-programming in only specific movements and leaves the user free to use both their hands as they do not need to press a button or move a joystick to operate the device. Stroke rehabilitation training with EMG initiated and controlled devices also offer advantages in terms of patient improvement as when the patient observes the intended movement occurring it mentally reinforces and establishes the correct neural pathways required for self-production of the movement (Hara, 2013; Vaca Benitez et al., 2013). For patients who lack the production of action potentials at the muscle site electroencephalography (EEG) can be used instead, although this typically means a longer set up time.

Functional electrical stimulation (FES) is the application of high frequency electrical pulses to the nerves in order to elicit contractions in the muscle. There are many commercially available FES devices which are used for stroke rehabilitation. FES may be used to provide assistive training where the FES is initiated in response to a patient pushing a button or by some other intention estimation method (e.g EMG) with the goal being to practice performing a specific movement. Alternatively, it may be used at a lower dose with the main aim being not to produce significant movement of the limb but rather to strengthen the muscle through small cyclically induced muscle contractions. For stroke patients cyclically induced FES is often used to prevent subluxation of the shoulder muscle (Chae, Sheffler, & Knutson, 2008). FES devices also often have preprogramed sessions available which can be initiated with the single press of a button and which allow the patient to take the device home to use without worrying that they might alter the settings accidently. FES can help reduce muscle atrophy and increase muscle strength but it is less useful for aid with ADL on its own, as FES is difficult to control precisely, and quickly results in muscle fatigue (Doucet, Lam, & Griffin, 2012; Maciejasz et al., 2014). More recently FES has been combined with passive orthoses and robotic devices (Bioness Inc, 2015; Loureiro et al., 2011).

Requirements and Current State of Stroke Rehabilitation Robotics

Despite the expected increase in demand for at-home rehabilitation systems and ADL supporting devices many currently available rehabilitation robotic devices are lacking in portability, complex to use, and extremely expensive. Therefore, they are only suitable to use in clinics under the direct supervision of physicians (Lu et al., 2011; Maciejasz et al., 2014). If a patient's access to at home rehabilitation devices increases then this should also lead to an increase in the frequency of their training (Lu et al., 2011). It has been shown repeatedly that an increased dosage of training is highly positively correlated with patient improvement. Thus a higher uptake of at-home rehabilitation devices should result in increases in patient improvement (Vaca Benitez et al., 2013). To encourage a higher uptake of at-home rehabilitation devices there needs to be a focus on designing rehabilitation devices, especially exoskeletons, which are cheap, portable, and easy to use.

A study (Biddiss & Chau, 2008) which surveyed a range of prosthetic hand users indicated that the most important requirement as defined by the users was a reduction in the weight of the device. A low-cost prosthetic was ranked as the third most important feature. Both cost and weight were considered more important than precise movement of the device. While this study was of prosthetic hand users and not exoskeleton users it is quite likely that a similar trend will occur for exoskeleton users. Patients are not likely to be tolerant of rehabilitation devices which must be constantly removed and re-attached, especially if the devices are time-consuming to attach or noticeable (heavy or bulky) when worn. It is therefore important that an exoskeleton designed for rehabilitation purposes must be lightweight as well as quick and easy to attach. A well-working exoskeleton is useless if patients will not use it.

Assessment tools currently used by physicians to monitor patient progress tend to be time-consuming, subjective, and lacking in sensitivity (Loureiro et al., 2011). Some of the more commonly used scales are only moderately reliable when patient improvements are small (Patton, Small, & Zev Rymer, 2008). Robotic devices offer new, more sensitive, and consistent options for monitoring of patient improvement but they need to be able to provide measures that can be easily understood by physicians and related to current methods. There is little consensus how to best evaluate the effect of robotic systems on patient progress and a wide variety of measures have been used throughout current research (Del-Ama et al., 2012; Lu et al., 2011; Maciejasz et al., 2014; Sivan, O'Connor, Makower, Levesley, & Bhakta, 2011). There is also a lack of wide-scale assessments of robotic devices. This can likely be contributed to the high cost to produce the devices (Loureiro et al., 2011). Some attempt has been made to assess and compare the pros and cons of current rehabilitation measures and create a standard for assessing the effect of robotic devices on patient improvement (Sivan et al., 2011). One of the more widely used scales in literature related to rehabilitation robots is the Fugl-Meyer scale (Sivan et al., 2011).

It is highly desirable in stroke rehabilitation robotics that a robot or exoskeleton be capable of performing assistance-as-needed. This way the patient is encouraged to make the effort to achieve movement rather than learning to rely on the robot to perform the movement (Jarrassé et al., 2014; Lu et al., 2011; Maciejasz et al., 2014). Appropriately timed action is more important than strength for functional gains, however repetitive practice which builds strength without specific functional application can still help to diminish impairment (Patton et al., 2008). The ability of rehabilitation robots to adapt to different users and even to the same user on different days or throughout the same session is also highly important with regards to minimising set-up time and cost of rehabilitation (Maciejasz et al., 2014). Compliance of assistive robots is another important factor to consider as the robot should aid and encourage but not limit the movement of the patient (Loureiro et al., 2011; Maciejasz et al., 2014). This is important for safety as well. The exception to this is robots which are deliberately used to temporarily constrain certain joints or for resistive training. However, consideration of the robot's interaction with the user in a safe manner is still the most important feature. Above all else the robot should pose no harm to the user or nearby individuals.

REVIEW OF UPPER LIMB HYBRID EXOSKELETONS

The Potential Benefits

In future, we may find that a hybrid of FES and robotics may be the most efficient for providing continuous locomotion or performance of vital activities of daily living in individuals with paralysis. (Doucet et al., 2012)

Electrically controllable actuated exoskeletons offer huge advantages in their ability to repetitively and precisely provide assistance or resistance to a user. However electrically controlled actuators which provide the required forces are often heavy in weight and have high power requirements which limits portability. In addition, external movement of a patient's limb without effort from the patient does nothing to combat muscle atrophy. Thus exoskeleton devices are more useful to patients who already have some ability to move their limb (Hara, 2013).

Functional Electrical Stimulation (FES) allows for a much lighter device which is also better suited to reducing muscle atrophy in patients with no, or extremely limited, movement. However, precise control of FES is extremely difficult and producing the specific, repetitive, and functional movement desired for effective stroke rehabilitation is not easily accomplished. Furthermore extended use of FES is limited by the introduction of muscle fatigue caused by the unnatural motor unit recruitment order (Doucet et al., 2012). The forces required for large movements, such as shoulder abduction, are often too great to be provided by the use of FES which is much better suited to smaller movements such as finger extension (Pylatiuk et al., 2009; Schill et al., 2011). Some patients also find the use of FES painful.

Combining the use of FES and an electrically controlled actuator within an exoskeleton can potentially result in a device which is overall lighter (and thus more portable) than a non-hybrid exoskeleton and which can produce greater forces and more precise control than FES on its' own. Using FES to provide some of the required forces allows the use of smaller actuators and a reduction in the energy requirements of the device. Likewise, as some of the forces are provided by the actuator the FES does not need to be as intense to produce the same movement thus reducing the amount of pain at the electrode sites and delaying muscle fatigue.

The combination of devices potentially offers a greater ability to precisely monitor patient progress. FES in combination with electromyography (EMG) can provide information on the muscle fatigue of the patient which, when combined with the consistency in movement provided by the electrically controlled actuators, could be useful for tracking patient progress across several sessions. With these features, hybrid exoskeletons have potentially great benefits for upper-limb stroke rehabilitation. However, there is little literature reporting the use of such devices. This chapter reviews the state of the field and proposes that there is a need for a portable, hybrid, assist-as-need exoskeleton for upper-limb stroke rehabilitation.

A review has previously been conducted on hybrid exoskeletons for lower limb rehabilitation of spinal injury patients (Del-Ama et al., 2012) and readers may refer to this if they wish to learn more about lower limb hybrid exoskeletons. It should be noted that in stroke rehabilitation "often, more severe impairments are found in the upper limb" (Lu et al., 2011) yet rehabilitation research focus is largely placed on recovery of function in the lower limb. All further discussion within this chapter will be with regards to the upper-limb and focus will mainly be from a stroke rehabilitation perspective.

A recent review was conducted on hybrid robotic systems for upper limb stroke rehabilitation (Resquín et al., 2016) however it predominately covered passive devices and actively actuated end-effector devices.

No actively actuated hybrid exoskeletons were reviewed. The main focus of this chapter will be on upper-limb hybrid exoskeletons which use an electrically controlled actuator in combination with functional electrical stimulation to provide active assistance or resistance to the user. The FES and actuator do not necessarily need to be operating on the same joint and the goal of the hybrid exoskeleton may be purely rehabilitative or it may be designed to aid with ADL during rehabilitation.

A search was conducted in the IEEE, PubMed, ScienceDirect, and Google Scholar databases for relevant papers added prior to October 2016. A total of 12 different exoskeletons were found which combine an electrically controlled actuator with FES to provide active assistance or resistance to the user. Table 1 presents a summary of these 12 reviewed upper limb hybrid exoskeletons.

The results are discussed in five subsections based on the key features of exoskeletons in stroke rehabilitation. The subsections are as follows: Firstly, the suitability of hybrid exoskeletons to aid movement

Table 1. Reviewed upper limb hybrid exoskeletons

System	Supported Movements	Portable	Stage of Development	FES	Actuator	Main Control Input	Control Methods	Feedback	Advantages	Possible Improvements
Orthojacket (Schill et al., 2011)	Shoulder, elbow (actuator only). Wrist and fingers (FES only).	Fixed to wheelchair. Elbow and wrist sections of the device can be made portable.	Investigative study on 3 healthy subjects (elbow only). Investigative study on 1 SCI subject (elbow only).	Motionstim 8 (wrist, fingers)	Stepper Motors (Shoulder). Fluidic actuators (elbow).	sEMG and/or joystick	sEMG proportional control.	Speed, position	Modularity. FES electrode array.Can be worn under clothing.	Full device has not yet been tested. Only testing of the modules have been done. No fatigue monitoring.
Haptic Robotic Glove (Hartopanu, Poboroniuc, Serea, Irimia, & Livint, 2015)	Fingers, thumb	No	Prototype	Motionstim 8	Firgelli linear actuators	Key/button press	Button proportional control.	Position	Allows independent control of fingers.	No fatigue monitoring. Response time is slow.
FES/Robot Hand (Wei Rong, Tong, Hu, & Ho, 2013)	Fingers	No	Pilot clinical study on 4 stroke subjects over 20 sessions.	2 channel stimulator	Firgelli linear actuators	sEMG	Model based. Open loop.	Position – for display only.	Shown to improve movement accuracy as well as voluntary effort and subject focus.	Some patients had trouble with the weight of the device. No fatigue monitoring.
Wearable Rehabilitation Robot (Tu, Huang, Yu, Xu, & He, 2012)	Shoulder, fingers	No	Prototype	Beurer EM41	Pneumatic muscles	Key/button press	PID control of a predefined trajectory.	Position	Demonstrates that hybrid exoskeletons can reduce the weight of the device with no performance loss.	Predefined trajectory. No fatigue monitoring.
EXOSLIM (Serea, Poboroniuc, Hartopanu, & Irimia, 2015)	Shoulder, elbow	No	Prototype	Motionstim 8	DC motors	Key/button press	Model based. Button proportional control.	Position, acceleration		No fatigue monitoring.
BCI-Controlled Exoskeleton (Elnady et al., 2015)	Elbow, wrist (actuator only). Fingers (FES only).	Yes	Investigative study on 9 stroke subjects over 2 sessions 7 days apart.	RehaStim 1 (fingers)	BLDC motor (elbow). Brushed DC motor (wrist).	EEG (Emotiv EPOC neuroheadset)	State machine with proportional control of predefined trajectories.	Position	Easy to put on (<30 s with help).	Predefined trajectory. No fatigue monitoring.

continued on following page

Table 1. Continued

System	Supported Movements	Portable	Stage of Development	FES	Actuator	Main Control Input	Control Methods	Feedback	Advantages	Possible Improvements
MUNDUS (Ambrosini et al., 2009; CROSTA, 2012; Pedrocchi et al., 2013)	Shoulder, elbow (FES only). Fingers (actuator only or FES only, not both)	No. Fixed to wheelchair. Restricted to table.	Investigative study on 1 SCI subject for the hybrid part. Test was unsuccessful.	RehaStim 1 (shoulder, elbow)	DC motors (fingers)	EEG (other control options available as well, e.g. sEMG or head/eye movement).	Intergral control, or Biomimetic feedforward control, or Adaptive feedforward control.	Position, force, object location	RFID tags for object recognization. FES electrode array. Modularity.	Trial was unsuccessful for the hybrid part.Takes 35-40 minutes to set up and calibrate.
BCI-Controlled Wearable Robot (Looned, Webb, Xiao, & Menon, 2014)	Elbow, wrist (actuator only). Fingers (FES only).	Yes	Investigative study on 5 healthy subjects.	EMPI300 (fingers)	BLDC motor (elbow, wrist)	EEG	State machine with proportional control of predefined trajectories.	Position, acceleration	Easy to put on (<30 s with help).	Predefined trajectory. Not backdrivable. No fatigue monitoring.
RUPERT (Tu et al., 2014)	Shoulder (actuator only). Elbow. Fingers (FES only)	Unknown (tethered?)	Investigative study on 3 healthy subjects.	RehaStim 2	Pneumatic muscles	sEMG	sEMG PID control.	Position, pressure, Kinect		No fatigue monitoring.
THINK2GRASP (Rohm, Schneiders, Kreilinger, Müller-Putz, & Rupp, 2012; Rohm et al., 2013; Rupp et al., 2013)	Elbow (actuator only and not all users). Fingers (FES only).	No	An investigative study was done on hybrid portion of the device with 2 SCI subjects.	Motionstim 8 (fingers)	Electric motor (elbow)	EEG + shoulder position sensors	State machine for type of movement (EEG). Shoulder position sensors control degree of movement. Open loop.	System state, and joystick angle – for display only	Neoprene sleeve with Velcro attached electrodes. Provides visual feedback to the patient. Modular.	Subject required trunk restraint in order to produce normal correct movement. Takes at least an hour to set up and the device caused mental fatigue due to complex control. Control is switched between elbow and hand.
Early Hybrid Exoskeleton (Varoto, Barbarini, & Cliquet Jr, 2008)	Elbow (actuator only). Fingers (FES only).	Fixed to wheelchair.	Investigative study on 5 quadriplegic subjects.	2 channel stimulator	Electric motor (elbow)	Voice	Neural network processes pretrained voice commands.	Force, position, 3D visual motion capture.	Force feedback is provided to users.	Requires movement in the shoulder.
ALEx (Crema et al., 2015)	Shoulder, elbow (actuator only). Fingers (FES only).	No	Investigative study on 1 healthy patient of the elbow and shoulder modules.	Rehastim	Electric motor (shoulder, elbow)	GUI - key press	Impedance control.	Position, impedance	Back-drivable. Adjustable assistance/ resistance.	No fatigue monitoring. Possible misalignments with the shoulder of the subject and not adaptable to different sized patients.

in the different joints is discussed. The second subsection discusses ability of hybrid exoskeletons to provide a portable aid and the impact of actuator selection. The third subsection addresses hybrid exoskeletons as a patient monitoring tool. The fourth subsection discusses the various control schemes used as well as intention estimation methods. The final subsection presents an evaluation of the usability of the current upper-limb hybrid exoskeletons and the clinical evidence regarding the benefits and costs of hybrid exoskeletons. Future directions are then discussed and several concluding remarks summarizing the chapter are made.

Supported Movements

Presently when FES is used on the upper-limb for stroke rehabilitation it is either used to prevent subluxation of the shoulder or to aid with grasping movements in the hand. Rarely, if ever, is FES used to actuate the shoulder or elbow joints. There are several reasons for this. FES induced fatigue is considerably less pronounced for movement of smaller limbs, such as the fingers. In addition, for some patients FES is only capable of producing limited forces and so is much less suited to larger movements. There is also greater difficulty in controlling FES-induced movements when they are very large. It is much easier for a patient to practice only opening and closing of the hand than it is for a patient to practice only shoulder or elbow movement. Any movement of the shoulder and elbow will also result in movement of the hand. When FES is used to help actuate shoulder or elbow movements a physiotherapist or other helper is generally required to hold the lower part of the patients' arm to keep it from flopping about uncontrollably.

The joints which electrically controllable actuated exoskeletons are used for in stroke rehabilitation depend greatly on where the weight distribution of the exoskeleton is predominately located. Exoskeletons and end-effectors which place weight through some medium other than the patient themselves (e.g. the floor, a table, a chair, a lower limb exoskeleton) also often tend to be used more for fingers and wrist movement. As mentioned previously, functional gains are predominately related to regaining movement in these joints and most patients tend to regain movement in the shoulder and elbow first so it is no surprise that the fingers and wrist joints receive so much attention. However, the movement of the arm as a whole should not be ignored. Many functional tasks require control of all of the joints in the upper-limb. The ability to reach out, grasp a cup, bring it to the mouth, take a drink, and return it to the table, is an example of a key functional movement that utilises all of the main joints in the arm. Evidence strongly indicates that rehabilitation which practices movement of the arm as a whole produces better clinical outcomes than rehabilitation which focuses only on one or two joints (Resquín et al., 2016).

For electrically controllable actuated exoskeletons which distribute their weight predominantly through the user, focus is typically on the more proximal joints. Actuator weight is a much bigger issue the further one gets from the body, especially when one considers that stroke patients are even less capable than healthy individuals of supporting added weight on the end of their arm. Even when placed close to the body's centre of gravity actuator weight can still be an issue.

Of the twelve upper-limb hybrid exoskeletons in Table 1 only five use FES and an electrically controlled actuator for the same joint. With one exception, in cases where FES and an electrically controlled actuator are used for different joints the FES is used for the fingers and the electrically controlled actuator is used for the elbow and shoulder. All of the exoskeletons with the exception of one, attempt to aid with movement of the fingers. In contrast, only a little over half of the exoskeletons attempt to aid with movement of the shoulder joint. The distribution of actuation type used in these exoskeletons is displayed in Table 2 and may be further divided up as shown in Table 3.

When grasping an object like a cup the fingers must first extend then flex while the wrist remains extended. Generally, using FES to flex the fingers will also result in flexion of the wrist. Some devices which aid with grasping hold the wrist in a constant position while FES is used to close the fingers. The hybrid exoskeletons which use FES only for grasping and an actuator only for shoulder/elbow movements vary in their treatment of the wrist joint. The wrist joint may be fixed, actuated, or neither. Despite the differences in wrist support, all four exoskeletons which report results on reach and grasp performance using this combination, enabled the users to achieve successful reach and hand-opening movements or reaching and grasping of objects such as a cup or pretzel (Elnady et al., 2015; Looned et al., 2014; Rohm et al., 2012; Rohm et al., 2013; Rupp et al., 2013; Varoto et al., 2008).

In the more proximal joints fatigue from FES is a much bigger issue, as is the case with actuator weight on more distal joints. MUNDUS (Ambrosini et al., 2009; CROSTA, 2012; Pedrocchi et al., 2013) is unique in its use of FES only for actuation of the shoulder and elbow joints in combination with use of an actuator only for finger movements. It is important to note that FES for the shoulder and elbow is not used entirely on its own for this exoskeleton but is done in combination with passive elements (springs and elastic wires) or electric lockable/unlockable breaks which provide support against gravity. The use of FES in combination with passive components for the elbow and shoulder was shown to result in a reduction in the energy required from the patient to perform tasks. The testing of the grasping component of this exoskeleton however, was not successful due to the weight of the actuators.

The size and weight of the actuators was also found to be an issue for two out of three of the exoskeletons using hybrid actuation for the fingers; the Haptic Robotic Glove and the FES/Robot Hand (Hartopanu, Poboroniuc, Serea, Irimia, & Livint, 2013; Hartopanu, Poboroniuc, Serea, & Livint, 2014; W Rong, Tong, Hu, & Ho, 2012; Wei Rong et al., 2013). Despite its' issues with actuator weight, the FES/Robot Hand was able to demonstrate a decrease in the muscular effort required to perform tasks and a decrease in unnecessary muscular activity. Artificial tendons proved to be a solution to the actuator weight issues for the Haptic Robotic Glove (Hartopanu et al., 2015) as they allowed placement of the actuators to be situated closer to the centre of mass of the patients' body. Initial validation of the Haptic Robotic Glove appears to have been successful with regards to performing the desired movement however the device was limited by its slow response time.

The Wearable Rehabilitation Robot (Tu et al., 2012) successfully demonstrated the ability of hybrid exoskeletons to improve the accuracy, and reduce the weight and energy requirements, when compared respectively to an electrically controlled actuator or FES alone for both grasping and shoulder movement. Actual weight reduction of the exoskeleton has yet to be done however. The other two exoskeletons

Table 2. Distribution of actuation type by joint location

	Number
Fingers/wrist – Electrically controlled actuator + FES	2
Shoulder/elbow – Electrically controlled actuator + FES	2
Shoulder/elbow and fingers/wrist – Electrically controlled actuator + FES	1
Fingers/wrist – FES, shoulder/elbow – Electrically controlled actuator	6
Fingers/wrist – EC actuator, shoulder/elbow – FES	1
Total	12

Table 3. Distribution of actuation type by joint

	Fingers	Wrist	Elbow	Shoulder
Electrically controlled actuator and FES	3	0	2	2
Electrically controlled actuator only	1	2	6	3
FES only	7	1	2	1

which use hybrid actuation for the shoulder and/or elbow have only been developed very recently. The EXOSLIM and RUPERT use hybrid actuation for the shoulder and elbow, and the elbow respectively (Serea et al., 2015; Serea, Poboroniuc, Hartopanu, & Olaru, 2014; Serea, Poboroniuc, Irimia, Hartopanu, & Olaru, 2013; Tu et al., 2014). So far only limited testing has been conducted of these two exoskeletons but initial results are promising.

Actuator Selection and Portability

Actuator selection for an exoskeleton greatly influences several key features of resulting device and in many cases selection of one type of actuator results in a trade-off somewhere. An ideal actuator is lightweight, powerful, fast, easy-to-control, durable, portable, cheap, and compliant. Portability of an exoskeleton may be divided into three key areas; ease of relocating the device, portability with a wheelchair, and full portability. Even if the device must be plugged into a wall or computer the ability to take it home enables the patient to perform rehabilitation away from the hospital, reducing travel costs, and increasing the frequency and repetitiveness of training. Most of these exoskeletons, with a few exceptions could likely be easily relocated to a different location. There are a couple of exceptions, the Wearable Rehabilitation Robot and RUPERT (Tu et al., 2014; Tu et al., 2012), which are quite bulky and require large air supplies.

Various types of pneumatic actuators are used by three hybrid exoskeletons in this review; the Wearable Rehabilitation Robot, RUPERT, and the OrthoJacket (Pylatiuk et al., 2009; Schill, Rupp, & Reischl, 2009; Schill et al., 2011; Schulz et al., 2009; Schulz, Schmitz, Wiegand, Pylatiuk, & Reischl, 2011; Tu et al., 2014; Tu et al., 2012; Wiegand, Schmitz, Pylatiuk, & Schulz, 2011). Pneumatics can be light weight and are very compliant. However, they are difficult to control due to a non-linear performance, they can be slow to operate, and are they only capable of forces in one direction so must work in pairs. Perhaps the greatest barrier to the use of pneumatics is that they require a nearby air supply which can be difficult to make fully portable without adding significant weight. While the pneumatics used for the Wearable Rehabilitation Robot and RUPERT are quite bulky and pose as a barrier to portability, the OrthoJacket offers a much more compact solution.

The OrthoJacket is unique in its use of flexible fluidic actuators which may be actuated either by air or by liquid. These are lightweight and compliant and when pneumatic operation is used the supply easily fits under the wheelchair seat such that the OrthoJacket is portable with the use of the wheelchair. Typically hydraulic actuation sees more use in military lower-limb performance enhancement exoskeletons, such Raytheon XOS 2 (Army Technology, 2015), due to its' ability to produce large forces. However hydraulic actuation is usually very heavy and very expensive. Thus it is generally not well suited to medical rehabilitation applications, especially for the upper limb. Once again however, the OrthoJacket offers a unique solution, and when hydraulic operation of its's actuators is used the exoskeleton is capable of being made fully portable for the elbow and wrist sections (Pylatiuk et al., 2009). For the shoulder section, which uses an electric motor for actuation, the OrthoJacket is portable with the use of a wheelchair (Schill et al., 2011).

It should be mentioned here that only about half of the reviewed exoskeletons are aimed at stroke patients. The rest are aimed primarily at spinal cord injury (SCI) patients with some accounting for other neuromuscular and neurodegenerative dieses as well. Portability for SCI patients beyond the attachment to the patient's wheelchair is not often considered a high priority, as without a second exoskeleton for the legs the patient is not able to locomote independently anyway. On the other hand, stroke patients, who may have sufficient lower limb function but be lacking in upper limb function, can desire assistance with activities of daily living (ADL) which demand a higher level of portability.

Of the reviewed hybrid exoskeletons only the BCI-controlled Exoskeleton and the BCI-controlled Wearable Robot are fully portable for all of their joints (Elnady et al., 2015; Looned et al., 2014). In addition to the OrthoJacket, MUNDUS and the Early Hybrid Exoskeleton are also portable with the use of a wheelchair (Ambrosini et al., 2009; CROSTA, 2012; Pedrocchi et al., 2013; Varoto et al., 2008). Many of these exoskeletons are not fully portable because they are tethered to either a power/air supply or to a computer. For most of these exoskeletons, it would be possible to improve portability with wireless communications or a battery pack fixed on the user/wheelchair and in many cases future developments are planned to do so.

Electric motors are the most commonly used electrically controlled actuator among the reviewed hybrid exoskeletons and among upper-limb exoskeletons in general (Lo & Xie, 2012; Maciejasz et al., 2014), especially for movement of wrist, elbow, and shoulder joints. They are cheap, easy to control, compact, and powerful. The greatest limitations of electric motors are their weight and lack of compliance. Combining an electric motor with FES in an exoskeleton should reduce the amount of force needed and thus also the weight of the required electric motor. It is possible compliance could be improved by using methods such as Series Elastic Actuators (SEAs) (Veneman, Ekkelenkamp, Kruidhof, van der Helm, & van der Kooij, 2006). Electric motors have had most success on the proximal joints and are the actuators of choice in both reviewed exoskeletons which achieved full portability.

Two hybrid exoskeleton, the Haptic Robotic Glove and MUNDUS, attempted to use electric motors for finger actuation and neither was successful (Ambrosini et al., 2009; CROSTA, 2012; Hartopanu et al., 2013; Hartopanu et al., 2014; Pedrocchi et al., 2013). Linear actuators were then used as a replacement in the Haptic Robotic Glove. They were also used by the FES/Robot Hand (W Rong et al., 2012; Wei Rong et al., 2013). Linear actuators are slightly more suited for actuation of the fingers, however even these can be too heavy as was seen with the FES/Robot Hand. As mentioned earlier one solution to this may be the use of artificial tendons as used by the Haptic Robotic Glove. Linear actuators can potentially offer more compliance and reduced weight than electric motors, however they can quickly become very bulky if used for larger movements.

Patient Monitoring and Feedback

The vast majority of these hybrid exoskeletons are aimed purely at ADL assistance, or a combination of ADL assistance and rehabilitation, rather than rehabilitation alone. Thus, for these hybrid exoskeletons, more focus has been placed on achieving functional movement and reducing energy requirements, rather than on patient monitoring and feedback. Many of the assessment measures currently used in stroke rehabilitation consist of the physiotherapist evaluating the ability of a patient to perform a movement and assigning a course grading to this. In the case of the commonly used Fugl-Meyer Assessment, movements are assessed as; can perform, can partially perform, or cannot perform. Obviously, each of these measures, especially 'can partially perform', covers a very large range of abilities, and small improvements are impossible to distinguish. Use of an exoskeleton which can measure position, speed, and/or the amount of assistance needed to help a patient achieve a movement greatly increases the precision and range which a patient ability can be measured across and removes the subjectivity of the measurement.

Among these reviewed hybrid exoskeletons, in cases where the speed of the patient is monitored, such as the BCI-controlled Exoskeleton and the BCI-controlled Wearable Robot, the timing is not done by the exoskeleton itself but rather by the observer (Elnady et al., 2015; Looned et al., 2014). Often where error measurement is done for tracking it is not fed back to the user but is rather used simply for testing of the exoskeleton. Only in two cases, The FES/Robot Hand and ALEx, is the tracking task feedback an inbuilt function designed to be presented to the patient and/or therapist as a means of patient monitoring (Crema et al., 2015; W Rong et al., 2012; Wei Rong et al., 2013). In the case of ALEx, the impedance of the exoskeleton can also be adjusted and monitored, which when compared with the tracking error feedback, can help the physiotherapist to assess and adjust the difficulty of the tasks. Currently, this assistance adjustment is done manually.

Using sEMG to simply show the user and physiotherapist when an attempt to move is made, can also be very useful. The THINK2GRASP allows the patient to see the activation level and current state of the exoskeleton which can improve their ability to control the exoskeleton (Rohm et al., 2012; Rohm et al., 2013; Rupp et al., 2013). The FES/Robot Hand also monitors muscle activation levels however it is not known as to whether this is just for testing purposes of the exoskeleton or if it is intended to provide patient monitoring in a clinical setting. Other types of feedback can also be very useful when combined with exoskeleton rehabilitation. The Early Hybrid Exoskeleton provides haptic force feedback to the user which helps them relearn the force required to pick up objects and can especially aid patients who lack proprioception (Varoto et al., 2008).

At present, measurements of patient ability are largely limited to a combination of strength, precision, and speed. The ability of fatigue monitoring to be performed in combination with tracking and control of patient movements offers many advantages if fed back to the physiotherapist and could be extremely useful when selecting training programs. In addition, an exoskeleton could also be designed to automatically adjust its assistance in response to a combination of patient performance and physiotherapist commands. Little focus has been placed on muscular fatigue monitoring so far, let alone incorporating it into the feedback and control of the exoskeleton. Two of the exoskeletons, MUNDUS and THINK-2GRASP, monitor the muscular fatigue of the subject in order to test the performance of the exoskeleton (Ambrosini et al., 2009; CROSTA, 2012; Pedrocchi et al., 2013; Rohm et al., 2012; Rohm et al., 2013; Rupp et al., 2013) but none incorporate fatigue monitoring feedback in the control or user interface of the exoskeleton. While some of the exoskeletons are modular and can be adapted to the different patients, none are capable of automatically adjusting their assistance in response to muscle fatigue. Much more research is required in the area of muscle fatigue monitoring. Indeed, much improvement is needed in the area of patient monitoring in general.

Intention Estimation and Control

Compatible control of both the electrically controlled actuator and the electrical stimulation is one of biggest challenges facing hybrid exoskeletons. Control of exoskeletons is typically focused mainly on achieving the correct the movement of the exoskeleton however in the case of hybrid exoskeletons and any device using FES, control of muscle fatigue is also an important consideration.

Prior to achieving any type of control however, the desired movement must first be specified. Typically, the initialisation of movement is always specified by the user in some way, and the actual movement itself can be either user-controlled or user-initialised. If the movement is user-initialised then a predefined trajectory is used and the user has limited control once the movement has begun. If the movement is user-controlled, then the exoskeleton will produce a movement which is continuously proportional to the user's intent. This should not be confused with a proportional controller, which is a type of controller with a feedback loop, where the control signal is proportional to the size of the error between desired output and measured output (Figure 1).

User-controlled movement is used in most of these exoskeletons and is preferable for two reasons. Firstly, it is more useful for the patient with regards to helping them with ADL, and secondly a movement which accurately reflects that which the patient is attempting to perform has been shown to better help with the formation and reestablishment of neural pathways (Maffiuletti, Minetto, Farina, & Bottinelli, 2011). This can speed up the recovery of the patient and help them to relearn self-movement. It is however, much more complex to implement than user-initialised movement.

As described earlier, the simplest methods used to estimate the intention of the user is the use of key/button presses or a joystick. These methods are used in a few of these hybrid exoskeletons both as the sole method of intention estimation and in combination with other methods. The downside to using key/button presses or a joystick is that often then the patient's movement is limited in some way. Other methods for obtaining user intention include the monitoring of speech or eye movement. One exoskeleton, the Early Hybrid Exoskeleton, used voice recognition as its method of intention estimation. This method was correct in recognizing a particular command 92% of the time (Varoto et al., 2008). However, it should be noted that often speech can be impaired following a stroke so this method may not be suitable to all users. The exoskeleton, MUNDUS, which monitors eye movement was also successful in obtaining correct user intention. However, the eye-tracking unit used is mounted to a table and this method is thus limited in its portability (Ambrosini et al., 2009; CROSTA, 2012; Pedrocchi et al., 2013).

The use of EMG or EEG is much more portable and poses no limitations on movement. EEG was used by three of the reviewed exoskeletons, the BCI-Controlled Exoskeleton, the BCI-Controlled Wearable

Figure 1. An example of closed loop PID exoskeleton control

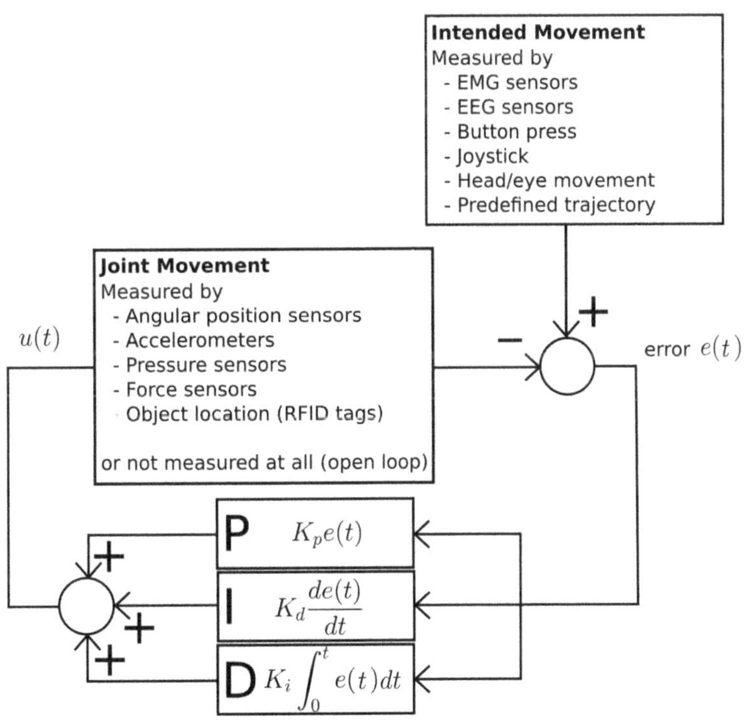

Robot, and THINK2GRASP. In all three of these exoskeletons EEG was used primarily to initialise movement which was either simply user-initialised or, in user-controlled exoskeletons, where the continuous control of the exoskeleton was done using a different method (Ambrosini et al., 2009; CROSTA, 2012; Elnady et al., 2015; Pedrocchi et al., 2013; Rohm et al., 2012; Rohm et al., 2013; Rupp et al., 2013). As described earlier, when EMG signals are present in a stroke patient, measurement of these is preferable over the use of EEG, as EMG is particularly useful for fatigue monitoring and does not require the training that EEG does. Three exoskeletons use EMG for user intention estimation, the OrthoJacket, the FESFES/Robot Hand, and RUPERT (Wei Rong et al., 2013; Schill et al., 2011; Tu et al., 2014).

Control of the reviewed exoskeletons is either open loop, or more predominately, proportional or a variation of proportional-integral-derivative (PID) control (Figure 1). Two exceptions are The Early Hybrid Exoskeleton, which uses a neural network to recognize and respond to predefined voice commands (Varoto et al., 2008), and ALEx, which uses impedance control to follow a predefined trajectory (Crema et al., 2015). In exoskeletons which are capable of multiple different movements state machines are often used to select which movement to perform. For example, THINK2GRASP uses a state machine based on EEG readings to select the type of movement and position sensors at the shoulder to control the degree of movement (Rohm et al., 2012; Rohm et al., 2013; Rupp et al., 2013).

The parameters typically used for FES vary depending on the specific muscle, patient, and other factors, and have been discussed elsewhere (Doucet et al., 2012). Little feedback control of FES parameters has been performed in these hybrid exoskeletons. Parameters are often adjusted until desired movement is achieved and then fixed at a set level. The FES/Robot Hand compared different combinations of FES and electrically controlled actuator support with the aim of reducing tracking error. They found the optimal combination was to provide equally balanced support from each (W Rong et al., 2012; Wei Rong et al., 2013). It is important to note that the FES/Robot Hand did not use feedback in the control of the exoskeleton however, and thus these results are specific only to open loop control. With feedback, different combinations of FES and electrically controlled actuator support may produce different results. An example of what the PID control loop of an exoskeleton may look like with feedback of muscle fatigue added is displayed in Figure 2.

Cycling pulses on and off has been shown to improve muscle recovery, reduce fatigue, and increase patient comfort (Doucet et al., 2012). It was noted earlier however, that minimal fatigue monitoring has been performed among hybrid exoskeletons so far and as such no automatic control or management of

Figure 2. An example of closed loop PID exoskeleton control with fatigue feedback

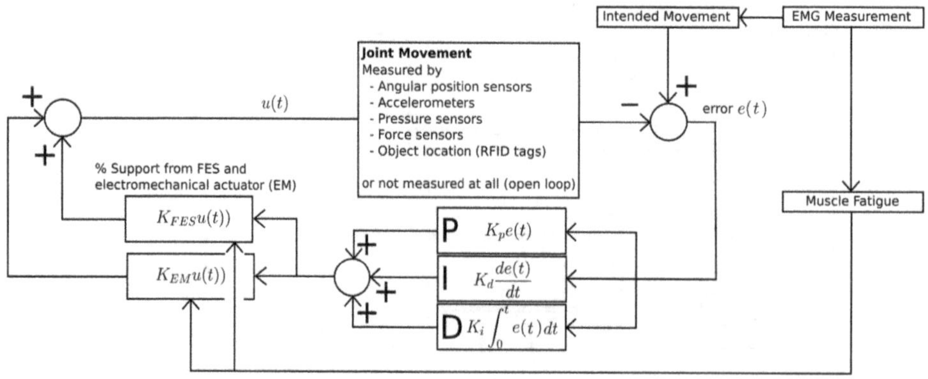

muscle fatigue has been done. In some hybrid exoskeletons, like THINK2GRASP, when muscle fatigue is noticed by the researcher the stimulation intensity is simply increased to compensate (Rohm et al., 2012; Rohm et al., 2013; Rupp et al., 2013). FES parameters such as frequency, pulse-width, ramp-time, and duty-cycle, can all affect onset of muscle fatigue. With the ability of the electrically controlled actuator to provide variable support the resulting movement of the hybrid exoskeleton need not alter with a variation in FES parameters. In such a case where multiple parameters of FES are varied it may be preferable to treat the muscle response from the FES as a disturbance and have the electrically controlled actuator compensate for the resulting output variation (Figure 3).

Usability and Trials

Usability is a key component in the ability of hybrid exoskeletons to improve patient outcomes. If patients and physiotherapist do not want to use the exoskeleton due to difficulties in using it or issues with comfort, then it does not matter how capable the exoskeleton is in improving patient outcomes.

The preparation time of the hybrid exoskeletons reviewed ranges from the very quick (<30 s with help) as seen with the BCI-controlled Exoskeleton and BCI-controlled Wearable Robot (Elnady et al., 2015; Looned et al., 2014), to the excessive (up to 1 hour) for the THINK2GRASP exoskeleton (Rohm et al., 2012; Rohm et al., 2013; Rupp et al., 2013). However, many of these exoskeletons do not detail the preparation time so it is difficult to fully evaluate the status of this area. A typical patient session is about an hour long and will sometimes be divided into one section for the upper-limb and one section for the lower-limb. For a hybrid exoskeleton to be useful it should not take up a significant portion of this time to achieve a ready-to-use state. In addition, if the patient is to use the device at home then they need to be able to attach it without help.

Some exoskeletons were difficult to use, caused mental fatigue, and/or required several training sessions prior to their use, as was the case with THINK2GRASP. In certain cases, like the Early Hybrid Exoskeleton, patients did not have sufficient limb movement to use the device (Varoto et al., 2008), or in the case of THINK2GRASP, were not sufficiently able to use the control interface. ALEx was only capable of fitting patients of a smaller and thinner stature and suffered from possible joint misalignment at the shoulder (Crema et al., 2015). On a more positive note some exoskeletons proved useful even

Figure 3. Closed loop PID exoskeleton control with fatigue modelled as a disturbance

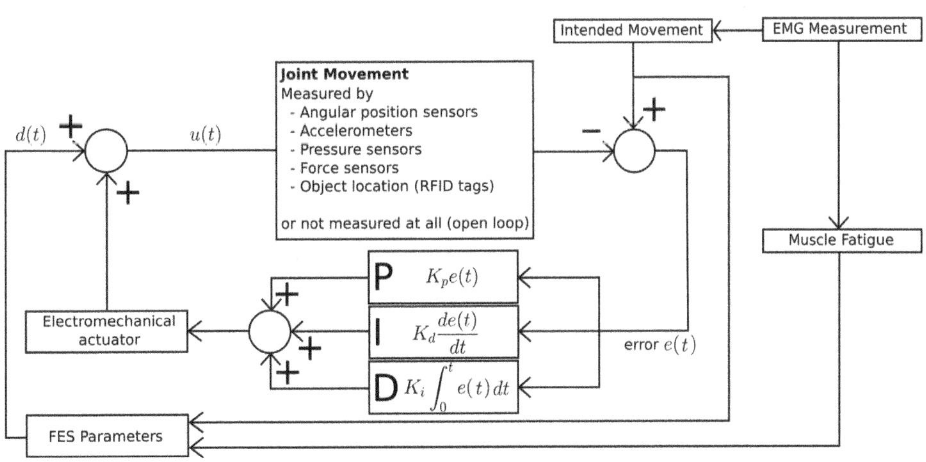

for patients with proprioception problems (Hartopanu et al., 2014), and others, such as the FES/Robot Hand, reduced the energy required of the user for a particular task (W Rong et al., 2012; Wei Rong et al., 2013). Some exoskeletons, such as the OrthoJacket, consisted of modular components and were capable of being adjusted to suit patients with different abilities.

The user interface of the hybrid exoskeleton needs to be very simple and intuitive to interact with, as cognitive ability and focus of a patient can be considerably impaired following a stroke. It is desirable that a patient have some control over the device without the need for direct observation. Control which a physiotherapist has over the exoskeleton should also be distinctively separate to the control which the patient has, much like the commercial FES devices which can have their parameters pre-set and then locked so the patient has no concerns about accidently changing physiotherapist settings. Modularity of an exoskeleton and the ability to fit different-sized or different-skilled patients is highly useful and desirable if adjustments are not overly time-consuming or complicated.

Hybrid exoskeletons are a very recent development and thus only limited testing has been done to date. In some cases, the entire exoskeleton has not even been tested as a whole, such as can be seen with the OrthoJacket, the Haptic Robotic Glove, and ALEx (Crema et al., 2015; Hartopanu et al., 2013; Hartopanu et al., 2014; Pylatiuk et al., 2009; Schill et al., 2009; Schill et al., 2011; Schulz et al., 2009; Schulz et al., 2011; Wiegand et al., 2011). In many cases the exoskeleton has only been tested on the designer themselves or a couple of healthy subjects. None of these exoskeletons have been tested on more than 10 subjects or over more than 20 sessions, and only one hybrid exoskeleton, the FES/Robot Hand, has been tested where established clinical measures (such as the Fugl-Meyer scale) have been performed on patients as an evaluative measure of the exoskeleton's ability to improve rehabilitation (W Rong et al., 2012; Wei Rong et al., 2013). In most cases only functional testing has been performed.

FUTURE RESEARCH DIRECTIONS

Overall most success towards achieving functional movement, such as a reach and grasp of an object, has been had with hybrid exoskeletons which do not utilize the combination of FES and electrically controlled actuation at the same joint, or which use a passive component in place of an active actuator. Indeed, both setups also result in a simplification of the control of the exoskeleton. The use of FES on its' own for grasping has also proven successful for allowing patients to achieve functional movement with sufficient precision. There is room for improvement with regards to increasing precision, however actuator weight is likely to remain a significant barrier towards achieving this using full hybrid actuation, as has been seen in the results of the hybrid exoskeletons which use full hybrid actuation for grasping. Other methods may be more useful here, such as electrode arrays which have previously demonstrated a significant increase in the precision of grasping movements (Malešević et al., 2012).

With regards to the shoulder and elbow joints there remains the issue that unless the patient has some movement in these joints already, using only electrically controlled actuation for these two joints can only help with neural repair but not with reducing muscle atrophy. This is especially true for patients early on in recovery who have very limited movement. It may be that exoskeletons with passive components are sufficiently suited in some cases to reducing exoskeleton weight, reducing muscle fatigue, and producing functional movement, however these exoskeletons still rely on the use of FES on its' own to produce the entire movement. In many cases forces produced from FES are insufficient even with the help of a passive component (Resquín et al., 2016). Precise control is also easier to achieve with the use of an active actuator. Hybrid exoskeleton development in this area is especially new. Initial results do indicate that hybrid exoskeletons may prove useful for helping to achieve functional movement and to improve clinical outcomes, especially for patients with limited movement, in these more proximal joints. Electric motors have had the most success so far although there is plenty of opportunity for research into novel methods of actuation. While hybrid exoskeletons have been shown capable of reducing the energy requirements of the exoskeleton, actuator weight remains a common issue. Placement of the actuator is an important consideration and cable-actuated exoskeletons offer a lot of flexibility here.

Hybrid exoskeletons present a unique opportunity to further investigate the effects of FES parameters on muscle fatigue and potentially to delay muscle fatigue onset. In all cases where the FES device has been specified a commercial device has been used instead of one being designed specifically for use in the hybrid exoskeleton. This makes it easier to incorporate hybrid exoskeletons into a clinical setting as most rehabilitation clinics already use FES, however, depending on the device used, it may limit the flexibility and controllability of the hybrid exoskeleton possibly reducing their potential advantages. While proportional, PID, impedance, and admittance control are simple to implement it may be that more robust control schemes, such as artificial neural networks (ANNs) or other types, are be better suited to take full advantage of the hybrid combination and to perform fatigue management, patient improvement monitoring, or assist-as-need.

There are currently no commercially available upper-limb hybrid exoskeletons and predicted prices for exoskeletons which do have a long term view of commercialization are not expected to be cheap (Pedrocchi et al., 2013). Due to cost of hybrid exoskeletons and indeed exoskeletons in general wide scale evaluation is difficult (Loureiro et al., 2011). Devices which can prove beneficial to a wider range of subjects will likely be easier to commercialize. Modularity, which has been attempted in some of these exoskeletons, may help to lessen the barriers to commercialization, however it is important that usability and comfort are not sacrificed.

CONCLUSION

At this stage, success has been most prominent among hybrid exoskeletons which are focused on the more proximal joints. Indeed, hybrid exoskeletons may be more useful for the shoulder and elbow joints while FES alone may be successfully used for grasping. Many of the reviewed hybrid exoskeletons are only in the initial stages of development with limited trials conducted thus far and the potential benefits of hybrid exoskeletons have yet to be fully taken advantage of. The control schemes used are typically simple in nature and none perform involved fatigue management. The assist-as-need paradigm has yet to be implemented in any hybrid exoskeleton. The usability of current upper-limb hybrid exoskeletons is mixed with many exoskeletons performing quite well in some areas but lacking in others. Additionally, predicted commercial prices are not expected to be cheap. Despite this, preliminary results are promising and the ability of hybrid exoskeletons to improve the energy efficiency of exoskeletons has been successfully demonstrated in prototypes.

REFERENCES

Ambrosini, E., Ferrante, S., Gfoeler, M., Reichenfelser, W., Karner, J., Schauer, T., . . . Pedrocchi, A. (2009). *A hybrid assistive system to support daily upper limb activities*. Academic Press.

Army Technology. (2015). *Raytheon XOS 2 Exoskeleton, Second-Generation Robotics Suit - Army Technology*. Retrieved from http://www.army-technology.com/projects/raytheon-xos-2-exoskeleton-us/

Biddiss, E., & Chau, T. (2008). Dielectric elastomers as actuators for upper limb prosthetics: Challenges and opportunities. *Medical Engineering & Physics*, *30*(4), 403–418. doi:10.1016/j.medengphy.2007.05.011 PMID:17632030

Bioness Inc. (2015). *H200 Wireless Hand*. Retrieved from http://www.bioness.com/Products/H200_for_Hand_Paralysis.php

Carpi, F., Mannini, A., & De-Rossi, D. (2008). *Elastomeric contractile actuators for hand rehabilitation splints*. Paper presented at the The 15th International Symposium on: Smart Structures and Materials & Nondestructive Evaluation and Health Monitoring. doi:10.1117/12.774644

Chae, J., Sheffler, L., & Knutson, J. (2008). Neuromuscular electrical stimulation for motor restoration in hemiplegia. *Topics in Stroke Rehabilitation*, *15*(5), 412–426. doi:10.1310/tsr1505-412 PMID:19008202

Chandrapal, M. (2012). *Intelligent assistive knee orthotic device utilizing pneumatic artificial muscles*. Academic Press.

Crema, A., Mancuso, M., Frisoli, A., Salsedo, F., Raschella, F., & Micera, S. (2015). *A hybrid NMES-exoskeleton for real objects interaction*. Paper presented at the Neural Engineering (NER), 2015 7th International IEEE/EMBS Conference on. doi:10.1109/NER.2015.7146710

Crosta, E. (2012). *An assistive device based on the detection of the user's intention from residual EMG to drive an upper limb neuroprosthesis*. Academic Press.

De Luca, C. J. (2002). Surface electromyography: Detection and recording. *DelSys Incorporated*, *10*, 2011.

Del-Ama, A. J., Koutsou, A. D., Moreno, J. C., De-Los-Reyes, A., Gil-Agudo, Á., & Pons, J. L. (2012). Review of hybrid exoskeletons to restore gait following spinal cord injury. *Journal of Rehabilitation Research and Development*, *49*(4), 497–514. doi:10.1682/JRRD.2011.03.0043 PMID:22773254

Doucet, B. M., Lam, A., & Griffin, L. (2012). Neuromuscular electrical stimulation for skeletal muscle function. *The Yale Journal of Biology and Medicine*, *85*(2), 201. PMID:22737049

Ellis, L., Jackson, S., Liu, C., Molloy, P., & Paterson, P. (2013). *Lower Limb Exoskeleton Final Report*. Academic Press.

Elnady, A. M., Zhang, X., Xiao, Z. G., Yong, X., Randhawa, B. K., Boyd, L., & Menon, C. (2015). A single-session preliminary evaluation of an affordable BCI-controlled arm exoskeleton and motor-proprioception platform. *Frontiers in Human Neuroscience*, 9. PMID:25870554

Gijbels, D., Lamers, I., Kerkhofs, L., Alders, G., Knippenberg, E., & Feys, P. (2011). The Armeo Spring as training tool to improve upper limb functionality in multiple sclerosis: A pilot study. *Journal of Neuroengineering and Rehabilitation*, *8*(5), 5. doi:10.1186/1743-0003-8-5 PMID:21261965

Hara, Y. (2013). Rehabilitation with Functional Electrical Stimulation in Stroke Patients. *International Journal of Physical Medicine & Rehabilitation*, *1*(147), 2.

Hartopanu, S., Poboroniuc, M., Serea, F., Irimia, D., & Livint, G. (2013). *Design of a hybrid FES-mechanical intelligent haptic robotic glove.* Paper presented at the System Theory, Control and Computing (ICSTCC), 2013 17th International Conference. doi:10.1109/ICSTCC.2013.6689040

Hartopanu, S., Poboroniuc, M., Serea, F., Irimia, D., & Livint, G. (2015). *New Issues on FES and Robotic Glove Device to Improve the Hand Rehabilitation in Stroke Patients.* Paper presented at the 6th International Conference on Modern Power System.

Hartopanu, S., Poboroniuc, M., Serea, F., & Livint, G. (2014). *Towards human arm rehabilitation in stroke patients by means of a hybrid FES&robotic glove.* Paper presented at the Electrical and Power Engineering (EPE), 2014 International Conference and Exposition on. doi:10.1109/ICEPE.2014.6969886

Jarrassé, N., Proietti, T., Crocher, V., Robertson, J., Sahbani, A., Morel, G., & Roby-Brami, A. (2014). Robotic exoskeletons: A perspective for the rehabilitation of arm coordination in stroke patients. *Frontiers in Human Neuroscience*, 8. PMID:25520638

Lo, H. S., & Xie, S. Q. (2012). Exoskeleton robots for upper-limb rehabilitation: State of the art and future prospects. *Medical Engineering & Physics*, *34*(3), 261–268. doi:10.1016/j.medengphy.2011.10.004 PMID:22051085

Looned, R., Webb, J., Xiao, Z. G., & Menon, C. (2014). Assisting drinking with an affordable BCI-controlled wearable robot and electrical stimulation: A preliminary investigation. *Journal of Neuroengineering and Rehabilitation*, *11*(51), 1–13. PMID:24708603

Loureiro, R. C., Harwin, W. S., Nagai, K., & Johnson, M. (2011). Advances in upper limb stroke rehabilitation: A technology push. *Medical & Biological Engineering & Computing*, *49*(10), 1103–1118. doi:10.1007/s11517-011-0797-0 PMID:21773806

Lu, E. C., Wang, R. H., Hebert, D., Boger, J., Galea, M. P., & Mihailidis, A. (2011). The development of an upper limb stroke rehabilitation robot: Identification of clinical practices and design requirements through a survey of therapists. *Disability and Rehabilitation. Assistive Technology*, *6*(5), 420–431. doi: 10.3109/17483107.2010.544370 PMID:21184626

Maciejasz, P., Eschweiler, J., Gerlach-Hahn, K., Jansen-Troy, A., & Leonhardt, S. (2014). A survey on robotic devices for upper limb rehabilitation. *J. Neuroeng. Rehabil*, *11*(3).

Maffiuletti, N. A., Minetto, M. A., Farina, D., & Bottinelli, R. (2011). Electrical stimulation for neuromuscular testing and training: State-of-the art and unresolved issues. *European Journal of Applied Physiology*, *111*(10), 2391–2397. doi:10.1007/s00421-011-2133-7 PMID:21866361

Malešević, N. M., Maneski, L. Z. P., Ilić, V., Jorgovanović, N., Bijelić, G., Keller, T., & Popović, D. B. (2012). A multi-pad electrode based functional electrical stimulation system for restoration of grasp. *Journal of Neuroengineering and Rehabilitation*, *9*(1), 1. doi:10.1186/1743-0003-9-66 PMID:23009589

Patton, J., Small, S. L., & Zev Rymer, W. (2008). Functional restoration for the stroke survivor: Informing the efforts of engineers. *Topics in Stroke Rehabilitation*, *15*(6), 521–541. doi:10.1310/tsr1506-521 PMID:19158061

Pedrocchi, A., Ferrante, S., Ambrosini, E., Gandolla, M., Casellato, C., Schauer, T., ... Gföhler, M. (2013). MUNDUS project: MUltimodal Neuroprosthesis for daily Upper limb Support. *Journal of Neuroengineering and Rehabilitation*, *10*(1), 66. doi:10.1186/1743-0003-10-66 PMID:23822118

Pylatiuk, C., Kargov, A., Gaiser, I., Werner, T., Schulz, S., & Bretthauer, G. (2009). *Design of a flexible fluidic actuation system for a hybrid elbow orthosis.* Paper presented at the Rehabilitation Robotics, 2009. ICORR 2009. IEEE International Conference on. doi:10.1109/ICORR.2009.5209540

Resquín, F., Gómez, A. C., Gonzalez-Vargas, J., Brunetti, F., Torricelli, D., Rueda, F. M., ... Pons, J. L. (2016). Hybrid robotic systems for upper limb rehabilitation after stroke: A review. *Medical Engineering & Physics*, *38*(11), 1279–1288. doi:10.1016/j.medengphy.2016.09.001 PMID:27692878

Rocon, E., Belda-Lois, J., Ruiz, A., Manto, M., Moreno, J. C., & Pons, J. (2007). Design and validation of a rehabilitation robotic exoskeleton for tremor assessment and suppression. *Neural Systems and Rehabilitation Engineering. IEEE Transactions on*, *15*(3), 367–378. PMID:17894269

Rohm, M., Schneiders, M., Kreilinger, A., Müller-Putz, G., & Rupp, R. (2012). First evaluation results of a BCI-controlled hybrid neuroprosthesis for restoration of grasping in a high spinal cord injured individual. *Proceedings of TOBI Workshop LII.*

Rohm, M., Schneiders, M., Müller, C., Kreilinger, A., Kaiser, V., Müller-Putz, G., & Rupp, R. (2013). Hybrid brain–computer interfaces and hybrid neuroprostheses for restoration of upper limb functions in individuals with high-level spinal cord injury. *Artificial Intelligence in Medicine*, *59*(2), 133–142. doi:10.1016/j.artmed.2013.07.004 PMID:24064256

Rong, W., Tong, K., Hu, X., & Ho, N. (2012). *Combined Electromyography (EMG)-driven robotic system with Functional Electrical Stimulation (FES) for rehabilitation.* Paper presented at the Bioengineering Conference (NEBEC), 2012 38th Annual Northeast.

Rong, W., Tong, K. Y., Hu, X. L., & Ho, S. K. (2013). Effects of electromyography-driven robot-aided hand training with neuromuscular electrical stimulation on hand control performance after chronic stroke. *Disability and Rehabilitation. Assistive Technology*, *10*(2), 149–159. doi:10.3109/17483107.2013.873491 PMID:24377757

Rupp, R., Rohm, M., Schneiders, M., Weidner, N., Kaiser, V., Kreilinger, A., & Müller-Putz, G. (2013). *Think2grasp-BCI-controlled neuroprosthesis for the upper extremity.* Biomedical Engineering/Biomedizinische Technik.

Schill, O., Rupp, R., & Reischl, M. (2009). *Signal processing concepts for optimal myoelectric sensor placement in a modular hybrid FES orthosis.* Paper presented at the 4th European Conference of the International Federation for Medical and Biological Engineering. doi:10.1007/978-3-540-89208-3_433

Schill, O., Wiegand, R., Schmitz, B., Matthies, R., Eck, U., Pylatiuk, C., ... Rupp, R. (2011). OrthoJacket: An active FES-hybrid orthosis for the paralysed upper extremity. *Biomedizinische Technik. Biomedical Engineering*, *56*(1), 35–44. doi:10.1515/bmt.2010.056 PMID:21210758

Schulz, S., Pylatiuk, C., Kargov, A., Gaiser, I., Schill, O., Reischl, M., . . . Rupp, R. (2009). *Design of a hybrid powered upper limb orthosis.* Paper presented at the World Congress on Medical Physics and Biomedical Engineering, Munich, Germany. doi:10.1007/978-3-642-03889-1_125

Schulz, S., Schmitz, B., Wiegand, R., Pylatiuk, C., & Reischl, M. (2011). *The Hybrid Fluidic Driven Upper Limb Orthosis-Orthojacket*. Academic Press.

Senelick, R. C. (2010). Technological Advances in Stroke Rehabilitation—High Tech Marries High Touch. *US Neurology*, *6*(2), 102–104. doi:10.17925/USN.2010.06.02.102

Serea, F., Poboroniuc, M., Hartopanu, S., & Irimia, D. (2015). *Towards Clinical Implementation of an FES&Exoskeleton to Rehabilitate the Upper Limb in Disabled Patients.* Paper presented at the Control Systems and Computer Science (CSCS), 2015 20th International Conference on. doi:10.1109/CSCS.2015.114

Serea, F., Poboroniuc, M., Hartopanu, S., & Olaru, R. (2014). *Preliminary tests on a hybrid upper arm exoskeleton for upper arm rehabilitation for disabled patients.* Paper presented at the Electrical and Power Engineering (EPE), 2014 International Conference and Exposition on. doi:10.1109/ICEPE.2014.6969887

Serea, F., Poboroniuc, M., Irimia, D., Hartopanu, S., & Olaru, R. (2013). *Preliminary results on a hybrid FES-exoskeleton system aiming to rehabilitate upper limb in disabled people.* Paper presented at the System Theory, Control and Computing (ICSTCC), 2013 17th International Conference. doi:10.1109/ICSTCC.2013.6689046

Sivan, M., O'Connor, R. J., Makower, S., Levesley, M., & Bhakta, B. (2011). Systematic review of outcome measures used in the evaluation of robot-assisted upper limb exercise in stroke. *Journal of Rehabilitation Medicine*, *43*(3), 181–189. doi:10.2340/16501977-0674 PMID:21305232

Tu, X., He, J., Wen, Y., Huang, J., Huang, X., Huang, H., . . . Yuan, Y. (2014). *Cooperation of electrically stimulated muscle and pneumatic muscle to realize RUPERT bi-directional motion for grasping.* Paper presented at the Engineering in Medicine and Biology Society (EMBC), 2014 36th Annual International Conference of the IEEE.

Tu, X., Huang, J., Yu, L., Xu, Q., & He, J. (2012). *Design of a wearable rehabilitation robot integrated with functional electrical stimulation.* Paper presented at the Biomedical Robotics and Biomechatronics (BioRob), 2012 4th IEEE RAS & EMBS International Conference on. doi:10.1109/BioRob.2012.6290720

Vaca Benitez, L. M., Tabie, M., Will, N., Schmidt, S., Jordan, M., & Kirchner, E. A. (2013). Exoskeleton technology in rehabilitation: Towards an EMG-based orthosis system for upper limb neuromotor rehabilitation. *Journal of Robotics*.

Varoto, R., Barbarini, E. S., & Cliquet, A. Jr. (2008). A hybrid system for upper limb movement restoration in quadriplegics. *Artificial Organs*, *32*(9), 725–729. doi:10.1111/j.1525-1594.2008.00597.x PMID:18684204

Veneman, J. F., Ekkelenkamp, R., Kruidhof, R., van der Helm, F. C., & van der Kooij, H. (2006). A series elastic-and bowden-cable-based actuation system for use as torque actuator in exoskeleton-type robots. *The International Journal of Robotics Research*, *25*(3), 261–281. doi:10.1177/0278364906063829

Wiegand, R., Schmitz, B., Pylatiuk, C., & Schulz, S. (2011). Mechanical Performance of Actuators in an Active Orthosis for the Upper Extremities. *Journal of Robotics*.

World Heart Federation. (2015). *Stroke*. Retrieved from http://www.world-heart-federation.org/cardiovascular-health/stroke/

World Stroke Campaign. (2015). *Facts and Figures*. Retrieved from http://www.worldstrokecampaign.org/learn/facts-and-figures.html

KEY TERMS AND DEFINITIONS

Activities of Daily Living (ADL): Activities conducted routinely during everyday life which most people can perform without assistance, for example, eating, bathing, dressing, toileting, transferring (walking), and continence.

Electroencephalography: The recording of the electrical activity in the brain.

Electromyography: The recording of electrical activity in muscle tissue.

Fugl-Meyer Scale: An index used to assess the sensorimotor impairment of people who have had a stroke.

Functional Electrical Stimulation (FES): Electrical stimulation applied to a muscle which results in functional movement.

Muscle Atrophy: A reduction in the mass of a muscle.

Muscle Fatigue: A reduction in the ability of a muscle to generate force.

Proprioception: The ability to sense the position and movement of the parts of one's own body.

Stroke: A stroke occurs when the blood supply to a section of the brain is blocked or reduced significantly, resulting in the deprivation of oxygen and nutrients.

Chapter 5
Towards Cost-Oriented User-Friendly Robotic Systems for Post-Stroke Rehabilitation

Iuliu Adrian Nadas
Technical University of Cluj-Napoca, Romania

Doina Pisla
Technical University of Cluj-Napoca, Romania

Calin Vaida
Technical University of Cluj-Napoca, Romania

Bogdan George Gherman
Technical University of Cluj-Napoca, Romania

Giuseppe Carbone
University of Cassino and South Latium, Italy

ABSTRACT

This chapter deals with robotics for assisting patients in their recovery after stroke. The first section outlines the key physiotherapist movements that are needed in post-stroke treatment. Sets of experimental data are discussed for defining the main specifications for post-stroke rehabilitation devices. Then, a detailed overview is provided on the existing key achievements and solutions concerning rehabilitation robotics for post-stroke treatment. This section is followed by a detailed critical analysis aiming at the definition of new concepts for cost-oriented user-friendly solutions.

INTRODUCTION

In the next two decades, the current evolution in average life span and medical progress, will force a paradigm shift in the treatment of many diseases. Such is the case of stroke, a cardiovascular disease, which affects mostly the elder population (over 65) and has, as the most common consequence, a nega-

DOI: 10.4018/978-1-5225-2993-4.ch005

tive effect on the patient's capability of performing activities of daily living (ADL). From a medical standpoint, specific protocols exist, which attempt to correct these disabilities by performing specific drug, neurologic and physical therapy. One of the most common disabilities, which affects over 50% of the stroke survivors is hemiparesis (one side weakness) or hemiplegia (one side paralysis), namely the partial or total loss of the ability to control half of the body. This disability is treated by a physical therapist by using different techniques in a process called rehabilitation, which spreads in time up to two years after the stroke.

The Royal College of Physicians defines rehabilitation as a collaboration with limited time between a person with disabilities and professionals, plus other relevant people, to produce sustained reduction in the impact of disease and disability on daily life. The interventions are focused on the human, on the social or physical implications, or a combination of these, (RCF 2010).

In the current approach one therapist can attend only one patient at a time which limits the number of procedures that one person can undergo daily, thus having a negative effect upon the recovery process and progress. Unfortunately, based on the demographic shift estimates, the number of elderly people will increase, making it impossible for the medical system to deal properly with all of them. One possible answer here is a shift in the rehabilitation paradigm which involves the use of specialized devices to perform specific exercises with the patients, changing the role of the therapist to a supervisor which monitors and adapts the rehabilitation process for each individual, based on his/her specific needs and progress. The specialized devices which can perform these tasks efficiently are the rehabilitation robots. Currently multiple devices exist for both upper and lower limb, balance, gait and so on but there is a need for further development in this field aiming to achieve not only efficient but also cost-oriented devices that could be afforded by the medical systems of lower income countries. Rehabilitation robotics has become one of the strategic fields of research promoted all over the world in an attempt to gain sufficient knowledge and reach a good maturity level before the moment when demographics will force the shift from one-to-one therapy to personalized robotic assisted rehabilitation. If compared with conventional therapy via physiotherapists, rehabilitation robots can offer repeatable, controlled, and interactive training at a much higher intensity and pace (Housman et al., 2009).

This chapter presents a detailed overview of the current problems and progress in patient rehabilitation after stroke, illustrating some of the commercially available technologies for the rehabilitation of the lower limb, upper limb and hand, followed by on-going developments in different maturity stages. Several new solutions for the rehabilitation of the upper limb are illustrated along with an overview of the most important challenges and future research directions.

Note: performing multiple movements significantly affects the rehabilitation of post-stroke patients, as highlighted for example in (Klein et al., 2008). However, research in fully understanding the relationship between limb training and recovery is still undergoing, as pointed out for example in (Yoo, D. H., Kim, S. K., 2015). Additionally, besides existing protocols, such as (Vaida, et al., 2017), significant research is carried out towards defining and understanding what is the most effective rehabilitation protocol, (Vaida, et al., 2017).

BACKGROUND

A stroke, as defined by the World Health Organization (WHO), is a "neurological deficit of cerebrovascular cause that persists beyond 24 hours or is interrupted by death within 24 hours". In simple terms, a stroke will cause either a burst or a blockage in a blood vessel inside the brain which will lead to the temporary or permanent loss of the brain functions from the affected region, the damage being influenced by the blood vessel size and medical reaction time. Some recent numbers published on the WHO page in September 2016, reveal that in 2012 there were 6.7 million deaths registered due to stroke with over 75% being registered in the low- and middle-income countries. However, a mortality study published by the Romanian government shows that Romania, which trails behind the advanced European countries, registered an average of 18.6% decrease in stroke deaths from 2006 and 2010. Quoting from the Multi-Annual Report (MAR, 2015):" It is estimated that in the EU the proportion of the population aged over 65 will rise from 17.1% in 2008 to 30% in 2060 and that the proportion of persons aged over 80 will rise from 4.4% to 12.1% over the same period (EUROSTAT population projections). Neurological conditions, especially stroke, are a major cause of disability among older people. Incidence of a first stroke in Europe is about 1.1 million and prevalence is about 6 million. Currently, about 75% of stroke sufferers survive one year after...approximately 80% of stroke patients experience long-term reduced manual dexterity and half of all patients with neurological conditions are unable to perform everyday tasks."

Another recent study published in Spain (Lopez-Espuela, 2016), reveals that out of the 175 stroke patients admitted in a medical center in acute post-stroke phase, 22 have died and 110 (62.9%) suffered hemiparesis or hemiplegia.

Stroke is the first cause of long-term disability in the United States, about 3% of men and 2% women were reported having a long-term disability after they suffered a stroke. (Benjamin, et al. 2017)

The main objectives of stroke rehabilitation are to regain independence and improve the quality of life by improving the capacity of the patient to perform activities of daily living (ADL). (Sullivan, Schmit &Fulk, 2014)

Stroke rehabilitation support is achieved by a professional team which targets multiple areas, starting with the improvement of the patient's physical impairments, and continuing with cognitive, emotional, and vocational skills. (Clarke & Forster, 2015)

Generally, when discussing about physical rehabilitation the motions that a physical therapist performs are complex, working on the mobilization of each joint which are sometimes difficult to replicate with a robotic device but, in some cases there are reports which emphasize the advantages of existing robotic devices. Such is the case of devices for gait rehabilitation, which can more easily support the weight of the patient in comparison to a person, and because of their repeatability, tirelessness and ability to quantitatively assess rehabilitation effectiveness with high accuracy, (Patton, Kovic, Mussa-Ivaldi, 2006) can provide better physiotherapy training than ordinary physiotherapists.

The market for rehabilitation robots is continuously increasing. The use of rehabilitation robots in hospitals and ADL is increasing in the US and European countries. Due to the high demand of rehabilitation robotic devices, the market is expected to increase to 24.27% by 2019 (Chen et al., 2015) Based on another market research, a rehabilitation robot market size value of 37.2 million euros was registered in 2014 and has expectations to grow up to 1.6 billion euros by 2020. (Winter Green Research [WGR]).

As an overview of the current trends at an international level reference is made to the documents which define the critical domains where the development of robotics represents a necessity in the next years. Healthcare is among the most important areas where robotics systems can enhance the level of medical care that can be provided for patients. With reference to neuro-rehabilitation, the Multi-Annual Roadmap (MAR, 2015) shows an increase in the implementation of robotic solutions which is however insufficient. Their report makes reference to statistical data which forecast very important demographic changes in the next decades. Nowadays the ratio between elderly, retired people and the active (working) population is around 26.8% (one elderly person for every four working persons) whereas in 2060 this ratio will increase to 52.6% (one elderly person for every two working persons). These demographic changes will generate automatic modifications to the expenses in population healthcare (a higher number of pensions and people requiring medical services, increased number of medical personnel and so on) which will put pressure upon the economic development and sustainability of economies worldwide. Furthermore, EUROSTAT studies show that in 2011 20.5% of the elderly European population (EU 27) live in poverty at the border of social exclusion (Bulgaria – 61.1%, Romania – 35.3%, Greece – 29.3%, Portugal – 24.5%, Italy – 24.2%) which shows that with a decrease in the funding allocated for healthcare or an increase in the healthcare costs, this category of people will suffer even more, leading to economic crises which will not be sustainable causing a domino effect with global implications. This data demonstrates once more that actions must be taken before it is too late. The same European document, the Multi-Annual Roadmap (MAR, 2015), points out the fact that the utilization of robotic systems in post-stroke rehabilitation along other neurologic pathologies (Parkinson, Multiple Sclerosis, ataxia) proved to have positive effects (in many cases higher than traditional therapies). Some of the necessary characteristics of these robotic systems are: sustainable production costs, clinical validation, proper protocols for the evaluation of patient progress, large scale implementation strategies.

From the medical point of view post-stroke patient rehabilitation covers three large periods of time, with the general consensus that progress in functional recovery is obtainable in a period of up to two years after which only progress preservation can be achieved. These time periods represent the acute phase which covers a period of up to three months after the stroke, sub-acute phase which extends up to six months post stroke and the chronic phase which spreads up to two years. In each phase, the patients have specific needs and rehabilitation is achieved differently being highly dependent on the patient progress.

Taking into account that currently over 75% of the stroke patients survive one year after the stroke and that the large majority of them (over 80%) experience a certain level of physical disability, the target population which could benefit from robotic rehabilitation is extremely high and will increase even more due to the continuous increase of average life expectancy. As in the next years the available number of physical therapists will no longer be able to provide proper medical care to this population segment, a general conclusion emphasis a big gap between the actual demands in rehabilitation and the available solutions, a gap which extends with time. This raises not just the opportunity but the necessity to act fast towards the development of new rehabilitation solutions, in an economically feasible environment capable of supporting stroke patients in each stage of rehabilitation starting from the acute post-stroke phase which takes place in hospitals, followed by longer stages which can take place in day care centers and home environment.

COMMERCIAL DEVICES

Upper Limb

InMotionArm (Duret, Courtial, Grosmaire & Hutin, 2015)

It is made by Bionik Laboratories Corp.

InMotion ARM Dimensions: Workstation: 1.12m(W) x 1.85m(D) x 1.2m(H) at lowest position. Allows 81cm for the chair. Weight 83kg. Electrical Requirements 100-240VAC, 50/60Hz, automatic < 1250VA.

This device is a 2 translational degrees-of-freedom planar robot that focuses on shoulder (flexion/extension) and elbow (flexion/extension) motions in the horizontal plane. It was planned to have low intrinsic end point impedance with a low inertia. It allows clinicians to efficiently deliver optimum intensive sensorimotor therapy to the shoulder and elbow to achieve the development of new neural pathways. The InMotion ARM is an interactive rehabilitation technology that senses patient movements and limitations, providing "assistance-as-needed" in real-time. This robotic device can provide a serious number of treatment types, including an adjustable (or assist-as-needed) program using a performance-based algorithm that adapts forces to assist or challenge the patient's motions according to his/her motor performance. Especially, if a function cannot be completed in a voluntary manner, then the robot helps the patient to reach the target. The point-to-point unrestricted reaching program is mostly used for the evaluation of kinematic motions.

Advantages: It is able to execute the following motions: shoulder protraction/retraction, shoulder internal/external rotation and elbow flexion/extension, it is an accurate device, it is able to improve the patient's range of motion, eye-hand coordination, strength, movement speed, movement smoothness. It is easy to use, it can be accessed directly from a wheel chair and patient progress reports can be printed directly from the robot.

Drawbacks: Being a planar robotic device it allows the human limb to achieve motions in a three-dimensional space but it is heavily limited by the structure.

Bi-Manu-Track (van Delden, Peper, Kwakkel, & Beek, 2012)

It is produced by Reha-Stim Medtec Gmbh, it is designed especially for both sides of the human musculoskeletal system. Lost movements are regained with the help of the healthy side. The stand-alone therapy permits full concentration on the pertinent brain areas in the affected treatment area.Passive and active therapy offers, two different training movements: forearm pronation and supination, wrist extension/flexion.

Power supply: 230V/50Hz, servo motor driven operating panel, has an individually adjustable resistance, amplitude and speed of arm movement, height adaptable table and arm rest (25x18x13cm).

Figure 1. InMotionArm prototype (Duret, Courtial, Grosmaire & Hutin, 2015); a) a general scheme b) a kinematic scheme; c) a kinematic scheme with arm attachments

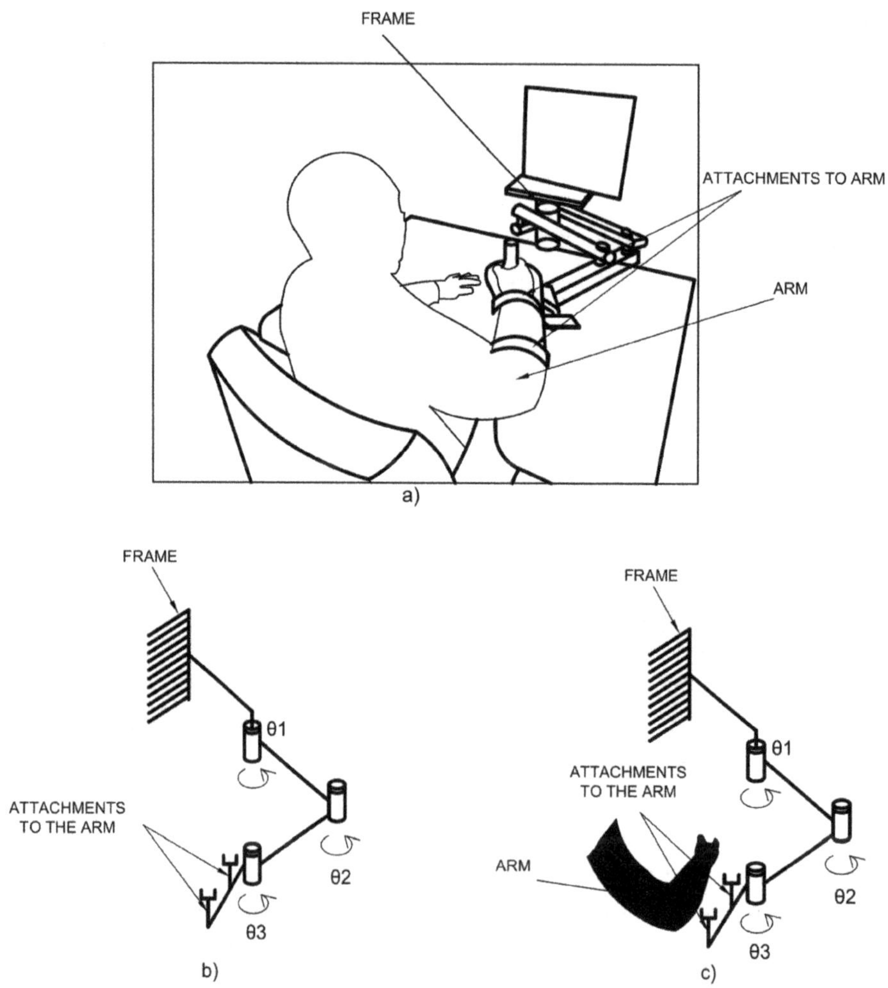

Dimensions: height:72-105cm, width: 85 cm, depth:90cm (including table), weight 51kg.

The Bi-Manu-Track is an automated motor-driven arm trainer that permits bilateral training of two movement patterns: forearm pronation and supination and wrist flexion and extension. To change between movement patterns the device can be rotated by 90o. The Bi-Manu-Track is suitable for three computer-controlled modes of practice. In the passive-passive mode the robot device commands both arms. In the active-passive mode the less affected upper limb actively moves the handle while the robot moves the most affected upper limb. In the active-active mode both arms execute the motions actively by overcoming an initial isometric resistance. Movements can be either mirror-symmetric (in-phase) or parallel (antiphase). Amplitude, speed, and resistances can be set individually.

Advantages: The robot device is simple and easy to use, low cost and permits bilateral exercises.
Drawbacks: Because of the involvement of the healthy limb, the projections to the paretic muscle can be affected.

ArmeoPower

ArmeoPower (Hocoma Web Page) [HWP] 2016) is made by Hocoma AG, Switzerland, and is a rehabilitative exercise device for upper extremity therapy. The main component of the Armeo Power consists of a motorized arm orthosis that is able to sustain the weight of a patient's arm, it can also assist the patient during specific training in a large 3D workspace. The Armeo Power does specific exercises for increasing the strength of muscles and the range of motion of joints in order to improve motor functions.

The motorized arm orthosis with 6 DOF reaches a large 3D workspace. Dimensions 2.7 m (l)x0.8 m (w) x 1.65 m (h), weight: approx. 205 kg. It can be adapted to patient sizes with an electric lifting column (range: 400 mm), forearm lengths (elbow to handgrip): 310-420mm, upper arm lengths (shoulder to elbow): 250-340mm, maximum weight of patient can reach 135 kg. It has an optional module named ManovoPower that is a mechanized hand module which supports the opening and closing of the hand in reach and grasps movements.

Advantages: Allows patients to get early treatment of motor disabilities, enabling patients to perform repetitive, intensive, and goal-oriented exercises.
Drawbacks: The price is high, about 250000 euros.

Figure 2. The Bi-Manu-Track prototype (van Delden, Peper, Kwakkel, & Beek, 2012); a) a general scheme b) a kinematic scheme for wrist flexion/extension; c) a kinematic scheme for forearm pronation/supination

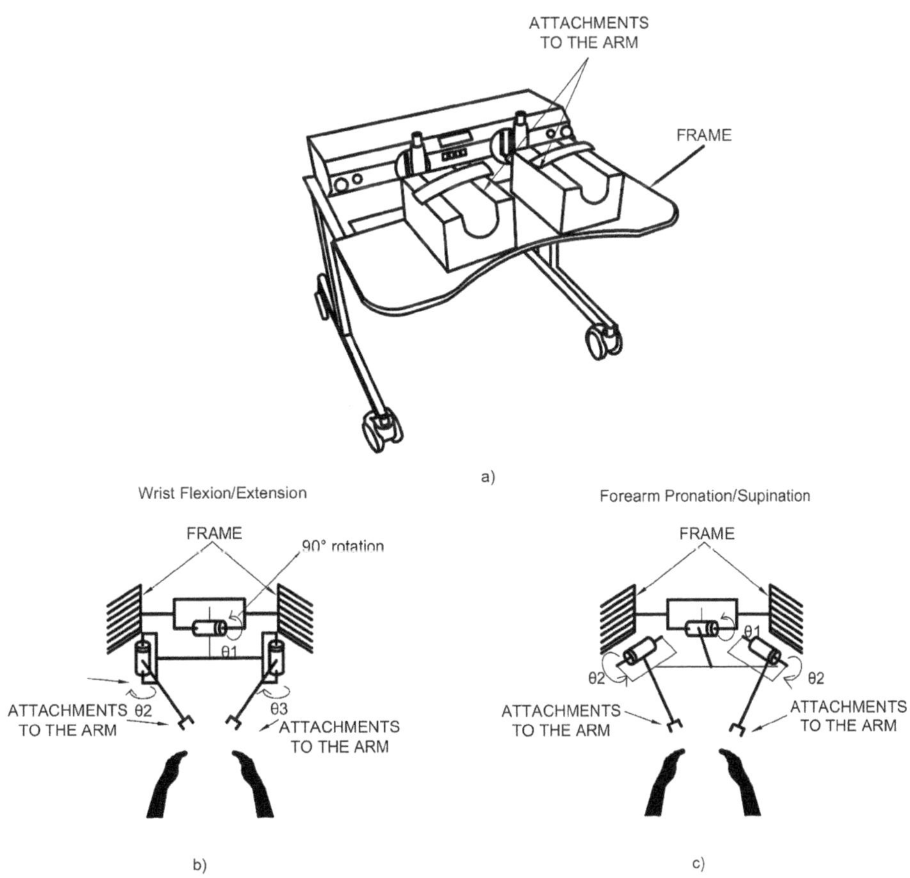

ArmeoSpring

ArmeoSpring (HWP, 2016) is made by Hocoma AG, Switzerland, it is an instrumented arm orthosis with a spring mechanism for an adjustable Arm Weight Support in a large 3D workspace that can be used as a real-time input device to the associated therapy software ArmeoControl. The aim of ArmeoSpring is to provide functional therapy for patients who have impaired upper limbs due to cerebral, neurogenic, spinal, muscular or bone-related disorders.

Device's dimensions: 81 cm (l) x 75 cm (w) x 85-125 cm (h), weight: max. 82 kg, room requirements: 3x3x2m (l x w x h), temperature: 10°C-30°C. Forearm lengths 290-310mm, upper arm lengths:220-310mm.

It has a 5 degree-of-freedom (3 in the shoulder, 1 in the elbow, 1 in the forearm) orthosis without robotic actuators, a so called passive system. The adaptable mechanical arm permits variable levels of gravity support by means of a spring mechanism. This allows patients, using residual upper limb function, to reach a bigger active range of motion (ROM) within a 3-dimensional work volume than is possible without support.

Figure 3. ArmeoPower prototype (van Delden, Peper, Kwakkel, & Beek, 2012); a) a general scheme b) a kinematic scheme; c) a kinematic scheme with detail of arm attachments

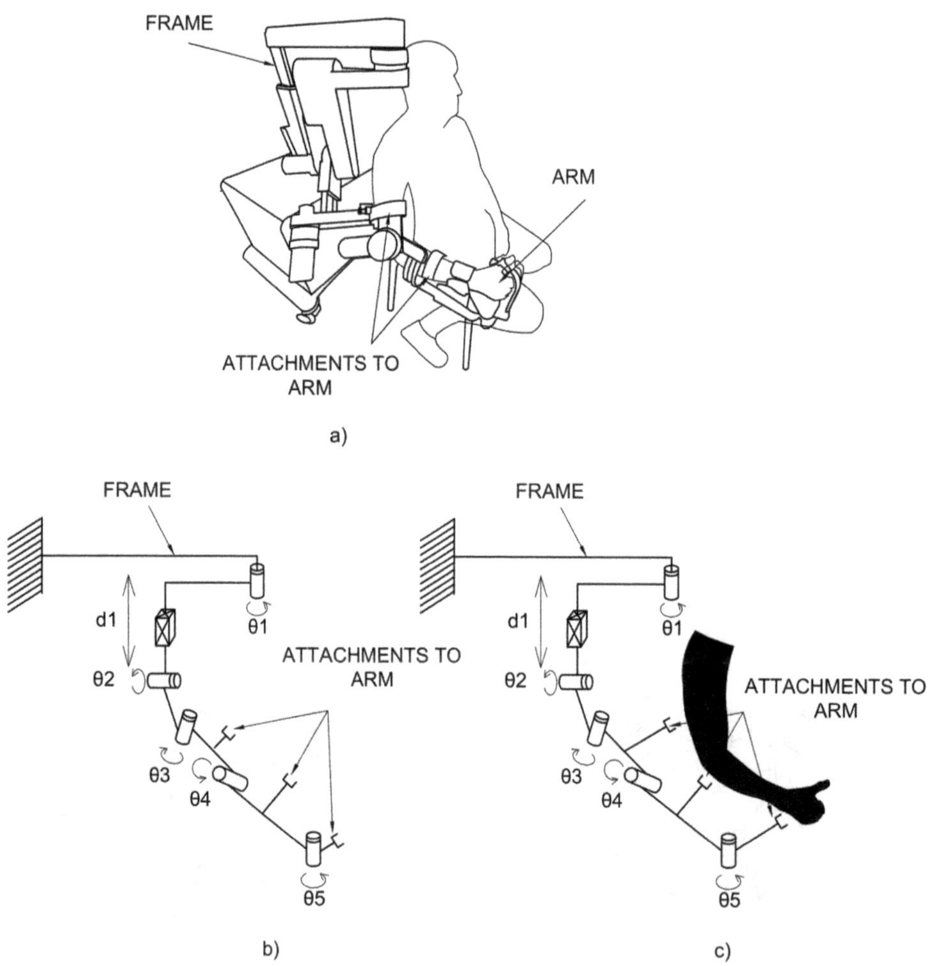

Advantages: Permits customized patient's therapy, feedback exercises have been designed to train core movement patterns that are used in activities of daily living.

Drawbacks: Being a passive system, the robotic devices address patients with mild moderate chronic impairment of the upper limb following a stroke, so this robotic system can't be used for patients with severe hemiparesis.

Lower Limb

G-EO-System (Hesse, Waldner, &Tomeller, 2010)

The robotic device is built on the end-effector principle, the trajectories of the foot plates are freely programmable allowing not only the practice of simulated floor walking but also up and down stair climbing.

Its design characteristics are as follows: Length: 406 cm, Width: 124 cm, Standard height: 280 cm, Reduced Height: 260 cm, Weight: 900 kg, power Supply: 110/230 V. The device adopted the end effector principle. The harness secures the patient that stands on two-foot plates, whose trajectories were completely programmable. The two-foot plates have each been connected via a pivoting arm to two moving sledges. The foot's forward motion was given by the movement of the main sledge, which was connected to the transmission belt of the linear guide. The transmission belt was driven by a 1.500 W

Figure 4. ArmeoSpring prototype (HWP, 2016); a) a general scheme b) a kinematic scheme; c) a kinematic scheme with detail of arm attachments

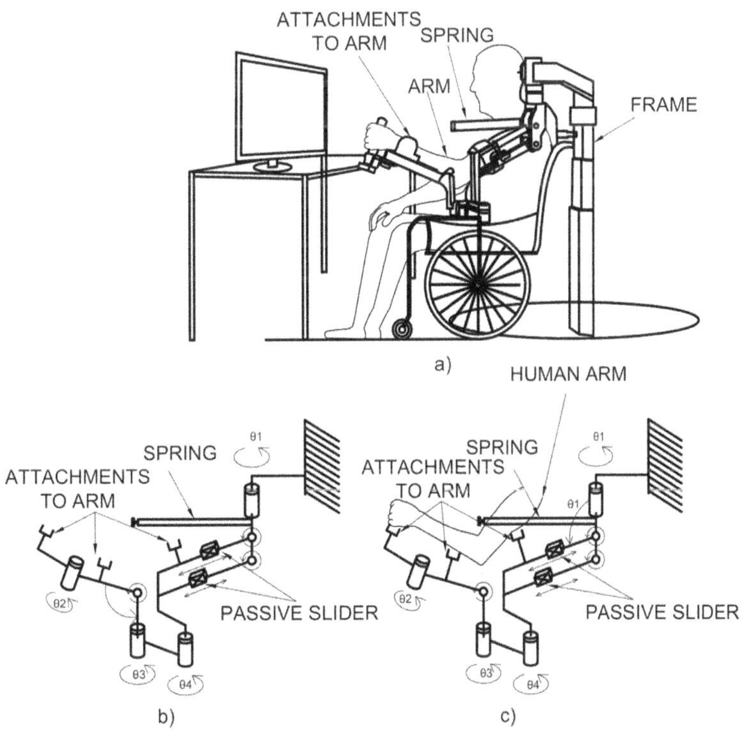

servomotor mounted in the back end of the linear guide. The robotic system is used to improve the floor walking and stair climbing of patients that suffered a stroke. The G-EO-System robot device patent is held by Reha-Technologies GmbH.

Advantages: Easy for patients to use, G-EO system can simulate walking on level ground and stair climbing, quick setup and easy patient handling.

Drawbacks: It doesn't have a virtual reality interactive training module; this detail could help patients increase their motivation during the therapy exercises.

HAL-5 (Type-B) (Hybrid Assistive Limb) (Sankay, 2007)

This exoskeleton was developed by the University of Tsukuba, Japan and it's produced by CYBERDYNE Inc. The exoskeleton helps in daily activities, heavy work and also entertainment.

The Robotic Suit HAL is a cyborg type robot which can enlarge, augment and sustain physical capacity. The robotic device HAL has two kinds of control systems such as "Cybernic Voluntary Control System" and "Cybernic Autonomous Control System". The application fields of HAL are medical well-being, heavy work support, entertainment etc.

The exoskeleton has the following specifications: Height 1,600mm, Weight Full Body Type approx.

23kg (Lower body approx. 15kg), Battery Drive Charged battery (AC100V), Continuous operating time Approximately 2 hours 40 minutes, Motions Daily Activities (standing up from a chair, walking, climbing up and down stairs), hold and lift heavy objects up to 70kg, Hybrid Control System (Cybernic Control: Cybernic Voluntary Control, Cybernic Autonomous Control), indoor and outdoor usage.

Advantages: It is a wearable and portable device that confers the patients significant aid in rehabilitation gait training.

Drawbacks: It has a limited torque, it is heavy and can't be used by paralyzed patients.

AlterG Bionic Leg (Li, Ding, Chen, Mao, Huang, Li 2015)

This is an advanced and dynamic lower extremity orthosis that is battery-powered and wearable. It intends to increase existing muscle strength, it gives sensory inputs (i.e. auditory and sensory feedback) and mobility support for users with disability at the lower-extremity functions during rehabilitation. The ROM and quantity of support provided by the Bionic Leg is personalized through the device settings. These parameters are determined by the clinician.

The robotic device has a weight of 3.5 kilograms and an internal sensor that detects the user's foot pressure. The orthosis provides forward propulsion when the users generate enough force on their paretic knee. The RLO (robotic leg orthosis) delivered support with extension (during standing posture, forward gait, and so on), controlled flexion (during standing to sitting posture), and free movement (during the swing period).

Three types of settings on the robotic device can be modified: threshold (the minimum force necessary to activate the device), assistance (the percentage of body weight provided through the limb during extension in the stance phase of the gait cycle), and resistance (level of resistance during flexion on the stance phase of the gait cycle). The robotic device's support physiotherapy training procedure was built

Figure 5. G-EO-System prototype (Hesse, Waldner & Tomeller, 2010); a) a general scheme b) a kinematic scheme; a kinematic scheme with detail of leg attachment

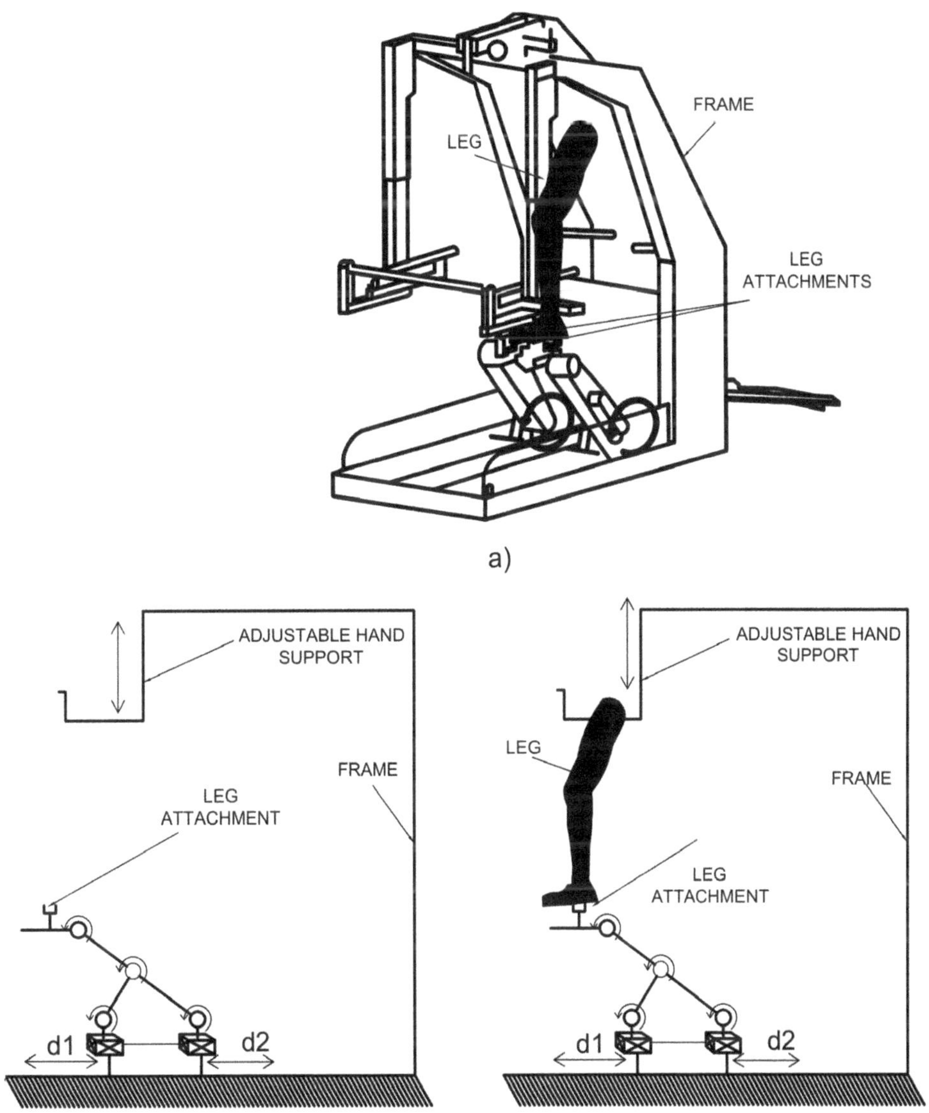

respecting a motor control learning theorem that included fifteen sessions held five times per week for three weeks

Advantages: It is a lightweight wearable device which assist the patient with the impaired leg by helping the user to relearn gait support, it also helps patients to place the body weight on the healthy leg.
Drawbacks: Battery life gives limited daily usage.

Figure 6. HAL-5 (Type-B) (Hybrid Assistive Limb) prototype (Sankay, 2007); a) a general scheme b) a kinematic scheme; c) a kinematic scheme with detail of leg attachment

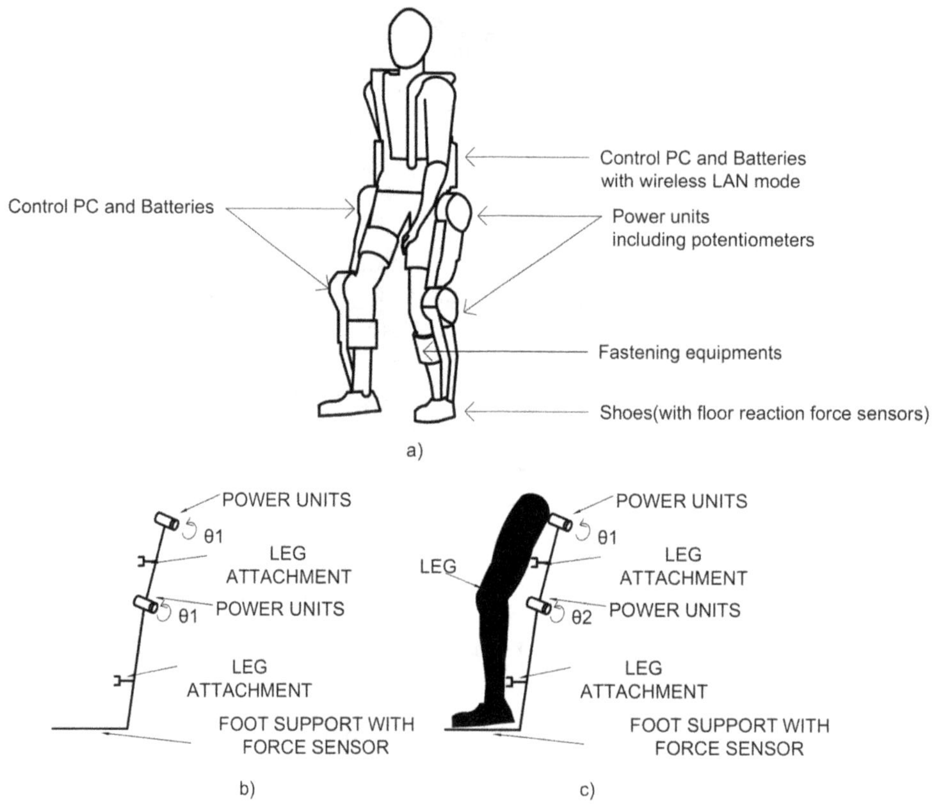

Hand

AMADEO Hand-Therapy System (Sale, Lombardi, Franceschini 2012)

AMADEO is made by TYROMOTION GmbH Austria, it is a modern mechatronic finger-hand therapy system for the rehabilitation of patients with motor disability in the distal upper extremity. The AMADEO system is made up of the hand unit, that contains the electrically powered movement mechanism, the hand-arm support, finger supports and finger plasters, a height-adjustable load-bearing frame with table surface, control panel and a PC-based control and operating unit for configuring therapy parameters.

The AMADEO system has the following specifications: power supply 110-240V a.c., supply frequency 50/60Hz, electricity/power consumption 1,83-4A/440W, supply voltage for motors 24V d.c,, max speed 210mm/sec, nominal power of motors 3watts/motor, max. power of motors 30N/Motor, max. permissible finger strength 30N, measurement range strength sensors ±20N, weight 60kg, dimension(WxLxH) 1160x754x1209 (mm) collapsed, 1160x754x1589 (mm) extended.

The Amadeo robot has got 5 DOFs and supplies the motion of one or all five fingers, due to a passive rotation joint placed between fingertip and an entity moving laterally; (the thumb has got two passive rotation joints). All five translational DOFs are independent and provide large coverage of the fingers

Figure 7. AlterG Bionic Leg prototype (Li, Ding, Chen, Mao, Huang, Li); a) a general scheme; b) a kinematic scheme; c) a kinematic scheme with leg attachment

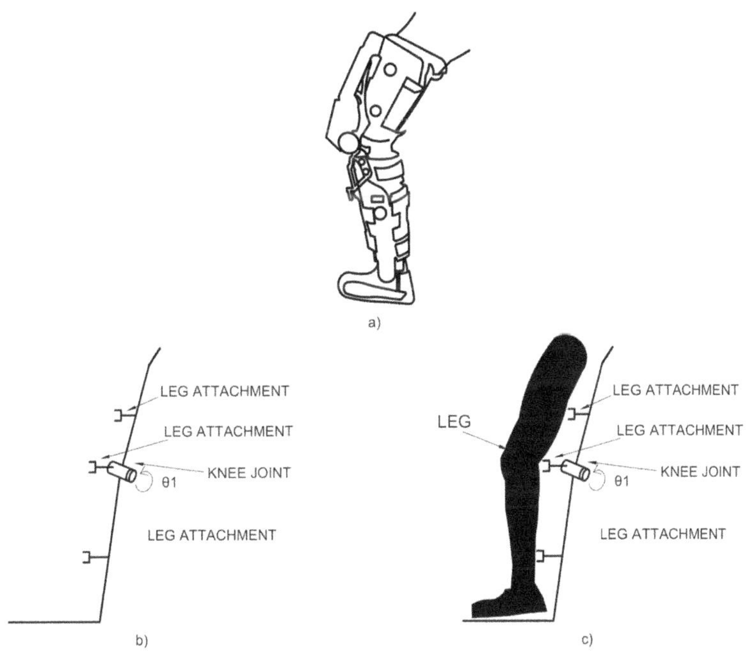

workspace (but not all of it is covered). The interface between the human hand and the machine is realized thanks to elastic bands or plasters, and the wrist is restrained from the movements by a Velcro strap.

Advantages: This robotic system covers the human hand fingers workspace and performs an intensive training having a high frequency of gripping movements. It can be used even for patients with severe hemiparesis.

Drawbacks: It is a fixed system it is not wearable, it cannot be complementary with a robotic system for upper limb rehabilitation.

YouGraber (Shuster-Amft et al., 2014)

YouGrabber is made by YouRehab, Switzerland, it helps patients with sensory-motor and cognitive impairments during rehabilitation. It delivers bi-manual training exercises with a focus on visuomotor finger, hand and arm coordination. The interactive physiotherapy is using Virtual Reality technology and can be done even in a three-dimensional environment.

At the beginning of every training session the system is adapted to the patient's current abilities. Particular settings in every training application allow the training difficulty and content to be quickly and easily adapted to the patient's progression and therapy needs. The Auto mode allows an automatic adjustment of the software's difficulty and always ensures an optimal challenge. The software is in accordance with state of-the-art therapy approaches and mirror therapy is also possible. The training software induces the patient to train intensively, using the following available training targets: Arm pronation and supination, arm lifting and lowering, hand and finger extension, grasping movements, selective fine

Figure 8. AMADEO Hand-Therapy System prototype (Sale, Lombardi, Franceschini 20012); a) a general scheme b) a kinematic scheme; c) a kinematic scheme with detail of hand attachments

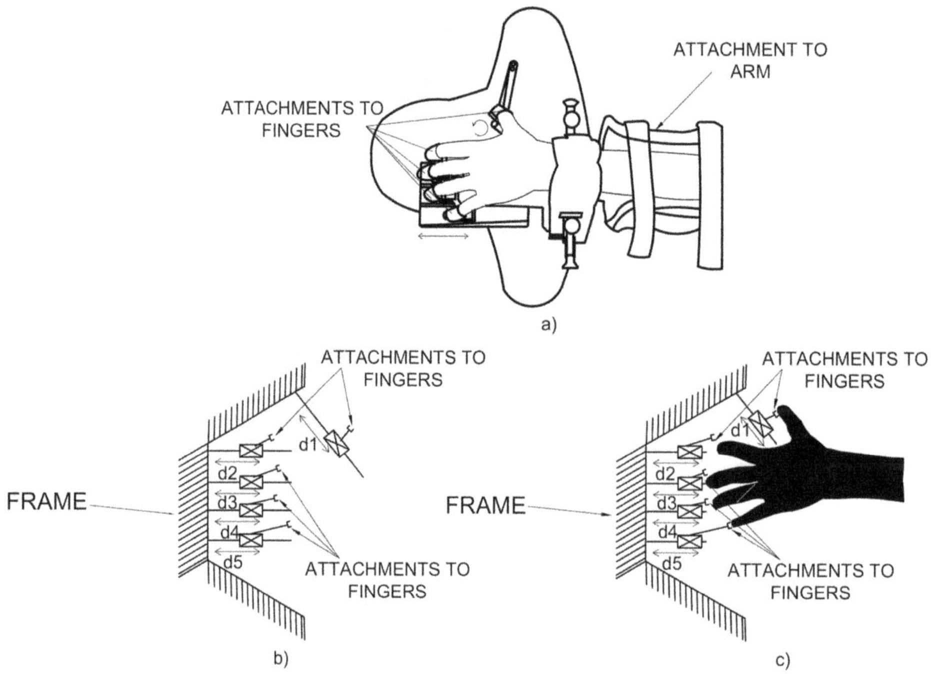

finger movements, wrist flexion and extension elbow flexion and extension, divided attention, reaction time, visual attention, concentration and visual information processing.

Advantages: The YouGrabber is a wearable, adjustable device that can be adapted in every training session based on patient's current abilities. The training software provided with the system can be used on an usual personal computer for interactive therapy training at home or in a clinic.
Drawbacks: It is designed to move only 3 fingers.

CyberGrasp (Cyber Glove System Web Page[CGSWP] 2014)

The CyberGrasp system, made by CyberGlove Systems LLC, San Jose, CA, USA, is an innovative force feedback system for the fingers and hand. It allows patients to "reach into a computer" and grasp computer-generated or tele-manipulated objects.

It has a force of 12 N per finger (max, continuous), Weight: 454 g (exoskeleton without CyberGlove system), Workspace: 1-meter spherical radius from the Actuator Module.

Grasp forces are generated by a network of tendons routed to the fingertips via the exoskeleton. It has five actuators, one actuator per finger, which can be one by one programmed to prevent the user's fingers from penetrating or crushing a virtual solid object.

The device exerts grasp forces that are roughly perpendicular to the fingertips throughout the range of motion, and forces can be specified individually. The CyberGrasp system allows full range-of-motion

Figure 9. YouGraber prototype (Shuster-Amft et al., 2014); a) a general scheme; b) a kinematic scheme; c) a kinematic scheme with hand attachment

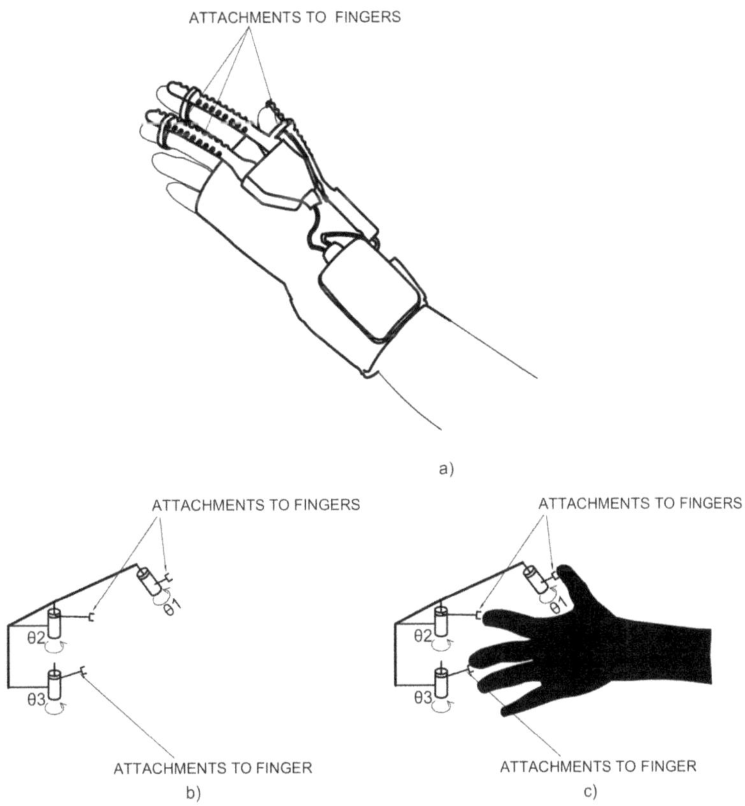

of the hand and does not obstruct the wearer's movements. The device is fully adjustable and designed to fit a wide variety of hands. The CyberGrasp system exerts grasp forces perpendicular to each fingertip, allowing full range of hand motion without restriction. The CyberGrasp system is fully adjustable and designed to accommodate a wide variety of hand sizes.

Advantages: Due to the cable driven actuation system the robot can reach a continuous force of 12N per finger. It can be used in different domains for example in haptic rehabilitation and in the training of the visually impaired.

Drawbacks: the system is pretty complex and has a limited workspace from the actuator module; the price is also an inconvenience, around 34000 euros. Also, with the weight of exoskeleton being 454 grams it is not easy to use the system for a certain amount of time.

CONCEPTUAL DEVICES

Upper Limb

Robotic systems for rehabilitation are mainly classified in two large categories: end-effector based and exoskeletons (Figure 11) (Loureriro, Harwin, Nagai, & Johnson, 2011) the end-effector based robotic systems for rehabilitation interact with a patient through a single lateral attachment point on the patient's

Figure 10. CyberGrasp prototype (Cyber Glove System Web Page[CGSWP] 2014); a) a general scheme b) a functional scheme

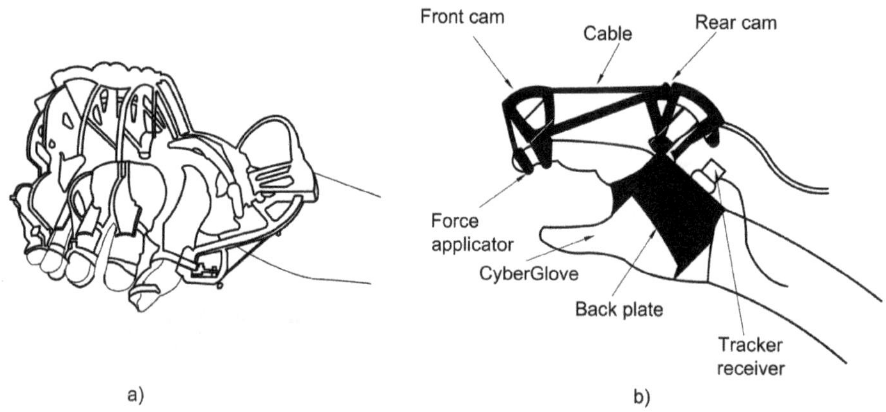

Figure 11. Rehabilitation robots categorized by end-effector and Exoskeleton; a) end-effector actioned unilateral; b) end-effector actioned bilateral; c) partial or full exoskeleton; d) wire-based robot; e) multi-robot

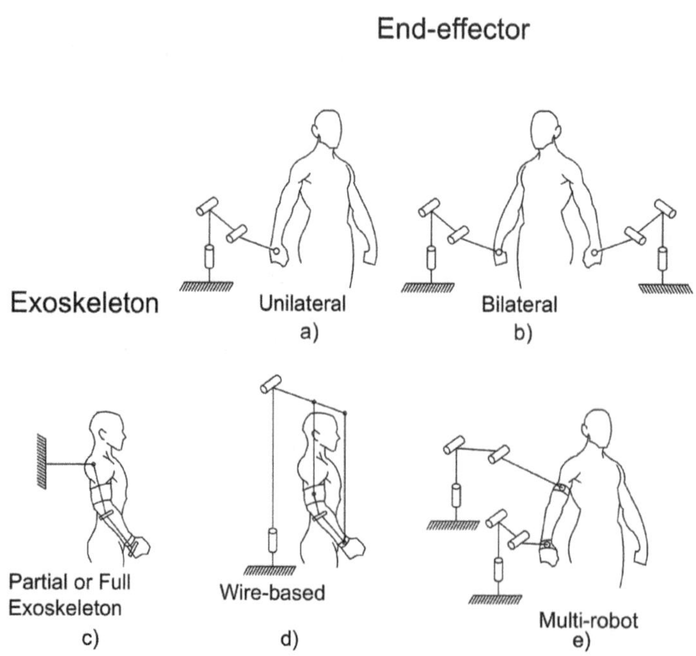

hand. These types of robots are rather simple and can be easily adjusted to different sizes and shapes of the patient's upper limb (e.g. MIT-MANUS (Hogan, Krebs, Charnnarong, Srikrishna, & Sharon, 1992) and ARM-Guide (Reinkensmeyer, Dewald, & Rymer, 1999). Because of their architecture which allows only the control of the end-effector tip, it is rather difficult to control a certain single joint of the upper limb of the patient, which is why at least another anchor point is usually needed for rehabilitation training.

If not mastered well, this deficiency could cause serious lesions of the patient's joints, which is a serious problem especially for 3D spatial robots.Two types of robots were designed: table-top robots that are able to perform 2D planar training: e.g. MIT-MANUS (Hogan et al, 1992) or Braccio di Ferro (Casadio, Sanguineti, Morasso, Arrichello 2006) and multi-robots that can perform the predesignated exercises of the wrist, elbow and shoulder at the same time or separately (e.g. REHAROB (Tóth, Arz, Fazekas, Bratanov, Zlatov, 2004) and iPAM (Jackson et al., 2007).

A number of end-effectors were developed to address the mirroring concept, namely to achieve the bilateral rehabilitation training (Figure 11), which supposes that the affected arm imitates the healthy or the less affected arm to perform synchronous training exercises assisted by the robots (e.g. MIME (Lum, Van der Loos, Shor, Burgar, 1999), Bi-Manu-Track (Hesse, Schulte-Tigges, Konrad, Bardeleben, Werner, 2003) and Driver SEAT (Johnson, Van der Loos, Burgar, Leifer,1999)).

An exoskeleton robot encloses the upper-limb into the bionic-construction to control its motion. Because of its enclosed mechanism, exoskeletons can completely control all joints of the upper- limb and determine how much torque is needed to actuate each upper-limb joint separately. (Sheng, Zhang, Meng, Deng, Xie, 2016) Moreover, exoskeletons require less working space compared to end-effectors when achieving the same therapeutic exercises. In any case, this capsule structure means that the joint axes of upper-limb are pre-determined, and mechanical singularity will happen if the robots' axes do not align correctly with the patients' upper-limb anatomical axes (Yang, Cao, Zhao, 2013). Both end-effector based robots and exoskeletons in Figure 11 can be actuated using wires, that are connected to a splint, becket or even the exoskeleton, to support the body weight, to control the height, torque and orientation of the upper-limb (e.g. Swedish Helparm (Kinsman Enterprises Product Catalog Web Page, [KEPCWP]) and NeReBot (Masiero, Celia, Rosati, Armani, M 2005)).

Lately, an increased number of robots were designed for bilateral exercise (mirroring), specifically for upper-limb stroke rehabilitation. Cauraugh (Cauraugh, Summers, 2004) and Stinear (Stinear, J., Byblow, 2002) further reported that the activation of the damaged hemisphere could be assured through simultaneous motions of the damaged arm and the less damaged/unaffected arm. (Sheng, et al., 2016)

Biomimetic Orthosis for the Neurorehabilitation of the Elbow and Shoulder (BONES) (Klein et al., 2008)

The authors have proposed a robotic structure having a light weight and a robust force control using a parallel robotic device with mechanically grounded actuators. This enables the use of large, direct-drive actuators since the weight of the actuators themselves is not important. The key feature in the design of BONES consists in the human forearm. The parallel device mechanism allows a 3 DOF shoulder motion, including shoulder internal/external rotation. To allow the elbow flexion/extension, the robot includes a serially-placed actuator for elbow flexion/extension.

The upper arm exoskeleton reproduces the upper arm motion via a spherical joint rotating around the shoulder. The arm is anchored at the elbow and moved by means of two passive sliding rods.

The authors have used pneumatic cylinder actuators to achieve the movements described above. In this study, the inverse kinematics has been used to determine the cylinder coordinates required to produce a desired motion of the end-effector.

The weight of the upper arm exoskeleton is 3438 g (including the elbow actuator). The weight of the forearm exoskeleton module is 921g leading to a total arm weight of 4359g. So, the mass of the exoskeleton is comparable with a human upper limb mass that is around 4220g, in average across sexes.

Advantages: Allows naturalistic motion of human arm using mechanically grounded, direct drive actuators at the shoulder.

Drawbacks: It is not wearable and can be used only in clinics thus not for ADL.

Rocker (APBT)

The Rocker is a rehabilitation device that allows active–passive bilateral motion exercise (APBT, Stinear and Byblow (Stinear, Byblow, 2004)), that is an upgraded version of Hand-Object-Hand system (H –O –H, Lum P. S. et al. (Lum, Reinkensmeyer, Lehman, 1993), (Reinkensmeyer, Lehman, Lum,1993)). The

Figure 12. BONES prototype (Klein et al., 2008); a) a general scheme; b) a kinematic scheme; c) a kinematic scheme with arm attachment attachment

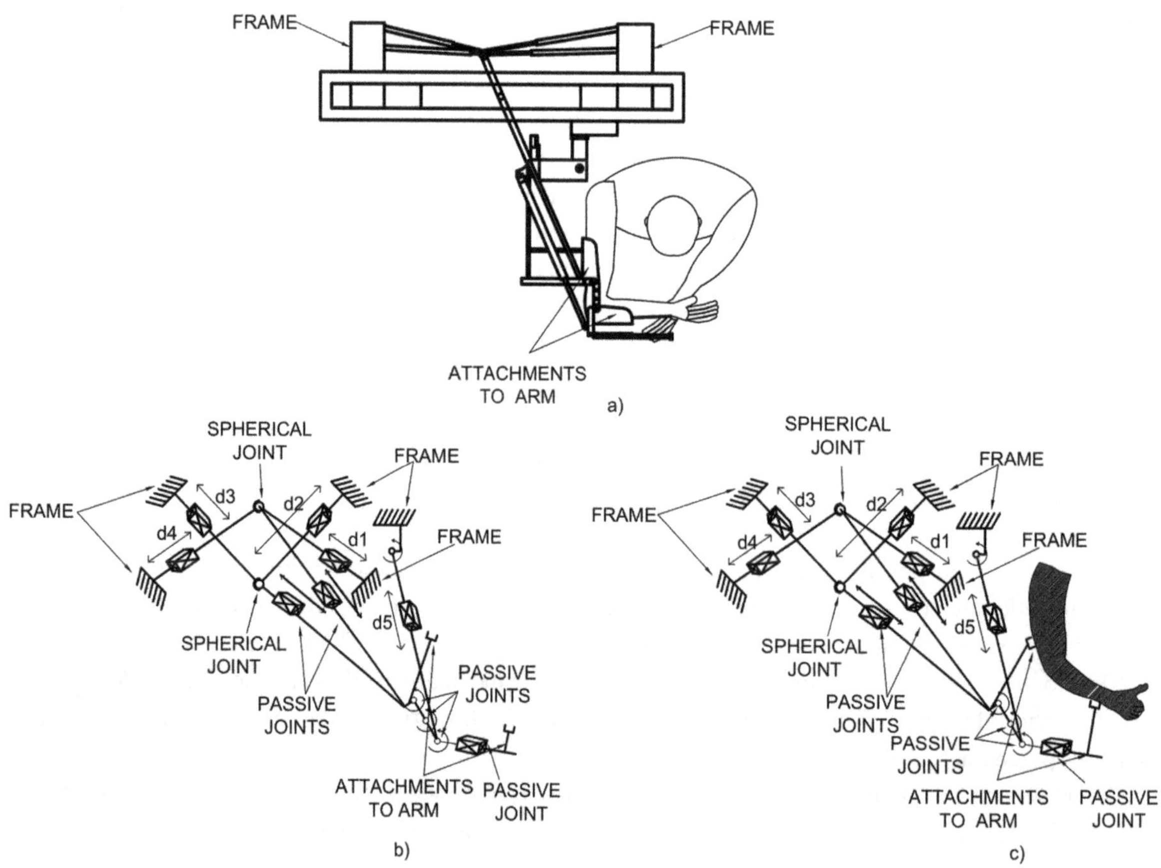

robotic device has two connected crankshafts which are placed on a table that connects to two manipulators (Stinear et al., 2004) and can provide symmetrical or asynchronous trainings (60 ° phase lag) for wrist flexion/extension movements in horizontal plane (Figure 13). This robot system allows actively moving the unimpaired upper limb and can passively move the impaired upper limb in either a synchronous or (60 ° phase lag) asynchronous mode. Compared with the Hand-Object-Hand system, this device supports active training mode, which is important for nerve and coordination recovery (Sheng, et al., 2016).

Advantages: It is a good device which helps patients to improve wrist flexion/extension. Based on mirror symmetrical coordination the less impaired limb can passively move the most impaired limb.
Drawbacks: It allows only planar flexion/extension of wrist.

Figure 13. Rocker prototype (APBT, Stinear and By- blow (Stinear, Byblow, 2004); a) a general scheme b) a kinematic scheme of one side; c) a kinematic scheme with detail of hand attachments

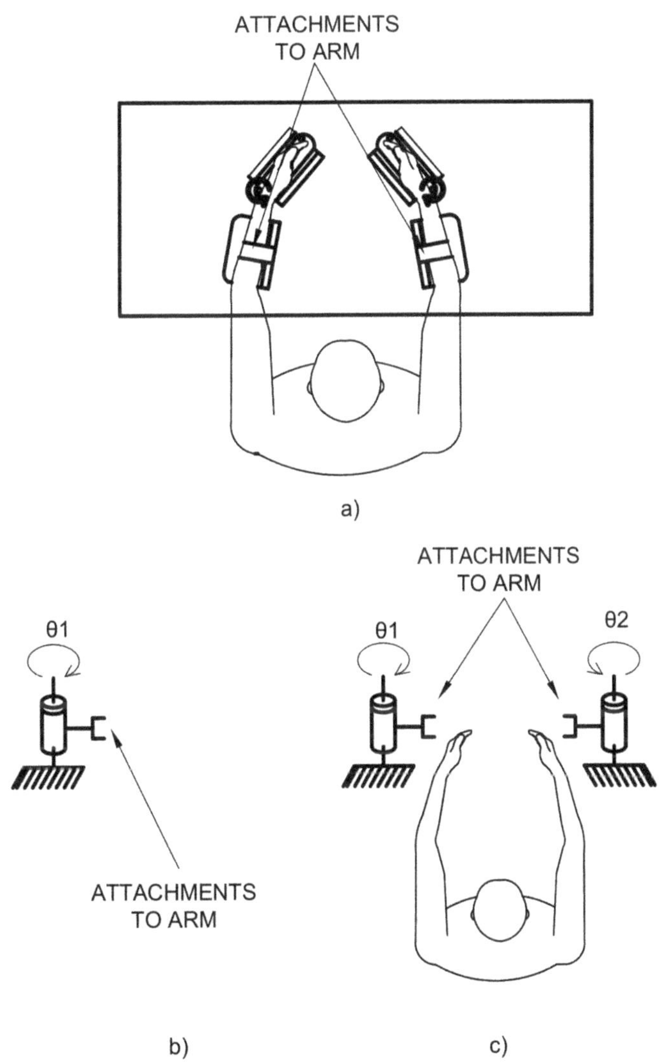

Self-Controlled Master-Slave Robot

The Self-controlled Master-Slave Robot was developed starting from the ideas of the Bimanual Lifting Rehabilitator (Lum, Lehman, Reinkensmeyer,1995) and Hand Robotic Rehabilitation Device (Rashedi, Mirbagheri, Taheri, Farahmand, Vossoughi, Parnianpour, 2009), which is composed of two identical terminals for elbow flexion/extension movements (SCMSR, Li C.G. et al. (Liu, Li, Inoue, Shibata, 2010). Two adjustable handles are coupled to the terminals with two identical motors. To measure the torque the authors have used two torque transducers and one strain gage signal amplifier (Figure 14). The motor on which a larger force is applied (by the healthy upper-limb) acts as the master and the other one as slave and guides the impaired limb. The device can support three training types: 1) passive-driven mode: the impaired upper-limb will be passively handled and moved by the robot; 2) active-assisted mode: the impaired upper-limb will be helped by the other (healthy) upper-limb to complete the predefined exercises; 3) active-resistive mode: a resistive force will be applied to the impaired upper-limb by the healthy (contralateral) upper-limb, for the rehabilitation training. (Sheng, et al., 2016)

Advantages: This robotic device has a great potential to be applied in rehabilitation, manipulation. The system provides bilateral arm coordinated training for patients that suffered unilateral paralysis.
Drawbacks: It performs only flexion/extension of elbow.

Figure 14. Self-controlled Master-Slave Robot prototype (Lum, Lehman, Reinkensmeyer,1995) ; a) a general scheme b) a kinematic scheme of one flexion; c) a kinematic scheme with detail of hand attachment

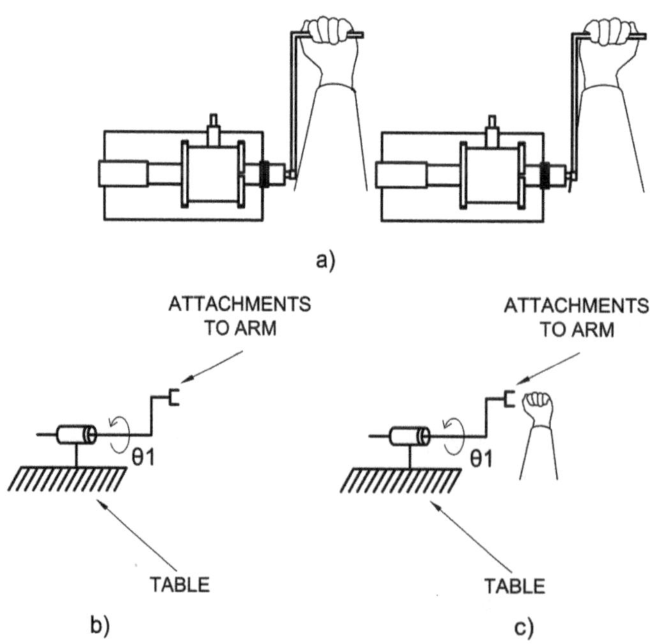

Supinator Extender (SUE): A Pneumatically Actuated Robot for post-stroke Forearm/ Wrist Rehabilitation (Allington, Spencer,Klein, Buell,Reinkensmeyer, Bobrow, 2011)

The robot presented in this study, SUE (Supinator extender), has been developed as an additional forearm/wrist rehabilitation device to the UCI BONES (Klein et al., 2008) exoskeleton robot leading to a 6 DOF robotic system. It can also be attached to the ArmeoSpring rehabilitation device. SUE has a 2-DOF serial chain and has two functions: to measure and to achieve the forearm supination-pronation and the wrist flexion-extension. It uses pneumatic actuators, having large power to weight ratio, thus allowing SUE to reach the required forces for the rehabilitation therapy. Due to its lightweight (0.56 kg) it can be carried by BONES (Klein et al., 2008) or ArmeoSpring (HWP, 2016). Each DOF has a range of 90 degrees and a nominal torque of 2.7 N-m.

It is a strong and compact back-drivable robot that provides forearm/wrist capability for two arm rehabilitation exoskeletons. One of the reasons for the use of pneumatic actuation consists in the low impedance of the cylinders and their lightweight. SUE uses two control types, a simple position control and a force control that uses valves to control the air pressure separately in each chamber. In both cases, a sliding control approach is used to take advantage of the fast on-off solenoid valves. The controller is implemented using xPCTarget in standalone mode running at 1000 Hz.

Advantages: The robot device is lightweight (0.56 kg), the system is complementary with the robot system UCI BONES (Klein et al., 2008) and the Armeo Spring (HWP, 2016), the robot system performs forearm supination/pronation and wrist flexion/extension.

Drawbacks: Non-linear forces that came from pneumatic cylinders can be overcome by good controls so adding complex controls is necessary.

Lower Limb

"Physioterabot" (Akdogan, Mehmet, 2011)

This study presents a robotic system with 3 DOF attached to a chair, as shown in Figure 16.

The system can perform all major active and passive rehabilitation exercises but it also has the capacity of learning specific exercise movements and performing these exercises in the absence of the physiotherapist. The implemented Human-Machine interface allows a larger adaptability of the system, based on the patient's reaction, so that the robot manipulator can change its position while performing exercises according to the feedback data about the patient's state. The remote manipulator can achieve: flexion-extension for the knee where only link 1 is used, flexion-extension for the hip where Link 1 and Link 2 are working together and abduction-adduction for the hip where link 0 works around the OZ axis. The remote manipulator is extremely versatile in terms of the patient's physical characteristics, generating torque for patients up to 100 kg. The robot is actuated using servomotors, equipped with gearboxes for the motions reported in Table 1

Advantages: It can almost reach the ROM of human limb as is shown in the Table 1 and it can be adapted to fit different weight of patients up to 100 kg.

Drawbacks: It doesn't perform rehabilitation exercises for ankle.

Figure 15. Supinator Extender (SUE) prototype (Allington, Spencer, Klein, Buell,Reinkensmeyer, Bobrow, 2011) a) a general scheme; b) a kinematic scheme; c) a kinematic scheme with arm attachment

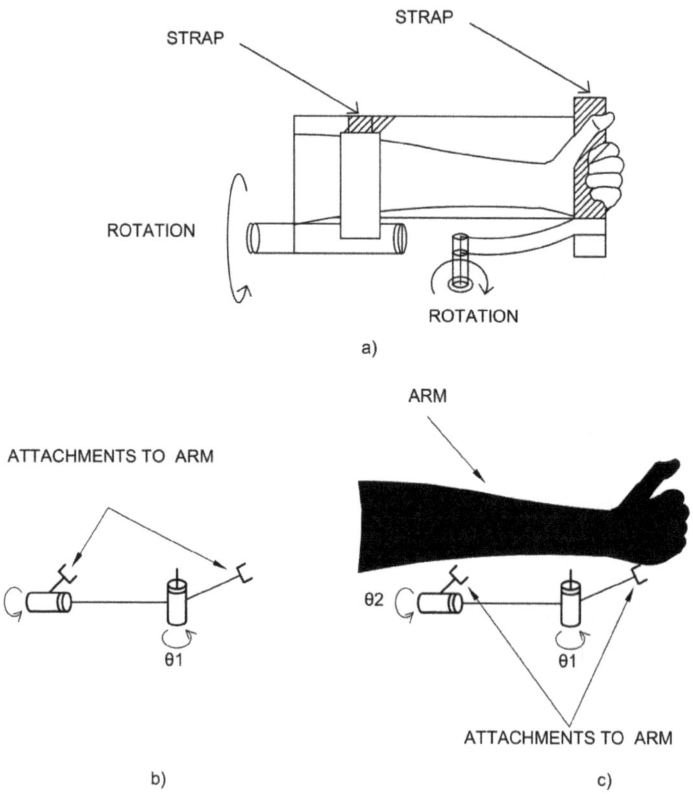

Figure 16. Physioterabot prototype
(Akdogan & Mehmet, 2011)

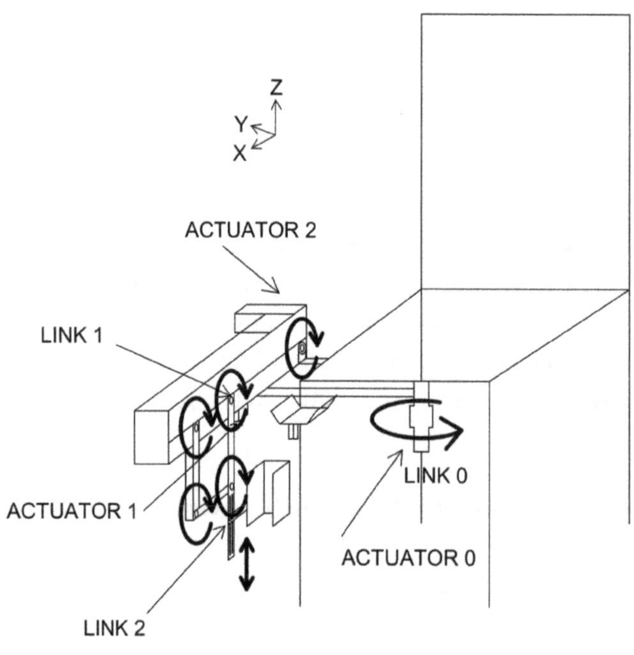

Table 1. Motion ranges as reported in Akdogan (2011)

Limb	Movement	ROM of Limb	ROM of Physiotherabot
Knee	Flexion-extension	-45°/90°	-35°/90°
Hip	Flexion-extension	-15°/45°	-15°/45°
	Abduction-adduction	0°/120°	0°/120°

Knee-Ankle-Foot Robot for Gait Rehabilitation (Cheng, Qi, Guo, Yu, 2016)

Knee-foot-ankle is a 2 DOF robot, featuring a modular and portable configuration for post-stroke gait training either in a clinical environment or at home. A new compact series elastic actuator (SEA) has been designed for safe human-robot interactions. For an improved portability, the robotic mechanism is optimized for the walking biomechanics and the mechanical construction has been built using lightweight materials.

The authors have performed an analysis of human walking and starting from this, the design specifications of the robotic system have been delineated. The authors have used Clinical Gait Analysis (CGA) data for the design of the exoskeleton and actuators, collected from the CGA normative gait database. To determine the bio-mechanic features in terms of torque, joint angle and power of the knee flexion/extension and ankle dorsiflexion/plantarflexion, studies have been achieved using a healthy subject (70 kg, 0,9 m leg-length) walking at 1m/s.

The authors have added a certain number of sensors to monitor human kinematics, muscle activity, interaction forces and the force feedback, to control the actuator. The angles in the robotic joints have been determined using rotary potentiometers. The authors used joint angles as control input for the conventional impedance-based control strategy. The robotic device has integrated EMG (electromyography) amplifiers to record muscle activation patterns. EMG signals are used to evaluate human intention or to rate human effort during rehabilitation training.

Advantages: It is a wearable exoskeleton that is a lightweight system of about 3.5 kg.
Drawbacks: It can be used for low hemiparetic patients but can't help the rehabilitation of hip joint.

LLRR: A 3 DOF Lower Limb Rehabilitation Robot (Wu, J., Gao, J., Song, Li, Li, Jiang, 2016)

The LLRR is an exoskeleton with an adaptive robot sub-controller. The design of the exoskeleton includes three joints which can perform the motions of the following human joints: hip joint, knee joint as well as ankle joint. The robotic mechanism provides an adaptable exoskeleton (the length of linkages can be adjusted per the patient's limbs dimensions) and a bracket. The exoskeleton has three modules: thigh, calf and foot, where the length of the thigh and calf are adjustable. The robot is actuated using DC motors.

Advantages: Several parameters must be selected and set prior to starting the training program: the exoskeleton length should be adapted per patient's height; the training program parameters must be selected: the maximum intensity, the training time, the rehabilitation training trajectory and the training joints. The control model includes the kinematics, joint friction and actuator characteristics.

Figure 17. Knee-foot-ankle robot a kinematic scheme
(Cheng, Qi, Guo, Yu, 2016)

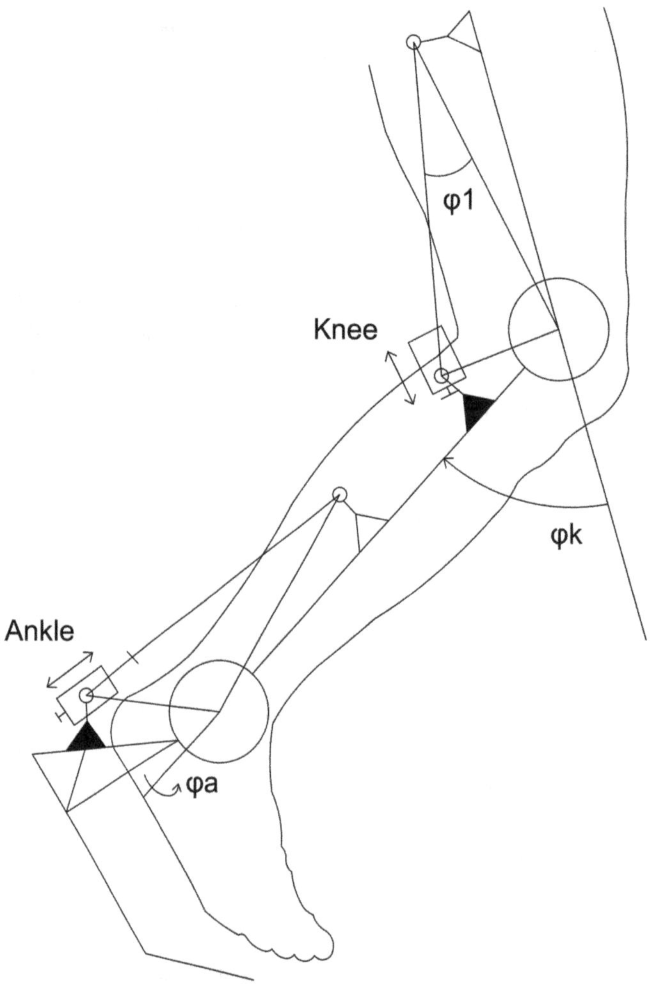

Advantages: The system can be used successfully in lower limb rehabilitation therapy involving all of the lower limb joints.

Drawbacks: It is a static system and it can't achieve the proper simulation of the human gait because the leg is suspended. The patient must be in a standing position therefore those who have serious health problems after a stroke can't use this system.

ROAD: Domestic Assistant and Rehabilitation Robot (Carrera, Moreno & Saltaren, 2011)

This study focuses on the development of a robotic system for the assistance and rehabilitation of patients in walking, balance and standing up activities. This robotic system is designed to interact with the user in the indoor environment both in hospitals or at home. The design is safety compliant, assuring the integrity and safety of the user and the surrounding elements. The robot device moves using tracks

Figure 18. The design and control of a 3DOF lower limb rehabilitation robot prototype (Wu, J., Gao, J., Song, Li, Li, Jiang, 2016); a) a general scheme; b) a kinematic scheme; c) a kinematic scheme with leg attachment

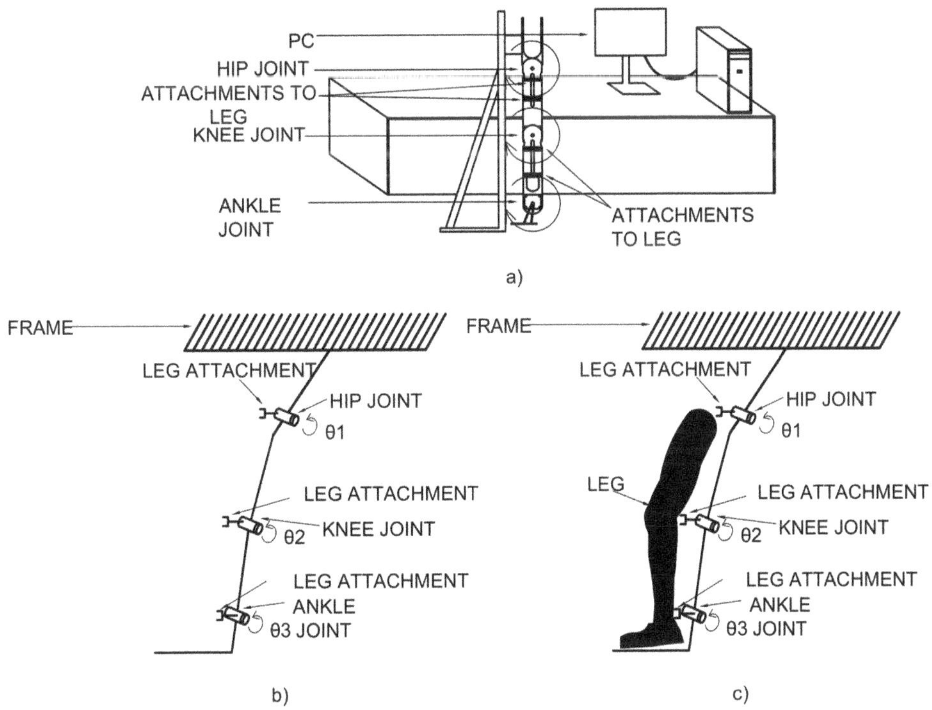

placed on the room ceiling see Figure 19, adding some advantages compared to other robots that move on the floor such as reduced risk of collision with the disabled person, better performance and balance support, better access to high places, no need of a navigation system for detecting objects on the floor, improved usage of the available space and adaptability to perform different tasks using different assembly tools placed on the robot.

Drawbacks: Because it is suspended on the celling the user's routes are established beforehand, so humans who used this type of robots could feel some kind of spatial restriction as they might feel surrounded, therefore the robotic system has limited mobility and workspace.

CALOWI: A Reconfigurable Cable-Based Parallel Manipulator as a Motion Aiding System (Ottaviano E., Ceccarelli M., Paone A., & Carbone G., 2005)

This study presents the CALOWI (Casino Wire Low-Cost Robot), a reconfigurable four-cable-based parallel manipulator developed as a guiding system for people that have motion disabilities.

The reconfigurability is the main idea behind the concept of this project, being able to achieve different tasks by simply modifying the position of the end-effector anchor points and by building two versions: the planar and the spatial version.

Figure 19. ROAD: domestic assistant and rehabilitation robot prototype (Carrera, Moreno & Saltaren, 2011); a) a general scheme; b) a kinematic scheme; c) a kinematic with human attachment

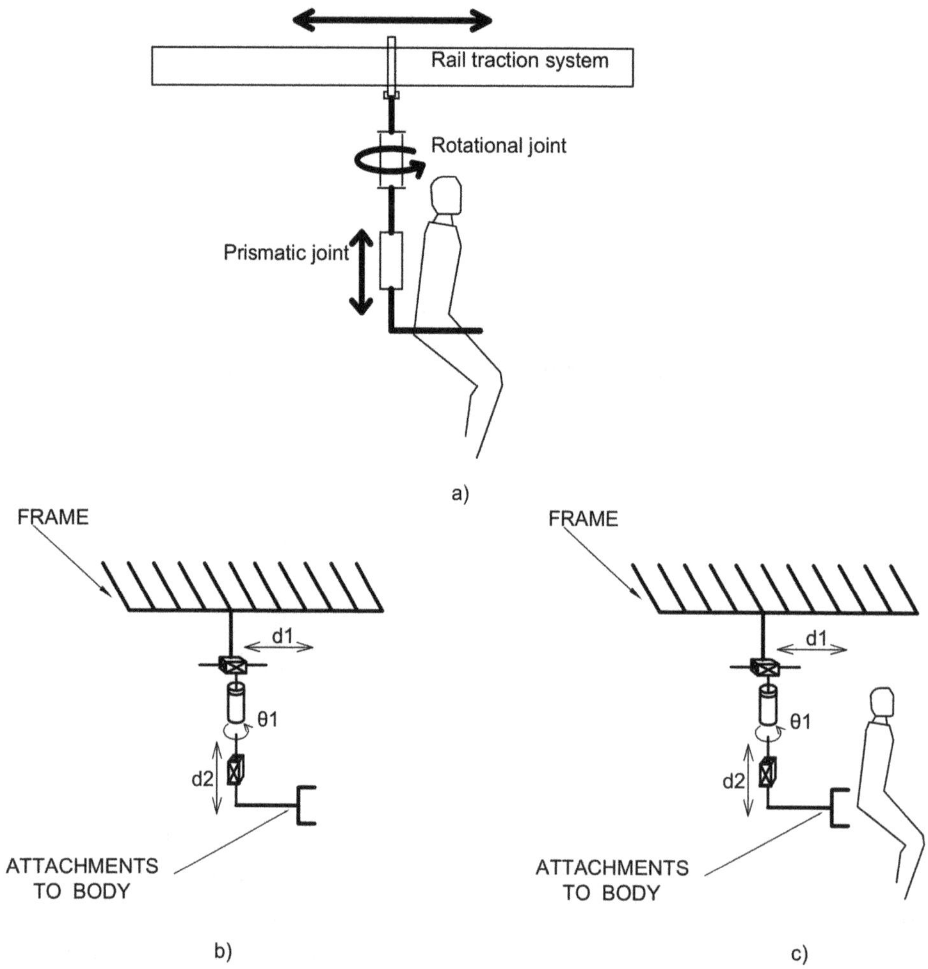

A model for the cable-based parallel manipulator and assisting application has been built. The overall robot dimensions are 2300×2700× 2300 mm, being modeled as a rigid body. The workspace of the robot is quite large, compared to its dimensions. The end-effector has also been modeled as a rigid body, having the overall dimensions: 300× 105 × 216 mm and weighing 3 kg. The selected cable for the simulation is made of steel with a diameter of 2 mm, a Young's modulus of 200 GPa, a breaking load of 3300 N and a mass of 0.02 kg/m.

The human-robot mechanical interface is achieved by using a sling attached to the upper torso of the patient, while four cables connect the sling to the fixed frame of the manipulator, as shown in Figure 20.

Advantages: It can be easy reconfigurable, it is able to do a variety of tasks by changing the position of the end-effector. It is not a complicated solution and also, the costs can't be high. It has a large workspace.

Drawbacks: This system needs a large space to be installed.

Figure 20. The CALOWI prototype
(Ottaviano E., Ceccarelli M., Paone A., & Carbone G., 2005)

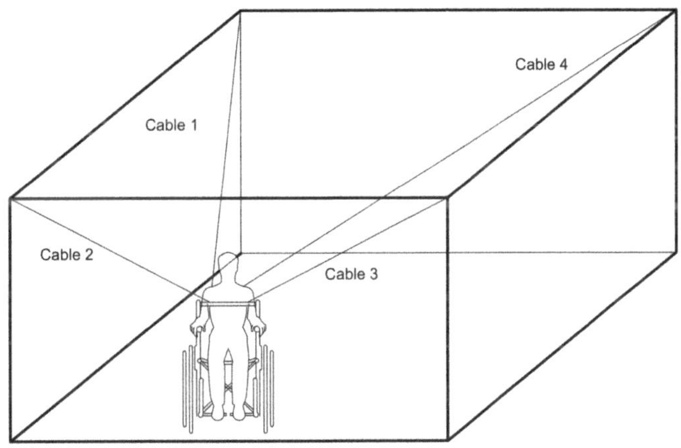

Hand

A Rehabilitation Device to Improve the Hand Grasp Function of Stroke Patients using a Patient-Driven Approach (Park, Jeong, Kwon, Kim, Y., H., Kim, L., 2013)

This device was developed to achieve the rehabilitation of stroke patients with the aim of recovering their hand grasp function in order to undertake activities of daily living (ADL). The device has a control unit, two small actuators, an infrared sensor, and pressure sensor in the grasp handle.

The presented system is comprised of a control unit, two actuators, an infrared (IR) sensor, and pressure sensor in the grasping handle. The actuators are LSA-3024SM by PoteNit. The stroke length is 24 mm and the maximum force is 30 N. These actuators assist the grasping (closing the hand) in both passive and patient-driven models.

In the passive training mode, the presented device drives the grasping movement for patients who do not have voluntary hand and finger movements. At this time, the patient's hand must be fixed to the grasping device, thus, a fixing bandage strap is included. The active training mode is utilized when the presented device does not need to assist the patient's motion. Stroke patients who suffer from light after-effects and are close to being recovered can train the grasping motions in this mode. The patient-driven (active-assistive) mode can be utilized using the movement intention of patients with minimal voluntary hand and finger movements. When a patient tries to move the handle for grasp training, the device detects the patient's attempt by way of pressure sensors that are positioned between the patient's hand and the handle.

Advantages: The device improves hand grasping, it can be used in passive training mode and in patient-driven mode. The device helps patients that do not have voluntary hand and finger movements. It has a small size, it is inexpensive, and available for use at home without specialist's supervision.

Drawbacks: the system needs to be improved to offer a higher level of comfort to patients.

A Soft Exoskeleton for Hand and Rehabilitation Application using Pneumatic Actuators with Variable Stiffness (Yap, Lim, Nasrallah, 2015)

The design implicates a soft wearable exoskeleton that is composed of a glove embedded with pneumatic actuators of variable stiffness for hand assistive and rehabilitation application. This device is light weight therefore it is easy to wear due to the usage of soft pneumatic actuators. The wearable robotic device is described by the authors as a soft wearable robotic device for assistive and rehabilitative exercises by providing customizable motion.

In the study, an actuator with variable stiffness is utilized, thus enabling the stiffness of the actuator to be adapted at different locations, according to the dimensions of the patient's fingers. The actuator concentrates most of the torque at the joints avoiding to bend the links. Moreover, the stiffness of the actuator can be adjusted based on different physiotherapy rehabilitation exercises required such as full fist, straight fist, hook fist and table top. The actuators with variable stiffness are made by using a two –part mold that was made on a 3D printer. The lower-part mold (channel mold) is used to make pneumatic channels into the actuator, which will inflate upon pressurization, and the upper-part mold (feature mold) is used to impose variable stiffness at different locations of the actuator, which decide the bending profile of the actuator.

The general construction of the device is in the form of a glove that has soft actuators with variable stiffness attached on the dorsal side. The ExoGlove prototype was constructed by attaching a total of five actuators on the glove. The glove is glued with five Velcro straps, as well as the bottom part of the actuators. Therefore, the actuators can achieve DON and DOFF motions when required. The total weight of the prototype is approximately 200g.Once the actuators are attached on the glove, the connector tubing is plugged in and connected to the air source.

Figure 21. A Rehabilitation Device to Improve the Hand Grasp Function of Stroke Patients using a Patient-Driven Approach. The Prototype. (Park, Jeong, Kwon, Kim, Y., H., Kim, L., 2013); a) a general scheme; b) a kinematic scheme; c) a kinematic scheme with hand attachment

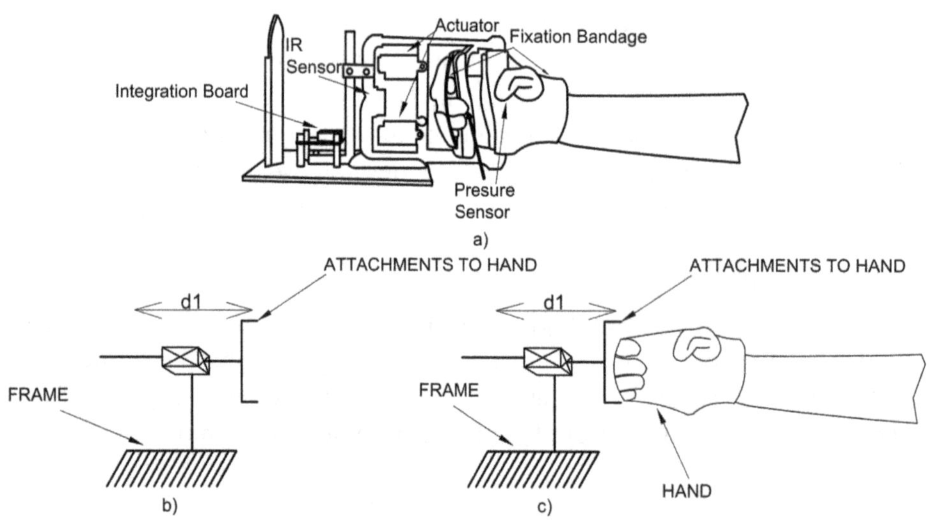

Advantages: It is a lightweight glove 200g, allows fingers to reach all necessary DOF by using pneumatic actuators with variable stiffness in different locations. It helps the human hand to perform motions like grasping and pinching.

Drawbacks: The device is too customizable, a single set of actuators is able to support only one type of exercises therefore, to reproduce a different type of exercise it is necessary to change the actuators, this could take a pretty long time.

The Design of Fine Motion Assistance Equipment for a Hand Oriented Robotic Rehabilitation System (Ito, Kawasaki, Ishigure, Natsume, Nishimoto, 2011)

This study presents hand rehabilitation equipment and the control system for operating it. A prototype of the hand rehabilitation equipment has been designed, which is able to achieve a total of 18 DOFs of motion, 3 DOFs for help each finger, 4 DOFs for assisting the thumb, and 2 DOFs for the wrist.

The motion assistance equipment that was developed is composed from three parts: mechanisms for the fingers and thumb, a mobile base of mechanism for the aforementioned, and a mechanism for the wrist. Two types of mechanisms are developed, for the four fingers and the thumb. Two closed link

Figure 22. A Soft Exoskeleton for Hand and Rehabilitation Application using Pneumatic Actuators with Variable Stiffness prototype (Yap, Lim, Nasrallah, 2015); a) a general scheme; b) a kinematic scheme; a kinematic scheme with hand attachment

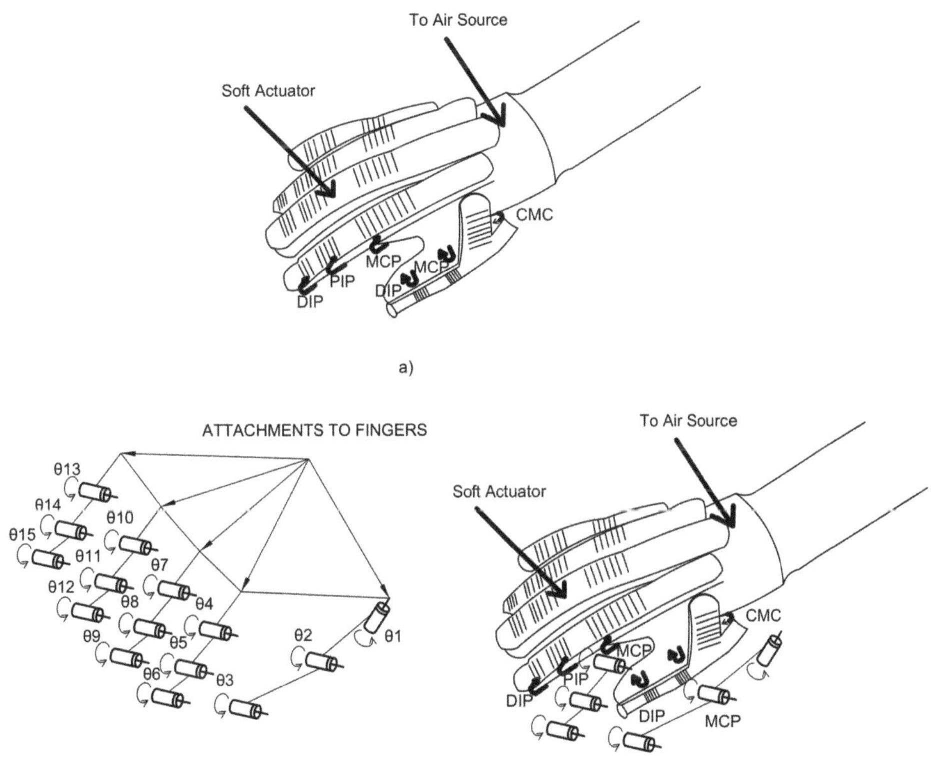

structures are established when the finger mounts the equipment. In the first test model of this equipment, the two closed link structures collided each other, which limited the motion of the link. The equipment presented in this study has been developed to solve this problem. The kinematic relation in the closed link structure allows authors to calculate the MP/PIP joint angles according to the active joint angles detected from the built-in rotary encoder, and vice versa.

A self-motion control strategy is adopted in which the impaired hand is controlled by commands from the healthy hand, so in this way bilateral symmetrical motions are realized. Regarding the motion of the unimpaired side, the angular displacements of the finger joints are observed using a data glove. Information that is coming from the unaffected hand's stance is also displayed for the disabled user using computer graphics, which allows the user to visually identify the hand's state of motion. Based on inverse kinematics, the reference angles for the active articulations of each movement assistance mechanisms are computed so that the impaired hand takes the same stance as the unimpaired one. These reference angles are sent to the controller of the motion assistance equipment through TCP/IP communication. To do the above tasks the authors used a personal computer operated by a Windows operating system.

Advantages: The robotic device covers 18 DOFs

Drawbacks: Being a prototype concept the authors will have carry out improvements based on experimental results.

CRITICAL DISCUSSIONS

In order to assess the current progress in robotic rehabilitation and to define some general directions which would enhance the impact of robotic devices in post-stroke rehabilitation, reference is made to a number of review studies which focus on multiple studies which evaluated the efficiency of the systems that reached maturity thus enabling patient testing. Thus, the major conclusions can be summarized in the following guidelines:

- There exists no preferred physical therapy or "more efficient" set of exercises, the selection of the proper rehabilitation plan depends on the physical therapist and the patient progress (Proietti, T., et al., 2016);
- Most available robotic devices focus on patient rehabilitation in the post-acute and chronic phase, emphasizing the need of more devices capable to accommodate the special requirements of the acute phase of post-stroke patients (Yadav, R.K., et al., 2016);
- The preferred modalities for rehabilitation exercises used in robotic devices are: active, active-assisted, passive and resistive control, which based on the existing definitions are the simpler human-robot interaction strategies (Vaida, et al., 2017);
- Unfortunately, some results point out that the progress registered by robotic devices based on consecutive patient assessments using classical scoring methods have no positive influence in the ADL (Yoo, et al., 2015);
- The limited transfer of the motor gains into ADL capability improvement is considered to be caused by the non-uniform distribution of the exercises on the different limb areas, even though medical studies proved that distal to proximal combined training is essential for efficient functional progress; (Proietti, T., et al., 2016);

Figure 23. A design of fine motion assist equipment for disabled hand in robotic rehabilitation system (Ito, Kawasaki, Ishigure, Natsume, Nishimoto, 2011) a) a kinematic scheme; b) a scheme of mechanism for fingers; c) a scheme of mechanism for thumb

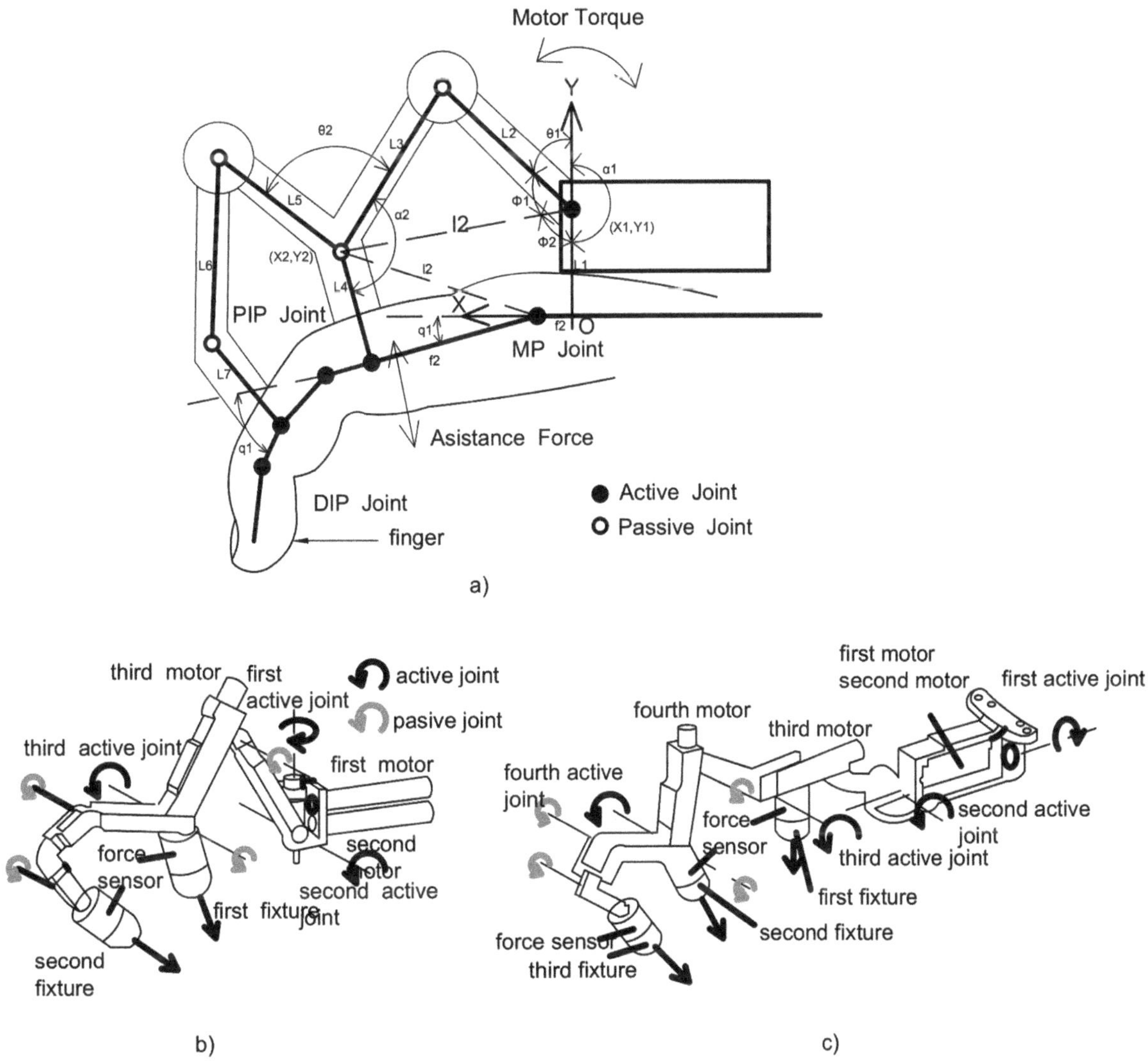

- The exclusive use of repetitive exercises hinders the patient progress due to a lack of physiological implication of the patient in the rehabilitation activities which would stimulate also cognitive and neuronal progress (Vaida, et al., 2017);
- There must be an increase in the human robot interaction process with the time interval from the stroke as it was shown that in acute phase simple exercises are efficient but later on, the best results are obtained through more complex strategies with motivational elements (Abdollahi F., et al, 2014);
- In the context of independent living and the use of such devices in multiple environments safely, ease-of-use and tele-monitoring are elements that the new devices must integrate.

From the social and economic point of view it has to be underlined that the developing economies face a larger number of stroke patients. As some of the social conditions leading to this situation cannot be solved, proper medical care has to be provided also for this population segment. Thus, the future robotic rehabilitation devices must not only provide efficient therapeutic options but must be cost – efficient, sustainable by average healthcare systems and user-friendly, a concept that must be seen beyond its classical definition, providing:

- The definition of standards in robotic rehabilitation, dedicated for this particular field – as an observation, such a standard is in work with the International Organization for Standardization, under the code IEC/CD 80601-2-78, with the latest reported development stage 30.20 (Committee Draft Study);
- The definition of comprehensive medical protocols adapted for robotic assisted rehabilitation (Vaida, et al., 2017);
- Adaptable devices which can cover different anthropometric patient parameters making a system suitable for multiple patients (Vaida, et al., 2017);
- The use of modular concepts where the robot can be easily adapted to certain needs;
- The integration of efficient but simple sensory systems that create efficient progress monitoring solutions allowing fast patient oriented exercise optimization (Proietti, T., et al., 2016);
- The implementation of multiple HRI concepts dependent on the patient degree of disability and applications that motivate the patient in multiple ways;
- Dedicated protocols for hospital, day care centers and own environment use;
- Safety functions which prevent unwanted injuries upon the affected limbs that can be configured during the periodic patient evaluation;
- The user-interface must be effective for both the patient, stimulating more than the physical motor gains and the physical therapist that can safely and quickly adapt the therapy plan for maximum patient progress;
- A cost oriented design approach should be followed to guarantee affordable solutions also for home use as proposed for example in (Cafolla & Carbone 2014; Carbone et al. 2017).

  Note: It is to note that it is still an open research issue to identify proper requirements and protocols for rehabilitation robots. Accordingly, literature shows a wide range and variety of rehabilitation robots. They can provide quite different features and characteristics while their use and real application still requires investigation to transfer them from laboratories to real end-users.

The above-mentioned aspects have been highlighted in the comparison Tables 2, 3, and 4 as well as they have led to the development of some novel concepts for rehabilitation devices where the main focus has been focused at upper limbs. Further concepts are under investigation for lower limbs and hands.

NEW DEVICES FOR THE REHABILITATION OF THE UPPER LIMB

An Exoskeleton-Based Rehabilitation Robot for Elbow and Wrist

Based on the results presented in (Major, K., et. al., 2016), the following motions were selected to be implemented in the rehabilitation process (Figure 24):

- **Elbow**: Flexion/extension and pronation/supination;
- **Wrist**: Adduction/abduction and the flexion/extension.

For each motion two contact areas (anchor points) will be used, one acting as the motion generator (the active anchor point – represented with dotted line in the figure) and one acting as joint stabilizer and motion isolator (the passive anchor point – represented with continuous line in the Figure 24).

For the achievement of the defined motions the ReExRob robotic device is proposed as a 4<u>R</u> robot having an anthropometric structure integrating only rotational joints. The kinematic scheme of ReExRob, presented in Figure 25. illustrates the 4 active joints $q_i, i = 1..4$ and the links that $l_i, i = 1..5$ connect them. Through efficient and modular design, the robot can be constructed with adjustable link lengths making it suitable for different anthropometric patient data.

The initial design of the ReExRob robot is illustrated in Figure 26. The main components are: the fixing elements, the three anchors: passive one and the active anchors (2 – for supination/pronation and 3 for the wrist motions). The flexion/extension of the elbow is actuated by Motor 1, the pronation /supination is designed using a spur gear mechanism actuated by Motor 2, while Motor 3 and Motor 4 achieve the motions for the wrist.

An Innovative Spherical Parallel Robot for Shoulder Rehabilitation

Moving to the proximal joint of the upper limb, the shoulder, the motions of interest are flexion/extension and abduction/adduction which, with respect to the shoulder joint, can be simply described as motions performed either in a horizontal plane or a vertical plane. Analyzing the motion from the shoulder joint

Figure 24. Shows motion anchor for elbow and wrist mobilization: a) flexion/extension; b) pronation/supination; c) wrist flexion/extension; d) adduction/abduction
(Major et al., 2016)

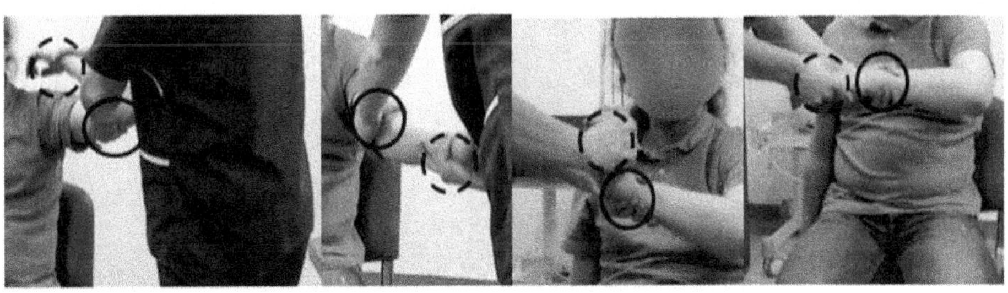

Figure 25. ReExRob robotic system kinematic scheme for elbow and wrist mobilization
(Plitea et al., 2017)

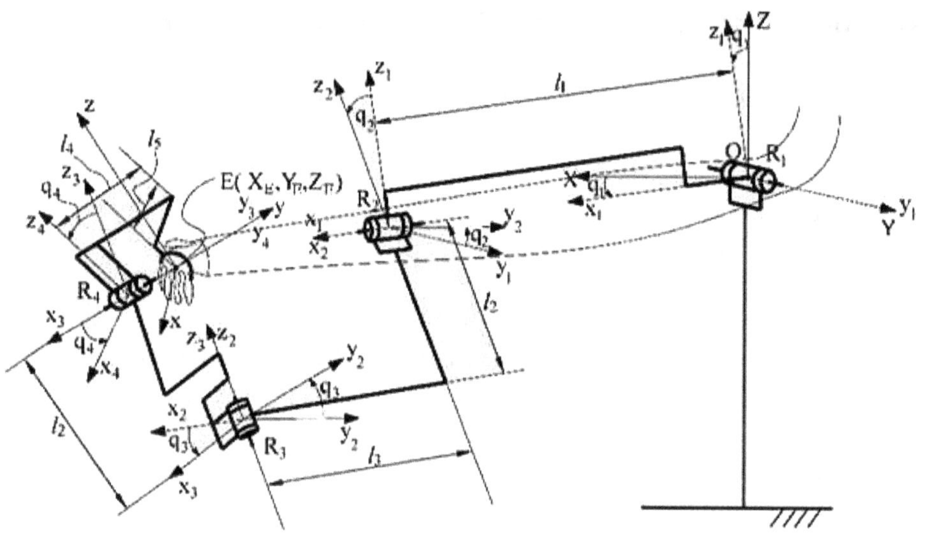

Figure 26. ReExRob robotic system scheme for elbow and wrist mobilization
(Plitea et al., 2017a)

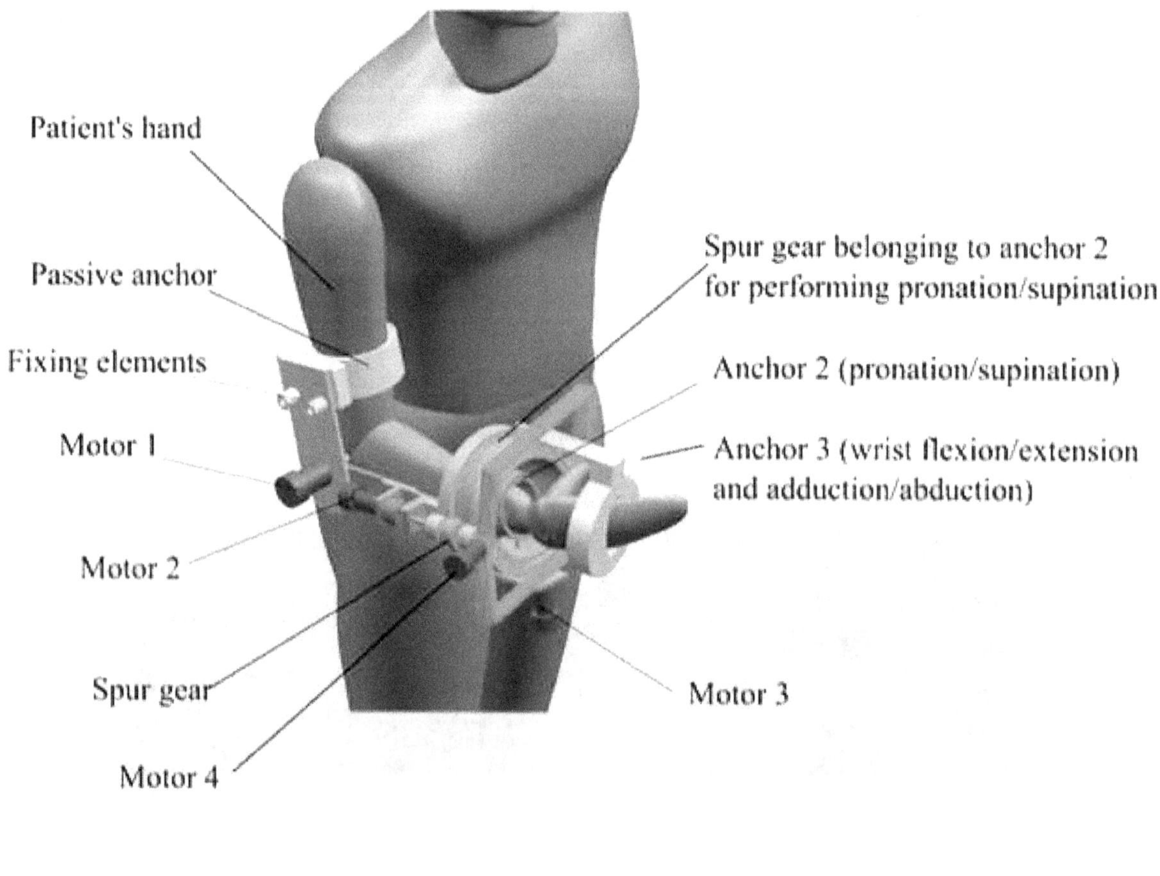

point of view the two elementary motions describe arcs with a common rotation center, in the joint itself. Thus, one simple mechanism that can replicate easily these two motions is a spherical robot (ASPIRE), whose kinematic scheme is illustrated in Figure 27.

The advantages of this spherical mechanism are the simple concept, wide range of motions, modular design allows the possibility to attach new modules. Due to its structure, the mechanism allows a generalized motion on the sphere which extends its generality degree allowing the definition of multiple exercises with large amplitudes working in a virtually unconstrained volume. One possible drawback could be considered the larger volume the structure occupies, but due to its design the mechanism itself does not hinder in any way the motions of the arm. A concept design of the mechanism is illustrated in Figure 28. (Plitea et al., 2017b)

Based of robotic rehabilitation systems presented above a comparison table (Table 2) is added to clearly see the main aspects of robotic systems analyzed in this chapter.

CONCLUSION

The general conclusion with results from the data presented in this chapter emphasis a big gap between the actual demands in rehabilitation and the available solutions, a gap which extends with time due to the ageing of the population. The current prognosis in world population evolution estimates that in 15 years the available number of caregivers will not be able to satisfy the population need in rehabilitation thus imposing a paradigm shift in the therapeutic approaches, where the physical therapist will become

Figure 27. Kinematic scheme of the spherical mechanism for shoulder rehabilitation
(Plitea et al., 2017b)

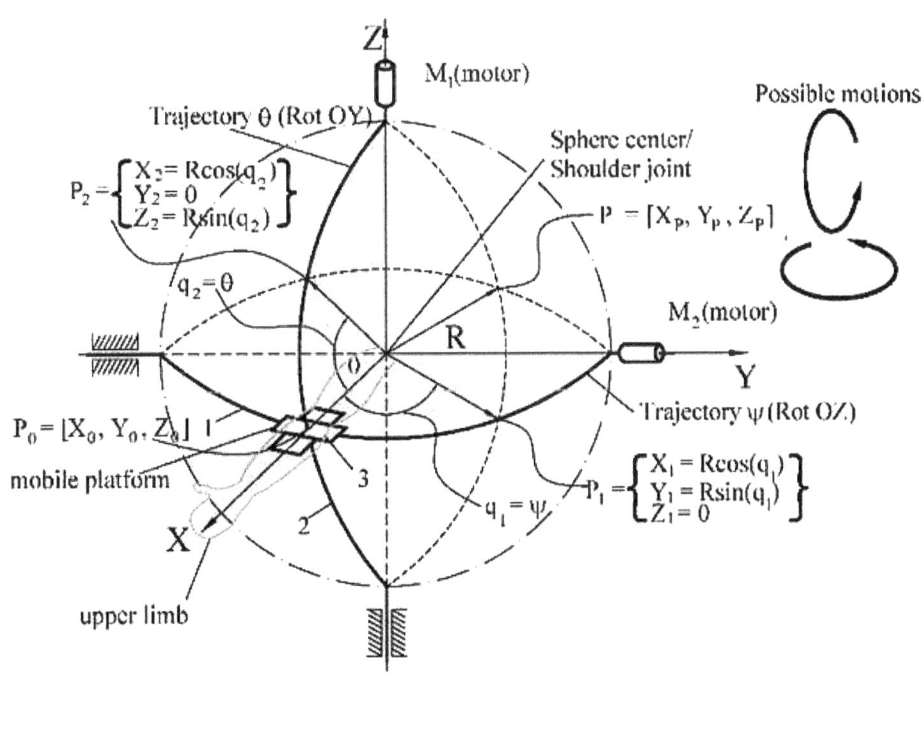

Figure 28. CAD model of the Aspire structure
(Plitea et al., 2017b)

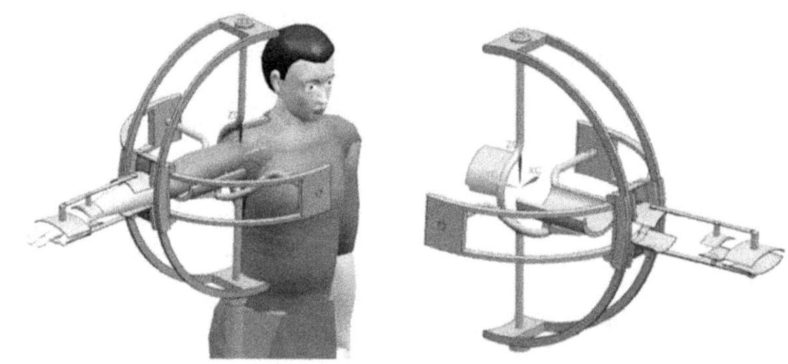

Table 2. Comparison of robotic rehabilitation systems for upper limbs

No.	Name	DOF	Actuation System	Evol. Stage	Advantages	Drawbacks	Source
1	InMotionArm	2DOF	DC brushless motors	commercial system	-improve patient's range of motion, eye hand coordination, easy to use -it can be accessed direct from wheelchair	being a planar robot device, it allows the human limb to achieve motions in a limited three-dimensional space.	(Duret et al., 2015)
2	Bi-Manu-Track	2DOF	servo motors	commercial system	-easy to use -low cost -permits bilateral exercises	because of the healthy limb's involvement, the projections to the paretic muscle can be affected.	(van Delden, et al., 2012)
3	ArmeoPower	6DOF	DC motors	commercial system	-allow patients to get early treatment of motor disabilities; enable patients to perform repetitive, intensive, and goal oriented exercises.	-high price about 250.000 euros	(HWP, 2016)
4	ArmeoSpring	5DOF	passive (springs)	commercial system	-permits to customize patient's therapy; - feedback exercises to train core movement patterns that are used in activities of daily living	-can't be used for patients with severe hemiparesis	(HWP, 2016)
5	BONES	4DOF	Pneumatic	prototype	-allows naturalistic motion of upper limb using mechanically grounded, direct drive actuators at the shoulder.	-It is not wearable and can be used only in clinics not for ADL	(Klein et al., 2008)
6	Rocker (APBT)	1DOF	DC motors	prototype	-helps patients to improve wrist flexion/extension - mirror symmetrical coordination	-it allows only planar flexion/ extension of wrist	(Stinear & Byblow, 2004)
7	SCMRSR (SelfCon. Master-Slave Robot)	1DOF	DC motors	prototype	- provides bilateral arm coordinated training for patients which suffered unilateral paralysis.	- it performs only flexion/ extension of elbow.	(Rashedi et al., 2009)
8	SUE (Supinator Extender)	2DOF	Pneumatic	prototype	-lightweight (0.56 kg) - performs forearm supination/ pronation and wrist flexion/extension	- non-linear forces induced by the pneumatic cylinders a complex control is necessary.	(Allington et al., 2011)
9	ReExRob	4DOF	servo motors	concept	-efficient and modular design -due to its adjustable link lengths, it is suitable for different anthropometric patient data.	-being a concept, the must be improved to meet a proper pHRI (physical human-robot interaction)	(Major et. al., 2016)
10	ASPIRE	2DOF	servo motors	concept	-simple concept; wide range of motions; modular design allows to attach new modules	-it occupies a large volume	(Plitea et al., 2017b)

Table 3. Comparison of robotic rehabilitation systems for lower limbs

1	G-EO-System	3DOF	electric motors	commercial system	-easy for patients to use - G-EO system can simulate walking on level ground and stairs climbing, quick setup and easy patient handling.	- No virtual reality interactive training module.	(Hesse et al., 2010)
2	HAL-5	2DOF	DC motors	commercial system	-It is wearable and portable device that confers to patient a good help in rehabilitation gait training.	-it has limited torque -is heavy and can't be used by paralyzed patients.	(Sankay, 2007)
3	AlterG Bionic Leg	1DOF	NXT motors	commercial system	lightweight wearable device which assist person with impaired leg, helps patients to place the body weight on healthy leg.	- battery life gives limited uses per day	(Li et al., 2015)
4	Physioterabot	3DOF	DC motors	prototype	-it can reach the range of motion of human limb as in Table 1; -can be adaptive for different weight of patients up to 100kg.	-it doesn't perform rehabilitation exercises for ankle.	(Akdogan & Mehmet, 2011)
5	Knee-ankle-foot robot	2DOF	SEA (series elastic actuator)	prototype	-lightweight wearable exoskeleton around 3.5 kg.	-it can be used for low hemi paretic patients -can't help rehabilitation joint.	(Cheng et al., 2016)
6	LLRR	3DOF	DC motors	prototype	- the system can be used successfully in lower limb rehabilitation therapy involving whole lower limb joints.	-static system, leg is suspended -the patient with serious health problems after stroke can't use it	(Wu et al., 2016)
7	ROAD	3DOF	electric motors	concept	- less risk of collision with the disabled person; good support for balance; no need of a navigation system for detecting objects on the floor; improved usage of the available space	- some limitations to the patient mobility, access to the surrounding environment is limited; users could feel some kind of spatial restriction.	(Carrera et al., 2011)
8	CALOWI (Cassino Low-Cost wire driven)	4DOF		concept	easy reconfigurable; -it is able to do a variety of tasks by changing the end-effector position; large workspace; low-cost; suitable also for upper limbs	-the system needs a large space, to be installed.	(Ottaviano, et al. 2005)

Table 4. Comparison of robotic rehabilitation systems for hands

1	AMADEO Hand-Therapy System	5DOF	DC motors	commercial system	- covers the human hand fingers workspace - performs an intensive training having a high frequency of gripping movements - even for patients with severe hemiparesis.	- it is a fixed system it is not wearable	(Sale et al., 2012)
2	YouGrabber	3DOF	bending sensors	commercial system	-it allows bilateral training aside from unliteral; the device can be adjusted based on patient's current abilities; the training software can be used on an ordinary PC at home or in a clinic.	-it is designed to move only 3 fingers.	(Shuster-Amft et al., 2014)
3	CyberGrasp	5DOF	stepper motors	commercial system	-as it is a cable driven system, the robot can maintain a continuous force of 12N per finger -it can be used in different domains such as haptic rehabilitation and training of visual impaired.	- a limited workspace -high price around 34000 euros -it is not lightweight (around 454 grams) for using the system for a prolonged period of time.	(CGSWP., 2014)
4	Rehabilitation Device to Improve the Hand Grasp Function of Stroke Patients	1DOF	linear servo actuator	prototype	-improves hand grasping; it can be used in passive training mode and also in patient-driven mode; helps patients that do not have voluntary hand and finger movements; it has a small size and it is inexpensive; it is available to be used at home	-the system needs to be improved to offer a higher level of comfort to patients.	(Park et al., 2013)
5	Soft Exoskeleton for Hand and Rehabilitation Application	15DOF	pneumatic actuators	prototype	-Lightweight glove 200g -allows fingers to reach all necessary DOF by using pneumatic actuators with variable stiffness in different locations. -it helps humans to perform motions like grasping and pinching.	- too many customizations; each actuator type is able to support only one type of exercises; long set up time for actuators changes.	(Yap et al., 2015)
6	Fine motion assist equipment for disabled hand	18DOF	servo motors	prototype	-the robotic devices cover 18 DOF's	-being a prototype concept, the authors will have to make improvements based on experimental results.	(Ito et al., 2011)

the planner and decision maker for the personalized assistance provided for a patient through dedicated robotic devices, supervising, monitoring and adjusting the therapeutic parameters. These conditions raise not just the opportunity but the necessity to act fast towards the development of new rehabilitation solutions, in an economically feasible environment capable of supporting stroke patients in each stage of rehabilitation starting from the acute post-stroke phase which takes place in hospitals, followed by longer stages which can take place in day care centre and home environment. The existing commercial solutions demonstrate that robotic rehabilitation is very promising and can be used for the rehabilitation of multiple disabilities caused by stroke. The results obtained during their first trials define for the entire community the current limitations and areas that have to be enhanced in parallel with the setup of a first set of standards in robotic rehabilitation. The first clinical trials point out the need for the development of HRI strategies that will integrate the patient within the robot control system as an active component to increase the collaboration between man and machine. This will enable an enhanced rehabilitation treatment with focus on multiple areas, cognitive, neural, motivational and physical, aiming to achieve a faster patient recovery improving, in the end, the life quality and contributing to the concept of independent living. The selection criteria have been based on the engineering significance of each solution. A classification has been made based on the limb which can be treated as well as on the availability of either just a concept/laboratory solution or a commercial product.

REFERENCES

Abdollahi, F., Case Lazarro, E. D., Listenberger, M., Kenyon, R. V., Kovic, M., Bogey, R. A., ... Patton, J. L. (2014). Error augmentation enhancing arm recovery in individuals with chronic stroke: A randomized crossover design. *Neurorehabilitation and Neural Repair*, *28*(2), 120–128. doi:10.1177/1545968313498649 PMID:23929692

Akdogan, E., & Mehmet, A. A. (2011). The design and control of a therapeutic exercise for lower limb rehabilitation: Physiotherabot. *Mechatronics*, *21*(3), 509–522. doi:10.1016/j.mechatronics.2011.01.005

Allington, S., Klein, J., Buell, M., Reinkensmeyer, D. J., & Bobrow, J. (2011). Supinator extender (SUE): A pneumatically actuated robot for forearm/wrist rehabilitation after stroke. *Proceedings of Annual International Conference of the IEEE Engineering in Medicine and Biology Society*. doi:10.1109/IEMBS.2011.6090459

Ates, S., Haarman, C. J. W., & Stienen, A. H. A. (2016). SCRIPT passive orthosis: Design of interactive hand and wrist exoskeleton for rehabilitation at home after stroke. *Autonomous Robots*, *41*(3), 711–723. doi:10.1007/s10514-016-9589-6

Basteris, A., Nijenhuis, S. M., Stienen, A. H., Buurke, J. H., Prange, G. B., & Amirabdollahian, F. (2014). Training modalities in robot-mediated upper limb rehabilitation in stroke: A framework for classification based on a systematic review. *Journal of Neuroengineering and Rehabilitation*, *11*(111), 15. PMID:25012864

Benjamin, E. J., Blaha, M. J., Chiuve, S. E., Cushman, M., Das, S. R., Deo, R., & Muntner, P. (2017). Hearth Disease Statistics – 2017 Update: A Report From the American Heart Association. In Heart Disease and Stroke Statistics—2017 Update. American Health Association.

Cafolla, D., & Carbone, G. (2014). A study of feasibility of a human finger exoskeleton. *Service Orientation in Holonic and Multi-Agent Manufacturing and Robotics, Studies in Computational Intelligence, Springer, 544*, 355–364. doi:10.1007/978-3-319-04735-5_24

Carbone, G., Cavero, C. A., Ceccarelli, M., & Altuzarra, O. (2017). A Study of Feasibility for a Limb Exercising Device. Advances in Italian Mechanism Science, 47, 11-21.

Carbone, G., Gherman, B., Ulinici, I., Vaida, C., & Pisla, D. (2017). Design Issues for an Inherently Safe Robotic Rehabilitation Device. *Proceedings of the 26th International Conference on Robotics in Alpe-Adria-Danube Region RAAD 2017*, 967-974.

Carrera, J., Moreno, H. A., Saltaren, R., Pérez, C., Puglisi, L., & Garcia, C. (2011, October). ROAD: Domestic assistant and rehabilitation robot. *Medical & Biological Engineering & Computing*, *49*(10), 1201–1211. doi:10.1007/s11517-011-0805-4 PMID:21789672

Casadio, M., Sanguineti, V., Morasso, P. G., & Arrichello, V. (2006). Bracio di Ferro: A new haptic workstation for neuromotor rehabilitation. *Technology and Health Care*, *14*(3), 123–142. PMID:16971753

Cauraugh, J., & Summers, J. (2004). Natural plasticity and bilateral movements: A rehabilitation approach for chronic stroke. *Progress in Neurobiology*, 5(75), 309–320.

Ceccarelli, M., & Romdhane, L. (2010). 2010 Design issues for human-machine platform interface in cable-based parallel manipulators for physiotherapy applications. *Journal of Zhejiang University. Science A*, *11*(4), 231–239. doi:10.1631/jzus.A1000027

Chen, B., Ma, H., Qin, L.-Y., Gao, F., Chan, K.-M., Law, S.-W., ... Liao, W.-H. (2006). Recent developments and challenges of lower extremity exoskeletons. *Journal Orthopaedic Translation*, *5*, 26–37. doi:10.1016/j.jot.2015.09.007

Cheng, G., Qi, P., Guo, Z., & Yu, H. (2016). Mechanical design and evaluation of a compact portable knee-ankle -foot robot for gait rehabilitation. *Mechanism and Machine Theory*, *103*, 51–64. doi:10.1016/j.mechmachtheory.2016.04.012

Clarke, D. J., & Forster, A. (2015). An Improving post-stroke recovery: The role of the multidisciplinary health care team. *Journal of Multidisciplinary Healthcare*, *8*, 433–442. doi:10.2147/JMDH.S68764 PMID:26445548

Colomer, C., Noe, E., & Llorens, R. (2016). Mirror therapy in chronic stroke survivors with severely impaired upper limb function: A randomized controlled trial. *European Journal of Physical and Rehabilitation Medicine*, *52*(3), 271–278. PMID:26923644

Cyber Glove Systems Web Page. (n.d.). Retrieved from http://www.cyberglovesystems.com/cybergrasp/#specs

Duret, C., Courtial, O., Grosmaire, A. G., & Hutin, E. (2015). Use of a Robotic Device for the Rehabilitation Severe Upper Limb Paresis in Subacute Stroke: Exploration of Patient/Robot Interaction and the Motor Recovery Process. *BioMed Research International*, *2015*, 482389. doi:10.1155/2015/482389 PMID:25821804

Hesse, S., Schulte-Tigges, G., Konrad, M., Bardeleben, A., & Werner, C. (2003). Robot-assisted arm trainer for the passive and active practice of bilateral forearm and wrist movements in hemiparetic subjects. Archives of Physical Medicine and Rehabilitation, 84(6), 915-920.

Hesse, S., Waldner, A., & Tomeller, C. (2010). Innovative gait robot for the repetitive practice of floor walking and stair climbing up and down in stroke patients. *Journal of Neuroengineering and Rehabilitation*, *7*(1), 30. doi:10.1186/1743-0003-7-30 PMID:20584307

HOCOMA Webpage. (n.d.). Retrieved from http://www.hocoma.com

Hogan, N., Krebs, H. I., Charnnarong, J., Srikrishna, P., & Sharon, A. (1992). MIT-MANUS:a workstation for manual therapy and training. Robot and Human Communication. *Proceedings., IEE International Workshop*. DOI: .1992.253895 doi:10.1109/ROMAN

Housman, S., Scott, K., & Reinkensmeyer, D. (2009). Randomized Controlled Trial of Gravity-Supported, Computer-Enhanced Arm Exercise for Individuals With Severe Hemiparesis. *Neurorehabil Neural Repair.* doi:10.1177/1545968308331148

Ito, S., Kawasaki, H., Ishigure, Y., Natsume, M., Mouri, T., & Nishimoto, Y. (2011). A design of fine motion assist equipment for disabled hand in robotic rehabilitation system. *Journal of the Franklin Institute*, *348*(1), 79–89. doi:10.1016/j.jfranklin.2009.02.009

Jackson, A., Holt, R., Culmer, P., Makower, S., Levesley, M., Richardson, R., & Bhakta, B. B. (2007). Dual robot system for upper limb rehabilitation after stroke: The design process. *Journal of Mechanical Engineering Science. Proceedings of the Institution of Mechanical Engineers, Part C.*, *221*(7), 845–857.

Johnson, M., Van der Loos, H., Burgar, C., & Leifer, L. (1999). Driver's SEAT: Simulation Environment for Arm Therapy. *Proceedings of 6th International Conference on Rehabilitation Robotics (ICORR'99)*, 227-234.

Johnson, W., Onuma, O., Owolabi, M., & Sachdev, S. (2016). Stroke: A global response is needed. *Bulletin of the World Health Organization*, *94*(9), 634–634A. doi:10.2471/BLT.16.181636 PMID:27708464

Kinsman Enterprises Product Catalog Web Page. (n.d.). Retrieved from http://www.kinsmanenterprises.com/

Klamroth-Marganska, V., Blanco, J., Campen, K., Curt, A., Dietz, V., Ettlin, T., ... Riener, R. (2014). Three-dimensional, task-specific robot therapy of the arm after stroke: A multicentre, parallel-group randomised trial. *Lancet Neurology*, *13*(2), 159–166. doi:10.1016/S1474-4422(13)70305-3 PMID:24382580

Klein, J., Spencer, S. J., Allington, J., Minakata, K., Wolbrecht, E. T., Smith, R., & Bobrow, J. E. (2008). Biomimetic orthosis for the neurorehabilitation of the elbow and shoulder (BONES) 2008. *Proceedings of the 2nd Biennial IEEE/RAS-EMBS International Conference on Biomedical Robotics and Biomechatronics*.

Li, L., Ding, L., Chen, N., Mao, Y., Huang, D., & Li, Le. (2015). Improved walking ability with wearable robot-assisted training in patients suffering chronic stroke. *Proceeding of the 4th International Conference on Biomedical Engineering and Biotehnology*. doi:10.3233/BME-151320

Liu, T., Li, C., Inoue, Y., & Shibata, K. (2010). A Self-controlled Master-Slave Robot and Its Application for Upper Limb Rehabilitation. Lecture Notes in Electrical Engineering, 75, 219-235. doi:10.1007/978-3-642-15687-8_12

Loureriro, R., Harwin, W., Nagai, K., & Johnson, M. (2011). Advantages in upper limb stroke rehabilitation: A technology push. *Medical & Biological Engineering & Computing*, *49*(10), 1103–1118. doi:10.1007/s11517-011-0797-0 PMID:21773806

Lum, P., Lehman, S., & Reinkensmeyer, D. (1995). The bimanual Lifting Rehabilitator: An Adaptive Machine for Therapy of Stroke Patients. *IEEE Transactions on Rehabilitation Engineering*, *3*(2), 166–174. doi:10.1109/86.392371

Lum, P., Reinkensmeyer, D., & Lehman, S. (1993). Robotic Assist Devices for Bimanual Physical Therapy: Preliminary Experiments. *IEEE Transactions on Rehabilitation Engineering, 1*(3), 185-191. DOI: 10.1109/86.279267

Lum, P., Van der Loos, M., Shor, P., & Burgar, C. (1999). A robotic system for upper-limb exercises to promote recovery of motor function following stroke. *Proceedings of Sixth International Conference on Rehabalition Robotics*, 235–9.

Major, K. A., Major, Z. Z., Carbone, G., Pîslă, A., Vaida, C., Gherman, B., & Pîslă, D. (2016). Ranges of motion as basis for robot-assisted post-stroke rehabilitation. *In. J Bioflux Society, Human & Veterinary Medicine*, *8*(4), 192–196.

Masiero, S., Celia, A., Rosati, G., & Armani, M. (2005). Robotic-Assisted Rehabilitation of the Upper Limb After Acute Stroke. Archives of Physical Medicine and Rehabilitation, 88(2), 142-149.

Ottaviano, E., Ceccarelli, M., Paone, A., & Carbone, G. (2005). A Low-Cost Easy-Operation 4-Cable Driven Parallel Manipulator. *IEEE International Conference on Robotics and Automation ICRA'05*, 4019-4024. doi:10.1109/ROBOT.2005.1570734

Park, W., Jeong, W., Kwon, G. H., Kim, Y. H., & Kim, L. (2013) A Rehabilitation Device to Improve the Hand Grasp Function of Stroke using a Patient-Driven Approach. *Proceeding of International Conference on Rehabilitation Robotics*. DOI: doi:10.1109/ICORR.2013.6650482

Patton, J. L., Kovic, M., & Mussa-Ivaldi, F. A. (2006). Custom-designed haptic training for restoring reaching ability. *Journal of Rehabilitation Research and Development*, *43*(5), 643–656. doi:10.1682/JRRD.2005.05.0088 PMID:17123205

Plitea, N., Gherman, B., Carbone, G., Ceccarelli, M., Vaida, C., Banica, A., & Pisla, D. (2017). Kynematic analysis of an exoskeleton-based robot for elbow and wrist rehabilitation. *6th International Symposium on Multibody Systems and Mechatronics-MuSMe*.

Plitea, N., Pisla, D., Carbone, G., Vaida, C., Gherman, B., & Banica, A. (2017). *Robotic system for elbow and wrist rehabilitation*. Patent pending.

Plitea, N., Pisla, D., Carbone, G., Vaida, C., Gherman, B., & Ulinici, I. (2017). *Spherical mechanism for shoulder rehabilitation*. Patent pending

Plitea, N., Vaida, C., Carbone, G., Pisla, A., Ulinici, I., & Pisla, D. (2017). On the Kinematics of an Innovative Spherical Parallel Robot for the Shoulder Rehabilitation. *6th International Symposium on Multibody Systems and Mechatronics-MuSMe.*

Proietti, T., Crocher, V., Roby-Brami, A., & Jarrasse, N. (2016). Upper-limb robotic exoskeletons for neurorehabilitation: A review on control strategies. *IEEE Reviews in Biomedical Engineering, 9*, 4–14. doi:10.1109/RBME.2016.2552201 PMID:27071194

Rashedi, E., Mirbagheri, B., Taheri, B., Farahmand, F., Vossoughi, G. R., & Parnianpour, M. (2009) Design and Development of a Hand Robotic Rehabilitation Device for Post Stroke Patients. In *Proceeding of the Annual International Conference of the IEEE Engineering in Medicine and Biology Society.* IEEE Engineering in Medicine and Biology Society. DOI: doi:10.1109/IEMBS.2009.5333827

Reinkensmeyer, D., Lehman, S., & Lum, P. (1993) A bimanual therapy robot: controller design and prototype experiments. In *Proceedings of the 15th Annual International Conference of the IEEE.* Engineering in Medicine and Biology Society. DOI: doi:10.1109/IEMBS.1993.978943

Reinkensmeyer, D. J., Dewald, J. P. A., & Rymer, W. Z. (1999). Guidance-based quantification of arm impairment following brain injury: A pilot study. *IEEE Transactions on Rehabilitation Engineering, 7*(1), 1–11. doi:10.1109/86.750543 PMID:10188602

Royal College of Physicians. (2010). *Medical rehabilitation in 2011 and beyond. Report of a working party*. London: RCP.

Sale, P., Lombardi, V., & Franceschini, M. (2012). Hand Robotics Rehabilitation: Fesability and Preliminary Results of a Robotic Treatment in Patients with Hemiparesis. *Stroke Research and Treatment.* 10.1155/2012/820931

Sankay, Y. (2007). HAL: Hybrid Assistive Limb based on Cybernics. New York, NY: Springer

Sheng, B., Zhang, Y., Meng, W., Deng, C., & Shengquan, X. (2016). Bilateral robots for upper-limb stroke rehabilitation: State of the art and future prospects. *Medical Engineering & Physics, 38*(7), 587–606. doi:10.1016/j.medengphy.2016.04.004 PMID:27117423

Shuster-Amft, C., Eng, K., Lehmann, I., Schmid, L., Kobashi, N., Thaler, I., … Kiper, D. (2014). Using mixed methods to evaluate efficacy and user expectations of a virtual reality–based training system for upper-limb recovery in patients after stroke: a study protocol for a randomized controlled. *Trial, 15*(3). DOI 1745-6215-15-350

Stinear, J., & Byblow, W. (2002). Disinhibition in the human motor cortex is enhanced by synchronous upper limb movements. *The Journal of Physiology, 543*(Pt 1), 307–316. doi:10.1113/jphysiol.2002.023986 PMID:12181301

Stinear, J., & Byblow, W. (2004). Rhythmic Bilateral Movement Training Modulates Corticomotor Excitability and Enhances Upper Limb Motricity Postroke: A Pilot Study. *Journal of Clinical Neurophysiology, 21*(2), 124–131. doi:10.1097/00004691-200403000-00008 PMID:15284604

Sullivan, S. B., Schmitz, T. J., & Fulk, G. D. (2014). *Physical rehabilitation* (6th ed.). Philadelphia, PA: F.A. Davis Co.

Tóth, A., Arz, G., Fazekas, G., Bratanov, D., & Zlatov, N. (2004). Post stroke shoulder-elbow physiotherapy with industrial robots. Advances in Rehabilitation Robotics, 391–411.

Vaida, C., Carbone, G., Major, K., Major, Z., Plitea, N., & Pisla, D. (2017). On Human Robot Interaction Modalities in the Upper Limb Rehabilitation After Stroke. *Acta Tehnica Napocensis, Applied Mathematics, Mechanics and Engineering*, *60*(1), 91–102.

van Delden, L., Peper, C. E., Kwakkel, G., & Beek, P. J. (2012, November). A Systematic Review of Bilateral Upper Limb Training Devices for Poststroke Rehabilitation. Stroke Research Treatment.

Winter Green Research. (2014). *Rehabilitation robots, active prostheses, and exoskeletons: Market shares, strategies, and forecasts, worldwide, 2014 to 2020*. Retrieved from http://wintergreenresearch.com/reports/RehabilitationRobots.html

Wu, J., Gao, J., Song, R., Li, R., Li, Y., & Jiang, L. (2016). The design and control of a 3DOF lower limb rehabilitation robot. *Mechatronics*, *33*, 13–22. doi:10.1016/j.mechatronics.2015.11.010

Yadav, R. K., Sharma, R., Borah, D., & Kothari, S. Y. (2016). Efficacy of Modified Constraint Induced Movement Therapy in the Treatment of Hemiparetic Upper Limb in Stroke Patients: A Randomized Controlled Trial. *Journal of Clinical and Diagnostic Research: JCDR*, *10*(11), YC01–YC05. PMID:28050492

Yang, Q., Cao, D., & Zhao, J. (2013). Analysis on State of the Art of Upper Limb Rehabilitation Robots. *Journal of Robotics*, *5*, 630–640.

Yap, H. K., Lim, J. H., & Nasrallah, F. (2015) A Soft Exoskeleton for Hand Assistive and Rehabilitation Application using Pneumatic Actuators with Variable Stiffness. In *Proceeding of International Conference on Robotics and Automation*. Seattle, WA: IEEE. DOI: doi:10.1109/ICRA.2015.7139889

Yoo, D. H., & Kim, S. K. (2015). Effects of upper limb robot-assisted therapy in the rehabilitation of stroke patients. *Journal of Physical Therapy Science*, *27*(3), 677–679. doi:10.1589/jpts.27.677 PMID:25931706

Chapter 6
Lithotripsy of Renal Stones With Avicenna Roboflex Robotic-Assisted Retrograde Intra-Renal Surgery (RA-RIRS)

Ahmet Sinan Kabakci
Hacettepe University, Turkey

Anup Patel
Spire Roding Hospital, UK

Aydan M. Erkmen
Middle East Technical University, Turkey

Ismet Erkmen
Middle East Technical University, Turkey

Özlem Gümüşkanat
Middle East Technical University, Turkey

Mehmet Çetinkaya
Middle East Technical University, Turkey

ABSTRACT

The role of retrograde intra-renal surgery (RIRS) with flexible ureterorenoscopy (FURS) in the management of kidney stone formation has garnered a prevalent use among urologists thanks to new improvements in medical equipment; however, it still remains a challenging technique. The purpose of this chapter is to overview robotic assistance for RIRS and mainly focus on ELMED's robotic device by thoroughly describing it and investigating its effects on surgical performance based on the IDEAL (idea, development, evaluation, assessment, long-term study) framework. The robotic device, Avicenna Roboflex, is a teleoperated robot consisting of a surgeon's console (master control console – MCC) and manipulator arm (MA) to house and couple with the FURS. It provides a suitable and safe platform for robotic-assisted (RA) RIRS with significant improvements in ergonomics and potentially prolongs equipment durability which may lead to its general acceptance and preference of use by many surgeons in various countries including Turkey, Germany, France, the UAE, and Romania.

DOI: 10.4018/978-1-5225-2993-4.ch006

INTRODUCTION

In this chapter, the authors overview the field of Robot Assisted Retrograde Intra-Renal Surgery (RA-RIRS) motivated by the state of the art in the field, and introduce an innovative robotic solution for flexible ureterorenoscopy, the Avicenna Roboflex (ELMED, Ankara, Turkey, 2012). It is the first commercially available robotic master control console (MCC) and manipulator arm (MA) for remote control of all commercial flexible ureterorenoscopes equipped with functions, such as rotation, deflection, insertion and retraction, used in Robotic Assisted Retrograde Intra-Renal Surgery (RA-RIRS). The reader is first presented with key descriptions and the status quo of existing robotic solution for flexible ureterorenoscopy prior to the arrival of Avicenna Roboflex in order to gain insight to surgical specifications robotic assistance needs to comply to, followed by an introduction into the IDEAL framework as a means of evaluating surgical innovations. Then, the focus turns into explaining the need for a robotic option for RIRS, with a thorough explanation of functionalities provided by the robot related to task specific requirements. Effectiveness of the device is later evaluated, introducing an application of the IDEAL framework to the reader. Lastly, certain additional features are presented, which can further enhance the effectiveness of the robot.

To briefly introduce the robot, Avicenna Roboflex is an interface between a surgeon and the surgical equipment needed for flexible ureterorenoscopy (FURS). It allows the surgeon to fine-tune the endoscope tip control by choosing low or high precision in movement scaling and enables more dexterity in tip rotation when compared to manual operation. Beyond that, the scope used can be selected by the surgeon among various EU- and US-types, providing greater versatility in the choice of the end effector. The scope can be fixed in position as well, allowing continuous access to the target and preventing surgeon fatigue by manipulating the scope in a comfortable seated position instead of a hard-to-maintain standing bent posture for the necessary scope tip orientation. It also keeps the surgeon outside the radiation field eliminating the need to wear any lead protection, which in turn improves surgeon performance through better comfort and dexterity and reduced physical fatigue. Avicenna Roboflex provides an interface for laser fibers that fire pulsed (or continuous wave) laser beams which create vaporization bubbles that in turn fragment renal stones, and incise or ablate soft tissue. The tracking of the laser beam nozzle tip and the precise remote advancement-retraction controls are key for preventing unintended laser firing near the scope tip; thus, accidental shots, and consequent damage to the scope tip are avoided, leading to instrument longevity. Precise pulsed laser stone fragmentation has made it possible to treat larger intra-renal stones, which was impractical before and required lengthy procedures of great discomfort. Additionally, video processing and computer measurements are overlaid, providing a better awareness and orientation without any need of looking to separate control screens. Being potentially adaptable as an interface for all endoscopic equipment, Avicenna Roboflex may also decrease required manpower to perform any surgical and diagnostic endoscope based procedures.

Thus, this chapter aims to provide the detailed description of this system, compare it to other commercially available robotic surgical systems and present details on the development process of the device designed and manufactured by ELMED (Ankara, Turkey), from prototype to commercialization. The authors will also provide surgical innovation analyses of Roboflex Avicenna (IDEAL stage 1), and summarize its early clinical experience with treatments performed by different experienced surgeons (IDEAL stage 2).

BACKGROUND

In this section, the authors provide the reader with an introduction to the field of Flexible Uretero-renoscopy (FURS), a technique, where a flexible endoscope is inserted through natural orifices into the ureter(s), allowing more versatility in diagnosis and treatment of ureteral and intra-renal collecting system disorders. The status quo of surgical telerobotics before the arrival of ELMED's robotic telemanipulator system, namely the Roboflex Avicenna, are also discussed. The section is finalized with the description of IDEAL, a framework for evaluating surgical innovation in the relevant field.

Ureteroscopy has evolved over almost 40 years following the introduction of first generation ureteroscopes whose initial general use was for diagnostic purposes only. This was mainly due to the size limitations of the large rigid ureteroscopes of those early days. With the introduction of smaller semi-rigid and later of flexible uretero-renoscopes, which provided greater dexterity as opposed to their rigid and semi-rigid counterparts thanks to the elimination of protective metallic sheaths, reduced size, and greater accessory instrument capability, they gained a foothold in urological practice in key centres. Increased scope active tip deflection, or equivalently higher scope manoeuvrability, together with smaller scope sizes, made it possible for a surgeon to navigate through the upper urinary tract more easily with a considerably reduced chance of urothelial damage. Increased instrumentation capability, on the other hand, enabled the scopes to be equipped with more advanced imaging devices and accessory equipment to be used for treatment purposes rather than diagnosis alone (Wetherell et al., 2014; Somani, Srivastava, Traxer, & Aboumarzouk, 2013).

Being able to reach the intra-renal collecting system, FURS has become a valid alternative to two other standard methods for lithotripsy, a medical procedure for the treatment of kidney stones, termed [extracorporeal] shock wave lithotripsy ([E]SWL) and percutaneous nephrolithotomy (PCNL). [E]SWL is often the suggested first-line treatment to be used thanks to its least possible amount of complications, but relies entirely on natural efflux of stone fragments once created. However, as the stones become more complex in shape, larger in number and volume, harder in terms of composition, and unfavourable in terms of collecting system location (lower calyx)-anatomy for spontaneous efflux (impeded in situations such as neurogenic disease which can affect urinary tract motility, congenital or acquired anomalies, or if there is any physical anatomical impediment to stone egress), the performance of ESWL degrades drastically. PCNL is more effective in achieving high stone-free results as fragments are actively removed. However, this better performance comes at a cost, which is: losing the advantages of minimally invasive operations and causing pain, bleeding, and urosepsis etc, which can lead to longer recovery times and critically increased morbidity rates. This is where RIRS provides a desirable impact. When equipped with laser fibers for lithotripsy where energy can be delivered directly to the stone under visual control, it is a more effective method than [E]SWL (stones can be broken and fragments can be dust left for spontaneous passage or actively removed), and is a safer procedure than PCNL by being natural orifice surgery, making it an optimum choice for certain types of stones, a fact that has been clinically verified (Breda, Ogunyemi, Leppert, & Schulam, 2008).

During classical manual Retrograde Intra-Renal Surgery (RIRS), stone fragmentation is usually performed under general anaesthesia. The surgeon inserts a uretero-renoscope into the bladder through urethra. The aim is then to move the ureteroscope forward through the ureter up to the intra-renal collecting system where the stone is located – note that stones formed do not have to be that deep inside the upper urinary tract but they may lay anywhere along that path. Moving forward is achieved by manually inserting the scope further into the urethra and relying on a certain level of shaft stiffness and accommodating ureteral lumen. Manoeuvring along the desired way requires two kinds of movements. The first movement is done manually by the surgeon as a combination of advancement and rotation. The second movement is achieved by controlling the tip deflection with a control lever placed near the other end of the scope (the handle) where the surgeon's deflection commands are translated into tip deflection by means of a cable-and-pulley system. There is continuous visual feedback to the surgeon from the camera placed at the scope eyepiece while fluoroscopic extrinsic visual feedback is available upon request. After reaching the target stone, the surgeon has to maintain the position of the endoscope tip stationary while a laser fiber is inserted through one of the working channels of the scope (i.e. hollow tube[s] within the scope for insertion of accessory equipment, which in a stone fragmentation case corresponds to laser fiber[s] and retrieval basket[s]). Following this, after targeting with an aiming beam, laser energy is deployed onto the stone at varying energy and frequency and, as needed, the area is flooded with fluid to help clean up stone fragments and to provide better visualization of the area with stone fragment removal. Fig. 1 illustrates a case where the uretero-renoscope is inserted up to the lower pole of the kidney for stone fragmentation. Note that stone fragmentation with the use of lasers is not the only use of uretero-renoscopes – they may be used for diagnostic purposes too as well as for treating soft tissue maladies – nor is the use of lasers along with the uretero-renoscopes the only means for stone treatment. One may use a variety of small accessories (stainless steel or Nitinol), inserted within the working channels of the uretero-renoscopes to capture and extract the small stones-fragments with a diameter less than 3-5mm (Rajamahanty & Grasso, 2008). However, the authors will only discuss the case where the stones are to be treated with RIRS and fragmented with laser beams to keep the discussion focused.

By advanced technological improvements and receding the horizon of engineering techniques, endoscopes are getting smaller in terms of cross-sectional size and their optical specifications are highly improved using Microelectromechanical systems (MEMS) technology (Xie, 2009). Newly innovated uretero-renoscopes offer greater accessibility to every point of an upper urinary tract - renal collecting system. Even in patients with more complicated situations, containing abundant variety of medical problems (obesity, elderly, high risk comorbidities, anticoagulation etc.), RIRS has become increasingly appropriate and easily applicable. (Wright, A. E., Rukin, N. J., & Somani, B. K., 2014). Thanks to these

Figure 1. A basic illustration of Flexible Uretero-renoscopy

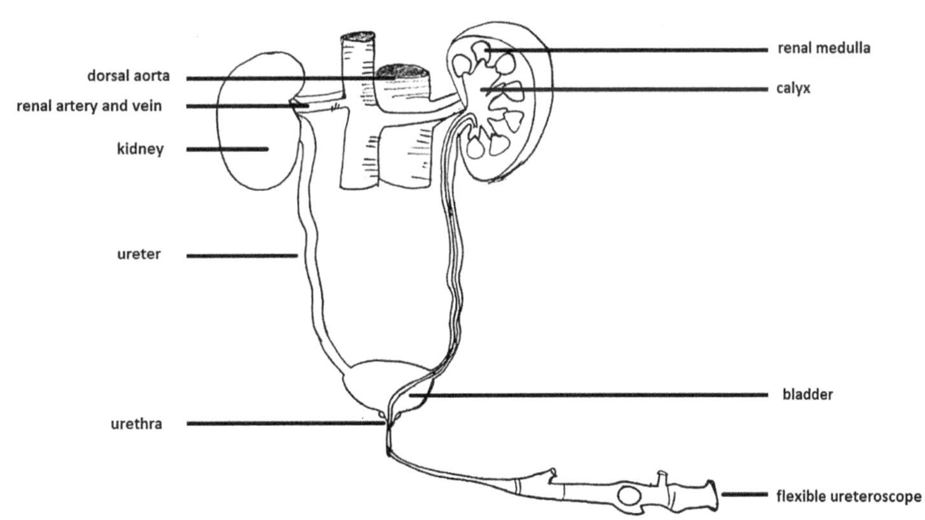

innovations and developments, robotic surgery with uretero-renoscopes also has taken on an important role in complex endourology. Another example of innovation in RIRS strategy is the use of peptide-coated iron oxide-based microparticles (Fernandez et al., 2012). These microparticles can recognize and coat calcium stone fragments. After coating with microparticles, a magnetic device provides a practical and suitable tool to gather fragments and extract them easily out of body. This operation is rapid compared to classical renal stone management methods which are reliant on spontaneous fragment passage.

Prior to the release of Roboflex Avicenna, there have been two clinically used telesurgical robot assistive systems, namely: ZEUS (Goleta, California, USA, 1999) and Da Vinci (Intuitive Surgical, Sunnyvale, CA, USA, 1997) (Pugin, F., Bucher, P., & Morel, P., 2011). Both robotic systems had been developed as generic tele-operated systems for minimally invasive operations and both had their first clinical usage for cardiovascular surgery. Thanks to their generic characteristics, their end effector can be equipped with various surgical tools, allowing them to be used in procedures ranging from targeted radiation therapy to Robotic-assisted laparoscopic operations. The ZEUS telesurgical system used for laparoscopic surgery consists of three (3 DoF – degrees of freedom) robotic arms roughly imitating surgeons' hand movements, mounted on carts moving along the operation table that adds one more DoF to each arm. One of the arms is used to place the endoscope positioner, or is used as a camera manipulator, to provide a 2D visual representation of the operative field. Having only the remaining two arms of 4 DoF for instrument placement, ZEUS resulted in a significant increase in surgery lengths and thus has been used for less critical and coarser surgeries such as gallbladder removal and aorto-coronary bypasses (Rassweiler, Safi, Subotic, Teber, & Frede, 2005). The Da Vinci system, on the other hand, can address the prominent agility problem encountered with ZEUS and its undesirable effect on increased surgical times and types of operations that can be performed. Equipped with a stereo-endoscopic system, it is able to provide the operator with depth perception. Orientation and placement of various surgical instruments can be easily and accurately achieved thanks to its 6 DoF arms, significantly decreasing operative times potentially compared to ZEUS. If Da Vinci, or even ZEUS, were to be equipped with an appropriate instrument actuator that could drive a uretero-renoscope, they could be able to perform the required telesurgical operation for RIRS. However, even if an additional actuator were not needed, high up-front investment and running costs of both systems plus regular disposable cost (excluding additional service contract costs), which might be justified for certain cancer operations, make them impractical and unaffordable for RIRS (Rassweiler et al., 2005). High costs of the Da Vinci system, the more advanced one among the duo, stem from the fact that it is a more complicated system designed as a generic system that can perform various surgeries such as cardiac and head surgeries together with general surgeries encompassing pancreas removal, gastric bypass and acid reflux surgeries to name a few. Higher complexity of the device naturally results in higher costs while some of the required functions of such surgeries are not even needed for RIRS.

As a feasibility trial for a RIRS robotic system, Desai et al. (2011) tested the Sensei system (Hansen Medical, Mountain View, CA, USA), a novel robotic catheter system designed for intra-cardiac electrophysiological applications. The system consisted of a surgeon console (for user interaction with displays, Master Input Device, etc.), steerable catheter system, remote catheter manipulator (a modified robotic arm to drive catheter system) and electronics rack (equipped with power supplies, computers and interfacing electronics). The steerable catheter subsystem was composed of an outer catheter sheath and an inner catheter guide, the flexible uretero-renoscope being glued to the latter. By manipulating the latter from the remote end, the ureteroscope tip was guided through the desired path for FURS, effectively reducing the manipulation of the flexible uretero-renoscope to a passive one. By performing in vivo experiments for the treatment of kidney stones in pigs and in human patients, they demonstrated that such a robotic system might be beneficial in clinical use (Desai M et al. 2008; Desai M et al. 2011). This work was just a feasibility study to see potential benefits of a tele-operated system for RIRS for commercial and wider usage, where any of their proposed generic solutions were more complex than needed. There was still a need for a simpler and cheaper solution better fulfilling the specific tasks required for RIRS. Avicenna Roboflex has been specifically designed for this need: it can work with any commercially available flexible uretero-renoscope, reaping the advantages they provide, while enhancing their performance with higher manoeuvrability and accuracy and potentially solving the problem of their fragility as a cost saving exercise. It also provides operators with all the advantages of robot assisted surgery: higher device safety, better ergonomics and shorter learning curve for beginners to name a few.

In order to underline the innovations and benefits provided to the sector by Avicenna Roboflex, which are stated at the end of the above paragraph, the authors hereby recapitulate the "Idea, Development, Exploration, Assessment and Long term study" framework (IDEAL), which provides a systematic basis and distinct development method for evaluations of surgical innovations. It was a critically needed framework that set out a structure for better quality of research in surgery and described the stages surgical innovations must go through to be accepted as successful treatment methods. The first stage named as "Idea" is a new planned method for the treatment of a disease, stemming from the expertise of a surgeon. "Development" is the second phase 2a, where complete successes and failures of any new treatment method were expected to be reported by the practitioner with full transparency. It is the phase when the new method is tested provided that required consent from all relevant parties were obtained. Collected data for the effectiveness of the method are sparse and no decisive conclusion can be drawn at this stage. During the "Exploration" stage, phase 2b, in contrast to early evaluation phase, patient data are expected to become more statistically meaningful with more and more surgeons using the same techniques on a wider variety of less carefully selected patients. At stage 3, assessment of the method has to be conducted over a series of randomized controlled trials, even if the exploration stage suggested a strong, definite conclusion where the possibility of bias may still have been present. It is at this stage, that possible vulnerabilities of the new method can be tested and that possible long term complications of the method should be observed (Mcculloch et al., 2013; Ergina et al. 2013). As an example to this stage, Avicenna Roboflex has enabled treatment of larger stones, which in turn, prolongs surgery times, impeding patient safety through longer exposure to radiation and laser energy and surgeon fatigue. Tissue damage caused by this last aspect may present serious threats for certain patients. In the remainder of this chapter, Avicenna Roboflex will be evaluated according to the first three stages of the IDEAL framework (1, 2a and 2b). Having only been used for a short period of time in limited centers, stage 3 of the evaluation of Avicenna Roboflex, is yet to be completed.

MAIN FOCUS OF THE CHAPTER

Issues, Controversies, Problems

Flexible uretero-renoscopy (FURS) has made substantial progress towards enabling better quality classical retrograde intra-renal surgery (c-RIRS) over recent decades thanks to innovations in uretero-renoscopic instrumentation and intracorporeal stone fragmentation techniques. RIRS provides a unique natural orifice solution to pathology in the intra-renal collecting system and is therefore associated with faster recovery, also almost eliminating post-operative pain. It is associated with predictable and reproducible outcomes in expert hands. In this manner, it has become a feasible and attractive alternative to [E]SWL ([Extracorporeal] Shock Wave Lithotripsy) for smaller stones, and PCNL (Percutaneous Nephrolithotomy) for larger renal stones, as introduced in the background section (Knoll T. et al., 2011; Rassweiler J. J. et al. 2011; Breda A. et al, 2009; Preminger G. M. et al., 2007; Beiko D.T. et al. 2007; Patel A. et al., 1997). However, c-RIRS requires very sensitive and specific adjustments by the operator, which can frequently be technically challenging, and hence the variability of outcomes at present. (Somani B. K. et al., 2013; Holden T. et al., 2008; Monga M. et al., 2002). Despite the advancements in the design of flexible uretero-renoscopes and ureteroscopic accessory instruments, c-RIRS is still reliant on a single surgeon supported by many other team members for different aspects of the procedure (Table 1, Saglam et al., 2014). Insertion of, and positioning of laser fibers or accessories suitable for capturing stones, starting and maintaining irrigation at adjustable rates of flow (controllably pumping fluid to facilitate optimal visibility) from the flexible uretero-renoscope, should all be conducted synchronously while the endoscope is focusing on stone position or orientation. Additionally, as needed, the endourologist has to use a foot pedal to activate laser lithotripsy. Most surgeons suffer from postural skeletal complaints such as neck, back, and hand disorders, as a result of doing all procedural aspects in a standing or seated position for long time periods with a suboptimal ergonomic posture while wearing a heavy lead apron and other lead protection to limit their personal radiation exposure to vital organs which worsen their complaints (Elkoushy M.A. et al., 2011; Healy K. A. et al., 2011). Taking into consideration all these inconveniences and c-RIRS procedural aspects, the Avicenna Roboflex was envisaged and designed to provide a suitable, ergonomic, and safe, robot-assisted RIRS (RA-RIRS). Fig. 2, compares a surgeon performing c-RIRS versus one performing RA-RIRS with the Avicenna Roboflex.

Table 1. Ergonomic requirements for classic flexible uretero-renoscopy and retrograde intrarenal surgery

Action related to operation	Activity Requirements	Perfomed by
Insertion of ureteroscope	Fingers of both hands(at instrument and the beginning of urinary tract)	Surgeon
Deflection of ureteroscope	Hand holding hand piece and thumb; fingers of the other hand at the beginning of urinary tract	Surgeon
Rotation of ureteroscope	Hand holding hand piece; fingers of the other hand at the beginning of urinary tract	Surgeon
Translation of ureteroscope	Hand holding hand piece; pulling or pushing	Surgeon
Fluoroscopy activation	Foot	Surgeon or radiotechnician
Movement of table or C-arm	Foot or hand	Surgeon or radiotechnician
Irrigation control	Observing related equipment	Nurse or assistant
Syringe	Hand	Nurse or assistant
Mechanical device	Foot	Surgeon or technician
Pump device	Finger activation	Nurse or assistant
Basket and grasper insertion	Finger and hand	Nurse or assistant
Basket and grasper activation	Finger and hand	Nurse or assistant
Laser fiber insertion	Finger	Nurse or assistant
Laser setting	Finger	Nurse or technician
Laser activation	Foot	Surgeon

The Avicenna Roboflex is a robotic device whose intended use is endoscopic lithotripsy (and precision treatment of urothelial soft tissue lesions-non-invasive malignancy). Since it is not an accessory to any device, it has no interaction with blood, tissue, or drug. It does not directly control any life supporting or life sustaining function or the delivery of potentially harmful energy that could result in death or serious injury. In addition, a device error or malfunction, were such a thing to occur, is unlikely to result in death or serious injury. It neither provides any diagnostic information, nor should it affect any vital signals or alarms for potentially life threatening situations. Consequently, the Avicenna Roboflex® (ELMED) is classified as a medical device with minor risk level according to the U.S. Food and Drug Administration (U.S. Food and Drug Administration, 2011).

During the Avicenna Roboflex® design process, a touch screen HMI (Human Machine Interface) was used as a master in conjunction with two Programmable Logic Controllers (PLC's) that work as slaves which meant that HMI had uni-directional control over the PLC's. The robot manipulator arm (MA) of the Avicenna Roboflex, which is responsible for the movements of flexible uretero-renoscope, when coupled by a customized adapter to the manipulator, is controlled by one of the PLCs, while the other PLC is embedded in the robot master control console (MCC). The Avicenna Roboflex MCC is mainly responsible for the transmission of orders, given from the HMI to the related parts. These two

Figure 2. The postures of a surgeon as he performs manual c-RIRS and RA-RIRS with the Avicenna Roboflex

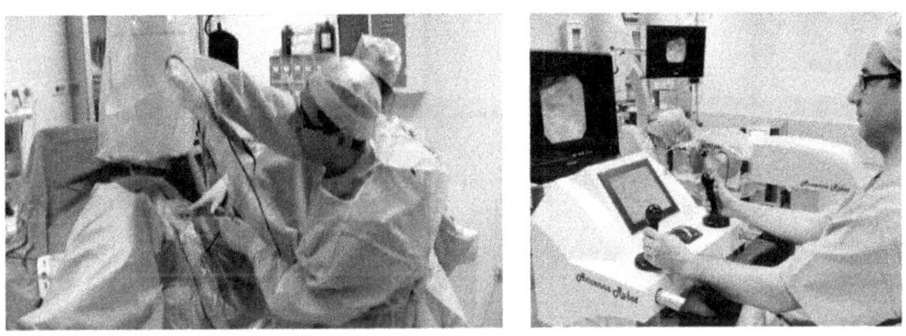

Figure 3. Block diagram of the system

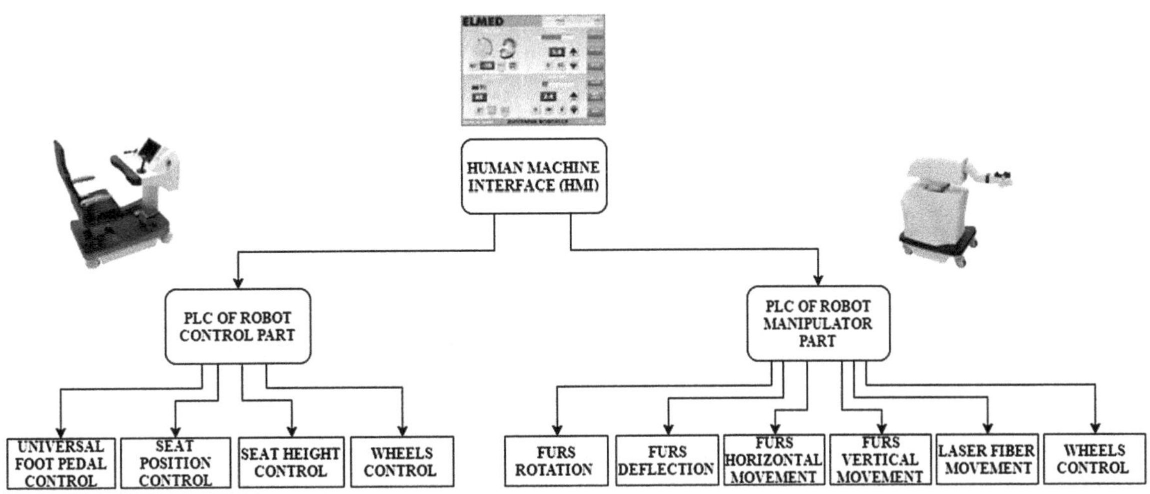

PLCs communicate with the HMI directly according to specified duties. The system block diagram and main tasks of PLCs can be seen in Fig. 3. Also a picture of the MCC-HMI screen is shown in Figure 4. The surgeon's Master Control Console (MCC) and Manipulator Arm (MA) are shown in Figures 5 and 6, respectively.

The PLC of the manipulator arm reads the rotational/horizontal motion data imparted by the control handle located on the left side of the master control console and calculates the rotation speed, the movement direction, together with the new position of the manipulator arm simultaneously, enabling the manipulator arm to move to a new position promptly with a required speed. Rotation angles are shown on the MCC-HMI screen both graphically and numerically in degree as located at top-left of MCC-HMI screen (grid#1 in Figure 4). The MA can also be manoeuvred forwards and backwards (in-and-out)

Figure 4. A picture of MCC-HMI screen

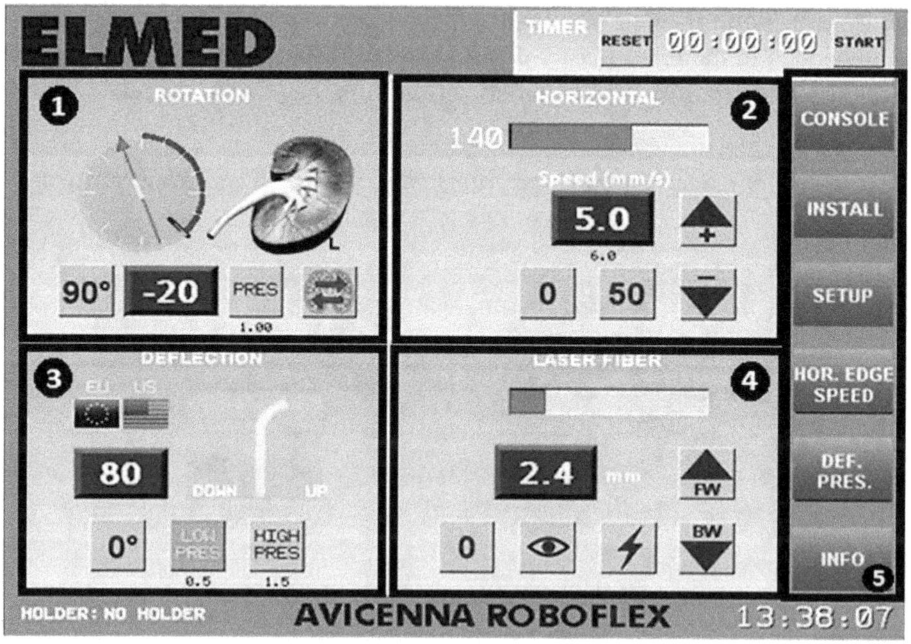

Figure 5. The surgeon's Master Control Console (MCC) of the Avicenna Roboflex

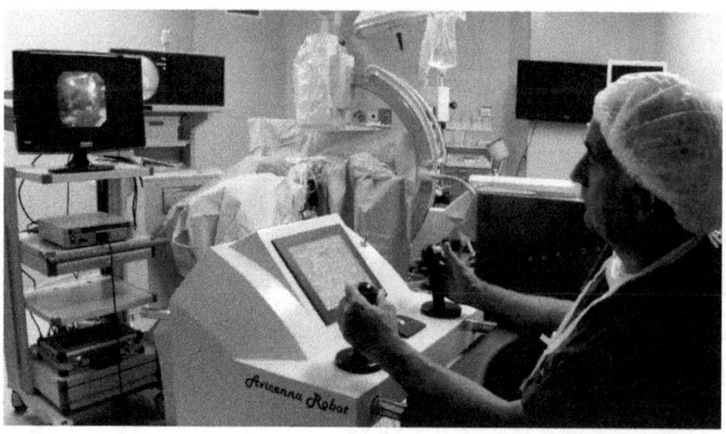

Figure 6. The manipulator arm (MA) of the Avicenna Roboflex

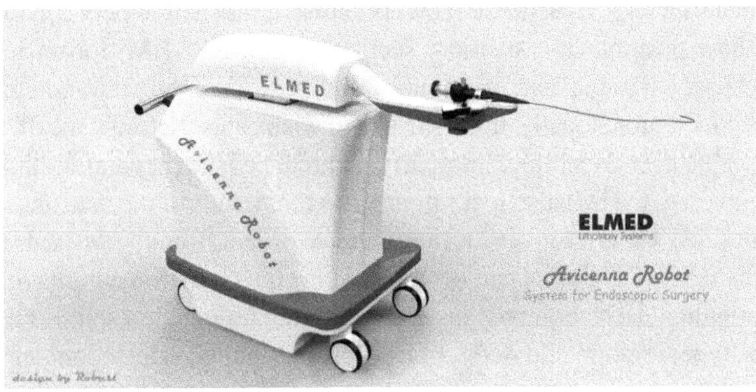

manually in the horizontal direction with a physical control handle. When MA moves in the horizontal direction, the position information is also provided on the right of rotation information part (grid#2 in Figure 4) both graphically and numerically in millimetres to aid the user in precision positioning. The user can also pre-select speed precision by order of 0.5, 1, 3, 5, 10 and 15 mm/sec, which are cut-sets of fuzzy variables as will be discussed further down in this section. The position and velocity information provide the user with accurate operational facility.

The manipulator controller reads the deflection amount of the control handle that is located on the right of the control console when the user activates it, and calculates the deflection speed of the deflection lever, movement direction, the new position of the deflection manipulator on the manipulator arm, to subsequently move it to its new position with the related speed. All information and settings buttons related to deflection are placed at the lower-left of MCC-HMI screen (grid#3 in Figure 4). The precision of the deflection can be manually fine-tuned on the MCC-HMI according to the application. User friendliness is obtained by precision values being selected from a predefined set of fuzzy variables. The fuzzy parameters define the number of turns the user should apply to the deflection wheel for full span of deflection: more specifically, if the cut-set of the fuzzy variable "LOW PRES" is selected as 0.5, it means that the relevant user should turn the deflection wheel half of one full turn for full span of deflection. If "HIGH PRES" value is selected as 1.5, the users should turn the deflection wheel 1.5 turns for full span of deflection. It means that "HIGH PRES" is three times more precise compared with "LOW PRES" according to selected values. An angle and an animated indicator are available on the MCC-HMI screen as a feedback information to the user. In addition, the user may select the FURS deflection type, as non-US type (thumb up / tip down or vice versa - illogical) or US type (thumb up / tip up or vice versa – logical) by touching the related MCC-HMI flag button. The laser fiber once fixed in the actuator device and zeroed, can also be moved forwards and backwards by pressing the related touchscreen buttons on the MCC-HMI. Positioning of the laser fiber bundle tip is shown on the lower-right part of MCC-HMI screen (grid#4 in Figure 4) as a bar graph and in millimetres to inform the user, simultaneously. Thereby, the user can always get feedback with regard to their commands without any need of being in the same room with the patient. The user can also give commands by simply touching the related buttons in order to move the laser nozzle until the minimum firing distance is achieved and fixing it there. These RA-RIRS abilities provide critical safety and efficacy in laser firing compared to c-RIRS. When the MCC-HMI info button (#5 in Figure 4) is pushed, important other menus appear. The MCC-HMI then provides the information and request data to control the horizontal and vertical position of the MCC console seat together with the height of the control unit for acquiring a suitable setting optimized to the individual user so that the MCC-HMI is kept at a safe height between the seat and console. The user may save these positions to a memory section in the MCC-HMI for retrieving these settings at a later date. If the user saves the seat and control unit positions, the seat and control unit takes the same position by the PLC automatically, upon the user's wish. This "settings recall" option as a default yields a very comfortable user-friendliness property and reduces the preparation time for the operation to begin. The PLC of the MCC-HMI also locks or unlocks the wheels of the console and the manipulator arm when the user desires. The user gives these respective orders from the MCC-HMI using the Lock/Unlock buttons on the HMI. The device can be prepared for the operation automatically for initial rotation, horizontal positioning, deflection, and laser fiber bundle tip position settings by pressing [Install] button. Ureteroscope type, rotation and deflection precisions can be selected under the "Setup" menu. Also a "Help" menu is provided so that the user can access a brief instruction about the use of device.

The system algorithm also checks the distance between the tip of the laser fiber bundle guiding the laser beam emitted, and the tip of the FURS, by the help of the mounted encoder. The flowchart of distance control is shown in Figure 7. Because of the high purchase and repair cost of flexible ureterorenoscopes and their potential fragility with improper use, the laser fiber tip should be positioned such that the camera chip/fiber optic bundle at the endoscope tip does not get damaged by cavitation bubbles created during the laser activation. When the laser fiber tip is at a distance of 3 mm or more from the FURS tip, cavitation bubbles cannot reach the optics of the scope. Also while the tip of laser fiber is kept at minimum 3 mm distance, the erosion of the fiber tip cannot occur thus freeing the system of such a risk. Thus, this position is established as "safe distance" (Talso, M. et al., 2016). If the distance is not safe, the Avicenna Roboflex HMI over-rides laser activation in order to protect the FURS optics. Thereby, any possible user fault that could damage the FURS optics at the tip is prevented. System analyses of the Avicenna Robotic assistance is detailed in the upcoming section together with discussions on superiorities and limitations.

SOLUTIONS AND RECOMMENDATIONS

In this section, performance and IDEAL framework evaluation of Avicenna Roboflex is presented, with the aim of providing evidence of its effectiveness as well as an example case study for the framework.

Phase 1 of the IDEAL framework, "Idea", starts with the expertise of a surgeon. Activities needed to be performed during a c-RIRS operation were compiled in a stepwise fashion with the help of a surgeon

Figure 7. Flowchart of controlling distance between laser fiber bundle tip and FURS tip

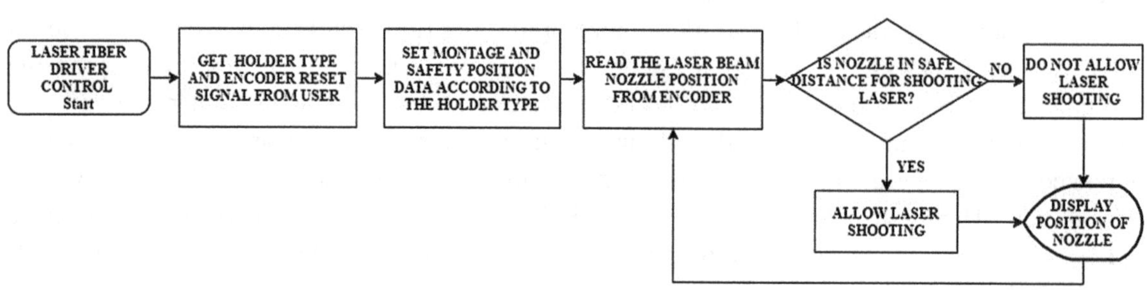

and are reported in Table 1. As listed in this table, there are various activities, some of which are quite rudimentary, that need to be performed by various operating room team personnel. On top of this, such activities result in surgery-related musculoskeletal complaints that are prevalent amongst practitioners as indicated under complaints related to c-RIRS in Table 2. These have brought about the idea of designing a robotic manipulator specifically targeted to RIRS. Phase 2a, the "Development" phase, started with the idea of imitating the surgeon's movements as well as automating the aforementioned activities as much as possible. The main difficulty faced by a surgeon is when the flexible uretero-renoscope is actively directed within the ureter or kidney, and easing this was the main goal during the design. With one prismatic joint and one revolute joint – along and about the axis of insertion directed towards the urethra, respectively – all crucial movements of the surgeon can be performed. Another actuator was also added to control the height of the tip to facilitate the operation. One last actuator was added to reflect the operator's commands to the coupled scope. There are other functions that are or will be developed according to emerging needs – covered under the other sections of this chapter – but the main development of the Avicenna Roboflex about its intended use is the insertion of the scope. That is why it is the only one covered in Phase 2a of the IDEAL framework.

One note on the design considerations regarding the support of different types of flexible uretero-renoscopes by the platform is that the Avicenna Roboflex has specific end effectors (couplers) designed specifically for each type of flexible uretero-renoscope. As each new scope may come in different shapes or sizes, there still is a need to redesign this *coupling* equipment, i.e., holder. From time to time, the resolution of the built-in camera and communication protocols used by the uretero-renoscope may also change. As of now, the user is to be presented with the option of choosing the right model of the scope used and each additional platform yet to be supported by the Avicenna Roboflex is to be integrated by the support of its parent company. A picture of two different couplers for different uretero-renoscopes can be found in Figure 8.

As for the "Evaluation" phase, Phase 2b, Saglam et al. (2014) reported performance results from 81 operations where the Avicenna Roboflex was used. There were no drastic changes in operation (mean 74 min.) and fragmentation times (mean 46 min.), which were comparable to c-RIRS. This is most likely explained by the learning curve of surgeons beginning with RA-RIRS (c-RIRS experts not as quick to adapt to the technology as younger novices), and the fact that larger stones required longer laser exposure times for complete stone disintegration (more dusting than fragmentation and extraction). It remains to be seen whether post-learning curve surgeons (especially of the younger generation unhampered by previous c-RIRS experience) can deliver more efficient fragmentation and operating times leading to cost savings when c-RIRS is compared to RA-RIRS for the same stone volume, hardness, composition, and location.

As a second evaluation metric, the Avicenna Roboflex's enhancement of ergonomics is discussed. Seven experienced surgeons, with classical RIRS experience ranging from 5 to 16 years, were presented

Figure 8. Coupler designs for different uretero-renoscopes

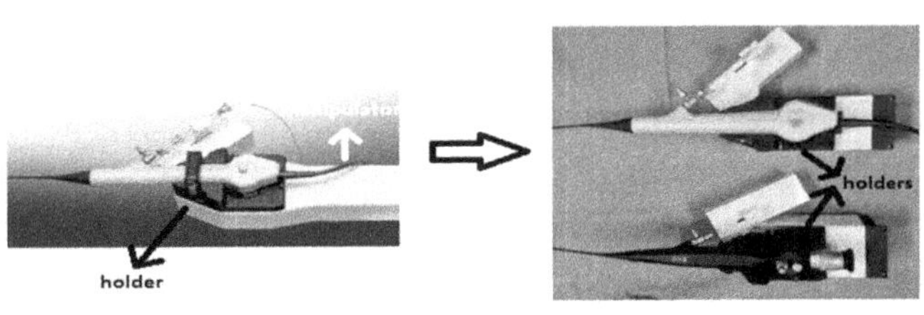

Table 2. Comparison of conventional and RA-RIRS using a validated questionnaire concerning ergonomics of minimally invasive surgeries

Surgeon	B.E.	A.Y.M.	K.S.	R.S.	Z.T	J.J.R.	O.T.	Mean
Age, yr	51	52	50	67	40	59	46	52.14
FURS experience, yr	16	7	10	5	5	16	15	10.57
FURS workload, h/wk	10	12	15	7	10	6	18	11.14
Classic FURS, complaints (0-5)								
Musculoskeletal pain	3	3	3	3	3	3	1	2.71
Neck pain	3	2	2	2	2	3	1	2.14
Shoulder stiffness	5	3	3	4	2	3	1	3.00
Arm pain	5	3	3	4	3	3	1	3.14
Forearm pain	5	4	4	3	3	4	1	3.14
Elbow stiffness	5	2	2	2	2	4	2	2.71
Hand pain	3	3	3	3	4	4	2	3.14
Wrist stiffness	3	4	4	3	3	4	1	3.14
Finger numbness	1	2	2	1	3	3	2	2.00
Back pain	3	1	1	2	3	3	1	2.00
Leg pain	2	2	2	2	3	4	2	2.43
Eye strain	0	2	2	2	3	2	1	1.71
Total score	38	30	30	31	34	40	16	31.3*
Robot-assisted FURS, complaints (0-5)								
Musculoskeletal pain	0	1	0	0	0	0	0	0.14
Neck pain	0	0	0	0	1	1	0	0.29
Shoulder stiffness	2	0	0	0	0	0	0	0.43
Arm pain	2	1	1	1	0	0	0	0.71
Forearm pain	2	1	1	1	0	0	0	0.57
Elbow stiffness	2	0	0	0	0	0	0	0.29
Hand pain	2	0	0	0	1	1	1	0.71
Wrist stiffness	0	0	0	0	1	0	1	0.29
Finger numbness	0	1	1	1	1	1	1	0.86
Back pain	0	0	0	0	0	1	0	0.14
Leg pain	0	0	0	0	0	2	0	0.29
Eye strain	0	1	1	1	2	1	1	0.86
Total score	10	5	4	4	6	7	4	5.6*

with a questionnaire following an hour of experience with the device. The questionnaire was based on previous work that compared the discomfort between traditional and robot-assisted minimally invasive laparoscopic surgery (Rassweiler et al., 2012). The surgeons were asked to scale their ergonomic complaints in the range from 0 to 5, 0 corresponding to no complaint at all and 5 to a serious disturbance. Their results as well as the type of complaints were reported (Table 2, Saglam et al., 2014), and showed improvement in each metric, with only some remaining unchanged. Although mean metrics for certain complaints did not diminish (e.g. hand pain and finger numbness), there was a great improvement when total questionnaire scores were compared: 31.3 versus 5.6 for c-RIRS and RA-RIRS.

The accuracy performance of the Avicenna Roboflex, results of which can be seen in Table 3, can be justified as successful since the measured values were within the identified tolerance limits, which were defined according to the requirements of the task at hand, and the ratio of success was greater than 80% in all tests. As can be seen in the table, console movements required coarser precision as the tolerance limits of the console and chair movements were not critical for the operation, but to the comfort of the user which should not be affected by such small errors. There was one outlier which performed not as well as anticipated: Manipulator Vertical Movement; it had a steady state error of 1% in 13.33 test cases. Although such errors are within tolerance limits thanks to the fact that vertical positioning of the manipulator was performed only once at the beginning of the surgery and residual errors can be fixed by the surgeon if needed, it is an issue that needs to and will be addressed in the near future.

Table 3. Results of accuracy performance tests

Performance Test Section		Target	Tolerance	#of measurements within tolerances	# of measurements not within tolerances	Ratio of success (%)
Seat Up/Down Movement		270 mm	± 5mm	15	0	100
Seat Forward/Backward Movement		145 mm	± 5mm	15	0	100
Console Up/Down Movement		295 mm	± 5mm	15	0	100
Manipulator Rotation		± 225þ	± 5°	15	0	100
Manipulator Deflection		Limit Positions	-	15	0	100
Manipulator Horizontal Movement		228 mm	± 2mm	15	0	100
		50 mm	± 2mm	15	0	100
Manipulator Vertical Movement		300 mm	± 3mm	13	2	86.67
Manipulator Horizontal Speed		0.5 mm/s	± 0.1mm/s	15	0	100
		1 mm/s	± 0.1mm/s	15	0	100
		3 mm/s	± 0.3mm/s	15	0	100
		5 mm/s	± 0.3mm/s	15	0	100
		10 mm/s	± 0.7mm/s	15	0	100
		15 mm/s	± 0.7mm/s	15	0	100
Laser Fiber Bundle Movement	for STORZ FLEX-XC	14 mm	± 0.5mm	15	0	100
	for STORZ FLEX-X2	14 mm	± 0.5mm	15	0	100
	for OLYMPUS URF-V2	14 mm	± 0.5mm	15	0	100
Laser Bundle Tip Position for Beam	for STORZ FLEX-XC	1.5 mm	± 0.5mm	15	0	100
	for STORZ FLEX-X2	1.5 mm	± 0.5mm	15	0	100
	for OLYMPUS URF-V2	1.5 mm	± 0.5mm	15	0	100

FUTURE RESEARCH DIRECTIONS

To discuss possible and surmountable limitations of the Avicenna Roboflex, absence of tactile feedback is a limitation. Although successful surgery is possible by the virtue of that every movement of endoscope is visually controllable by a urologist and Avicenna Roboflex is compatible with many digital endoscopes with greater image quality, incorporation of tactile feedback is in the list of future work. For this purpose, two main developments with regard to control handles have been worked on to warn the surgeon with the aim of comfortable and efficient usage. The first improvement is to prevent the uretero-renoscope from making more deflection than its present deflection angle when it has reached its limit such that it contacts with collecting system tissue or a renal stone before fragmentation begins. The second is to give the operator a vibration response via the control handle in a case of inadvertent touching of a vital structure.

Further developments such as automated control of laser fiber tip positioning at the optimal distance from uretero-renoscope tip and stone surface assisted by image processing are work in progress. Thanks to image processing, the lateral scatter movements and length of laser fiber emerging from the uretero-renoscope tip can be also detected simultaneously. By this means, an extra control mechanism could be constituted to perform safe laser firing while protecting the uretero-renoscope tip by preventing shooting if the safe distance is not present. A short cross-section related to real-time recognition of the laser fiber tip is shown in Fig. 9. In addition to these, ELMED is working on the 3D reconstruction of the patient's relevant anatomy from CT images. Thus, the most efficient sequence and range of movements could be pre-planned and programmed into the device pre-operatively for each patient, but would be subject to real time adjustment as deemed necessary and safe by the console surgeon during the procedure itself. This visual opportunity would optimize patient safety and shorten intervention time leading to further cost savings for healthcare providers to offset costs of the device. An illustration of the beginning of this work can be seen in Fig. 10. In this example, the patient obviously has several renal stones on the right kidney. By this means, the surgeon can plan and run simulated analyses before the intervention according to this output.

CONCLUSION

The Avicenna Roboflex is the first robotic interface between a surgeon and a flexible uretero-renoscope and establishes a new platform for the future natural orifice treatment of upper urinary tract collecting system pathology. The concept of the device can however be extended to any flexible endoscope used

Figure 9. Detection of laser fiber tip

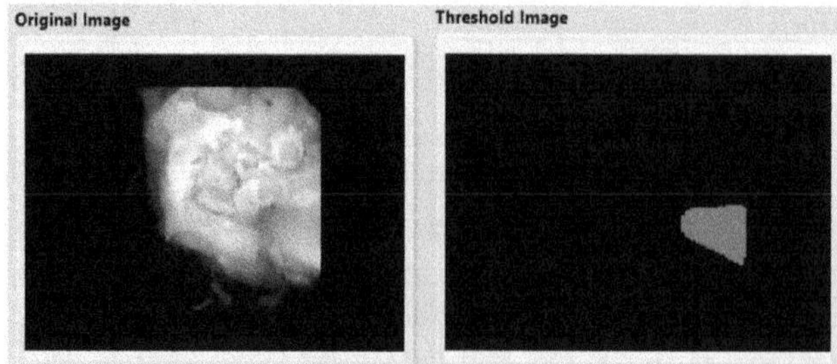

by any specialty. As such, it has enhanced operator control of key surgical procedural aspects by at least 50% as was previously described. It has shortened the learning curve for novice surgeons (Saglam et al., 2014), although it required some re-adaptation time for some classical RIRS experts. The latter, however, can be justified by experienced surgeons' preference of this system thanks to increased safety, functionality and comfort, as well as elimination of radiation safety protocols for the operating surgeon regarding. Mean surgery time beyond the short learning curve has generally decreased, and the cases where it has "apparently increased" were those corresponding to the previously impractical tiring treatment of larger stones, which invariably required more than 1 stage with classical RIRS. The range of stone volume, hardness, and chemical-composition that can be treated, which are of interest to RIRS expert surgeons in the past decade, has also been expanded.

In the Turkish Technology Contest, the Avicenna Roboflex was awarded First Prize and has also won the Best Scientific Invention Prize. Thanks to its innovative features, it has rapidly gathered the attention of the urological fraternity. The Avicenna Roboflex has been mentioned in many international and national congresses, some of which were WCE 2013 New Orleans, EAU 2014 Stockholm and AUA 2015 New Orleans. The 1st live Avicenna Roboflex surgery was performed in Karolinska University Hospital, via EAU Congress 2014, followed by live surgeries in congresses which included ERUS 2014 Amsterdam, AUA Segura Qatar 2015 Doha, IAE 2015 Milano and the 8thInt. Course on FlexURS 2015 Rome. Furthermore, the Avicenna Roboflex has been used by many renowned urologists; it has been installed at Medicana International Hospital in Ankara-Turkey operated by Remzi Saglam, at SLK Kliniken in Heilbronn-Germany by Jens Rassweiler, at Tenon Hospital in Paris-France by Olivier Traxer, at Sheikh Khalifa General Hospital – UAQ, U.A.E by AbdulQadir Al-Zarouni and Yasser Farahat and at Sanador Hospital in Bucharest-Romania by Petrisor Geavlette.

As we enter the 4th Industrial revolution, from the viewpoint of technological advances, the limitations of c-RIRS are reduced every day as Robotic Assisted surgery takes up an important role in complex RIRS, and may have important applications in many other areas of medicine. Together, these advancements are likely to change training of young apprentice surgeons, and treatment methods, leading to

Figure 10. 3D reconstruction of patient body and kidney stone(s)

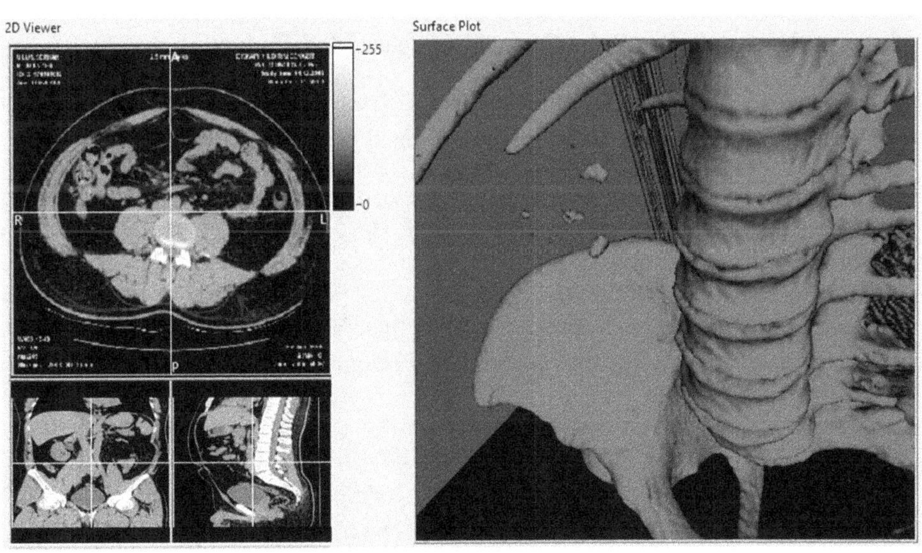

better patient safety, surgeon comfort and precision, while delivering potentially significant cost savings to increasingly stressed healthcare systems, as demand rises from ageing and or obese populations with multiple co-morbidities globally.

REFERENCES

Bagley, D. H., & Cubler-Goodman, A. (1990). Radiation exposure during ureteroscopy. *The Journal of Urology*, *144*(6), 1356–1358. doi:10.1016/S0022-5347(17)39739-2 PMID:2231925

Beiko, D. T., & Denstedt, J. D. (2007). Advances in ureterorenoscopy. *The Urologic Clinics of North America*, *34*(3), 397–408. doi:10.1016/j.ucl.2007.05.003 PMID:17678989

Breda, A., Ogunyemi, O., Leppert, J. T., & Schulam, P. G. (2009). Flexible Ureteroscopy and Laser Lithotripsy for Multiple Unilateral Intrarenal Stones. *European Urology*, *55*(5), 1190–1197. doi:10.1016/j.eururo.2008.06.019 PMID:18571315

Desai, M. M., Aron, M., Gill, I. S., Pascal-Haber, G., Ukimura, O., Kaouk, J. H., ... Moll, F. (2008). Flexible Robotic Retrograde Renoscopy: Description of Novel Robotic Device and Preliminary Laboratory Experience. *Urology*, *72*(1), 42–46. doi:10.1016/j.urology.2008.01.076 PMID:18372029

Desai, M. M., Grover, R., Aron, M., Ganpule, A., Joshi, S. S., Desai, M. R., & Gill, I. S. (2011). Robotic Flexible Ureteroscopy for Renal Calculi: Initial Clinical Experience. *The Journal of Urology*, *186*(2), 563–568. doi:10.1016/j.juro.2011.03.128 PMID:21683380

Elkoushy, M. A., & Andonian, S. (2011). Prevalence of Orthopedic Complaints Among Endourologists and Their Compliance with Radiation Safety Measures. *Journal of Endourology*, *25*(10), 1609–1613. doi:10.1089/end.2011.0109 PMID:21819221

Ergina, P. L., Barkun, J. S., Mcculloch, P., Cook, J. A., & Altman, D. G. (2013). IDEAL framework for surgical innovation 2: Observational studies in the exploration and assessment stages. *BMJ (Clinical Research Ed.)*, *346*. doi:10.1136/bmj.f3011 PMID:23778426

FernandezR.TanY. K.KaberleW.BestS. L.OlwenyE. O.PearleM. S.CadedduJ. A. (2012).

Guidance for the Content of Premarket Submissions for Software Contained in Medical Devices. (2015, May 11). U.S. Department of Health and Human Services, Food and Drug Administration.

Hellawell, G., Mutch, S., Thevendran, G., Wells, E., & Morgan, R. (2005). Radiation Exposure and The Urologist: What Are The Risks? *The Journal of Urology*, *174*(3), 948–952. doi:10.1097/01.ju.0000170232.58930.8f PMID:16094003

Holden, T., Pedro, R. N., Hendlin, K., Durfee, W., & Monga, M. (2008). Evidence-Based Instrumentation for Flexible Ureteroscopy: A Review. *Journal of Endourology*, *22*(7), 1423–1426. doi:10.1089/end.2007.0327 PMID:18613778

Kim, K. P., Miller, D. L., Gonzalez, A. B., Balter, S., Kleinerman, R. A., Ostroumova, E., ... Linet, M. S. (2012). Occupational Radiation Doses to Operators Performing Fluoroscopically-Guided Procedures. *Health Physics*, *103*(1), 80–99. doi:10.1097/HP.0b013e31824dae76 PMID:22647920

Knoll, T., Jessen, J. P., Honeck, P., & Wendt-Nordahl, G. (2011). Flexible ureterorenoscopy versus miniaturized PNL for solitary renal calculi of 10–30 mm size. *World Journal of Urology*, *29*(6), 755–759. doi:10.1007/s00345-011-0784-y PMID:22037633

Mcculloch, P., Cook, J. A., Altman, D. G., Heneghan, C., & Diener, M. K. (2013). IDEAL framework for surgical innovation 1: The idea and development stages. *BMJ (Clinical Research Ed.)*, *346*. doi:10.1136/bmj.f3012 PMID:23778427

Monga, M., Dretler, S. P., Landman, J., Slaton, J. W., Conradie, M. C., & Clayman, R. V. (2002). Maximizing ureteroscope deflection: "play it straight. *Urology*, *60*(5), 902–905. doi:10.1016/s0090-4295(02)01891-5

Patel, A., & Fuchs, G. J. (1997). Expanding the Horizons of SWL Through Adjunctive Use of Retrograde Intrarenal Surgery: New Techniques and Indications. *Journal of Endourology*, *11*(1), 33–36. doi:10.1089/end.1997.11.33

Preminger G. M., Tiselius H. G., Assimos D. G., Alken P., Buck C., . . . Wolf J. S. (2007) Guideline for the Management of Ureteral Calculi. *American Urological Association, 178*(6), 2418-2434. doi:10.1016/j.juro.2007.09.107

Pugin, F., Bucher, P., & Morel, P. (2011). History of robotic surgery: From AESOP® and ZEUS® to da Vinci®. *Journal of Visceral Surgery*, *148*(5), e3–e8. doi:10.1016/j.jviscsurg.2011.04.007 PMID:21974854

Rajamahanty, S., & Grasso, M. (2008). Flexible ureteroscopy update: Indications, instrumentation and technical advances. *Indian Journal of Urology*, *24*(4), 532. doi:10.4103/0970-1591.44263 PMID:19468513

Rassweiler, J., Safi, K. C., Subotic, S., Teber, D., & Frede, T. (2005). Robotics and telesurgery – an update on their position in laparoscopic radical prostatectomy. *Minimally Invasive Therapy & Allied Technologies*, *14*(2), 109–122. doi:10.1080/13645700510010908 PMID:16754625

Rassweiler, J. J., Goezen, A. S., Jalal, A. A., Schulze, M., Pansadoro, V., Pini, G., ... Turner, C. (2012). A New Platform Improving the Ergonomics of Laparoscopic Surgery: Initial Clinical Evaluation of the Prototype. *European Urology*, *61*(1), 226–229. doi:10.1016/j.eururo.2011.09.018 PMID:21978421

Saglam, R., Muslumanoglu, A. Y., Tokatlı, Z., Çaşkurlu, T., Sarica, K., Taşçi, A. İ., ... Traxer, O. (2014). A new robot for flexible ureteroscopy: Development and early clinical results (IDEAL stage 1–2b). *European Urology*, *66*(6), 1092–1100. doi:10.1016/j.eururo.2014.06.047 PMID:25059998

Somani, B. K., Srivastava, A., Traxer, O., & Aboumarzouk, O. (2013). Flexible ureterorenoscopy: Tips and tricks. *Urology Annals*, *5*(1). doi:10.4103/0974-7796.106869

Talso, M., Emiliani, E., Haddad, M., Berthe, L., Baghdadi, M., Montanari, E., & Traxer, O. (2016). Laser Fiber and Flexible Ureterorenoscopy: The Safety Distance Concept. *Journal of Endourology*, *30*(12), 1269–1274. doi:10.1089/end.2016.0209

Wright, A. E., Rukin, N. J., & Somani, B. K. (2014). Ureteroscopy and stones: Current status and future expectations. *World Journal of Nephrology*, *3*(4), 243. doi:10.5527/wjn.v3.i4.243 PMID:25374818

Xie, H. (2009, September). MEMS-based 3D optical microendoscopy. In *Engineering in Medicine and Biology Society, 2009. EMBC 2009. Annual International Conference of the IEEE* (pp. 6703-6705). IEEE.

KEY TERMS AND DEFINITIONS

Endoscopic Lithotripsy: The visualization of a calculus in the urinary tract and the simultaneous application of energy to fragment the stone or stones into either extractable or passable pieces.

Extracorporeal Shock Wave Lithotripsy: The least invasive surgical stone treatment using high frequency sound waves from an external source (outside the body) to break up kidney stones into smaller pieces and allow them to pass out through the urinary tract.

Lithotripsy: Medical procedure of destroying or fragmenting stones.

Percutaneous Nephrolithotomy: A minimally invasive procedure to remove stones from the kidney by a small puncture wound (up to about 1 cm) through skin.

Renal: Related to or located within kidneys.

Retrograde Intra-Renal Surgery: Surgery within the kidneys where a viewing tube called a fiber-optic endoscope is used for visual guidance.

Uretero-Renoscopy: Procedure used to diagnose or treat problems in kidneys using a small scope (uretero-renoscope) inserted through ureters.

Chapter 7
The "Arm" Line of Devices for Neurological Rehabilitation

Alessandro Scano
Institute of Industrial Technologies and Automation, Italy

Andrea Chiavenna
Institute of Industrial Technologies and Automation, Italy

Tito Dinon
Institute of Industrial Technologies and Automation, Italy

Alessio Prini
Institute of Industrial Technologies and Automation, Italy

Giulio Spagnuolo
Institute of Industrial Technologies and Automation, Italy

Matteo Malosio
Institute of Industrial Technologies and Automation, Italy

Lorenzo Molinari Tosatti
Institute of Industrial Technologies and Automation, Italy

ABSTRACT

In the modern scenario of neurological rehabilitation, which requires affordable solutions oriented toward promoting home training, the Institute of Industrial Technologies and Automation (ITIA) of the Italian National Research Council (CNR) developed a line of prototypal devices for the rehabilitation of the upper limb, called "Arm." Arm devices were conceived to promote rehabilitation at affordable prices by capturing all the main features of the state-of-the-art devices. In fact, Arm devices focus on the main features requested by a robot therapist: mechanical adaptation to the patient, ranging from passive motion to high transparency, assist-as-needed and resistive modalities; proper use of sensors for performance monitoring; easy-to-use, modular, and adaptable design. These desirable features are combined with low-cost, additive manufacturing procedures, with the purpose of meeting the requirements coming from research on neuro-motor rehabilitation and motor control and coupling them with the recent breakthrough innovations in design and manufacturing.

DOI: 10.4018/978-1-5225-2993-4.ch007

INTRODUCTION

The use of robotic devices for upper-limb neuro-motor rehabilitation is usual practice in clinical centers. In respect to conventional therapies, robots allow to increase training intensity and help patients to promote their active contribution. Furthermore, robots can act as measurers of patients' performances and adapt their interaction modalities to the emerging needs during the rehabilitation course. Robots like ARMin, MIT Manus, Armeo Spring, Braccio di Ferro, represent the state of the art devices for rehabilitation of the upper-limb and for promoting motor recovery. According to the available assessments and studies in the literature, their efficacy is slightly/moderately higher than the one of conventional therapies. Furthermore, robots are used in research to learn more about physiological and pathological motor control and neuromuscular diseases. Unfortunately, while being the state of the art devices for neuro-motor stimulation and training, such robots are very expensive and not compliant to user-friendly requirements that are needed for semi-autonomous home use. Consequently, they can be used only in clinical environments, under the supervision of medical personnel. Furthermore, sanitary costs related to rehabilitation are increasing and clinical centers can hardly support their burden. The possibility of delocalizing rehabilitation from clinical centers opens the chance for training performed in home environment, with time and costs savings for both the sanitary system and patients. In this scenario, which requires affordable solutions oriented toward promoting home training, the Institute of Industrial Technologies and Automation (ITIA) of the Italian National Research Council (CNR) developed a line of prototypal devices for the rehabilitation of the upper-limb, called *-Arm*. *Arm* devices were conceived to test the possibility of promoting rehabilitation at affordable prices but capturing all the main features of the state of the art devices. In fact, *Arm* devices focus on the main features requested by a robot therapist: mechanical adaptation to the patient, ranging from passive motion to high transparency, assist-as-needed and resistive modalities; proper use of sensors for performance monitoring; easy-to-use, modular and adaptable design. These desirable features are combined with low-cost, additive manufacturing procedures, with the purpose of meeting the requirements coming from research on neuro-motor rehabilitation and motor control and coupling them with the recent breakthrough innovations in design and manufacturing. *Arm* devices cover both clinical and home-oriented training and are designed for adaptation to patients with different motor impairment.

The *Arm* prototypes are:

- **LINarm**: linear device, freely orientable in space, suitable for functional movements. It features a variable stiffness actuation, allowing to adapt the mechanical behavior of the device to patients' needs. Functional Electrical Stimulation, simple Virtual Environments and a Patient Model, gathering data from integrated sensors and modulating the level of assistance, are integrated in the set-up. The LINarm++ Echord++ Project ended in October 2016 and guided the development of a second, more refined prototype, enhancing the original concept.
- **PLANarm**: planar device, freely orientable in space, suitable for planar functional movements. The state of the art planar robots used in literature for motor control and motor learning research inspired PLANarm. It features a variable stiffness actuation, allowing adapting the mechanical behavior of the device depending on patients' needs.
- **DUALarm**: Low-Cost device for bimanual rehabilitation, exploiting the capability of the less affected limb to provide rehabilitation to the more affected limb. DUALarm is completely realized in 3D printing technology and aims at being an easy-to-use, low-cost, open-source project.

Currently, reaching movements can be trained, but the device is conceived to be suitable for training of other functional gestures.

- **LIGHTarm**: Exoskeleton for the rehabilitation of the upper-limb, designed in two versions: LIGHTarm, not actuated, and conceived to support the weight of the impaired limb. The mechanical design includes high backdrivability, focusing on shoulder rhythm and elbow singular configurations.
- **VIRTUALarm**: Kinect One-based platform for motor monitoring, including body and limb tracking and a biomechanical evaluation of the performance in relation to databases of healthy subjects. Assessments include range of motion, motion dynamics, effort, motor control indexes, body segments barycenter tracking.

In this chapter, the main mechanical and design features of the devices will be described, along with the preliminary trials executed with the devices on healthy people and neurological patients. Instrumental assessments during the use of the devices, related to kinematics, EMG and human-robot interaction, will be discussed. Finally, a discussion will describe potential, appealing scenarios of the use of *Arm* devices in clinical and home environment.

BACKGROUND

Neurological Diseases and Stroke: Epidemiology Neurological Diseases are one of the main sources of disability, especially in Western countries. A wide variety of pathologies and symptoms can lead to partial or complete disability, which influences people life up to preventing self-autonomy. Stroke is one of the most noticeable sources of motor impairment. The World Health Organization (WHO) defines stroke as the clinical syndrome of rapid onset of focal (or global) cerebral deficit, lasting more than 24 hours with no apparent cause other than a vascular one (Hsieh, Chen, Li, & Lai, 2015). According to the recently published Global Burden of Disease 2010 Study (GBD 2010), stroke is the second leading cause of death globally and the third leading cause of premature death and disability as measured in DALY (Daily-affected living years). It is estimated that interventions to reduce mortality from stroke and ischemic heart disease by 10% are projected to reduce economic burden by an average of USD $25 billion annually with a threefold return on investment (Hsieh et al., 2015). Ischemic heart disease, sharing many risk factors of stroke, is the leading cause of both death and premature death and disability. The GBD 2010 clusters DALY for cerebrovascular disease in the category "Cardiovascular and circulatory diseases." Within the combined burden of neurologic disorders and cerebrovascular disease as measured in the GBD 2010, hemorrhagic stroke (35.7%) and ischemic stroke (22.4%) contribute the largest proportions of DALY (from GBD 2010 Results by Cause 1990–2010) (Chin & Vora, 2014).

Robots and Devices for Rehabilitation in Clinical Environment: An Overview

The Schema Theory of motor learning (Schmidt, 1975) (Sherwood & Lee, 2003) provides a framework for understanding the way human Central Nervous System (CNS) produces voluntary movement. Such issue has been addressed with the generalized motor program (GMP) concept proposed by Schmidt. In his theory, Schmidt argued that motor output is achieved through neural representations of the movements, called motor programs (Muratori, Lamberg, Quinn, & Duff, 2014). Motor programs are not related to specific tasks; rather, each motor program accounts for a large class of similar actions. Such framework not only is compliant with storage limit capabilities connected to human motor memory, but also explains that new skills can be performed by drawing from already stored similar classes of movements. Furthermore, Schmidt argues that stored rules of each GMP rely on invariant features that the program specifies (Muratori et al., 2014). These invariants help define classes of movement and help our CNS in minimizing the amount of information that must be stored.

The use of robotic devices for upper-limb neuro-motor rehabilitation is usual practice in clinical centers. It can be seen as the use of robotic therapists that help to reshape GMPs that have been lost due to stroke and, consequently, lists of motor commands that the CNS specifies to produce purposeful movement. Consequently, the robot should act as an intelligent care-giver, by triggering assistance and promoting the process of motor learning and have a biomimetic behavior (Habib, 2011). Assist-as-needed control systems have been developed in order to minimize the intervention of the robot that, according to such logics, intervenes only in case of failure in executing the motor task (Morasso et al., 2009). Thus, the requirements for rehabilitation robots include high compliance and softness: a little exerted force should guarantee motion. Backdriveability is requested to allow patients to interact and experience the results of their muscular activation patterns (Maciejasz, Eschweiler, Gerlach-Hahn, Jansen-Troy, & Leonhardt, 2014). Force fields created by the robot can be used to assist the execution of the motor task, or to resist patients' capabilities to provide challenging exercises. Furthermore, robots can act as objective measurers of patients' performances. Such capability allows adapting their interaction modalities to the emerging needs during the rehabilitation course.

Furthermore, in respect to conventional therapies, robots allow to increase training intensity and help patients to promote their active contribution. Resilience can be monitored and the therapy can be made more repeatable.

Two basic approaches to robotic therapy for the upper-limb can be identified. The first one includes end-effector robots, the second one exoskeletons. End-effector robots require patients to hold the handle and displace the robot. Some state-of-the-art examples are the ARMin, MIT Manus, Braccio di Ferro. They include a handle that can be displaced in a planar workspace, together with simple games or virtual environments that address patients' efforts towards specific tasks and targets. During the interaction, the robot, by providing haptic feedbacks and force fields, can ease the execution or making it more challenging. Exoskeletons, instead, mimic the kinematics of the limb and wrap the affected limb, such as Armeo Spring. They are designed to support the weight of the limb and execute 3D movements, usually in interaction with games or virtual environments that ask to perform specific tasks. According to the available assessments and studies in the literature, robots efficacy is slightly/moderately higher than the one of conventional therapies (Lo et al., 2010). Furthermore, robots are used in research to learn more about physiological and pathological motor control and neuromuscular diseases.

The Emerging Need for Home-Oriented, Cost-Effective Solutions

Unfortunately, while being the state of the art devices for neuro-motor stimulation and training, previously mentioned robots are very expensive and not compliant to user-friendly requirements that are needed for semi-autonomous home use. Consequently, they can be used only in clinical environments, under the supervision of medical personnel. Furthermore, sanitary costs related to rehabilitation are increasing and clinical centers can hardly support their burden. The possibility of delocalizing rehabilitation from clinical centers opens the chance for training performed in home environment, with time and costs savings for both the sanitary system and patients. The neuro-motor rehabilitation process involves training exercises whose intensity and quality, along with the amount of damage, are related to functional recovery. Conventional approaches to therapy, robotic or supported by virtual interactive environments are standard practice in the clinic as well as the possibility of surgical and pharmacological support. In the United States, the costs related to the provision of these therapies are expected to double by 2030, going to affect 4% of the population. There are also scientific evidence that the benefits related to rehabilitation are preserved over time if the partially restored features are constantly kept under workout and used continuously in daily life. In Western countries, it is therefore necessary to identify new strategies to promote the rehabilitation process as an intensive and low-cost. The previous conditions indicate as a natural consequence the transfer of neurological therapies at the household level. To the benefits listed above, are added objective factors of decrease of the crowding of hospital centers, the possibility to make daily motor training without having to reach the reference structure, with savings of time not only for the health system but also for the patient.

The ability to run training within the family context, generally more comfortable for the patient, is an added value. It should however be noted that the delivery of treatments at the clinical level takes advantage of: 1) technologies that represent the state for training oriented neuro-motor recovery, 2) equipment and sensors for the assessment of therapy, and 3) medical and physiotherapy that can support the patient. These factors may not be present at home; their absence poses considerable challenges for the effective implementation of motor training, although the home solution seems to be the way of the natural and preferred out of the problems highlighted above.

The first requirement to be analyzed is the absence of a direct supervision to therapy. For this reason, in addition to adequate patient and family education on the proper execution of the training, rehabilitation is based on computer applications, which require the presence of virtual assistants, equipped with artificial intelligence, which direct the activities and show the patient how to rehabilitate, just as would a physical therapist at the clinic structure.

The type of rehabilitation activities proposed at the household level can occur in two main types of approaches. The first is based on the use of motor learning strategies centered on the teaching of specific functional motor tasks related to activities of daily living (coaching); during the training, the patient focuses on the repetitive execution of one or more specific gestures, maximizing the intensity of the workout in order to regain that specific functionality (Cameirão et al., 2016). The second approach, generally associated with the presence of virtual environments, is otherwise to promote physical activity in conjunction with the execution of fun and motivating activities. According to the Cochrane review of Laver (K. E. Laver, George, Thomas, Deutsch, & Crotty, 2015; K. Laver, George, Thomas, Deutsch, & Crotty, 2012), the adoption of virtual environments to support the rehabilitation does not offer substantial improvements compared to conventional therapy and may present cognitively complex challenges and require coordination of multiple motor districts. However, if provided with appropriate logical and appropriate to the patient's ability, the approach related to gaming can enhance motivation by offering a variety of activities and engaging, and more regular training and long-lasting. An element that promotes this aspect is the possible presence of the storyboard, which is a narrative progression motor experience, more emotionally engaging. Functional involvement is also the promotion of appropriately balanced logic, enabling the patient to co-operate or "compete on equal terms" with other people.

Applications should guide the patient in therapy and adapt to his motor skills, by calibrating the difficulty of the exercise in terms of gestures (stimulating the execution of specific motor tasks) and in terms of required skills (within the same gesture stimulating higher execution speed or accuracy).

Scoring systems and performance evaluation are useful for both the monitoring of the training distance, for immediate feedback to patients who may be aware of the trend of their motor performance. These applications are based on low-cost technologies, including optical tracking sensors or accelerometers. These sensors make it possible to track the movement of the body and then bind to the patient's motor gestures or an interaction with a virtual environment or performance evaluation. Some examples of this type of approach are Neuro@Home, and VirtualRehab Jintronix, rehabilitative platforms that use Kinect sensors, derived from industry of video games, for the integration of exercises related to functional motors objectives drawn directly from clinical staff.

The domestic therapy can also make use of electromechanical devices dedicated to functional recovery. These devices act as a support to the movement, replicating the core functionality of the clinical counterparts, with less complexity and extremely lower price solutions. One of the state of the art device for home rehabilitation is Tailwind, a commercial device that allows you to perform bimanual movements of the upper limbs and rehabilitate them to motion.

THE "ARM" LINE OF DEVICES

In this scenario, which requires affordable solutions oriented toward promoting home training, the Institute of Industrial Technologies and Automation (ITIA) of the Italian National Research Council (CNR) developed a line of prototypal devices for the rehabilitation of the upper-limb, called *"Arm"*. *Arm* devices were conceived to test the possibility of promoting rehabilitation at affordable prices but capturing all the main features of the state of the art devices. In fact, *Arm* devices focus on the main features requested by a robot therapist: mechanical adaptation to the patient, ranging from passive motion to high transparency, assist-as-needed and resistive modalities; proper use of sensors for performance monitoring; easy-to-use, modular and adaptable design. These desirable features are combined with low-cost, additive manufacturing procedures, with the purpose of meeting the requirements coming from research on neuro-motor rehabilitation and motor control and coupling them with the recent breakthrough innovations in design and manufacturing. *Arm* devices cover both clinical and home-oriented training and are designed for adaptation to patients with different motor impairment.

Variable Stiffness Actuation

From the invention of the MIT-Manus (Krebs et al., 1999), force feedbacks and force-based controls are considered important features for neurorehabilitation devices (Marchal-Crespo & Reinkensmeyer, 2009), enabling them to react to the actual force exchanged with the patient, modifying, in most cases, the stiffness of the device used by the patient. Within this context, Variable Stiffness Actuators (VSA) (Vanderborght et al., 2013) are considered a promising technology, thanks to their inherent adjustment of the mechanical stiffness, force estimation and robustness to external perturbations and model errors in physical human-robot interacting scenarios (Jafari, 2014). In particular, this type of actuation has been adopted to realize some prototypes specifically dedicated to neurorehabilitation (Vitiello et al., 2013; Zhang, Guo, Pang, Gao, & Guo, 2014). Among the VSA, it is worth to mention the Agonist-Antagonist VSA (AAVSA), made up of two Series Elastic Actuators (SEA), arranged in parallel and exerting opposite forces. The principle of operation of AAVSAs are at the basis of two VSA devices described hereafter, LINarm and PLANarm.

Figure 1. Architecture of an Agonist-Antagonist Variable Stiffness Actuator: two non-linear Series Elastic Actuators arranged in parallel control the position and the stiffness of a mobile mass m

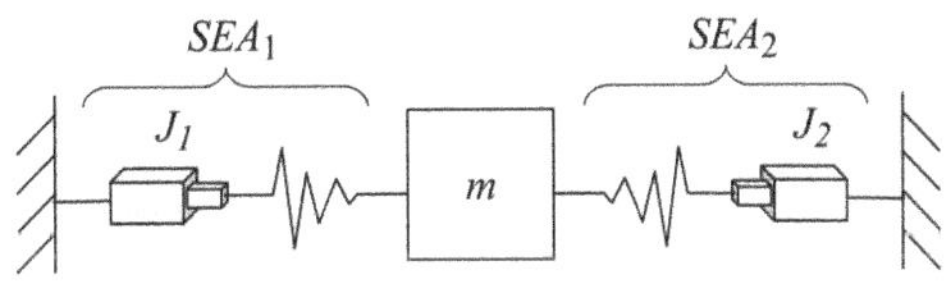

LINarm

The upper limb is an incredible adaptive organ that allows performing high-level interactive movements and is capable of accomplishing numerous functional tasks in an infinitive number of kinematic solutions. It is a sophisticated multi-joint system designed for multitasking functioning and capable of performing from heavy jobs to tasks requiring fine movements, not to mention exploration and manipulation of objects (Rau, Disselhorst-Klug, & Schmidt, 2000). Certainly, not all movements may be trained and this variability makes the selection of a reduced number of them necessary, a selection that, preferably, should be done based on key determinants of the movement control.

As reported above, two functional gesture only are sufficient to study shoulder and elbow compound movement that are frequently used in Activities of Daily Living (ADLs): 1) the Reaching Movement against gravity and 2) the Hand-to- Mouth Movement (Caimmi et al., 2015), Figure 2. These two gestures are representative of important compound movements like shoulder flexion (against gravity) and elbow extension, and, elbow flexion (against gravity) and shoulder flexion.

Considering the relevance of RM and HtMM and promising results obtained so far, a solution to enable patients to be trained by a robotic device with one these two movements is considered beneficial and constituted the origin of the LINarm device development. Furthermore, for both RM and HtMM, linear movement represents a good approximation. For these reasons, the development of a modular solution enabling differently oriented movements, approximating them as linear, is considered a good compromise and strategy to be included in a low-cost device for home rehabilitation.

Focusing on the mechatronic actuation schemes of rehabilitation devices, rigid mechanical actuations coupled with force sensors or back-drivable transmissions are typically employed to infer the interaction force between the device and the patient. However, in the last years, the effectiveness of adding mechanical compliance to the actuation system is being explored.

LINarm is a low-cost VSA robotic device for the rehabilitation of the upper limb. It is characterized by a more compact and optimized design, and embeds a novel VSA architecture exploiting a wire or bend-based mechanical transmission. The peculiar novel VSA scheme, enabling a compact and light design, is illustrated in Figure 3 embedding the principle of VSA actuation (G Spagnuolo, Malosio, Dinon, Tosatti, & Legnani, 2017a).

As depicted in the Figure 4 LINarm is characterized by two motors (Pololu), two motor-driven drums which wrap the wires that are connected with two non-linear spring embedded in the handle. The two motors permit to control the stiffness and the position of the handle. The handle slides on a linear guide composed by two rounds bar contained by two mechanical end-stroke.

The controller is composed of an Arduino DUE board and two Pololu driver required to control the motors. These are low-cost components, chosen in order to contain the overall cost of the device, but are in any case sufficient to ensure the proper functioning of the device. The controller is also connected to

Figure 2. The 3D Reaching Movement (RM) and the Hand-to-Mouth Movement (HtMM)

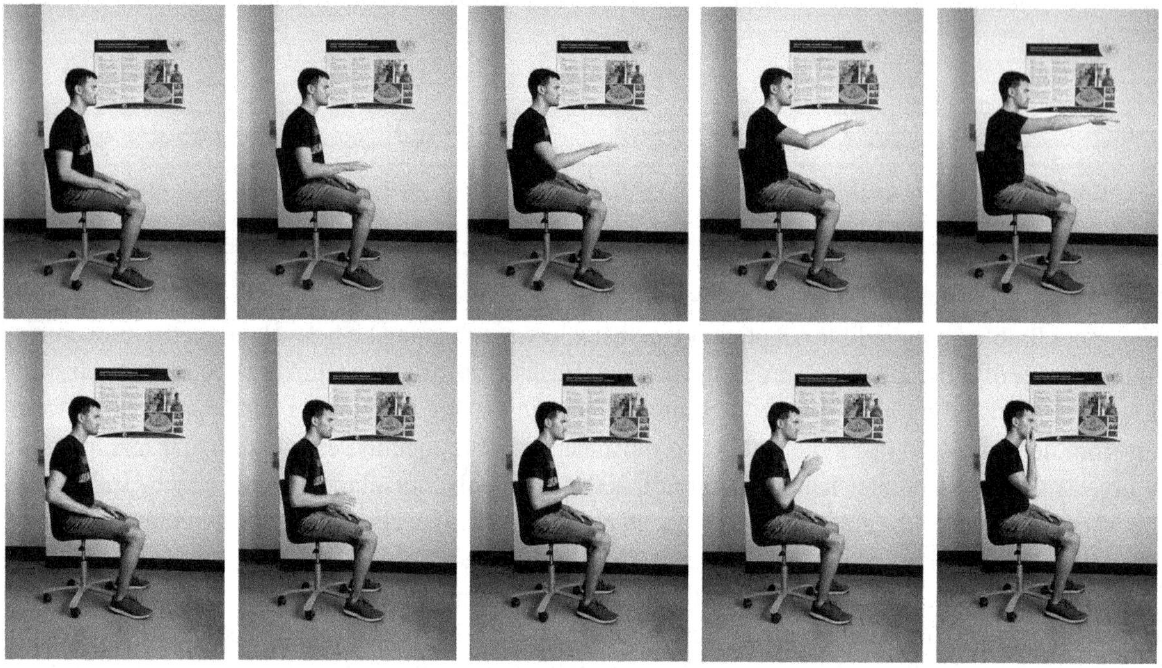

Figure 3. VSA scheme of LINarm

Figure 4. LINarm mechanical design: Render and demonstrator

the handle encoder and to two mechanical switch that represent the electrical end-stroke. The controller software developed in C++ is loaded in the Arduino board.

The control software must be able to control the robot in different modalities that represent the state of art of the control strategies in rehabilitation. For this kind of device we can divide the control strategies in two main group. The assistive control group and the challenge-based control group. In the first one, the robot help the patient to reach a pre-determinate target. In the latter, a generic resistance can be added to the movement made by the patient. Gravity in the anti-gravity movement can be considered as a resistance and for this reason LINarm is equipped with a gravity compensation algorithm. The principal control strategies provided in the LINarm robot are:

- **Passive Control**: In this control strategy the patient is guided along the path to reach the target. With passive control the patient could remain totally passive and still reach the expected target. Anyway, studies have shown that also in this case the patient tends to actively participate in the exercise. At mechatronic level this kind of control is characterized by a position control closed on the motors position feedback. A scheme of the real-time passive control is depicted in Figure 5.
- **Admittance Control**: This is a challenged-based control strategy. In this kind of control the interaction force between the patient and the robot handle is converted in a velocity command. So in this kind of control strategy the velocity depends proportionally to the force that the patient is able to apply to the handle. The admittance control is described by the formula:

$$v = a \cdot F$$

Where v represent the velocity command, F represent all the external force while a is the admittance value. It easy to understand that a higher admittance result in a lower perceived resistance during the robot movement. As said previously during the anti-gravity movement the gravity can be felt as a resistance. It is possible to add a virtual force, able to reduce or eliminate the gravity force component. In this control strategy, no help is given to the patient in order to achieve a possible target. The target can be shown to the patient using a virtual reality or other type of visual input.

Assistance control: as said in the passive control the great part of the work is done by the robot and the patient simply follow the robot during along the path. During this kind of exercise the correct movement is assured by the robot but a negative effect can be represented by the slacking hypothesis. In this case, due to the excessive assistance provided by the robot, the patient stop to actively follow the robot during its movement. On the other hand a pure admittance control in patient with severe motor disease can bring to incomplete exercise due to difficulties faced by the patient in reaching the target. A hybrid

Figure 5. LINarm position control loop

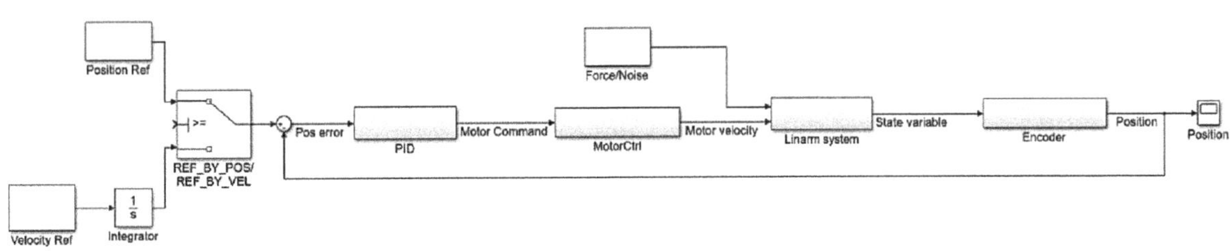

control strategies can be considered, and this strategy is represent by the assistance control. In this control paradigm, a virtual force could be added to the admittance input force in order to reach the target:

$$v = a \cdot (F_e + F_t)$$

The above formula represent the concept behind the assistance control. The F_t represent the assistance force, while F_e represent all the external force, including interaction force. F_t force can be varied according the particular needs.

Patients is helped by a force linearly proportional to the target distance. The value k_t is representative of this proportionality and could be considered as the stiffness of a virtual spring connected to the handle and to the target.

All modalities can be selected by a PC through a serial communication with the LINarm controller. It's also possible access to the LINarm robot data (like motor and handle position, interaction force ecc.) and command all the device functions. This characteristic has been exploited in the LINarm++ platform in order to expand LINarm functionalities by integrating additional modules and augment application scenarios and applicable clinical techniques.

The LINarm++ (Matteo Malosio et al., 2016) rehabilitation platform is a multisensory and multimodal system made up of the following set of optional modules interacting with the patient:

- LINarm device, a low-cost unobtrusive sensory system for measuring the patient's physical activity and his physiological state, in order to obtain and constantly update a comprehensive state of the patient;
- An easily wearable FES system, allowing selective and effective stimulations of upper-limb muscles;
- Virtual reality (VR) in order to represent target exercise. Different level of difficult can be selected during the exercise.

The *patient model*, a software that starting from the kinematic, physiological and VR data is able to select the best VR difficult level, and LINarm and FES assistance level.

These modules are virtually connected by different software communication channel realized exploiting the ROS framework functionalities. Hardware and software modularity meets the need of satisfying the requirements and budget availability of different clinical centers. Its affordability and semi-autonomy in targeting patient's needs can enable more than one patient to be treated simultaneously, increasing labor productivity and improving the overall services for the patients. These same characteristics facilitate its use in a domestic environment, giving the patients the chance of being treated directly at home with challenging and personalized exercises, exploiting device adaptability to conditions and improvements of each specific patient. The platform features self-adaptation of rehabilitation parameters during its functioning and includes auto-tuning procedures during the setup, as the one to identify the muscles to be stimulated by the NMES system, allowing a rough positioning of the electrode pseudo matrixes on the arm, without requiring deep medical knowledge. The system aims at satisfying the increasingly larger request of more and more effective and affordable technologies complementary to traditional rehabilitation techniques, to face population ageing and the related increase of neurological diseases. The first LINarm++ prototype is currently available, technical and its clinical assessment is planned in the near future.

PLANarm

Despite the existence of one-DOF end-effector devices and exoskeletons embedding VSAs (Vitiello et al., 2013; Zhang et al., 2014), no end-effector VSA planar device has been so far developed and exploited for rehabilitation purposes, even if different planar-based force-feedback devices has been developed. Only one general-purpose prototype featuring planar movements and embedding a VSA architecture was developed so far, at authors' knowledge. It is a cable-driven device arranged on a triangular framework, featuring three non-linear SEAs (Zhou, Jun, & Krovi, 2014). Because of the presence of only three tendons, its capability of adjusting the stiffness of the load along different directions is reduced. Each tendon is, moreover, controlled independently by two actuators, for a total requirement of six actuators.

In order to take a step forward with respect to this previously existing prototype, PLANarm is a planar device (Di, Industriale, & Informazione, 2015) embedding a novel architecture conceived with the aim of favoring the isotropy in terms of stiffness and decoupling of dofs, by an orthogonal two-dof planar tendon-driven VSA actuation scheme. Its peculiar architecture allows to tune, in parallel, both the equilibrium position and the stiffness of a mobile platform along two orthogonal directions, in order to favor isotropy and to tune stiffness ellipsis along two independent axis (G. Legnani, Tosi, Fassi, Giberti, & Cinquemani, 2010; Giovanni Legnani, Casalo, Righettini, & Zappa, 1996; Molinari-Tosatti, Fassi, & Legnani, 2003).

The kinematic architecture of the submodules making up PLANarm is depicted in Fig. 6. The orthogonality of agonist-antagonist tendons connected to the handle is guaranteed by a proper mechanical transmission. For a deeper understanding of geometrical, analytical and numerical aspects refer to (M Malosio, Corbetta, & Ramirez, 2016).

Figure 6. Simplified representations of the two main submechanisms embedded in PLANarm. Continuous lines refer to the two horizontal antagonist actuators, dotted lines refer to the two vertical antagonist actuators. (Left) The tendons subsystem includes two orthogonal pairs of antagonistic non-linear series elastic actuators. The direction of the tendons is defined by four carriages c_{ij} which embeds deflecting pulleys. (Right) The carriages subsytem depicts the principle of operation of the tendon-based system which guarantees that the tendons are constantly orthogonal if no external force is applied to m.

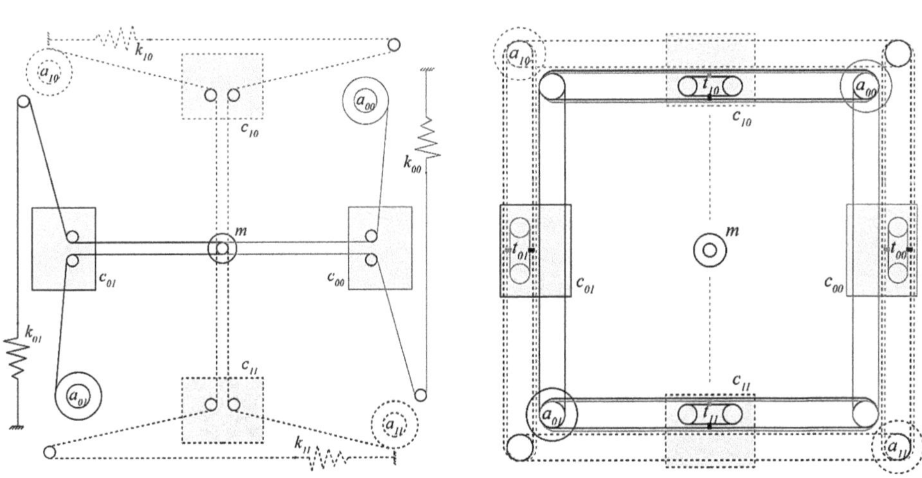

The handle is free to slide on the rectangular ferromagnetic surface of the device. The bilateral constraint is guaranteed by a set of rollers and a magnet, which generates an attractive force between the handle and the surface, guaranteeing that the handle remains vertical, even if an external force is applied to it (Figure 7).

The prototype (Figure 8) is actuated by four rotational DC motors Pololu 30:1 Metal Gearmotor 37Dx68Lmm with an embedded encoder, assembled at the corners of the PLANarm (a_{ij} in 6). The non-linear elastic elements have been designed taking inspiration of the principle of operation described in (G Spagnuolo, Malosio, Dinon, Tosatti, & Legnani, 2017b), designing proper cam profiles. The electronics includes an Arduino Due micro-controller, in charge of executing low-level and real-time control algorithms to directly control the motors, and two motor drivers to power the four actuators. The Arduino board is connected to a computer through a micro-USB cable to communicate with a PC, in charge of executing the high-level software application, interfacing the user and the device, with graphical and numerical feedback, providing required information to correctly command the control algorithms at low level. The control structure and algorithms executable by PLANarm are very similar to the ones illustrated for the LINarm device, applied to planar movements instead merely linear.

Figure 7. Design of the PLANarm prototype which embeds a two-dof planar tendon-driven VSA actuation scheme and cam-based non-linear springs

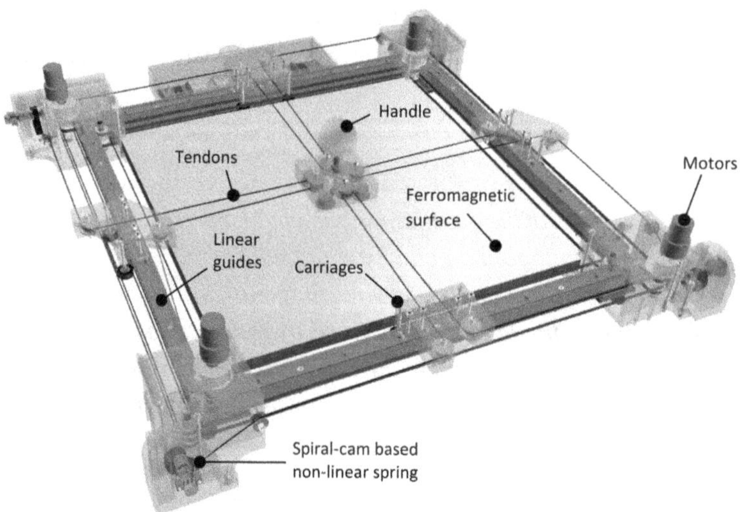

Figure 8. The PLANarm prototype and handle

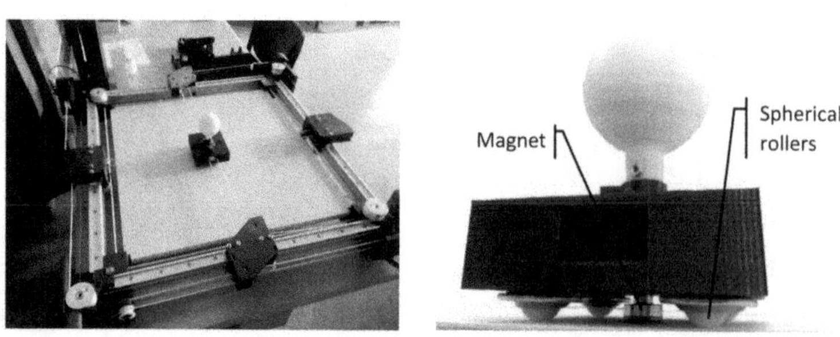

DUALarm

The global requirement of technologies able to facilitate the execution of neurorehabilitation therapies at patients' home and the diffusion of low-cost and home-oriented rehabilitation devices are the basis and source of inspiration of DUALarm (Figure 9), a low-cost, open-source and 3Dprintable device for bilateral upper-limb rehabilitation, whose core mechanism could be exploited in different installation configurations (Figure 10).

Bilateral UL training is a relatively new form of stroke rehabilitation. It is based on the premise that, during simultaneous movements, the non-paretic UL may support the movement of the paretic UL at brain level. The interest in this therapy, which is becoming more and more common, arose partly accidentally and partly from insights, gained from the motor control literature. In this literature, rhythmic inter-limb-coordination studies investigated extensively coupling and interaction effects between the two ULs in healthy subjects (Van Delden, Peper, Kwakkel, & Beek, 2012). It was well proved that human beings do show a basic tendency towards in-phase (i.e. symmetrical movements) or anti-phase (i.e. alternating movements) coordination, with a prevalent 1:1 frequency locking mode for UL bilateral movements (Swinnen, 2002).The tendency towards these patterns reflects the coupling between the ULs. In bilateral UL training, this coupling is exploited using interactions between both sides of the central nervous system through intact connecting structures, such as the corpus callosum (Cauraugh & Summers, 2005) In this work, the authors are presenting the current version of DUALarm a low-cost, open-source and 3D-printable rehabilitation device based on a geared mechanism (Fig. 11) enabling in-phase and anti-phase movements and developed to support reaching rehabilitation movements of the UL.

Figure 9. The DUALarm prototype

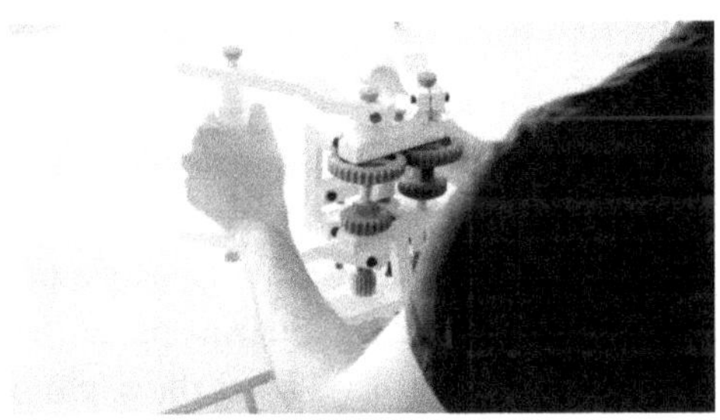

Figure 10. DUALarm designed set-ups

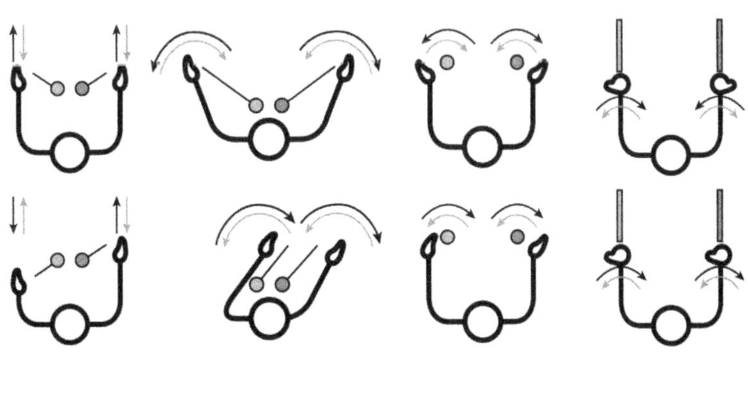

DUALarm has been designed in order to take the most of 3Dprinting technology, limiting as much as possible the number of commercial components. The version presented in Fig. 11 is an almost-completely 3D-printed plastic (PLA) device, equipped by two handles (one for each side) held by the subject and standing upon a regular table, with an optional clamp to fix it. The core mechanism is made up of: a properly supported gearbox, a mode-selection mechanism and a signal acquisition system. The pivotal component of the device is the mechanism of the gearbox, based on a plastic structure and containing five wheel-gears and three aluminum shafts (10 mm in diameter). Referring to Fig. 11, three groups of gears are recognizable: the fixed (hereafter f) group rotating about *af* and made up of gears *gfu, gfm, gfl* stationary constrained to an aluminum shaft, the inversion gear *gim* rotating about *am* and the mobile (hereafter *m*) group rotating about am and made up of gears *gmu, gmm, gml*. All the gears are mechanically constrained to three aluminum shafts, and secured by steel screws and bolts to avoid free spinning. The gap between *gfu* and *gfm* is different from the one existing between *gmu* and *gmm*. Shaft am is free to move axially in order to obtain two different configurations of the mechanism and let gears engage differently. Consequently, sliding up and down the m group, two different configurations can be selected: if *gfu* engages *gmu, af* and *am* perform anti-phase rotations; if *gfm* engages *gmm, af* and *am* perform in-phase rotations, thanks to the presence of *gim*. All the gears in the mechanism are 2 mm in module to guarantee a good mechanical resistance and have a double chamfer in order to ease the engagement of the gears. A selection mechanism, made up of a 3D-printed lever, is placed above the gears. If it is engaged, it works as a spacer that causes *gmu* to engage *gfu*, leading to the anti-phase rotation of *af* and *am*. If it is not engaged, *gmm* is engaged with *gfm* through *gim*, leading to an in-phase rotation of shafts. An aluminum pin p (with a 3D-printed cup) locks the position of s for safety. To track the movement performed by the patient, the mechanism is equipped with two potentiometers, measuring *af* and am rotations through two pairs of gears (*gfl, gfk and gml, gmk*) to compensate the vertical sliding of *am*: the first element of each couple is spliced and secured to the rod by a steel screw and bolt, the second one is simply spliced onto the slotted-shaft of the potentiometer. On the back-side of the device an Arduino Uno R3 board is located while two potentiometers *gfp* and *gmp* are accommodated into the base.

In the DUALarm configuration shown in Figure 11, two arms are connected to axis *af* and *am*. Each arm is made up of one handle, one aluminum rod (about which the handle is free to rotate) and two links connecting the shaft of the wheelgears to the handle. Because of the slight sliding movement of am to select the required configuration, every arm includes a spacer that can be placed above or below the handle in order to maintain the left and right handles at the same height. Links are constrained to the shafts of the core mechanism.

The device was manufactured thanks to the Fused Deposition Modeling (FDM) technology, an Additive Manufacturing process consisting in laying tracks of molten thermoplastic polymer onto a

Figure 11. DUALarm design

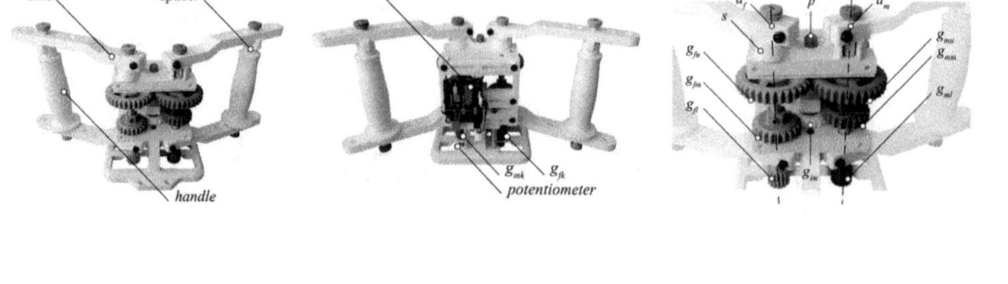

platform to obtain a layer of material on the XY plane of the machine (3D-printer). Once the molten layer has solidified, other subsequent layers are deposited in a bottom to top manner, in order to create the complete 3D object in the Z direction of the machine. All the 3D-printed parts of the device were made by the low-cost FDM 3D-printer Sharebot NG. The plastic used to build the entire prototype was PLA with a wire diameter of 1.75 mm. Custom parameters for the aforementioned printer have been created and main parameters used are the following: 3 layers of bottom, 3 layers of top, 0.2 mm of layer height, 4 perimeters, 30% grid infill percentage, 45_/ 45_ raster orientation for bottom and top solid layers. These parameters were used to guarantee a suitable mechanical resistance in order to test the first prototype in several conditions, from normal to heavy use, and to ensure the presence of enough material, especially around holes, to perform subsequent operations, like boring or countersinking, to remove potential imperfections coming from the 3D-printing process. In order to avoid the typical high friction between two PLA 3D-printed parts, a sheet of a very smooth polypropylene (1.5 mm in height) was positioned onto critical surfaces.

Even though DUALarm is a completely mechanical device and does not require any additional mechatronical component to perform its rehabilitative tasks, it is equipped with some basic electronical components in order to monitor the activity of patients during rehabilitation sessions. Two potentiometers (100k @ 300_) are in charge of measuring and recording the angular rotation of the DUALarm's links and, therefore, of estimating the patient's limbs positions. The data acquisition system is implemented in an Arduino UNO microcontroller, acquiring the angular position of each of the two handles in real time. Thanks to this acquisition system, it is possible to convert the analog signal coming from the potentiometers to a digital signal that can be transmitted through a serial connection to a common PC. The high-level PC program is implemented in Python, an easily customizable, open-source and cross-platform language, allowing the program to run on different operating system, even on Linux-based open-source and free distributions, contributing to lower overall use costs of the device. A graphic user interface of the device is implemented too. It is characterized by four main panels, allowing to initialize the device, displaying a tridimensional feedback for the user, a real-time updated log of measured angular positions and the actual position of the handles. Through customized python scripts it is possible to program exercises, showing to the patient different target points to be reached, tuning the required range of motion and target distances according to the actual patient capabilities.

DUALarm was tested in preliminary experimental trials to characterize the interaction with the device of healthy subjects and a patient, in three modalities:_

Monolateral (M), executed alternately with the left and the right limb), Bilateral (B) In Phase (IP) and Anti-Phase (AP). 5 right-handed healthy subjects and 1 patient (post-stroke, left hemiplegia) were enrolled to participate to the preliminary trials.

Kinematics of the upper-limbs was measured with BTS Smart Elite system to detect movement phases; EMG of Deltoid Anterior (DA) and Posterior (DP), Biceps (BIC) and Subspinatus (SS) on both limbs. Mean Muscular Activation (MMA) of each muscle was computed as:

$$MMA = \frac{1}{n} \cdot \sum_{i=1}^{n} \int_{t_b}^{t^e} EMG_i(t)dt$$

The following results were observed on Healthy Subjects (Fig. 12):

Figure 12. EMG muscles activations during DUALarm trials

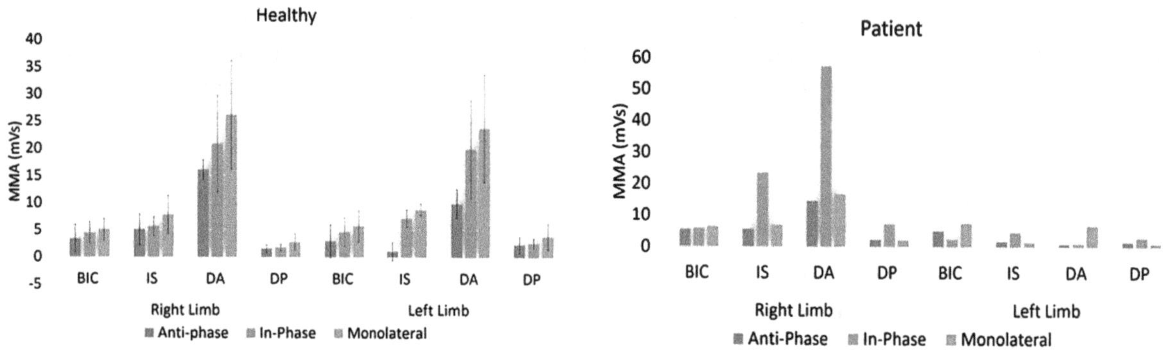

- DA is the main agonist muscle
- DA EMG is higher in (M) trials than in (B) ones.
- trials show higher activations than AP ones.
- Muscle EMGs are symmetrical between in M and IP trials, while AP trials show higher activity in the dominant limb.
- In AP trials, clear predominance of the dominant side.
- In IP trials, instead, DA of the right and left limb activate equally, and higher than in AP trials.
- The following results were observed on the Patient (Figure 12):
- Weakness of the more affected limb.
- Infact, as expected, the less affected limb strongly helps the more affected one by dragging it.
- During the bilateral IP trial, the MMAs of the less affected limb is comparable to the (M) one.
- MMAs are spotted also in the more affected limb.

LIGHTarm

Support limb weight due to gravity is a valuable strategy to enable upper limb rehabilitation in neurological patients. Raising the upper-limb is one of the more challenging tasks for neurological patients, since neurological diseases may affect the capability of performing functional movements and interacting with the environment (Krabben et al., 2012). Weight support partially discharge the muscles from the burden of supporting the limb. The effectiveness of weight-support approaches was observed in exoskeleton-assisted studies that showed benefits in terms of motor function recovery (Reinkensmeyer, 2012) and reduction of the influence of abnormal coupling (flexural-synergies) between shoulder and elbow. Starting from (Matteo Malosio, Pedrocchi, Vicentini, & Tosatti, 2011) and by a patent pending kinematic scheme (Malosio, M., Pedrocchi, N., & Tosatti, L. M. (2014)) a passively gravity-compensated exoskeleton for the upper limb neuro-rehabilitation was developed. Its peculiar kinematic structure has been specifically conceived to deal with shoulder rhythm and elbow singularity issues, achieving high backdrivability (transparency) throughout a large exploitable range of motion. An introduction describes the upper-limb and the exoskeleton kinematic models which where the basis for the development of the device (Giulio Spagnuolo et al., 2015). Then, the passively gravity-balanced exoskeleton is described.

Afterwards, the results of an experimental campaign designed to test gravity support features and transparency in static and dynamic conditions, are reported (Scano, Spagnuolo, et al., 2015).

A model of the upper limb, resembling the one presented in (Malosio, M., Pedrocchi, N., & Tosatti, L. M. (2014)) is depicted in Figure 13(a) and Figure 13(c). It is a serial kinematic chain characterized by five rotational joints $J_{h,i}$ (i=1,…,5). $J_{h,1}$ denotes the horizontal add/abduction, $J_{h,2}$ the arm elevation, $J_{h,3}$ the arm internal/external rotation, $J_{h,4}$ the elbow flexo/extension and $J_{h,5}$ the forearm prono-supination. The set $\{J_{h,1}, J_{h,2}, J_{h,3}\}$ models, as a whole, the gleno-humeral joint. The three degrees of freedom translational joint $J_{h,sg}$ models the humeral head movement according to the shoulder rhythm. Moreover, let us denote by:

- $\{t\}$ and O_t a generic coordinate frame and its center, respectively; in particular, $\{hs\}$ and $\{he\}$ denote the reference frames centered in the shoulder and elbow human articulations, respectively. $\{hs\}$ translates with respect to the chest $\{w\}$ according to the shoulder girdle movement.
- $\{\mathbf{e}_{t,x}, \mathbf{e}_{t,y}, \mathbf{e}_{t,z}\}$ the unit vectors of $\{t\}$ axes;
- $\pi_{w,xy}$, $\pi_{w,yz}$ and $\pi_{w,xz}$ the coronal, transverse and sagittal planes, respectively;
- $\mathbf{u}_{h,i}$ the unit vector, respectively, of the i-th rotational axis.

The exoskeleton kinematics consists of a hybrid mechanism, composed by four distinct serial chains, denoted by S_s, S_u, S_e and S_f - nearby the shoulder, the upper arm, the elbow and the forearm, respectively, as depicted in Figure 13(c).

This peculiar kinematic structure a) enables human shoulder to move complying with the physiological shoulder rhythm and b) avoids the occurrence of any singularity in the whole elbow range of motion (ROM) (Matteo Malosio et al., 2011). These features, in conjunction with low inertia and low frictions, guarantee high transparency, allowing the exoskeleton to follow accurately arm movement in a large exploitable range of motion. Even though the exoskeleton kinematics allows decoupling exoskeleton and human upper limb shoulder and elbow, hereafter, without loss of generality, the analysis is performed considering the limb shoulder and elbow centres coincident with the exoskeleton shoulder and elbow centres.

Figure 13. Human upper-limb modeling and LIGHTarm kinematic design. (a) Human-arm and (b) exoskeleton kinematic models.(c) Joint connection graph: joint are revolute (R), prismatic (P), 3dof translational (T), gimbal (G).

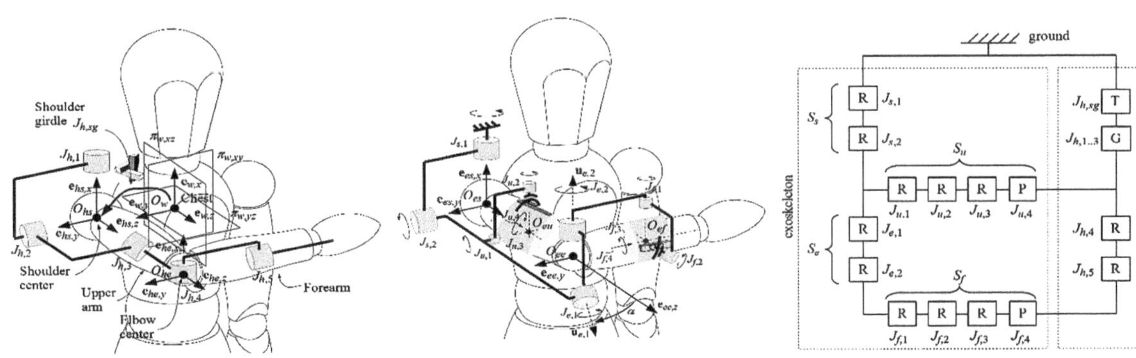

An articulated mechanism is *gravity-balanced* if it is in indifferent equilibrium without requiring joint actuator forces or torques. Physically, this means that the total potential energy of the device is invariant whatever its configuration is. This condition can be achieved in different ways, using counter masses, springs and/or auxiliary parallelograms (Lu, Ortega, & Ma, 2011). The gravity-balancing system of the unactuated version of LIGHTarm is made up of a combination of these three mechanical elements, in order to partially decouple the gravitational load contributions of the upper-arm and the forearm segments (Figure 14(a)). The gravity-balancing contribution has been partially decoupled between a counterweight W_s, in charge of supporting the arm and exoskeleton masses, and a spring s in charge of supporting the elbow mass. It is important to note that, no torque is transferred from the human upper limb and the exoskeleton, through the interface cuffs (S_u, S_f, Figure 14(c)) because of the presence of the spherical joints. The counterweight W_s has to counterbalance all the arm and forearm masses. For convenience, referring to Figure 14(b) and Figure 14(c), let us denote by:

- $\mathbf{Q}_{\{t\}}$ the vector connecting the origin O_t of frame *{t}* to point Q;
- $\mathbf{Q}_{\{t\},i}$ the component of vector $\mathbf{Q}_{\{t\}}$ along the direction i (i.e. x,y,z) of frame *{t}*;
- $\mathbf{G}_a$ and m_a the center of mass and the value of the mass of a, respectively;
- $\mathbf{G}_f$ and m_f the center of mass and the value of the mass of f, respectively.

Auxiliary Parallelogram Counter-Balance Mechanism

The overall mass of the exoskeleton and of the upper limb is counterbalanced by a mass W_s, rigidly connected to the exoskeleton link parallel to the human upper arm. This link has been realized through an auxiliary-parallelogram mechanism (Figure 14(b)): referring to the plane of elevation, the planar hinged-parallelogram produces 1-DoF circular translation. It consents to the reference frame *{ee}* centered in the elbow to translate with respect to (w.r.t.) the shoulder reference frame *{es}* (Fig. 14(b)), instead of rotating about $\mathbf{e}_{es,y}$, (Figure 14(b)). This leads the point A to be above O_{ee}, having the line $\overline{OA} // \boldsymbol{e}_{es,x}$ for any configuration, guaranteeing that the forces generated by the spring s supporting the forearm is unaffected by the upper-arm orientation. Referring to Figure 14(b), let us consider the fixed reference frame

Figure 14. Gravitational load contributions of the upper-arm and the forearm segments; Auxiliary Parallelogram Counter-Balance Mechanism; Forearm Spring-Balancing Mechanism

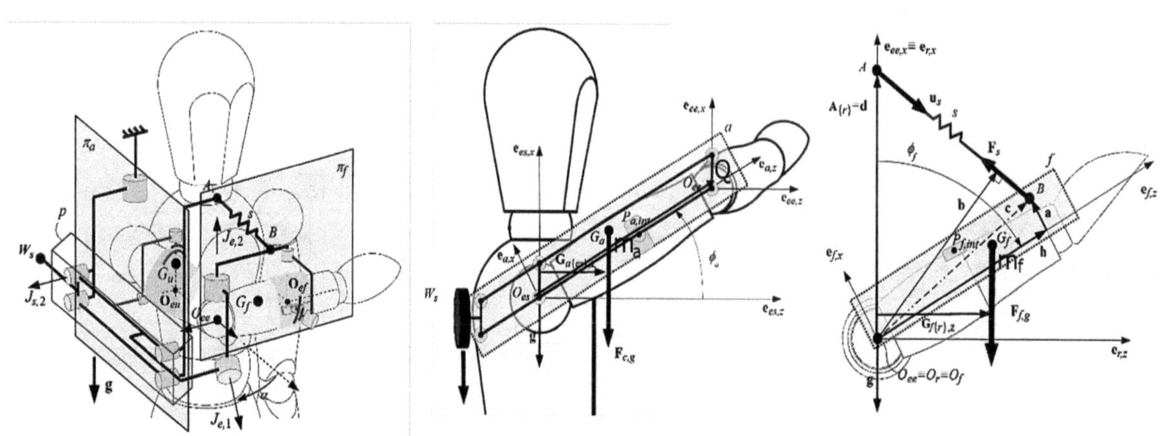

{es} w.r.t. which the body *a* rotates, and a reference frame *{a}* rigidly connected to *a*. In depth, let us analyze each contribution acting on the shoulder joints. No torque is transferred from the human upper limb and the exoskeleton, through the interface cuffs (S_u, S_f, Figure 13(c)) because of the presence of the spherical joints. The counterweight W_s has to counterbalance all the arm and forearm masses. The auxiliary parallelogram allows the mass balance in respect to O_{es} to be not influenced by the actual configuration of the masses downline the parallelogram.

The forearm mass is suspended by tension spring, generating a tension force between point *A* of the exoskeleton structure and point *B* along the forearm (Figure 14(a)). Let us refer to the model depicted in Figure 14(c), schematizing the spring balancing system, supporting a body *f*, modelling the forearm, hinged to a point O_f. Referring to Figure 14(a) it is a representation of the forearm kinematics on the plane π_f. Let us consider a fixed reference frame *{r}* w.r.t. which the body *f* rotates, and a reference frame *{f}* rigidly connected to *f*. A tension spring *s* is connected between points *A* and *B*, respectively fixed w.r.t. *{r}* and *{f}*.

Several trials of the first LIGHTarm prototype were performed. Firstly, to test LIGHTarm weight support characteristics in different configurations; EMG activity of the Deltoid Anterior (DA), the main shoulder flexor muscle in the sagittal plane, was used as assessment parameter. Second, to investigate LIGHTarm transparency during the execution of significant functional motor tasks; an analysis of the differences of the neuromuscular activation patterns of the relevant agonist and antagonist muscles during free movements and exoskeleton-assisted ones was used for assessment.

Three healthy subjects were enrolled for this study: age 32.7±11.9, height 181±2 cm, weight 77±8 kg, 3M, all right-handed. The following equipment was used: the prototype of the LIGHTarm exoskeleton; 8-channel EMG acquisition system (FreeEMG, BITIS, Italy); 6-TVC Marker-based Motion Capture system (Smart-D, BITIS, Italy). s-EMG were recorded to monitor muscular activation patterns of the following muscles: Deltoids Anterior, Lateralis and Posterior, Trapezius Superior, Pectoralis Major,

Figure 15. LIGHTarm CAD render

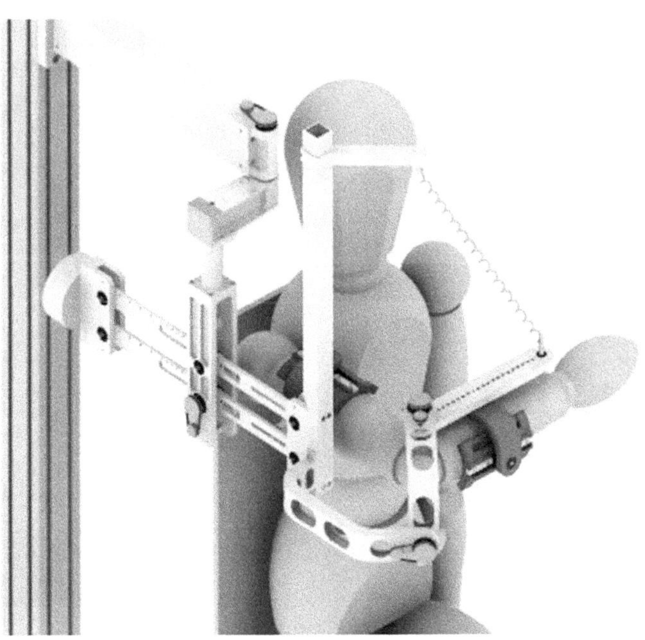

Triceps Lateral Head, Biceps Brachii Caput Longus and Brachioradialis. All subjects were right-handed and the recordings were performed on the right limb. Kinematics of the right limb was recorded too. Retroreflective Markers were positioned on: C7 and D5 vertebras, acromion, right epicondyle of the elbow, and styloid process of the ulna (Caimmi et al., 2015). To face the fact that the elbow marker is obscured by the exoskeleton in some configurations of the upper limb, a set of four markers was used to reconstruct elbow location. The procedure is not reported for the sake of brevity.

Two experimental conditions were investigated (Figure 16). *Static* trials were conceived to evaluate the efficacy of the gravity-support features in different configurations. *Dynamic* trials were designed to assess transparency. In both cases, the subjects' performances with LIGHTarm were compared to free movements. We denoted each trial with a 2-cyphers alphanumeric id. The letter identifies the task while the progressive number identifies different trials conditions. Static trials (Tab. II) were conducted with the exoskeleton and without exoskeleton in three different configurations (Figure 3): 1) *Begin (B)* position, where subjects held their hand on the thigh and the arm was relaxed; 2) In the *End (E)* position, where subjects held the arm raised at 90° of shoulder flexion and the elbow was extended; 3) In the *Intermediate (I)* position, subjects held the arm raised at 90° of shoulder flexion and the elbow flexed at 90° in the horizontal plane. In the Free Positioning trials, *Not Supp (0)* denoted the subjects had to keep their posture without external support, while *Supp (1)* indicated that a physiotherapist was statically supporting the upper-limb weight. In the exoskeleton trials, *Not Supp (2)* denotes that only the exoskeleton weight was supported, while *Supp (3)* indicates that both the weight of the exoskeleton and of the upper-limb were compensated with the gravity-support mechanism.

In order to investigate the perceived *transparency* during motion, three dynamic tasks were designed (Tab. III): the 3D Reaching (*R*, Figure 16), the Hand-to-Mouth (*M*, Figure 16) and the Hand-to-Nape (*N*, Figure 16). Their combination allows a daily-life purposeful interaction with the environment.

Figure 16. LIGHTarm trials: Static postures and dynamic gestures

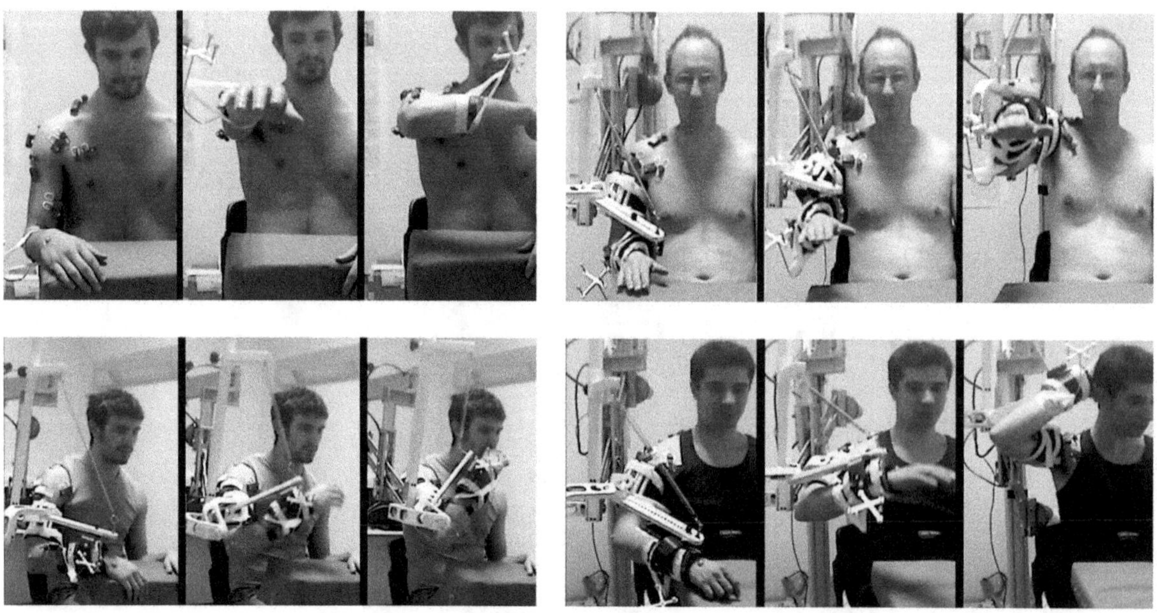

The Deltoid Anterior (DA) EMG activity, related to 8 seconds of the three static postures holding (B, I and E), in the four different trial conditions.

In the upper-left panel, the free positioning subject 1' trials are shown. The DA EMG activity is negligible in the Begin configuration (B0). This is expected because no torque is required to support the limb as the hand lies on the thigh. In the End and Intermediate configurations, the DA activity is quite high as the whole upper-limb weight is supported. The DA EMG activity is slightly superior in the End position (E0) consistently with the increase in shoulder torque due to elbow extension.

In the upper-right panel, the DA EMG activity during *exoskeleton-assisted not-supported* trials is shown for the same subject. The EMG activity in the three different configurations is highly comparable to the ones in the corresponding Free static postures. This indicates that the exoskeleton weight was supported. A slightly lower EMG activity in trial E2 could indicate a slight, though observable, over compensation of LIGHTarm weight.

The comparison between the EMG activity of the DA when the upper-limb is supported by the therapist (trials B1, E1, I1) and when supported by the exoskeleton (trials B3, E3, I3), is reported.

Considering as gold standard the DA EMG activity when the physiotherapist was supporting the arm (left-lower panel), the following considerations follow:

- Being the weight of the upper limb fully supported by the therapist (trials 1), no difference in DA EMG activity would be expected among the End, Intermediate and Begin configurations; residual DA EMG activity in the End position is due to the inability of the subject to fully relax;
- The DA EMG activity at End position are comparable between physiotherapist (B1, I1, E1) and exoskeleton support (B3, I3, E3) trials, indicating satisfactory weight support provided by LIGHtarm.
- Having obtained a perfect support in the configuration that requires more support (End), the slightly higher DA EMG activity in the exoskeleton supported trials at the Intermediate position, could be due to the inability of the subject to completely relax. Authors cannot exclude that LIGHtarm cannot provide perfectly uniform weigh-support being the cause of the residual DA EMG activity in trial I; however, the EMG activity in supported trials is considerably lower than in free ones.

Movement phases were calculated analyzing the vertical displacement of the wrist. Dynamic trials refer to subject 1, as representative of all the three subjects. As a first result, it is remarked that, while not being trained in the use of the device, subjects were able to move at self-selected speed without incurring in any mechanical lock.

Kinematic data indicated that both free movements and exoskeleton-assisted ones are characterized by similar range of motions. The main difference lies in the fact that exoskeleton-assisted trials (1 and 3 referring to Table. III) were executed at a spontaneously slower velocity than the free ones. Such results are likely due to inertial properties of the device (new parts with lighter materials will be realized for next trials) and to motor control adaptation (fully supported trials were executed lastly).

RCH: The comparison between trials R0 and R2 shows a slightly lower activity of the DA muscle to elevate the arm. Such result can be explained by: 1) only a partial compensation of the limb weight and, 2) lower torques required by slower movements. The Deltoid Posterior (DP) contraction is strong in both trials. In R0 a slight EMG activity is shown also during shoulder flexion (DP co-contraction), indicating the physiological need of stabilizing the upper-limb. R3 instead shows a remarkable decrease of the activity both of the DA and the DP, indicating that the weight support system is reducing the amount of torque required to elevate and stabilize the arm.

HTM: The comparison between trials M0 and M2 shows similar activation patterns, for both biceps (BIC) and triceps (TRIC). The triceps activity is slightly reduced in the exoskeleton-assisted trial, again, probably indicating less need to stabilize the arm. In the M3 trial, the exoskeleton supports the biceps in flexing the elbow against gravity and the EMG activity decreases.

HTN: The comparison between trials N0 and N2 shows that the trapezius (TRAP) activates with a very similar pattern, while the DA has a lower peak value, probably due to LIGHTarm slight over weight-support action. The DP shows similar activation patterns and amplitudes.

Authors underline that EMG activity suggests that transparency features are generally achieved; the exoskeleton does not alter significantly the EMG patterns of the main agonists and antagonists muscles. The shape of the EMG activity is generally preserved, even if sometimes amplitudes are partially altered. All subjects referred their comfort when assisted by the exoskeleton. Some discomfort was noticed when moving in vertical direction in full-compensated trials. The inertial load probably accounts for the problem, as healthy subjects perform movements at high velocity. This will be not the case of the majority of pathological subjects, for whom achievable velocities are significantly lower.

VIRTUALarm

According to the World Health Organization (WHO), the essential components of successful neurorehabilitation include expert multidisciplinary assessment, goal-oriented programs and evaluation of impact on patient and goal achievement through the use of clinically appropriate, scientifically sound outcome measures incorporating the patient's perspective (World Health Organization, 2002). VIRTUALarm is a Kinect One-based platform for motor performance monitoring, including body and limb tracking and a biomechanical evaluation of the performance in relation to databases of healthy subjects. Assessments include range of motion, motion dynamics, effort, motor control indexes, body segments barycenter tracking. Thus, at the current state, VIRTUALarm does not implement feedback during the execution of games, but provides an assessment similar to the ones made with marker-based systems (Cappozzo, Catani, Leardini, Benedetti, & Della Croce, 1996; Petuskey, Bagley, Abdala, James, & Rab, 2007; Roux, Bouilland, Godillon-Maquinghen, & Bouttens, 1999).

VIRTUALarm exploits Kinect One technology that allows that tracking of human body segments as shown in Fig. 17. Preliminary tests were performed on healthy subjects and patients, for a twofold reason: 1) to build database of healthy people performances, to be used as reference when evaluating neurological patients; 2) to assess Kinect One capabilities of assessing motor performances and discriminate different levels of motion capability. In a recently published study (Scano et al., 2017), 78 healthy subjects were enrolled to perform the Reaching Against Gravity Movement (RM) and Hand-To-Mouth Movement (HtMM) according to the protocol described in (Caimmi et al., 2015). Subjects stood comfortably on a chair, with their back straight. They were asked not to move their torso during the experiment and to perform the motor tasks only by limb motion. A target for the RM was set at shoulder height, indicating the target point. The target of the HtMM was the mouth itself. Subjects were requested to perform the repetitions of the gestures at natural, comfortable, self-selected speed, with no pauses between one repetition and the following one. The starting position was with the elbow flexed of about 90°, with pronated hand leaning on the thigh. Four acquisitions was performed by each subject: RM and HtMM

Figure 17. Kinect One graphical interface, visual feedback and graphs

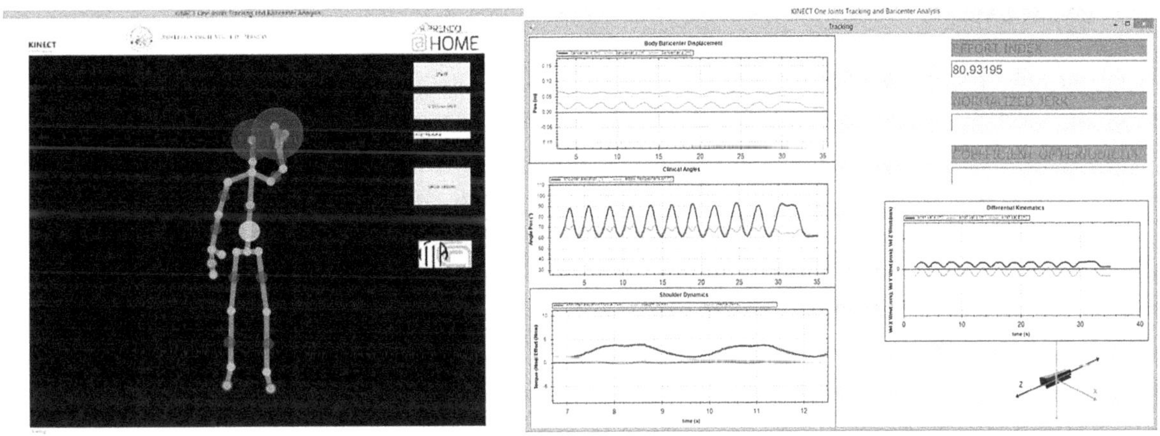

gestures, both performed with their dominant and non-dominant limbs. Twelve repetitions of the motor task was performed per acquisition.

Microsoft Kinect One can track human movements but does not allow defining other reference points, except for the articular joint centers. Consequently, a fine characterization of the interaction with an object, as with marker-based systems, is not possible. However, as stated by (Schaefer & Hengge, 2016) motor control analysis and knowledge acquired in tests on "standardized" reaching movements, can be considered as representative of "natural" reaching movements, involving physical interaction with real objects. Consistently, the use of the Microsoft Kinect One sensor for the analysis of reaching movements is a reasonable (and cost-effective) choice.

During the trials, the experimental set-up was composed of:

- A Microsoft Kinect One sensor version 2.0, mounted on an easel and placed at about 2.0 m distance from the torso of the subject;
- A PC with Microsoft Windows 8.1, USB 3.0, and Microsoft Kinect One S.D.K. version 2.0;

Figure 18. Kinect One set-up

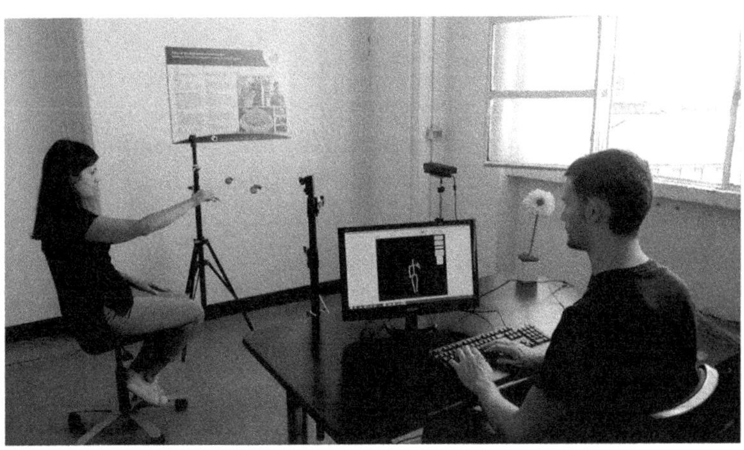

- A screen providing on-line visual feedback to the operator to visually monitor the correctness of the kinematic tracking.

Data were recorded and logged with an in-house C# developed software for visual feedback and data acquisition. Offline analysis was performed by the means of an in-house developed Matlab software.

3D joints tracking data of shoulder, elbow, wrist and time labels were recorded and logged for off-line analysis. Tracking data were low-pass Butterworth filtered, 3rd order, cut-off frequency 6Hz. Data were used to identify the forward and backward phases of the movements, and to compute the following parameters, according to the protocol presented and described in detail in previous studies (Scano, Caimmi, Chiavenna, Malosio, & Tosatti, 2015):

- Execution Time (T, s)
- Shoulder Elevation angle (SE, °)
- Elbow Flexion and Extension angle (EF, °)

The above listed parameters account for movement kinematics (timing and range of motion) at the end of the forward phase of the movement.

- Shoulder Elevation Torque (ST, Nm)
- Shoulder Effort Index (SEI, Nms)
- Elbow Elevation Torque (ET, Nm)
- Elbow Effort Index (EEI, Nms)

The above listed parameters account for movement dynamics (torques at the end of the movement, and cumulative torque during the forward phase). They were obtained by the means of a simple biomechanical model of the upper-limb accounting gravitational and inertial contributions to articular torques.

- Normalized Jerk (NJ, a-dimensional number)
- Coefficient of Periodicity (ACP, a-dimensional number)

The above listed parameters account for movement quality of execution, representing respectively the smoothness (Normalized Jerk) and the Repeatability of the Acceleration Profile (Coefficient of Periodicity) of the forward phase.

The protocol was designed to be a simplified, synthetic version, made of a selection of the parameters computed in clinical environments during robotic therapies.

In general, Shoulder evaluations mainly characterize the RM, while Elbow evaluations, instead, are related to the HtMM. This is relevant especially for healthy subjects that execute "correctly" the motor tasks. However, patients might show impaired motor functions and compensatory motor strategies (for example, to bring the hand to the mouth, they abnormally elevate the shoulder). As reference values for the assessment of abnormal motor strategies, both Shoulder and Elbow Evaluations were computed for RM and HtMM.

In the preliminary investigation, only motor control indexes were considered (NJ and ACP) since they are more sensible to measurement because they involve derivatives of position (see Discussion for details). A previous study (Caimmi et al., 2015) performed the same analysis on healthy people using a "gold standard" marker-based system, and was used as reference to evaluate Kinect One performances. Subjects were divided into four subsets (young, middle-aged, senior, and all). Regression curves were analyzed for both RM and HtMM, NJ and ACP, and for each age subset and considering all the subjects together, for a total of 32 regression curves (2 gestures x 2 limbs x 2 motor indexes x 4 subsets each). Regression curves were compared to the above mentioned study.

In the main investigation, subjects were clustered into three groups of data subsets: data subset 1) Dominancy: dominant/non dominant limb; 2) Gender: males/females; 3) Age: young/middle-aged/senior. For each of the three group, each subset was tested in comparison to others on each variable of the biomechanical assessment. Means and standard deviations were computed along with p-values of the comparisons.

Since the study is still undergoing, results will be reported briefly in the following paragraph. Most notable remarks are:

- Regressions (t, NJ) and (t, ACP) indicate that, the higher the execution time, the lower the repeatability and the lower the smoothness. The same results were found in (Caimmi et al., 2015) this indicates that, will losing sensibility, Kinect One can be considered as a valuable tool for assessing motor control indexes. In general, senior people have higher performance decline (higher regression slopes) in comparison to young adults and middle aged-people.
- Limb dominancy does not introduce remarkable different, probably because the investigated gestures are of easy execution;
- Gender affected the execution of HtMM gestures; especially with the non-dominant limb, women proved to be more "precise" in motor execution because they elevated less the shoulder, flexed more the forearm, with best motor control and less effort. Such a result required to distinguish male and female performances when creating references for neurological evaluations;
- Age affected many parameters and was the major source of differentiation. In fact, senior people are in general slower and execute less refined gestures. Young adults and middle-aged people are usually similar in performances.

Summarizing the results, Kinect V2 can be considered as a cost-effective solution for motor evaluation. Furthermore, in order to analyze patients' performances, healthy people reference data should not be considered as a unique group, but should be clustered at least considering gender and age.

Future Research Directions

The –arm line will be exploited in future experimental works for two main objectives. First, the devices will be tested on healthy people to verify their usability and applicability and therefore planning eventual modification of the initial prototype concept. Furthermore, some clinical studies will be conducted to assess the efficacy of the devices in the clinical environment with neurological impaired patients.

In particular, some future application have already be planned:

- LINarm application on post-stroke subjects. The final version produced at the end of Echord++ project already allows the possibility of different interaction modes which can be suitable for different level of impairment. The sensors placed on the handle and the patient model assure a good monitoring of user's performances reducing risks of unwanted reactions. The possibility to orientate the device grants, in addiction, to propose different movements in the workspace like a hand-to-mouth movement, already tested on healthy subjects, and therefore customize even more the therapy.
- PLANarm application on healthy subjects. The interaction of healthy subjects with the device in different control algorithms could test the mechanical design proposed leading to eventual modification of the concept.
- Development of new handles for the DUALarm device. The positive results on healthy subjects and patients have verified the applicability of the system and the usefulness of a bimanual approach. Other configurations could possibly expand the range of possibilities and movements available for a more tailored rehabilitation protocol, maintaining the core feature of easily switch from in-phase to anti-phase movements.
- An active version of the LIGHTarm with actuators for targeted joints and force sensors to detect the intention of movement or for assist-as-needed interactions. The muscular activation profiles acquired during preliminary sessions on healthy subjects could be exploited to generate a reference for patients' usage. The passive version, with some modifications for safety and easy wearing, could undergo a clinical trial with patients to verify the actual benefit of the weight compensation.
- A Kinect-based environment for the monitoring of the patient. The human tracking with low-cost devices widens a broad list of solutions for clinical evaluation either at hospital or at home. Coupling two or more Kinect sensors placed inside the house could allow the clinician to acquire data of the patients' lifestyle and therefore program future interventions.

CONCLUSION

The devices described in this chapter share the common feature of being low-cost, while capturing many of the basic functionalities needed to promote motor recovery. The expected outcome of the arm line of device is to support the development of reliable but affordable products that can be placed in almost every clinic, even in low and middle income countries, and reach also the houses of the people, where usually a robot device is too expensive to be bought. The smart design and the exploiting of emerging technologies such as additive manufacturing may be also an example for manufacturers and developers of solutions for home rehabilitation.

REFERENCES

Caimmi, M., Guanziroli, E., Malosio, M., Pedrocchi, N., Vicentini, F., Tosatti, L. M., & Molteni, F. (2015). Normative Data for an Instrumental Assessment of the Upper-Limb Functionality. *BioMed Research International*, *2015*, 1–23. doi:10.1155/2015/484131 PMID:26539500

Cameirão, M. S., Smailagic, A., Miao, G., Siewiorek, D. P., Sander, M., Oxlund, B., ... Brooke, J. (2016). Coaching or gaming? Implications of strategy choice for home based stroke rehabilitation. *Journal of Neuroengineering and Rehabilitation*, *13*(1), 18. doi:10.1186/s12984-016-0127-8 PMID:26921185

Cappozzo, A., Catani, F., Leardini, A., Benedetti, M. G., & Della Croce, U. (1996). Position and orientation in space of bones during movement: Experimental artefacts. *Clinical Biomechanics (Bristol, Avon)*, *11*(2), 90–100. doi:10.1016/0268-0033(95)00046-1 PMID:11415604

Cauraugh, J. H., & Summers, J. J. (2005). Neural plasticity and bilateral movements: A rehabilitation approach for chronic stroke. *Progress in Neurobiology*, *75*(5), 309–320. doi:10.1016/j.pneurobio.2005.04.001 PMID:15885874

Chin, J. H., & Vora, N. (2014). The global burden of neurologic diseases. *Neurology*, *83*(4), 349–351. doi:10.1212/WNL.0000000000000610 PMID:25049303

Di, S., Industriale, I., & Informazione, E. D. (2015). *Design and development of a planar cable-driven VSA based robot for upper-limb*. Academic Press.

Habib, M. K. (2011). Article. *Biomimetics: Innovations and Robotics, 4*(2), 113–134.

Hsieh, C.-Y., Chen, C.-H., Li, C.-Y., & Lai, M.-L. (2015). Validating the diagnosis of acute ischemic stroke in a National Health Insurance claims database. *Journal of the Formosan Medical Association, 114*(3), 254–9. 10.1016/j.jfma.2013.09.009

Jafari, A. (2014). Coupling between the Output Force and Stiffness in Different Variable Stiffness Actuators. *Actuators*, *3*(3), 270–284. doi:10.3390/act3030270

Krabben, T., Prange, G. B., Molier, B. I., Stienen, A. H., Jannink, M. J., Buurke, J. H., & Rietman, J. S. (2012). Influence of gravity compensation training on synergistic movement patterns of the upper extremity after stroke, a pilot study. *Journal of Neuroengineering and Rehabilitation*, *9*(1), 44. doi:10.1186/1743-0003-9-44 PMID:22824488

Krebs, H., Hogan, N., Volpe, B. T., Aisen, M. L., Edelstein, L., & Diels, C. (1999). Overview of clinical trials with MIT-MANUS: A robot-aided neuro- rehabilitation facility. *Technology and Health Care*, *7*, 419–423. Retrieved from http://www.scopus.com/inward/record.url?eid=2-s2.0-0033400738&partnerID=40&md5=1cfad3a95596bfa282229f338a7fb624 PMID:10665675

Laver, K., George, S., Thomas, S., Deutsch, J., & Crotty, M. (2012). Virtual reality for stroke rehabilitation. *Stroke*, *91*, 1–16. doi:10.1097/PHM.0b013e31826bce79.Comparison PMID:22713539

Laver, K. E., George, S., Thomas, S., Deutsch, J. E., & Crotty, M. (2015). Virtual reality for stroke rehabilitation. *Cochrane Database of Systematic Reviews*, (2): CD008349. doi:10.1002/14651858.CD008349. PMID:25927099

Legnani, G., Casalo, F., Righettini, P., & Zappa, B. (1996). A homogeneous matrix approach to 3D kinematics and dynamics - II. Applications to chains of rigid bodies and serial manipulators. *Mechanism and Machine Theory*, *31*(5), 589–605. doi:10.1016/0094-114X(95)00101-4

Legnani, G., Tosi, D., Fassi, I., Giberti, H., & Cinquemani, S. (2010). The "point of isotropy" and other properties of serial and parallel manipulators. *Mechanism and Machine Theory*, *45*(10), 1407–1423. doi:10.1016/j.mechmachtheory.2010.05.007

Lo, A. C., Guarino, P. D., Richards, L. G., Haselkorn, J. K., Wittenberg, G. F., Federman, D. G., ... Peduzzi, P. (2010). Robot-Assisted Therapy for Long-Term Upper-Limb Impairment after Stroke. *The New England Journal of Medicine*, *362*(19), 1772–1783. doi:10.1056/NEJMoa0911341 PMID:20400552

Lu, Q., Ortega, C., & Ma, O. (2011). Passive Gravity Compensation Mechanisms: Technologies and Applications. *Recent Patents on Engineering, 5*, 32–44. Retrieved from http://cat.inist.fr/?aModele=afficheN&cpsidt=24159482

Malosio, M., Caimmi, M., Cotti Cottini, M., Crema, A., Dinon, T., Mihelj, M., ... Spagnuolo, G. (2016). An affordable, adaptable, and hybrid assistive device for upper-limb neurorehabilitation. *Journal of Rehabilitation and Assistive Technologies Engineering*, *3*, 205566831668098. doi:10.1177/2055668316680980

Malosio, M., Corbetta, F., & Ram, F. (n.d.). *The kinematic architecture of a two-dof parallel and orthogonal variable stiffness actuator*. Academic Press.

Malosio, M., Pedrocchi, N., Vicentini, F., & Tosatti, L. M. (2011). Analysis of elbow-joints misalignment in upper-limb exoskeleton. *IEEE International Conference on Rehabilitation Robotics*, 1–6. doi:10.1109/ICORR.2011.5975393

Marchal-Crespo, L., & Reinkensmeyer, D. J. (2009). Review of control strategies for robotic movement training after neurologic injury. *Journal of Neuroengineering and Rehabilitation*, *6*(1), 20. doi:10.1186/1743-0003-6-20 PMID:19531254

Molinari-Tosatti, L., Fassi, I., & Legnani, G. (2003). Kineto-static optimisation of PKMs. *CIRP Annals - Manufacturing Technology, 52*(1), 337–341. 10.1016/S0007-8506(07)60596-4

Morasso, P., Casadio, M., Giannoni, P., Masia, L., Sanguineti, V., Squeri, V., & Vergaro, E. (2009). Desirable features of a "Humanoid" robot-therapist. *Proceedings of the 31st Annual International Conference of the IEEE Engineering in Medicine and Biology Society: Engineering the Future of Biomedicine, EMBC 2009*, 2418–2421. doi:10.1109/IEMBS.2009.5334954

Muratori, L. M., Lamberg, E. M., Quinn, L., & Duff, S. V. (2014). Article. *Extremity Rehabilitation*, *26*(2), 94–103. doi:10.1016/j.jht.2012.12.007.Applying

Petuskey, K., Bagley, A., Abdala, E., James, M. A., & Rab, G. (2007). Upper extremity kinematics during functional activities: Three-dimensional studies in a normal pediatric population. *Gait & Posture*, *25*(4), 573–579. doi:10.1016/j.gaitpost.2006.06.006 PMID:16875821

Rau, G., Disselhorst-Klug, C., & Schmidt, R. (2000). Movement biomechanics goes upwards: From the leg to the arm. *Journal of Biomechanics*, *33*(10), 1207–1216. doi:10.1016/S0021-9290(00)00062-2 PMID:10899329

Roux, E., Bouilland, S., Godillon-Maquinghen, A.-P., & Bouttens, D. (1999).. . *Evaluation of the GO Method Within the Upper Limb Kinematics Analysis*, *35*(1), 1–14.

Scano, A., Caimmi, M., Chiavenna, A., Malosio, M., & Tosatti, L. M. (2015). Kinect One-based biomechanical assessment of upper-limb performance compared to clinical scales in post-stroke patients. *Proceedings of the Annual International Conference of the IEEE Engineering in Medicine and Biology Society, EMBS, 2015*, 5720–5723. doi:10.1109/EMBC.2015.7319691

Scano, A., Caimmi, M., Chiavenna, A., Malosio, M., & Tosatti, L. M. (2017). A Kinect-Based Biomechanical Assessment of Neurological Patients' Motor Performances for Domestic Rehabilitation. *Virtual Reality Enhanced Robotic Systems for Disability Rehabilitation*, 252–279. 10.4018/978-1-4666-9740-9.ch013

Scano, A., Caimmi, M., Malosio, M., & Tosatti, L. M. (2014). Using Kinect for upper-limb functional evaluation in home rehabilitation: A comparison with a 3D stereoscopic passive marker system. *5th IEEE RAS/EMBS International Conference on Biomedical Robotics and Biomechatronics*, 561–566. doi:10.1109/BIOROB.2014.6913837

Alessandro Scano, Andrea Chiavenna, Matteo Malosio, and Lorenzo Molinari Tosatti, "Kinect V2 Performance Assessment in Daily-Life Gestures: Cohort Study on Healthy Subjects for a Reference Database for Automated Instrumental Evaluations on Neurological Patients," Applied Bionics and Biomechanics, vol. 2017, Article ID 8567084, 16 pages, 2017. doi:10.1155/2017/8567084

Scano, A., Spagnuolo, G., Caimmi, M., Chiavenna, A., Malosio, M., Legnani, G., & Tosatti, L. M. (2015). Static and dynamic characterization of the LIGHTarm exoskeleton for rehabilitation. *IEEE International Conference on Rehabilitation Robotics,* (1), 428–433. doi:10.1109/ICORR.2015.7281237

Schaefer, S. Y., & Hengge, C. R. (2016). Testing the concurrent validity of a naturalistic upper extremity reaching task. *Experimental Brain Research, 234*(1), 229–240. doi:10.1007/s00221-015-4454-y PMID:26438508

Schmidt, R. (1975). A Schema Theory of Discrete Motor Skill Learning. *Psychological Review, 82*(4), 225–260. doi:10.1037/h0076770

Sherwood, D. E., & Lee, T. D. (2003). Schema theory: Critical review and implications for the role of cognition in a new theory of motor learning. *Research Quarterly for Exercise and Sport, 74*(4), 376–382. doi:10.1080/02701367.2003.10609107 PMID:14768838

Spagnuolo, G., Malosio, M., Dinon, T., Tosatti, L. M., & Legnani, G. (2017). Analysis and synthesis of LinWWC-VSA, a Variable Stiffness Actuator for linear motion. *Mechanism and Machine Theory, 110*, 85–99. 10.1016/j.mechmachtheory.2016.11.002

Spagnuolo, G., Malosio, M., Scano, A., Caimmi, M., Legnani, G., & Tosatti, L. M. (2015). Passive and active gravity-compensation of LIGHTarm, an exoskeleton for the upper-limb rehabilitation. *IEEE International Conference on Rehabilitation Robotics, 2015*, 440–445. doi:10.1109/ICORR.2015.7281239

Swinnen, S. P. (2002). Intermanual Coordination: From Behavioural Principles To Neural-Network Interactions. *Nature Reviews. Neuroscience, 3*(5), 348–359. doi:10.1038/nrn807 PMID:11988774

Van Delden, A. E. Q., Peper, C. E., Kwakkel, G., & Beek, P. J. (2012). A systematic review of bilateral upper limb training devices for poststroke rehabilitation. *Stroke Research and Treatment, 2012*, 1–17. doi:10.1155/2012/972069 PMID:23251833

Vanderborght, B., Albu-Schaeffer, A., Bicchi, A., Burdet, E., Caldwell, D. G., Carloni, R., ... Wolf, S. (2013). Variable impedance actuators: A review. *Robotics and Autonomous Systems*, *61*(12), 1601–1614. doi:10.1016/j.robot.2013.06.009

Vitiello, N., Lenzi, T., Roccella, S., De Rossi, S. M. M., Cattin, E., Giovacchini, F., ... Carrozza, M. C. (2013). NEUROExos: A powered elbow exoskeleton for physical rehabilitation. *IEEE Transactions on Robotics*, *29*(1), 220–235. doi:10.1109/TRO.2012.2211492

World Health Organization. (2002). Towards a Common Language for Functioning, Disability and Health ICF. *International Classification, 1149*, 1–22.

Zhang, S., Guo, S., Pang, M., Gao, B., & Guo, P. (2014). *Mechanical Design and Control Method for SEA and VSA-based Exoskeleton Devices for Elbow Joint Rehabilitation*. Academic Press.

Zhou, X., Jun, S. K., & Krovi, V. (2014). Stiffness modulation exploiting configuration redundancy in mobile cable robots. *Proceedings - IEEE International Conference on Robotics and Automation*, 5934–5939. doi:10.1109/ICRA.2014.6907733

KEY TERMS AND DEFINITIONS

Arm: Line of devices for rehabilitation developed by ITIA.

Assistance: Paradigms of interaction that robot implement to facilitate motor recovery.

Biomechanical Evaluation: Evaluation of patients' motor capability with the techniques of biomedical and mechanical engineering.

End-Effector: Device for rehabilitation that requires the holding of a handle for interaction.

Exoskeleton: Device for rehabilitation that wraps patients' body.

Neurological Patients: Patients with injury or impairment due to neurological diseases.

Rehabilitation: Process for the recovery of motor function.

Robots: Devices for promoting the neuro-motor rehabilitation process.

Variable Stiffness Actuators: Actuators that can change their stiffness to meet patients' rehabilitative needs.

Chapter 8
Brachytherapy Needle Steering Guidance Using Image Overlay

Carlos Rossa
University of Alberta, Canada

McNiel Inyani Keri
University of Alberta, Canada

Mahdi Tavakoli
University of Alberta, Canada

ABSTRACT

This chapter presents a physical simulator for needle steering in brachytherapy. As the user inserts the needle in a phantom tissue, images of the needle and prostate shape reconstructed from 2D transverse ultrasound images are displayed online in a semi-transparent mirror. During insertion, the user sees the images as if they were floating inside the phantom accounting for scale and orientation. The ultrasound images of the needle are combined with a needle-tissue interaction model that predicts the needle deflection further along the insertion process. The necessary manoeuvres that bring the needle towards its intended target location are displayed to the user along with the actual needle location. This platform allows the user to test different manual and robotic-assisted needle steering techniques. Reported experimental results confirm the accuracy of the system in reconstructing and overlaying images onto the phantom.

1. MOTIVATION

Percutaneous needle insertion is a widespread minimally invasive surgical intervention. Applications of needle insertion include brachytherapy cancer treatment, tissue biopsy sampling, neurosurgery, radio-frequency ablation, and drug delivery. In brachytherapy, for instance, needles loaded with tiny radioactive seeds are inserted in the patient's prostate. Once the needles are fully inserted, the needles are pulled back to permanently leave the seeds in their desired destinations. As a consequence, controlling the radiation dose that will destroy cancer cells without affecting adjacent healthy tissues critically depends on the accurate guidance of the needle towards a body target.

DOI: 10.4018/978-1-5225-2993-4.ch008

Due to the lack of sufficient feedback during insertion and the limited steerability of needles, needle placement during brachytherapy is usually planned with needles assumed to follow a straight-line trajectory. In practice, this assumption does not hold true as the needle deviates from this path. For a comprehensive survey on issues related to needle steering, see Rossa (2017). In turn, inserting and retracting needles in tissue causes the tissue to move and deform, posing a strong risk of inadvertent target misses. To deal with these challenges, ultrasound images of the needle within the tissue are typically used to guide the needle towards the target. Ultrasound images often contain artifacts that are hard to identity and distinguish from targets and needles. In addition, to view the ultrasound images, the surgeon must look away from the patient and towards a monitor while manoeuvring the needle, introducing an additional challenge in interpreting the images.

In order to improve the visualization of needles and targets during clinical interventions, augmented reality (AR) has been the focus of significant research. Image processing combined with augmented reality has the ability to project onto real world images, reconstructed images of the inner body acquired from any medical imaging modality Weiss (2011). An augmented reality display adds an extra layer of virtual information on top of the perception of the real world in real time, making many surgical tasks simpler and safer for the surgeon. The simplest AR system consists of a semitransparent mirror through which the viewer looks directly at the patient Nikou (2000). An image is projected onto the mirror by a standard monitor so that the viewer simultaneously sees a reflection of the computer display that appears with correct orientation and scale and the real world image Blackwell (2000). In brachytherapy, this can be used as a training environment to help user develop needle steering skills based on the ability to visualize needles and anatomical structures within the tissue in an intuitive way.

Related applications of AR for needle insertion guidance include arthrography Fichtinger (2005), Westwood (2006), ultrasound guided needle placement training Magee (2007), Zhu (2006), surgical laparoscopy Wacker (2005), magnetic resonance image guided biopsy Wacker (2006), liver puncture Nicolau (2005) and ablation Nicolau (2009), Teber (2009), and computed tomography Stetten (2001), Stetten (2001b), Sauer (2002). In Fischer (2007), the AR system identifies the optimal needle insertion path such that the surgeon can follow a desired needle insertion trajectory under the assumption that the needle will remain unbent during insertion. This chapter introduces the first implementation of an AR system for skills development for brachytherapy needle steering. The system displays reconstructed images of the actual shape of a needle in the tissue, as well as the location of the prostate phantom and the target locations in real time. In addition to measuring the actual needle deflection, ultrasound image processing is combined with a needle-tissue physical model Rossa (2016) that predicts the needle deflection further along the insertion process. The necessary manoeuvres that bring the needle towards its intended target location are then displayed to the user online.

This chapter is structured as follows. After we get acquainted with the proposed physical system in Section II, we will see how several transverse ultrasound images are combined to reconstruct the actual 3D shape of a prostate phantom embedded in a synthetic tissue. Next, the needle- tissue model presented in Rossa (2016) is used to predict needle deflection during the insertion process. Once we have reconstructed the prostate geometry and the needle trajectory, we will proceed with image registration to ensure the prostate and needle will appear to the user with correct orientation and scale floating inside the patient. Finally, in Section IV, experimental results confirm the ability of the proposed system in overlaying the reconstructed images.

2. THE AUGMENTED REALITY SYSTEM

The augmented reality system for needle steering training is schematically depicted in Figure 1. The system is composed of a tissue phantom that embeds a synthetic prostate phantom. Whilst the user inserts an 18-gauge, 200 mm long, brachytherapy needle through a standard brachytherapy template grid, an ultrasound probe acquires 2D images of the tissue, in a plane normal to the straight needle (the (x, y) plane of Figure 1). A linear stage connected to a stepper motor by a belt-pulley mechanism controls the position of the ultrasound probe such that it slides back and forth above the tissue (along z). Template grids consist of equidistant rows and columns of holes spaced 5 mm apart. In prostate brachytherapy, the needle is inserted through a hole and the corresponding position is marked on an overly in the calibrated ultrasound image, which corresponds to a target location in the prostate. These transverse ultrasound images show a cross section of an object inside the tissue. By combining several transverse images acquired at different positions, the 3D volume of the objects can be reconstructed. This simulates an operating room setup where a transrectal ultrasound probe is placed below the pelvic tissues of inter-

Figure 1. The training system. As the user inserts a needle into a phantom, an ultrasound probe moves above the tissue to acquire 2D images. The reconstructed images of the needle and of a prostate phantom embedded in the tissue are displayed in the monitor and reflected in the semi-transparent mirror. The images will then appear to the user to be floating inside the tissue with proper size and location. A motion tracker measures the position of the mirror, the monitor, the needle base, and the user's eyes.

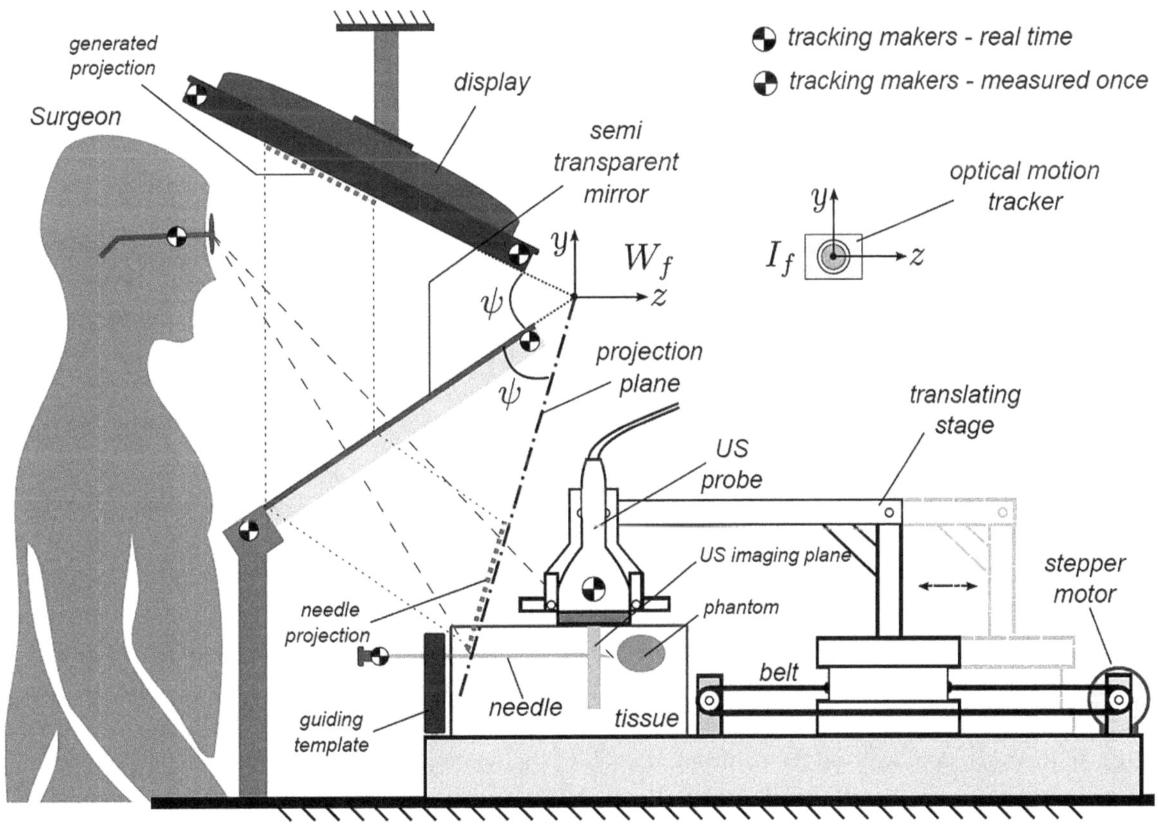

est. The techniques presented here can be employed in the clinical context, as they only require the current position of the ultrasound probe.

As the user inserts the needle, we combine the 2D ultrasound images of the needle and of the target and reconstruct their 3D images, adjust their magnification, appropriately rotate the images and then render them on a monitor placed above the tissue phantom. These images are reflected in the semitransparent mirror located between the user and the phantom. The user sees the images generated by the monitor as if they were floating behind the mirror in the so-called projection plane. The goal of the system is to match the projected image and the real objects inside the phantom with the user's viewpoint such that they appear to the user to be inside the phantom with proper location and size.

To this end, one needs to generate the images to be displayed to the user, and then calculate their position in the monitor. We begin by placing the images obtained in the ultrasound frame in the real world frame. In other words, we need to calculate their relative position in the setup. Next, we calculate the image points in the projection plane by tracing a straight line from the user's eye to a particular point in the image, called a perspective line. Notice that as shown in Figure 1, the image in the projection plane has the same size as it appears in the monitor. Therefore, the corresponding points in the monitor image are found when the perspective lines intercept the projection plane.

To perform these steps, one needs to measure the position of the user's eyes, of the needle base, and of the ultrasound probe online. As well, the orientation of the mirror and of the display must be known. Tracking markers are placed in each of the above referenced objects whose position is measured at by an optical motion tracker placed on the side of the setup. In order to measure the position of the user's eyes, the user wears glasses without lenses that are equipped with two tracking markers on the right side.

In the following, we describe the three parts that constitute the augmented reality system, i.e., 1) the ultrasound-based reconstruction of the target and prostate phantom, 2) estimation of future needle deflections, and 3) the rigid body transformation that generates the images of 1) and 2) in the display.

2.1 Segmentation of Prostate Phantom

The first step in running the augmented reality system is to scan the tissue phantom with the ultrasound probe. The tissue phantom used in the setup is made of plastisol gel (M-F Manufacturing, Fort Worth, USA) with a volume ratio of four parts of plastisol to one part of plastic softener (see Figure 2(a)). To simulate the target location, we embedded in this tissue a prostate phantom fabricated with a custom made 3D printed mould (see Figure 2(b)), using a mixture with a higher concentration of plastisol i.e., 6 to 1. In order to mimic the ultrasound imaging characteristics of a human prostate, 10 grams of silica gel are added into the 0.5 litre mixture during fabrication of the prostate phantom.

The ultrasound probe is controlled to slide above the tissue phantom while acquiring successive transverse images at every 0.15 millimetres increment. Each image will show a cross section of the prostate phantom. The image processing sequence used to find the contour of the prostate for each of the acquired transverse images is shown in Figure 3. First, the original image shown in (a) is filtered using a 2D Gaussian low pass filter and results in the image (b). The generated kernel is then convolved with the original image in order to sharpen the edges of the object in the image. After stretching the pixel intensity range in the gray scale to 0 to 255, a 45% image thresholding generates the binary image seen in (c). In the next step, we find the centroid of each of the white regions shown in (c) using the Matlab function *regionprops*. An elliptically shaped morphological structuring element is then created to fit each of these regions in order for one to calculate their approximate area. The biggest area is assumed

Figure 2. Tissue phantoms. The prostate phantom in (right), fabricated using a 3D printed mould is embedded in the phantom tissue shown in the right. Both tissues are made of plastisol gel with different stiffness.

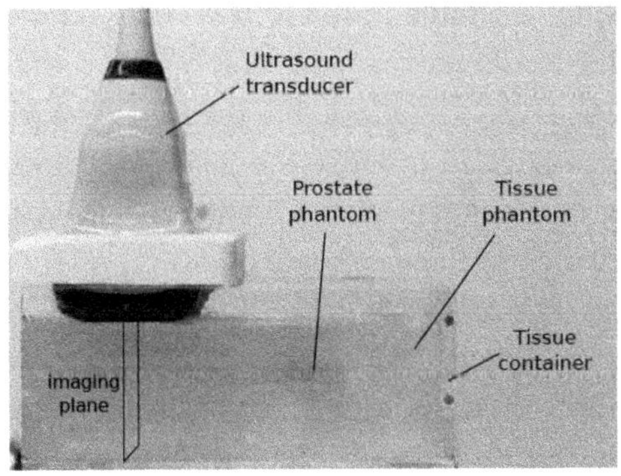

Figure 3. Image processing steps used to segment the prostate phantom. (a) Original transverse image, (b) filtered image using a 2D Gaussian low pass filter, (c) binary image after convolution and thresholding, (e) obtained contours after filling the small white regions in the image, (e) extracted contour, and (f) interpolated contour on the original image (only used for visual verification).

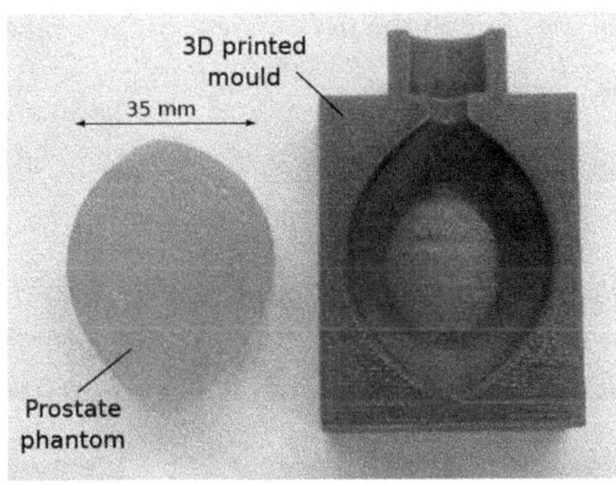

to correspond to the prostate shape and the other regions are discarded. The result is shown in (d). Next, the Matlab function *bwperim* returns the contour of the retained image (see (e)), which is converted into polar coordinates and interpolated in order to close the contour. The final result overlaid with the original image is shown in (f). On a 6 GB of RAM with Intel Core i3 3.3 GHz computer, the processing time for each of the ultrasound images is 85 milliseconds.

By combining the successive prostate contours with the corresponding measured position of the ultrasound probe, the 3D volume of the prostate phantom is reconstructed.

2.2. Estimating Future Needle Deflection

Once the prostate phantom is found using the steps described in the previous section, the ultrasound probe moves back to the point where the needle enters the tissue and is controlled to follow the position of the needle tip as the needle is inserted. In each of the ultrasound images, the needle tip appears as a bright spot along with extraneous background objects. To remove the majority of these objects, we first define a region of interest (ROI) that limits the search for the needle to a small section of the image (see Figure 4(a)), initially located at the needle's entry point in the tissue. The ROI consists of a square with a width of 5 times the needle radius. In frame f, the ROI is centred on the position of the needle found in frame $f-1$. Therefore, as the needle moves transversely in successive ultrasound images, the ROI moves such that the needle tip will always be located in the ROI. The Matlab function *imadjust* applies an intensity transformation to the image to improve the visibility of bright points. Next, we use an intensity threshold to obtain a set of candidate pixels for the needle within the axial image. Finally, a Kalman filter is used to predict which of the candidate pixels represent the actual needle tip within the ROI, given the needle's historical trajectory, similar to Zhao (2012).

The measured needle deflection is then used in a mechanical model of the needle-tissue system that calculates the future needle tip deflection, given the past positions of the needle tip Rossa (2016). Two cases are considered. In the first case, the model calculates the needle deflection as the user inserts it until a desired depth in the tissue. In the second case, the deflection is calculated considering that the user rotates the needle base by 180 around its shaft, at the current insertion depth, with the aim of reversing the direction of the needle deflection. Both cases are then displayed to the user. An example is shown in Figure 4(b). Combined, the image processing and the needle deflection estimation are performed in less than 60 milliseconds on a 6 GB of RAM with Intel Core i3 3.3 GHz computer.

Now that images have been acquired and processed, the objective is to find their position in the monitor such that their projection will match the user's view of the real objects. This is described in the next section.

Figure 4. Image processing and needle deflection estimation. In the first figure, the different steps used to find the needle tip in each transverse image. In second figure, the obtained deflection is used in a model to predicted future deflections of the needle tip as the needle is inserted and rotated at the current depth.

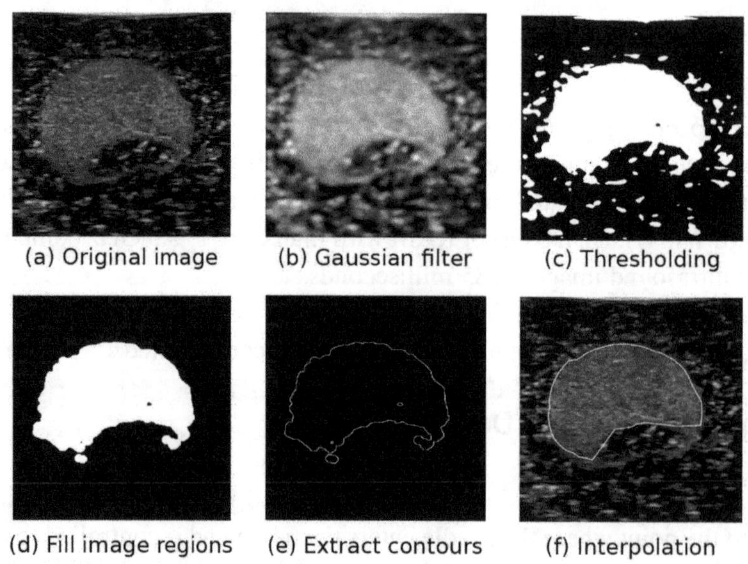

3. IMAGE REGISTRATION

In the following, M^N defines a column vector M, which has coordinates m_x^N, m_y^N and m_z^N along the x, y, and z axes, respectively, as expressed in the frame N. The frame I_f is the frame of the optical tracker. To determine the relative position of the semi-transparent mirror and of the display shown in Figure 1, we place i tracking markers on the side of each object whose position P_i is measured by the motion tracker in the frame I. The point P_1 will be considered to be the origin of the display frame, called D_f (see Figure 5).

The 3D equations of the semi-transparent mirror and of the display planes can be written as $y = m_m^I z + b_m^I, \forall x$ and $y = m_d^I z + b_d^I, \forall x$, respectively, where m_d^I and m_m^I are the plane slopes, and b_m^I and b_d^I are the intersection points of each plane with the y^I axis. As functions of the tracking markers position p_i^I, m and b are computed as follows:

$$\begin{aligned} m_d^I &= \frac{p_{2y}^I - p_{1y}^I}{p_{1z}^I - p_{2z}^I},\ b_d^I = p_{2y}^I - p_{2z}^I m_d, \\ m_m^I &= \frac{p_{4y}^I - p_{3y}^I}{p_{3z}^I - p_{4z}^I},\ b_m^I = p_{4y}^I - p_{4z}^I m_m, \end{aligned} \tag{1}$$

For convenience, we will define a second frame, called the world frame W_f, whose origin coincides with the intersection of the mirror and display planes, as shown in Figure 5. We assume that the two planes are not rotated with respect to each other around z z and y y, and that the coordinate of the world frame along x is arbitrarily chosen to be p_{1x}^I. By setting the equations of the mirror and display planes to be equal, one finds the position of the world frame, as expressed in the motion tracker frame, as

$$W_f^I = \begin{bmatrix} p_{1x}^I & \frac{b_m - b_d}{m_b - m_m} & \frac{m_d b_m - m_m b_d}{m_d - m_m} \end{bmatrix}^T. \tag{2}$$

Figure 5. Rigid transformations. The point $\boldsymbol{S^W}$ $\boldsymbol{S^W}$ *obtained from ultrasound images is found in the projection plane as a point* $\boldsymbol{J^W}$ $\boldsymbol{J^W}$, *which corresponds to the intersection of a line connecting the observer's eye* $\boldsymbol{E^W}$ $\boldsymbol{E^W}$ *to* $\boldsymbol{S^W}$ $\boldsymbol{S^W}$. *Dashed crosses indicate known points measured by the optical tracker. If and Wf are the origins of the motion tracker frame and world frame, respectively.*

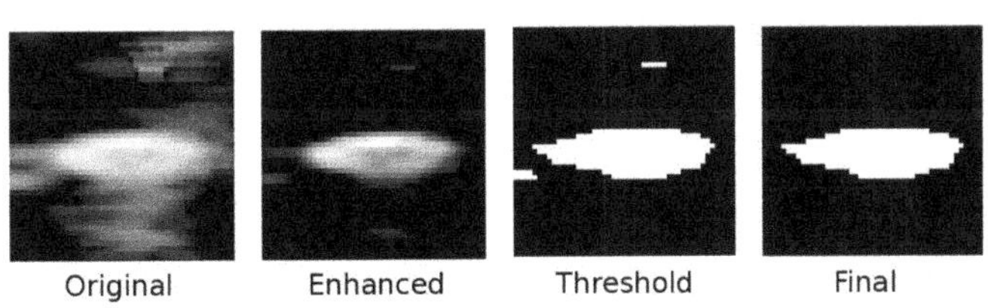

Thus, the coordinates of the tracking markers in the world frame are calculated as $P_1^W = P_1^I - W_f^I$, $P_2^W = P_2^I - W_f^I$, $P_3^W = P_3^I - W_f^I$, and $P_4^W = P_4^I - W_f^I$. With the world frame and the reference points defined, we can proceed with the registration of the processed ultrasound images.

The ultrasound probe acquires 2D images (in the (x, y) plane) that are to be express in the 3D world frame. Let us begin the image registration by scaling the ultrasound images. The 2D ultrasound image frame U_f is located at the upper-right corner of the ultrasound imaging plane such that it is always parallel to the (x, y) plane in W_f. A point S^U in an ultrasound image has coordinates s_x^U and s_y^U in pixels along the x and y axes, respectively. The number of pixels in each axis is px_{Ux} and px_{Uy}. Let v_{Ux} and v_{Uy} be the ultrasound image field of view in meters along each axis, respectively. The number of pixels and the field of view size are constant and are known from the ultrasound machine settings. The scaling factor in x and y that converts a point from pixels to meters is given by $f_{Ux} = v_{Ux} / px_{Ux}$ and $f_{Uy} = v_{Uy} / px_{Uy}$.

Now that the magnification ratios of the image are known, the points in the ultrasound frame can be translated to the world frame. The 3D position of the ultrasound probe measured by the optical tracker is $R^I = \left[r_x^I \; r_y^I \; r_z^I\right]^T$, which can be expressed in the world frame as $R^W = R^I - W_f$. In the setup, R^I is only measured once. When the probe moves, its location along z with respect to p_z^W is measured by a potentiometer in real time, and it is called p_0. Finally, the position S^W in the world frame of a point S^U is given by:

$$S^W = \begin{bmatrix} f_{Ux} & 0 & 0 \\ 0 & f_{Uy} & 0 \\ 0 & 0 & 0 \end{bmatrix} . \begin{bmatrix} s_x^U \\ s_y^U \\ 0 \end{bmatrix} + \begin{bmatrix} r_x^W \\ r_y^W \\ r_z^W \end{bmatrix} + \begin{bmatrix} 0 \\ 0 \\ p_0 \end{bmatrix} \tag{3}$$

Having converted the ultrasound point to the world frame, the next step is to calculate where the projection plane is. Let us call y the angle between the display and mirror planes with respect to x^W. The projection plane will appear to the user to be floating behind the mirror, rotated from the mirror plane around x^W by ψ. The angle ψ is:

$$\psi = \arctan\left[\frac{m_d^W - m_m^W}{1 + m_d^W m_m^W}\right] \tag{4}$$

where m_d^W, and m_m^W are calculated by inserting the reference points P_i^W in (1). Now, the projection plane is obtained by rotating the display plane by 2ψ degrees clockwise about the axis x^W. The rotation matrix is

$$R_x = \begin{bmatrix} 1 & 0 & 0 \\ 0 & \cos(2\psi) & \sin(2\psi) \\ 0 & 0 - \sin(2\psi) & \cos(2\psi) \end{bmatrix} \tag{5}$$

The coordinates of the rotated points are then given by $R_x P_1^W$ and $R_x P_2^W$. As for the mirror and for the display, the projection plane can be expressed as a first order polynomial in the form $y = m_p^W z + b_p^W$ where m_p^W and b_p^W $\forall x$, are given as functions of the rotated points as

$$m_p^W = \frac{\sin(2\psi)(P_{2y}^W - P_{1y}^W)}{\cos(2\psi)(P_{1y}^W - P_{2y}^W) + \sin(2\psi)(P_{1z}^W - P_{2z}^W)}, \tag{6}$$

$$b_p^W = -\sin(2\psi)P_{1y}^W + \cos(2\psi)P_{1z}^W - m_p^W(\cos(2\psi)P_{1y}^W + \sin(2\psi)P_{1z}^W)$$

The parametric equation of the perspective line $\ell(t)^W$ that connects the observer's eye E^W to a point S^W in the ultrasound image is:

$$\ell(t)^W = \begin{bmatrix} te_x + s_x(1-t) & te_y + s_y(1-t) & te_z + s_z(1-t) \end{bmatrix}^T \tag{7}$$

After straightforward manipulation, we find that the perspective line ℓ^W intercepts the projection plane $y = m_p^W z + b_p^W$ when $t = t_p$, where t_p is

$$t_p = \frac{s_y^W - m_p^W s_z^W - b_p^W}{m_p^W(e_z^W - s_z^W) + s_y^W - e_y^W} \tag{8}$$

Thus, the point that must be generated in the projection plane in order to match the user's angle of view with the actual point seen in the ultrasound image is $J^W = \ell(t_p)^W$.

The last transformation will convert the point J^W found in the projection plane (world frame) to the display frame. As for the ultrasound images, let px_{Dx} and px_{Dz} be the number of pixels in the x and z axes of the display, and v_{Dx} and v_{Dz} be the display image field of view in meters along each axis. The scaling factor in x and z is given by $f_{Ux} = v_{Ux} / px_{Ux}$ and $f_{Uz} = v_{Uz} / px_{Uz}$. The display and monitor are only rotated around x, and therefore have the same scaling. In order to convert a point from the projection plane to the display frame, one simply accounts for the distance between the display and the origin of the world frame. As is has been defined, this distance is $J^W - R_x P_1^W$. Thus, the coordinates of a point J^W in pixels, in the display frame is:

$$J^D = \begin{bmatrix} 1/f_{Dx} & 0 & 0 \\ 0 & 0 & 0 \\ 0 & 0 & 1/f_{Dz} \end{bmatrix} . (J^W - R_x P_1^W) \tag{9}$$

4. HARDWARE IMPLEMENTATION AND RESULTS

The prototype of the system is shown in Figure 6(a). A 4DL14-5/38 linear 4D ultrasound probe connected to a Sonix Touch ultrasound machine (Ultrasonix, Richmond, CA) is linked to a linear stage motorized by a stepper motor. A linear potentiometer (LP-250FJ from Midori Precisions, Tokyo, JP) measures the absolute position of the ultrasound probe in real time. The semi-transparent mirror is made of a 5 mm thick standard piece of glass covered by a 40% reflective tint film. Two computers running Matlab in xPC real-time mode are used in the setup. Computer 2 receives measurements via UDP of the current position of the ultrasound probe, and uses a digital PID regulator in order to control the horizontal position of the ultrasound-imaging plane. The first computer runs the image processing and rigid transformation software. As the first computer moves the ultrasound probe; images generated in the ultrasound machine are captured by a frame grabber and sent to Computer 1 (host), where they are processed. In addition, Computer 1 receives real-time measurements of tracking markers from the optical motion tracker (BB2-

Figure 6. Prototype of the augmented reality system for needle insertion. In the top figure, the implementation of the system is shown with our existing needle insertion robot. The needle and the ultrasound probe are connected to two independent linear stages. Two computers are used to run the setup as depicted in the second figure. The user looks at the phantom tissue through the mirror from the left side and controls the needle insertion robot based on the provided visual feedback.

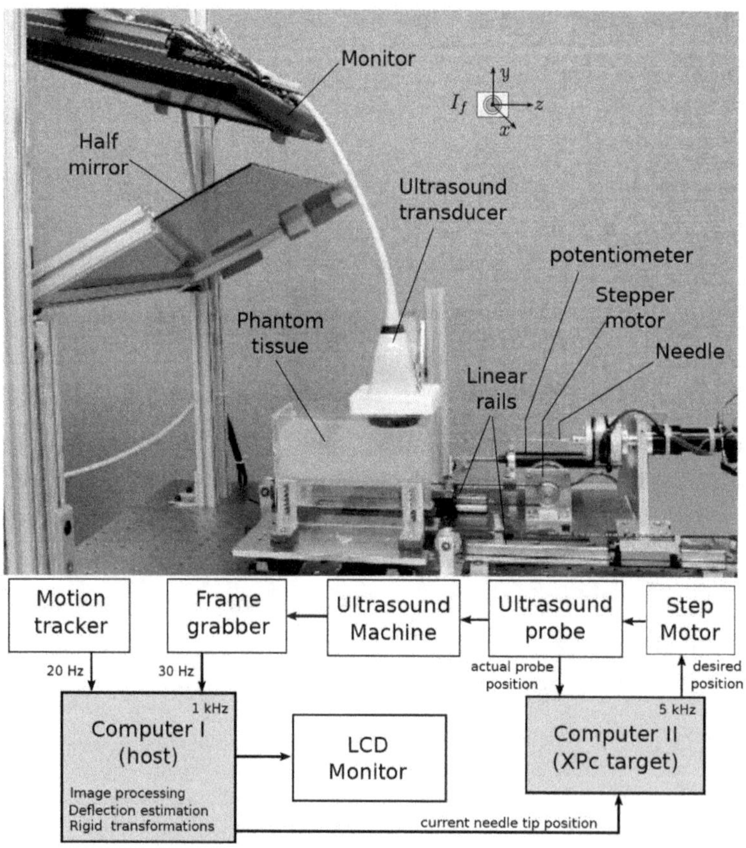

BW-Hx60 from Claron Tech, Toronto, CA) and generates the desired projection of the objects that are displayed in the monitor (see 6(b)).

For simplicity, the needle is attached to a second linear stage that rotates the needle base such that the deflection only occurs on the (z, y) plane. Notice that the needle enters the tissue on the right side of the container as shown in Figure 6(a), while the user looks at the tissue from the left side. In a different configuration, tracker markers connected to the needle base would allow the user to be in charge of the needle insertion. Notice also that the image registration routine is independent from the needle's entry point in tissue. Thus, inserting the needle on the same side as the mirror can be performed with only minor changes in the needle physical model. In the current configuration, the user is able to manually control the needle insertion robot based on the visual feedback received from the augmented reality display.

4.1 Experimental Results

An example of the images seen by the user through the mirror is shown in Figure 7(a). After the ultrasound probe has scanned the tissue and the image of the prostate is reconstructed, a horizontal cross section of the phantom volume, in the (x, z) plane placed at the same height along y y as the needle's base, is displayed to the user (see Figure 7(a)). The needle is then inserted to 5 mm in the phantom tissue. Based on the measured needle deflection by the ultrasound probe, future needle deflections are calculated using Rossa (2016). In Figure 7(b), the line projected to the left shows the predicted future deflection of the needle tip if it is further inserted to reach a depth of 150 mm. The user can reverse the direction of the deflection by rotating the needle base by 180° at the current depth. The curved line projected to the right indicates the resultant path. In Figure 7(c), the insertion is carried out until the needle reaches 40% of the desired insertion depth. The updated estimation of the needle deflection, based on the historical position of the tip and of the online-estimated local mechanical properties of the tissue, indicates to the user that the needle must be rotated at the current depth in order for the needle to reach the target. Please see the enclosed video for the complete insertion.

During insertion, the position of the user's head might change. In order for the images to match the user's viewpoint, the projected images are recalculated given the current position of tracker markers connected to glasses that the user wears. Figure 7(d) shows the adjusted images when the user's head moves to the right by approximately 200 mm as compared to the position of Figure 7(b). This demonstrates that the system is able to automatically compensate for the position of the user's head online.

5. DISCUSSION AND CONCLUDING REMARKS

In this paper we present the first implementation of an augmented reality system for skills development in needle steering during brachytherapy. As the needle is inserted, an ultrasound probe follows the needle tip to measure its deflection. This information is used in a needle-tissue interaction model that calculates the future needle tip deflection given the current and past measurements, and updated local mechanical properties of the phantom Rossa (2016). The system then displays the predicted needle tip deflection considering two cases. In the first case, the user keeps inserting the needle without performing any steering manoeuvres. The second case calculates the future deflection of the needle assuming the user rotates the needle base around its shaft by 180 degrees at the current insertion depth. As both

Figure 7. Images as seen by the user. In (a) a cross section of the phantom is shown in the plane that contains the needle. In (b) the needle is inserted by 5 mm in the tissue and the future deflections are calculated by the needle-tissue model. In (c) the needle is inserted to 40% of the maximum insertion depth and the estimated deflection is updated. In (d), the user moves his/her head to the right and the images are readjusted.

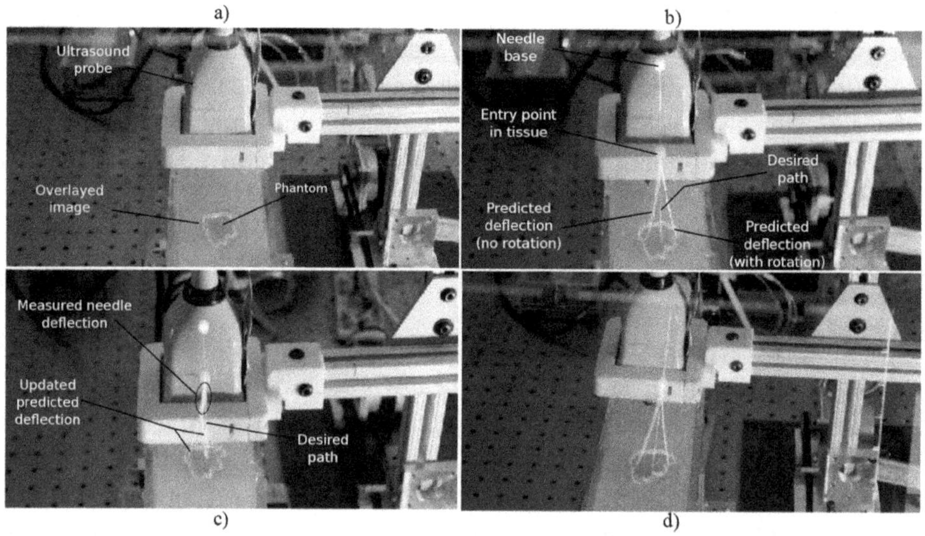

cases are displayed simultaneously to the user, the optimal depth at which the needle must be rotated in order to reach the target can be easily identified during the insertion process.

In the current configuration, the needle is connected to an existing needle insertion robot. This provides the user with a training environment where he can control the insertion robot actions, i.e., insertion depth and needle axial rotation, based on the provided visual feedback. If the user is manually inserting the needle, one simply needs to measure the position of the user's hand. This can be done in various different ways, including a wearable wristband with an embedded inertial measurement unit, or using additional optical trackers. Inserting the needle on the same side as the mirror can be done with only minor changes in the needle model.

The system is easy to implement, and is cost-effective. The added information supplied by this device allows the use to develop needle steering skills in either manual or robotic-assisted brachytherapy. Future efforts will focus on tracking the movements of implanted seeds to provide an accurate map of the seeds distribution and to evaluate the efficiency of the proposed system as a training environment for prostate brachytherapy.

REFERENCES

Blackwell, M., Nikou, C., DiGioia, A. M., & Kanade, T. (2000). An image overlay system for medical data visualization. *Medical Image Analysis*, *4*(1), 67–72. doi:10.1016/S1361-8415(00)00007-4 PMID:10972322

Fichtinger, Deguet, Masamune, Balogh, Fischer, Mathieu, … Fayad. (2005, August). Image overlay guidance for needle insertion in CT scanner. *Biomedical Engineering. IEEE Transactions on*, *52*(8), 1415–1424.

Fischer. (2007). MRI image overlay: Application to arthrography needle insertion. *Computer Aided Surgery*, *12*(1), 2–14. PMID:17364654

Magee, D., Zhu, Y., Ratnalingam, R., Gardner, P., & Kessel, D. (2007). An augmented reality simulator for ultrasound guided needle placement training. *Medical & Biological Engineering & Computing*, *45*(10), 957–967. doi:10.1007/s11517-007-0231-9 PMID:17653784

Nicolau, S., Soler, L., Mutter, D., & Marescaux, J. (2011). Augmented reality in laparoscopic surgical oncology. *Surgical Oncology*, *20*(3), 189–201. doi:10.1016/j.suronc.2011.07.002 PMID:21802281

Nicolau, S. A., Pennec, X., Soler, L., Buy, X., Gangi, A., Ayache, N., & Marescaux, J. (2009). An augmented reality system for liver thermal ablation: Design and evaluation on clinical cases. *Medical Image Analysis*, *13*(3), 494–506. doi:10.1016/j.media.2009.02.003 PMID:19282234

Nicolau. (2005). A complete augmented reality guidance system for liver punctures: First clinical evaluation. *Medical Image Computing and Computer-Assisted Intervention*, 539–547.

Nikou, C., Digioia, A. M. III, Blackwell, M., Jaramaz, B., & Kanade, T. (2000). Augmented reality imaging technology for orthopaedic surgery. *Operative Techniques in Orthopaedics*, *10*(1), 82–86. doi:10.1016/S1048-6666(00)80047-6

Rossa, C., & Tavakoli, M. (2017). Issues in closed-loop needle steering. *Control Engineering Practice*, *62*, 55–69. doi:10.1016/j.conengprac.2017.03.004

Rossa. (2016). Adaptive quasi-static modelling of needle deflection during steering in soft tissue. *IEEE Robotics and Automation Letters*.

Sauer. (2002). An augmented reality navigation system with a single-camera tracker: System design and needle biopsy phantom trial. *Medical Image Computing and Computer-Assisted intervention*, 116–124.

Stetten, G. D., & Chib, V. S. (2001). Overlaying ultrasonographic images on direct vision. *Journal of Ultrasound in Medicine*, *20*(3), 235–240. doi:10.7863/jum.2001.20.3.235 PMID:11270527

Stetten. (2001b). Real time tomographic reflection: phantoms for calibration and biopsy. *Augmented Reality, 2001. Proceedings. IEEE and ACM International Symposium on*, 11–19.

Teber, D., Guven, S., Simpfendörfer, T., Baumhauer, M., Güven, E. O., Yencilek, F., ... Rassweiler, J. (2009). Augmented reality: A new tool to improve surgical accuracy during laparoscopic partial nephrectomy? Preliminary in vitro and in vivo results. *European Urology*, *56*(2), 332–338. doi:10.1016/j.eururo.2009.05.017 PMID:19477580

Wacker, F. K., Vogt, S., Khamene, A., Jesberger, J. A., Nour, S. G., Elgort, D. R., ... Lewin, J. S. (2006). An augmented reality system for MR image–guided needle biopsy: Initial results in a swine model 1. *Radiology*, *238*(2), 497–504. doi:10.1148/radiol.2382041441 PMID:16436814

Weiss, C. R., Marker, D. R., Fischer, G. S., Fichtinger, G., Machado, A. J., & Carrino, J. A. (2011). Augmented reality visualization using image overlay for MR-guided interventions: System description, feasibility, and initial evaluation in a spine phantom. *AJR. American Journal of Roentgenology*, *196*(3), W305–W307. doi:10.2214/AJR.10.5038 PMID:21343479

Westwood. (2006). MRI image overlay: applications to arthrography needle insertion. In *Medicine Meets Virtual Reality 14: Accelerating Change in Healthcare: Next Medical Toolkit*, (vol. 119, p. 150). Academic Press.

Zhao. (2012). Tracking micro tool in a dynamic 3D ultrasound situation using kalman filter and ransac algorithm. *Biomedical Imaging (ISBI), IEEE International Symposium on. IEEE*, 1076–1079.

Zhu. (2006). A virtual ultrasound imaging system for the simulation of ultrasound-guided needle insertion procedures. *Proceedings of Medical Image Understanding and Analysis*, 61–65.

Chapter 9
Neural Networks to Solve Nonlinear Inverse Kinematic Problems

Fusaomi Nataga
Tokyo University of Science, Japan

Maki K. Habib
The American University in Cairo, Egypt

Keigo Watanabe
Okayama University, Japan

ABSTRACT

In making neural networks learn nonlinear relations effectively, it is desired to have appropriate training sets. In the proposed method, after a certain number of iterations, input-output pairs having worse errors are extracted from the original training set and form a new temporary set. From the following iteration, the temporary set is applied to the neural networks instead of the original set. In this case, only pairs with worse errors are used for updating the weights until the mean value of errors decreases to a desired level. Once the learning is conducted using the temporary set, the original set is applied again instead of the temporary set. The effectiveness of the proposed approach is demonstrated through simulations using kinematic models of a leg module with a serial link structure and an industrial robot.

INTRODUCTION

For example, when designing a leg module with a serial link structure as shown in Fig. 1 for a multi-legged robot used in mechatronics education, first of all, its inverse kinematics problem must be solved. Up to now, biomimetic artificial neural network approaches have been devoted to robotic inverse kinematics problems. The neural network technique mimicking the human grain is known as one of the important results in biomedical engineering. In this first section, several related researches that challenged to solve inverse kinematics problems of manipulators or mechatronics systems are surveyed. Mao and

DOI: 10.4018/978-1-5225-2993-4.ch009

Hsia proposed a neural network approach to solve the inverse kinematics of redundant robot manipulators in an environment with obstacles (Mao 1997), in which only the knowledge of the robot forward kinematics functions was required and the neural network was trained in the inverse modeling manner. Tejomurtula and Kak (1999) proposed a solution of inverse kinematics concerning a simple two link manipulator (Tejomurtula 1999). It was reported that the proposed neural network method yielded multiple and precise solutions suitable for real-time applications. Xia and Wang proposed a new recurrent neural network called the dual network (Xia, 2001). It was shown that the dual network was capable of asymptotic tracking for the motion control of kinematically redundant manipulators.

Koker et al. proposed a neural network based inverse kinematics solution for a three joints manipulator, in which the designed neural networks gave reliable three angles according to a given position in Cartesian coordinates (Koker 2004). Then, neural networks and genetic algorithms were fused together to solve the inverse kinematics problem of a Stanford robotic manipulator with six-DOFs while minimize the error at the end effector (Koker 2013). Hoang et al. proposed a differential kinematics algorithm to generate omnidirectional walking trajectory of a leg based on backstepping control using Lyapunov stability. Simulation results for walking motion of one leg of the 6LR were shown to prove the effectiveness and applicability of the proposed controller (Hoang 2014). Aggarwal et al. presented a non-conventional technique for solving the inverse kinematics problem using artificial neural networks. A feed forward multi-layer neural network was applied to this research. The inverse kinematic solution for a PUMA 560 robot was developed by training the neural network with the end-effector's Cartesian co-ordinates and its corresponding joint angles.

As for other types of approaches, Duka designed a feedforwad neural network to solve the inverse kinematics problem of a three-link planar manipulator (Duka 2014), and then presented adaptive neuro-fuzzy inference system (ANFIS) that could produce the solution to the inverse kinematics problem for a planar robot, in which the ANFIS learned the mapping between the end-effector's position/orientation and three joint angles with acceptable accuracy (Duka 2015). Furthermore, Maeda et al. designed a position control scheme for actual robot systems using high dimensional neural networks, in which complex-valued neural network and quaternion neural network learned the inverse kinematics of the robot systems. Two dimensional SCARA robot and three-dimensional robot were well controlled using the inverse kinematics (Maeda 2014).

An expected effect of artificial neural networks in the field of biomedical engineering is to be able to simulate the structure and functionalities of biological neural networks. As introduced above, many researches using neural networks have been conducted to cope with robotic inverse kinematics problem, however, efforts on the improvement of learning efficiency, i.e., weights updating times, have been hardly seen. The main contribution of this chapter is a proposal of a weight updating method. Generally, in making neural networks learn nonlinear relations suitably, desired training set prepared in advance is used. The training set consists of multiple pairs of an input vector and an output one. Each input vector is given to the input layer for forward calculation and the paired output vector is compared with the vector yielded from the output layer. Also, backward calculation means updating the weights using a back propagation algorithm. One cycle consists of one forward calculation and backward one. The time required for the learning process of the neural networks depends on the number of total weights in the neural networks and the one of the input-output pairs in the training set. This chapter describes neural networks with efficient weights update ability in order to effectively learn the inverse kinematics of a leg kit with multi-DOFs (Nagata 2015). In the proposed learning process, after certain number of iteration, input-output pairs having the worse errors are extracted from the original training set and form a new temporary set. Note that an iteration of learning uses all pairs in the training set. Then, from the following iteration, the temporary set is applied instead of the original set. In this case, only pairs with worse errors are used for updating weights until the mean value of errors reduces to a desired level. Once the learning is conducted using the temporary training set, the original set is applied again instead of the temporary set. It is expected by alternately giving the two kinds of training sets the convergence time can be efficiently reduced. The effectiveness is proved through simulation experiments using a kinematic model of a leg with four-DOFs. Additionally, when the tip of the leg with a serial link structure is controlled kinematically in Cartesian space, an inverse Jacobian matrix is needed to generate velocities in joint space. However, the calculation of the inverse Jacobian is not easy. Hence, neural networks based inverse kinematics is used to implement the function of Jacobian to overcome the complexity.

Then, the proposed learning method is applied to the inverse kinematics problem of an industrial robot with six-DOFs (Nagata 2016). The effectiveness is further proved through simulation experiments using the kinematic model of the industrial robot.

LEG WITH FOUR-DOFS AND ITS FORWARD KINEMATICS

Generally, multi-legged robot has multiple legs with a serial link structure. Figure 2 shows an example of a leg with four-DOFs, which is used for graduation study of undergraduate students (Nagata 2013). Five coordinate systems $\Sigma_k - x_k y_k z_k (0 \le k \le 4)$ are assigned at each joint. The position of the arm tip in base frame $\Sigma_0 - x_0 y_0 z_0$ is defined with $\mathbf{x} = [x \ \ y \ \ z]^T$. Table 1 tabulates the Denavit-Hartenberg (DH) notation extracted from the leg with four joints. a is the link length which is the distance between two adjacent z-axes measured along x-axis, α is the link twist angle between two adjacent z-axes measured around x-axis, d is the link offset which is the distance between two adjacent x-axes measured along z-axis and θ is the joint angle between two adjacent x-axes measured around z-axis. Actually, Fig. 2 shows the initial pose of the leg with the angles of $\boldsymbol{\theta} = [\theta_1 \ \theta_2 \ \theta_3 \ \theta_4]^T = [0 \ 0 \ 0 \ 0]^T$. The homogeneous transform ${}^{k-1}\mathbf{T}_k$ using the four parameters is written by

$$ {}^{k-1}\mathbf{T}_k = \mathrm{Rot}(z,\theta)\mathrm{Trans}(0,0,d)\mathrm{Trans}(a,0,0)\mathrm{Rot}(x,\alpha) = \begin{pmatrix} C_\theta & -S_\theta C_\alpha & S_\theta S_\alpha & aC_\theta \\ S_\theta & C_\theta C_\alpha & -C_\theta S_\alpha & aS_\theta \\ 0 & S_\alpha & C_\alpha & d \\ 0 & 0 & 0 & 1 \end{pmatrix} \tag{1} $$

where S_θ and C_θ denote sinθ and cosθ, respectively. ${}^0\mathbf{T}_4$ is obtained by

$$ {}^0\mathbf{T}_4 = {}^0\mathbf{T}_1 {}^1\mathbf{T}_2 {}^2\mathbf{T}_3 {}^3\mathbf{T}_4 = \begin{pmatrix} R_{11} & R_{12} & -S_{\theta 1} & x \\ R_{21} & R_{22} & C_{\theta 1} & y \\ R_{31} & R_{32} & 0 & z \\ 0 & 0 & 0 & 1 \end{pmatrix} \tag{2} $$

where

$$R_{11} = C_{\theta 4}(C_{\theta 1}C_{\theta 2}C_{\theta 3} - C_{\theta 1}S_{\theta 2}S_{\theta 3}) + S_{\theta 4}(C_{\theta 1}S_{\theta 2}C_{\theta 3} - C_{\theta 1}S_{\theta 3}S_{\theta 4}) \tag{3}$$

$$R_{12} = S_{\theta 4}(C_{\theta 1}S_{\theta 2}S_{\theta 3} - C_{\theta 1}C_{\theta 2}C_{\theta 3}) - C_{\theta 4}(C_{\theta 1}S_{\theta 2}C_{\theta 3} + C_{\theta 1}C_{\theta 2}S_{\theta 3}) \tag{4}$$

$$R_{21} = C_{\theta 4}(S_{\theta 1}C_{\theta 2}C_{\theta 3} - S_{\theta 1}S_{\theta 2}S_{\theta 3}) - S_{\theta 4}(S_{\theta 1}S_{\theta 2}C_{\theta 3} + S_{\theta 1}C_{\theta 2}S_{\theta 3}) \tag{5}$$

$$R_{22} = S_{\theta 4}(S_{\theta 1}S_{\theta 2}S_{\theta 3} - S_{\theta 1}C_{\theta 2}C_{\theta 3}) - C_{\theta 4}(S_{\theta 1}S_{\theta 2}C_{\theta 3} + S_{\theta 1}C_{\theta 2}S_{\theta 3}) \tag{6}$$

$$R_{31} = S_{\theta 4}(S_{\theta 2}S_{\theta 3} - C_{\theta 2}C_{\theta 3}) - C_{\theta 4}(S_{\theta 2}C_{\theta 3} + C_{\theta 2}S_{\theta 3}) \tag{7}$$

$$R_{32} = S_{\theta 4}(S_{\theta 2}C_{\theta 3} + C_{\theta 2}S_{\theta 3}) + C_{\theta 4}(S_{\theta 2}S_{\theta 3} - C_{\theta 2}C_{\theta 3}) \tag{8}$$

$$\begin{aligned} x = {} & l_1C_{\theta 1} + l_2C_{\theta 1}C_{\theta 2} + l_3C_{\theta 1}(C_{\theta 2}C_{\theta 3} - S_{\theta 2}S_{\theta 3}) + \\ & l_4C_{\theta 1}(C_{\theta 2}C_{\theta 3}C_{\theta 4} - S_{\theta 2}S_{\theta 3}C_{\theta 4} - S_{\theta 2}C_{\theta 3}S_{\theta 4} - C_{\theta 2}S_{\theta 3}S_{\theta 4}) \end{aligned} \tag{9}$$

$$\begin{aligned} y = {} & l_1S_{\theta 1} + l_2S_{\theta 1}C_{\theta 2} + l_3S_{\theta 1}(C_{\theta 2}C_{\theta 3} - S_{\theta 2}S_{\theta 3}) + \\ & l_4S_{\theta 1}(C_{\theta 2}C_{\theta 3}C_{\theta 4} - S_{\theta 2}S_{\theta 3}C_{\theta 4} - S_{\theta 2}C_{\theta 3}S_{\theta 4} - C_{\theta 2}S_{\theta 3}S_{\theta 4}) \end{aligned} \tag{10}$$

$$\begin{aligned} z = {} & -l_2S_{\theta 2} - l_3(S_{\theta 2}C_{\theta 3} + C_{\theta 2}S_{\theta 3}) - \\ & l_4C_{\theta 4}(S_{\theta 2}C_{\theta 3} + C_{\theta 2}S_{\theta 3}) + l_4S_{\theta 4}(S_{\theta 2}S_{\theta 3} - C_{\theta 2}C_{\theta 3}) \end{aligned} \tag{11}$$

DESIGN OF NEURAL NETWORK-BASED INVERSE KINEMATICS

An original mobile robot for mechatronics education can be designed with the leg module shown in Figure 1. In this chapter, Eq. (2) for $\theta \in \Re^4 \rightarrow \mathbf{x} \in \Re^3$ is called the forward kinematics $\mathbf{x} = fkine(\theta)$ and can be analytically calculated. On the contrary $\mathbf{x} \in \Re^3 \rightarrow \theta \in \Re^4$ is to be the inverse kinematics

Figure 1. Leg module with a four-DOFs serial link structure for mechatronics education

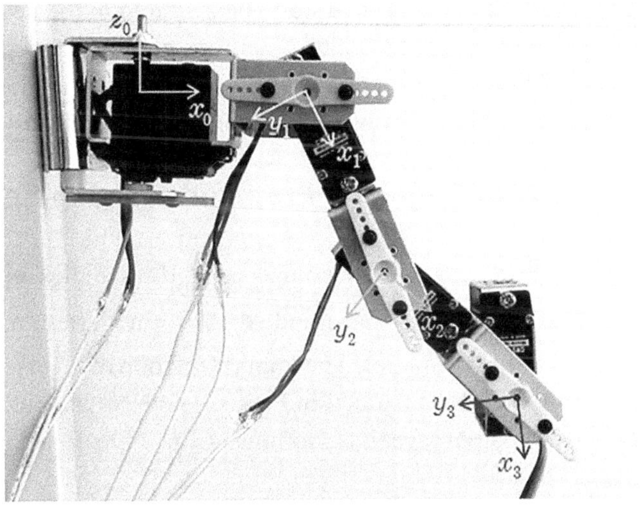

Figure 2. Kinematics model of the leg with four-DOFs

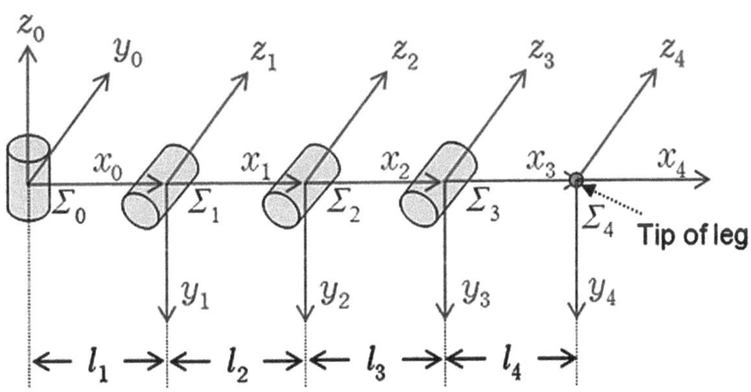

Table 1. Denabit-Hartenberg notation designed for the leg with four joints

k	a	α	d	θ
1	l_1	$-\frac{\pi}{2}$	0	θ_1
2	l_2	0	0	θ_2
3	l_3	0	0	θ_3
4	l_4	0	0	θ_4

$\theta = ikine(\mathbf{x})$. It is not easy but complex to obtain the analytical solutions of the inverse kinematics. In this section, the mapping of inverse kinematics is tried to be acquired in the neural networks shown in Figure 3. The number of the hidden layers and that of neurons are not important for researching the weights update method proposed in the next section. This means that they are out of evaluation, so that two hidden layers and 30 neurons were tentatively set. After here, the neural networks are called the NN.

Neural Networks Leaned With Regularly Prepared Training Set

Figure 4 shows training set $\mathbf{x}_j \in \Re^3$ $(1 \le j \le 100)$ composed of 100 samples representing the relations $\mathbf{x}_j = fkine\ (\theta_j)$ between $\theta_j \in \Re^4$ in joint space and $\mathbf{x}_j \in \Re^3$ in Cartesian space, which are randomly generated by using the forward kinematics. The implicit relation of inverse kinematics that Figure 4 has, is tried to be learned with the NN illustrated in Fig. 3. The NN has four layers where the input layer has three units of $\mathbf{x}_j$ and the output layer has four units of θ_j. The first and second hidden layers have thirty units, respectively.

After passing enough learning process using the training set, the performance of the trained NN was evaluated using the training test set along the rectangle path in Figure 4. Figure 5 shows the result, in which markers '+' are the training test set $\mathbf{x}_{dj} \in \Re^3$ $(1 \le j \le 120)$ for the input layer of the NN. Also 'x' are the output of forward kinematics $fkine(\theta_{nnj})$. Note that θ_{nnj} is the output from the NN in response to $\mathbf{x}_{dj}$. It is recognized from the results that the NN learned with the training set shown in Figure 4 performs a passable generalization for inverse kinematics but some small errors are observed.

Figure 3. Neural networks with three inputs of $\mathbf{x} \in \Re^3$ and four outputs $\theta_{nn} \in \Re^4$

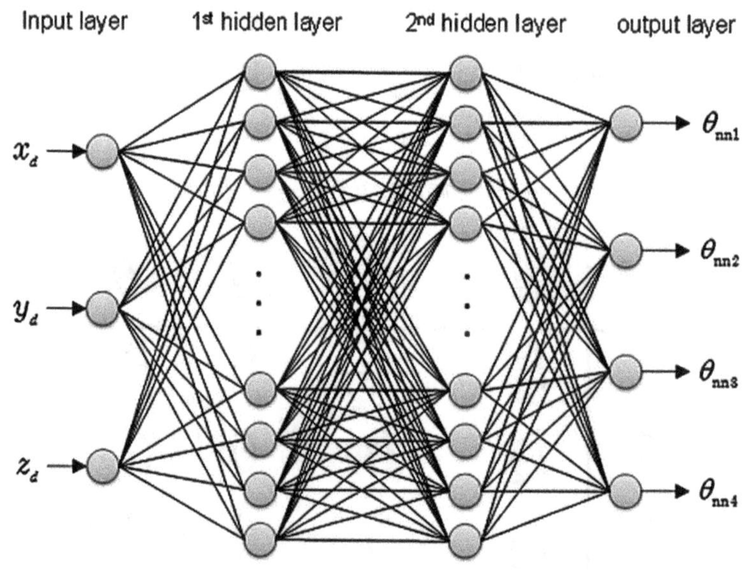

Figure 4. Training set $\mathbf{x}_j \in \Re^3$ *representing the relation* $\mathbf{x}_j = fkine\ (\theta_j)$ *between* $\theta_j \in \Re^4$ *in joint space and* $\mathbf{x}_j \in \Re^3$ *in Cartesian space, which are prepared randomly.*

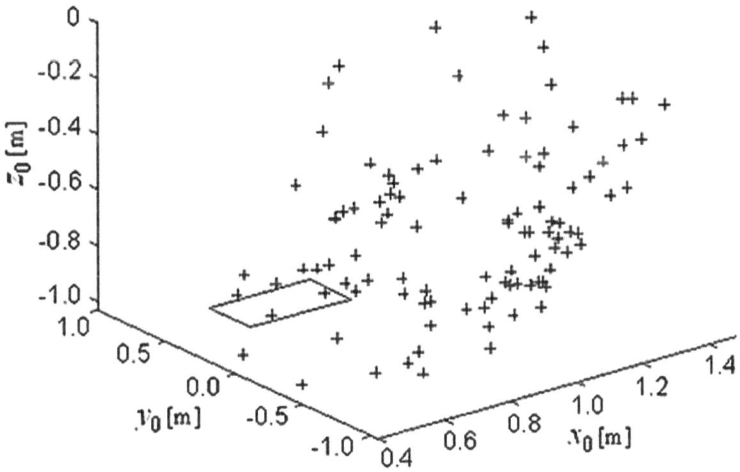

Figure 5. Markers '+' are the training test set $\mathbf{x}_{dj}$ *for learning of the NN. Also'x' are the output of forward kinematics* $fkine(\theta_{nnj})$. *Note that* θ_{nnj} *is the output from the forward calculation of the NN when* $\mathbf{x}_{dj}$ *is given to the input layer.*

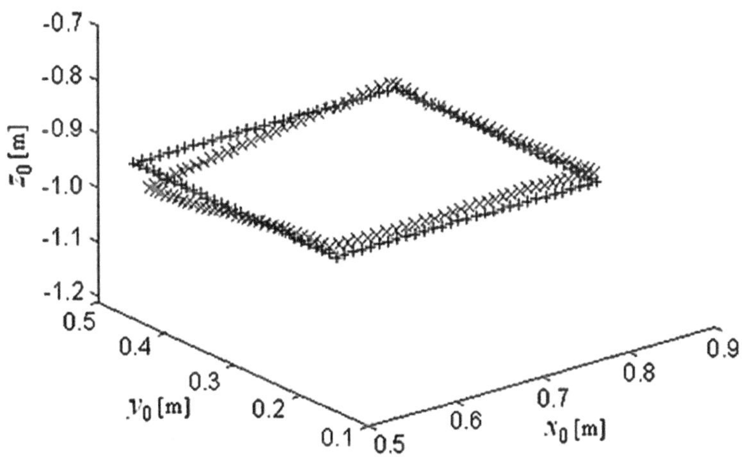

Neural Networks Learned With Randomly Prepared Training Set

In this subsection, the performance of the NN learned with regularly prepared training set (60 pairs of $\mathbf{x}_j$ and θ_j) is evaluated. Figure 6 shows the training set representing a path consisting of $\mathbf{x}_j$ $(1 \le j \le 60)$, i.e., the sampled number for training set is 60. The NN shown in Figure 3 was trained with the set until the error satisfactorily converged. Figure 7 shows the performance result of the trained NN. Marker '+' in the upper figure are training test set $\mathbf{x}_{dj}$ $(1 \le j \le 300)$ for the NN. Also, 'x' in the lower figure are

Figure 6. Training set representing the relation between $\boldsymbol{\theta}_j \in \Re^4$ *and* $\mathbf{x}_j \in \Re^3$ *, which are prepared regularly*

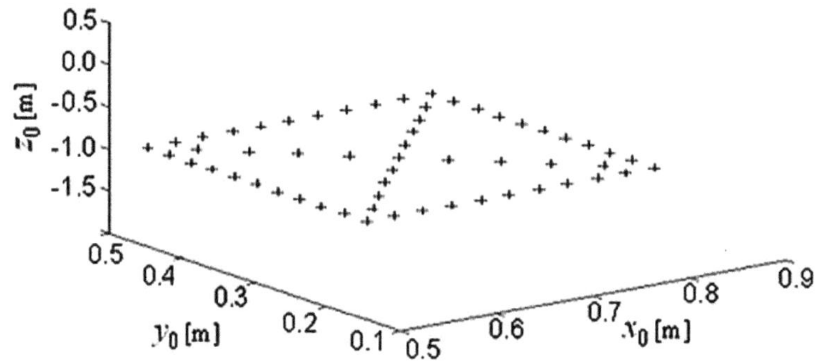

Figure 7. Marker '+' in the upper figure are the training test set $\mathbf{x}_{dj}$ *for NN. Also 'x' in the lower figure are the output* $\mathbf{x}_j$ *of forward kinematics* $fkine(\theta_{nnj})$*. Note that* θ_{nnj} *is the output from the forward calculation of the NN when* $\mathbf{x}_{dj}$ *is given to the input layer*

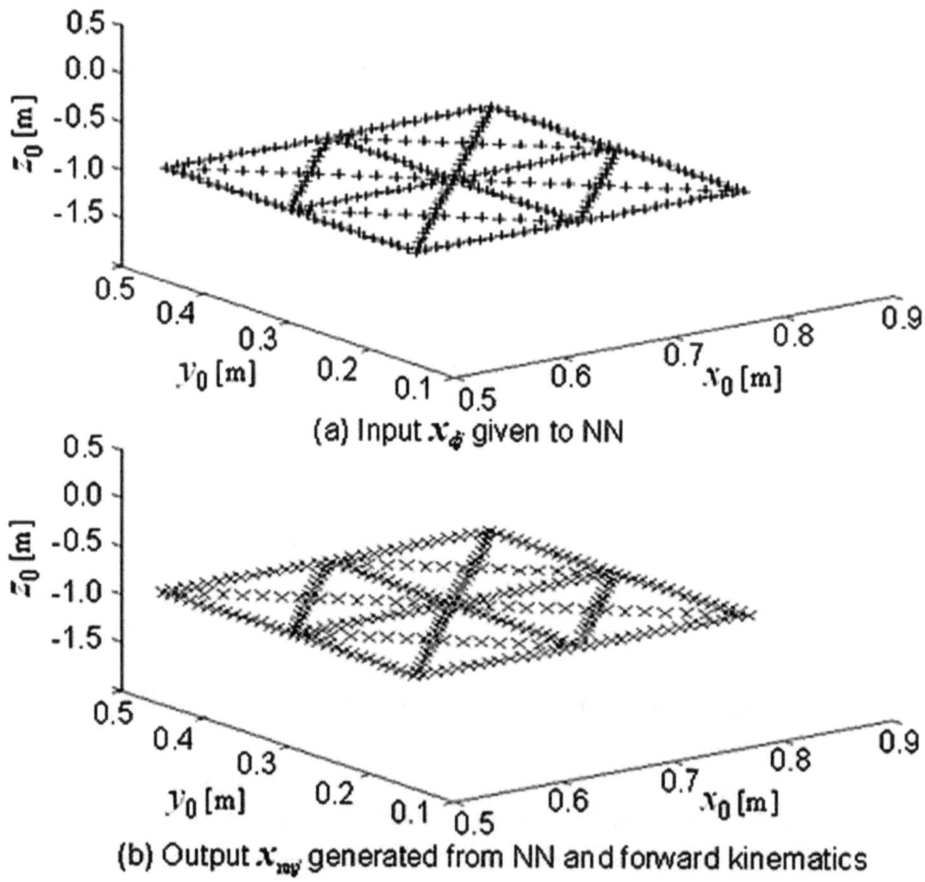

the output of forward kinematics $fkine(\theta_{nnj})$. Note that θ_{nnj} is the output from the forward calculation of the NN when $\mathbf{x}_{dj}$ $(1 \le j \le 300)$ is given to the input layer. It is recognized from the result that the NN learned with the training set shown in Fig. 6 performs a desirable generalization for inverse kinematics.

EFFICIENT LEARNING ABILITY

In the Case of Conventional Learning Method

When the back propagation algorithm is used for adjusting weights in neural networks, it is serious problem that much time is required for satisfactory convergence. Here, an efficient training method is introduced by using the case shown in Figure 6. First of all, error E_i for criterion at the i-th learning procedure was defined by

$$E_i = \bar{e} = \frac{1}{60}\sum_{j=1}^{60} e_j \tag{12}$$

where e_j is the error in case that the j-th sample, i.e., a pair of $\mathbf{x}_j$ and θ_j in training set is given to input and output layers, which is obtained by

$$e_j = \sqrt{\sum_{k=1}^{4} (\theta_{jk} - \theta_{nnjk})^2} \tag{13}$$

where $\theta_j \in \Re^4$ and $\theta_{nnj} \in \Re^4$ are the training vector for the output layer and the actual output from the NN, respectively. k $(1 \le k \le 4)$ is the number of each joint. j $(1 \le j \le 60)$ is the sampled number in the training set. The learning, i.e., the tuning of weights, was continued until the following condition is satisfied.

$$E_i < E_d \tag{14}$$

where E_d is the maximum allowable error to be set in advance. In the conventional learning method, all samples in the training set, i.e., sixty pairs, were sequentially given for updating weights. Figure 8 shows the learning result, in which sixty times of weights updating by using a back propagation algorithm were conducted in one learning procedure on the horizontal axis. In other words, one learning procedure means one iteration using sixty pairs. Also note that one weights updating procedure consists of a pair of a forward propagation to calculate the error and a back propagation to update all weights based on the error. It was confirmed that totally 22,810 × 60 = 1,368,600 times of weights updating were conducted and the total calculation time was 8,351 seconds. In this case, the desired maximum allowable error E_d was set to 0.02. The specification of CPU was Intel(R) Core(TM) i3 560 3.33 GHz.

Figure 8. An example of learning result using a conventional method, in which the desired maximum allowable error E_d is set to 0.02

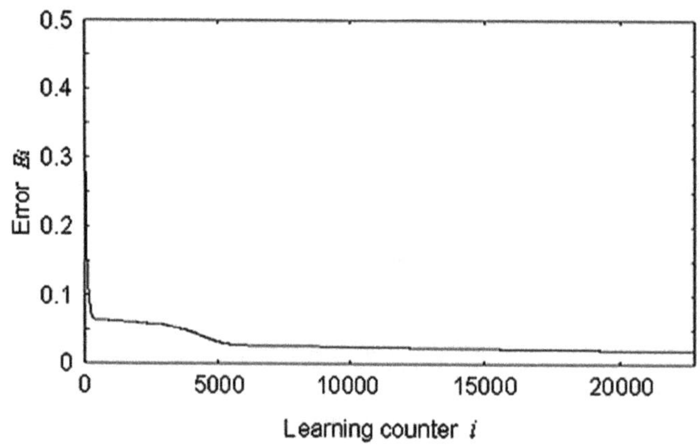

In the Case of Proposed Efficient Learning Method

On the other hand, in our proposed method, a new training set consisting of pairs of input $\mathbf{x}_l$ and output θ_l, that have not been well trained, is extracted from the original training set. The condition of the extraction is given by

$$e_l > \bar{e} + e_{\mathrm{sd}} \tag{15}$$

where

$$e_l = \sqrt{\sum_{k=1}^{4} (\theta_{lk} - \theta_{\mathrm{nn}lk})^2} \tag{16}$$

$$e_{\mathrm{sd}} = \sqrt{\frac{1}{60} \sum_{j=1}^{60} (\bar{e} - e_j)^2} \tag{17}$$

where e_{sd} is the standard deviation of e_j $(1 \le j \le 60)$. The pairs of $\mathbf{x}_l$ and θ_l satisfying Eq. (15) are extracted and form a new training set with less number of pairs than that of the original training set, i.e., 60. When this new training set is applied instead of the original set, the error $\tilde{E}_i$ at the i-th learning procedure, i.e., iteration, is calculated by

$$\tilde{E}_i = \frac{1}{m} \sum_{l=1}^{m} e_l \tag{18}$$

where m is the number of the pairs extracted from the original set based on Eq. (15). In our proposed method, E_i and $\tilde{E}_i$ are alternately used for the evaluation of convergence at the i-th learning procedure. Note that the extracted training set does not increases the mean errors E_i and $\tilde{E}_i$. E_i and $\tilde{E}_i$ are criteria to know the progress of training. Figure 9 illustrates the concept of the proposed efficient learning process, in which the original training set and the extracted one are alternately applied for 200 times of iterative learning, respectively. The vertical axis means the number of the pairs in the training set. The learning process using the extracted training set works as mini-batch gradient descent used in deep learning techniques. In the mini-batch gradient descent, samples randomly extracted from original training set are used for weights updating in each iteration. The number of samples is designated as mini-batch size. On the other hand, in the proposed learning process, m samples with larger errors are extracted and forms a temporary subset for weights updating.

Then, the proposed method was applied to the same problem explained in the previous section to evaluate the effectiveness. Figure 10 shows the learning result until E_i reached to 0.02, in which se-

Figure 9. Image of the proposed learning process, in which original training set and extracted one are alternately applied to the NN for 200 times of iterative learning, respectively

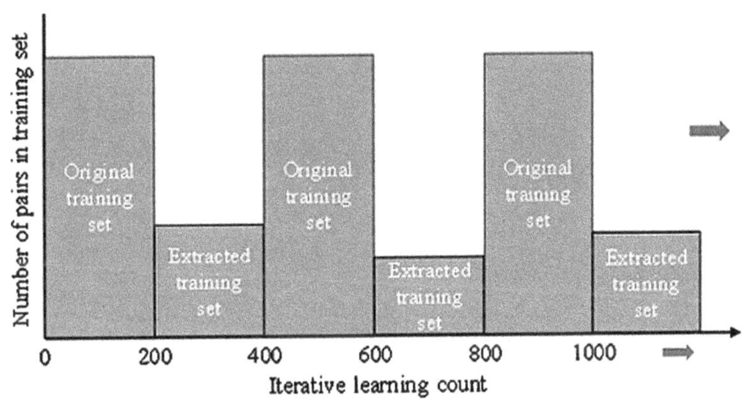

Figure 10. Learning result using the proposed method, in which original training set and extracted one are alternately applied to the NN for 100 and 1000 times of iterative learning, respectively

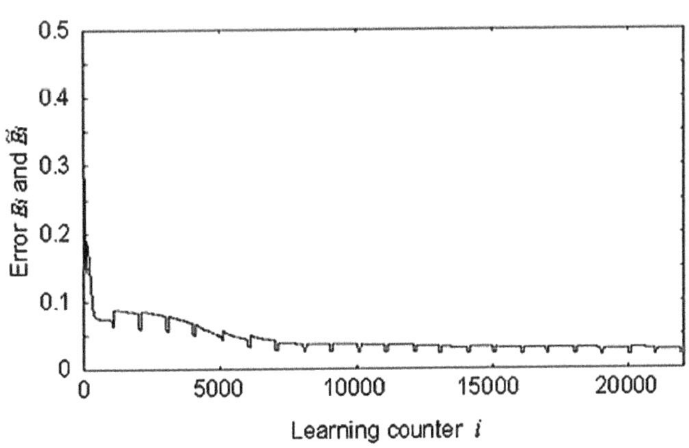

quential 100 times of E_i and 1,000 times of $\tilde{E}_i$ are used alternately and repeatedly. It is observed that although the iterative learning counter is almost the same as 22,000 times shown in Figure 8, total times of weights updating were reduced to 650,220 and the calculation time largely decreased to 4,096 seconds.

KINEMATIC CONTROL OF THE LEG MODULE

Jacobian of the Leg Module

When the tip of the leg is controlled kinematically in Cartesian space, Jacobian matrix is needed. By differentiating Eqs. (9), (10) and (11) by time, the following relation is obtained.

$$\begin{pmatrix} \dot{x} \\ \dot{y} \\ \dot{z} \end{pmatrix} = \begin{pmatrix} J_{11} & J_{12} & J_{13} & J_{14} \\ J_{21} & J_{22} & J_{23} & J_{24} \\ 0 & J_{32} & J_{33} & J_{34} \end{pmatrix} \begin{pmatrix} \dot{\theta}_1 \\ \dot{\theta}_2 \\ \dot{\theta}_3 \\ \dot{\theta}_4 \end{pmatrix} \tag{19}$$

where

$$\begin{aligned} J_{11} = & -l_1 S_{\theta1} - l_2 S_{\theta1} S_{\theta2} - l_3 S_{\theta1}(C_{\theta2}C_{\theta3} - S_{\theta2}S_{\theta3}) \\ & - l_4(S_{\theta1}C_{\theta2}C_{\theta3}C_{\theta4} - S_{\theta1}S_{\theta2}S_{\theta3}C_{\theta4} - S_{\theta1}S_{\theta2}C_{\theta3}S_{\theta4} - S_{\theta1}C_{\theta2}S_{\theta3}S_{\theta4}) \end{aligned} \tag{20}$$

$$\begin{aligned} J_{12} = & -l_2 C_{\theta1} S_{\theta2} - l_3 C_{\theta1}(S_{\theta2}C_{\theta3} + C_{\theta2}S_{\theta3}) \\ & - l_4(C_{\theta1}S_{\theta2}C_{\theta3}C_{\theta4} + C_{\theta1}C_{\theta2}S_{\theta3}C_{\theta4} + C_{\theta1}C_{\theta2}C_{\theta3}S_{\theta4} - C_{\theta1}S_{\theta2}S_{\theta3}S_{\theta4}) \end{aligned} \tag{21}$$

$$\begin{aligned} J_{13} = & -l_3 C_{\theta1}(S_{\theta2}C_{\theta3} + C_{\theta2}S_{\theta3}) \\ & - l_4(C_{\theta1}S_{\theta2}C_{\theta3}C_{\theta4} + C_{\theta1}C_{\theta2}S_{\theta3}C_{\theta4} + C_{\theta1}C_{\theta2}C_{\theta3}S_{\theta4} - C_{\theta1}S_{\theta2}S_{\theta3}S_{\theta4}) \end{aligned} \tag{22}$$

$$J_{14} = -l_4(C_{\theta1}S_{\theta2}C_{\theta3}C_{\theta4} + C_{\theta1}C_{\theta2}S_{\theta3}C_{\theta4} + C_{\theta1}C_{\theta2}C_{\theta3}S_{\theta4} - C_{\theta1}S_{\theta2}S_{\theta3}S_{\theta4}) \tag{23}$$

$$\begin{aligned} J_{21} = & \, l_1 C_{\theta1} + l_2 C_{\theta1} C_{\theta2} + l_3 C_{\theta1}(C_{\theta2}C_{\theta3} - S_{\theta2}S_{\theta3}) \\ & + l_4(C_{\theta1}C_{\theta2}C_{\theta3}C_{\theta4} - C_{\theta1}S_{\theta2}S_{\theta3}C_{\theta4} - C_{\theta1}S_{\theta2}C_{\theta3}S_{\theta4} - C_{\theta1}C_{\theta2}S_{\theta3}S_{\theta4}) \end{aligned} \tag{24}$$

$$\begin{aligned} J_{22} = &-l_2 S_{\theta 1} S_{\theta 2} - l_3 S_{\theta 1}(S_{\theta 2} C_{\theta 3} + C_{\theta 2} S_{\theta 3}) \\ &- l_4(S_{\theta 1} S_{\theta 2} C_{\theta 3} C_{\theta 4} + S_{\theta 1} C_{\theta 2} S_{\theta 3} C_{\theta 4} + S_{\theta 1} C_{\theta 2} C_{\theta 3} S_{\theta 4} - S_{\theta 1} S_{\theta 2} S_{\theta 3} S_{\theta 4}) \end{aligned} \tag{25}$$

$$\begin{aligned} J_{23} = &-l_3 S_{\theta 1}(S_{\theta 2} C_{\theta 3} + C_{\theta 2} S_{\theta 3}) \\ &- l_4(S_{\theta 1} S_{\theta 2} C_{\theta 3} C_{\theta 4} + S_{\theta 1} C_{\theta 2} S_{\theta 3} C_{\theta 4} + S_{\theta 1} C_{\theta 2} C_{\theta 3} S_{\theta 4} - S_{\theta 1} S_{\theta 2} S_{\theta 3} S_{\theta 4}) \end{aligned} \tag{26}$$

$$J_{24} = -l_4(S_{\theta 1} S_{\theta 2} C_{\theta 3} C_{\theta 4} + S_{\theta 1} C_{\theta 2} S_{\theta 3} C_{\theta 4} + S_{\theta 1} C_{\theta 2} C_{\theta 3} S_{\theta 4} - S_{\theta 1} S_{\theta 2} S_{\theta 3} S_{\theta 4}) \tag{27}$$

$$J_{32} = -l_2 C_{\theta 2} - l_3(C_{\theta 2} C_{\theta 3} - S_{\theta 2} S_{\theta 3}) + l_4(S_{\theta 2} C_{\theta 3} S_{\theta 4} + C_{\theta 2} S_{\theta 3} S_{\theta 4} - C_{\theta 2} C_{\theta 3} C_{\theta 4} + S_{\theta 2} S_{\theta 3} C_{\theta 4}) \tag{28}$$

$$J_{33} = -l_3(C_{\theta 2} C_{\theta 3} - S_{\theta 2} S_{\theta 3}) + l_4(S_{\theta 2} C_{\theta 3} S_{\theta 4} + C_{\theta 2} S_{\theta 3} S_{\theta 4} - C_{\theta 2} C_{\theta 3} C_{\theta 4} + S_{\theta 2} S_{\theta 3} C_{\theta 4}) \tag{29}$$

$$J_{34} = -l_4(C_{\theta 2} C_{\theta 3} C_{\theta 4} - S_{\theta 2} S_{\theta 3} C_{\theta 4} - S_{\theta 2} C_{\theta 3} S_{\theta 4} - C_{\theta 2} S_{\theta 3} S_{\theta 4}) \tag{30}$$

Here, for example, a constraint condition is considered. The condition is that the orientation of the leg tip is fixed to the ground as given by

$$\theta_2 + \theta_3 + \theta_4 = c \quad (c = \text{const.}) \tag{31}$$

so that

$$\dot{\theta}_2 + \dot{\theta}_3 + \dot{\theta}_4 = 0 \tag{32}$$

By including Eq. (32) into Eq. (19),

$$\begin{pmatrix} \dot{x} \\ \dot{y} \\ \dot{z} \\ 0 \end{pmatrix} = \begin{pmatrix} J_{11} & J_{12} & J_{13} & J_{14} \\ J_{21} & J_{22} & J_{23} & J_{24} \\ 0 & J_{32} & J_{33} & J_{34} \\ 0 & 1 & 1 & 1 \end{pmatrix} \begin{pmatrix} \dot{\theta}_1 \\ \dot{\theta}_2 \\ \dot{\theta}_3 \\ \dot{\theta}_4 \end{pmatrix} \tag{33}$$

is obtained, in which Jacobian $\mathbf{J}(\theta)$ is defined as

$$\mathbf{J}(\theta) = \begin{pmatrix} J_{11} & J_{12} & J_{13} & J_{14} \\ J_{21} & J_{22} & J_{23} & J_{24} \\ 0 & J_{32} & J_{33} & J_{34} \\ 0 & 1 & 1 & 1 \end{pmatrix} \tag{34}$$

In order to conduct the kinematic control, desired relative position $\Delta\mathbf{x}$ should be designed in Cartesian space and be transformed into relative angle $\Delta\boldsymbol{\theta}$ in joint space with $\mathbf{J}^{-1}(\theta)$. However, since the calculation of $\mathbf{J}^{-1}(\theta)$ is considerably complicated, an easy method to generate $\Delta\boldsymbol{\theta}$ with the NN shown in Figure 3 is introduced.

Kinematic Control Without Inverse Jacobian

It is assumed that $\theta(k)$ and $\mathbf{x}(k)$ are the position vectors in joint and Cartesian spaces, respectively, so that the relation is represented with the forward kinematics function as

$$\mathbf{x}(k) = fkine\{\theta(k)\} \tag{35}$$

where k denotes the discrete time. The desired movement $\Delta\mathbf{x}_d(k)$ in Cartesian space is written by

$$\Delta\mathbf{x}_d(k) = \mathbf{x}_d(k) - \mathbf{x}_d(k-1) \tag{36}$$

To conduct the above motion, the following movement in joint space can be generated using the trained NN.

$$\Delta\theta_d(k) = ikine_{\_nn}\{\mathbf{x}_d(k)\} - ikine_{\_nn}\{\mathbf{x}_d(k-1)\} \tag{37}$$

where the function $\theta_{nn}(k) = ikine_{\_nn}\{\mathbf{x}_d(k)\}$ means the mapping of inverse kinematics through the NN shown in Fig. 3. Through the proposed procedure, the desired movement $\Delta\theta_d(k)$ for kinematic control can be generated without using $\mathbf{J}^{-1}(\theta)$.

INDUSTRIAL ROBOT MOTOMAN WITH SIX-DOFS

Figure 11 shows an industrial robot Motoman with six-DOFs. Six coordinate systems $\Sigma_k - x_k y_k z_k$ $(0 \le k \le 5)$ are assigned at each joint. The position of the arm tip viewed in base frame $\Sigma_0 - x_0 y_0 z_0$ is defined with $\boldsymbol{x} = [x\ y\ z]^T$. Table 2 tabulates the Denavit-Hartenberg notation (DH) extracted from the robot (Gan 2013). a is the link length which is the distance between two adjacent z-axes measured along x-axis, α is the link twist angle between two adjacent z-axes measured around x-axis, d is the link offset which is the distance between two adjacent x-axes measured along z-axis and θ is the joint angle between two adjacent x-axes measured around z-axis.

Figure 11. Six-DOFs Industrial robot Yaskawa Motoman

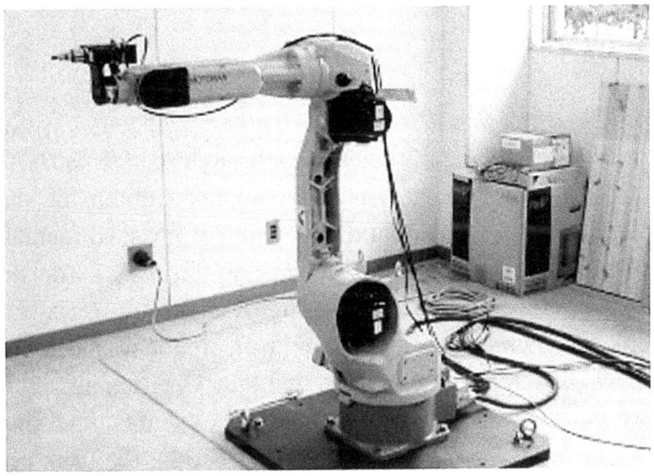

$$
{}^{k-1}\mathbf{T}_k = \mathrm{Rot}(z,\theta)\mathrm{Trans}(0,0,d)\mathrm{Trans}(a,0,0)\mathrm{Rot}(x,\alpha) = \begin{bmatrix} C_\theta & -S_\theta C_\alpha & S_\theta S_\alpha & aC_\theta \\ S_\theta & C_\theta C_\alpha & -C_\theta S_\alpha & aS_\theta \\ 0 & S_\alpha & C_\alpha & d \\ 0 & 0 & 0 & 1 \end{bmatrix} \tag{38}
$$

where, e.g., the homogeneous transform Rot(z, θ) conducts the rotation of θ [rad] to right screw direction around z-axis; Trans(0, 0, d) translates to z-direction with d [mm]; S_θ and C_θ denote sinθ and cosθ, respectively. Hence, ${}^0T_6$ is obtained by

$$
{}^0\mathbf{T}_6 = {}^0\mathbf{T}_1\,{}^1\mathbf{T}_2\,{}^2\mathbf{T}_3\,{}^3\mathbf{T}_4\,{}^4\mathbf{T}_5\,{}^5\mathbf{T}_6 = \begin{bmatrix} R_{11} & R_{12} & R_{13} & x \\ R_{21} & R_{22} & R_{23} & y \\ R_{31} & R_{32} & R_{33} & z \\ 0 & 0 & 0 & 1 \end{bmatrix} \tag{39}
$$

3×3 rotational matrix $\boldsymbol{R}$ in ${}^0\boldsymbol{T}_6$ has to be transformed into roll, pitch and yaw angles in kinematic simulation (Corke 1996). As is known, the inverse kinematics problem of industrial robots with a serial link structure becomes more complex with the increase of the DOFs.

NEURAL NETWORKED-BASED INVERSE KINEMATICS

Design of Neural Networks

In this section, the joint angle vector $\boldsymbol{\theta} = [\theta_1 \ldots \theta_6] \Rightarrow$ the homogeneous transform matrix ${}^0\boldsymbol{T}_6$ is called the forward kinematics and can be analytically calculated through ${}^0\boldsymbol{T}_6 = fkine(\boldsymbol{\theta})$ (Corke 1996). Reversely, ${}^0\boldsymbol{T}_6 \Rightarrow \boldsymbol{\theta}$ is to be the inverse kinematics. It is not easy but complex to obtain the analytical solutions of the inverse kinematics. In this section, it is tried that the mapping of inverse kinematics ${}^0\boldsymbol{T}_6 \Rightarrow \boldsymbol{\theta}$ is acquired in the neural network shown in Figure 12, in which $\boldsymbol{x}_d = [x_d \, y_d \, z_d \, \alpha_d \, \beta_d]^T$ is transformed from ${}^0\boldsymbol{T}_6$; α_d and β_d are the roll and pitch angles, respectively. The yaw angle can be always fixed to 0 because of the axis symmetry of the sixth axis. The number of the hidden layers and that of neurons are not important for researching the proposed weights update method, so that they were tentatively set to 2 and 20, respectively. The neurons in two adjacent layers are connected with weights randomly initialized within the range from -1 to 1. In the learning process shown in this section, the back propagation algorithm is applied for training of the weights.

The performance of the proposed efficient learning method is evaluated using the regularly prepared training set consisting of multiple pairs of $\boldsymbol{x}_{dj} = [x_{dj} \, y_{dj} \, z_{dj} \, \alpha_{dj} \, \beta_{dj}]^T$ and $\boldsymbol{\theta}_{dj}$ $(1 \le j \le 481)$, i.e., the sampled number in the training set is 481. The desired points $[x_{dj} \, y_{dj} \, z_{dj}]^T$ included in $\boldsymbol{x}_{dj}$ generated along a spiral path can be viewed as shown in Fig. 13. Roll angle α_{dj} and pitch angle β_{dj} are converted from the normal vector at the respective points. The joint angle $\boldsymbol{\theta}_{dj}$ in the training set was made by using an inverse kinematics function $\boldsymbol{\theta}_{dj} = ikine(\boldsymbol{x}_{dj})$ available on MATLAB (Corke 1996).

Figure 12. Neural network with five inputs of $\mathbf{x}_d$ and six outputs of $\boldsymbol{\theta}_{nn}$. In simulations, the number of the hidden layers and that of neurons are tentatively set to 2 and 20, respectively

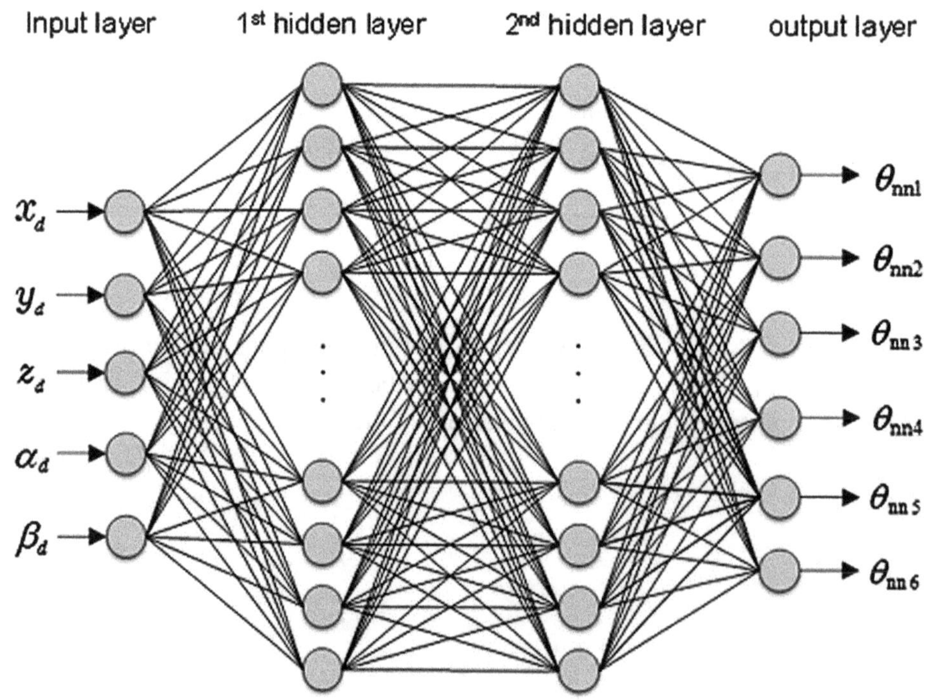

Figure 13. Desired points along a spiral path prepared for training set

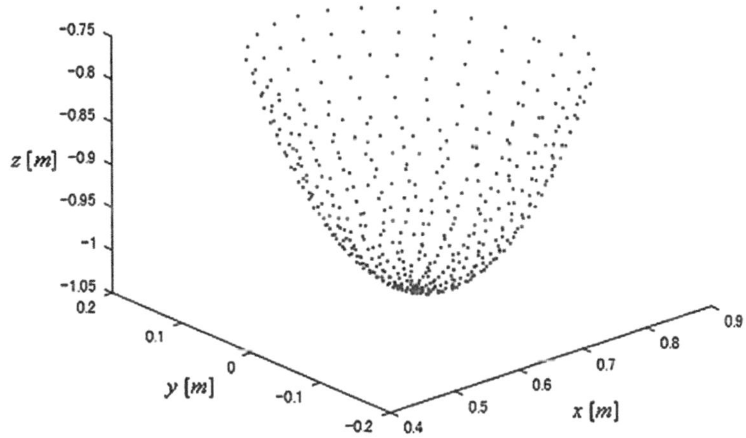

When the back propagation algorithm is used for adjusting weights in neural network, it is serious problem that much time is required for satisfactory convergence. Especially, undesirable stagnation of learning always depresses us. Here, an efficient training method is introduced by using the case shown in Fig. 13. In this method, when the learning falls into the situation of stagnation, a stimulating extraction from original training set is applied and temporally creates another training set with worse evaluation for more intensive training.

Conventional and Orthodox Learning Process

First of all, the conventional and orthodox learning process is described. Error E_i for criterion at the i-th learning process is defined by

$$E_i = \bar{e} = \frac{1}{481}\sum_{j=1}^{481} e_j \tag{40}$$

where e_j is the error in case that the j-th sample in training set is given to input and output layers, which is obtained by

$$e_j = \sqrt{\sum_{k=1}^{6} (\theta_{djk} - \theta_{\mathrm{nn}jk})^2} \tag{41}$$

where $\boldsymbol{\theta}_{dj} \in \Re^6$ and $\boldsymbol{\theta}_{\mathrm{nn}j} \in \Re^6$ are the training vector for the output layer and the actual output from the NN, respectively. k is the number of each joint. j is the sampled number in training set. The learning, i.e., the updating of weights, is continued until the following condition is satisfied.

$$E_i < E_d \tag{42}$$

where E_d is the maximum allowable error. The stagnation of training process can be judged by

$$E_{i-1} - E_i < E_s \tag{43}$$

where E_s is the stagnation index. In the conventional learning method used in this section, all samples in the training set, i.e., 481 pairs, are sequentially given for updating 620 weights.

Figure 14 shows the learning result, in which the updating of weights by using a back propagation algorithm was conducted 481 times in one learning process on the horizontal axis. In other words, one learning process means 481 times iterative computations using 481 input and output pairs. Also, note that one computation for updating the weights consists of a pair of a forward propagation to calculate the error and a back propagation to update 620 weights based on the error. Figure 14 has the abscissa with a logarithmic plot to clearly show the progress of training at the early stage. The structure of the industrial robot with six-DOFs is more complex than that of the serial link with four-DOFs, so that it also takes more time for convergence. That is the reason why the logarithmic plot is used in the horizontal axis in Figures 14 and 16.

It was confirmed that totally the updating of weights was conducted 81,702×481= 39,298,662 times. In this case, the desired maximum allowable error E_d was set to 0.03. The specification of CPU was Intel(R) Core(TM) i7 4790 3.6 GHz. In this research, the calculation time is not used for criterion because the value of the time depends on the performance of CPU built in PC.

Proposed Efficient Learning Process

On the other hand, in our proposed method, a new temporary training set consisting of pairs of input $\boldsymbol{x}_l$ and output $\boldsymbol{\theta}_l$, that have not been well trained, is extracted from the original training set. Three conditions of extraction are compared in the latter experiments, which are given by

$$e_l > \bar{e} \tag{44}$$

$$e_l > \bar{e} + e_{sd} \tag{45}$$

Figure 14. Learning result using a conventional method

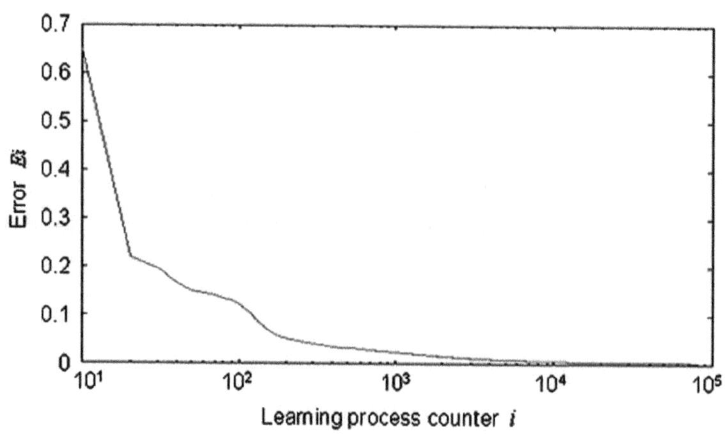

$$e_l > \bar{e} + 2e_{sd} \tag{46}$$

Where

$$e_{sd} = \sqrt{\frac{1}{481}\sum_{j=1}^{481}(\bar{e} - e_j)^2} \tag{47}$$

where e_{sd} is the standard deviation of e_j $(1 \le j \le 481)$. At the present time, some standards and criterions to systematically decide the condition have not been considered.

The pairs of $\boldsymbol{x}_l$ and $\boldsymbol{\theta}_l$ satisfying Eq. (44), (45) or (46) are extracted and form a new training set with less number of pairs than the original training set. When this new training set is applied instead of the original set, the error $\tilde{E}_i$ at the i-th learning process is calculated by

$$\tilde{E}_i = \frac{1}{m}\sum_{l=1}^{m} e_l \tag{48}$$

where m is the number of the pairs in the training set extracted from the original set based on Eq. (44), (45) or (46). E_i and $\tilde{E}_i$ are alternately used for the criterion in the back propagation algorithm in our proposed method. Figure 15 illustrates the conceptual comparison between the conventional and proposed learning methods. In the proposed method, the original training set and the extracted one are alternately applied for the iterative learning 1000 and 4000 times, respectively. Note that the times of iterative learning for original and extracted sets can be arbitrarily changed according to actual problem settings.

Figure 15. Concept of the proposed learning method, where original training set and extracted one are alternately applied

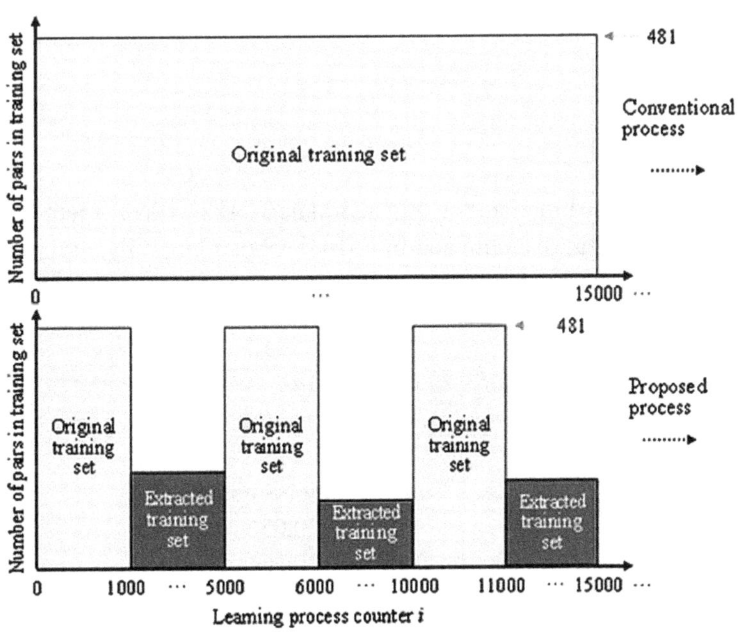

Table 2. Denavit-Hartenberg notation for Motoman HP20

k	a	α	d	θ
1	150	$-90°$	0	θ_1
2	760	0	0	θ_2
3	140	$-90°$	0	θ_3
4	0	$90°$	795	θ_4
5	0	$90°$	0	θ_5
6	0	0	-105	θ_6

In this experiment, 1000 and 4000 times are found experientially. The vertical axis means the number of the pairs in the training set.

Finally, the proposed method was applied to the same problem explained in the previous section to evaluate the effectiveness. As an example, Table 3 shows three results of the total times of weights updating in case that the numbers of iterative learning of original and extracted sets were set to 1,000 and 4,000, respectively. Figure 16 shows the learning result until E_i reached to 0.03 in the case of Eq. (44) in Table 3, in which sequential 1000 times of E_i and 4,000 times of $\tilde{E}_i$ are used alternately and repeatedly. It is observed that although the iterative learning counter is almost the same, total times of weights updating were improved by about 31%, i.e., from 39,298,662 to 27,050,036. It was experimentally observed that the numbers of sequential times for evaluating E_i and $\tilde{E}_i$ occurred different learning results, so that some method will be needed to systematically determine the numbers of times.

CONCLUSION

In this chapter, learning of inverse kinematics using neural networks with efficient weights update ability has been described for a serial link structure and industrial robot. Generally, in making neural networks learn a relation among multi inputs and outputs, a desired training set prepared in advance is used. The

Table 3. Three results of the total times of weights updating, in which numbers of iterative learning of original and extracted sets were 1,000 and 4,000, respectively

Extraction	Times of weights updating
Eq. (44)	27,050,036
Eq. (45)	29,130,464
Eq. (46)	31,901,186

Figure 16. Learning result in the case of using the extraction condition given by Eq. (44) in Table 3

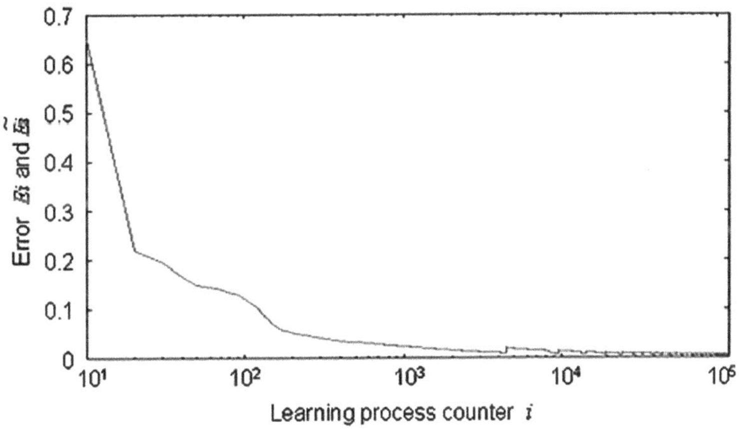

training set consists of multiple pairs of input and output vectors. Each input vector is entered into the input layer for forward computation and the paired output vector is compared with the yielded vector from the output layer in terns. The time required for the learning process of the neural networks depends on the number of total weights in the neural networks and that of the input-output pairs in the training set.

In the proposed learning process, input-output pairs, which have had the worse errors in learning process, are extracted from the original training set and form a new temporary training set. From the following iteration of learning, the temporary training set is applied to the neural network instead of the original one. That means only pairs with worse errors are used for updating weights until the mean value of errors decreases to a desired level. Once the learning process is conducted using the temporary training set, the original training set is applied again instead of the temporary set. By repeatedly using these two training sets alternately, the convergence time could be efficiently reduced. The effectiveness was proved through simulation experiments using a kinematic model of a leg with four-DOFs. Additionally, when the tip of the leg with a serial link structure is controlled kinematically in Cartesian space, an inverse Jacobian matrix is needed to generate velocities in joint space. However, the calculation of the inverse Jacobian is not easy. Hence, neural networks based inverse kinematics was introduced to overcome the complexity.

Then, neural networks with the efficient weights update ability have been also applied to the inverse kinematics of an industrial robot with six-DOFs. In the proposed learning process, after the learning is progressed, e.g., 1000 iterations, input-output pairs with worse errors are extracted from the original training set and form a new temporary set. From the next iteration, the temporary set is applied instead of the original set. After the learning conducted using the temporary set, the original set is applied again instead of the temporary set. The effectiveness of the proposed method was also proved through simulation experiments using the kinematic model of an industrial robot MOTOMAN HP20, i.e., the convergence time could be efficiently reduced by alternately applying the two kinds of sets.

In future work, better conditions for extraction from original training set will be investigated. Also, how to systematically determine the times of iterative learning for original training set and extracted one will be considered in order to reduce the load of weights updating.

REFERENCES

Aggarwal, L., Aggarwal, K., & Urbanic, R. J. (2014). Use of artificial neural networks for the development of an inverse kinematic solution and visual identification of singularity zone(s). *Procedia CIRP*, *17*, 812–817. doi:10.1016/j.procir.2014.01.107

Corke, P. (1996). A robotics toolbox for MATLAB. *IEEE Robotics & Automation Magazine*, *3*(1), 24–32. doi:10.1109/100.486658

Duka, A. V. (2014). Neural network based inverse kinematics solution for trajectory tracking of a robotic arm. *Procedia Technology*, *12*, 20–27. doi:10.1016/j.protcy.2013.12.451

Duka, A. V. (2015). ANFIS Based Solution to the Inverse Kinematics of a 3DOF Planar Manipulator. *Procedia Technology*, *19*, 526–533. doi:10.1016/j.protcy.2015.02.075

Gan, Y., Dai, X., & Li, D. (2013). Off-line programming techniques for multi-robot Cooperation System. *International Journal of Advanced Robotic Systems*, *10*(7), 1–17. doi:10.5772/56506

Hoang, G., Min, J. H., Lee, G. M., Jun, B. H., Kim, H. K., & Kim, S. B. (2014, October), Omni-directional walking control for a six-legged robot using differential kinematics algorithm, In *Proceedings of 2014 14th International Conference on Control, Automation and Systems (ICCAS 2014)* (pp. 1163–1168). Academic Press. doi:10.1109/ICCAS.2014.6987735

Koker, R. (2013, February). A genetic algorithm approach to a neuralnetwork-based inverse kinematics solution of robotic manipulators based on error minimization. *Information Sciences*, *222*, 528–543. doi:10.1016/j.ins.2012.07.051

Koker, R., Oz, C., Çakar, T., & Ekiz, H. (2004, December). A study of neural network based inverse kinematics solution for a three-joint robot. *Robotics and Autonomous Systems*, *49*(3), 227–234. doi:10.1016/j.robot.2004.09.010

Maeda, Y., Fujiwara, T., & Ito, H. (2014, September). Robot control using high dimensional neural networks. In *Proceedings of SICE Annual Conference 2014 -International Conference on Instrumentation, Control, Information Technology and System Integration-* (pp. 738–743). Academic Press. doi:10.1109/SICE.2014.6935220

Mao, Z., & Hsia, T. C. (1997, January). Obstacle avoidance inverse kinematics solution of redundant robots by neural networks. *Robotica*, *15*(1), 3–10. doi:10.1017/S0263574797000027

Nagata, F., Inoue, S., Fujii, S., Otsuka, A., Watanabe, K., & Habib, M. K. (2015, August). *Learning of inverse kinematics using a neural network and its application to kinematic control of position-based servo motor*. Paper presented at The 2015 World Congress on Advances in Aeronautics, Nano, Bio, Robotics, and Energy (ANBRE15), Incheon, South Korea.

Nagata, F., Otsuka, A., Sakakibara, K., Watanabe, K., & Habib, M. K. (2013, September), Experiment Systems Using Three Types of Motors for Biomimetic Machine Research, In *Proceedings of SICE Annual Conference 2013 -International Conference on Instrumentation, Control, Information Technology and System Integration* (pp. 2711–2717). Academic Press.

Nagata, F., & Watanabe, K. (2016). Neural network-based inverse kinematics for Motoman HS20 and its efficient learning method. *Journal of the Institute of Industrial Applications Engineers*, *4*(4), 166–171. doi:10.12792/JIIAE.4.166

Tejomurtula, S., & Kak, S. (1999). Inverse kinematics in robotics using neural networks. *Information Sciences*, *116*(2-4), 147–164. doi:10.1016/S0020-0255(98)10098-1

Xia, Y., & Wang, J. (2001, February). A dual neural network for kinematic control of redundant robot manipulators. *IEEE Transactions on Systems, Man, and Cybernetics. Part B, Cybernetics*, *31*(1), 147–154. doi:10.1109/3477.907574 PMID:18244777

KEY TERMS AND DEFINITIONS

Back Propagation Algorithm: The most famous and reliable algorithm based on the concept of the steepest decent method and can train weights between each neuron in neural networks based on errors. The errors are defined as the quantity between the output of neural network and teaching signal prepared in advance.

Industrial Robots: Progressed remarkably as one of the powerful automation machines and been applied to many tasks such as welding, paining, handling, assembling, and so on. General industrial robots have a serial link structure with five, six, or seven DOFs pursuing the function of a human arm, so that more flexible motion than NC machine tools can be realized.

Kinematics: Quantitatively handles the relation between joint angles in joint coordinate system and the corresponding position/orientation elements of the arm tip in Cartesian coordinate system without considering the masses of those parts nor the torques and forces that drive the joints.

Neural Networks: Typically composed of multiple layers and the signal traverses from the input layer to the output layer of neurons. Trained neural networks can approximate an arbitrary nonlinear function.

Serial Link Structure: The typical configuration of articulated-type industrial robots. Each link of the industrial robot is connected with a servo motor-driven joint for rotational motion. Flexible position and orientation motion with multi-DOFs is characterized.

Weights Update Method: Multi-layered neural networks have to be trained to acquire the non-linear relation between input and output teaching signals. The training means weights update the process to enable the neural networks to produce the output teaching signals according to the input teaching signals. In this chapter, a novel efficient weights update method is proposed.

Chapter 10
Ant-Like Walking Behavior of MEMS Microrobot With Artificial Neural Networks IC

Ken Saito
University of California – Berkeley, USA & Nihon University, Japan

Minami Kaneko
Nihon University, Japan

Fumio Uchikoba
Nihon University, Japan

ABSTRACT

This chapter explains how the MEMS microrobot system could perform the walking behavior of ants. MEMS microrobot system consists of micro-mechanical systems and micro-electro systems. The micro-mechanical systems mimic the alternating tripod gait of an ant by the shape memory alloy-type rotary actuator and the link mechanism. The micro-electro systems mimic the electrical activity of biological neural networks using the artificial neural networks IC. The artificial neural networks IC generates the driving pulses of shape memory alloy-type rotary actuator without using software programs. The micro-mechanical systems and micro-electro systems are integrated as a robot system. As a result, the authors show that the MEMS microrobot system could perform the ant-like walking behavior with a speed of 20 mm/min. The MEMS microrobot system was 0.079 g in weight, 4 mm width, 4 mm length, and 5 mm height in size. The robot system needs only the electrical power source as an external device.

INTRODUCTION

Biomimetics has been studied in many research fields because organisms have useful functions to develop technology (e.g., Habib, 2011; Seok et al., 2015). In particular, the locomotion mechanisms of insects attract the attention of researchers to develop microrobots. Hoover et. al. (2008) proposed the crawling hexapod robot RoACH, and Wood et. al. (2012) proposed the flying microrobot RoboBee. Mimicking the locomotion mechanisms of insects is useful to realize small sized actuators for millimeter scale microrobots. If the microrobot system could perform such as insects, microrobots can be used for agriculture,

DOI: 10.4018/978-1-5225-2993-4.ch010

manufacturing, medicine, informational technology and so on. However, further miniaturization of the microrobot is required. In the further miniaturizations, some researchers use micro fabrication technology to fabricate the small sized actuators of the microrobot (e.g., Donald, Levey, McGray, Paprotny, & Rus, 2006; Edqvist, Snis, Mohr, Scholz, Corradi, Gao, ... Johansson, 2009). The development of the small sized actuator was categorized into some groups (e.g., Fearing, 1998). For example, electric field driven actuators, magnetic field driven actuators, piezoelectric driven actuators, shape memory alloy driven actuators, electrostatic driven actuators, ion-exchange polymer driven actuators, and so on (e.g., Tang, Nguyen, & Howe, 1989; Sniegowski, & Garcia, 1996; Asada, Matsuki, Minami, & Esashi, 1994; Suzuki, Tani, & Sakuhara, 1999; Surbled, Clerc, Pioufle, Ataka, & Fujita, 2001; Yeh, Hollar, & Pister, 2002; Bell, Lu, Fleck, & Spearing, 2005; Ryu, Jeong, Tak, Kim, Kim, & Park 2002). The actuators using field forces are advantageous for miniaturization because the power source and the controller of the actuator do not need to be implemented on the robot body. However, the robot would not be able to locomote without the external controller. Thus, the actuators have to generate the driving force by themselves and be integrated inside the robot body to locomote the robot independently. The researchers have to choose the actuators because each actuator has different advantages, such as power consumption, switching speed, force generation, displacement, fabrication difficulty, and so on. Also, actuators can only generate rotary motion or linear motion; therefore, mechanical design is necessary to create microrobots that locomote using a step pattern.

Besides further miniaturization, microrobot systems require higher functionalization. Insects can control the movement according to the environmental situation because the brain can process the sensory information. A small size autonomous intelligent system realized by the brain can be a good example for the microrobot system. However, it is difficult to program the autonomous intelligent system on a small microcontroller. Some advanced studies have been done for applying artificial neural networks to robot control. Many studies have reported on both software models and hardware models (e.g., Matsuoka, 1987; Ikemoto, Nagashino, kinouchi, & Yoshinaga, 1997; Nakada, Asai, & Amemiya, 2003). The software model is useful because it is easy to program on a microcontroller. The program could change or update under different conditions. In contrast, using the hardware model is difficult to reconstruct the structure of the circuit. The insects use small sized biological neural networks which process information by oscillatory patterns. Hardware models which can generate an oscillatory pattern such as a biological neuron have possibilities to construct a new information processing method in a simple and compact system.

The authors are studying applying biomimetics to the millimeter-sized robot system. A microrobot of 4.0 mm width, 4.0 mm length, and 3.5 mm height in size and artificial neural networks constructed by a discrete circuit board were reported (e.g., Okazaki, Ogiwara, Yang, Sakata, Saito, Sekine, & Uchikoba, 2011). The microrobot had been downsized to 4.0 mm in width, 2.7 mm in length and 2.5 mm in height in a later work (e.g., Saito, Takato, Sekine, & Uchikoba, 2012). In addition, artificial neural networks were constructed by integrated circuit (IC) for the purpose of minimizing the circuit size. However, a peripheral circuit was needed to drive the shape memory alloy-type rotary actuator in the previous IC chip of artificial neural networks. The peripheral circuit consisted of the transistors to amplify the electric current and the operational amplifier for the buffer. The weight of the MEMS microrobot system was 0.33 g (IC with peripheral circuit: 0.31 g; MEMS microrobot: 0.02 g). The MEMS microrobot system could not locomote because the IC with the peripheral circuit was too heavy. Therefore, a reduction in size and weight for the IC with a peripheral circuit was done (e.g., Saito, Maezumi, Naito, Hidaka, Iwata, Okane, Oku, Takato, & Uchikoba, 2014). The authors made the optimal design to the actuator according to the result of the thermal analysis. As a result, MEMS microrobot system could perform the continuous and high speed walking (e.g., Saito, Sugita, Ishihara, Iwata, Asano, Okane, Ono, Chiba, Takato, & Uchikoba, 2017).

In this chapter, authors compile the results of previous research to explain how MEMS microrobot system could perform the walking behavior of ants. Firstly, the micro-mechanical systems of the MEMS microrobot are shown. Secondly, micro-electro systems of the MEMS microrobot using artificial neural networks are discussed. Finally, we combine the micro-electro systems and micro-mechanical systems as a MEMS microrobot system.

MICRO-MECHANICAL SYSTEMS

Design and Fabrication

The authors used microfabrication technology to construct the micro-mechanical systems of the MEMS microrobot. The micro-mechanical systems consisted of shape memory alloy-type rotary actuators, body frame, link mechanisms and six legs.

Figure 1 shows the silicon parts of the micro-mechanical system of the MEMS microrobot. All mechanical parts were made from a silicon wafer except the shaft and the shape memory alloy. 100 μm, 200 μm, 560 μm thickness silicon wafers were used (See Table 1 for specifications of the silicon wafer). The shapes of the silicon parts are machined by inductively coupled plasma (ICP) dry etching with photolithography technology (Bhardwaj, & Ashraf. 1995). The body frame consists of a center frame, a front frame, a rear frame and a top frame. The rotor is part of the shape memory alloy-type rotary actuator. Link mechanisms and six legs consist of two types of leg and two types of link bars. In addition, 0.1 ± 0.002 mm diameter cemented carbide was used for a shaft. The washer had used to the end of a shaft to fixed the silicon parts such as legs and link bars. All silicon parts are clearance-fitted to the other parts by 0.01 mm gap.

Figure 2 shows the shape memory alloy-type rotary actuator of the MEMS microrobot. Figure 2 (a) shows the exploded view and Figure 2 (b) shows the picture after being assembled. The body frames are fixed to each other using cyanoacrylate. The rotary actuator consists four pieces of shape memory alloy, shaft, GND wire, and a rotor. The rotary actuator uses the shape memory alloy to generate the rotational movement which is necessary for the locomotion of the microrobot. The shape memory alloy can change its length according to temperature (Over 70 °C: Shrink. Under 70 °C: Extend.). Therefore, our constructed actuator could not rotate if the temperature of the shape memory alloy could not cool less than 70 °C. In this study, the length of the shape memory alloy was changed by supplying or stopping the current flowing. One side of the shape memory alloy is connected to the rotor and the other side is connected to the copper wires by using solder paste (Figure 2 (b)). The GND wire is fixing to the rotor directly. The rotor of the rotary actuator is connected by the shape memory alloy to the body frame. The authors use the BioMetal® Helix BMX50 for the shape memory alloy (Available online at: http://www.toki.co.jp/biometal/english/contents.php). The basic characteristics of the shape memory alloy were as follows. The standard coil diameter of the shape memory alloy was 0.2 mm where wire diameter was 0.05 mm. The practical maximum force produced from 3 gf to 5 gf where kinetic displacement was 50%. The standard drive current was from 50 mA to 100 mA where standard electric resistance was 3600 ohm/m. The authors used the shape memory alloy because the generating force was enough to generate the rotary movement of the rotor. Previously, the authors proposed the piezoelectric-type rotary actuator (Takato,

Figure 1. Silicon parts of the micro-mechanical system

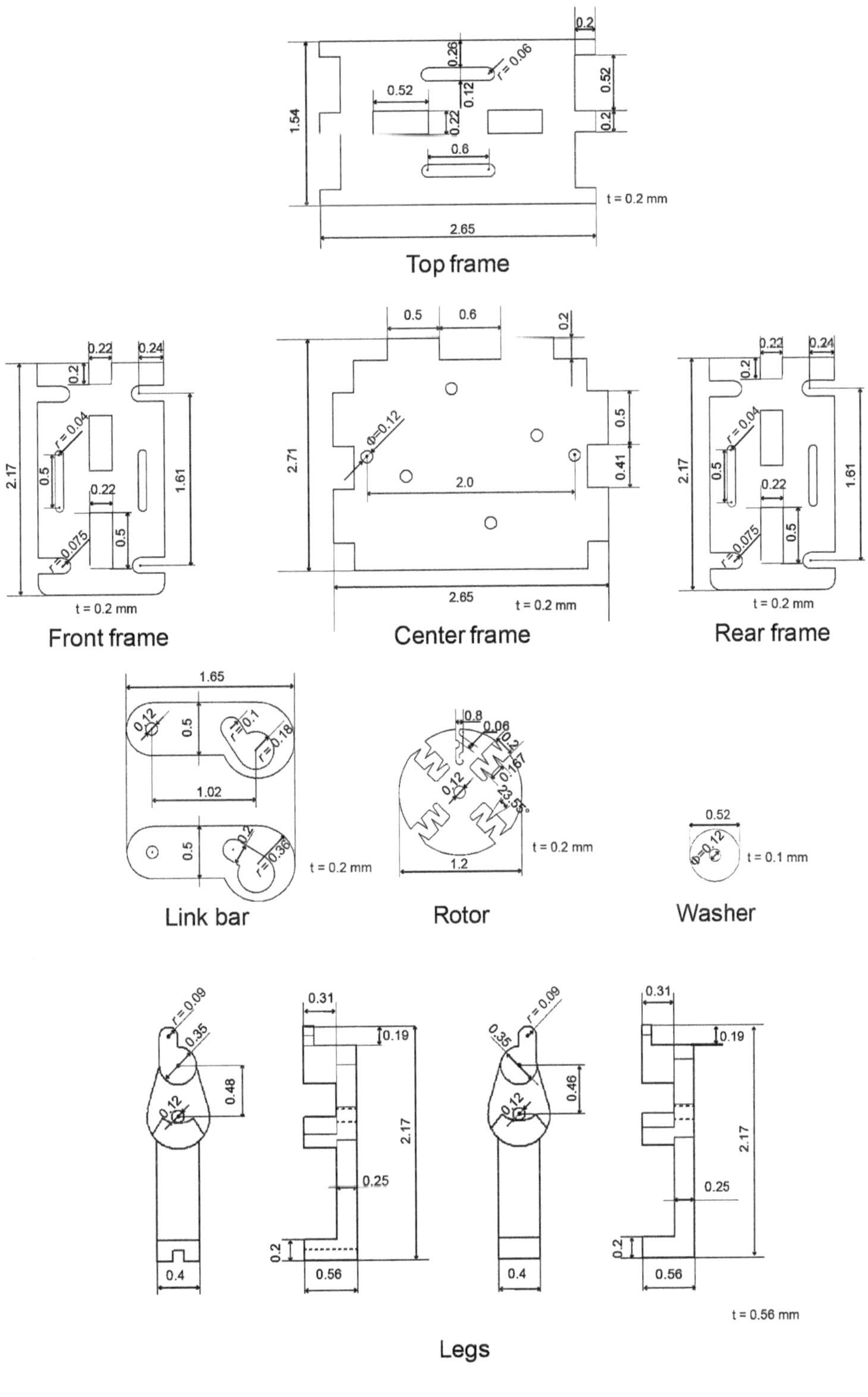

Table 1. Specifications of the silicon wafer

	Specifications
Diameter	100 ± 0.5 mm.
Type	p-type or n-type.
Crystal plane orientation	(100) ± 1°.
Thickness	100 ± 20 μm 200 ± 20 μm, and 560 ± 20 μm.
Resistivity	From 0.5 Ω • m to 100 Ω • m.
Orientation Flat	(100) ± 1°, 32.5 ± 2.5 mm.
Condition of the surface	Double side polished.
Particle count	Equal to or less than 10 (Larger than 3 μm particle).

Figure 2. Shape memory alloy-type rotary actuator

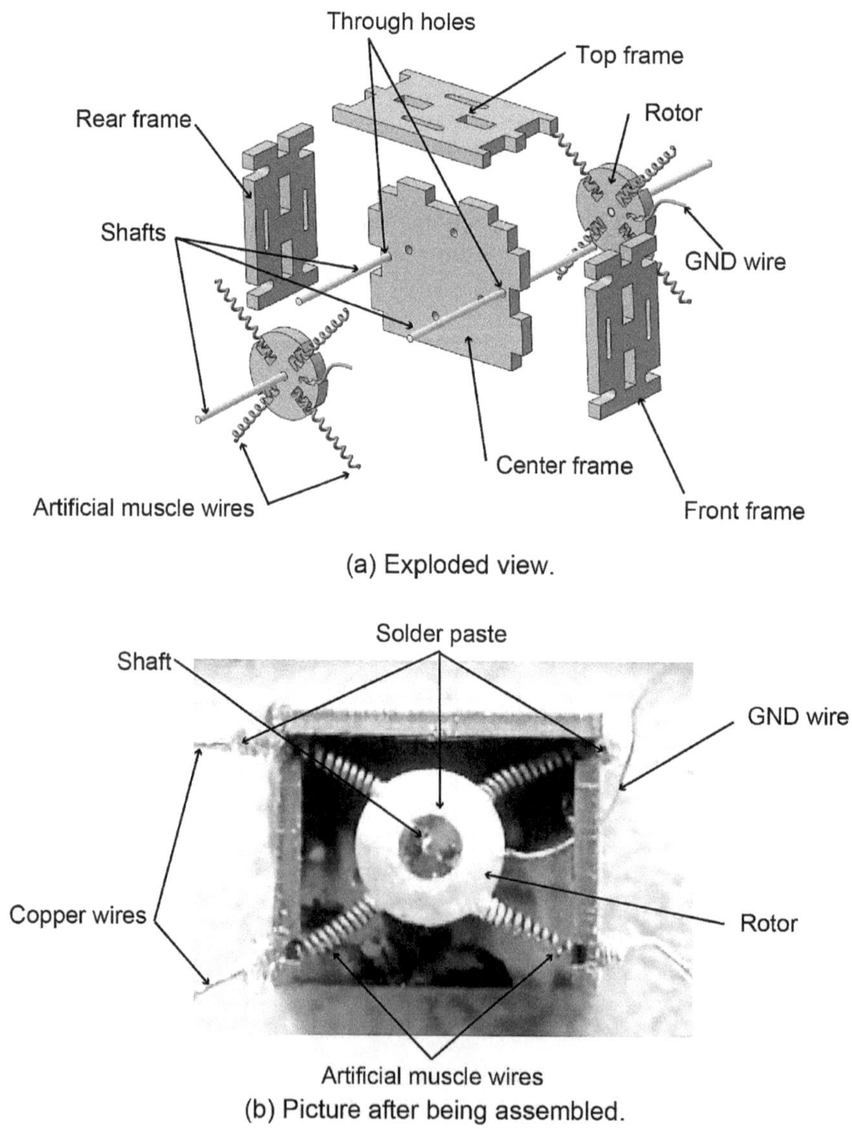

Naito, Maezumi, Ishihara, Okane, Oku, Tatani, Saito, & Uchikoba. 2015). The energy consumption was low compard with the shape memory alloy. However, the generating force was not enough to generate the rotary movement of the rotor.

Figure 3 shows the micro-mechanical system of the fabricated MEMS microrobot. Figure 3 (a) shows the exploded view and Figure 3 (b) shows the picture after being assembled. The link mechanism consists of three legs, shafts, and link bars. The middle legs are attached to the rotor using the shafts; therefore, the middle legs and the rotors have the same rotational phases. By contrast, the front legs and the rear legs are attached to the middle legs by link bars using shafts; therefore, the front legs and the rear legs can generate ± 90 ° phase shift comparing with middle legs. For attaching each part not to come off

Figure 3. Micro-mechanical system of the fabricated MEMS microrobot

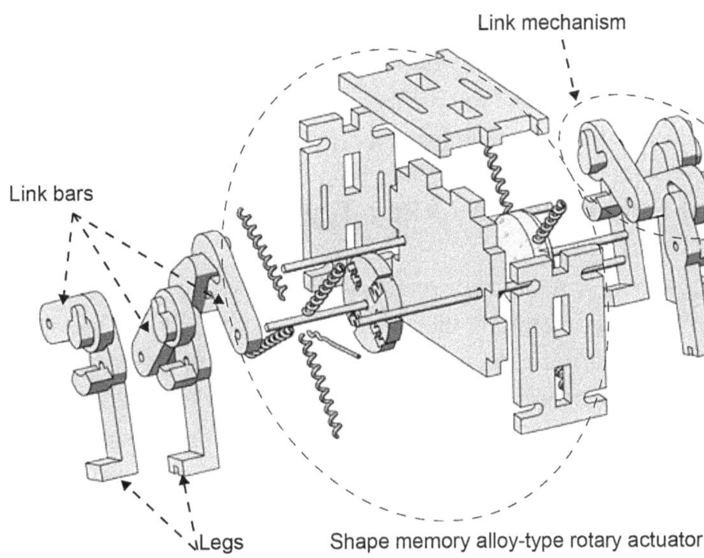

(a) Exploded view

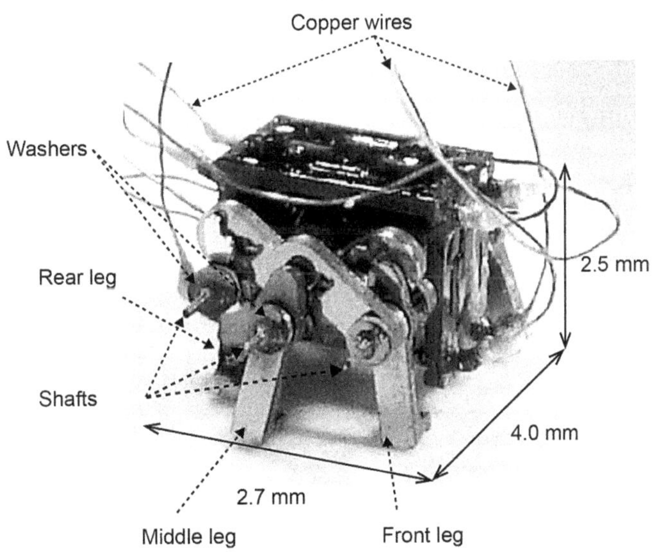

(b) Picture after being assembled

from a shaft, the washer and the shaft fixed using cyanoacrylate. The MEMS microrobot measures 4.0 mm in width, 2.7 mm in length, and 2.5 mm in height. The copper wires of figure 2 (b) show the eight signal wires and two GND wires, the MEMS microrobot can walk when connecting the signal wires to the artificial neural networks IC.

Mechanisms of Ant-Like Walking Behavior

Authors replicate the locomotion mechanisms of an ant using link mechanisms and shape memory alloy-type rotary actuators.

Figure 4 shows the mechanism of ant-like walking behavior. Figure 4 (a) shows the alternating tripod gait of an ant. The alternating tripod gait is very stable because three legs are always touching a ground. Figure 4 (b) shows the alternating tripod gait of the MEMS microrobot. The MEMS microrobot can perform the alternating tripod gait by the rotation of the rotary actuator. In the case of heating the shape memory alloy wires from A to D, the microrobot performs forward locomotion. In contrast, heating the shape memory alloy from D to A, the microrobot performs backward locomotion. The locomotion pattern of the left side legs and the right side legs has 180 ° phase shift.

Heat Conduction of Shape Memory Alloy-Type Rotary Actuator

Figure 5 shows the heat conduction of the shape memory alloy-type rotary actuator. In this figure, sections over 70 °C are indicated by white, therefore, the area of sections over 70 °C increases as time proceeds. The heat of the shape memory alloy conducts to the mechanical components of the MEMS microrobot. The shape memory alloy shrinks over 70 °C and extends under 70 °C. The thermal difference needs to be extended. Running for more than 20s raises the temperature of every shape memory alloy actuator above 70C, because of thermal conductivity. This conductivity does not allow the actuators to cool down, slowing the locomotion. To solve this problem, authors constructed the new type rotary actuator to reduce the heat conduction. The thermal conductivity of each material at room temperature is as follows; solder paste: 10.0 W / (m • K), shape memory alloy: 12.1 W / (m • K) Si: 149 W / (m • K). According to the thermal conductivity of each material, solder paste keeps the heat of the shape memory alloy.

Figure 4. Mechanism of ant-like walking behavior

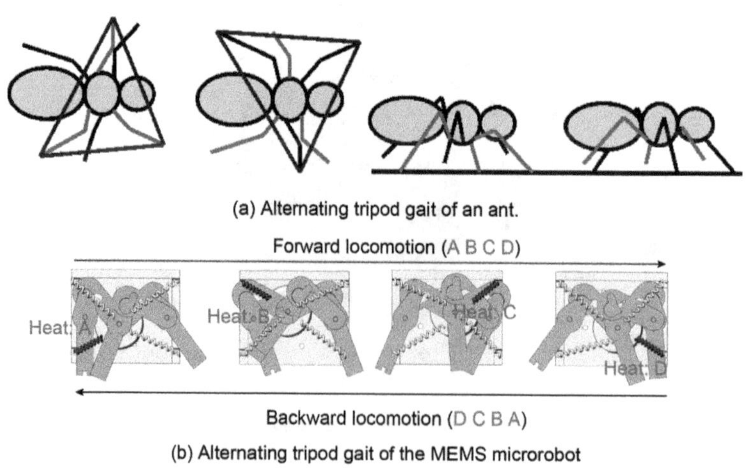

Figure 5. Heat conduction of the shape memory alloy-type rotary actuator

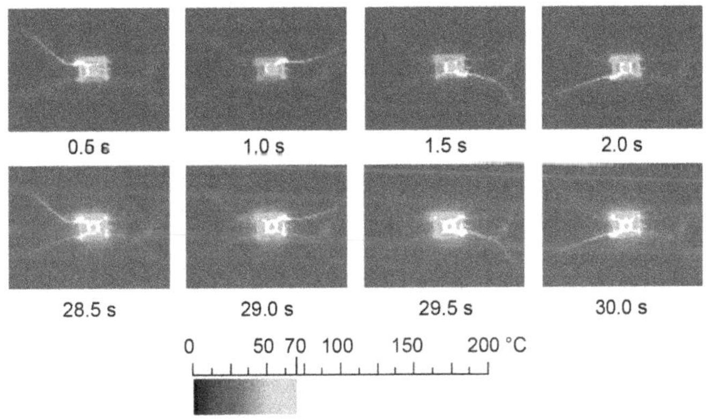

Improvement of Shape Memory Alloy-Type Rotary Actuator

According to the results of thermal conductivity, the authors designed the small rotor to reduce the uses of the solder paste. Figure 6 shows the shape memory alloy-type rotary actuator with a small sized rotor. Figure 6 (a) shows the exploded view of the rotary actuator and Figure 6 (b) shows the picture after being assembled. The differences between the two types of the shape memory alloy-type rotary actuators are the shape of the rotor and the number of the shape memory alloys (The differences between Figure 2 and Figure 6). The rotary actuator of Figure 2 uses four pieces of shape memory alloy while the rotary actuator of Figure 6 uses two pieces of shape memory alloy. Both types of the rotary actuators are assembled with exactly the same frame parts.

Figure 7 shows the heat conduction comparison of the shape memory alloy-type rotary actuator. Figure 7 (a) shows the heat conduction of rotary actuator of Figure 2 where Figure 7 (b) shows the heat conduction of rotary actuator of Figure 6. The driving pulse width is set to 0.2 s and the pulse period is set to 0.8 s. It is shown that the rotary actuator of Figure 2 cannot radiate under 70 °C; the shape memory alloy cannot generate the continuous rotational movement. By contrast, the rotary actuator of Figure 6 can keep the temperature at under 70 °C; the shape memory alloy can generate the continuous rotational movement.

Figure 8 shows the comparison of movements of the rotor of the rotary actuator of Figure 2 and Figure 6. Figure 8 (a) shows the rotor movements of the rotary actuator of Figure 2 and Figure 8 (b) shows the rotor movements of the rotary actuator of Figure 6. In these figures, solid circles indicate the center of the rotor before 10 s and open circles indicate the center of the rotor after 30 s. The solid lines indicate the trajectory of the solid circles and the broken lines indicate the trajectory of the open circles. It is shown that the rotary actuator of Figure 6 can generate continuous rotational movements after 30 s. The displacement of the shape memory alloy depends on its temperature. The displacement versus time of the shape memory alloy is about 0.63 mm/s (measurement value of Figure 8, where the pulse amplitude is 4 V, the pulse width is 0.2 s and the pulse period is 0.8 s.). The rotary rate of the rotor depends on the pulse period. For example, it is 30 rpm in the case of a pulse period that is 2 s and it is 75 rpm in the case of a pulse period that is 0.8 s. The maximum force produced by the shape memory alloy is 5 gf. The power consumption of the shape memory alloy-type rotary actuator depends on its temperature. The power consumption is about 80–400 mWh.

Figure 6. Shape memory alloy-type rotary actuator with small size rotor

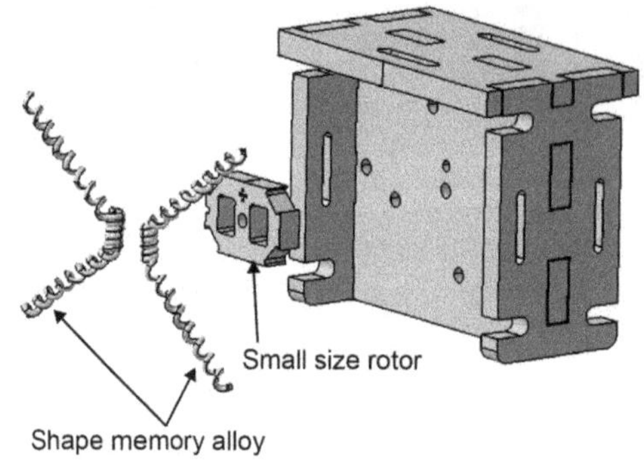

(a) Exploded view.

(b) Picture after being assembled.

Figure 7. Heat conduction comparison of the shape memory alloy-type rotary actuator

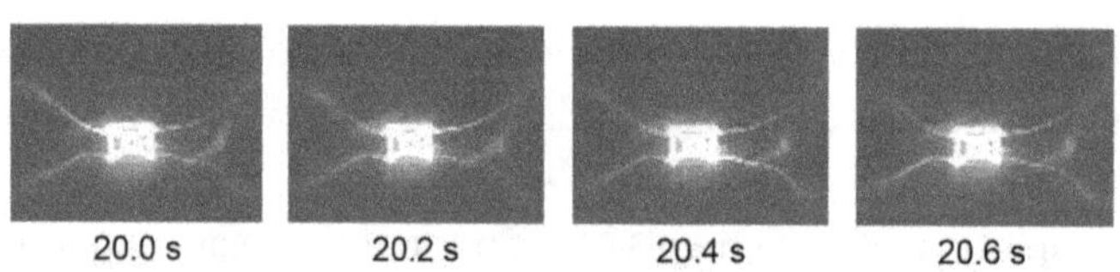

(a) Heat conduction of Figure 2

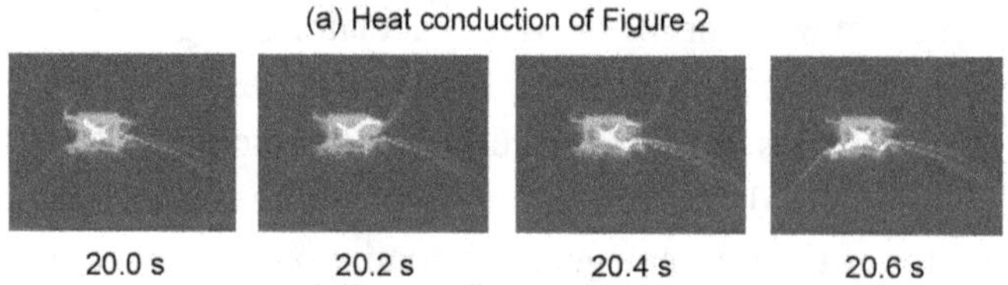

(b) Heat conduction of Figure 6

Figure 8. Comparison of movements of the rotor of the rotary actuator

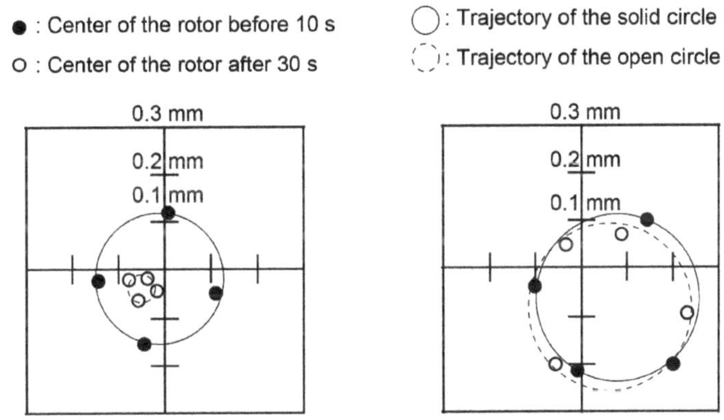

(a) Rotary actuator of the Figure 2 (b) Rotary actuator of the Figure 6

MICRO-ELECTRO SYSTEM

Design and Fabrication

The authors constructed the artificial neural networks using the pulse-type hardware neuron model as the basic component. The pulse-type hardware neuron model is a class II neuron model. The class II neuron model has the same features as biological neurons such as spatio-temporal summation characteristics, refractory period and threshold. Therefore, pulse-type hardware neuron model could generate the continuous pulse waves such as biological neurons.

Figure 9 shows the circuit diagram of the pulse-type hardware neuron model. The pulse-type hardware neuron model consists of a synaptic model and a cell body model. Figure 9 (a) and Figure 9 (b) shows the circuit diagrams of the inhibitory synaptic model and the cell body model, respectively. The synchronization phenomena of the cell body model could change by the coupling of the synaptic model. Coupling by the excitatory synaptic model and coupling by the inhibitory synaptic model causes the in-phase synchronization and anti-phase synchronization, respectively. The circuit parameters of the synaptic model are as follows: C_{IS}=1 pF, $M_{IS1\text{-}5}$: W/L=1 (where W is an effective channel width and L is an effective channel length of the MOSFET). The voltage source V_{ISDD} was set to 5 V. The circuit parameters of the cell body model are as follows: C_G=39 μF, C_M=270 nF, M_{C1}, M_{C2}: W/L=10, M_{C3}: W/

Figure 9. Circuit diagram of the pulse-type hardware neuron model

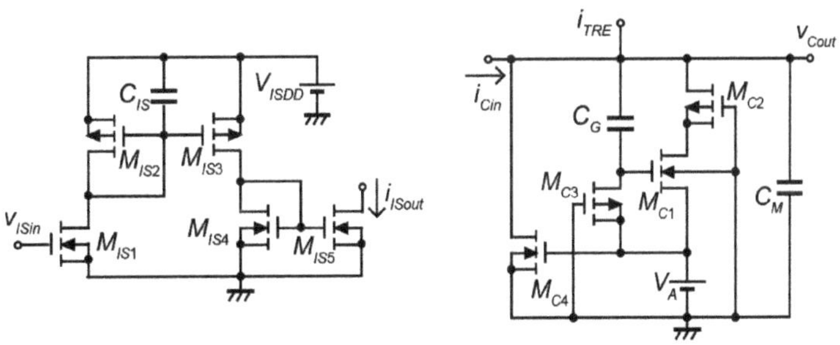

(a) Inhibitory synaptic model (b) Cell body model

L=0.1, M_{C4}: W/L=0.3. The voltage source V_A was set to 3.3 V. The input voltage v_{ISin} of the synaptic model are output voltage v_{Cout} of the other cell body model. The input current i_{Cin} of the cell body model is output current i_{ISout} of the synaptic model.

The authors designed the artificial neural networks by the inhibitory mutual coupling which could generate the anti-phase synchronization. The constructed artificial neural networks are shown in Figure 10. Figure 10 (a) shows the connection diagram and Figure 10 (b) shows the layout pattern of the artificial neural networks, respectively. The four cell body models are mutual coupled by 12 inhibitory synaptic models. In Figure 10 (a), v_{CoutA}, v_{CoutB}, v_{CoutC} and v_{CoutD} indicates the output ports of the artificial neural networks. In addition, i_{TREA}, i_{TREB}, i_{TREC} and i_{TRED} indicate the trigger pulse input ports of the artificial neural networks. The artificial neural networks can change the sequence of driving pulses according to the input timing of single external trigger pulse. The design rule of the IC is four metal two poly CMOS 0.35 μm. The cell body models output the voltage waveform. However, the shape memory alloy-type rotary actuators need electrical current to generate the rotational movements. Therefore, we convert the voltage waveform to current waveform.

Figure 11 (a) shows the circuit diagram of the current mirror circuit. The current mirror circuit generates the output current waveform i_{Oout} according to the voltage waveform v_{Cout} (v_{Oin}). The circuit parameters of the current mirror circuit are as follows: M_1: W/L=40, M_2: W/L=1, M_{On}: W/L=66.7. In

Figure 10. Design of artificial neural networks

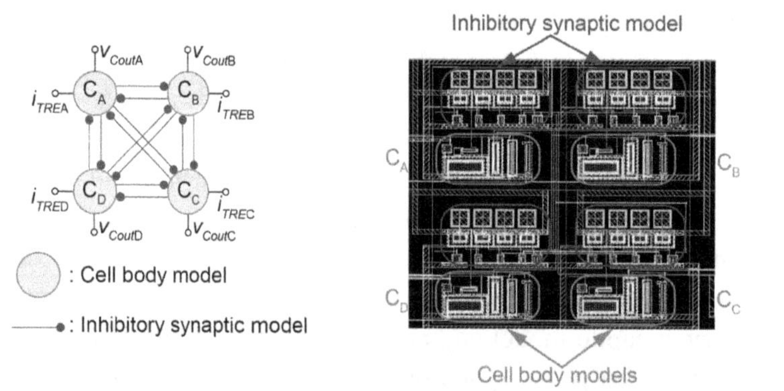

(a) Connection diagram (b) Layout pattern of the artificial neural networks

Figure 11. Circuit diagram of the current mirror circuit

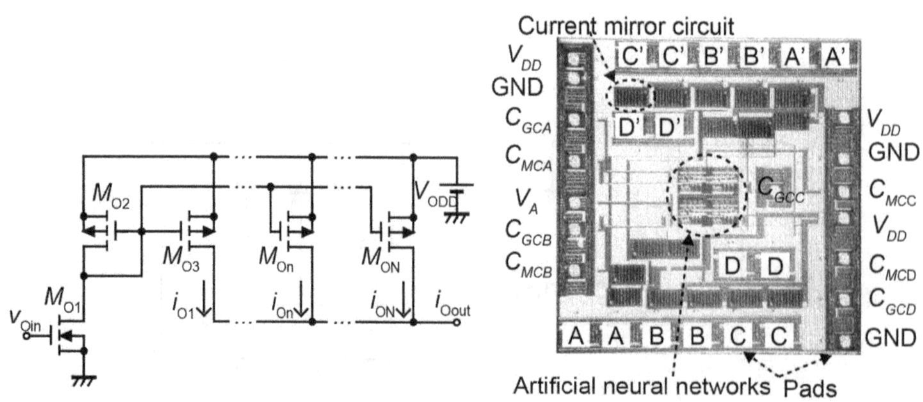

(a) Current mirror circuit (b) Layout pattern of the IC

this chapter, we set as N=11 ($3 \leq n \leq N$) because the wire bonding between the bare chip IC and the robot can only output 50 mA. The authors constructed the bare chip IC of artificial neural networks with current mirror circuit. The bare chip IC is fixed to the cavity of the flame retardant 4 (FR4) substrate. Each pad of the bare chip IC is connected to pad pattern of the FR4 substrate. We use the ultrasonic wire bonding to fix the aluminum wire. Figure 11 (b) shows the layout pattern of the IC of artificial neural networks with current mirror circuit. In Figure 11 (b), several pads are arranged to fix the wire bonding. In addition, two pads are placed for each port of microrobot. This is because shape memory alloy-type rotary actuator needs large current to actuate.

Figure 12 shows the bare chip IC with peripheral circuits. Figure 12 (a) and (b) show the top view and bottom view of the bare chip IC with FR4 substrate, respectively. The bare chip IC and FR4 substrate are connected by wire bonding. The top side and bottom side are electrically connected by through hole. Figure 12 (c) shows the output waveform of the bare chip IC (measured value). The capacitors of the cell body model were impossible to implement inside the bare chip IC because of the size. Therefore, the capacitors C_G and C_M are mounted externally of the bare chip IC as shown in Figure 12 (b). Artificial neural networks are coupled neural networks systems which can generate the same locomotion rhythms as living organisms. To heat the shape memory alloy, the input of the pulse of 0.2 s - 0.5 s in width, 0.8 s - 2 s in a period and 75 mA in amplitude is required. According to the Figure 12 (c), the output waveform is enough to actuate the rotary actuator because the resistance of the shape memory alloy-type rotary actuator is about 50 Ω.

Figure 12. Bare chip IC of artificial neural networks with current output circuit

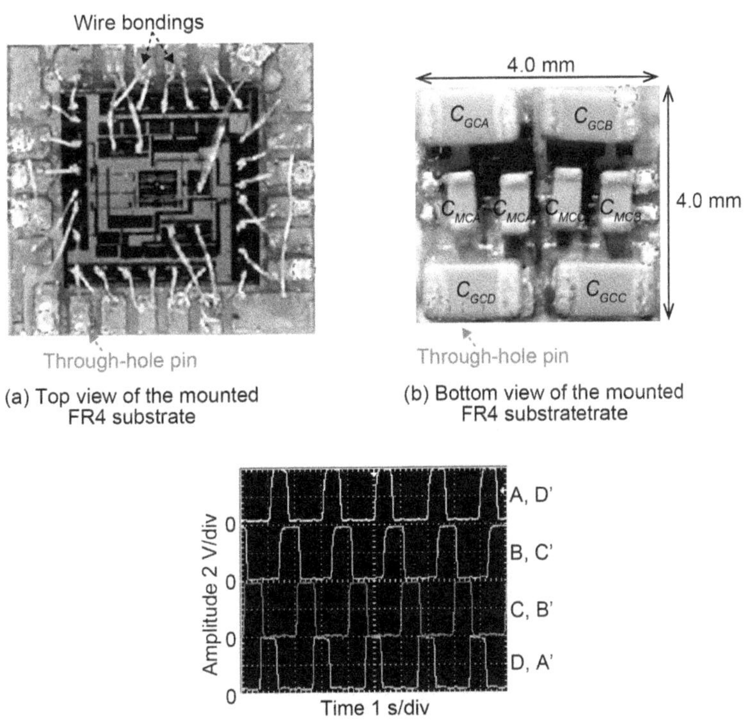

(a) Top view of the mounted FR4 substrate

(b) Bottom view of the mounted FR4 substratetrate

(c) Output waveform of the bare chip IC (measured value)

MEMS MICROROBOT SYSTEM

Figure 13 shows the assembled MEMS microrobot system. We mount the bare chip IC with the FR4 substrate on top of the microrobot. The whole system is 0.079 g in weight and less than 5.0 mm in size (Micro-mechanical system: 0.02 g in weight; 4.0 mm in width, 2.7 mm in length and 2.5 mm in height in size. Micro-electro system: 0.059 g in weight; 4.0 mm in width, 4.0 mm in length and 2.5 mm in height in size).

ANT-LIKE WALKING BEHAVIOR OF MEMS MICROROBOT

Figure 14 shows the example of locomotion of the MEMS microrobot system. Artificial neural networks can generate the driving pulse which is necessary to locomote the MEMS microrobot. The locomotion speed of the MEMS microrobot system is 4 mm/min. The MEMS microrobot system without mounting the bare chip IC with FR4 substrate was 26.4 mm/min [21]. The locomotion of the MEMS microrobot system is much slower when mounted with the bare chip IC with FR4 substrate. One of the reasons for slow locomotion speed is the heat conduction of the shape memory alloy-type rotary actuator. This can be solved by changing the rotor of the shape memory alloy-type rotary actuator to small size. The other reasons is the copper wire around the MEMS microrobot system. The copper wire limits the locomotion of the robot.

To solve the slow locomotion, the authors redesigned the body frame for internal wiring.

Figure 15 shows the internal wired MEMS microrobot system. Figure 15 (a) shows the body frames of the internal wired MEMS microrobot system. The other mechanical parts are exactly same as Figure 1. The copper wire connecting the microrobot and FR4 substrate was wired as shown in Figure 15 (b).

Figure 13. MEMS microrobot system

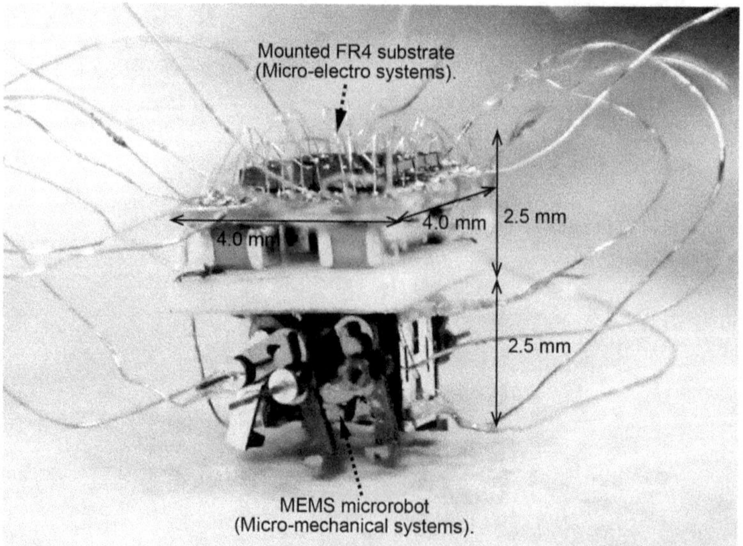

Figure 14. Example of locomotion of the MEMS microrobot system

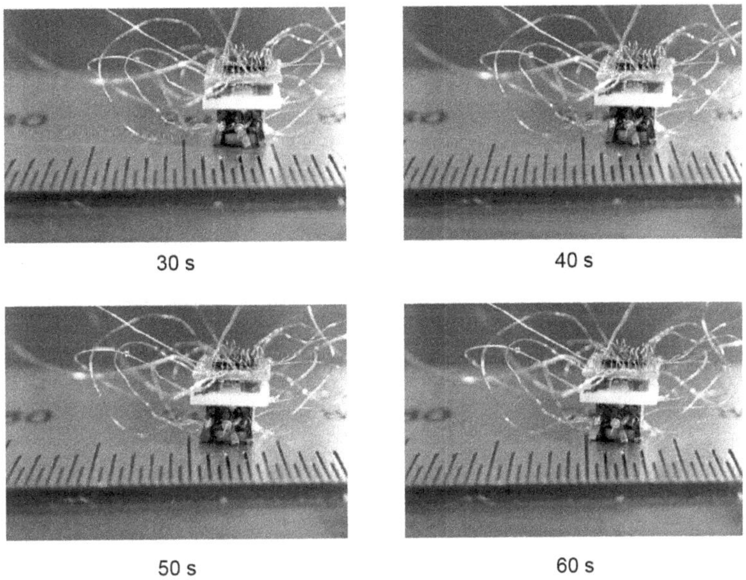

Figure 15 (c) shows the internal wired MEMS microrobot system after assembly. It is shown that the copper wire around the MEMS microrobot system has been improved.

Figure 16 shows the example of locomotion of the internal wired MEMS microrobot system. The locomotion speed of the MEMS microrobot system is 20 mm/min.

CONCLUSION

This chapter explained how the MEMS microrobot system could perform the walking behavior of ants.

As a result, the authors show that the MEMS microrobot system could perform the ant-like walking behavior at a speed of 20 mm/min. The MEMS microrobot system was 0.079 g in weight; 4 mm width, 4 mm length, and 5 mm height in size. The robot system needs only the electrical power source as an external device.

Now the authors are looking into using the electrostatic actuator to replace the shape memory alloy-type rotary actuator. The electrostatic actuator could actuate using lower power. Also, the high-voltage silicon photovoltaic cells will add to the microrobot system to power the electrostatic actuator. In the future, the artificial neural networks, the actuator, the power source system and the sensory system will be directly integrated with the silicon frame parts of the microrobot.

ACKNOWLEDGMENT

Authors would like to thank Prof. Liwei Lin and the members of Lin laboratory. Authors would also like to thank Daniel Contreras for proofreading. The fabrication of the MEMS microrobot was supported by Research Center for Micro Functional Devices, Nihon University. The VLSI chip in this study has been

Figure 15. Internal wired MEMS microrobot system

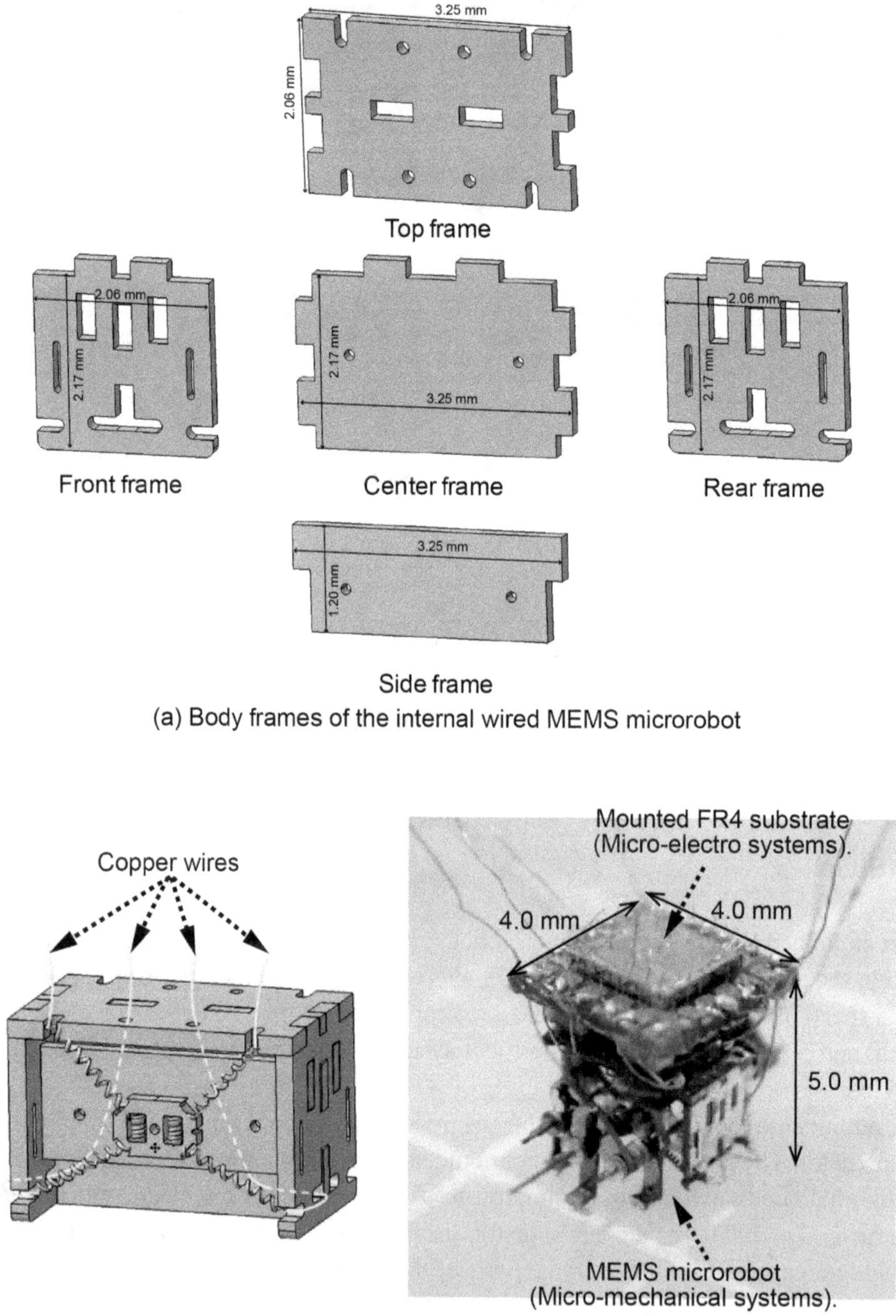

(a) Body frames of the internal wired MEMS microrobot

(b) Copper wire connection

(c) Internal wired MEMS microrobot system

fabricated by Digian Technology, Inc. This work is supported by VLSI Design and Education Center (VDEC), the University of Tokyo in collaboration with Synopsys, Inc., Cadence Design Systems, Inc. and Mentor Graphics, Inc.

Figure 16. Example of locomotion of the internal wired MEMS microrobot system

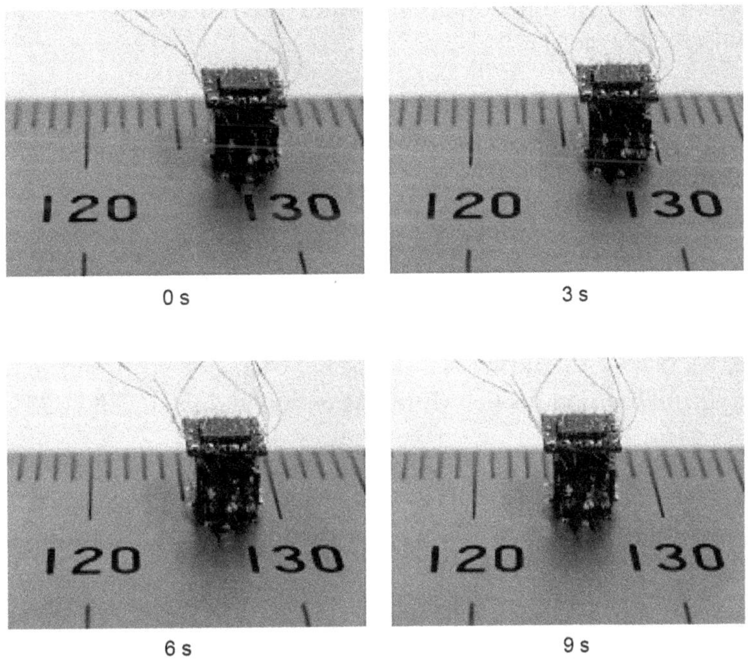

REFERENCES

Asada, N., Matsuki, H., Minami, K., & Esashi, M. (1994). Silicone micromachined two-dimensional galvano optical scanner. *IEEE Transactions on Magnetics*, *30*(6), 4647–4649. doi:10.1109/20.334177

Bell, D. J., Lu, T. J., Fleck, N. A., & Spearing, S. M. (2005). MEMS actuators and sensors: Observations on their performance and selection for purpose. *Journal of Micromechanics and Microengineering*, *15*(7), S153–S164. doi:10.1088/0960-1317/15/7/022

Bhardwaj, J. K., & Ashraf, H. (1995). Advanced silicon etching using high-density plasmas. *Proc. of SPIE Micromachining and Microfabrication Process Technology, 2639*, 224-233. doi:10.1117/12.221279

Donald, B. R., Levey, C. G., McGray, C. D., Paprotny, I., & Rus, D. (2006). An untethered, electrostatic, globally controllable MEMS micro-robot. *Journal of Microelectromechanical Systems*, *15*(1), 1–15. doi:10.1109/JMEMS.2005.863697

Edqvist, E., Snis, N., Mohr, R. C., Scholz, O., Corradi, P., Gao, J., ... Johansson, S. (2009). Evaluation of building technology for mass producible millimeter-sized robots using flexible printed circuit boards. *Journal of Micromechanics and Microengineering*, *19*(7), 1–11. doi:10.1088/0960-1317/19/7/075011 PMID:20161118

Fearing, R.S. (1998). Powering 3 Dimensional Microrobots: Power Density Limitations. *IEEE Int. Conf. Robotics and Automation*.

Habib, M. K. (2011). Biomimetcs: Innovations and Robotics. *International Journal of Mechatronics and Manufacturing Systems*, *4*(2), 113–134. doi:10.1504/IJMMS.2011.039263

Hoover, M. A., Steltz, E., & Fearing, S. R. (2008). RoACH: An autonomous 2.4 g crawling hexapod robot. *Proceedings of the 2008 IEEE/RSJ International Conference on Intelligent Robots and Systems*, 26–33. doi:10.1109/IROS.2008.4651149

Okazaki, K., Ogiwara, T., Yang, D., Sakata, K., Saito, K., Sekine, Y., & Uchikoba, F. (2011). Development of pulse control type MEMS micro robot with hardware neural network. *Artificial Life and Robotics, 16*(2), 229–233. doi:10.1007/s10015-011-0925-9

Ryu, J., Jeong, Y., Tap, Y., Kim, B., Kim, B., & Park, J. (2002). A Ciliary Motion Based 8-legged Walking Micro Robot Using Cast IPMC Actuators. *Proc. of 2002 International Symposium on Micromechatronics and Human Science.*

Saito, K., Maezumi, K., Naito, Y., Hidaka, T., Iwata, K., Okane, Y., ... Uchikoba, F. (2014). Neural Networks Integrated Circuit for Biomimetics MEMS Microrobot. *Robotics, 3*(3), 235–246. doi:10.3390/robotics3030235

Saito, K., Sugita, K., Ishihara, Y., Iwata, K., Asano, Y., Okane, Y., ... Uchikoba, F. (2017). Insect-Type MEMS Microrobot with Mountable Bare Chip IC of Artificial Neural Networks. *Artificial Life and Robotics, Springer, 22*(1), 118–124. doi:10.1007/s10015-016-0324-3

Saito, K., Takato, M., Sekine, Y., & Uchikoba, F. (2012). Biomimetics Micro Robot with Active Hardware Neural Networks Locomotion Control and Insect-Like Switching Behaviour. *International Journal of Advanced Robotic Systems, In Tech, 9*(226), 1–6.

Seok, S., Wang, A., Chuah, M. Y., Hyun, D. J., Lee, J., Otten, D. M., ... Kim, S. (2015). Design Principles for Energy-Efficient Legged Locomotion and Implementation on the MIT Cheetah Robot. *IEEE/ASME Transactions on Mechatronics, 20*(3), 1117–1129. doi:10.1109/TMECH.2014.2339013

Sniegowski, J. J., & Garcia, E. J. (1996). Surface-micromachined gear trains driven by an on-chip electrostatic microengine. *IEEE Electron Device Letters, 17*(7), 366–368. doi:10.1109/55.506369

Surbled, P., Clerc, C., Pioufle, B. L., Ataka, M., & Fujita, H. (2001). Effect of the composition and thermal annealing on the transformation temperature sputtered TiNi shape memory alloy thin films. *Thin Solid Films, 401*(1-2), 52–59. doi:10.1016/S0040-6090(01)01634-0

Suzuki, Y., Tani, K., & Sakuhara, T. (1999). Development of a new type piezo electric Micromotor. *Proc. of Transduceres, 99*, 1748–1751.

Takato, M., Naito, Y., Maezumi, K., Ishihara, Y., Okane, Y., Oku, H., ... Uchikoba, F. (2015). Development of impact-type rotary actuator and application for MEMS microrobot by bare chip IC of hardware neuron model. *Artificial Life and Robotics, 20*(4), 359–365. doi:10.1007/s10015-015-0239-4

Tang, W. C., Nguyen, T. H., & Howe, R. T. (1989). Laterally driven poly silicon resonant microstructure. *Proc. of IEEE Micro Electro Mechanical Systems. An Investigation of Micro Structures, Sensors, Actuators, Machines and Robots*, 53-59.

Wood, R. J., Finio, B., Karpelson, M., Ma, K., Pérez-Arancibia, N. O., Sreetharan, P. S., ... Whitney, J. P. (2016). Robotics Research. In H. I. Christensen & O. Khatib (Eds.), *Progress on "Pico" Air Vehicles* (pp. 3–19). Springer.

Yeh, R., Hollar, S., & Pister, K. S. J. (2002). Single Mask, Large Force, and Large Displacement Electrostatic Linear Inchworm Motors. *Journal of Microelectromechanical Systems, 11*(4), 330–336. doi:10.1109/JMEMS.2002.800937

KEY TERMS AND DEFINITIONS

Alternating Tripod Gait: The insect has six legs. Alternating tripod gait is the basic gait pattern of insects. It is very stable because three legs are always touching the ground.

Complementary Metal Oxide Semiconductor (CMOS): The complementary metal oxide semiconductor is one of the technologies for constructing the IC. CMOS technology also used for several analog circuits and digital circuits.

Integrated Circuit (IC): Integrated circuit is a set of multiple electronic circuits. Usually, an IC production line uses micro-fabrication technology to construct the small size electronic circuits. Bipolar transistor or field effect transistor are used to construct the electronic circuits of IC chip.

Micro-Electro Mechanical Systems (MEMS): Micro-electro mechanical systems consist of micro-mechanical systems and micro-electro systems. The micro-mechanical systems are fabricated by micro-fabrication technology based on integrated circuit production lines. Small size controlling circuit of the micro-mechanical systems could be constructed as an integrated circuit. The integrated circuit is the micro-electro system. Both micro-mechanical systems and micro-electro systems can be constructed on a silicon wafer. Therefore, micro-mechanical systems and micro-electro systems could be fabricated on the same silicon wafer.

Pulse-Type Hardware Neuron Model (P-HNM): Neuron model has two types: mathematical model and hardware model. In this chapter, hardware neuron model was used.

Shape Memory Alloy: In this chapter, shape memory alloy type artificial muscle wire was used. Artificial muscle wire displaces using shape-memory effect. Artificial muscle wire had two-way shape memory effect: shrank at high temperature and extended at low temperature.

Synchronization Phenomena: In engineering fields, synchronization phenomena are often used for signal processing. Also, a living organism often uses the synchronization phenomena for information processing. For example, locomotion rhythms of living organisms are generated by central pattern generator using synchronization phenomena.

Chapter 11
Bio-Inspired Snake Robots:
Design, Modelling, and Control

Mohammadali Javaheri Koopaee
University of Canterbury, New Zealand

Cid Gilani
University of Canterbury, New Zealand

Callum Scott
University of Canterbury, New Zealand

XiaoQi Chen
University of Canterbury, New Zealand

ABSTRACT

This chapter concerns modelling and control of snake robots. Specifically, the authors' main goal is introducing some of the fundamental design, modelling, and control approaches introduced for efficient snake robot locomotion in cluttered environments. This is a critical topic because, unlike locomotion in flat surfaces, where pre-specified gait equations can be employed, for locomotion in unstructured environment more sophisticated control approaches should be used to achieve intelligent and efficient mobility. To reach this goal, shape-based modelling approaches and a number of available control schemes for operation in unknown environments are presented, which hopefully motivates more scholars to start working on snake robots. Some ideas about future research plans are also proposed, which can be helpful for fabricating a snake robot equipped with the necessary features for operation in a real-world environment.

INTRODUCTION

Learning from natural organisms is conceived to be one of the key methods for developing innovative mechanisms. There are many biological organisms with smart structure, incredibly efficient locomotion mechanisms or intelligent sensing and processing capabilities, which are natural sources of inspiration for engineering designs. For instance, the amazing ability of some animals such as birds and turtles to

DOI: 10.4018/978-1-5225-2993-4.ch011

use geomagnetic information for navigation is imitated to develop a navigation system for autonomous underwater vehicles in (Liu, Liu, Zhang, & Li, 2016).

One of the fascinating topics in the field of bio-inspired design is snake locomotion. Snake's locomotion is a relatively efficient type of locomotion, which despite frictional opposition, consumes less energy compared to other biological forms with similar sizes, weights, and speeds. Also, small cross section of snake body is ideal for locomotion in narrow and remote environments, such as pipes and tankers. Snake robots have also been used as a flexible robot manipulators in Minimally Invasive Surgeries (MIS) by (Tully and Choset, 2016), who used highly articulated snake robots for performing surgeries with fewer incisions and consequently decreases the recovery time of patients

Snake's movement has always been a peculiar phenomenon for scientist. The elegant locomotion of snakes on smooth and uneven surfaces, under water and through narrow channels have motivated many scientists to study biological snake locomotion, years before considering it as an alternative way of locomotion for robots. However, it was not until twentieth century that the first accurate explanation of snake locomotion mechanism revealed.

The first discussion about snake's movement goes back to ancient times, when Aristotle studied snake locomotion as a part of a book about animal locomotion capabilities. However, for centuries, humans understanding of snake's unique movement mechanisms was very limited. The general belief even among zoologist was that snakes use their scales or tips of their ribs in a similar manner that other animals use their legs as means of locomotion.

(Hutchinson, 1879) challenged this theory. By observing a snake, which could easily ascend to the top of a glass jar, he argued that there should be some other justifications for snake-like movement. To explain this peculiar behavior, obviously contradicting the existing theory, (Fokker, 1927) came up with an alternative explanation, which can be considered as the first accurate description of snake movement mechanism. In this work, originally published in Dutch, the snake is modeled as a thin elastic body confined in a groove curved in a board and it is argued that snakes move in such a way that the total potential energy decreases. (Mosauer, 1932), without mentioning Fokker's results, categorized snake locomotion into three different categories, namely Horizontal Undulation, Sidewinding and Caterpillar movement, however did not discuss mechanical modelling or mathematical foundation of each types of movements. Later on, (Jones, 1933) motivated by a simplified Fokker's model, modeled the snake body as serially connected spools and mathematically showed that Mosauer results can indeed be supported theoretically, made Mosauer's work the foundation of the almost all published works after that.

In the 1940s, (Gray, 1946) conducted a comprehensive study on snake locomotion by observing a common grass snake. Motivated by Fokker and Mosauer's work, he searched for the answer to the basic question that how axial muscular activity generate propulsive force. To address this question, he considered the snake as a serially connected rigid rods hinged together, where axial muscles are regarded as elastic elements operating laterally to the hinges. Using this model, he empirically explained the mechanism behind each types of snake locomotion. One of the most important consequence of Gray's experiments was that each part of the body of the snake should be in contact with external obstacles preventing it to move in the direction normal to that segment of the snake body to enable the snake move forward. (Gans, 1962) further discussed limbless locomotion and described the dynamics of four major types of locomotion pattern qualitatively.

Although the studies on biological snakes continued, the critical findings of Gray and his predecessors enabled Hirose to further analyze bio-snakes movement and fabricate the world's first snake robot mimicking biological snake's movement in 1972 (Hirose, 1993). Hirose modeled the snake body assuming that the whole body is comprised of serially connected joints with infinitesimal length and then approximated the acting forces in tangential and normal direction to the ground, resulted from snake muscle activity. Based on these equations and considering continuity of muscle activity, he proposed *Serpenoid curve* for describing the shape of the snake performing lateral undulation. This was the beginning of major studies on bio-inspired snake locomotion.

BACKGROUND

Development of a snake robot by Hirose was the beginning of a new era in snake-inspired locomotion studies. This major development inspired Engineers to study snake locomotion with close collaboration with Biologist. Because of the relatively different nature of each, it is worthwhile to separate the available literature on snake like mechanism into two groups. The first one can be considered as the bio-mechanical studies of snake-like movement and eventually obtaining governing equations of motion, identifying key characteristic of motion and other contributing factors in snake locomotion. The second group of studies is related to bio-inspired snake robot's design and fabrication, mainly concerning modelling, control and operation of snake-like robots.

Studies on basic characteristics of snake locomotion through mathematical modeling is critical for mimicking these marvelous creatures. Early researchers, such as (Gray, 1946) and (Jones, 1933) proposed a segmented mechanical model of snakes to explain the fundamentals of their locomotion patterns. (Hirose, 1993), also proposed a segmented model and approximated the contributing forces in lateral undulation and identified the so called sinus-lifting behavior, which showed that snake locomotion is indeed should be performed in lateral and dorsal plane. Another modelling methods, such as the one by proposed by (Kuznetsov et al., 1967) mainly motivated by (Lavrentiev and Lavrentiev, 1962), consider a snake as an elastic road confined in a hard wall without friction. With the use of this model the movement of snake based on elastic energy of the snake body can be explained easily. This model is recently modified by (Dal Corso et al., 2017) to also encompass the case when just part of the snake body is in contact with the wall. Similar methods such as (Hu et al., 2009) has also been used, which instead of considering a wall around the snake body, the snake is modeled as a planar curve, incorporating frictional anisotropy.

In addition to efforts in mathematical modelling, experiment on real snakes also continued since the early studies of Gray. (Bennet et al., 1974) showed that for lateral undulation the space between the pegs directly affect the speed of the snake. (Jayne, 1986) further analyzed the kinematic of snake locomotion using cinematographic films mainly focusing the sidewinding and lateral undulation mode of locomotion. (Moon & Gans, 1998) studied lateral undulation mainly focusing on muscle activity during the locomotion and suggested that the bending of snake body around the peg generate the translation force and modeled it as a type of cam-follower. In 2012, Marvi recognized that the aforementioned experiments have been done in an artificial environment and conducted series of studies by observing snakes behavior in simulated real world conditions, such as studies on concertina locomotion, when the snake moves in a channel with a varying inclination and width (Marvi & Hu, 2012). The mathematical model of rectilinear motion also proposed by (Marvi, Bridges, & Hu, 2013), which successfully employed to predict the speed of movement. Sidewinding of snakes further studied in sandy slopes and revealed the behavior of snake in varying environmental conditions (Marvi et al., 2014).

As a consequence of mathematical and experimental results of snake locomotion studies, many engineers aimed to develop a locomotion mechanism similar to snakes. To mimic snake locomotion, early works such as (Hirose, 1993), considered sideslip constraints, which is implemented by installing passive wheels on the sideways of the robot. Although this design is not in harmony with biological snakes locomotion, many recent studies such as (Manzoor and Choi, 2016; Ryu, Chong, You, & Christensen, 2009) considered snake robots with sideslip constraint as well. In other works such as (Liljebäck, Pettersen, Stavdahl, & Gravdahl, 2012), sideslip constraints are no longer imposed, however it is assumed that sufficient numbers of vertical pegs exist around the robot, which could be used as push points. Other snake robots such as the one by (Rollinson et al., 2014), are not equipped with wheels and do not have anisotropic ground friction properties and are designed for performing 3D snake inspired locomotion, such as pole climbing and rolling.

Although creating a snake like mechanism is critical for mimicking snake like locomotion, modelling and control of such mechanism are of paramount importance. However, snake robots belong to the family of hyper redundant mobile mechanism, which their kinematic and dynamical models are not easy to obtain using conventional methods. To model this type of systems, one can obtain the kinematic and dynamic equations of the system using conventional methods such as Denavite-Hartenberg (D-H) (Denavit, 1955) and Newton-Euler formulation respectively, however the structure of these models heavily depend on the mechanism of the designed robot. For example, wheeled snake robot with side slip constraints has been treated as a non-holonomic system and the relationship between its internal shape changes and net motion has been obtained by (Krishnaprasad & Tsakiris, 1994). On the other hand, this method cannot be employed for wheel-less snake robots and dynamical models should be obtained to describe the motion of these robots because the forward motion of the robot heavily depend on the ground friction.

Searching for a computationally less expensive method for kinematic analysis of hyper redundant mechanisms, (Chirikjian, 1992) proposed Back Bone Curve approach as a general method for kinematic analysis of continuum and hyper redundant mechanism such as snakes. Although, Hirose Serpenoid curve was similar to this newly presented method, the proposed method by (Chirikjian & Burdick, 1995) and (Chirikjian, 1994) for kinematic and dynamic modelling respectively, was powerful enough to model variety of mechanisms, especially to achieve the ultimate goal of bio mechanic, which is designing a robot exhibiting all the fascinating characteristic of a real snake.

To control hyper redundant robot mechanism, such as snakes most of the related works are based on nonlinear oscillators for generating actuator displacements. These oscillators, usually named Gait Patterns, reduce the number of control parameters considerably and leads to a simpler system to control (Transeth, Pettersen, & Liljebäck, 2009). Alternative methods, such as (Date and Takita, 2007) are also proposed, based on torque control for adaptive locomotion. Recently, this method has received more attention by (Rollinson, Alwala, Zevallos, & Choset, 2014), who introduced compliant control approaches for a snake robot that includes series elastic actuator for sensitive torque control.

The rest of this chapter is organized as follows. First, the biological snake anatomy and modes of locomotion will be briefly discussed. Additionally, available approaches for modelling snake-like mechanisms will be introduced. Then, design specifications of snake robot Konda-II as an example of operating snake robot developed at the University of Canterbury will be presented. Moreover, some results obtained from implementation of presented approaches on the simulated model of Konda-II will be provided, to better explain the fundamentals of these approaches.

BIOLOGICAL SNAKE LOCOMOTION

The first step toward designing an efficient and smart snake robot is the study of biological snake itself. Many experiments have been reported regarding the snake's anatomy, behavior and locomotion, however for the purpose of this chapter, skeleton structure of common snakes will be discussed then different types of snake locomotion will be introduced.

The snake's skeleton structure consist of flexible vertebrae connected to each other with a ball and socket mechanism to form the snake backbone (Figure 1). Each vertebrae is also connected to two ribs on its sides similar to fish, which also use same methods for locomotion under the water. Vertebra and ribs are connected together with the help of three types of muscle, which their contraction and relaxation play the most important role in snake locomotion. It should be noted that the relative movement of vertebra caused by snake muscles activities are limited, however the number of vertebra is large enough, to result enough net motion.

As the result of close observation of snakes in their natural habitat, different types of snake locomotion have been identified, which each of them is efficient for a specific environment. In other words, snakes show different behavior, depending on their habitat, hence it is not possible to choose a specific locomotion pattern, suitable for all environments. Among all types of snake locomotion four of them are more common, not all snakes can perform all of these types of locomotion, however the upcoming locomotion types are dominant.

Serpentine Locomotion

Serpentine locomotion or lateral undulation is the most common type of locomotion among snakes and has been studied extensively. In early studies such as (Gray, 1946), serpentine mode of locomotion is studied considering the existence of some projections to push against and move forward, Figure 2 shows this type of locomotion. However, the most important aspect of lateral undulation is the existence of anisotropic ground friction force, which enables the snake to move in relatively smooth surfaces. In fact everything stems from the special structure of ventral scales of snake, which causes anisotropic friction between snake belly and the surface beneath the snake. This means that friction coefficient in the direction tangential to body is smaller than the friction coefficient in the perpendicular direction. This characteristic of snake skin is investigated in an experiment using an inclined plane in (Hu et al., 2009).

Figure 1. Snake skeleton structure

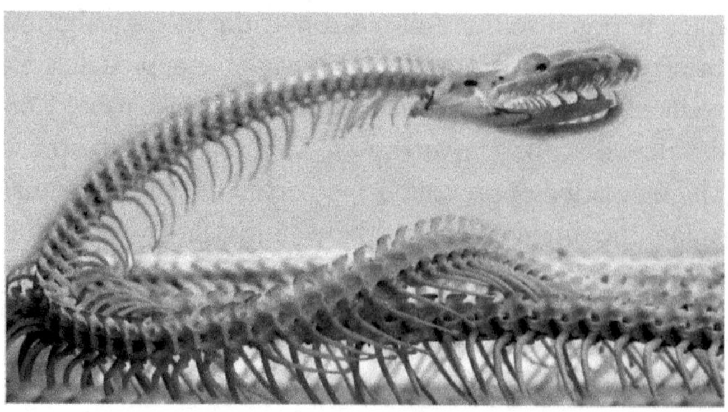

Figure 2. Serpentine locomotion of snakes

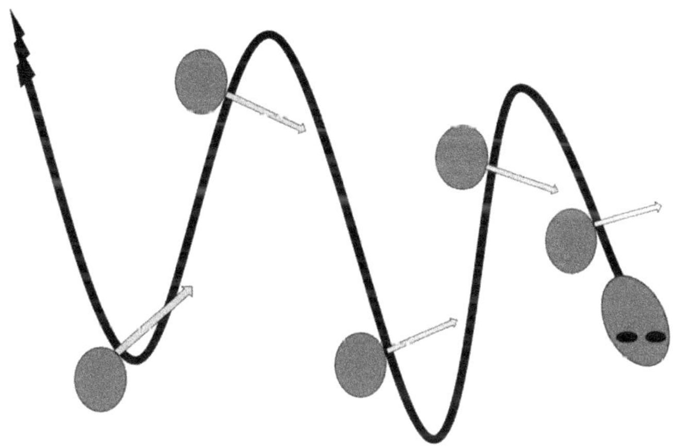

In lateral undulation a periodic wave travel backwards from head to tail, which cause the snake to move forward, so all part of the snake follow faithfully the path taken by the snake head (Mosauer, 1932). Also it should be noted that the weight distribution of the snake is not uniform during this type of locomotion, hence this type of locomotion cannot be performed precisely on 2D plane only.

Concertina Locomotion

Concertina locomotion, which is demonstrated in Figure 3, is among the first type of snake locomotion that has been studied because of its simple mechanism compared to other types of locomotion. In this type of locomotion, usually observed among snakes travelling through narrow channels, some parts of the body are in static contact with the wall of the channel, similar to anchor, to push against the wall in order to translate the rest of the body forward.

Side-Winding Locomotion

Side-winding locomotion, firstly studied in (Hutchinson, 1879) is another type of snake locomotion, which usually can be observed among the snakes locomotion on soft and slippery surfaces such as sand deserts. In this type of locomotion the snake use two separate parts of the body as static contact points to lift and bend the remaining parts (Mosauer, 1930), which is exhibited in Figure 4. The best way to imagine sidewinding in snakes is considering sidewise rolling of a spring coil, which leaves a series of disconnected tracks like snake tracks on sand.

Rectilinear Locomotion

This type of locomotion, which is also named caterpillar locomotion in (Mosauer, 1932) can be exhibited mostly in heavy snakes. Former studies on this type of locomotion suggested that, in rectilinear locomotion snake ribs act as legs, similar to walking, however snake does not ''travel on its ribs'', (Bogert, 1947) or in the other word the snake ribs is not the main means of locomotion.

Figure 3. Concertina locomotion of snakes

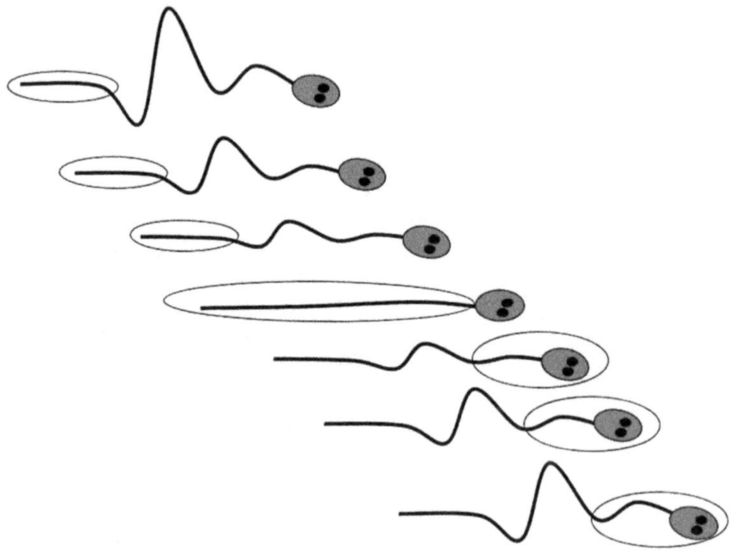

Figure 4. Side-winding locomotion of snakes

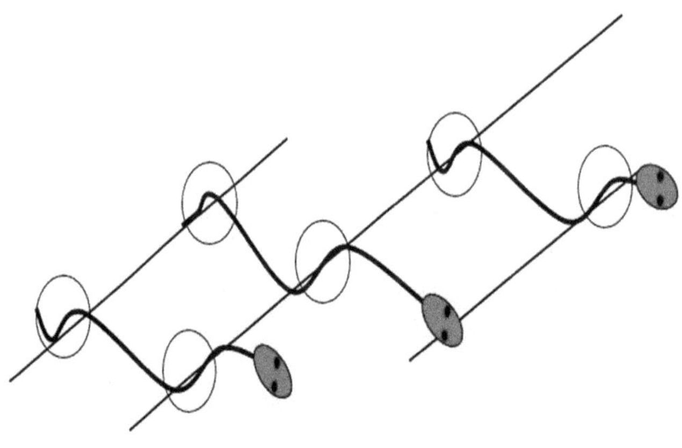

In rectilinear motion, similar to earthworms the snake travels on a straight line and unlike the other type of locomotion, sideways interaction with the environment is not essential. Instead contraction and relaxation wave pass over ventral muscles along with lifting body parts is the main method of locomotion (Lissmann, 1950; Marvi et al., 2013). Figure 5 displays the fundamental features of rectilinear locomotion, which the grounded body parts are shown inside the circles.

Contributing Factors in Snake Locomotion

Another aspects of snake locomotion that needs attention is gait selection mechanism in snakes (Guo & Mahadevan, 2008). It is not still clear that how and when they change their locomotion pattern. Studies on the efficiency of snake locomotion could be a key to find the answer to this question. Naturally,

Figure 5. Rectilinear Locomotion

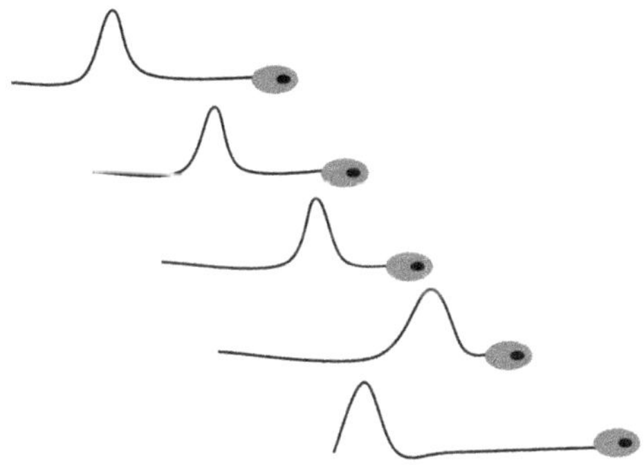

snakes have evolved in such a way to minimize their energy usage during the locomotion, however, the number of works in this field is quiet limited and restricted only to lab environment without considering possible defensive or offensive behavior of snakes.

(Walton et al., 1990) studied energetic cost of snake locomotion, by measuring the oxygen consumption of a real snake during the locomotion and concluded that energetic cost of lateral undulation in limbless species is equal to running of limbed animals with same weight, hence there might be other explanation for evolution of limbless species. (Secor et al., 1992) investigated the sidewinding energy cost and argued that sidewinding is more efficient than lateral undulation, however necessarily not all snake species are good sidewinders. In (Ariizumi & Matsuno, 2013), based on dynamical analysis on a snake-like robot, it is stated that under some assumption, it might be concluded that, even in biological snakes side-winding is more efficient than lateral undulation, however sinus-lifting is not considered.

The abovementioned results on snake efficiency are not conclusive, however from the available results and by observing snakes in their habitat, it can be concluded that no specific type of locomotion is efficient enough for all situations. For example, sidewinding is best suitable in a sandy desert but not on flat surfaces. However, whether the snake suddenly changes its locomotion pattern or just some parts of the body is also not investigated comprehensively. For example, lateral undulation is subject to existence of external push points, however it is not clear that what will happen when the number of push points are not enough to generate forward motion.

Another aspect of biological snake, closely related to its interesting locomotion capability is the tribological behavior of its skin. (Hazel et al., 1999) showed that the friction properties of snake skin in tangential and normal direction is not similar, which is also mathematically modeled by (Hu et al., 2009). It is also showed by (Marvi & Hu, 2012) that snake can actively control the friction between its ventral scales and the ground. (Benz et al., 2012) showed that the microstructure of snake skin is not the only contributing factor in anisotropic friction and indeed the underlying structure of snake skin is also a critical factor. (Cuervo et al., 2016) developed a snake-like surface by modifying the surface of metallic pins made of Titanium alloy. Despite these major advancements, our understanding of the function of snake skin is still in its infancy and to best of our knowledge it has never been implemented for the purpose of snake-inspired mechanism locomotion.

SNAKE-LIKE MECHANISM MODELLING

Generally speaking, there exist two methods to model a snake-like mechanism, considering a highly articulated body or a continuous curve to capture only the macroscopic shape of the snake. A segmented modelling approach can be really useful for analysis of muscle activities of snakes, however the number of degrees of freedom of typical snake-like mechanisms makes analysis based on this method rather complicated. Even a simplified segmented model such as the one proposed by (Marvi et al., 2013), is only useful for an specific motion pattern and relatively insufficient for effective description of others.

On the other hand, shape based modelling approaches are more flexible and there is almost no limitation exist for the type of the locomotion that can be modeled within this framework. It can easily be extended for modeling 3D locomotion patterns, such as sidewinding and can be also used to study muscle activity of a real snake by considering extensible curves. In addition, related to (Hu et al., 2009), within this framework, it is possible to model anisotropic ground friction or analyze the lateral undulation based on elastic energy release similar to what have been suggested by (Dal Corso et al., 2017). Hence, as a general approach for snake-like mechanism modeling, Back Bone Curve approach will be introduced here.

Backbone Curve Approach

Theoretically, the major contributing factor in snake locomotion is the macroscopic shape of the snake body and the change in response to environmental constraint. For this purpose, a flexible shape based modeling method like Backbone Curve approach (Chirikjian & Burdick, 1995) seems really effective. This modelling framework is initially introduced for hyper-redundant and continuum robots, however can easily be adapted for modeling and then control of modular snake robots using discretization methods, such as anneal chain fitting. Although dynamical modelling of such system using this approach will not be discussed here, using this modeling framework and based on continuum mechanics, it is also possible to dynamically analysis the locomotion (see (Chirikjian, 1994; Hu et al., 2009).

In the following section, we will describe the Backbone Curve approach for Kinematic modelling of snake-like mechanism. In addition, we will briefly explain the Annealed Chain Fitting algorithm, which provide us with a powerful tool to be employed for designing gait equation for efficient snake locomotion.

Considering a curve as a substance of R^3, a definition of differentiable curve could be given as follows:

Definition. 1(Do Carmo, 1976): A parametrized differentiable curve is a differentiable map $\alpha : \mathrm{I} \to R^3$ of an open interval $I = (a, b)$ of the real line R into R^3.

This means that if we consider

$$\alpha(s) = (x(s), y(s), z(s))$$

the functions $x(s), y(s)$ and $z(s)$ are differentiable functions, where s is the curve parameter, to be chosen latter.

This definition can easily be employed for capturing macroscopic shape of snake's backbone, however first, we need a method for parametrization of the curve. The most common method for parametrization of such curves is choosing s to be the curve arc length. Although other parameters can be used, choosing s to be the curve arc length will make its analysis more convenient.

By definition, see (Do Carmo, 1976) for more details, the arc length of the curve, from an initial point s_0 to an arbitrary point s, would be:

$$L(s) = \int_{s_0}^{s} \left|\alpha'(\sigma)\right| d\sigma$$

Where $\left|\alpha'(s)\right| = \sqrt{\left(\frac{dx}{ds}\right)^2 + \left(\frac{dy}{ds}\right)^2 + \left(\frac{dz}{ds}\right)^2}$.

As we chose the curve parameter s to be the arc length, then

$$L(s) = s - s_0$$

which results that:

$$\left|\alpha'(s)\right| = 1$$

This important conclusion, results that, if we choose the curve parameter to be the curve arc length, the magnitude of $\alpha'(s)$ is always equal to one. This also allows to define Curvature and Torsion functions and assign the well-known Frenet-Serret frames to each point on the curve, which uniquely determine the curve, however this method has some practical limitations (Chirikjian, 1992), hence in this chapter we discuss the simplified Backbone Curve approach, which is more suitable for practical applications.

After introducing the basics of curve geometry, now it is possible to introduce the Backbone Curve approach. Using this method, the position of each point on a snake relative to the origin of a frame attached to the head of the snake specified in Cartesian coordinate is as follows:

$$\vec{x}(s,t) = \int_{s_0}^{s} \vec{u}(\sigma,t)\, d\sigma$$

where t stands for time, $\vec{u}(\sigma,t) = \frac{\partial \alpha(\sigma,t)}{\partial \sigma}$ and $s \in I = (0,1)$, is the normalized length of the snake, which will not change with respect to time as the modular snakes are considered to be inextensible.

An intuitive way to understand is to consider the integral over s of $\vec{u}(\sigma,t)$ as sum of all tangent vector from base to the specific point s, hence the resultant position vector would be $\vec{x}(s,t)$.

To proceed, we need to use a parametrization method to parametrize the vector $\vec{u}(s,t)$. We also use Euler parametrization, similar to (Chirikjian & Burdick, 1995), which is the simplest type of parametrization. The definition of $K(\mathrm{s},t)$ and $T(\mathrm{s},t)$ is given in Figure 7.

Using Euler parametrization, the vector $\vec{u}(s,t)$ $\vec{u}(s,t)$ can be represented as follows:

$$\vec{u}(s,t) = \left[\int_{s_0}^{s} \sin(K)\cos(T)\,d\sigma \quad \int_{s_0}^{s} \cos(K)\cos(T)\,d\sigma \quad \int_{s_0}^{s} \sin(T)\,d\sigma \right]^{\mathrm{T}}$$

where $K(.)$ and $T(.)$ are functions of t and s. We also assume that $K(0,t) = T(0,t) = 0$, so $\frac{\partial\alpha(\sigma,t)}{\partial\sigma} = [0,1,0]$. In the other word, we assume that the attached frame to the base is chosen carefully so the unit tangent vector to the curve at the base is always in positive Y direction.

Figure 6. Konda-II

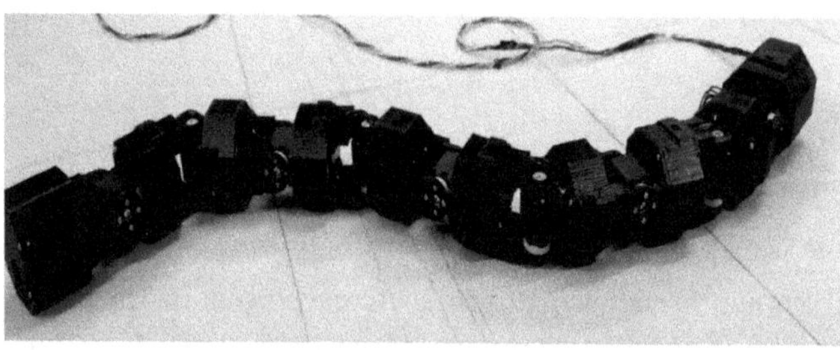

Figure 7. Backbone Curve and its set of reference frame

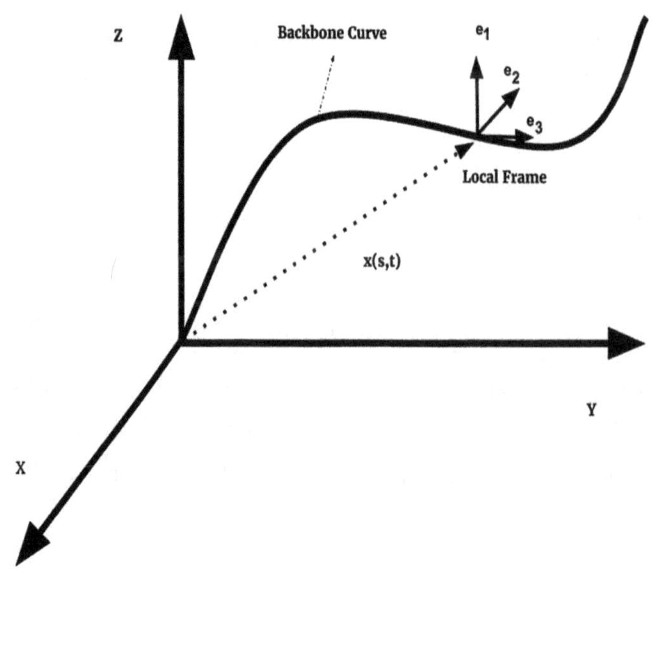

The advantage of this curve modelling framework is that just by determining the functions $K(\mathrm{s},t)$ and $T(\mathrm{s},t)$, one can determine the curve uniquely or in other word, assign a unique reference frame to each point on the curve. It is worth mentioning that for more general cases, two other functions *roll distribution* and *extensibility* is also required to determine the curve uniquely. These functions are mostly required for continuum manipulators and not modular snake models, which the length of each link is assumed to be constant and they cannot twist.

Annealed Chain Fitting

The Backbone Curve approach or similar methods such as (Yamada & Hirose, 2006), are very useful, when the designer is interested in designing shape-based controllers. This allows the designer to specify the desirable shape of the snake robot by determining $K_d(s,t)$, $T_d(s,t)$ and then use a fitting methods to obtain the actual rotation or displacement of actuators. One of the methods for approximating joint reference angles from a desired shape is Annealed Chain Fitting algorithm (Hatton & Choset, 2010), which will be described here.

At each iteration of Annealed Chain Fitting algorithm, using a moving active window, the joints should be grouped into three types, fixed joints, active joints and free joints. The only group of joints, which should be fitted to the curve are active joints, which their joint angles will be obtained to minimize the following cost function:

$$D_i\left(\alpha_{active}\right) = \sum_{active} d_j^{\,2} + \kappa\left(\alpha_j\right),$$

where d_j is the distance between the free end of the joint j and the curve and the penalty function $\kappa(\alpha)$ is defined as follows:

$$\kappa\left(\alpha_j\right) = max\left(0 \quad \left(abs\left(\alpha_j\right) - {}^2\right)^3\right)$$

where ² is a design parameter, which should be specified to make sure that the joint angles remain in an acceptable interval.To better understand above procedure, consider a planar snake robot, hence $K(s,t)$ alone is enough to describe the shape of the robot. To visualize the desired curve assume that:

$$K_d(s,t) = \cos\left(t + 6s\right),$$

which specifies a family of curves, depicted in Figure 8. It should be noted that $K_d(s,t)$ may not be known in prior, however $K_d(s,t)$ can be obtained uniquely by knowing the Cartesian position of each point on the desired curve.

Also assume that the joint angles $\alpha_j, j = 1,2,\ldots,9$, must satisfy the following condition due to practical limitation:

Figure 8. Representation of $K\left(s,t\right)$ and $T\left(s,t\right)$

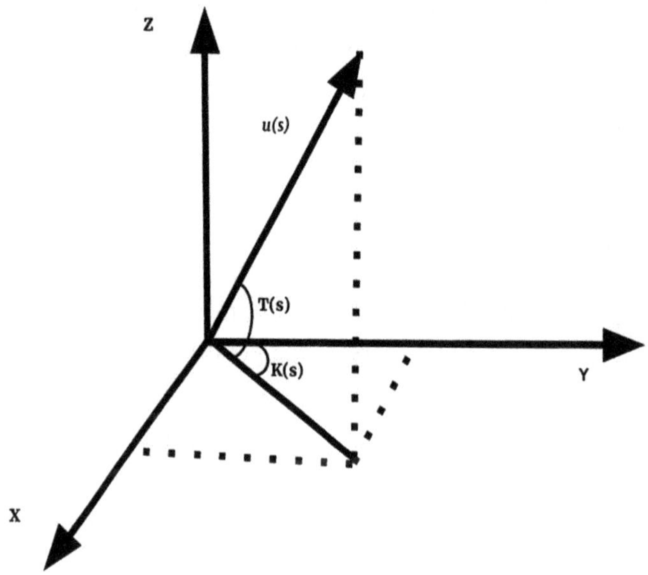

$$\left|\alpha_j\right| < 0.5$$

The problem to be solved is to obtain α_j to minimize the objective function, where the length of each link is equal to 0.1. To achieve this goal, Annealed Chain Fitting algorithm has been employed to obtain the joint angles. Figure 9 shows the joint angles fitted to the curve.

In case the designer aims to obtain the gait pattern generating same shape as, the above procedure should be carried out for fixed or variable time step. Then, a secondary fitting algorithm should be used to approximate the nonlinear equation generating these joint angles.

To use Annealed Chain fitting Algorithm there are a number of factors that one should keep in mind. The most important one is the initial condition of the optimization algorithm, which in fact is the joint angle of the first link. This is critical, mostly because placing the first link along the tangent line to the

Figure 9. Trace of the family of curves specified by (9)

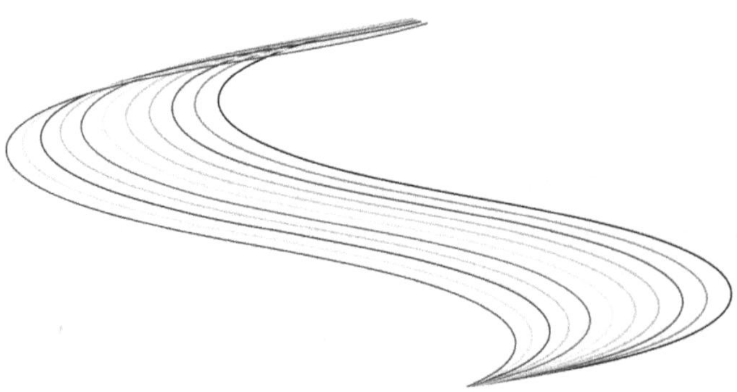

curve is not always the best choice (see (Hatton & Choset, 2010) for more details). Also the number of active joints and the step size of the algorithm can change the results considerably, hence one should carefully choose these parameters.

DESIGN OF SNAKE ROBOTS

Snake robots, pose special features that make them ideal for locomotion in unknown and challenging environments. These characteristics cause snake locomotion to be more robust and stable, compared to legged or wheeled locomotion, which makes snake locomotion ideal for many different purposes, including pipe inspection, maintenance purposes in industrial plants and exploration.

Currently, existing bio-inspired snake robots can be divided into two major groups, namely modular and Continuum snake robots. Snake-inspired continuum robots do not have distinct joints, however modular snake robots consist of several joint modules (bends), which are connected serially together. Continuum snake-like robots are mostly designed as highly flexible manipulators for medical and industrial purposes and few number of available prototypes are designed specifically for efficient locomotion in cluttered environment (see (Walker, 2013; Walker, Choset, & Chirikjian, 2016) for further details). On the other hand, modular snake robots, which are the main focus of this chapter, are more common for locomotion purposes.

The world's first modular snake robots was developed in 1972 by Prof. Shigeo Hirose at Tokyo Institute of Technology. This robot, which was equipped with passive wheels and active joints could move forward by controlling the joints to mimic a periodical motion; similar to biological snakes. The passive wheels are installed on each joint to mimic the idealistic anisotropic friction property of snake belly scales, meaning that the sideways velocity of snake joints, i.e., the velocity component normal to the links are always equal to zero. Similar design can also be found in (Tanaka, Nakajima, & Tanaka, 2016), which passive wheels have been used.

Over time, snake robots with or without wheels and with different structures and capabilities have been developed. The survey in (Hopkins, Spranklin, & Gupta, 2009) is a great source to gain some insight about different types of snake-inspired robot designs, where different types of snake robots are categorized based on their main characteristics.

Snake robots without anisotropic ground friction property or wheels, such as (Rollinson et al., 2014) or Konda-II, which is developed at the University of Canterbury are initially designed for different type of applications, such as pipe network inspection (Rollinson & Choset, 2016) and locomotion in cluttered environments (Kano, Sato, Kobayashi, & Ishiguro, 2012). In the other word, the initial motivation for development of these robots is locomotion in cluttered and unstructured environment and not locomotion on smooth surfaces.

The design of snake robot Konda-II, which is depicted in Figure 10 will be briefly described in the next section, to explain basic design procedures for developing a snake robot for lab experiments.

Mechanical Design

Konda-II consists of ten serially connected joints. These joints are equipped with HerkuleX DRS-0101 Smart Robot Servo motors as shown in Figure. 10, which are installed alternately in lateral and dorsal plane, which means that every consecutive module rotates in the axis 90o out of phase to the previous

Figure 10. Obtained results using Annealed Chain Fitting algorithm for t=0

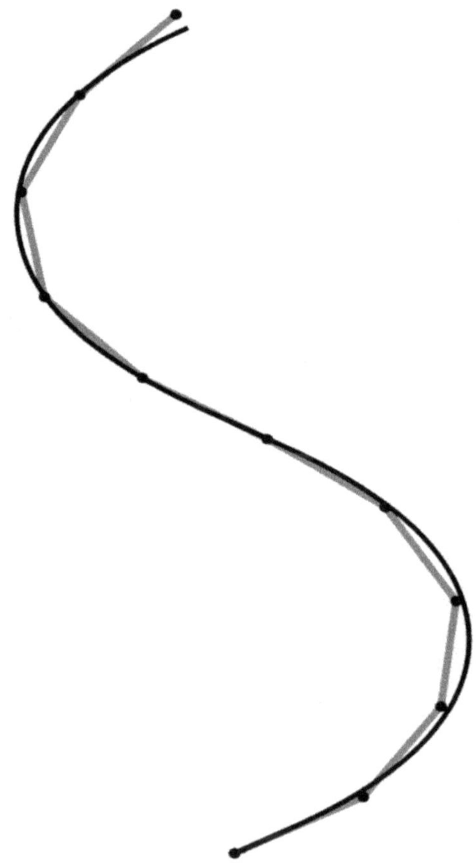

one. This specific design enables the robot to perform 3D movements without using two servo motors for each joint, which simplify the implementation considerably. It also worth mentioning that the pitch and the yaw modules have slightly different designs to one another to enable a neater cable management, hence the cable, which connects one module to another, is maintained at the bottom of the main frame and the joint of the next module.

A head module is also designed and manufactured, which is assembled at the end of the modular assembly. The head module of the snake is passive, which means that it has no actuation, and is designed specifically to accommodate a camera similar to the snake robot head module designed in (Wright et al., 2012). The head module is lighter than the rest of the modules weighing only 68.1gr, because it only accommodate a microcontroller and a camera.

The last module of the snake robot is the tail, which has a similar design as the rest of the modules, but with an extended part designed to hold a slip ring and some electronics elements. This module is the first module, which receive power from the power supply and also communicate with the computer. The slip ring allows wires to rotate while preventing them from getting twisted during the robot locomotion, especially when the robot performs a rolling gait. Based on the test conducted in the lab, the use of the slip ring had no effect on the communication and power of the snake. It also should be noted that the tail module was slightly heavier than the rest of the module weighing 108g.

The difference in the mass of the head and tail module and other modules is the problem that should be addressed. The easiest solution would be adding more weights to other modules to make their masses similar to that of the tail module. However, this requires redesigning all the modules to be large enough so that there would be more spaces to add weights to them. Though this difference might affect the locomotion of the snake, it would be an interesting effect to be investigated in future.

Electronic Design

As mentioned earlier, the actuator used in each module of the snake robot is a Herkulex smart servo motor, which can be connected to two other smart servos to form a parallel bus connection. This parallel bus connection has a pair of wires for communication and another pair wires for power. Also the smart servo is capable of feeding back the joint angle and power consumption.

The smart servo can be powered up to 12V with optimized voltage of 7.4 V, which maintains the consistency of the smart servo properties at any voltage level. The motor is capable of rotating at ±160o with the accuracy of 0.325o and stall torque 1.2 Nm with the free load speed 60 rpm. The maximum driving torque and maximum speed would respectively be 0.6 Nm and 30 rpm with the assumption that the smart servo has ideal DC motor characteristics.

The electronic design for Konda-II uses a parallel bus connection. The Teensy 3.2 communicates with the servo motors using a TTL serial method. In the Konda-II design each module is equipped with a microcontroller (Teensy LC), RS485 board and an IMU. Both the Teensy 3.2 and the Teensy LC are 3.3V boards while the servo motors require a 5 V single, as such a logical level convertor was used for the TTL signals sent to the servo motors. The Teensy LC reads the data from the IMU and transmits it to the Teensy 3.2 via RS485 communication. The RS485 board plays an important role in converting the UART signal from the Teensy LC to a RS485 signal. The advantage of using RS485 over I2C was that it can transmit signal more reliably at a longer length.

Software

Robot Operating System (ROS) (Quigley et al., 2009) is a framework used to write a robot application to operate the snake robot Konda II. The main advantage of ROS for the snake robot is its capability to have multiple processes computing and sharing data together. The processes which perform specific tasks for the snake robot control are known as nodes. All data communication between the Nodes are managed by the ROS Master, which enables the nodes to locate each.

The nodes enable computation to be performed modularly which increase the fault tolerance of the overall software system because the error would only be localized to a node. Because of this modularization, one can easily use existing open source codes developed for accomplishing complex tasks, such as image processing or state estimation. In Konda II, tasks such as providing each joint with desired relative joint angles, which with the help of PID controllers inside the servomotors will ensure that the measured value of joint angles match with the referenced angles.

The ROS node on the Teensy 3.2 subscribes to the target angle topic and translates them into TTL level communication for the smart servos. The Teensy 3.2s other function is to control the feedback from the Teensy LCs on each module through means of a single master, multiple slaves system over RS485. The Teensy LCs use Madgwick algorithm for each module, which are then converted into Euler angles and sent back to the Teensy 3.2 on request.

CONTROL STRATEGIES

Unlike many processes and simple mechanical systems, which the traditional model based control approaches have been used successfully, robotic engineers usually try to avoid direct usage of dynamical model because of its complexity, high nonlinearity and uncertainty. Especially, for periodical motions such as snake like locomotion or legged locomotion, the main focus of control engineers is on proposing efficient Gait patterns, instead of model based control approaches, such as Virtual Holonomic constraint (Mohammadi, Rezapour, Maggiore, & Pettersen, 2016). To better understand the general control structure proposed for control of snake robots, Figure 11 is advantageous.

The first layer, namely the Low-Level, makes sure that the desired angles of each joints matches with the measured angles with the use of an ordinary PID or a fractional order PID for more sophisticated control objectives like robust trajectory tracking (Rahmani, Ghanbari, & Ettefagh, 2016). Although some recent papers, such as (Rollinson, Alwala, Zevallos, & Choset, 2014) have considered the torque control strategies, most of the available literature is based on the position control of the servos and not the actual torque of each motor. Obviously, this control layer is as a closed-loop controller, because the PID controller use the measured value of relative joints angles to make sure that the errors converge to zero.

The second layer, namely the Mid-level, is responsible for generating the desired joint angles like a repetitive sequence of commands. This layer, which is usually called the Gait Pattern, is inspired by nature. Researchers, by observing the locomotion of animals, have concluded that to achieve rhythmic

Figure 11. Snake robot control structure

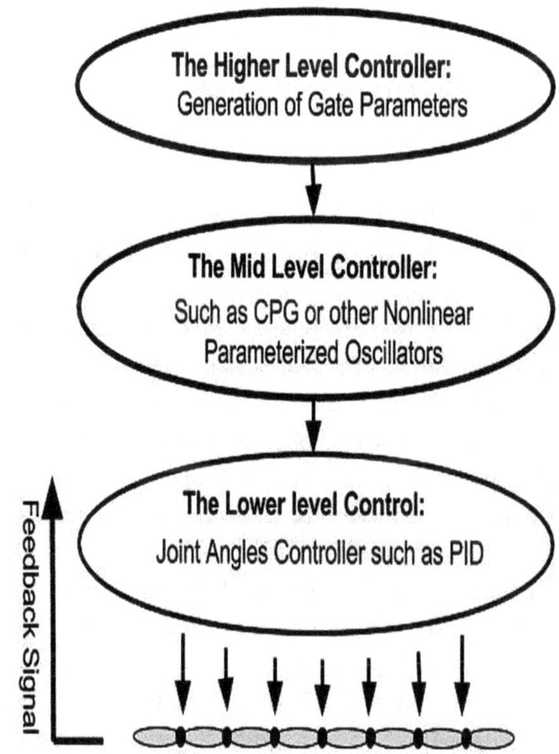

motor patterns, such as walking, the commands do not come from the brain directly, instead a local mechanism is responsible for activating the muscles. Similar mechanism is considered in snake robots too. Some paper such as (Hatton & Choset, 2010) and more recently (Hasanzadeh & Tootoonchi, 2010) have proposed their unique Gaits pattern with different properties, such adaptability or robustness. However, most of the Gait patterns have a similar structure. For example, in flat surfaces, most of the designed controllers are based on the following sinusoidal Gait pattern (Liljebäck, Pettersen, Stavdahl, & Gravdahl, 2012)

$$\phi_{i,ref} = A sin\left(\omega t + \left(i - 1\right)\delta\right) + \phi_0$$

where $\phi_{i,ref}$ is the reference angle of ith joint, δ is the phase shift and ϕ_0 is added to control the direction of the movement.

This Gait pattern and other similar ones, have been used extensively for snake robot control in flat surfaces. However, intelligent and efficient snake robot locomotion in unknown and unstructured environments requires that the snake robot could sense its environment and adapt its body shape and movements accordingly. Environment sensing and adaptation is not necessary when the surface beneath the snake robot is flat, since predetermined gait patterns, such as can be performed in open-loop without sensing the external forces (A.A. Transeth, Leine, Glocker, Pettersen, & Liljebäck, 2008).

It should be mentioned that, instead of using non-meaningful oscillators such as (1), a bio-inspired oscillator, namely Central Pattern Generator (CPG) is introduced (Ijspeert, 2008), which is a network of connected neurons capable of producing rhythmic motor patterns without any external input. In this type of methods, the rhythmic reference for each joint will be generated by a separate CPG network, hence with the help of series of successive rhythmic signals with certain phase difference one can generate joint angles for achieving snake-like locomotion. In some of the recent papers such as (Nor & Ma, 2014) smooth switching between different snake locomotion is investigated and in (Wu, 2011) a CPG based controller is designed for snake robot locomotion control and the relation between CPG parameters and the snake body shape is investigated.

Finally, the High-Level controller can be considered similar to the brain of animals, which most of the existing works do not take it into account. This Layer can also be considered as a human operator, who controls the snake robot using a remote controller. Feeding back the sensory information to this layer, the Gait parameters can be adopted to reach advanced control objectives such as trajectory tracking (Jin, Zhu, & Chen, 2016) and obstacle avoidance (Tanaka, Kon, & Tanaka, 2015). Also, instead of using force sensors, an alternative approach in (Rollinson & Choset, 2016) have been proposed to approximate the value of Gait parameters from actual joint angles to achieve more adaptive behavior in snake robot. This method, which is called compliant control method will be discussed with more details in following sections.

As it was mentioned earlier, snake robots should be able to move in cluttered and unstructured environments, however most of the existing works have only considered locomotion in flat surfaces, which is not in line with the initial motivation for designing snake robots. During the next section we will discuss some newly presented methods for control of snake robot in cluttered environment.

SNAKE ROBOT LOCOMOTION IN UNSTRUCTURED ENVIRONMENT

Recent works in the field of snake robot locomotion in cluttered environment can be categorized into three separate groups. In the next section we will discuss each group in more details.

Adaptive Gait Patterns

Consider the following Gait Pattern:

$$\phi_{i,ref} = A sin\left(\omega t + \left(i - 1\right)\delta\right) + ¥$$

The parameters of this nonlinear oscillator, i.e., $A,\ \omega,\ \ \delta$ are gait parameters that can be manipulated during the snake locomotion to achieve desired motion. This idea is regarded as adaptive locomotion, which its concept will be explained with an example taken from (Meister, Stepanenko, & Kernbach, 2011).

Consider a snake robot, which is moving in an environment with number of obstacles. To design an obstacle avoidance controller, an easy solution would be choosing Y to be equal to:

$$¥ = sign\left(phase\right) f_0 e^{-kd}$$

where $sign\left(phase\right)$ determine that the robot should turn right or left, *d* is the distance between snake head and obstacle, f_0 is a constant and *k* is a turning parameter. Using equation, the value of Y increases, when *d* decrease and vice versa, hence the robot turns whenever it is close to an obstacle in order to avoid collision.

Similar methods can also be employed with the use of CPGs. As discussed earlier, the fundamental function of CPG is producing rhythmic behaviors such as locomotion, respiration, sucking, licking, and mastication (Yu et al. 2014). One of the most important properties of CPG based controllers is adaptability, which can be achieved by incorporating sensory feedback signals. The CPG parameters can be tuned for achieving a desired locomotion pattern by changing the frequency, amplitude, phase lags, or waveforms of the GPC output. This high adaptively of CPGs make them ideal to be used for robot locomotion in cluttered environments. Employing CPGs, the parameters of oscillator can be changed to adapt to the new situation and generate different Gait patterns such as (Jae-Kwan Ryu, Chong, You, & Christensen, 2009), where an adaptive neural oscillator is designed, which the CPG parameters adapt to varying coefficients of body-ground friction.

One of the problems associated with general adaptive methods is computation burden of optimization algorithm, which make the real time control of snake robot very hard. So instead of online optimization, offline adaptive methods, such as (Gong, Tesch, Rollinson, & Choset, 2014) can be used, where an offline optimization should be carried out to obtain the desired Gait parameters for different situation, then a controller should be designed to select between these pre-obtained set of Gait parameters based on online sensing of the environment.

Obstacle Assisted Locomotion

The main idea of Obstacle Assisted locomotion of snake robots, which is employed in (Bayraktaroglu, Butel, Pasqui, & Blazevic, 2000), is that to propel forward with the use of Lateral Undulation or serpentine Gait, the sides of the snake should push against irregularities in the environments. In Obstacle Assisted locomotion, the joint angles should be specified in such a way that the resultant force from the obstacles that the robot is in contact with be in forward direction. Consequently, the snake robots can move in cluttered and unstructured environments by using external objects as push points, which aid the propulsion. Hence, this type of locomotion is also referred to Obstacle Aided locomotion in (A.A. Transeth et al., 2008). In the other words, instead of designing an obstacle avoidance controller, obstacles are surprisingly useful. To achieve this type of locomotion two existing methods will be explained during this sub section.

Firstly, in (Bayraktaroglu et al., 2000), using the information from tactile sensors mounted on the sideways of the joints, an algorithm for achieving Obstacle Assisted locomotion is proposed. In this paper, based on the observation of real snakes during the progressive motion, each section of the snake body is enforced to faithfully follow the path taken by the head of the animal. Hence, when the robot first module makes contact with a new push point, a curve fitting algorithm followed by a discretization method compute the reference angle for the rest of the joints during the future step times until a new obstacle comes with contact with the head module.

Secondly, in (Liljebäck, Pettersen, Stavdahl, & Gravdahl, 2010), employing hybrid dynamical system modeling framework the interaction between the robot and obstacles is modelled. The main idea of this work is that because the interactions between the robot and obstacles are discrete phenomena, it can be modeled as an instantaneous state jump.

The obstacle contact force acting on link i, consists of two components, depicted in the Figure 12, namely constraint force $f_{c,i}$ and friction force $f_{f,i}$.

Figure 12. Contact force components
(Liljebäck et al., 2010)

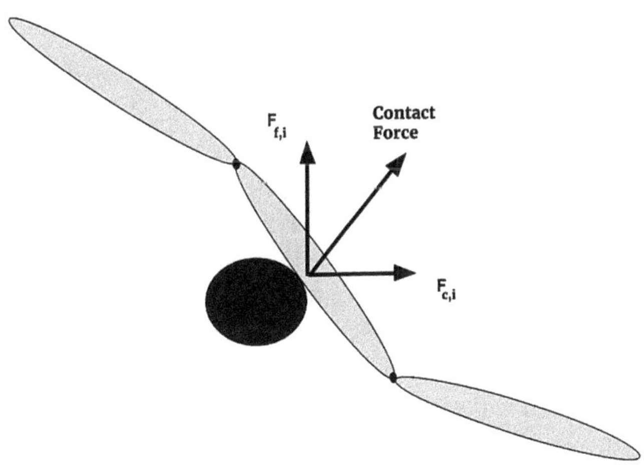

Considering the unilateral velocity constraint, which is imposed on link i to avoid sideway link motion toward the obstacle, the resulting contact force can be calculated by solving a Linear Complementary Problem (LCP), hence the continuous dynamic of the robot, when the robot is in contact with fixed set of obstacles will be obtained. Moreover, when the robot comes into contact with a new obstacle, the velocity vector of the robot jumps to new values. Considering these, the hybrid dynamical model of the snake robot, can be written as follows:

$$\begin{aligned} \dot{x} &= F_p\left(x,u\right) && if\, x \in C_p \\ x^+ &= G_p\left(x^-\right) && if\, x \in D_p \end{aligned}$$

where x is the state vector, x^+ and x^- denote the state values after and before touching a new push point, F_p is the continuous dynamic of the robot when the set of obstacle in the contact with the robot's link is fixed, C_p denotes the flow set, D_p denotes the jump set, and G_p is the jump map, obtained by solving another LCP. In this model, when the state vector belongs to the set C_p, it evolves according to the flow map F_p, however when a link comes into contact with a new obstacle, it jumps instantaneously to the new value x^+ according to the jump map G_p. The proposed hybrid can easily be used to model obstacle assisted locomotion of snake robots, however this model is more suitable for simulation purpose rather than controller design.

In (Liljebäck et al., 2010) in addition to the proposed hybrid model, a hybrid controller is proposed, which consist of two separate control strategy. The first one, which is called leader-follower scheme is responsible for producing oscillatory motion of the snake robot links. Based on the presented ideas, the body joint angles should follow the motion of the head module, hence the reference angle of link i is chosen as follows:

$$\phi_{i,ref} = \phi_{N-1,ref}\left(t - \left(N - i - 1\right)\Delta t\right)$$

where $\phi_{N-1,ref}$ will be switched between two values according to some pre-defined switching law to generate oscillatory body motion.

The second strategy, i.e., the jam resolution scheme, will be activated, when the robot is jammed between obstacles, which is a very common situation. In this situation, the links affected by contact forces will be rotated to increase the propulsive component of each contact force, which means that the adaptive component of hybrid controller activate, when the robot is jammed. Combination of these two schemes and a switching law, which select the appropriate controller at each time is shown to be very effective in simulation studies (Liljebäck, Pettersen, Stavdahl, & Gravdahl, 2011).

The most important part of this controller is the adaptation strategy, however this adaption law is only employed when the robot is jammed between the obstacles. Consequently, the proposed hybrid controller in (Liljebäck et al., 2010), do not use the obtained analytical hybrid model for stability analysis of the proposed controller and compared to the model, the controller itself is simpler.

Compliant Control Method

Another method proposed for snake locomotion in unstructured environment is compliant control, introduced in (Rollinson & Choset, 2016). Compliant robots or actuators have been studied extensively among the robotic community. In these types of studies, compliance is usually achieved using compliant actuators such as (Luo et al., 2015) or high precision torque and force control methods such as (D. Rollinson, Alwala, et al., 2014). However, the key points of the proposed method in (Rollinson & Choset, 2016) is using the estimated value of Gait parameters to represent the state of the robot.

Under typical operation, trajectories should be generated for the snake robot's joints based on the parametrized gait functions, however for some reasons such as actuator limitations or environmental constraints, the actual joint angles rarely match with these commands. Based on compliant control method, one can approximately describe the robot's actual shape using estimated Gait parameters in a more meaningful manner.

Consider the snake robot Konda-II described earlier. Suppose that the robot has to move forward autonomously through a narrow corridor, which the distance between the walls is unknown. To conduct an experiment a simulated model of the snake robot Konda-II is developed in V-rep environment (Rohmer, Singh, & Freese, 2013).

Once again for simplicity it is assumed that the motion of the robot is performed in a plane, hence only yaw joints of the snake robot will be commanded. Now we use the complaint control method to design an adaptive controller for the snake robot, based on the following Gait:

$$\phi_{i,ref} = A sin\left(\omega t + i\delta\right)$$

where i=2,4,6,8.

The first issue that needs to be considered is obtaining an estimation of the gait parameters based on the measured value of the joint angles. In (Rollinson & Choset, 2016) an Extended Kalman Filter (EKF) is used to estimate these parameters, however for the sake of brevity the details of this method will not be discussed here. However, it worth mentioning that to estimate the Gait parameters more suitable for real time control, the idea of Virtual Chassis (Rollinson, Buchan, & Choset, 2012) is found to be very useful.

After obtaining the estimated values of Gait parameters, which will be denoted by $\tilde{A}$,$\tilde{\omega}$ and $\tilde{\delta}$, the controller itself is very simple, it is just enough to command the new Gait parameters to be an offset of actual Gait parameters. For example for locomotion through a corridor, we command:

$$A - \tilde{A} \pm \rho$$

where $\rho$is a positive value whenever A should increase because the distance between the walls is large and the snake body cannot touch the walls. This situation can easily be detected whenever the commanded A become equal to $\tilde{A}$. Also ρwill be chosen to be negative, when $\tilde{A}$ becomes very larger than A.

This control scheme is implemented in a simulation environment when the distance between the walls is unknown. At first the snake will start moving by setting A to a small value, then the value of A

will be increased until the difference between commanded value and estimated value of A becomes 0.2. Figure 13 shows the obtained results in V-rep simulation environment.

It should be noted that the given example in here is just for demonstration. For more sophisticated control objective such as pole climbing, ω and δ are also should be controlled in a similar way.

In conclusion, the compliant control method is very useful for sensing the environment, which the robot is operating in. However, the phenomenon, which should be carefully addressed when using this method is delay. Addressing this issue is very important because the measured value of joint angles in step-time k is usually available at time $k+d$, where d is the sum of the actuator and sensor delays. Hence, one should carefully choose the step-time to avoid incorrect measurement.

FUTURE RESEARCH DIRECTIONS

Considering numerous practical applications of snake robots and amazing abilities of bio-snakes, it seems that there are still many aspects of snake robots that need to be investigated in more details. Researches on snake robots, as efficient mobile robots for undertaking variety of tasks, such as rescue operation, pipe inspection and operation in remote and harsh environments will not be thorough, unless scholars from different fields such as Biology, Mechanics, Control and Computer science work together. During this section, some of the possible future research plans, related to these fields will be presented to hopefully motivate more scholars to start working on bio-inspired snake locomotion.

Firstly, still there is a long way toward developing a skin with characteristic similar to biological snake's skin. Development of such skin with anisotropic friction property is necessary for developing a snake robot, similar to biological snakes. Moreover, the skin of snakes are the main means of environmental sensing, which the available prototype of snake robots should certainly have to be able move intelligently. Recently, in (Greiner & Schäfer, 2015) a laser-created texture similar to snake skin has been developed, which can be used for snake robots as well.

Obtaining dynamical model of snake robot, considering the external forces acting on the body of the snake is another area that should be carefully considered. Although some results on dynamical modeling of snake robots for the purpose of control design are available, these studies are only based on planar movement of snake robot and not 3D motions

Figure 13. Snake Robot locomotion between two walls with unknown distance

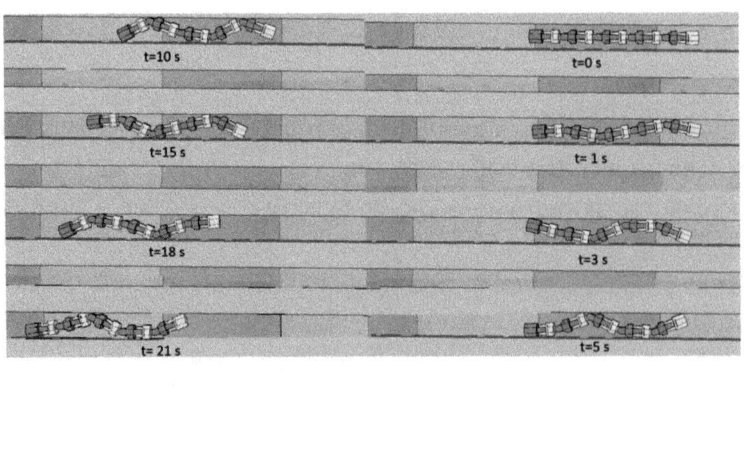

From control perspective, wheel-less modular snake robots belong to the family of under actuated robot mechanisms, as the number of DOFs of the system is N+3 but the number of control inputs is N-1, when N is the number of links. This means that the number of control inputs is fewer than the number of degrees of freedom of the robot, hence conventional control approaches such as Feedback Linearization or Fuzzy control cannot be employed directly. For model based control of planar snake robots VHC (Maggiore & Consolini, 2013) or Partial Feedback Linearization (Liljebäck et al., 2012) can be employed, however for Gait patterns designed for performing in 3D space, such as Rolling and Side-Winding the mentioned approaches should be extended to encounter 3D movement as well.

Snake robots have the ability to operate in environments, where human cannot. This amazing feature of snake robots, stemming from their small cross section, can be employed for undertaking variety of tasks. However, a main issue that also need to be considered is placing a camera on the snake robot, which is necessary for operation in remote and hazardous environments. Based on the discussions of this chapter, it is obvious that internal change of the snake body shape is the major means of locomotion, hence placing a camera on a robot, which not only move relative to global coordinate but also move periodically relative to its body frame is a challenging task. To study recent results see (Mutlu, Melo, Vespignani, Bernardino, & Ijspeert, 2016).

Another aspects of snake robots that should be investigated is fault tolerant locomotion. One of the major advantages of snake robots is the hyper redundancy, which is essential for locomotion in harsh environment, where maintenance is impossible or costly. To make advantages from this property of snake robot more research should be done on the effect of fault of snake locomotion and possible methods for reducing adverse effects on snake locomotion. There are some available results in this area such as (Irani et al., 2015), however there is still a long way toward designing Fault Tolerant Gaits.

CONCLUSION

One of the most important aspect of biological snakes is locomotion in unstructured environment, hence in this chapter, the authors aimed to introduce some of the available approaches to mimic this ability of biological snakes. Firstly, a shape based modelling framework, namely Backbone Curve model introduced. Using this modelling framework or similar ones, a desired shape of the robot for each time-step can be specified to be used for developing efficient Gait patterns, which are the most efficient way for control of snake robots so far.

After obtaining a suitable Gait pattern, which might be different for different prototype of snake robots, the next step is designing a controller for locomotion in unstructured environment. Three general methods introduced, which each one of them is suitable for a specific application.

Adaptive control methods depend extensively on sensing the environment. For example, for adaptive locomotion on surfaces with varying friction coefficient or on a slope, suitable sensing devices should be available to detect changes in environment. Also online optimization is always an essential component of any adaptive control approaches, hence the behavior of the system will not adapt instantaneously to environmental changes, which may not be acceptable for some applications.

Obstacle-Assisted locomotion was another approach that proposed in this chapter. The idea of this method is using the obstacles as push points for the robot to push against and move forward. Although this method seems very interesting, it cannot be employed as the main method of locomotion, because there are often not enough obstacles to be used as push points. Hence, it is more reasonable to use this idea to increase the efficiency of snake robot locomotion, instead of using it as the main means of locomotion.

Compliant control method, which was the last discussed method, uses a feedback of the actual Gait parameters estimated from measured values of joint angles to gain some information about the environment. This method works very well, as the simulation results show. However delay of sensor and actuators can considerably affect the results. Also the proposed controller, which was only an offset from estimated value of Gait parameters, seems in efficient for some applications. Nevertheless this idea can be further explored for designing more sophisticated control methods.

REFERENCES

Ariizumi, R., & Matsuno, F. (2013). Dynamical analysis of sidewinding locomotion by a snake-like robot. *Proceedings - IEEE International Conference on Robotics and Automation*, 5149–5154. doi:10.1109/ICRA.2013.6631313

Bayraktaroglu, Z. Y., Butel, F., Pasqui, V., & Blazevic, P. (2000). Snake-like locomotion: Integration of geometry and kineto-statics. *Advanced Robotics*, *14*(6), 447–458. doi:10.1163/156855300741942

Bennet, S., McConnell, T., & Trubatch, S. L. (1974). Quantitative analysis of the speed of snakes as a function of peg spacing. *The Journal of Experimental Biology*, *60*(1), 161–165.

Benz, M. J., Kovalev, A. E., & Gorb, S. N. (2012). Anisotropic frictional properties in snakes. *SPIE Smart Structures and Materials+ Nondestructive Evaluation and Health Monitoring*.

Bogert, C. M. (1947). Rectilinear locomotion in snakes. *Copeia*, *1947*(4), 253–254. doi:10.2307/1438921

Chirikjian, G. S. (1992). *Theory and applications of hyper-redundant robotic manipulators*. California Institute of Technology.

Chirikjian, G. S. (1994). Hyper-redundant manipulator dynamics: A continuum approximation. *Advanced Robotics*, *9*(3), 217–243. doi:10.1163/156855395X00175

Chirikjian, G. S., & Burdick, J. W. (1995). The Kinematics of Hyper-Redundant Robot Locomotion. *IEEE Transactions on Robotics and Automation*, *11*(6), 781–793. doi:10.1109/70.478426

Cuervo, P., López, D. A., Cano, J. P., Sánchez, J. C., Rudas, S., Estupiñán, H., ... Abdel-Aal, H. A. (2016). Development of low friction snake-inspired deterministic textured surfaces. *Surface Topography: Metrology and Properties*, *4*(2), 24013. doi:10.1088/2051-672X/4/2/024013

Dal Corso, F., Misseroni, D., Pugno, N. M., Movchan, A. B., Movchan, N. V., & Bigoni, D. (2017). Serpentine locomotion through elastic energy release. *Journal of the Royal Society, Interface*, *14*(130), 20170055. doi:10.1098/rsif.2017.0055 PMID:28566512

Date, H., & Takita, Y. (2007). Adaptive locomotion of a snake like robot based on curvature derivatives. *IEEE International Conference on Intelligent Robots and Systems*. doi:10.1109/IROS.2007.4399635

Denavit, J. (1955). A kinematic notation for lower-pair mechanisms based on matrices. *Trans. of the ASME. Journal of Applied Mechanics*, *22*, 215–221.

Do Carmo, M. P., & Do Carmo, M. P. (1976). *Differential geometry of curves and surfaces* (Vol. 2). Prentice-hall Englewood Cliffs.

Fokker, A. D. (1927). De Voortbeweging Der Slangen. *Phisica, 7*(65).

Gans, C. (1962). Terrestrial locomotion without limbs. *American Zoologist, 2*(2), 167–182. doi:10.1093/icb/2.2.167

Gong, C., Tesch, M., Rollinson, D., & Choset, H. (2014). Snakes on an inclined plane: Learning an adaptive sidewinding motion for changing slopes. *IEEE International Conference on Intelligent Robots and Systems*. doi:10.1109/IROS.2014.6942697

Gray, J. (1946). The mechanism of locomotion in snakes. *The Journal of Experimental Biology, 23*(2). PMID:20281580

Greiner, C., & Schäfer, M. (2015). Bio-inspired scale-like surface textures and their tribological properties. *Bioinspiration & Biomimetics, 10*(4), 044001. doi:10.1088/1748-3190/10/4/044001 PMID:26125522

Guo, Z. V., & Mahadevan, L. (2008). Limbless undulatory propulsion on land. *Proceedings of the National Academy of Sciences of the United States of America, 105*(9), 3179–3184. doi:10.1073/pnas.0705442105 PMID:18308928

Hasanzadeh, S., & Tootoonchi, A. A. (2010). Ground adaptive and optimized locomotion of snake robot moving with a novel gait. *Autonomous Robots, 28*(4), 457–470. doi:10.1007/s10514-010-9179-y

Hatton, R. L., & Choset, H. (2010). Generating gaits for snake robots: Annealed chain fitting and keyframe wave extraction. *Autonomous Robots, 28*(3), 271–281. doi:10.1007/s10514-009-9175-2

Hazel, J., Stone, M., Grace, M. S., & Tsukruk, V. V. (1999). Nanoscale design of snake skin for reptation locomotions via friction anisotropy. *Journal of Biomechanics, 32*(5), 477–484. doi:10.1016/S0021-9290(99)00013-5 PMID:10327001

Hirose, S. (1993). *Biologically inspired robot*. Oxford University Press.

Hopkins, J. K., Spranklin, B. W., & Gupta, S. K. (2009). A survey of snake-inspired robot designs. *Bioinspiration & Biomimetics, 4*(2), 21001. doi:10.1088/1748-3182/4/2/021001 PMID:19158415

Hu, D. L., Nirody, J., Scott, T., & Shelley, M. J. (2009). The mechanics of slithering locomotion. *Proceedings of the National Academy of Sciences of the United States of America, 106*(25), 10081–10085. doi:10.1073/pnas.0812533106 PMID:19506255

Hutchinson, H. (1879). About Snakes. *Nature*. Retrieved from http://adsabs.harvard.edu/abs/1879Natur.20.528H

Hutchinson, H. F. (1879). About Snakes. *Nature, 20*(518), 528–530. doi:10.1038/020528c0

Ijspeert, A. J. (2008). Central pattern generators for locomotion control in animals and robots: A review. *Neural Networks, 21*(4), 642–653. doi:10.1016/j.neunet.2008.03.014 PMID:18555958

Irani, R., Bauer, R., North, L., Nicholson, M., Nolan, D., & West, B. (2015). Analysis of joint failures on the lateral undulation gait of a robotic snake. *Transactions of the Canadian Society for Mechanical Engineering*, *39*(2).

Jayne, B. C. (1986). Kinematics of terrestrial snake locomotion. *Copeia*, *1986*(4), 915–927. doi:10.2307/1445288

Jin, H., Zhu, L., & Chen, Z. (2016). Trajectory Control of Snake-Like Robots in Natural Oscillation. *Asian Journal of Control*. 10.1002/asjc.1271

Jones, A. T. (1933). The gliding of a snake. *Physics*, *4*(4), 164–165. doi:10.1063/1.1745173

Kano, T., Sato, T., Kobayashi, R., & Ishiguro, A. (2012). Local reflexive mechanisms essential for snakes' scaffold-based locomotion. *Bioinspiration & Biomimetics*, *7*(4), 46008. doi:10.1088/1748-3182/7/4/046008 PMID:22918023

Krishnaprasad, P. S., & Tsakiris, D. P. (1994). G-snakes: Nonholonomic kinematic chains on Lie groups. In *Decision and Control, 1994., Proceedings of the 33rd IEEE Conference on* (*Vol. 3*, pp. 2955–2960). IEEE.

Kuznetsov, V. M., Lugovtsov, B. A., & Sher, Y. N. (1967). On the motive mechanism of snakes and fish. *Archive for Rational Mechanics and Analysis*, *25*(5), 367–387. doi:10.1007/BF00291937

Lavrentiev, M.A., & M. M. L. (1962). On a principle for creating a tractive force of motion. *Journal of Applied Mechanics and Technical Physics*, 4.

Liljebäck, P., Pettersen, K. Y., Stavdahl, O., & Gravdahl, J. T. (2010). Hybrid modelling and control of obstacle-aided snake robot locomotion. *IEEE Transactions on Robotics*, *26*(5), 781–799. doi:10.1109/TRO.2010.2056211

Liljebäck, P., Pettersen, K. Y., Stavdahl, O., & Gravdahl, J. T. (2011). Experimental investigation of obstacle-aided locomotion with a snake robot. *IEEE Transactions on Robotics*, *27*(4), 792–800. doi:10.1109/TRO.2011.2134150

Liljebäck, P., Pettersen, K. Y., Stavdahl, O., & Gravdahl, J. T. (2012). *Snake robots: modelling, mechatronics, and control*. Springer Science & Business Media.

Lissmann, H. W. (1950). Rectilinear locomotion in a snake (Boa occidentalis). *The Journal of Experimental Biology*, *26*(4), 368–379. Retrieved from http://jeb.biologists.org/content/26/4/368.short

Liu, K., Liu, M., Zhang, X., & Li, H. (2016). A bio-inspired geomagnetic navigation model based on course constraint strategy under anomalies field disturbing for AUV. OCEANS 2016 - Shanghai. doi:10.1109/OCEANSAP.2016.7485469

Luo, M., Pan, Y., Skorina, E. H., Tao, W., Chen, F., Ozel, S., & Onal, C. D. (2015). Slithering towards autonomy: A self-contained soft robotic snake platform with integrated curvature sensing. *Bioinspiration & Biomimetics*, *10*(5), 55001. doi:10.1088/1748-3190/10/5/055001 PMID:26335857

Ma, S., Araya, H., & Li, L. (2001). Development of a creeping snake-robot. *Proceedings of IEEE International Symposium on Computational Intelligence in Robotics and Automation, CIRA*. doi:10.1109/CIRA.2001.1013176

Maggiore, M., & Consolini, L. (2013). Virtual Holonomic Constraints for Euler–Lagrange Systems. *IEEE Transactions on Automatic Control*, *58*(4), 1001–1008. doi:10.1109/TAC.2012.2215538

Manzoor, S., & Choi, Y. (2016). A unified neural oscillator model for various rhythmic locomotions of snake-like robot. *Neurocomputing*, *173*, 1112–1123. doi:10.1016/j.neucom.2015.08.069

Marvi, H., Bridges, J., & Hu, D. L. (2013). Snakes mimic earthworms: Propulsion using rectilinear travelling waves. *Journal of the Royal Society, Interface*, *10*(84), 20130188. doi:10.1098/rsif.2013.0188 PMID:23635494

Marvi, H., Bridges, J., Hu, D. L., Wright, C., Johnson, A., Peck, A., … Quillin, K. (2013). Snakes mimic earthworms: propulsion using rectilinear travelling waves. *Journal of the Royal Society, Interface / the Royal Society, 10*(84), 20130188. 10.1098/rsif.2013.0188

Marvi, H., Gong, C., Gravish, N., Astley, H., Travers, M., Hatton, R. L., ... Goldman, D. I. (2014). Sidewinding with minimal slip: Snake and robot ascent of sandy slopes. *Science*, *346*(6206), 224–229. doi:10.1126/science.1255718 PMID:25301625

Marvi, H., & Hu, D. L. (2012). Friction enhancement in concertina locomotion of snakes. *Journal of The Royal Society Interface*.

Meister, E., Stepanenko, S., & Kernbach, S. (2011). *Adaptive locomotion of multibody snake-like robot.* arXiv Preprint arXiv:1108.6175

Mohammadi, A., Rezapour, E., Maggiore, M., & Pettersen, K. Y. (2016). Maneuvering Control of Planar Snake Robots Using Virtual Holonomic Constraints. *IEEE Transactions on Control Systems Technology*, *24*(3), 884–899. doi:10.1109/TCST.2015.2467208

Moon, B. R., & Gans, C. (1998). Kinematics, muscular activity and propulsion in gopher snakes. *The Journal of Experimental Biology*, *201*(19), 2669–2684. PMID:9732322

Mosauer, W. (1930). A note on the sidewinding locomotion of snakes. *American Naturalist*, *64*(691), 179–183. doi:10.1086/280308

Mosauer, W. (1932). On the locomotion of snakes. *Science, 76*(1982), 583–585.

Mutlu, M., Melo, K., Vespignani, M., Bernardino, A., & Ijspeert, A. J. (2016). Where to place cameras on a snake robot: Focus on camera trajectory and motion blur. *IEEE International Symposium on Safety, Security, and Rescue Robotics*. doi:10.1109/SSRR.2015.7442948

Nor, N. M., & Ma, S. (2014). Smooth transition for CPG-based body shape control of a snake-like robot. *Bioinspiration & Biomimetics*, *9*(1), 16003. doi:10.1088/1748-3182/9/1/016003 PMID:24343201

Quigley, M., Conley, K., Gerkey, B., Faust, J., Foote, T., & Leibs, J. … Ng, A. Y. (2009). ROS: an open-source Robot Operating System. In ICRA workshop on open source software (Vol. 3, p. 5). Kobe, Japan: ICRA.

Rahmani, M., Ghanbari, A., & Ettefagh, M. M. (2016). Robust adaptive control of a bio-inspired robot manipulator using bat algorithm. *Expert Systems with Applications*, *56*, 164–176. doi:10.1016/j.eswa.2016.03.006

Rohmer, E., Singh, S. P. N., & Freese, M. (2013). V-REP: A versatile and scalable robot simulation framework. In *Intelligent Robots and Systems (IROS), 2013 IEEE/RSJ International Conference on* (pp. 1321–1326). IEEE. doi:10.1109/IROS.2013.6696520

Rollinson, D., Alwala, K. V., Zevallos, N., & Choset, H. (2014). Torque control strategies for snake robots. *IEEE International Conference on Intelligent Robots and Systems*. doi:10.1109/IROS.2014.6942694

Rollinson, D., Bilgen, Y., Brown, B., Enner, F., Ford, S., & Layton, C., … Choset, H. (2014). Design and architecture of a series elastic snake robot. *IEEE International Conference on Intelligent Robots and Systems*. doi:10.1109/IROS.2014.6943219

Rollinson, D., Buchan, A., & Choset, H. (2012). Virtual Chassis for Snake Robots: Definition and Applications. *Advanced Robotics*, *26*(17), 2043–2064. doi:10.1080/01691864.2012.728695

Rollinson, D., & Choset, H. (2016). Pipe Network Locomotion with a Snake Robot. *Journal of Field Robotics*, *33*(3), 322–336. doi:10.1002/rob.21549

Ryu, J.-K., Chong, N. Y., You, B. J., & Christensen, H. I. (2009). Locomotion of snake-like robots using adaptive neural oscillators. *Intelligent Service Robotics*, *3*(1). doi:10.1007/s11370-009-0049-4

Ryu, J.-K., Chong, N. Y., You, B. J., & Christensen, H. I. (2009). Locomotion of snake-like robots using adaptive neural oscillators. *Intelligent Service Robotics*, *3*(1), 1–10. doi:10.1007/s11370-009-0049-4

Secor, S. M., Jayne, B. C., & Bennett, A. F. (1992). Locomotor performance and energetic cost of sidewinding by the snake Crotalus cerastes. *The Journal of Experimental Biology*, *163*(1), 1–14.

Tanaka, M., Kon, K., & Tanaka, K. (2015). Range-Sensor-Based Semiautonomous Whole-Body Collision Avoidance of a Snake Robot. *IEEE Transactions on Control Systems Technology*, *23*(5), 1927–1934. doi:10.1109/TCST.2014.2382578

Tanaka, M., Nakajima, M., & Tanaka, K. (2016). Smooth control of an articulated mobile robot with switching constraints. *Advanced Robotics*, *30*(1), 29–40. doi:10.1080/01691864.2015.1102646

Transeth, A. A., Leine, R. I., Glocker, C., Pettersen, K. Y., & Liljebäck, P. (2008). Snake robot obstacle-aided locomotion: Modeling, simulations, and experiments. *IEEE Transactions on Robotics*, *24*(1), 88–104. doi:10.1109/TRO.2007.914849

Transeth, A. A., Pettersen, K. Y., & Liljebäck, P. (2009). A survey on snake robot modeling and locomotion. *Robotica*, *27*(7), 999–1015. doi:10.1017/S0263574709005414

Tully, S., & Choset, H. (2016). A filtering approach for image-guided surgery with a highly articulated surgical Snake Robot. *IEEE Transactions on Biomedical Engineering*, *63*(2), 392–402. doi:10.1109/TBME.2015.2461531 PMID:26241966

Walker, I. D. (2013). Continuous backbone "continuum" robot manipulators. *ISRN Robotics, 2013*.

Walker, I. D., Choset, H., & Chirikjian, G. S. (2016). Snake-Like and Continuum Robots. In *Springer Handbook of Robotics* (pp. 481–498). Springer. doi:10.1007/978-3-319-32552-1_20

Walton, M., Jayne, B. C., & Bennett, A. F. (1990). The energetic cost of limbless locomotion. *Science*, *249*(4968), 524–527. doi:10.1126/science.249.4968.524 PMID:17735283

Wright, C., Buchan, A., Brown, B., Geist, J., Schwerin, M., & Rollinson, D. … Choset, H. (2012). Design and architecture of the unified modular snake robot. *Proceedings - IEEE International Conference on Robotics and Automation*. doi:10.1109/ICRA.2012.6225255

Wu, X. (2011). *Sensor-driven Neural Controller for Self-adaptive Collision-free Behavior of a Snake-like Robot*. Academic Press.

Yamada, H., & Hirose, S. (2006). Study on the 3D shape of active cord mechanism. In *Proceedings - IEEE International Conference on Robotics and Automation* (Vol. 2006). IEEE. doi:10.1109/ROBOT.2006.1642140

KEY TERMS AND DEFINITIONS

Adaptive Locomotion: An important characteristic of the robot to behave differently according to change in the environmental features.

Back-Bone Curve: A shape-based modeling framework to capture the macroscopic body shape of snake based on a curve.

Concertina Locomotion: The first studied type of snake locomotion in which the snake body is in static contact with the wall of the channel, similar to anchor, to push against the wall in order to translate the rest of the body forward.

Gait Pattern: A periodical sequence of commands generated for actuators.

Hyper-Redundant Mechanism: A mechanism in which the number of control inputs is fewer than the number of degrees of freedom of the robot.

Lateral Undulation: The most common type of snake locomotion in which a periodic sinusoidal wave travel backwards from head to tail, which causes the snake to move forward.

Obstacle-Assisted Locomotion: An idea for snake robot locomotion using the obstacles as push points to push against and move forward.

Rectilinear Locomotion: Or caterpillar locomotion is a type of snake locomotion in which the snake travels in a straight line without the need to have sideways interaction; instead, contraction and relaxation waves passing over ventral muscles is the essential method for the locomotion.

Side-Winding Locomotion: The fastest type of snake locomotion usually observed in sans snake in which two parts of the snake body act as static contact with ground to lift and move the rest of it sideways.

Chapter 12
Review of Friction and Surface Properties of Snakeskin

Hisham A. Abdel-Aal
Drexel University, USA

ABSTRACT

Locomotion is an essential activity in the biological world. Friction affects the energy invested in motion. Natural systems tend to conserve energy. As such, it is necessary to mitigate the effects of friction. There are many intriguing examples of creatures where dynamics of motion and surface geometry (macro and micro) combine to optimize the tribological profile. One example is that of snakes. In snakes, texturing geometries and topologies specifically contribute to frictional mitigation in a variety of situations. Snakes are found everywhere on earth. Their diverse habitat presents a broad range of tribological environments. This requires customized surface design features that potentially can inspire deterministic solutions for many technical problems. Therefore, they have attracted attention lately. Many, from surface engineering to robotics, have started to study their tribological response and frictional characteristics. This chapter attempts to review the major findings about the tribology of the legless reptiles obtained through the last two decades.

INTRODUCTION

Allah has created every creature from water. And of them are those that move on their bellies, and of them are those that walk on two legs, and of them are those that walk on four. Allah creates what He wills. Indeed, Allah is over all things competent. (Quran 24:45)

Snakes have a wide distribution all over the planet. They are found almost everywhere on earth except where factors that limit their survivability are present (e.g., higher altitudes where very cold temperatures are predominant year round). Their ecological diversity, and thereby their diverse habitat, presents a broad range of tribological environments many of which are hostile in terms of sliding and contact conditions. Hostility and diversity require specific tribological response that manifests itself in functional practices

DOI: 10.4018/978-1-5225-2993-4.ch012

and surface design features. As such, they offer a great resource that can be mined for viable surface design inspirations, which address a wide spectrum of technical problems, faced in the technical world.

The motion of a snake is a delicate balance between the propulsive forces generated by the muscles and the friction tractions due to contact with the substratum. In some cases, the snake makes use of friction to generate thrust. However, for economy of effort, the COF needs to be minimized (especially in rectilinear locomotion) since friction opposes motion. As such, a self-regulating mechanism to control frictional tractions should exist in the snake. The texture of the surface (i.e., the micron sized fibril structures or denticulations) are a major component of such a mechanism. The geometry and topology of the fibril structures allow the snake to modify the frictional profiles in response to changes in contact situations. The presence of the fibrils contributes to the dynamic control of the real area of contact between the skin and the substratum upon sliding. The function of the denticulations (micron-sized fibrils) in this sense is similar to that of the deterministic textures used to regulate friction.

Frictional performance of snakes was a subject of several investigations that confirmed the unique features of the ventral skin of snakes and their optimized tribological response with respect to energy losses and resistance to abrasion and wear (Abdel-aal, 2013; Abdel-aal, et. Al, 2011, 2012; Klein & Gorb, 2012; Benz et. al, 2012; Schmidt & Gorb, 2012) Other results (Abdel-aal, et. al 2011; Schmidt & Gorb, 2012; Stewart & Daniel, 1973; Abdel-aal & El Mansory, 2011) attribute optimization of tribological function to the geometrical patterns and metrology of ventral scales. Frictional response in snakes is the culmination of functional specialization in terms of surface microstructure and metrology.

Ventral scales in snakes have hierarchical micro-geometry in that the scales contain arrays of micron-sized fibril structures (Abdel-aal et. al. 2011, 2012; Benz et. al., 2012; Abdel-aal, 2015; Berthe et al, 2009). The geometry and topology of these fibrils allow the snake to modify the frictional profiles in response to changes in contact situations (Abdel-aal, 2013,; Abdel-aal & El Mansory, 2011). The presence of the fibrils contributes to the dynamic control of the real area of contact between the skin and the substratum upon sliding. The function of the denticulations (micron-sized fibrils) in this sense is similar to that of the deterministic textures used to regulate friction. This similarity raised the curiosity of several researchers (Stewart & Daniel, 1973; Abdel-aal & El Mansory, 2011, 2014). The results pointed many common features between man textured surfaces and micro-ornamentation in ventral scales. Namely, the research pointed at the deterministic nature of the topology of the micro-texture of the scales. The building block of the micro-texture is a 3-D feature of primitive geometry. The feature repeats on the surface in a particular topological pattern so that the metrological description of the surface reduces to several deterministic attributes almost identical to those used for textured surfaces (Marvi, H., Gong, C., et al, 2014; Cicconofri et al, 2015; Rezapour, et. al, 2014, Toyoshima, et. al, 2014, Liljeback et. al., 2014; Virgala, et. al., 2014; Bogue et. al. 2014).

There are several similarities between a textured engineering surfaces TES and the ventral side of a snake, both in metrology and topology. The main similarity is the dependence of the metrological features on the scale of observation. That is both of the snakeskin and the LTS belong to the so-called multi-scale surfaces. In addition, the basic building block in each case (snakeskin and TES) is a textural element that is repeated in an array distribution. The snake's ventral side is composed of identical micron-sized fibrils (denticulations) distributed over the skin area in a particular pattern. Spacing, length, orientation and shape of denticulation is in general common to a particular family of snakes. TTS, on the other hand, by their very definition comprise an individual textural building block (cone, dimple, chevron etc.,) that also is distributed on the surface area in array form. Therefore, both types of surfaces share a common constructal origin. A snake, however, has to be self-sustaining over a wide spectrum of sliding terrains and conditions. The ventral texturing, therefore, has to be efficient over the spectrum of contact conditions that the species experiences. Such a feature offers an advantage to the natural surface over an engineered one where the dimples satisfy efficient performance over a narrow domain within possible contact conditions. It is no surprise that multi-disciplinary groups, from robotics to biology and passing by tribology and material science, have generated a considerable amount of data over the past three decades or so, especially when surface structure and frictional performance are concerned. During this period a detailed review of the newly found facts about this class of reptiles haven't emerged. Here, therefore we attempt to review findings concerning tribological behavior.

The entry point to this review is exploring the general physical features of the species. We emphasize their global distribution and allometry. Allometry in the biological sense serves to relate size to form. Linking allometry to global distribution, therefore, helps linking habitat to surface design through size. We follow the exposition of allometry by an exposition of locomotion modes adopted by various snakes.

Locomotion in snakes is a function of surfaces design, size, and more importantly a balance between muscle-generated tractions and frictional induced tractions. The later, while contributes to motion resistance, in the conventional mechanical sense, also is essential for the propagation of the reptile. As such, understanding locomotion provides an inside look into the interaction between surface structure, body form, and muscle kinematics. It forms the basis for understanding the tribological aspects of the kinetic efficiency of the species.

The second part considers surface characterization through focus on three major aspects. The first emphasizes the physical attributes of the skin including general appearance and architecture. The second entails a review of metrological aspects. This includes both dimensional and surface metrology of the surface. The focus within this part is engineering characterization of the topographical build-up of the surface. That is, characterizing the surface in terms of industrial protocols to establish feasibility of replication or mimicry. The third is tribology of the species. This entails reviewing efforts to characterize friction and wear behavior of snakeskin.

The third part of the review deals with the feasibility of using snakeskin as a vehicle for developing bio-inspired solutions to technical surface problems within the domain of tribology and robotics.

GENERAL BACKGROUND

Lizards and snakes (Squamate-scaled-reptiles) constitute a diverse group of terrestrial vertebrates. Conventional classification (figure 1) divides Squamata into three major taxa: the Amphisbaenia (worm lizards), Lacertilia (lizards) and Serpentes (snakes). Squamata comprise almost 6,000 species. Around 2,900 of these species are classified as snakes (*Sepkoski, 2002*). Snakes, suborder Serpentes, are elongate, legless, and carnivorous. They are distinguishable from legless lizards through the absence of eyelids and external ears. Squamates are cold blooded-ectothermic- amniote vertebrates. Their body is covered with overlapping scales.

They are grouped into eighteen families with each family divided into genera, and these are further divided into species. The number of species per genera depends on biological similarities between candidates.

Figure 1. Conventional classification of Squamata

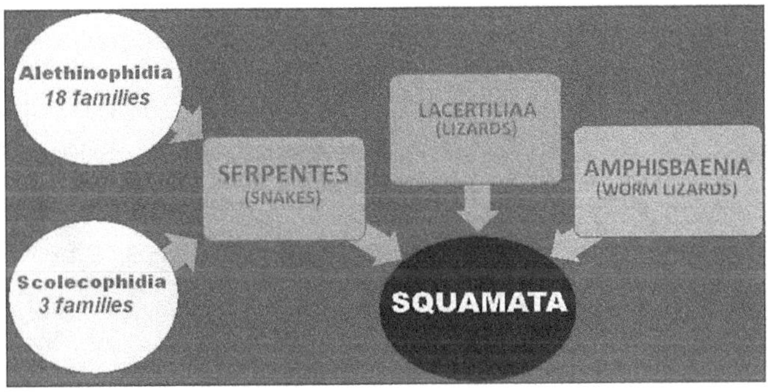

Size Distribution and Allometry

Snakes, have different body shapes (Figure 2). Difference in shape cause variation in body proportions. As such, two snakes may have the same length, while they considerably differ in mass. Vipers, for example, are generally more robust than other snakes (*Pough and Groves 1983*), and so are constrictors. Moreover, life history traits and ecological characteristics influence body shape whence mass-length relationships (Pough & Groves, 1983; Shine R., 1994; Martins et. al. 2001).

The cross section of the snake body is another factor that influences mass. The shape of the cross section, while influenced by habitat, is likely to affect the length-circumference relationship and also the ratio of length to maximum circumference. Pough and Groves (1983) studied the relationship between length and body circumference. Their results pointed that mass is not a unique function of body length. Feldman and Meiri *(2013)* suggested that SVL is a better predictor for mass. Based on their equations we may develop a mass-length map that ranks the order of the overall size of the different snake families. Figure 3 presents such a map. Ranking of species in the map observes the bounds of mass and length of a given snake family. Thus, the minimum length and mass and the maximum length and mass of a given snake family are plotted. The plots reflect the interference between sizes. Almost all families contain members that share a common length and body mass with other snakes belonging to different families. This implies that while mass and length are important proxies, they are not unique traits that can precisely identify or characterize a particular family (or individual species).

LIMBLESS LOCOMOTION

Snakes utilize a varicty of locomotion modes. The choice of a particular mode depends on speed requirements, the type of surface they contact (Vidal, Blair, & Hedges, 2005; Kumazawa, 2007; Vitt, et al, 2003; Gray, 1946; Gray & Lissman, 1950), body weight (Bogert, 1947), and vertebral number (Kelley, Arnold, Gladstone, 1997). Individual snakes often appear to use two locomotion modes simultaneously (Kumazawa, 2007), so that it is not always obvious whether different regions within a snake simultaneously combines characteristics of two modes. When a snake encounters a tunnel, for example, it adopts concertina locomotion since its body has to conform to the spacing of the walls in the tunnel. The kinematics of motion changes, in this case, when the width of the tunnel changes (Cowles, 1956-a).

Figure 2. Common encountered shapes of snake bodies. Slender (a) Ribbon Snake (Thamnophis sauritis), (b) Moderate, Northern pine snake (Pituophis melanoleucus) and (c) Stocky Brown Watersnake (Nerodia taxis pilota). Photographs of snakes are reproduced under the Wikipedia Foundation free licensing policy http://creativecommons.org/licenses/by-sa/3.0/.

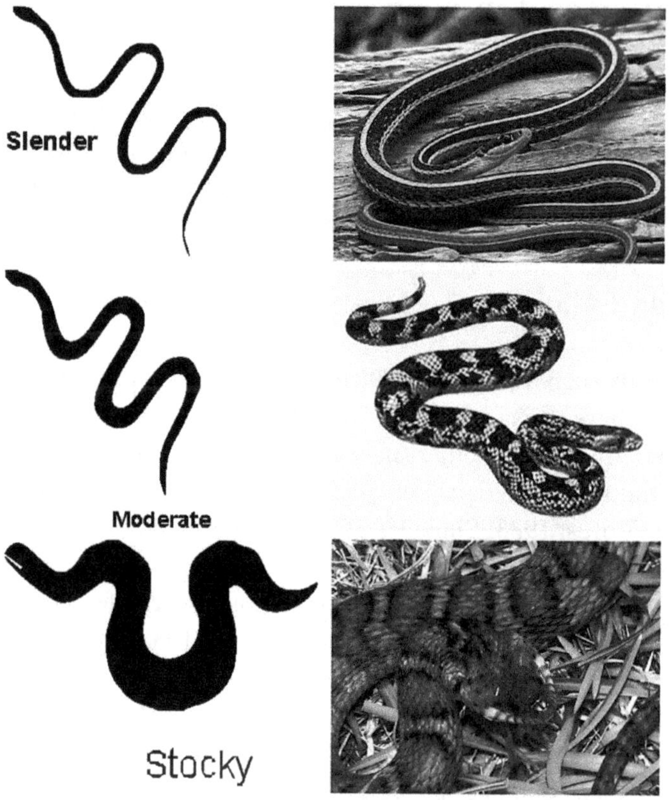

Snakes start their movement from a stationary position in one of two patterns associated with the speed of initiation of movement (slow or high speed) (Cowles, 1956-a, b). The slow pattern involves progressive acceleration of the body with the acceleration beginning with the head and proceeding towards the tail. The initiation of the slow speed pattern, generally, involves a variation of concertina propulsion. Once acceleration is initiated, the reptile tends to shift to lateral undulation. During the transition, both concertina and lateral undulation take place simultaneously.

The high-speed pattern of movement entails simultaneous acceleration of the entire trunk. In this pattern, lateral undulations are initiated directly from rest. In the absence of contact with push points, rapid movement entails considerable postero-lateral slippage and the reptile transitions into slide pushing mode. When movement starts from a stationary position in which the snake contacts a push point, the entire trunk undulates laterally in the absence of lateral slippage. Initiation of motion, and sustainment of propulsion, therefore, does not take place through a unique procedure.

Lateral Undulation

The pattern of lateral undulation in snakes entails the generation of alternate muscular contraction waves along the body. The size of the bends is established at the time of initial formation. The amplitude of the bends is a function of whether the contacting substratum includes irregularities. Contact and application of force also is intermittent and not continuous. Application of force is confined to particular pivoting (bearing) points on the ventral side of the reptile.

Lateral undulation in snakes is unique in that whenever a bend contacts a surface object, such as a rock or stick, the snake exerts force against it and deforms locally around it. Postural adjustment around each object tunes the control of the snake over the direction of force exertion. Force exertion against each object is inversely proportional to the number of objects being pushed against simultaneously by the snake, but total force is roughly constant for a given speed and substrate (Jayne, 1988; Stephen, 1992).

Sidewinding

Locomotion by sidewinding permits travel across relatively smooth substrata with higher velocities than those achieved by lateral undulations especially for thick and short bodied snakes. Locomotion usually comprises two essential movements, bearing and lifting. The snake starts from an extended position

Figure 3. Comparative allometry of various snake families. The map, developed based on the equations of Feldman and Meiri (2013), presents the maximum and minimum bounds of length and mass of the principle snake families.

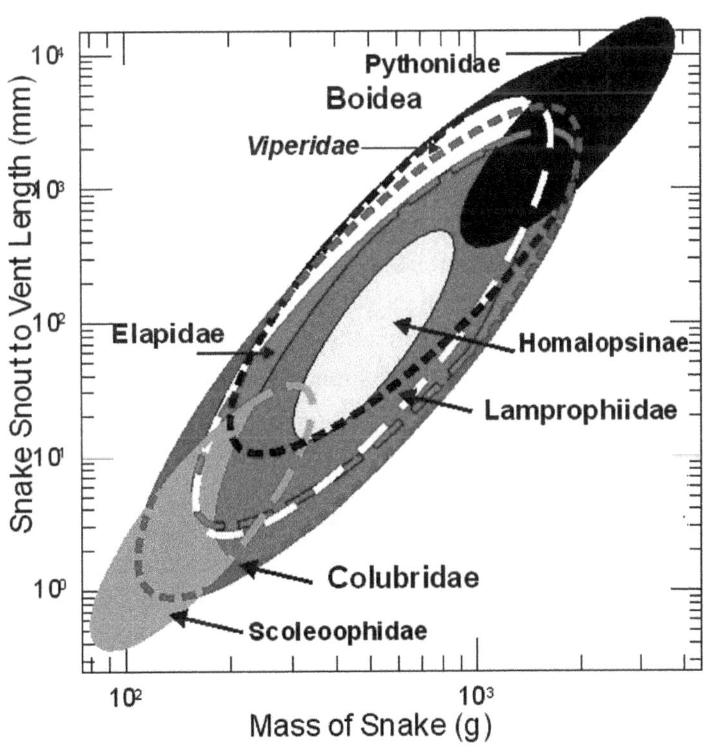

that is in full contact with the substratum. The snake then lifts its head, neck and anterior body. The reptile follows this lifting by swinging the head and anterior to right or left depending on whether right-handed or left-handed locomotion takes place. Typical swing angles may reach 90° or 120° (Nursall, 1962). Locomotion by sidewinding involves repetitive cyclic movements of body parts through which the reptile alternates flexion and extension of the vertebrate muscles on either side of the body. Through such a motion, a snake makes use of static frictional forces by establishing ground contact segments and ground contact tracks. Net snake displacement is the resultant of the sequential motion of the ground contact tracks.

Sidewinding bears some similarities to motion by lateral undulation in the pattern of bending. However, it differs in three distinct aspects: First, each point along the body sequentially rubs the substrate in a static, rather than dynamic, friction mode. Second, the reptile lifts above the ground those segments of the body that lie between the regions in static contact with the ground.

Thus, the body sequentially rolls along the ground from neck to tail, forming a characteristic track (that is proportional to body length). The front part of the body begins a new track after going through lifting off the ground and placement down again a short distance away. The path of the rear part of the body, meanwhile, traces the loci of the old track. Third, because of the combined effects of static friction, which acts upon bearing on the ground, and lifting of the body, the snake travels roughly diagonally relative to the tracks it forms on the ground (Figure 4).

Concertina Locomotion

Propulsion through the concertina mode involves flexion of the body against the walls of the tunnels. In this manner, the snake may adapt its body to the geometry of the tunnel by using the static friction forces that act upon contact. Propulsion in this mode is rather intermittent. Motion takes place through a series of stands and stops (Gans, 1994; Marvi & HU, 2012).

Concertina motion involves pulling the body of the snake into a series of S-shaped coils or bends (Table 1) and then straightening the moving portions of the body forward. The front part of the body then comes to rest on the surface and the back part of the body is pulled into bends again. This cycle of stretch-drag and re-bend repeats until the animal clears the tunnel. The process of bending and stretching, moreover, makes use of the differences between static and dynamic sliding friction. The posture of the reptile during concertina locomotion in tunnels changes in response to the width of the passage. The numbers of body segments that establish contact with the walls of the tunnel vary with the difference between the tunnel width and the body width. In addition, the angle between the tunnel and particular contacting segments is directly proportional to the tunnel width.

Rectilinear Locomotion

Rectilinear locomotion is used mainly by large snakes (e.g. large vipers, boas, and pythons) (Table 1). Rectilinear locomotion is cyclic. The snake alternately lifts ventral scales slightly from the ground then pulls forward then downward and backward. However, because the scales "*stick*" against the ground, the body moves forward over them. Once the body has moved far enough forward to stretch the scales, the cycle repeats. This cycle occurs simultaneously at several points along the body. During rectilinear propulsion, the distance between any two neighboring points on the body varies with the frequency and direction of the contraction-relaxation waves. Waves of alternating longitudinal contraction and

Table 1. Possible modes of locomotion in snakes

Locomotion Mode	*Schematic*
Serpentine locomotion	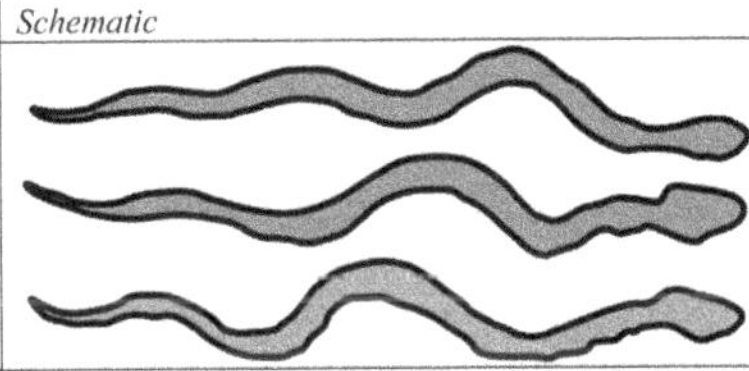
side winding	
Concertina locomotion	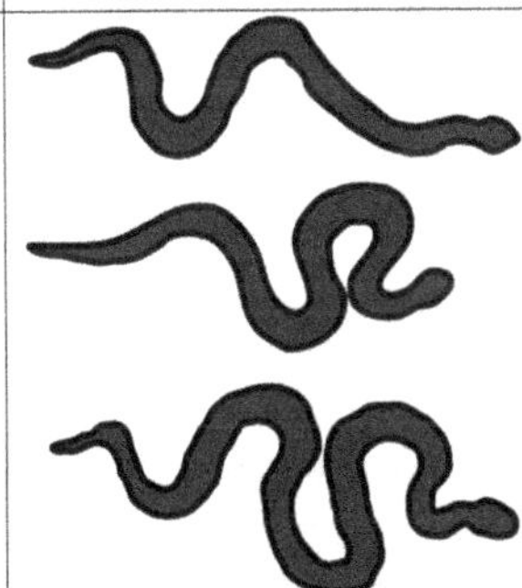
Rectilinear locomotion	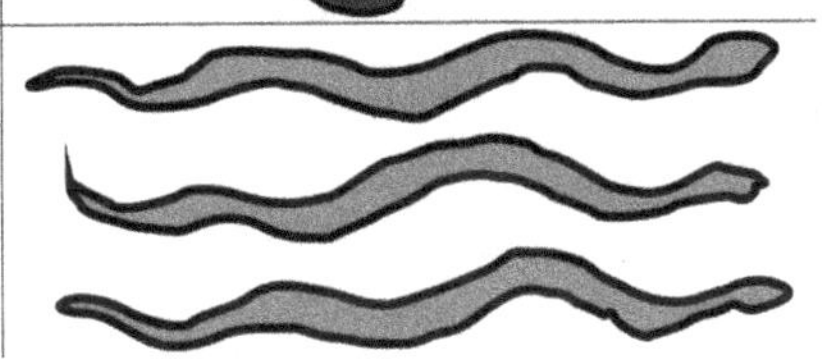

elongation pass axially in the caudal direction. As waves pass along, each point on the ventral surface progresses in rhythmic steps. The muscle waves may be irregular, and may stop short from reaching the tail. The head of the snake in this mode of locomotion moves with a constant speed. Points located on the ventral side undergo cycles of forward movement and rest. Points on the dorsal side, meanwhile, translate forward at a constant speed.

Therefore, all points on the dorsal side remain equidistant with respect to each other, whereas distances between ventral points do not necessarily remain constant. The distance between ventral points remain constant as long as the original spacing between them is one complete wavelength (or a multiple of such a distance). Ventral points that lie half a wavelength apart manifest maximum change in length between segments. In the absence of backward slip, the length of a step is a function of the difference between maximum elongation and maximum shortening between these points. The speed of propagation of the animal is the length of a step multiplied by the frequency of waves, which pass over the body.

Slide Pushing

Slide-pushing involves vigorous undulations of the body that slide widely over the surface. This gait is used when a snake on a smooth surface is startled and tries to escape quickly, but slips over the surface.

Figure 4. Schematic of the pathways established due to motion. The illustration depicts the portions of the body where static contact is established (solid marked parts) and the variables used to quantify force and velocity vectors for motion analysis.

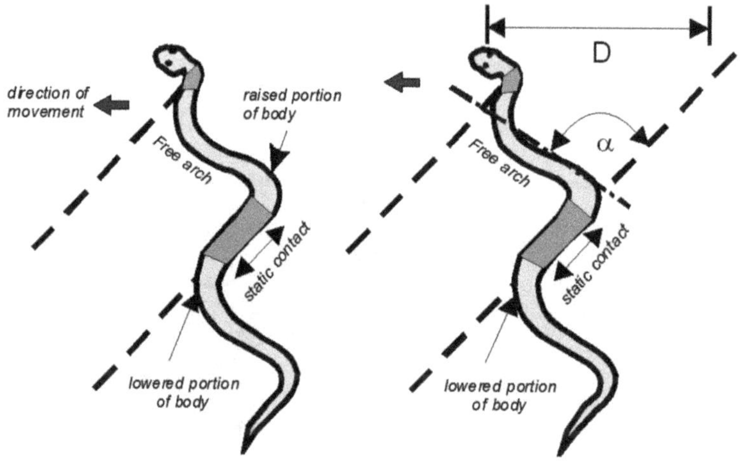

In slide-pushing, irregular bends of the body and tail press vertically on the surface at different points; although the body slips on the surface, it pushes down with enough force to move the center of mass in a quasi-regular, often step-like, pattern. Thus, the snake progresses irregularly by slipping along the ground. Sliding friction is most important in slide-pushing, although there may be occasional moments of static contact.

In slide pushing, the head is depressed, the neck often arches vertically and the snout presses against the substratum. The center of gravity of the animal progresses slowly and sometimes irregularly. Propulsive slide-pushing sequences are considerably complex. The animal may slide the terminal portions of the anterior and middle portions of the body. In the former, the anterior portion of the body is narrowly curved and slides anteriorly. The posterior portion, meanwhile, swings vigorously through wide curves first toward one side of the body then the reverse. The downward deflection of the trunk generates a contact zone at the more posteriorly placed segments of the fast swinging trunk in order to use sliding friction to generated propulsive thrust (Gans & Gasc, 1990; Dobrolyubov 1986).

MORPHOLOGICAL STRUCTURE OF SNAKESKIN

The epidermis of snakes is particularly complex (*Figure 5*). The functional scale consists of many layers of cells. The outer layers are heavily keratinized, but the deepest several layers are living. All these cells constitute a mature outer epidermal generation (Landman, 1986). An immature inner epidermal generation of cells, which has been produced more recently by the stratum germinativum, lies beneath. The same keratinized and living layers that characterize the outer generation begin to differentiate in the inner generation prior to skin shedding (Landmann, 1979). A new layer of scales is ready to take over for the old layer before the old layer is lost.

The epidermal layer in a snake entails seven layers. These are organized in plies of cells with tight packing. The epidermal layer (sub-layers included) encases the body of the retile to form an outer shield.

Figure 5. General structure of the epidermis of a squamate reptile. Figure, a, depicts the "outer" generation layer which is the layer about to be shed; and the "inner" generation layer which is the new replacement skin layer. Figure 5-b, Brigh Field TEM image showing the whole thickness of the skin of P. regius and B. Constrictor skins. O: Oberhautchen, m: mesos, l:lacunar tissue, c:clear layer.

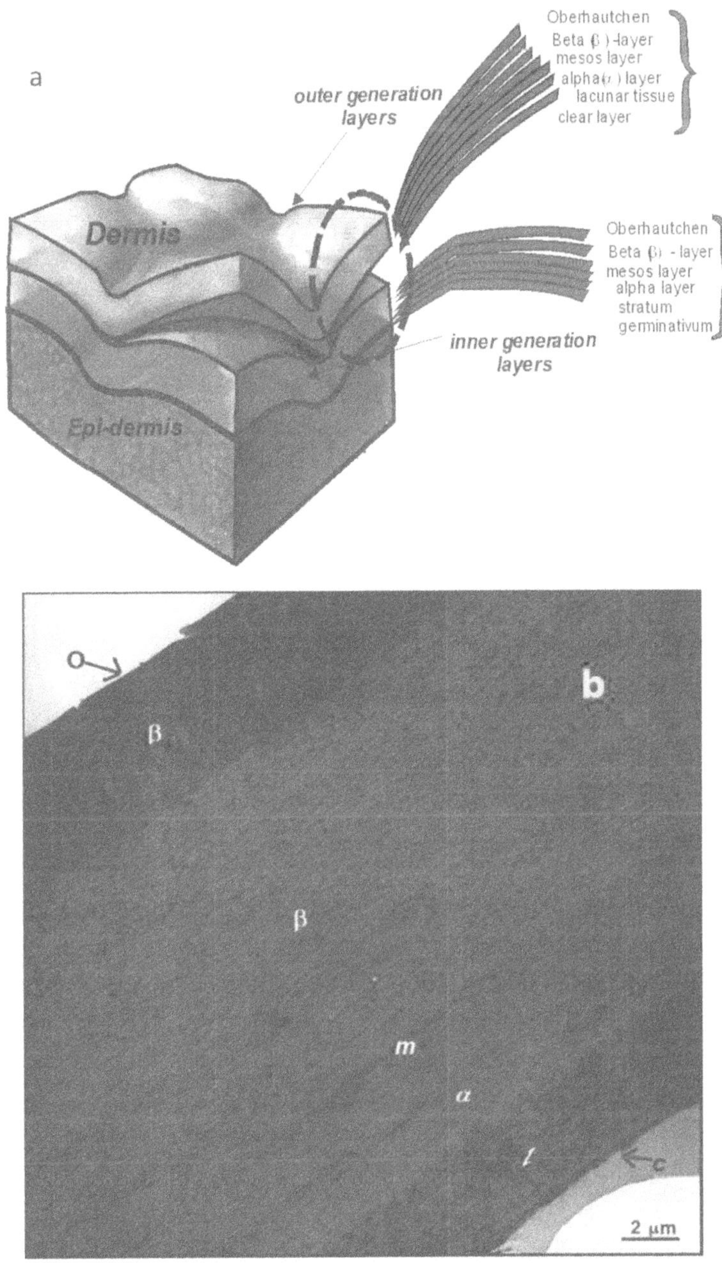

Figure 5 details the seven layers present within the epidermis. Described from the inside of the skin, the first layer is the "stratum germinativum". This is the deepest layer of the skin. It is lined with rapidly dividing cells. Six additional sub-layers, again from the inside of the skin, follow. Together the six sub-layers form a so-called "epidermal generation" (old and new skin layers). Thus, stacked above the stratum germinativum, there exists: the clear layer and the lacunar layer. The lacunar layer matures in the old skin layer as the new skin is growing beneath. Following there is the α–layer, the mesos layer and the β-layer. The mesos layer is similar to the human stratum corneum (Lillywhite & Maderson 1982) and contains several layers of flat and extremely thin cells surrounded by intercellular lipids (Maderson 1965, 1966, 1970, 1972; Maderson & Licht, 1967). These three layers consist of cells that become keratinized with the production of two types of keratin: α (hair-like) and β (feather like). Keratinisation continuously transforms these cells into a hard protective layer. Finally, there is the "oberhautchen" layer, which forms the toughest outer most layer of keratinized dead skin cells.

The oberhautchen layer contains the micro-ornamentation (Alibardi, 1996, 1998). Before the molt, a new layer of epidermis forms under the currently existing one; the two layers are zipped together by a spinulae structure (Alibardi & Toni,2007; Toni et. al., 2007). During the molt, the reptile sheds the outer (older) layer of the epidermis. The principle constituents of snakeskin are keratin fibers (Maderson et. al, 1998). Two types of keratins form the epidermis: the α-keratin (which in a snake is acid or neutral) and the β keratin (which in a snake is basic). The β-layer consists mainly of β-keratin; this type of keratin is not present in other layers of the skin. Alpha α-keratin constitutes most of the epidermal layers and it contributes to the mechanical properties of skin cells (Maderson & Licht, 1967; Alibardi, 1996; Maderson et. al, 1998; Alexander & Gans. 1966).

SCALES AND THEIR STRUCTURE

The Architecture of Snake Scales

A snake may have three or more types of scales on different locations of its body. Depending on the side, within which the scales are locate on the body, scales may be arranged in overlapping or successive rows. This is specially the case for dorsal scales (located on the back of the snake). Dorsal scales may be keeled (Figure 6-a) or smooth (Figure 6-b). Ventral scales, which are located on the underside of the snake (belly side), are apparently smooth, and hexagonal. They typically extend across the underside of the snake with one scale per row. They may be staggered side by side (Figure 6-c) or overlapping (Figure 6-d).

A unit building block of the skin is a so-called scale-hinge arrangement. Each scale has an outer and an inner surface (Figure 7) with the skin from the inner surface hinging back to form an area which overlaps the base of the next scale which emerges below the scale. The skin of the hinge is thin and stretchable. It allows the snake to expand its girth. The scales, meanwhile, are mostly transparent and rigid compared to the hinge.

Arrangement of Scales on Snake Species

A snake hatches with a fixed number of scales (both ventral and dorsal). As the body of the snake grows, the number of scales remains constant (within limits) but the size of individual scales changes (increases) to accommodate the increase in volume. Such a fact has been noticed by biologists and has been used to characterize the different species. Table 2 lists the variation in scale counts for several representative species. The data manifests a wide variation in scale counts among species.

The number of ventral scales in a species influences locomotion and may determine whether movement of the reptile is optimized (Mattison, 2006; Guyer & Donnelly 1990). Figure 8 compares variation

Figure 6. Examples of Snake scales, a smooth Dorsal Scales Blue-bellied Black Snake (Pseudechis guttatus), b Dorsal keeled Scales (bush Viper- Atheris squamigera), c- Ventral Scales Slender Coral Snake (Calliophis melanurus), and d- ventral scale Texas indigo snake- Drymarchon melanurus erebennus)

Figure 7. The details of dorsal scales from the inside of shed skin of a P. regius. The terminology used is: scale to denote the major area of the scale and hinge to denote the raised part forming the circumference of the scale. Scale marker is 1mm.

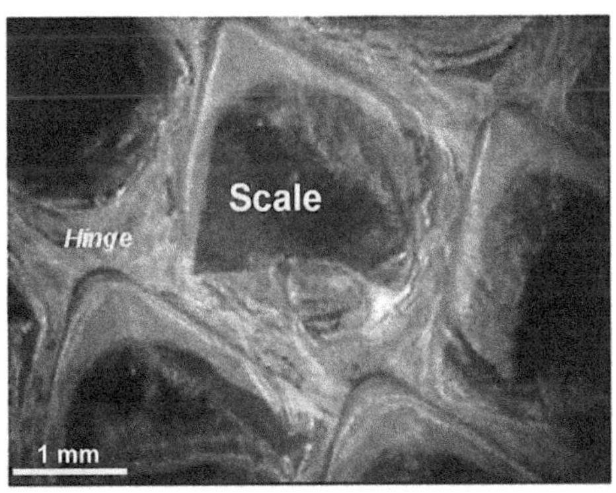

in ventral scale parameters (number and area) to length and mass. Figure 8-a, plots the SVL raw length of approximately seven hundred species versus the median ventral scale counts. The plot shows that the number of ventral scales directly varies in proportion to the SVL. The same trend emerges when plotting mass against the number of scales. This is shown in figure 8-b, which examines the relationship between mass and number of ventrals. The mass in the plot is a product of applying the model of Pough *(1980)*. The trend supports the notion that increased mass requires higher number of ventral scales. Higher scale counts reduce the loads on muscles and in principal reduce frictional resistance in locomotion.

Table 2. Scale counts for several snake species belonging to three snake families

SNAKE FAMILY	Scale counts		
Pythonidae	dorsal	Subcaudal	ventrals
Liasis Fuscus	45-55	60-90	270-300
L Mackloti	46-49	72-89	271-286
Morelia Amethistina	35-57	80-122	220-345
Boidea			
Boa Constrictor	81-87	49-60	234-243
Charina Bottae	41	31	184
Epicrates Cenchria Cenchria	88	45-68	246-279
Eunectes Murinus	--	55-78	239-266
Eunectes Notaeus	--	44-61	213-237
Sanzinia Madagascariensis	--	35-46	199-232
Elapinae			
Naja Naja	29-37	50-67	171-197
Naja Samarensis	17-19	41-52	161-184
Bungarus Fasciatus	15	23-39	200-234
Colubridae			
Heterodon nasicus gloydi	20	41-63	114-152
(Heterodon simus	23	28-50	112-132
Lamprophis Guttatus	21-25	46-72	186-230

Figure 8. Variation in ventral scale parameters (scale counts, area, and loading) against the allometric parameters mass and length

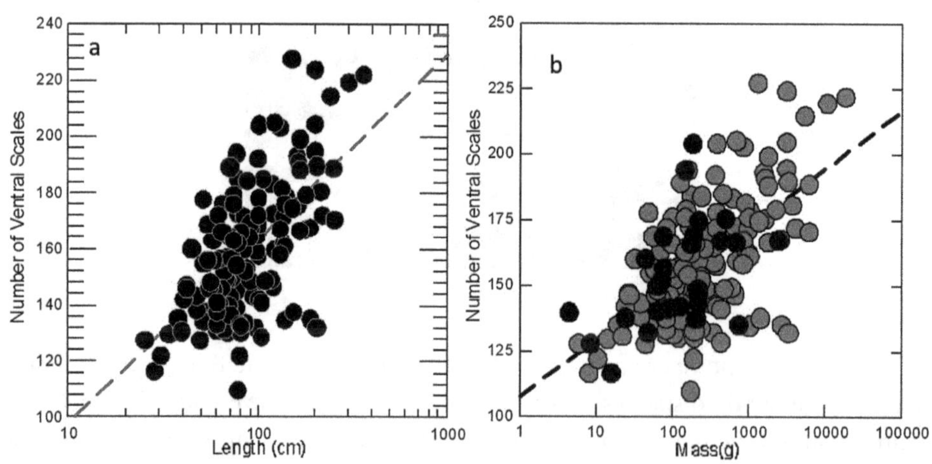

Dimensional Metrology of Ventral Scales

The general outline of a snake ventral scale is a hexagon (*Figure 9*). The hexagon, however, is not uniform. Rather, it is elongated along the lateral axis (labeled LL-RL in figure). The general outline of the scale is a curve that convexes toward the head of the reptile. The overall distribution of the areas of the ventral scales along the AP-axis of the body, and thereby the individual dimensions, is not uniform. The overall dimensions of individual ventral scales vary in proportion to the local variation in body girth.

Figure 10 (a, b) depicts the variation in the area of the ventral scales of two species as a function of the position of the individual scale along the body of the reptile.

The figure shows that area of individual scales vary according to position on the body. Close to the head (first ventral), the area is at a minimum. The area increases until it reaches a maximum value within the midsection of the trunk, thereafter, the area decreases until it reaches a minimum value at the last ventral scale. The minimum value may be less than, or equal to, the initial value of the first ventral scale (depending on the species). The distribution of the ventral scale area is asymmetric, in most species, with the scale of maximum area shifted away from the midsection of the reptile.

Distribution of the area of the ventral scales along the Snout-to-Vent Length SVL is not random. It relates to the general form of the body (*Figure 11*). The plotted data show that the area and the girth are linearly proportional to each other.

Figure 9. Details of the ventral scales on a P. regius. The boundaries of the scales are curves that convex toward the head of the reptile. The middle figure depicts the details of the junction between the scale and hinge regions, whereas the bottom picture illustrates the geometrical descriptors of the ventral scale. The last figure shows the actual geometry of a single ventral scale. Note the elongation of the ventral scale along the lateral axis.

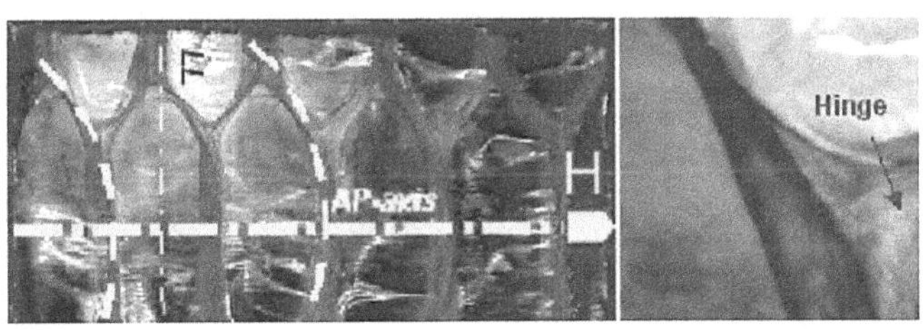

Figure 10. Distribution of the rea of the ventral scales along the AP-axis for several "stocky" snakes

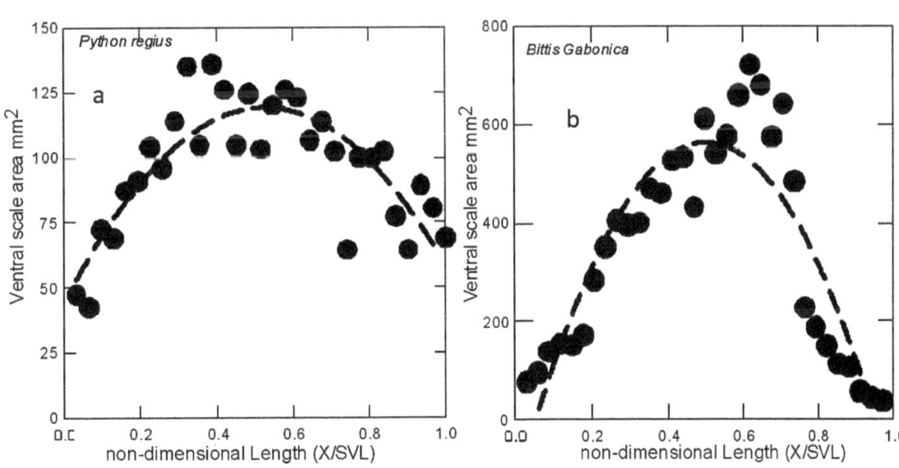

Figure 11. Variation in body girth in relation to the change in the area of ventral scales along the AP-axis of representative snake species. All data are normalized by the overall maximum value

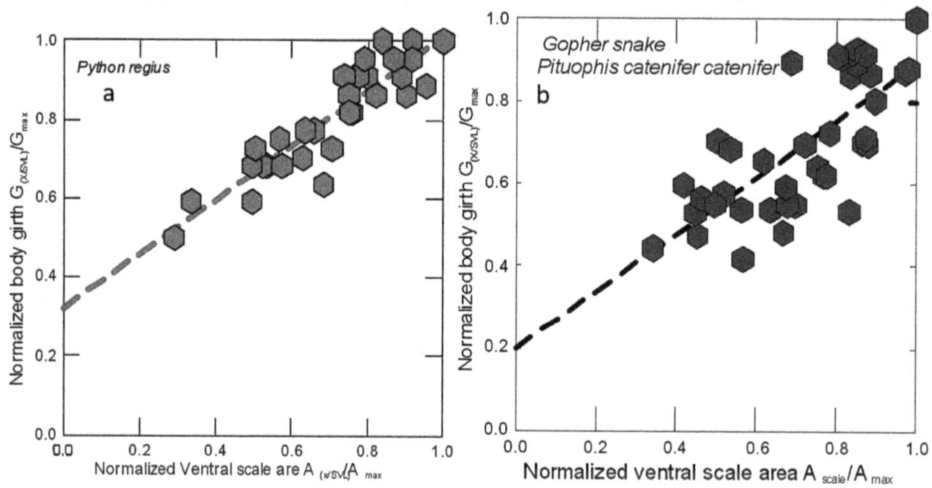

METROLOGY OF SKIN-TEXTURE

Scales in snakes comprises textural elements of primitive geometry (Figure 12). One approach to study the sense of micro-texturing is to view the topography and topology of the macrotexture in reference to the friction behavior of the species.

Several works (Abdel-Aal et. al, 2012; Abdel-Aal, 2013; Zahouani et al, 2008) pointed at the importance of four roughness parameters to frictional behavior. These are the mean arithmetic values of roughness, S_a, the root mean square of roughness S_q, the surface Kurtosis S_{Ku}, and the surface Skewness S_{sk}.

Figure 13-a presents a plot of the average roughness parameter S_a two species B. Gabonica and P. regius. For each snake, the plot shows values for the ventral as well as the dorsal side of the skin. Upon comparison, one notes that protrusions within the Bitis ventral skin are higher than protrusions on the Python. The dorsal side manifests the same trend in roughness height. However, in this case the difference in the heights is clearly visible (roughness of the Bitis species almost twice that of the Python). The RMS roughness parameter S_q reflects a similar trend.

Examination of the kurtosis parameter S_{ku}, Figure 13-, indicates that the distribution of heights does not depart considerably from being Gaussian (or random). From a statistical point of view, the skewness of the surface complements the physical description of the roughness. Figure 13-d presents the skewness data for the species. The figure compares the skewness parameter S_{sk} for the dorsal and ventral sides of the examined skin. In contrast to the Kurtosis the Bitis ventral side displays opposite skewness to the python (S_{sk}<0 for the former and <0 for the latter). Meanwhile, for both snakes the dorsal skin is of positive skewness.

Figure 12. Sample SEM-Images depicting the variation in micro-structure in the ventral scales of snake species

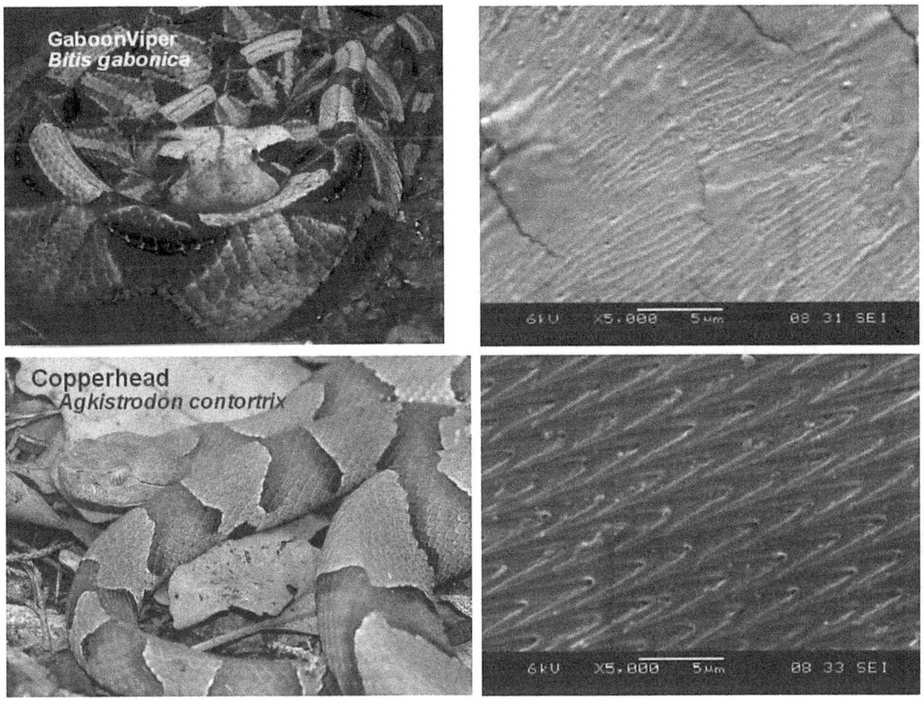

TRIBOLOGICAL BEHAVIOR

Frictional Behavior

Interest in characterization of friction response in snakes has its origin in the study of locomotion.

Hazel and co-workers *(1999)* used atomic force microscopy for qualitative frictional analysis at the nano-scale of three species (*Boa constrictor constrictor, Morelia spilotes, and P. regius*). Their results suggested that the micro-fibrils, recognized as part of the micro-ornamentation in ventral scales, were responsible for observed anisotropic frictional properties. Berthe *(Berthe et al, 2009)* examined the difference of skin friction as a function of body location. The study aimed to analyze the relationship between the microstructure of scales, at different regions of the body, and their frictional properties. She experimented with an Amazon tree boa *Corallus hortulanus* (Squamata, Boidae), which is a specialized arboreal snake. Static friction properties of the skin were evaluated on smooth and rough substrata in different directions relative to the body axis of the snake. The results showed that the COF is sensitive to the roughness of the substratum. SEM analysis revealed a correlation between the microstructural pattern of the ventral scales and frictional anisotropy. Dorsal and lateral scales showed higher friction coefficients compared to ventral scales (*Figure 14*). Since the dorsal and lateral scales are hardly involved in surface contact, their high friction coefficients implied functional irrelevance in locomotion. No detailed mechanism for the directional interaction of fibrils, however, was suggested.

Figure 13. Comparison between the surface roughness and aerial parameters of the examined reptilian surfaces; a- surface average roughness parameter, Sa, for ventral and dorsal sides of the Bitis and the Python species; b- root mean square roughness parameter Sq; c- surface kurtosis parameter Sku; and d- surface skewness parameter Ssk.

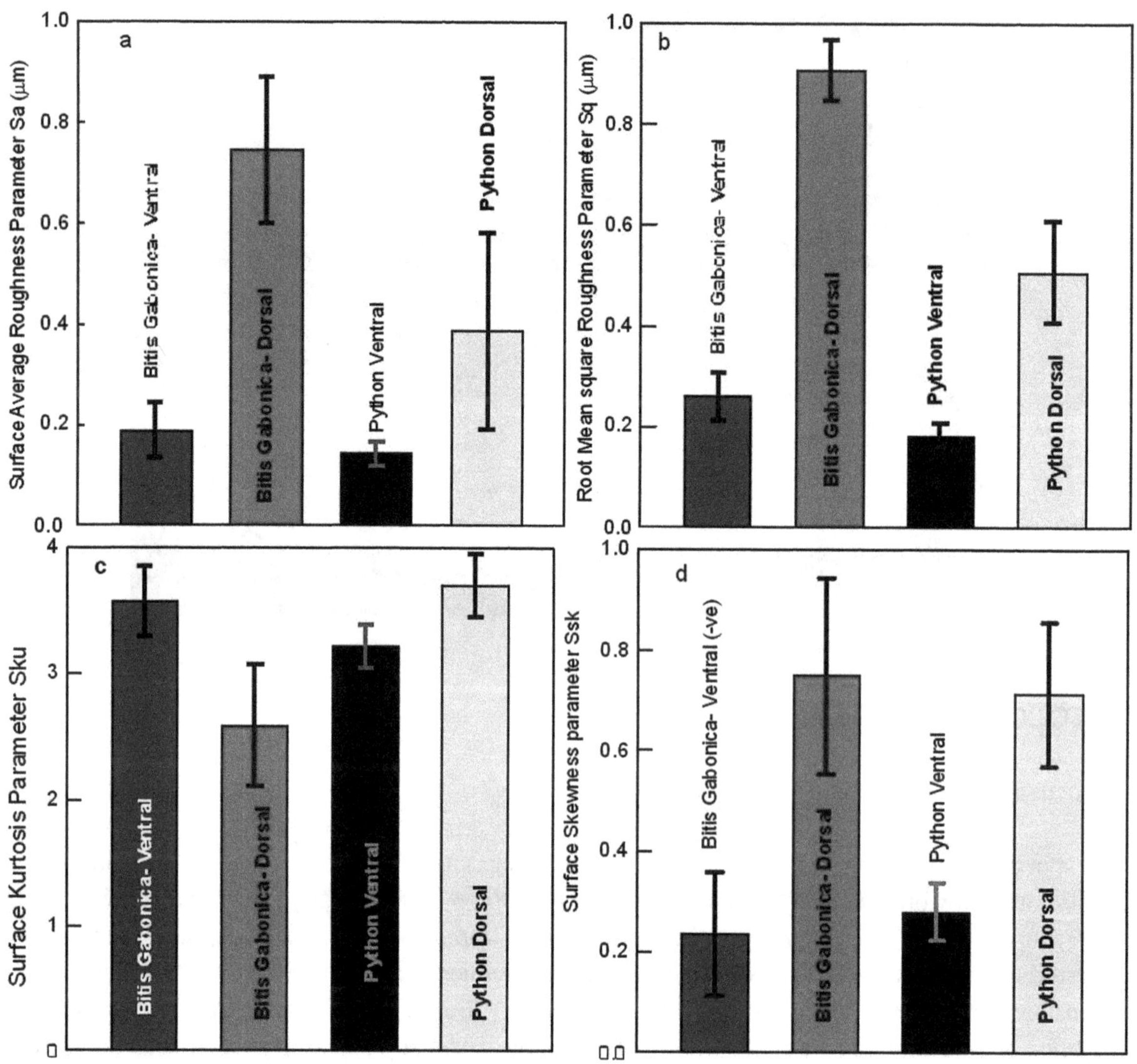

Locomotion also was the focus of Marvi and Hu (2012) who recognized that forces exerted by the snake on the sliding substratum trigger friction. Consistent with previous work, their results confirmed the anisotropy of the COF. An interesting find was the effect of animal consciousness. The COF in the unconscious state was consistently lower than that for the conscious state. In a later work (Marvi et al, 2014) asserted that muscle contractions can increase the COF depending on the slope of the plane a snake is moving through, and on the roughness of the substratum. Figure 15 plots the values of the dynamic friction coefficients measured in the unconscious and the conscious states. Values for unconscious snakes are greater than or equal to those of conscious snakes (static friction coefficients for conscious snakes

Figure 14. Friction coefficients of scales. Columns show mean values of corresponding friction measurements for different roughnesses and directions. Error bars indicate standard deviations. The percentages indicated over column are set to 100% for dorsal scales and the percentages within the columns are set to 100% for rostrad direction within one scale location. *This is an open access article distributed under the terms of the Creative Commons Attribution Noncommercial License, which permits any noncommercial use, distribution, and reproduction in any medium, provided the original author(s) and source are credited.*

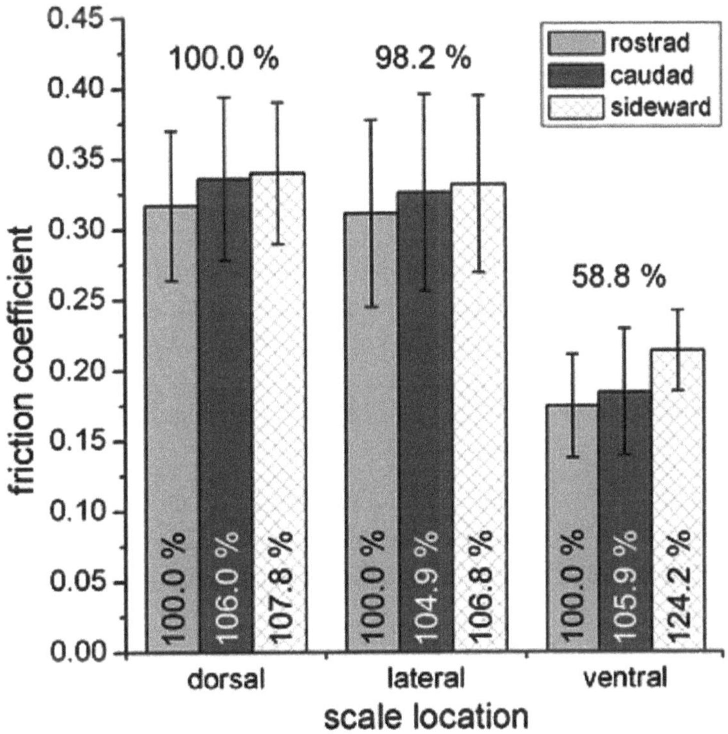

Figure 15. The dynamic and static friction coefficients for both conscious and unconscious snakes

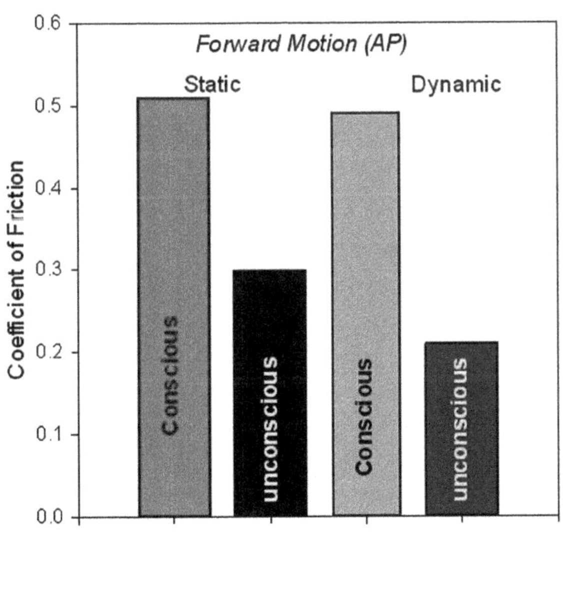

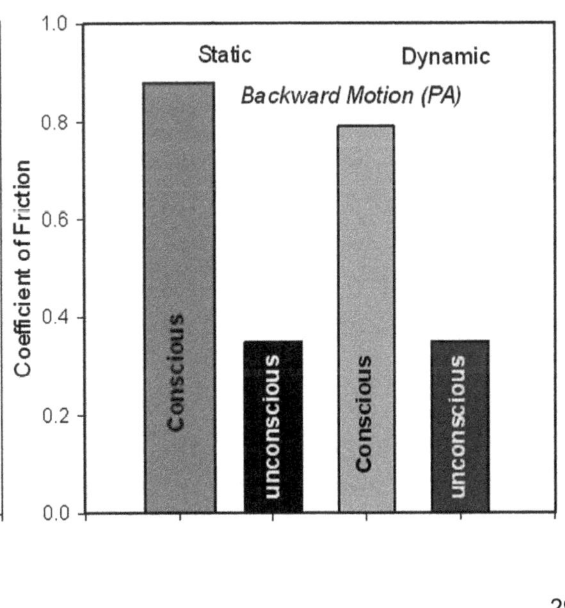

are double the corresponding coefficients of unconscious snakes). Differences in dynamic friction coefficients are not as pronounced. Coefficients for the conscious snakes approximately are 30% higher.

Abdel-aal (2013) measured the average kinematic friction coefficient for a *P. regius*. The experiments involved passing a probe over a stationary skin patch of length 15 cm. Using patches from different locations allowed a comparison between friction behavior of the leading and trailing halves of the reptile. Figure 16, presents a summary plot of the COF obtained. The data reveal the anisotropy of friction.

This anisotropy is not uniform (Figure 17), anisotropy along the anterior-posterior axis (AE-PE) is less than that within the trailing half. In the lateral direction, however, the anisotropy is almost invariable.

For diagonal motion, the highest anisotropy pertains to measurements taken along the dextro-caudal direction (and its converse). (Figure 18-b). The COF in forward motion for the trailing half of the skin was higher than that within the leading half. The inverse was noted for the COF in backward motion.

Comparison of Available Data

Studies confirm that the anisotropic frictional behavior of snakeskin not only is a function of the direction of motion, but also depends on the position on the body (trailing or leading half). Table 4 presents a compilation of available coefficient of friction data. Wide variation of data makes direct comparison likely not feasible. Several factors contribute to this situation: differences in measurement techniques, variation of micro geometry of species, and measurements performed on different parts of the reptile

Figure 16. Distribution of COF values for a P. regius a) measurement in all directions for leading and trailing halves of the skin., b-Mapping of the ratio of the frictional anisotropy (μB/μF). Printed with permission from: H. A, Abdel-Aal, On surface structure and friction regulation in reptilian limbless locomotion, JMBBM, 22, 2013, 115-135, doi /10.1016/j.jmbbm.2012.09.014.

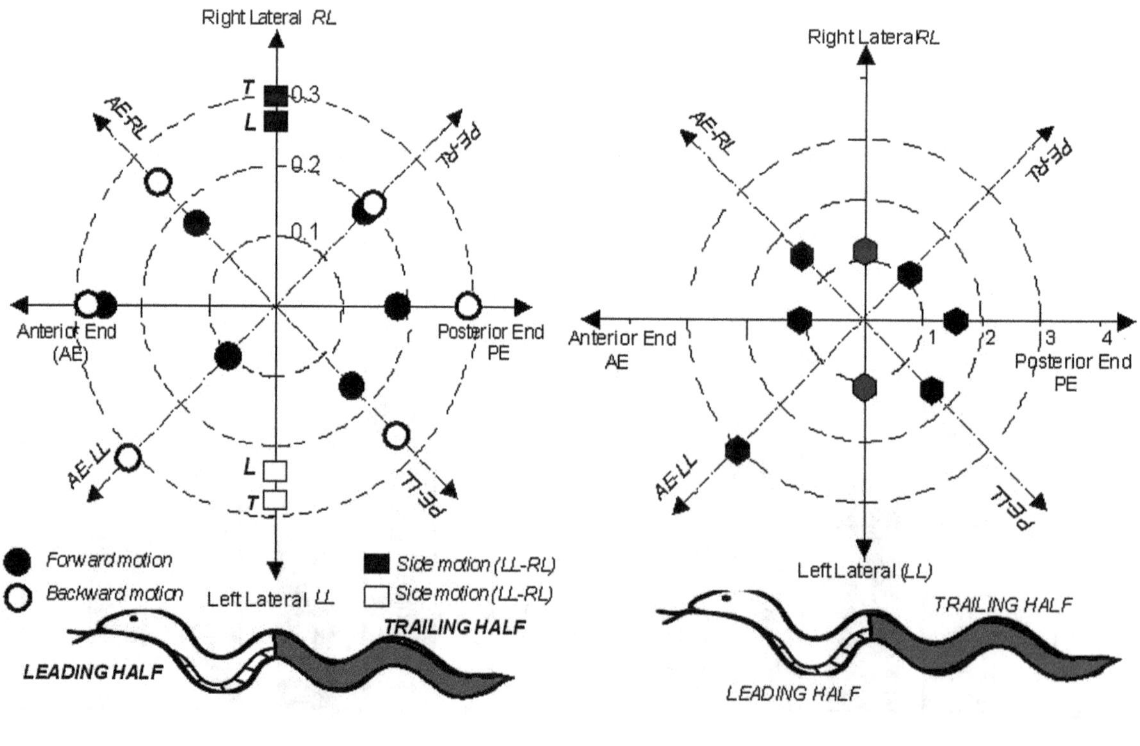

Figure 17. Comparison of COF values within the trailing half to its corresponding values of the leading half ($\mu_{T.H}/\mu_{L.H}$). Printed with permission from: H. A, Abdel-Aal, On surface structure and friction regulation in reptilian limbless locomotion, JMBBM, 22, 2013, 115-135, doi /10.1016/j.jmbbm.2012.09.014.

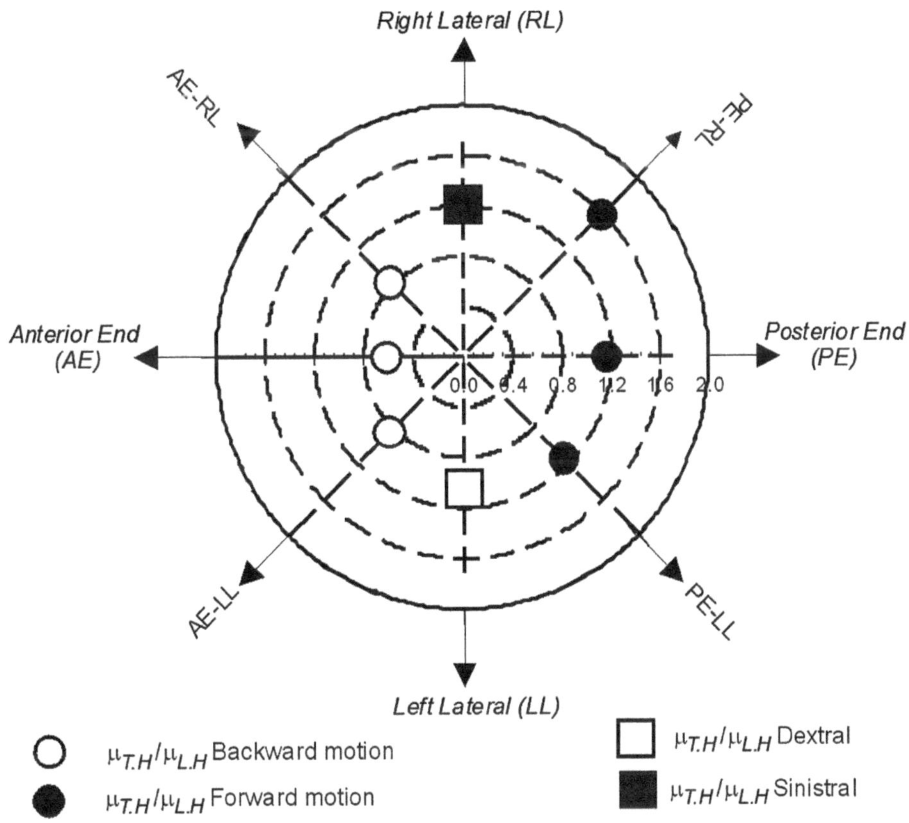

body. Evaluation of the effect of using different measurement techniques is still lacking. Abdel-aal (2016) compared values of the COF of a *P. regius* obtained by two methods: an acoustic probe (Abdel-aal, et al 2012) and a force gauge setup according to ASTM D1894-14. Figure 18-a compares the RFA in static and kinematic friction for P. regius.

The plot shows that the acoustic probe method yields higher values in static friction compared to the same value obtained by the force gauge method. The difference, however, is not pronounced ($RFA_{A.p} \approx 1.15\ RFA_{F.G}$). For kinematic friction, the RFA remains unchanged. Comparison between the static and kinematic COF for two species, P. regius and B. gabonicaGabonica, is provided in Figure 18-b. The order of magnitude of the COF follows common convention. Namely, the static COF is greater than the kinematic one.

DETERMINISTIC DESCRIPTION OF MICRO-TEXTURING FEATURES

Micro-texture of snake surfaces is deterministic (Abdel-aal & El Mansori, 2014). Topography of the surface comprises a basic textural feature of simple geometry that repeats on the surface in an array. Friction for such surfaces depends on several characteristic parameters that manifest ratios between texture attributes (i.e., lateral spacing, distance between rows of features, etc. see Table 4). Table 5 *(Abdel-aal, 2015)* provides a summary of some basic textural attributes for representative species. Table 6-meanwhile lists some of the average characteristic parameters of the species listed in table 5. The data demonstrate the variation between the parameters for the different species and even along the body of the same species.

Correlation of Frictional Behavior to Topography

The shape of fibril tips was considered important by Hazel *et.al, (1999)*, in invoking isotropy of the frictional response of snakeskin. *Abdel-aal (2013)* extracted the shape of the fibril tips in the ventral side of a P. regius using white light interferometry scans, WLI, of selected regions on the ventral scales. He further related the geometrical asymmetry of the fibril tips to the observed frictional anisotropy (Figure 19).

The figure contains plots that represent the RFA for the leading half of the skin (circular symbols), the trailing half of the skin (hexagonal symbols), and the average value of the surface (triangular symbols). Data trend reflects a linear relationship between the RFA and the FTAR, which indicates that frictional anisotropy of the skin is directly proportional to the asymmetric geometry of the fibril-tips. This sup-

Table 3. Coefficient of friction for various snake species

Substratum	μ_F	μ_B	μ_B/μ_F	μ_{Lat}	*Species*
					Grass snake (Natrix-Tropidonotus Natrix)
Metal					
dry	*0.33-0.4*	*0.33*	*1*		
oil film	*0.86*				
Glass					
(dry)	*0.40-0.42*				
(oil film)	*1.30*				
Wood					
(dry; smooth)	*0.34*				
Cardboard (smooth)	*0.24*				
Sandpaper					
(fine)	*0.65*	*0.75*			
medium	*0.65*	*0.88*			
(rough)	*0.44*	*1.31*			
Epoxy Resin					*Amazon Boa (C. hortulanos)*
Ra (μm)					
0.08	*0.207*	*0.227*	*1.09*	*0.23*	
Styrofoam					*Corn snake (Pantherophis guttatus guttatus)*
static					
Conscious	*0.51*	*0.88*	*1.72*		

continued on following page

Table 3. Continued

Substratum	μ_F	μ_B	μ_B/μ_F	μ_{Lat}	Species
Unconscious	0.3	0.35	1.17		
(dynamic)					
Conscious	0.49	0.79	1.61		
Unconscious	0.21	0.35	1.67		
Glass Ball (1mm diameter)					California King Snake L.G Californiae
Ra (0.006) μm					
Soft Cushioned	0.121	0.154	1.27	0.14	
Hard Cushioned	0.081	0.151	1.85	0.08	
Nitrocellulose-Ball					P. regius
Ra =4 μm					
Leading half	0.256	0.281	1.09	0.26	
Trailing half	0.185	0.293	1.58	0.27	
(Static Friction)					B. Constrictor
Gripping Tape	0.385-0.395				
Insulated Wire	0.347-0.353				
Insulated Wire	0.315-0.325				
Wood	0.246-0.256				
Bare Wire	0.174-0.186				
PVC	0.204-0.216				
Gripping Tape					
	0.98				B. Constrictor
Static					Mohave Shovel-nosed Snake
0.3 mm glass particles					Chionactis occipitalis occipitalis
Ventral	0.109	0.137			
Dorsal	0.161	0.236			
Dynamic					
Ventral	0.11±0.02	0.14			

Figure 18. Correlation of the ratio of frictional anisotropy μB/μF and the lateral COF μ_{Lat} reported in literature

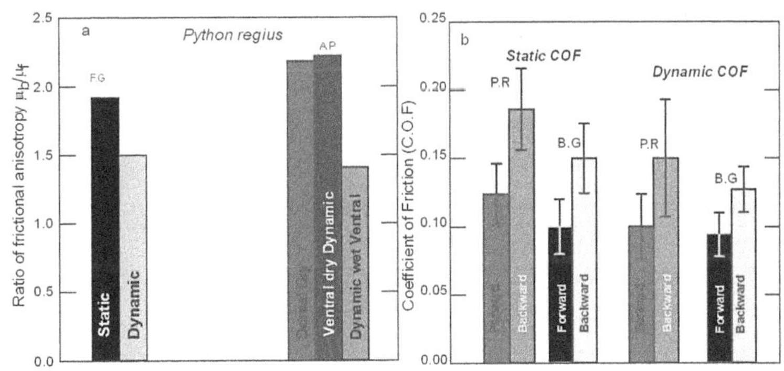

Table 4. Summary of formulas used to calculate parameters used in comparing LTS and python skin

Parameter	*Snake skin*
	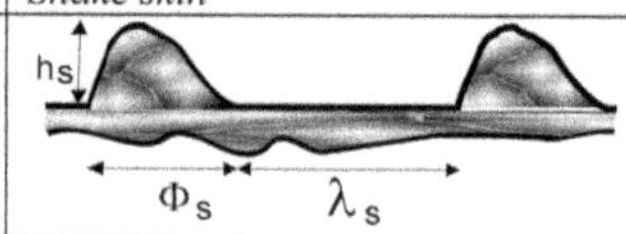
Total Area Ratio (TAR)	$\frac{\text{Total area of fibril tips}}{\text{total surface area}}$
Dimple Slenderness Ratio (DSR)	$DSR_s = \frac{h_s}{\Phi_s} \approx \frac{R_a}{\Phi_s}$
Surface Aspect Ratio (SAR)	$SAR_s = \frac{\lambda_s + \Phi_s}{h_s}$

(Abdel-Aal and El Mansori 2015)

Table 5. Summary of the primitive metrological dimensions needed for computing the performance metrics for a group of sake species

Species	*Lf(Fs)*	ϕ_{sd}	$\lambda s+\Phi s$	$\lambda s+\Phi_{slat}$	D_{rows}
Bittis Gabonica	*0.75*	*1.10*	*4.50*	*4.85*	*1.60*
Agkistrodon piscivorus	*1.75*	*1.00*	*4.85*	*4.10*	*0.33*
*Agikstron Taylori_*	*2.50*	*0.50*	*6.05*	*4.05*	*0.66*
Cerrophidion_Godmani	*1.00*	*0.85*	*4.00*	*3.85*	*0.74*
Echis_Carinatus	*5.50*	*5.50*	*20.50*	*20.50*	*0.26*
Pseudechis_Austarlis	*1.30*	*0.75*	*4.95*	*4.40*	*0.76*
Agkistrodon Contortrix	*3.90*	*2.12*	*7.75*	*5.97*	*4.40*
Pituophis Melanoleucus	*1.75*	*1.00*	*5.40*	*4.65*	*0.18*
P. regius	*1.00*	*0.50*	*5.50*	*5.00*	

(Abdel-Aal, 2015)

Table 6. Summary of performance indicators for the species listed in Table 5

	Texture Primitive Ratio							
Species	*Ra*	*TAR*	SAR_{min}	SAR_{max}	SAR_{av}	DSR_{max}	DSR_{min}	DSR_{AV}
Bittis Gabonica	*0.33*	*0.04*	*13.47*	*14.52*	*14.00*	*0.45*	*0.30*	*0.37*
Agkistrodon piscivorus	*0.04*	*0.28*	*110.45*	*93.37*	*101.91*	*0.03*	*0.04*	*0.03*
P. regius	*0.08*	*0.18*	*72.75*	*66.14*	*69.44*	*0.08*	*0.15*	*0.11*
Agikstron Taylori	*0.11*	*0.20*	*54.10*	*36.22*	*45.16*	*0.04*	*0.22*	*0.13*
Cerroph Godmani	*0.14*	*0.13*	*28.06*	*27.00*	*27.53*	*0.14*	*0.17*	*0.16*
Echis Carinatus	*0.04*	*0.54*	*467.18*	*467.18*	*467.18*	*0.01*	*0.01*	*0.01*
Pseudechis Austarlis	*0.14*	*0.18*	*34.46*	*30.63*	*32.54*	*0.11*	*0.19*	*0.15*
Agkist Contortrix	*0.10*	*0.68*	*77.50*	*59.74*	*68.62*	*0.03*	*0.05*	*0.04*
Pituophis Melanoleucus	*0.03*	*0.22*	*201.72*	*173.70*	*187.71*	*0.02*	*0.03*	*0.02*

(Abdel-Aal-2015)

Figure 19. Correlation of fibril tip asymmetry to frictional anisotropy in a P. regius. Printed with permission from: H. A, Abdel-Aal, On surface structure and friction regulation in reptilian limbless locomotion, JMBBM, 22, 2013, 115-135, doi /10.1016/j.jmbbm.2012.09.014.

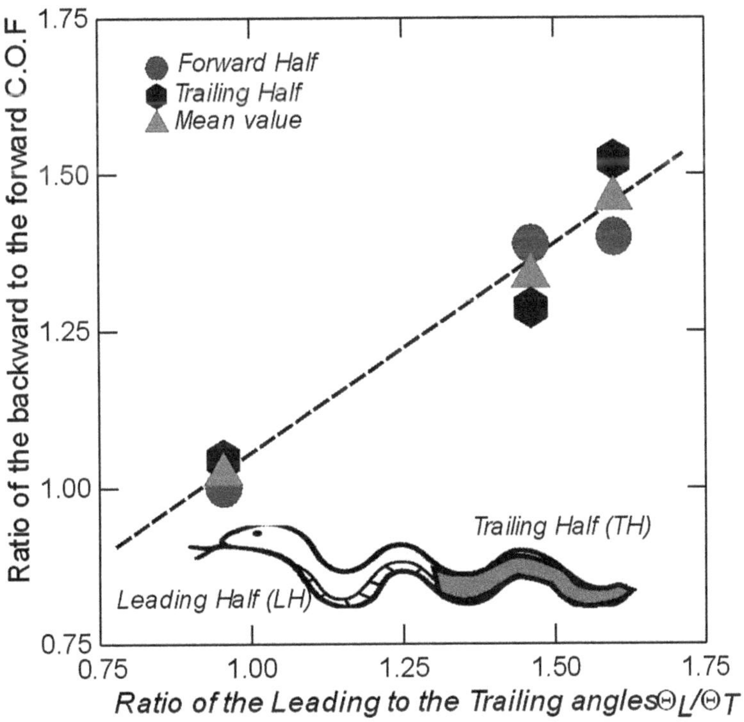

ports the suggestions of Hazel et al *(1999)* that the geometry of fibrils within the scales help the reptile to condition its frictional response through ratcheting. Other work *(Abdel-Aal et al, 2012)* have reported that frictional anisotropy is a manifestation of stick-slip sliding motion induced by the geometry of the fibril-tips. In this process, significant energy dissipation takes place through friction-induced acoustical emission. Presence of the fibrils, leads to the emission of quasi uniform acoustical waves when the reptile moves forward. However, upon attempting to move backwards, motion is not continuous and the fibrils appear to collide with the roughness of the substrate due to asymmetry.

Abdel-aal (2013) used Tabors analysis of the fiction force *(Bowden and Tabor 1964)* to study the effect of fibrils in modifying the friction force. The friction force between the ventral side of the reptile and the sliding substrate has two components a shear component, Fs, and a ploughing component, F_{pl}. These are given by

$$F_s = \tau A_{real} \quad (1\text{-a})$$

$$F_{pl} = H A_{pl} \quad (1\text{-b})$$

Where τ is the shear strength of the skin (assumed to be softer than the substrate), A_{real} is the real area of contact between the contacting region of the reptile and the substrate, H is the hardness of the softer material (the skin), and A_{pl} is the cross sectional area that the counter face material establishes upon indenting the skin. The total friction force, i.e., the sum of the two components, is given by:

$$F_f = F_s + F_{pl} = \tau A_{real} + H A_{pl} \tag{2}$$

Snakes move under the influence of their own body weight and the forces exerted by the muscle groups involved in locomotion. The contact pressures resulting in this case are relatively low. Under such conditions the real area, A_{real}, and the indentation area, A_{pl}, may be considered equal, which simplifies the expression of the friction force to,

$$F_f = \left(\tau + H\right) A_{real} \tag{3}$$

Equation (3) implies that when the shear strength and hardness remain unchanged, only the real area, A_{real}, will influence the friction force. This allows a comparison between the friction force developed in the ventral scale due to the presence of fibrils, and that developed in case of a non-textured surface, sliding on the same substrate. The resulting ratio of the coefficients of friction, under the same contact pressure, is expressed in the form,

$$\frac{\mu^{F}}{\mu^{NF}} = \frac{A^{F}_{real}}{A^{NF}_{real}} \tag{4}$$

The superscripts "*F*" and "*NF*" denote the presence of fibrils on the ventral scale and the absence of fibrils respectively. Equation (4) implies that area of contact for a ventral scale containing fibrils will be lower than that established in the case of a ventral scale with no fibrils present. This leads to lower COF for the former compared to that for the later. The equation further highlights the role of fibril distribution along the Anterior-Posterior axis of the reptile in controlling friction. This non-uniform distribution affects the variation of the COF along the body. Baum et al *(2014)* proposed that the geometry of the tips compliments ultrastructure of the skin to induce frictional anisotropy. Baum compared her frictional measurements, on the California king snake *(L. g. Californae)*, to the investigation of ultrastructure of the same snake (Klein & Gorb 2012; Klein et. al, 2010), who observed keratinous fibers oriented some 15° relative to the ventral body surface. Baum concluded that a combination of anisotropic geometry of surface microstructure and strategic fiber orientation within the epidermis architecture result in the anisotropic frictional behavior.

ENGINEERING OF CUSTOM-FRICTION SURFACES

Surface Engineering Techniques

Modification of surface texture is a relatively old concept. It is a subject of interest which is shared among many industries. Traditionally, therefore, many technologies have evolved to meet the individual demands of the particular industry. Differences in precision, quality of resulting surfaces, minimum size of the resulting textural feature, etc. do exist between the various technologies. One feasible classification of the various technologies is according to the physical principle. Broadly speaking, therefore, texturing methods may fall under one of the four broad categories shown in Figure 20.

The difference in the quality of the outcome, governing dynamics, or any generalized attributes may form the basis of a classification for techniques used to engineer a surface. Thus, one may develop a classification based on chemical aspects of the process, size and accuracy, cost, etc. One feasible classification is to distinguish between techniques according to the direction of material within the surface during the modification process itself. Broadly speaking, therefore, texturing methods may fall under one of the four categories shown in Figure 20. For some cases, modification of a surface may take place by adding a thin layer of a wear resistant coating, or ultra-lubricating film. In such instance modification is achieved by adding material, not originally present, to the surface. Carving texture out of the surface layer (e.g. by machining, Ion beam etching, or laser ablation) involves removal of original material from the surface (i.e. subtraction of material). Other modification methods may involve shifting of original material within the surface to enhance mechanical properties or to introduce intentional deformation (e.g in shot peening). Here the mechanism involves displacement of surface-layer constituents to manipulate properties. Finally, in some cases, modification of texture by manipulating the organization or bonding

Figure 20. Main classification categories of surface engineering techniques

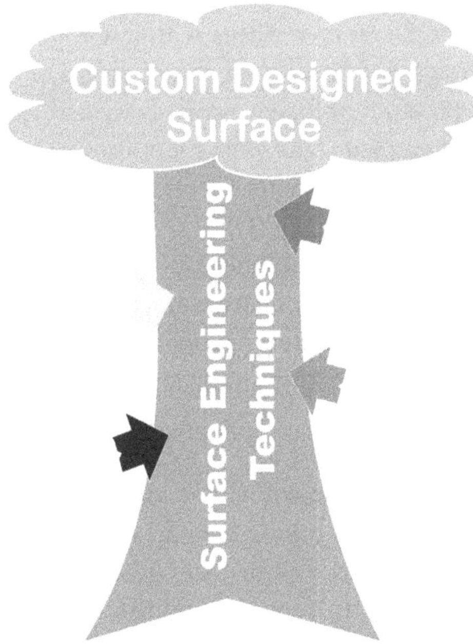

between molecules (e.g. through radiation) takes place for modifications of a surface. In such case, surface engineering takes place through inducing self-organization of the target interfacial layers. Table 7 presents a comprehensive review of the available texturing/surface modification techniques categorized according to the definitions of Figure 20.

The data presented imply that for each conceptual pathway to modification of texture several technological routes enable material manipulation. These routes provide unprecedented opportunities to manipulate surface material, virtually, across any scale, phase, thickness, or compositional threshold. The rapid evolution of material manipulation technologies, however, is not paralleled by viable surface design algorithms leading to true designer surfaces. In addition, even at the conceptual level a viable strategy to address basic technical challenges brought forward by futuristic needs haven't taken shape yet *(Abdel-aal, 2016)*. Summary of available material manipulation techniques for texture engineering

One challenge within the field of customized surface engineering is the design and implementation of adaptive gripping mechanisms for robots.

Surface Engineering for Legged Robots

There are two approaches that allow legged robots to traverse difficult terrains. The first entails application of general control approaches whereas the second comprises of implementing special robot structures and materials with specific control strategies. The second approach has come to be known as *morphological computation* (Pfiefer et al, 2006, 2007). The second approach, it is to be noted, is of recent origin and has not matured as of yet. However, there are several attempts for efficient implementation. The approaches are either bio-inspired or simply increasing "gripping" through using a high friction material. Kim *et al.(2008)* developed gecko-inspired adhesive materials and implemented them on a four-legged robot. Using the adhesive materials with a specific control mechanism for peeling toe pads of the surface allowed the robot to climb up a smooth wall. A similar strategy has been also applied to a leg-wheel hybrid robot (Daltorio 2006). Voigt *et al.* (2012) used foamy rubber materials as a gripper of the small robot Ratnic, with specific gripper movement control, for climbing on a pipe. These robots operate mostly on glass, smooth surfaces, or surfaces with low roughness. Extending the operational range to other surfaces of higher roughness (carpets or other felt-like or rough/dusty substrates) has been attempted as well. Spenko *et al.* (2008) developed special body and leg structures with compliant feet and embedded micro spines for a hexapod robot. The robot uses specific climbing control with force feedback to generate proper leg movements for climbing up different rough surfaces. Palmer *et al.* (2010) developed so-called Distributed Inward Gripping (DIG) with torsion springs (i.e., passive compliance) at the legs of a six-legged robot. Using this mechanism with specific leg movement control, the robot can generate adhesive forces and walk vertically on a mesh screen. Bretl (2006) developed a four-legged robot with single peg legs wrapped in high-friction rubber and used multi-step motion planning control to enable the robot to climb vertical rock. These strategies require specialized control, structure, and material designs to handle rough surfaces.

Achieving efficient locomotion on rough surfaces is a challenging task. However, it can be achieved by inducing frictional anisotropy and mechanical interlocking between surfaces at the microscale. IN other words it can be achieved by mimicking those creatures that display anisotropic friction (such as the subject of this review: snakes). In essence, strong mechanical interlocking in one direction should enhance the grip of a robot on the surface, thereby preventing slipping or backward sliding. Meanwhile almost no mechanical interlocking in another direction should allow releasing the robot from the sur-

Table 7.

Adding material		
Chemical processes	*Conversion coatings*	*Chromating*
		Phosphating
	Deposition coatings	*Patterned chemical deposition (CVD)*
		autocatalytic plating
		Electrolytic coating
		Anodising
		Electro-deposition
Physical deposition	*precipitation coating*	
	Inkjet printing	
	Patterned curing	
	Focused Ion Beam	
Removing material		
High temperature	*Laser methods*	*Laser texturing (LT)*
		Masked excimer Laser
		CNC focused lasers
		Femto-second lasers
	Electrical discharge machining (EDM)	*Electrical discharge texturing (EDT)*
		Micro EDM
	Ion beam texturing (IBT)	
Chemical etching	*Masking methods*	*Chemical texturing*
		Electrochemical texturing
	Non-masking methods	*Self-assembling*
		Maskless electrochemical texturing
		Anisotropic etching
Mechanical	*CNC ultrasonic machining*	
	Mechanical honing	
	Grinding	
	Free abrasive machining	
	Microcutting	
	Patterned erosion	
Moving material		
Mechanical	*Shot blasting*	
	Embossing	*Patterned embossing tools*
		Vibro-rolling
Chemical	*Molecular migration*	
Self-forming		
	UV contraction	
Localized diffusion	*Chemical conversion, thermochemical treatments*	
	Printing + heating	
hard and soft phases	*Thermal implantation of hard phases*	
	Embedding of soft phases	

face while moving forward. Implementing strong frictional isotropy should also allow the robot to grip to the surface, yet it will induce difficult release from the surface. Combined frictional anisotropy and mechanical interlocking, inspired Marvi *et al.* (2011) to develop active scales and their control to generate the frictional anisotropy for a snake-inspired robot (Scalybot). This combination allowed the robot to climb inclines up to 45°.

Shark skin in lieu of active scales was used by Voigt et al *(2012)* to enhance the efficiency of legged robot locomotion on inclines. The idea in essence is not new. Shark skin was used in ancient times by fishermen to make shoes to enhance their grip on the wet wooden decks of their ships. Moreover, shark skin was also used in various tools such as in wooden rasps *(*McQuirty 2011*)*, in grips and sheath of swords, and also to sand wood and ceramics (Vannuccini, 1999).

Lately, both approaches were adopted. Currently, there are reports of successful surfaces where direct replication took place. In addition attempts of building a so-called "snake-inspired" surface, where observing the metrological proportions of the geometrical features and not direct replication is the design claim, have been reported to be successful.

Muhlbergeret al *(*2015*)* applied powder injection molding combined to ultra violet light assisted nano-imprint lithography to replicate the delicate features of a Cobra (Najanigrocollis). Combining both techniques allowed direct copying of the biological surface and then imprinting unit textural features that are three times the length of actual features of the snake. Pin-on-disc experiments in parallel and anti-parallel sliding directions, with respect to the orientation of texturization, verified the benefits of the snake ornamentation. Snake skin textured samples running in the parallel direction gave a COF that is approximately 30% lower than that of the polished non-textured sample, and also lower than samples sliding in anti-parallel direction.

Another form of direct replication was endeavored by Greiner and Schäfer *(*2015*)* who attempted to mimic the overlapping structure of the sub-caudal scales (and not the micro structure within the scales) of a P. regius using laser ablation. Benefits of texturization were realized, however, only in dry sliding where, compared to the un-texturized sample, the biaxial pattern showed some 20% reduction in the COF, whereas the row arrangement was the most efficient with some 40% reduction.

An alternative to direct replication of bio-texturing is to design surfaces with variable "controlled" contributions of each of the contributing effective factors. This was attempted by Baum et al *(*2015*)* who produced a deterministic, wave like texturization embossed on epoxy resin. The biological analogue used in her work was the California King Snake (L. Getuta Californae). Frictional characterization resulted in a tradeoff between dimensions and structure on the one hand and the reduction in the real contact area and interlocking between surfaces on the other hand took place. This tradeoff manifests the interference between surface energy represented by the adhesion and the mechanical contribution of friction represented by the true contact area. When the surface roughness (texturization features) are of minimal size and the adhesion is eliminated due to orientation of sliding (parallel to the friction force), maximum reduction in friction is possible.

Adaptation of the primitive geometrical ratios influential to performance of texturized surfaces found on the skin of a snake was attempted by Cuervo et al *(*2016*)*. In this work, the authors used Photo-Lithography to texturize the surface of Titanium (Ti6Al4V) pins with elliptical patterns. The patterns were placed according to the ratio observed in the ventral scales of a P. regius thus mimicking the spatial distribution of the snake texture rather than the size or shape of the features.

The results confirmed the feasibility of mimicking the spatial ratios found on the snake surface. Consistent with other studies, friction anisotropy with the lateral COF being greater than the COF in parallel sliding was observed.

It has been seen in prvious sections that the essence of superior frictional performance of a snake is rooted in surface structure and in active control exerted by the reptile during motion. The same idea could be applied in robotics. While the control part is fairly advanced the surface engineering part is still lacking. In principle manufacturing of a custom friction surface reduces to manipulation of material at the desired level. Such a technology as seen before is present. The main obstacle currently is the design conceptualization of the needed surface.

Abdel-aal *(2016)* reviewed the state of the art of biological inspired surfaces. He asserted that there exist many common features between Custom Technological Surfaces, CTS, and the ventral side of a snake (where anisotropy of friction is manifested for locomotion). The principal similarity is that both, snakeskin and CTS, are multi-scale entities. The basic building block in each case is a textural element that is repeated in an array distribution. The snake's ventral side is composed of identical micron-sized fibrils (denticulations) distributed over the skin area in a particular pattern. Spacing, length, orientation and shape of denticulation is in general common to a particular family of snakes. CTS, on the other hand, by their very definition comprise an individual textural building block (cone, dimple, chevron etc.,) that also is distributed on the surface area in array formation. Based on similarities, a fundamental question always arises. Namely which is viable strict mimicry or application of design rules. An answer is complex due to the interdependence of the factors involved. A simple answer is to mimic, or replicate, the textural designs present on the reptile in a technological surface. However, the success of this approach depends on the existence of a one-to-one match between the function and perhaps, the material, of the target surface and that of a reptile. This match will limit the replication process to polymeric surfaces (because the skin is elastomeric). However, bio-inspired surface texturing has a broader goal than merely replication. In essence, it seeks to extend the potential tribological benefits of reptilian surfaces to the domain of man-engineered surfaces. Two features are important in this context: the quality of frictional performance of the snake and matching of the deterministic surface parameters of the biological target to those recommended for the technological surface whether it is a robotic surface or a surface with generalized tribological function.

CONCLUSION

There are many examples in nature where mitigation and control of friction is essential. Interaction with frictional effects in the natural world takes place through the surface. The surface, in turn, represents a true smart system that matches the modern views of technical surfaces. The paper provided a review of one of a true smart biological "systems" were the source of successful function is a integration of surface design and dynamic control of muscles, and sensory.

These surfaces of snakes are fit to interact with many hostile tribological situations, and although of different appearance they are linked through a unique fractal structure. This unifies the method of accommodating contact loads.

The theme of friction in snakes is anisotropy with respect to direction. There is a difference between forward motion and backward motion. The anisotropy also extends to lateral and diagonal motion. This anisotropy is partially a function of the geometry of the surface texture features. The ratio of forward to backward COF is on one-to-one correspondence with the ratio of leading to trailing angle of attack of the surface texture.

The efficient interaction between surface micro-texture and friction have prompted many efforts in surface design. The future of engineered surfaces is based on the "smart system view". That is a surface is considered as a smart entity that performs a combination of sensory, control, feedback and structural functions to name a few. Central to achieving this view is the ability to manipulate the textural features of a surface (geometry and topology). Many emerging surface designs directly replicate or mimics designs within snakes.

REFERENCES

Abdel-Aal, H. A. (2013). On Surface Structure and Friction Regulation in Reptilian Locomotion. *Journal of the Mechanical Behavior of Biomedical Materials*, *22*, 115–135. doi:10.1016/j.jmbbm.2012.09.014 PMID:23582565

Abdel-aal, H. A. (2015). The structure of Ventral Scale Textures in Snakes in comparison to Texturing of Deterministic Tribological Surfaces. In J. P. Davim (Ed.), *Processing Techniques and Tribological Behavior of Composite Materials, Rajnesh Tyagi* (pp. 268–316). IGI Global Publication. doi:10.4018/978-1-4666-7530-8.ch010

Abdel-Aal, H. A. (under consideration). Comparison of friction profiles of two snakes. *Journal of the Royal Society, Interface*.

Abdel-Aal, H. A., & El Mansori, M. (2011). Python Regius (Ball Python) shed skin: Biomimetic analogue for function-targeted design of tribo-surfaces In Biomimetics - Materials Structures and Processes Examples Ideas and Case Studies. Springer.

Abdel-Aal, H. A., & El Mansori, M. (2013). Tribological analysis of ventral scale structure in a python regius in relation to laser textured surfaces IOP-J. *Surface Topography: Metrology and Properties.*, *1*(1), 015001. doi:10.1088/2051-672X/1/1/015001

Abdel-Aal, H. A., & El Mansori, M. (2014). Characterization of Load Bearing Metrological Parameters in Reptilian Exuviae In Comparison To Plateau Honed Surfaces. *IOP-J. Surf. Top. Metrol. Prop.*, *2*, 045002. doi:10.1088/2051-672X/2/4/045002

Abdel-Aal, H. A., & El Mansori, M. (2015). The Fractal Structure of the Ventral Scales in Legless Reptiles. In *15th International Conference on Metrology and Properties of Engineering Surfaces*. UNC Charlotte.

Abdel-Aal, H. A., & El Mansori, M. (2011). A study on the frictional response of reptilian skin for green tribology. *Proceedings of 3rd European Conference on Tribology*.

Abdel-Aal, H. A., & El Mansori, M. (2011). *Metrological Characterization of Reptilian skin for Green Tribo Surfaces*. ASPE Spring Topical Meeting, Freeform Surfaces, Charlotte, NC.

Abdel-Aal, H. A., Vargiolu, R., Zahouani, H., & El Mansori, M. (2011). A study on the frictional response of reptilian shed skin. *Journal of Physics: Conference Series*, *311*, 012016. doi:10.1088/1742-6596/311/1/012016

Abdel-Aal, H. A., Vargiolu, R., Zahouani, H., & El Mansori, M. (2012). Preliminary investigation of the frictional response of reptilian shed skin. *Wear*, *290–291*, 51–60. doi:10.1016/j.wear.2012.05.015

Alexander, A. A., & Gans, C. (1966). The pattern of dermal-vertebral correlation in snakes and amphisbaenians. *Zoölogische Mededeelingen*, *41*(11), 171–190.

Alibardi, L. (1996). Scale morphogenesis during embryonic development in the lizard Anolis lineatopus. *Journal of Anatomy*, *188*, 713–725. PMID:8763488

Alibardi, L. (1998). Differentiation of the epidermis during scale formation in embryos of lizard. *Journal of Anatomy*, *192*(2), 173–186. doi:10.1046/j.1469-7580.1998.19220173.x PMID:9643418

Alibardi, L., & Toni, M. (2007). Alpha and beta keratins of snake epidermis. *Zoology (Jena)*, *110*(1), 41–47. doi:10.1016/j.zool.2006.07.001 PMID:17169542

Amarello, M., Novak, E. M., Taylor, E. N., Schuett, G. W., Repp, R. A., Rosen, P. C., & Hardy, D. L. (2010). Potential environmental influence on variation in body size and sexual size dimorphism among Arizona populations of the western diamond-backed rattlesnake (Crotalus atrox). *Journal of Arid Environments*, *74*(11), 1443–1449. doi:10.1016/j.jaridenv.2010.05.019

Arnold, S. (1988). Quantitative genetics and selection in natural populations: Microevolution of vertebral numbers in the garter snake Thamnophis eleguns. *Proc. 2nd Intl. Conf. Quant. Genet.*, *2*, 619-636.

Arnold, S., & Bennett, A. (1988). Behavioral variation in natural populations. V. Morphological correlations of locomotion in the garter snake. Thamnophis radix. *Biological Journal of the Linnean Society. Linnean Society of London*, *34*(2), 175–190. doi:10.1111/j.1095-8312.1988.tb01955.x

Arnold, S. J. (1983). Morphology, performance and fitness. *American Zoologist*, *23*(2), 347–361. doi:10.1093/icb/23.2.347

Ashton, K. G., & Feldman, C. R. (2003). Bergmann's rule in nonavian reptiles: Turtles follow it, lizards and snakes reverse it. *Evolution; International Journal of Organic Evolution*, *57*(5), 1151–1163. doi:10.1111/j.0014-3820.2003.tb00324.x PMID:12836831

Aubret, F., & Shine, R. (2009). The origin of evolutionary innovations: Locomotor consequences of tail shape in aquatic snakes. *Functional Ecology*, *22*(2), 317–322. doi:10.1111/j.1365-2435.2007.01359.x

Bauchot, R. (n.d.). *Snakes: A Natural History*. New York, NY: Sterling Publishing Co., Inc.

Baum, M. J., Heepe, L., Fadeeva, E., & Gorb, S. N. (2014). Dry friction of microstructured polymer surfaces inspired by snake skin. Beilstein Journal of Nanotechnology, 5, 1091-1103. doi:10.3762/bjnano.5.122

Baum, M. J., Kovalev, A. E., Michels, J., & Gorb, S. N. (2014). Anisotropic Friction of the Ventral Scales in the Snake Lampropeltis getula californiae. *Tribology Letters*, *54*(2), 139–150. doi:10.1007/s11249-014-0319-y

Benz, M. J., Kovalev, A. E., & Gorb, S. N. (2012). Anisotropic frictional properties in snakes. *Proc. of SPIE*, 8339. doi:10.1117/12.916972

Berthé, R. A., Westhoff, G., Bleckmann, H., & Gorb, S. N. (2009). Surface structure and frictional properties of the skin of the Amazon tree boa Corallus hortulanus (Squamata Boidae). *J. Comp. Phys. A*, *195*(3), 311–318. doi:10.1007/s00359-008-0408-1 PMID:19137315

Boback, S. M. (2003). Body size evolution in snakes: Evidence from island population. *Copeia*, *2003*(1), 81–94. doi:10.1643/0045-8511(2003)003[0081:BSEISE]2.0.CO;2

Boback, S. M., & Guyer, C. (2003). Empirical evidence for an optimal body size in snakes. *Evolution; International Journal of Organic Evolution*, *3*, 195–211. PMID:12683530

Bogue, R. (2014). Snake robots: A review of research, products and applications. Industrial Robot. *International Journal (Toronto, Ont.)*, *41*(3), 253–258.

Bowden, F. P., & Tabor, D. (1964). *The friction and lubrication of solids. Part II*. Oxford, UK: Clarendon Press.

Bretl, T. (2006). Motion planning of multi-limbed robots subject to equilibrium constraints: The free-climbing robot problem. *The International Journal of Robotics Research*, *25*(4), 317–342. doi:10.1177/0278364906063979

Brown, J. H., Gillooly, J. F., Allen, A. P., Savage, V. M., & West, G. B. (2004). Toward a metabolic theory of ecology. *Ecology*, *85*(7), 1771–1789. doi:10.1890/03-9000

Cicconofri, G., & DeSimone, A. (2015). Limbless Locomotion on Solid Surfaces: A Case Study in Soft Bio-Inspired Robotics. *Mathematical Modelling*, *8*(1), 827–828.

Cowles, R. B. (1956b). Sidewinding locomotion in snakes. *Copeia*, *4*(4), 211–214. doi:10.2307/1440272

Cowles, R. B. (1956a). Some of the activities of the side winder. *Copeia, 4*, 220-224.

Cox, C. L., Boback, S. M., & Guyer, C. (2011). Spatial dynamics of body size frequency distributions for North American squamates. *Evolutionary Biology*, *38*(4), 453–464. doi:10.1007/s11692-011-9132-9

Cuervo, P., López, D., Cano, J., Sanchez, J., Rudas, S., Estupiñan, H., . . . Abdel-Aal, H. A. (2015). Topographic characterization of snake skin ventral scales for tribological purposes. *VIII Congress of Materials*.

Daltorio, K. A. (2006). A robot that climbs walls using micro-structured polymer feet. Proc. of Climbing and Walking Robots, 131–138. doi:10.1007/3-540-26415-9_15

Dobrolyubov, A. I. (1986). The mechanism of locomotion of some terrestrial animals by travelling waves of deformation. In *Journal Theor* (pp. 457–466). Oxford, UK: Biology. doi:10.1016/S0022-5193(86)80195-3

Feldman, A., & Meiri, S. (2013). Length-mass allometry in snakes. *Biological Journal of the Linnean Society. Linnean Society of London*, *108*(1), 161–172. doi:10.1111/j.1095-8312.2012.02001.x

Forsich, Gorb, Einwögerer, Trappl, Heim, Schift, & Bergmair. (2015). UV-NIL fabricated bio-inspired inlays for injection molding to influence the friction behavior of ceramic surfaces. *Microelectronic Engineering, 141*, 140-144. doi:10.1016/j.mee.2015.02.051

Gans, B. C., & Gasc, J. P. (1990). Kinematics and performance capacity for the concertina locomotion of a Snake (Coluber Constrictor). *The Journal of Experimental Biology*, *1991*(156), 539–556.

Gans, C. (1962). Terrestrial locomotion without limbs. *American Zoologist*, *2*(2), 167–182. doi:10.1093/icb/2.2.167

Gans, C. (1994). Approaches to the evolution of limbless locomotion. *Cusdemos de Herpetologia*, *8*, 12–17.

Gans, C., & Gasc, J. P. (1990). Tests on the locomotion of the elongate and limbless reptile Ophisaurus apodus (Sauria: Anguidae). *Journal of Zoology (London, England)*, *220*(4), 517–536. doi:10.1111/j.1469-7998.1990.tb04731.x

Gans, C., & Mendelssohn, H. (1972). Sidewinding and jumping progression of vipers. In Toxins of Animal and Plant Origin (pp. 17–38). London: Gordon & Breach Science Publications.

Gorb, S. N., Beutel, R. G., Gorb, E. V., Jiao, Y., Kastner, V., Niederegger, S., ... Vötsch, W. (2002). Structural design and biomechanics of friction-based releasable attachment devices in insects. *Integrative and Comparative Biology*, *42*(6), 1127–1139. doi:10.1093/icb/42.6.1127 PMID:21680397

Gray, J. (1946). The mechanism of locomotion in snakes. *The Journal of Experimental Biology*, *23*, 101–120. PMID:20281580

Gray, J., & Lissmann, H. W. (1950). The kinetics of locomotion of the grass-snake. *The Journal of Experimental Biology*, *26*, 354–367.

Greenwood, J. A., & Tripp, J. H. (1971). The Contact of Two Nominally Flat Rough Surfaces. *Proceedings - Institution of Mechanical Engineers*, *185*(1), 625–633. doi:10.1243/PIME_PROC_1970_185_069_02

Guyer, C., & Donnelly, M. A. (1990). Length–mass relationships among an assemblage of tropical snakes in Costa Rica. *Journal of Tropical Ecology*, *6*(1), 65–76. doi:10.1017/S0266467400004041

Hazel, J., Stone, M., Grace, M. S., & Tsukruk, V. V. (1999). Nanoscale design of snakeskin for reptation locomotions via friction anisotropy. *Journal of Biomechanics*, *32*(5), 477–484. doi:10.1016/S0021-9290(99)00013-5 PMID:10327001

Jayne, B. C. (1988). Muscular mechanisms of snake locomotion: An electromyographic study of lateral undulation of the Florida banded water snake (Nerodia fasciata) and the yellow rat snake (Elaphe obsoleta). *Journal of Morphology*, *197*(2), 159–181. doi:10.1002/jmor.1051970204 PMID:3184194

Jayne, B. C., & Herrmann, M. P. (2011). Perch size and structure have species-dependent effects on the arboreal locomotion of rat snakes and boa constrictors. *The Journal of Experimental Biology*, *214*(13), 2189–2201. doi:10.1242/jeb.055293 PMID:21653813

Kalauni, K. (2014). *K. M. Gupta, and Isha Bharti*. Reptiles Inspired Biomimetic Materials and Their Novel Applications.

Kelley, K. C., Arnold, S. J., & Gladstone, J. (1997). The effects of substrate and vertebral number on locomotion in the grater snake Thamnophis elegans. *Functional Ecology*, *11*(2), 189–198. doi:10.1046/j.1365-2435.1997.00077.x

Kim, S. (2008). Smooth vertical surface climbing with directional adhesion. *IEEE Transactions on Robotics*, *24*(1), 65–74. doi:10.1109/TRO.2007.909786

Kim, S., Asbeck, A., Cutkosky, M., & Provancher, W. (2005). SpinybotII: Climbing hard walls with compliant microspines. *Proc. of the IEEE Int. Conf. on Advanced Robotics (ICAR)*, 601–606.

Klein & Gorb. (2014). Ultrastructure and wear patterns of the ventral epidermis of four snake species (Squamata, Serpentes). *Zoology, 117*(5), 295-314. doi:10.1016/j.zool.2014.01.004

Klein, M.-C. G., Deuschle, J., & Gorb, S. N. (2010). Material properties of the skin of the Kenyan sand boa Gongylophis colubrinus (Squamata, Boidae). *J. of Comp. Phys. A: Neuroethology, Sensory, Neural, and Behav. Phys.*, *196*, 659–668.

Klein, M-C. G., & Gorb, S. N. (2012). Epidermis architecture and material properties of the skin of four snake species. *J. R. Soc. Interface*. doi:1742-566210.1098/rsif.2012.0479

Kumazawa, Y. (2007). Mitochondrial genomes from major lizard families suggest their phylogenetic relationships and ancient radiations. *Gene*, *388*(1-2), 1–2, 19–26. doi:10.1016/j.gene.2006.09.026 PMID:17118581

Lancaster, J. K. (1968). The effect of carbon fibre reinforcement on the friction and wear of polymers. *Journal of Physics. D, Applied Physics*, *1*(5), 549–560. doi:10.1088/0022-3727/1/5/303

Landmann, L. (1986). The skin of Reptiles: epidermis and dermis. In Biology of the integument: Vertebrate. New York: Springer Verlag.

Landmann, L. (1979). Keratin formation and barrier mechanisms in the epidermis of Natrix natrix (Reptilia: Serpentes): An ultrastructural study. *J. Morphol., 162*, 93-126.

Liljeback, Stavdahl, Pettersen, & Gravdahl. (2014). Mamba-A waterproof snake robot with tactile sensing. *Intelligent Robots and Systems (IROS 2014), 2014 IEEE/RSJ International Conference on* (pp. 294-301). IEEE.

Lillywhite, H. B., & Maderson, P. F. A. (1982). Skin structure and permeability. In Gans, Lillywhite HB, Henderson RW. 1993. Behavioral and functional ecology of arboreal snakes. In R. A. Seigel & J. T. Collins (Eds.), *Snakes: Ecology and behavior* (pp. 1–48). New York: McGraw-Hill.

MacDonald, H. N. (1983). *Trees and networks in biological models*. Wiley.

MacQuitty, M. (2011). *Shark*. Dorling Kindersley Ltd.

Maderson, P. F. A. (1965). Histological changes in the epidermis of snakes during the sloughing cycle. *Journal of Zoology*, *146*, 98–113.

Maderson, P. F. A. (1966). Histological changes in the epidermis of the Tokay (Gekko gecko) during the sloughing cycle. *Journal of Morphology*, *119*(1), 39–50. doi:10.1002/jmor.1051190105

Maderson, P. F. A. (1970). Lizard hands and lizard glands: Models for evolutionary study. *Forma. Et. Functio.*, *3*, 179–204.

Maderson, P. F. A. (1972). When? Why? and How? Some speculations on the evolution of the vertebrate integument. *American Zoologist*, *12*(1), 159–171. doi:10.1093/icb/12.1.159

Maderson, P. F. A., & Licht, P. (1967). The epidermal morphology and sloughing frequency in normal and prolactin injected Anolis carolizensis (Iguanidae, Lacertilia). *Journal of Morphology*, *123*(2), 157–172. doi:10.1002/jmor.1051230205 PMID:6073684

Maderson, P. F. A., Rabinowitz, T., Tandler, B., & Alibardi, L. (1998). Ultrastructural contributions to an understanding of the cellular mechanisms involved in lizard skin shedding with comments on the function and evolution of a unique Lepidosaurian phenomenon. *Journal of Morphology*, *236*(1), 1–24. doi:10.1002/(SICI)1097-4687(199804)236:1<1::AID-JMOR1>3.0.CO;2-B

Martins, M., Araujo, M. S., Sawaya, R. J., & Nunes, R. (2001). Diversity and evolution of macrohabitat use, body size and morphology in a monophyletic group of Neotropical pitvipers (Bothrops). *Journal of Zoology*, *254*(4), 529–538. doi:10.1017/S0952836901001030

Marvi, H., Bridges, J., Meyers, G., Russell, G., & Hu, D. L. (2011). Scalybot: A snake-inspired robot with active control of friction. Proc. Of ASME Dynamic Systems and Control Conf., 443–450. doi:10.1115/DSCC2011-6174

Marvi, H., Gong, C., Gravish, N., Astley, H., Travers, M., Hatton, R. L., & Goldman, D. I. (2014). Sidewinding with minimal slip: Snake and robot ascent of sandy slopes. *Science*, *346*(6206), 224–229. doi:10.1126/science.1255718 PMID:25301625

Mattison, C. (2006). *The Encyclopedia of Snakes*. HarperCollins Publishers Limited.

Palmer, L. R., Diller, E. D., & Quinn, R. D. (2010). Toward a rapid and robust attachment strategy for vertical climbing. *Proc. of the IEEE Int. Conf. on Robotics and Automation*, 2810–2815. doi:10.1109/ROBOT.2010.5509680

Pfeifer, R., Iida, F., & Gomez, G. (2006). Morphological computation for adaptive behavior and cognition. *International Congress Series*, *1291*, 22–29. doi:10.1016/j.ics.2005.12.080

Pfeifer, R., Lungarella, M., & Iida, F. (2007). Self-organization, embodiment, and biologically inspired robotics. *Science*, *318*(5853), 1088–1093. doi:10.1126/science.1145803 PMID:18006736

Pough, F. (1980). The advantage of ectothermy for tetrapods. *American Naturalist*, *115*(1), 92–112. doi:10.1086/283547

Pough, F. H. (1980). The advantage of ectothermy for tetrapods. *American Naturalist*, *115*(1), 92–112. doi:10.1086/283547

Pough, F. H., & Groves, J. D. (1983). Specialization of the body form and food habits of snakes. *American Zoologist*, *23*(2), 443–454. doi:10.1093/icb/23.2.443

Rezapour, E., Pettersen, K. Y., Gravdahl, J. T., & Liljeback, P. (2014). Body shape and orientation control for locomotion of biologically-inspired snake robots. In *Biomedical Robotics and Biomechatronics 2014 5th IEEE RAS & EMBS International Conference on* (pp. 993-1000). IEEE. doi:10.1109/BIOROB.2014.6913910

Rivera, G., Savitzky, A. H., & Hinkley, J. A. (2005). Mechanical properties of the integument of the common gartersnake, Thamnophis sirtalis (Serpentes: Colubridae). *The Journal of Experimental Biology, 208*(15), 2913–2922. doi:10.1242/jeb.01715 PMID:16043596

Roderick, H. (1988). Myths and Symbols in Pagan Europe: Early Scandinavian and Celtic Religions. Manchester University Press.

Schmidt, C. V., & Gorb, S. N. (2012). Snake Scale Microstructure: Phylogenetic Significance and Functional Adaptations. *Zoologica, 187*, 1–106.

Secor, M. S., & Bruce, C. (1992). Jayne Albert F. Bennett, Locomotor performance and energetic cost of sidewinding by the snake crotalus cerastes. *The Journal of Experimental Biology, 163*, 1–14.

Sepkoski, J. (2002). A compendium of fossil marine animal genera (Reptilia entry). *Bulletins of American Paleontology, 364*, 560.

Sharpe, S. S., Koehler, S. A., Kuckuk, R. M., Serrano, M., Vela, P. A., Mendelson, J. III, & Goldman, D. I. (2015, February 1). Locomotor benefits of being a slender and slick sand swimmer. *The Journal of Experimental Biology, 218*(Pt 3), 440–450. doi:10.1242/jeb.108357 PMID:25524983

Shine, R. (1994). Allometric patterns in the ecology of Australian snakes. *Copeia, 1994*(4), 851–867. doi:10.2307/1446709

Spenko, M., Haynes, G. C., Saunders, J. A., Cutkosky, M. R., Rizzi, A. A., Full, R. J., & Koditschek, D. E. (2008). Biologically inspired climbing with a hexapedal robot. *Journal of Field Robotics, 25*(4-5), 223–242. doi:10.1002/rob.20238

Stewart, G. R., & Daniel, R. S. (1973). Scanning electron microscopy of scales from different body regions of three lizard species. *Journal of Morphology, 139*(4), 377–388. doi:10.1002/jmor.1051390402

Terrible, L. C., Diniz-Filho, J. A. F., Rodríguez, M. A., & Rangel, T. F. L. V. B. (2009). Ecological and evolutionary components of body size: Geographic variation of venomous snakes at the global scale. *Biological Journal of the Linnean Society. Linnean Society of London, 98*(1), 94–109. doi:10.1111/j.1095-8312.2009.01262.x

Toni, M., Dalla, L. V., & Alibardi, L. (2007). Hard (Beta-) Keratins in the Epidermis of Reptiles: Composition Sequence and Molecular Organization. *Journal of Proteome Research, 6*(9), 3377–3392. doi:10.1021/pr0702619 PMID:17705524

Toyoshima, S., Tanaka, M., & Matsuno, F. (2014). A study on sinus-lifting motion of a snake robot with sequential optimization of a hybrid system. Automation Science and Engineering. *IEEE Transactions on, 11*(1), 139–144.

Vannuccini, S. (1999). Shark utilization, marketing, and trade vol. 389. Food and Agriculture Organization of the United Nations (FAO) Fisheries Technical Papers.

Vidal N., S. (2005). The phylogeny of squamate reptiles (lizards, snakes, and amphisbaenians) inferred from nine nuclear protein-coding genes. *Comptes Rendus Biologies, 328*(10-11), 1000-1008. DOI: 10.1016/j.crvi.2005.10.001

Virgala, I., Dovica, M., Kelemen, M., Prada, E., & Bobovský, Z. (2014). Snake Robot Movement in the Pipe Using Concertina Locomotion. *Applied Mechanics and Materials*, *611*, 121–129. doi:10.4028/www.scientific.net/AMM.611.121

Vitt, L. J., Pianka, E. R., Cooper, W. E. Jr, & Schwenk, K. (2003). History and the global ecology of squamate reptiles. *American Naturalist*, *162*(1), 44–60. doi:10.1086/375172 PMID:12856236

Voigt, D., Karguth, A., & Gorb, S. (2012). Shoe soles for the gripping robot: Searching for polymer-based materials maximising friction. *Robotics and Autonomous Systems*, *60*(8), 1046–1055. doi:10.1016/j.robot.2012.05.012

Voigt, D., Karguth, A., & Gorb, S. (2012). Shoe soles for the gripping robot: Searching for polymer based materials maximising friction. *Robot. Auton. Syst.*, *60*(8), 1046-1055. doi:10.1016/j.robot.2012.05.012

APPENDIX

Nomenclature

A_{real}: Real area of contact between the contacting region of the reptile and the substrate
F_f: Friction force
F_s: Shear component of friction force
F_{pl} : Ploughing component of friction force
M: Mass
Ra: Mean arithmetic value of roughness (μm)
R_{ku} : Profile Kurtosis parameter
Rq : Root mean square average of the roughness profile ordinates (μm)
R_{sk} : Profile skewness parameter
R_T : Radius of curvature in transverse direction

Directions

AE-PE: Anterior Posterior
RL-LL: Lateral Axis
LR: Lateral right hand side
LL: Lateral left hand side
LF: Lateral forward,
LB: Lateral backward.
SB: Straight backward
SF: Straight forward

Greek symbols

τ: Shear strength of the skin
μ_B: Coefficient of friction in backward motion
μ_F: Coefficient of friction in forward motion
$\mu_{T.H}$: Coefficient of friction for the trailing half of the skin
$\mu_{L.H}$: Coefficient of friction for leading half of the skin

Abbreviations

COF: Coefficients of friction
DFE: Differential Friction Effect
FTAR: Fibril Tip Asymmetry Ratio,
FAR: Fibril Aspect Ratio

LBH: Leading Body Half
LCR: Length to Circumference Ratio
RFA: Ratio of frictional anisotropy
SVL: Snout-to-vent length
TBH: Trailing Body Half
TL: Total snake length
VSAR: Ventral Scale Aspect Ratio

Chapter 13
Modeling a Predictive Control of Human Locomotion Based on the Dynamic Behavior

Joao Mauricio Rosario
State University of Campinas, Brazil

Leonimer Flavio de Melo
State University of Londrina, Brazil

Didier Dumur
CentraleSupélec, France

Maria Makarov
CentraleSupélec, France

Jessica Fernanda Pereira Zamaia
State University of Londrina, Brazil

Gabriel Fillipe Centini Campos
State University of Londrina, Brazil

ABSTRACT

This chapter presents the development of a lower limb orthosis based on the continuous dynamic behavior and on the events presented on the human locomotion, when the legs alternate between different functions. A computational model was developed to approach the different functioning models related to the bipedal anthropomorphic gait. Lagrange modeling was used for events modeling the non-holonomic dynamics of the system. This chapter combines the comparison of the use of the predictive control based on dynamical study and the decoupling of the dynamical model, with auxiliary parallelograms, for locating the center of mass of the mechanism using springs in order to achieve the balancing of each leg. Virtual model was implemented and its kinematic and dynamic motion analyzed through simulation of an exoskeleton, aimed at lower limbs, for training and rehabilitation of the human gait, in which the dynamic model of anthropomorphic mechanism and predictive control architecture with robust control is already developed.

DOI: 10.4018/978-1-5225-2993-4.ch013

INTRODUCTION

The growing number of legged impaired in the world is a reality that should be confronted. The available passive mechanic orthosis presents locomotion difficulties when walking on inclined terrain with obstacles, such as slopes and stairs (Floriano-Batista & Rosario 2013).

Modeling of a system is a fundamental resource on the study, comprehension and simulation of its dynamic behavior. Real systems are rich in complexity, making it a difficult task reducing the dynamic behavior for a mathematic approach (de Melo, Rodrigues, & Rosário, 2015). By approaching robotics, with the objective of optimizing the system, there is an attempt to emulate the biomechanical behavior present in the human locomotion in which a trajectory assumed by the lower limbs is denominated gait (Gutiérrez-Carvajal, de Melo, Rosário, & Machado, 2016). This gait presents discontinuities and mechanical impacts that make modeling a complex and laborious task by means of the conventional mechanics formalism, and frequently limits the extension of the studies. This becomes critical when the interest is mimicking, i.e. the perfect assimilation of dynamic behavior on this system, intended objective when approaching the design of exoskeletons (mechatronics devices integrated with the human body), or parts of the body, intended to map and/or amplify its motor functions, and that interact in a physical and cognitive way with its user (Pons, Ceres, & Calderón, 2008).

The development of a study platform that simulates the locomotion of human gait and that permits the dynamic integration of the continuous behavior with the event dynamic, in order to create a model that can be applied to a gait, aiding exoskeleton, is the purpose of this paper. The main objectives of this are the development of a physical model of the human locomotion system, taking the gait in the sagittal plane as a first step and, then, connecting the control system, previously developed for the human leg, integrated by the human gait modeled system.

This chapter presents the study and the simulation of an exoskeletics orthosis for reproducing the human gait movement and integrating the patient within today society and a proposal for development of a lower limb orthosis based on the continuous dynamic behavior and on the events presented on the human locomotion, when the legs alternate between different functions. Computational model was developed to approach the different functioning models related to the bipedal anthropomorphic gait. One of the investigated factors is the gait bilateralism, that provides balance to the center of mass of the human body during locomotion. The authors combines the comparison of the use of the predictive control based on dynamical study and the decoupling of the dynamical model, with auxiliary parallelograms, for locating the center of mass of the mechanism using springs in order to achieve the balancing of each leg. Virtual model is implemented and its kinematic and dynamic motion analyzed through simulation of an exoskeleton, designed to lower limbs, for training and rehabilitation of the human gait. For this, a strategy was developed for dynamic model of anthropomorphic mechanism and predictive control architecture with robust control.

BACKGROUND

Many researchers have tried to accurately measure and discriminate the biological signals generated by gestures using various types of sensors and discriminators. Others presents bio-inspired morphologies, sensors, and actuators. The results clearly indicate that the gestures can be used for purposes such as automatic sign-language recognition, and operation of machine control interfaces. Some authors introduces biomimetics as an emerging discipline that studies and examines nature, its models, elements, processes, systems, structures, mechanisms, etc. to take inspiration from, or emulate, nature's best biological ideas in order to solve scientific and engineering problems. Biomimetics incorporates materials, concepts and techniques drawn from naturally made substances, and resembles the aspiration of biological systems in structure, mechanism and/or function. Biologically inspired approaches will create new reality with great potential. It has been utilized to advance the development of innovative robotics through the integration of bioinspired ideas and solutions into engineered systems (Habib, 2011). Some researches describes bioinspired morphologies, sensors, and actuators with their control architecture that, beyond mere reflexes, implement cognitive abilities like memory or planning, or adaptive processes like learning, evolution and development (Meyer & Guillot 2008). Whether approached from an engineering or a biological perspective, there is no doubt that by whatever term one chooses to characterize it, bioinspired engineering, biorobotics, biological inspiration, or biomimetics, the fusion of biology and engineering is emerging as a discipline in its own right (Delcomyn, 2007). Robotic engineering inspired by biology—biomimetics—has many potential applications: robot snakes can be used for rescue operations in disasters, snake-like endoscopes can be used in medical diagnosis, and artificial muscles can replace damaged muscles (Liu & Sun 2011). Biologically inspired robotics has enabled today's robots to operate in a variety of unstructured and dynamically changing environments in addition to traditional structured environments (Shi, Habib, Xiao, & Hu, 2015).

In the presence of fairly well defined kinematic patterns in human walking there was considerable variability at the kinetic level. Intra-subject variability of joint moment patterns over the stride period was high at the knee and hip, but low at the ankle and in a recently defined total limb pattern, called support moment. A similar profile of variability was evident for inter-subject trials at slow, natural and fast cadences, with the percentage variability at the knee and hip decreasing as cadence increases. (Winter, 1984). This works presents the study and the simulation of an exoskeletics orthosis for reproducing the human gait movement and integrating the patient within today society and a proposal for development of a lower limb orthosis based on the continuous dynamic behavior and on the events presented on the human locomotion.

BIOMECHANICAL APPROACH

Orthosis are devices that improve the performance of the body. Its purpose is align, prevent or correct deformities, of the mobile parts of the body. The therapeutic benefits are associated to minimization of the limitation of the movements. The orthosis allow assisting movement, transference of force and protection of parts of the body (Floriano-Batista & Rosario 2013). The exoskeleton is a device that can be used for improving people biomechanics such, as walk, "run faster", carry some equipment, among others.

Human Gait

The human gait is performed while being on the biped position. The gait cycle is initiated when the heel contacts the ground and is finished when the same heel touches the ground again; it is divided in two phases: the support and the rocking phase (Figure 1).

Figure 1. Events during the human gait cycle
Source: (Floriano-Batista & Rosario 2013).

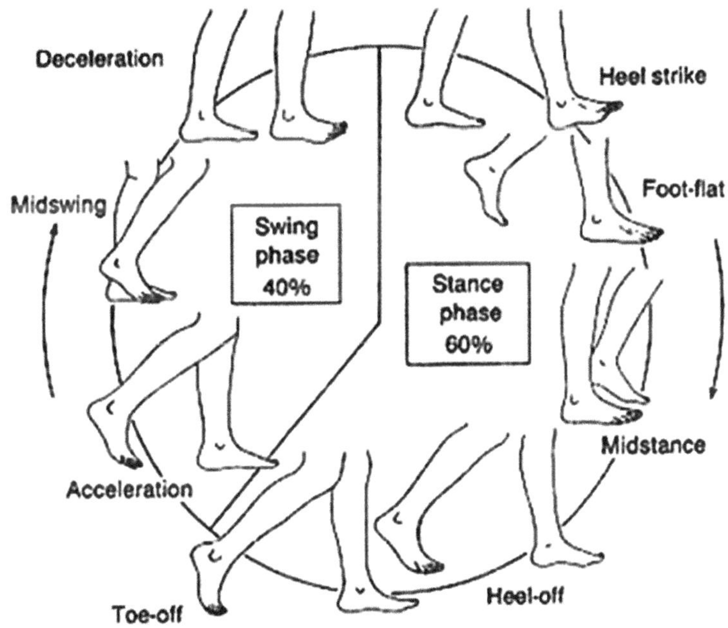

Proposed Mechanism

The designed exoeskeletic orthosis has to reproduce human gait as with much realism as possible. The joints are responsible for transmitting the angular motion between each member. The joint between the foot and the leg, is the ankle, between the leg and the thigh is the knee and finally, between the thigh and the pelvis is me hip (Floriano-Batista & Rosario 2013). Each one of these joints allows three basic rotations: flexion-extension, abduction-adduction and internal-external rotation.

The flexion-extension corresponds to the movement executed by the member in the sagittal plane. The abduction/adduction corresponds to the movement of the member executed in the frontal plane. Finally, the internal/external rotation corresponds to the movement executed in the direction of its own axis (Low, 2011).

DIRECT AND INVERSE KINEMATICS MODEL

From experimental data from a biomechanical study of the human gait, was adopted as the angular position reference signal, the curves obtained by the corresponding iterative model the angular variation of the knee joint angle, leg and foot of a person developing the speed 5 km/h (Abernethy, Kippers, Hanrahan, Pandy, McManus, & Mackinnon, 2013). From the kinematic model obtained, linear movement files relating to the corresponding trajectories of these three degrees of freedom (DOF) were established, and they are used as position reference signal during the calculation of the inverse kinematic model of the system under study. For dynamic effects, the foot was considered as a concentrated mass. From the kinematic motion analysis, were obtained the complete representation model gait motion for the phases of support and balance, which will be used as a reference trajectory in the implementation of dynamic control proposed in this study (Figures 2 and 3).

Kinematic Control Architecture

Figure 4 depicts the architecture implemented to obtain the angular reference position for each of the joints (3 DOF), corresponding to the knee, leg and foot movements.

Kinematic model allows the calculation of the current Cartesian position corresponding to each DOF (hip, leg and foot) from its geometric model and associated parameters.

Inverse kinematic model is realized doing the inversion of the Jacobian matrix, and multiplying its terms by the increase of Cartesian position (Delt_X). It allows the calculation of the angular position increment, and when is added to the current position of the joints (angular reference input), allows the calculation of the new angular position to be used as reference signal to the controller.

DYNAMIC STUDY

The dynamic study proposed in this chapter was developed for the creation of a drive and control system that presents three types of mechanisms to be put in the knee: serial, parallelogram and hybrid (parallelogram with efforts retaining springs). These mechanisms are arranged on the bars that connect the

Figure 2. Full motion gate: Leg-support
Source: (Abernethy et al., 2013).

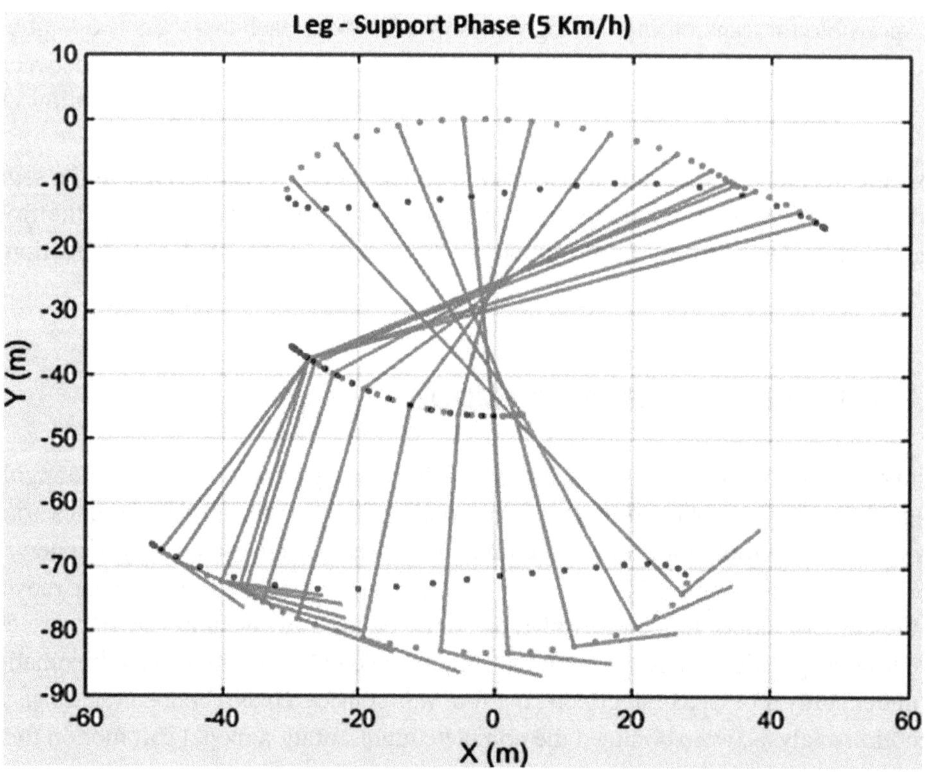

Figure 3. Full motion gate: Leg-balance
Source: (Abernethy et al., 2013).

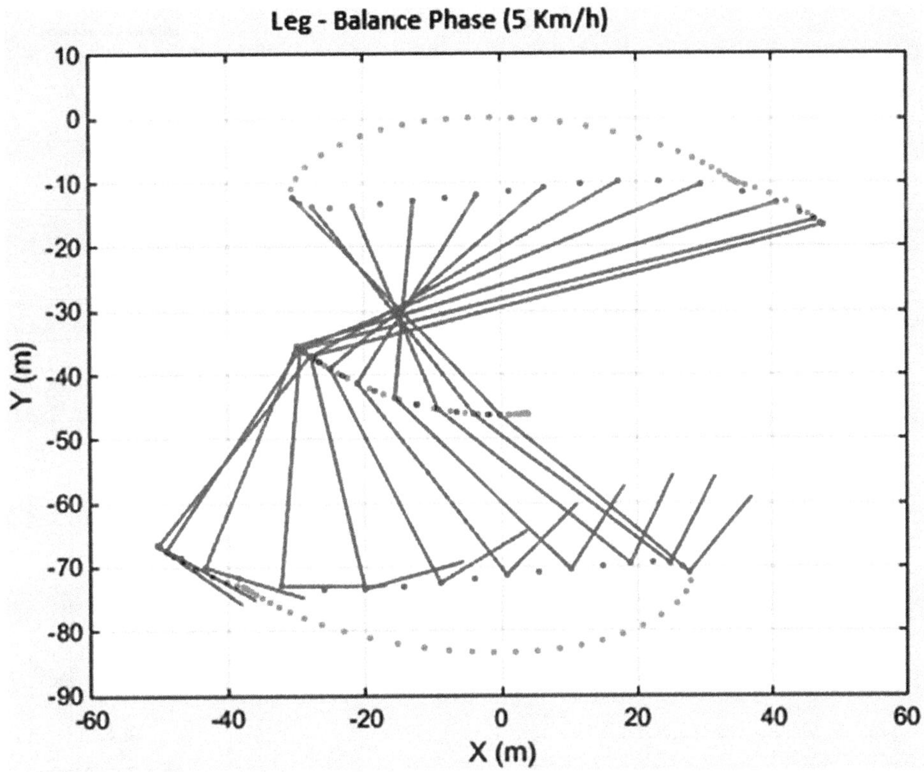

hip to the leg and the leg to the heel, in the way to not damaging the kinematics associated with joints (Figure 5). Through this study, it can be noted that the use of a parallel mechanism, decoupling the complete dynamic model, simplifies the control architecture.

CONTROL ARCHITECTURE

Data for the dynamic model and for the implemented controller, made in Matlab-Simulink, were parameterized, allowing the development of an open architecture control, based on direct and inverse kinematic model, and a complete dynamic model as well as PID controller and GPC (Generalized Predictive Controller). The gains of these controllers was previously obtained, allowing complete simulation results of the study exoskeleton device to be achieved (Cassemiro, Rosario, & Dumur, 2005). Figure 6 depicts a block diagram with full model control architecture implemented.

For systems with dynamic modeling control it is noted a decoupling of the dynamic model when a parallelogram mechanism or a hybrid mechanism (with spring) was used with respect to the serial mechanism (Banala, Agrawal, Fattah, Rudolph, & Scholz, 2004). It uses a reducer with a small transmission rate for the highest coupling joint (hip joint), enabling the driver to manage the effect on a dynamic system that is liable to perturbations caused by changes of its load.

Figure 4. Kinematic direct and inverse model

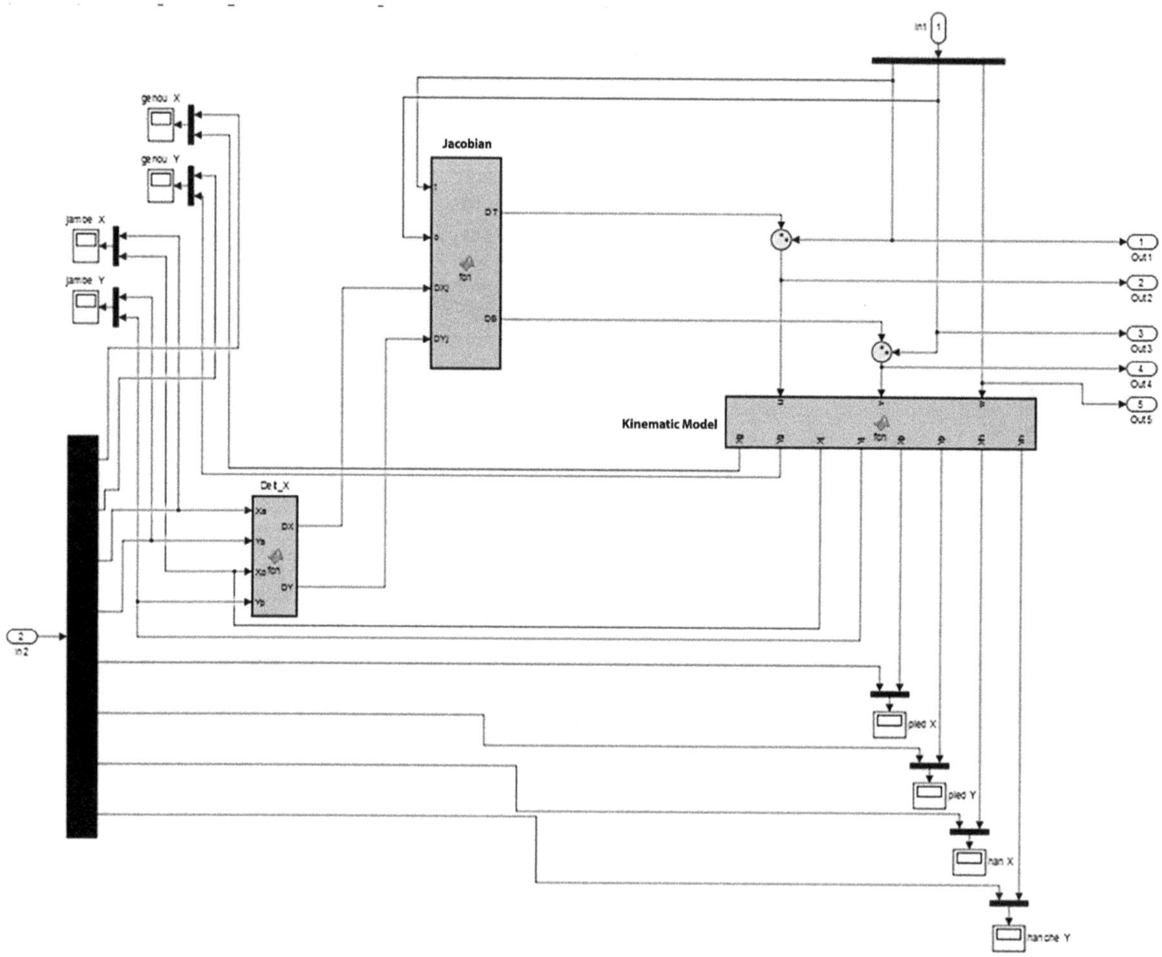

Figure 5. Serial leg model with parallelogram

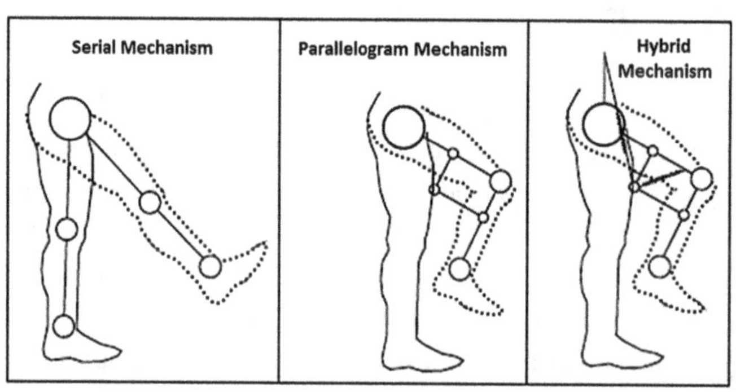

Figure 6. Dynamic model and coupling joints

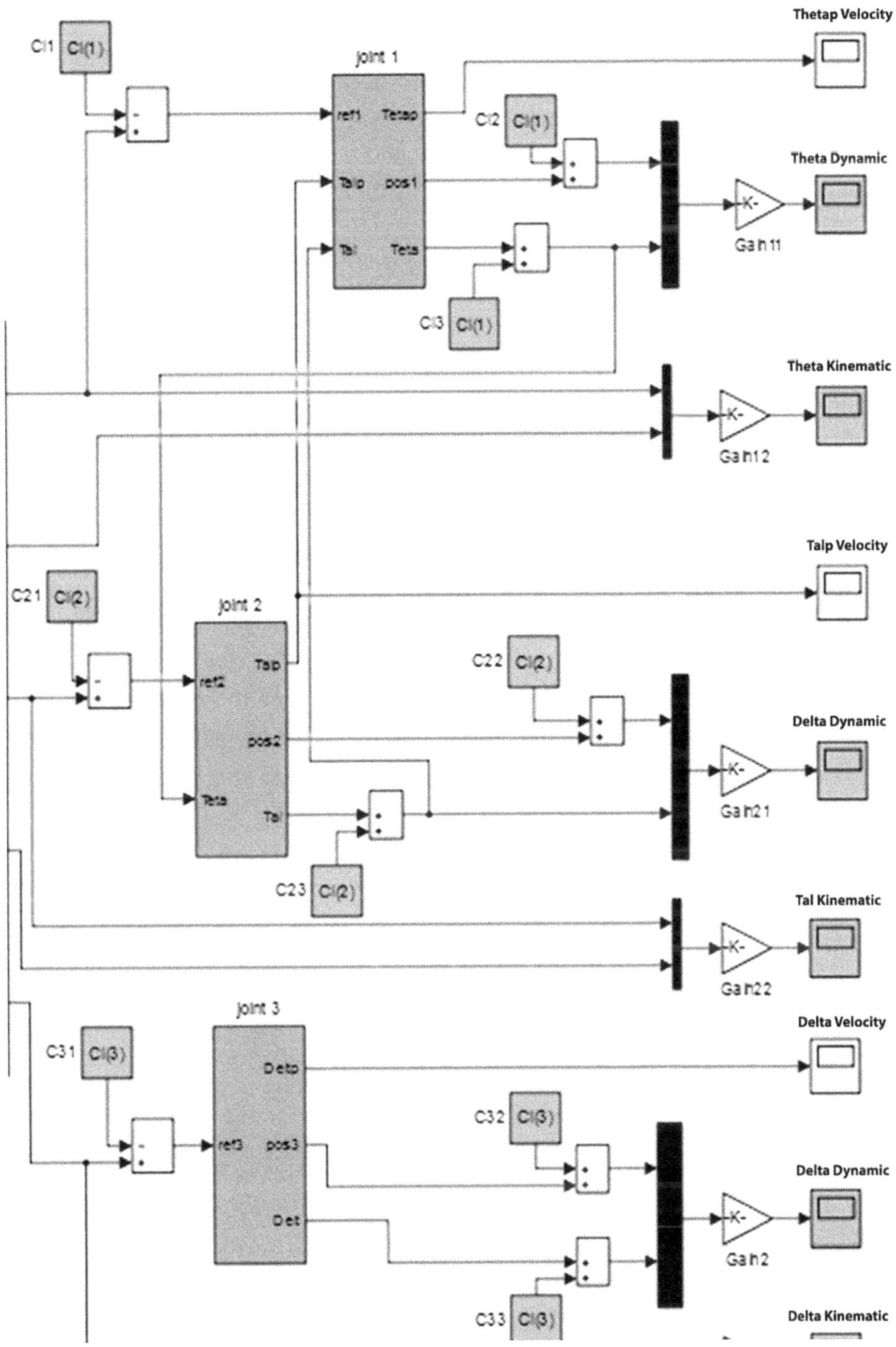

As expected, a self-tuning PID controller presented, for joints with strong coupling effect (hip), the occurrence of instabilities that can be eliminated only with the significant increase in the gain of the charge, reducing the mechanical transmission, conducting a comparative study based on controllers that has GPC, presents later.

Position Controller Implementation

The control architecture proposed for the system (3 DOF) was implemented with joint position controllers dynamically coupled (hip, leg and foot). For the dynamic model, the foot was regarded as a concentrated mass applied to the end of the leg. The parameters used were obtained from actual measurements of a mechanical design optimized (Kazerooni, 2008).

Figure 7 depicts joint control architecture used to simulate the previously described controllers, considering dynamically decoupled.

A Comparative Study: Criteria

A comparative study of the dynamic effect on the coupling of the hip-leg joints mechanism (2 DOF) was realized using serial, parallel, and hybrid mechanism (Banala et al., 2004), where can be saw the effect of dynamic decoupling in the use of parallelogram mechanism and the addition of the inertial compensation effect (or mass), when a spring was added (hybrid mechanism).

The parallelogram mechanism provides decoupling of the dynamic equations of the model, that permits simplify the dynamic model, and decrease the mechanical transmission rate as well as the possibility to use of a simple position controller (PID). In this study it was initially implemented a robust predictive controller for the case of a serial architecture, and it was used two criteria for comportment analysis:

- **Direct Sensitivity Function**: Direct influence of disturbance on the system output;
- **Complementary Sensitivity Function**: Influence of the measurement noise on the system output.

For a correct analysis of Robust Predictive Controller implemented the following adjustment parameters were used:

- **Delay Range**: More than one sample period;

Figure 7. Joint control architecture

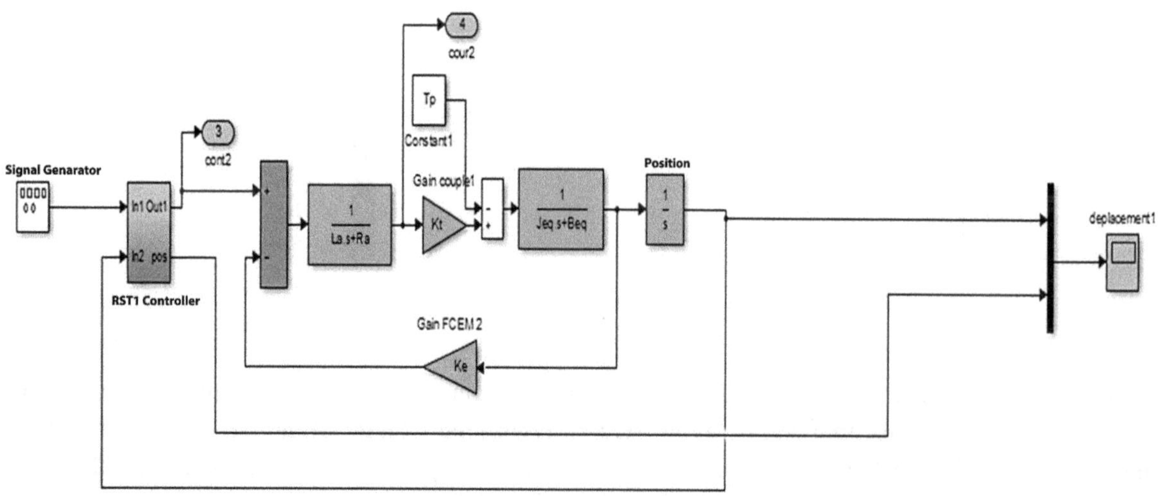

- **Direct Sensitivity Function**: Modulo less than 6 dB;
- **Complementary Sensitivity Function**: Modulo less than 3 dB;
- **Noise Power**: About 10% of the control signal.

Implementation of Control Based on Dynamic Model

In this study, was initially used a self-tuning PID controller in its classical form, which showed the expected correction behavior only when instability increases significantly to reducing the transmission ratio. At the same model was implemented a PID controller described as RST form. In this case, the joint corresponding to the hip presented itself very much sensitive (due to the lower number of DOF) for stabilization, then the controller needs to increase the constant discrete time.

This study also implemented Robust GPC controller considering the existence of a possible noise in the control signal, and/or the existence of disturbance in the system, which can run in the case of dynamic coupling joints. The joints transfer functions (hip, leg and foot) are described by equations 1, 2 and 3.

$$H_{Hip}(s) = \frac{1.37}{6,509e^{-7}s^{3} + 1,979e^{-4}s^{2} + 3,838e^{-2}s} \quad (1)$$

$$H_{Leg}(s) = \frac{1.37}{1,091e^{-7}s^{3} + 3,367e^{-4}s^{2} + 3,731e^{-2}s} \quad (2)$$

$$H_{Foot}(s) = \frac{1.37}{6,747e^{-8}s^{3} + 2,106e^{-4}s^{2} + 3,731e^{-2}s} \quad (3)$$

Controller Study Comparative uncoupled joints (3 DOF)

Figure 8, 9 and 10 presents a comparative study of the dynamic behavior of the corresponding uncoupled joins (hip, leg and foot) using a self-tuning PID controller, PID is in the form RST, and robust predictive controller (GPC) with a noise sensitive in the signal control and in the disturbance in the system. For comparative study of the dynamic behavior of the complete system (hip, leg and foot), it was used the corresponding transfer functions for each DOF of the coupled dynamic model. A comparative study is presenteded, of the dynamic behavior of joins, using for each of the DOF a predictive controller described as RST form (GPC) and a robust predictive controller: sensitive to noise in the control signal (GPC–R1), and robust noise-sensitive predictive in signal control witch disturbance in the system (GPC-R2), subjected a step input

Figure 8. Comparative study of controller (hip)

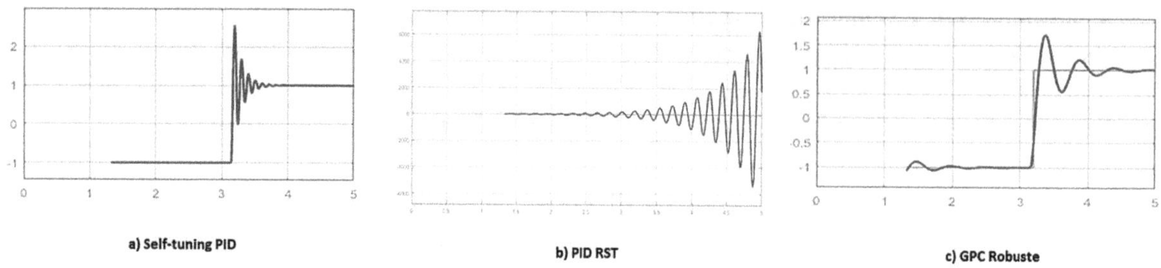

Figure 9. Comparative study of controller (leg)

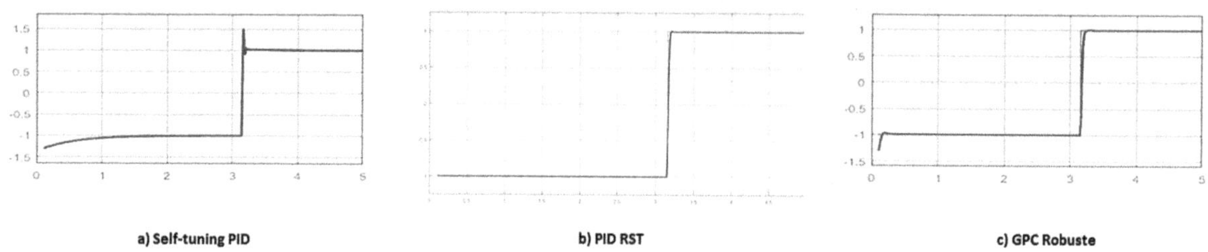

Figure 10. Comparative study of controller (foot)

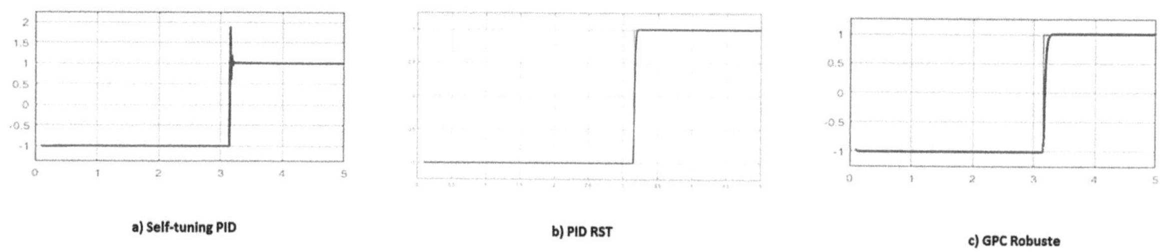

POSITION CONTROLLER IMPLEMENTATION

Figure 11 shows the complete structure control using the dynamic model implemented. For comparative study of effects was compared the 3 DOF mechanism in study using the control architecture proposed and the controllers GPC, GPC-R1 and GPC-R2.

For each joint (3 DDL) was implemented a position controller (GPC, GPC-R1 and GPC-R2), as shown at Figure 12, with the block diagram corresponding to each degree of freedom associated (hip, leg and foot). These blocks contain sub blocks concerning the position of control architecture used in each degree of freedom (GPC controller in the RST form), and the function block MATLAB corresponding to the dynamic coupled model of the joint (motor and associate load).

Figures 13, 14 and 15 show the results of the simulation that were obtained by the comparative study between the reference position and calculating the inverse kinematic model, and the system response analysis with dynamical coupling using structure Robust GPC controller, with disturbance in output and noise (coupled dynamic model).

Figure 11. Proposed control architecture for 3 DOF

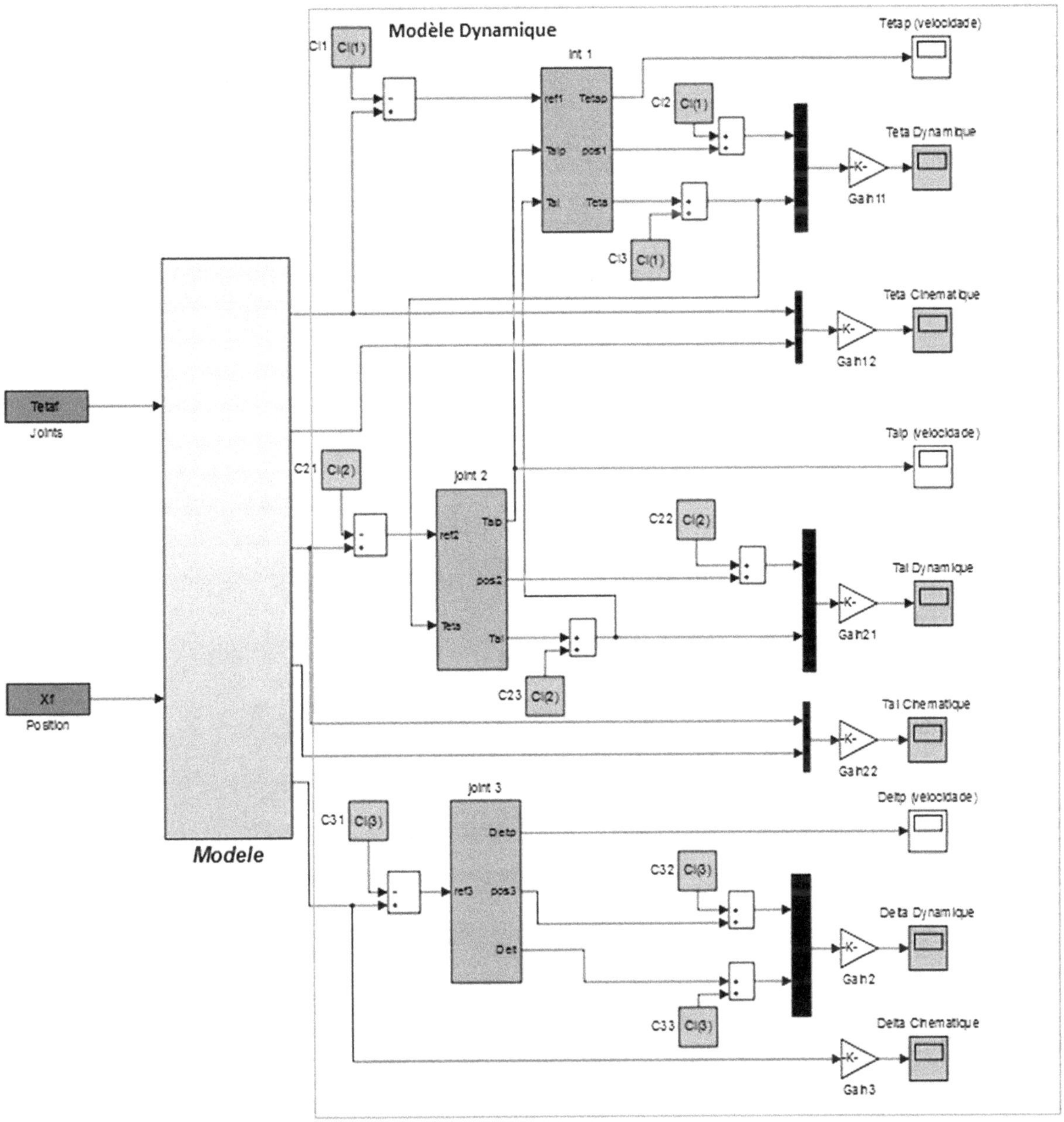

MAIN RESULTS AND COMMENTS

The self-tuning regulator PID presents the expected correction comportment, but need stabilization with respect to a Robust GPC controller. The PID-RST controller does not stabilize with respect to PID controller with constant gain, while the GPC controller is more sensitive to disturbances compared to controllers GPC-R1 and R2.

Figure 12. DC motor drive and joint dynamic controller

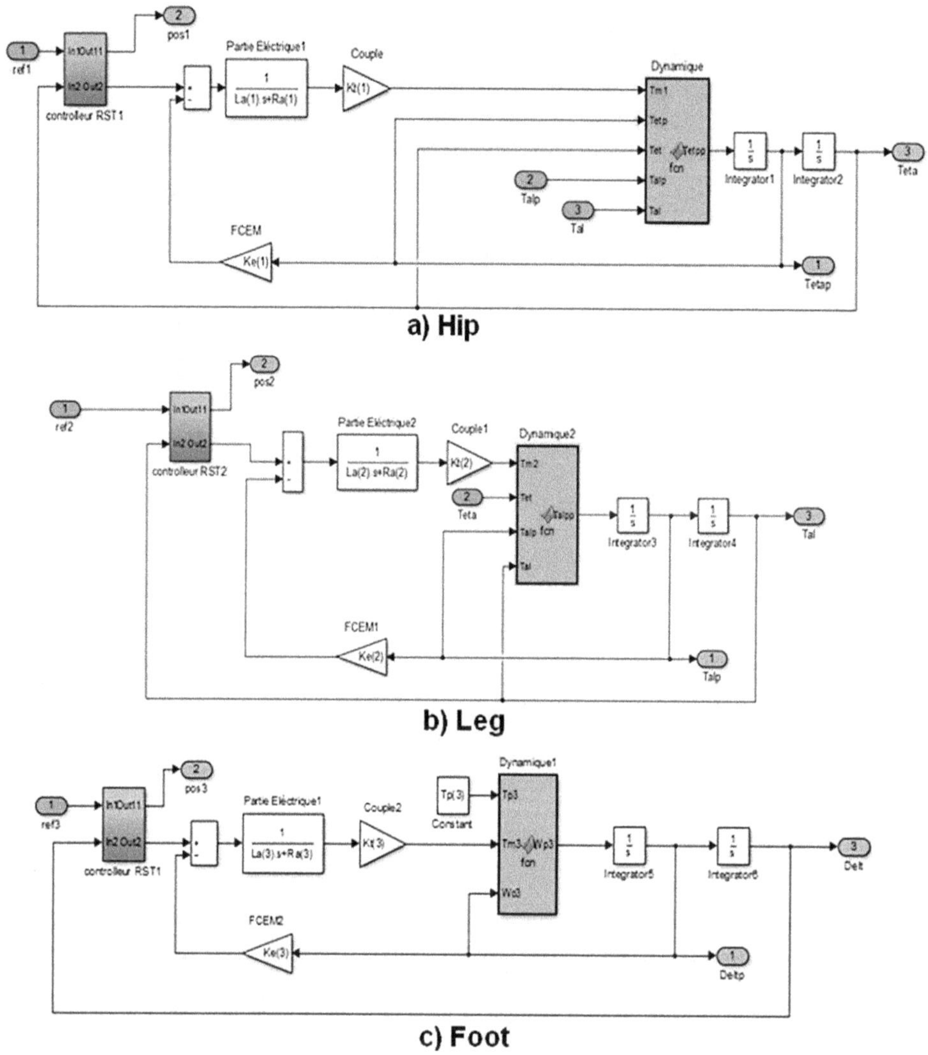

Motor + Dynamical Coupling Joint (Serial Mechanism)

The use of a mechanical transmission is essential to compensate for dynamic effects due to coupling joints. For correction effects, a comparative analysis was made with the different controller proposal in this study:

- **Self-Tuning PID Controller (Classical Form):** Shows the expected correction of behavior when the mechanical transmission (gear ratio) was increased, but it necessary that be stabilized in relation to a Robust GPC controller. The PID controller written in the RST form, in the case of joint corresponding to the hip, is very sensitive (due to the lower number of DOF) and does not stabilize easily.

Figure 13. Kinematics and dynamics comportment: Hip (Joint 1)

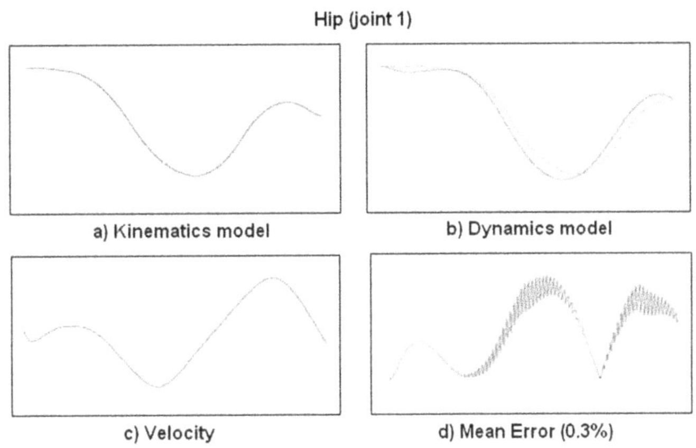

Figure 14. Kinematics and dynamics comportment: Leg (Joint 2)

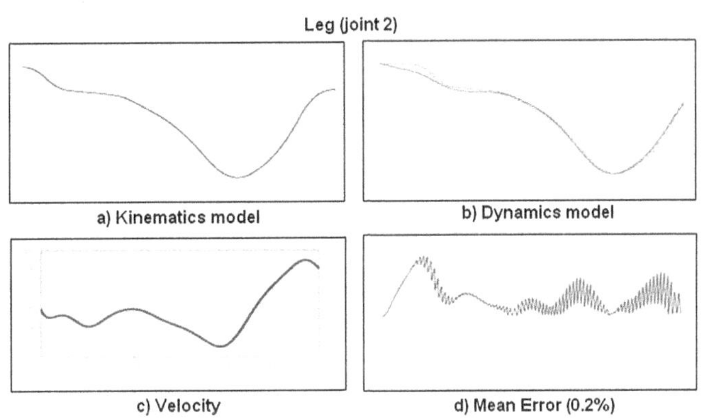

Figure 15. Kinematics and dynamics comportment: Foot (Joint 3)

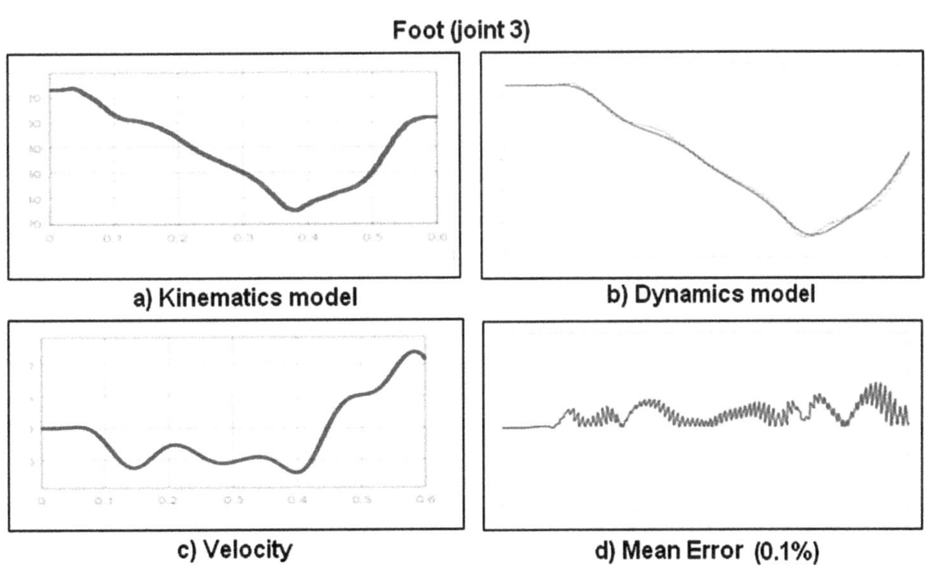

- **GPC (RST Form) Controller:** The same as the self-tuning PID controller is very sensitive to disturbance, especially if it was looked at the coupled joint (hip), not stabilizing it if not lift the mechanical transmission (gear ratio), also showing a considerable delay in relation to the reference input.
- **GPC Robust (R1) Controller:** Presents a less sensitive dynamic behavior disturbance in relation to the controller GPC-RST, especially if it was looked at the more coupled joint (hip), also presenting a significant delay in relation to the reference input, where that can be attenuated, if the mechanical transmission (gear ratio) was increased about the joint (occurs a decrease of its inertial coupling).
- **GPC Robust (R2) Controller:** Presents a dynamic comportment even less sensitive to disturbance with respect to the controller GPC-R1, particularly for the more dynamical coupled joint (hip), also presenting a significant delay in relation to the reference input. In the same way for the regulator GPC–R1, it can be attenuated, if increased the mechanical transmission (gear ratio) of the joint (when occurs a decrease of its inertial coupling).

CONCLUSION

From the development of this research work can be verified that solutions of problems of mobility of lower members using the hybrid method cause great impact the areas of Engineering of Rehabilitation. Orthosis exoskeletics of lower members was designed aiming at to the reproduction of the movements of the gait human being, with functions next to a natural member, for the implementation in patients with motor difficulty, having one of the weak legs.

The comparison of the use of the predictive control based on dynamical study and the decoupling of the dynamical model with auxiliary parallelograms for locating the center of mass of the mechanism with springs, with the objective of achieve the balancing of each leg, is obtained from a virtual model implemented and the kinematic and dynamic motion analyzed through simulation of an exoskeleton aimed at lower limbs for training and rehabilitation of the human gait, in which it is already developed the dynamic model of anthropomorphic mechanism and predictive control architecture with robust control.

The controller GPC-R2 presents a convenient way for use in exoskeleton, with high join dynamic coupling, however it presents the possibility of a delay in relation to the reference input, and the same can be corrected in two forms:

- By eliminating the dynamic coupling between the joints, from a hybrid compensation mechanism (spring + parallelogram) that will dynamically decouple the system under study; and/or
- Increased mechanical transmission (gear ratio).

Finally, the main results of this work can be used to increase the number of degrees of freedom, to extend the balance study of the gravity for the two legs.

REFERENCES

Abernethy, B., Kippers, V., Hanrahan, S., Pandy, M., McManus, A., & Mackinnon, L. (2013). *The Biophysical foundations of Human Movement* (3rd ed.). Champaign: Human Kinetics.

Banala, S. K., Agrawal, S. K., Fattah, A., Rudolph, K., & Scholz, J. P. (2004). A gravity balancing leg orthosis for robotic rehabilitation. In *Robotics and Automation, 2004. Proceedings. ICRA'04. 2004 IEEE International Conference on* (Vol. 3, pp. 2474-2479). IEEE. doi:10.1109/ROBOT.2004.1307432

Cassemiro, E. R., Rosario, J. M., & Dumur, D. (2005). Robot axis dynamics control using a virtual robotics environment. *2005 IEEE Conference on Emerging Technologies and Factory Automation*. doi:10.1109/ETFA.2005.1612539

de Melo, L. F., Rodrigues, E. J., & Rosário, J. M. (2015). Wheelchair Secure Navigation with RF Signal Triangulation and Genetic Algorithm Optimization. In *Handbook of Research on Advancements in Robotics and Mechatronics* (pp. 750–792). IGI Global. doi:10.4018/978-1-4666-7387-8.ch024

Delcomyn, F. (2007). Biologically inspired robots. In *Bioinspiration and Robotics Walking and Climbing Robots*. InTech. doi:10.5772/5506

Floriano-Batista, R., & Rosario, J. M. (2013). *Control Architecture Proposal for Robotic Exoskeleton to Human Gait in Assistive Technology*. Campinas, São Paulo: Unicamp.

Gutiérrez-Carvajal, R. E., de Melo, L. F., Rosário, J. M., & Machado, J. T. (2016). Condition-based diagnosis of mechatronic systems using a fractional calculus approach. *International Journal of Systems Science*, *47*(9), 2169–2177. doi:10.1080/00207721.2014.978833

Habib, M. K. (2011). Biomimetics: Innovations and robotics. *International Journal of Mechatronics and Manufacturing Systems*, *4*(2), 113–134. doi:10.1504/IJMMS.2011.039263

Kazerooni, H. (2008). Exoskeletons for human performance augmentation. In *Springer handbook of robotics* (pp. 773–793). Springer Berlin Heidelberg. doi:10.1007/978-3-540-30301-5_34

Y. Liu, & D. Sun (Eds.). (2011). *Biologically inspired robotics*. CRC Press. doi:10.1201/b11365

Low, K. H. (2011, August). Robot-assisted gait rehabilitation: From exoskeletons to gait systems. In *Defense Science Research Conference and Expo (DSR)*, 2011 (pp. 1-10). IEEE. doi:10.1109/DSR.2011.6026886

Meyer, J. A., & Guillot, A. (2008). Biologically inspired robots. In *Springer Handbook of Robotics* (pp. 1395–1422). Springer Berlin Heidelberg. doi:10.1007/978-3-540-30301-5_61

Pons, J. L., Ceres, R., & Calderón, L. (2008). Introduction to Wearable Robotics. In J. L. Pons (Ed.), *Wearable Robots: Biomechatronic Exoskeletons*. Chichester, UK: John Wiley & Sons, Ltd; doi:10.1002/9780470987667.ch1

Shi, L., Habib, M. K., Xiao, N., & Hu, H. (2015). Biologically inspired robotics. *Journal of Robotics*.

Winter, D. A. (1984). Kinematic and kinetic patterns in human gait: Variability and compensating effects. *Human Movement Science*, *3*(1), 51–76. doi:10.1016/0167-9457(84)90005-8

Chapter 14
Kinodynamic Motion Planning for a Two-Wheel-Drive Mobile Robot

Kimiko Motonaka
Kansai University, Japan

ABSTRACT

Since a nonholonomic system such as a robot with two independent driving wheels includes complicated nonlinear terms generally, it is hard to realize a stable and tractable controller design. However, about a dynamic control method for the motion planning, it is guaranteed that a nonholonomic-controlled object can always be converged to an arbitrary point using a control method based on an invariant manifold. Based on it, the method called "kinodynamic motion planning" was proposed to converge the states of the two-wheeled mobile robot to the arbitrary target position while avoiding obstacles by combining the control based on the invariant manifold and the HPF. In this chapter, how to combine the invariant manifold control and the concept of the HPF is explained in detail, and the usefulness of the proposed approach is verified through some simulations.

INTRODUCTION

Recently, many robots are working in our living environment. Especially, a mobile robot with two independently driven wheels has been commonly used for transportation or giving service to human. For these robots, autonomous locomotion is important, and there are many kinds of researches about an obstacle avoidance or a path planning for it.

For the smooth moving of the robot in a living environment, it is desirable to consider the kinematic constraint (generated from the constraint based on an obstacle or target position) and the dynamic constraints (generated from the dynamic characteristics of the controlled object). Therefore, the method called "kinodynamic motion planning" was proposed by Donald, Xavier, Canny, and Reif (1993) to solve this problem. The kinodynamic motion planning aims to design the control input so that the kinematic constraints generated from the environment and the dynamic constraints generated from the dynamic characteristics of the controlled object are both considered simultaneously. Now, there are many re-

DOI: 10.4018/978-1-5225-2993-4.ch014

searches about kinodynamics (Rimon, 1998; Hsu, 2002; Peng, 2005). However, most of these methods do not generate the control input in real time reflecting the current state of the robot but calculate the all of the control input in advance. This is unfavorable when the overshoot occurs and it needs to correct the state of the robot in an actual environment. Masoud (2010) realized the kinodynamic motion planning for a point mass by combining two control inputs: one is generated by considering the dynamics of the controlled object and the other is based on the gradient information of the potential field called harmonic potential field (HPF) which is generated from the boundary information of the environment. In this chapter, the method to extend and apply the kinodynamic motion planning based on the HPF to the two-wheeled mobile robot is described.

The kinodynamic motion planning using the HPF is achieved by adding the gradient information of an HPF to the control input that has considered the dynamic characteristics of the controlled object. To apply the kinodynamic motion planning based on an HPF to some controlled object, at first, it needs the control input that considers the dynamic model of the controlled object. Here, the stabilizing control to the origin using an invariant manifold is given as one of the methods to control the two-wheeled mobile robot. The controller based on the invariant manifold can reliably stabilize the controlled object to the origin kinematically and dynamically (Tsiotras, Corless, & Longuski, 1993). However, that method can stabilize the controlled object to the origin only. In the case of the robot works in the actual environment, the robot is required to move onto not only the origin but also the arbitral target position. To converge the controlled object to a non-zero desired point, Goto, Watanabe, and Maeyama (2014) extended the controller based on the invariant manifold by using an error model (2014). This extended method is suitable to combine the controller based on the HPF. By combining these controllers, the kinodynamic control can be achieved and the controlled object can reach the desired target position with the desired state while avoiding obstacles. In the following sections, the model of the two-wheeled mobile robot, the controller using an invariant manifold, the controller using the HPF, how to combine them, and some experiment and results are described.

A Model of Two-Wheeled Mobile Robot

Figure 1 shows the schematic structure of the two-wheeled drive mobile robot when assuming that the world coordinate system is the X-Y coordinate system. Note that, $2L$, R, and θ in Figure 1 mean the tread, the radius of the wheel, and the direction angle from the X-axis respectively. The controlled point of the two-wheeled drive mobile robot, $O(x, y)$ exists at the center of the tread. v means the forward velocity, and omega means the angular velocity around the point O. Hereafter, the authors call the two-wheeled drive mobile robot "the robot." The dynamic model of the robot with two-inputs and five-states is shown as:

$$\begin{bmatrix} \dot{x} \\ \dot{y} \\ \dot{\theta} \\ \dot{v} \\ \dot{\omega} \end{bmatrix} = \begin{bmatrix} v\cos x \\ v\sin x \\ \omega \\ 0 \\ 0 \end{bmatrix} = \begin{bmatrix} 0 \\ 0 \\ 0 \\ \frac{1}{M} \\ 0 \end{bmatrix} F + \begin{bmatrix} 0 \\ 0 \\ 0 \\ 0 \\ \frac{1}{I} \end{bmatrix} \tau \tag{1}$$

Figure 1. The two-wheeled mobile robot

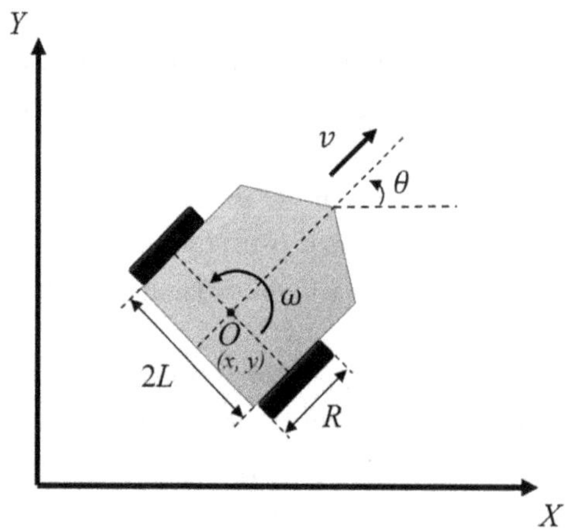

Here, M , I , F and τ mean the mass of the robot, the moment of inertia, the force and the torque, respectively. Assuming that the motor torque of the right and left wheels is given by (τ_1, τ_2), it follows that:

$$F = \frac{1}{R}(\tau_1 + \tau_2)$$

$$\tau = \frac{L}{R}(\tau_1 - \tau_2) \tag{2}$$

CONTROL USING THE INVARIANT MANIFOLD

The invariant manifold means a manifold constructed by the scalar functions whose differential value is 0 with respect to time. There are two types of control using the invariant manifold. One is the switching control that is able to divide the first-step (until on the invariant manifold) and the second step (on the invariant manifold), and the other is the pseudo-continuous exponential stabilization control that excludes the switching control by combining the first and second steps. It is known that the robot can converge to the origin by using these control methods. Generally, a nonholonomic system includes several complex nonlinear terms, and it is difficult to design the controller from the original model. On the other hand, by using the canonical model, such as the chained form, it becomes easy to make a control system, which assures a globally asymptotic stability. In the following subsections, we describe the process of the chained form transformation, the chained form transformation that considers the error for the stabilization to the arbitrary point, the switching control using the invariant manifold, and the pseudo-continuous exponential stabilization control.

Chained Form Transformation of the Dynamic Model

The chained form transformation is useful to present a complex equation simply. When converging the robot to the origin, the chained form transformation of the dynamical model is shown as:

$$\begin{bmatrix} z_1 \\ z_2 \\ z_3 \\ z_4 \\ z_5 \end{bmatrix} = \begin{bmatrix} \theta \\ x\cos\theta + y\sin\theta \\ x\sin\theta - y\cos\theta \\ \omega \\ v - \left(x\sin\theta - y\cos\theta \right)\omega \end{bmatrix} \tag{3}$$

$$\dot{z} = \begin{bmatrix} \dot{z}_1 \\ \dot{z}_2 \\ \dot{z}_3 \\ \dot{z}_4 \\ \dot{z}_5 \end{bmatrix} = \begin{bmatrix} z_4 \\ z_5 \\ z_2 z_4 \\ u_1 \\ u_2 \end{bmatrix} \tag{4}$$

$$u_1 = \frac{\tau}{I}$$

$$u_2 = -z_4^2 z_2 - \frac{\tau}{I} z_3 + \frac{F}{M} \tag{5}$$

The transformation from Eq. (1), which is an origin dynamical model, to Eq. (4) is called "chained form transformation." When transforming the two-input and five-output system of the robot by a chained form, the dynamical model of the robot becomes Eq. (4). Equation (3) shows the state $\left[z_1 z_2 z_3 z_4 z_5 \right]^T$ after providing the coordinate transformation to Eq. (1). Moreover, u_1 and u_2 denote the control inputs on the chained form, which are shown as Eq. (5). Here, again, M, I, F and τ mean the mass of the robot, the moment of inertia, the force and the torque, respectively.

Chained Form Transformation That Considered the Error

It needs to consider the error from the origin to an arbitrary point to converge the robot state value to the arbitrary point. The chained form transformation, which takes account of the error, is called an "error model." The error model is shown as:

$$\dot{e} = \begin{bmatrix} \dot{e}_1 \\ \dot{e}_2 \\ \dot{e}_3 \\ \dot{e}_4 \\ \dot{e}_5 \end{bmatrix} = \begin{bmatrix} e_4 \\ e_5 \\ e_2 e_4 \\ u_1 \\ u_2 \end{bmatrix} \tag{6}$$

Then, the error state $\left[e_1 e_2 e_3 e_4 e_5\right]^T$ from the desired target state $\left[z_{1d} z_{2d} z_{3d} z_{4d} z_{5d}\right]^T$ is represented by:

$$\begin{bmatrix} e_1 \\ e_2 \\ e_3 \\ e_4 \\ e_5 \end{bmatrix} = \begin{bmatrix} z_1 - z_{1d} + \alpha \\ z_2 - z_{2d} + \beta \\ z_3 - z_{3d} + \gamma \\ z_4 - z_{4d} \\ z_5 - z_{5d} \end{bmatrix} \tag{7}$$

Note that, z_{1d}, z_{2d}, z_{3d}, z_{4d}, and z_{5d} are the constant values which can be obtained in Eq. (3), when given the arbitrary target values x_d, y_d, θ_d, z_{1d}, and z_{1d} to x, y, θ, v and ω. Then, the differential equation of e_4 and e_5 shown in Eq. (6) may be easily delivered from Eq. (7). Here, α, β, and γ include z_{4d} or z_{5d} generally, but the final target states of v and ω are assumed to be set to 0. Thus, we derive the error equation assuming $z_{4d} = z_{5d} = 0$. In that case, α and β become $\alpha = \beta = 0$ (the detailed derivation is omitted). Next, we describe the derivation of γ. The time derivative of e_3 is denoted, from Eq. (7), as follows:

$$\dot{e}_3 = \dot{z}_3 + \dot{\gamma} = z_2 z_4 + \dot{\gamma} \tag{8}$$

In addition, $\dot{e}_3$ can be delivered from Eq. (7) as follows:

$$\dot{e}_3 = e_2 e_4 = \left(z_2 - z_2\right)\left(z_4 - z_{4d}\right) = z_2 z_4 - z_{2d} z_4 \tag{9}$$

To hold relating of Eq. (8) and Eq. (9), $\dot{\gamma}$ becomes $\dot{\gamma} = -z_{2d} z_4$. Hence, γ is reduced to $\gamma = -z_{2d} z_1$ from Eq. (4). Strictly speaking, an integration constant should be added, but this parameter is assumed to 0 in this method. Finally, the error state becomes:

$$\begin{bmatrix} e_1 \\ e_2 \\ e_3 \\ e_4 \\ e_5 \end{bmatrix} = \begin{bmatrix} z_1 - z_{1d} \\ z_2 - z_{2d} \\ z_3 - z_{3d} - z_{2d} z_1 \\ z_4 \\ z_5 \end{bmatrix} \quad (10)$$

As mentioned above, this is the result for the case where z_{4d} and z_{5d} are 0.

Derivation of the Invariant Manifold and the Switching Control

The switching control using the invariant manifold is able to be divided into the first stage which occupies the time domain until reaching the invariant manifold, and the second stage after reaching the invariant manifold. In the following sections, the invariant manifold is first derived, the control on the invariant manifold is described, and the control needed until reaching the invariant manifold is stated.

Assuming that the desired invariant manifold is $S(e(t))$. In the previous research, the invariant manifold was written using the chained form transformation z, but in this paper, we use the error model $\mathbf{e}$. To begin with, the following feedback control, if the gain $k > 0$, is provided to Eq. (6):

$$u_1 = -2ke_4 - k^2 e_1$$

$$u_2 = -2ke_5 - k^2 e_2 \quad (11)$$

Then, from the initial value of time response of the state e, the invariant manifold is given by the following equation:

$$S(e(t)) = e_3(t) - \frac{1}{2} e_1(t) e_2(t) + \frac{1}{4k}\left(e_2(t) e_4(t) e_5(t)\right) \quad (12)$$

Let us consider the case of using the control input written in Eq. (11) on the derived invariant manifold. If $S(e(t)) = 0$ is true at the time $t = T$, then, the $S(e(t)) = 0$ is also true in the following time *t*. In addition, from the time response of the state ***e***, if $t \to \infty$, then it is true that $e_1 \to 0$, $e_2 \to 0$, $e_3 \to 0$, $e_4 \to 0$, and $e_5 \to 0$. Therefore, the controller can converge the controlled object to the target value by providing the control input shown in Eq. (11) on the invariant manifold.

The control input in the first step for being attracted to the invariant manifold is derived. To derive the control input for reaching the state on the invariant manifold, i.e., for making $S(e) = 0$, the Lyapunov function $V(e)$ of the $S(e)$ is chosen as follows:

$$V(e) = \frac{1}{2} S^2(e) \quad (13)$$

For this, selecting the control input which satisfies the condition $\dot{V}(e) \le 0$ the control input can satisfy $S(e) = 0$. In this case, if the control input is selected as Eq. (14):

$$u_1 = -2ke_4$$

$$u_2 = \frac{fS(e)}{e_1} - 2ke_5 \tag{14}$$

then, $\dot{V}(e)$ becomes like the following equations:

$$\begin{aligned} \dot{V}(e) &= S\dot{S} \\ &= S\left[\dot{e}_3 - \frac{1}{2}(\dot{e}_1 e_2 + e_1\dot{e}_2) + \frac{1}{4k}(\dot{e}_2 e_4 + e_2\dot{e}_4 - \dot{e}_1 e_5 - e_1\dot{e}_5)\right] \\ &= S\left[e_2 e_4 - \frac{1}{2}(e_2 e_4 + e_1 e_5) + \frac{1}{4k}(e_4 e_5 + e_2 u_1 - e_4 e_5 - e_1 u_2)\right] \\ &= S\left[\frac{1}{2}e_2 e_4 - \frac{1}{2}e_1 e_5 + \frac{1}{4k}(-2ke_2 e_4 - fS + 2ke_1 e_5)\right] \\ &= -\frac{f}{4k}S^2 \end{aligned} \tag{15}$$

Therefore, $\dot{V}(e)$ becomes negative definite and $S(e(t))$ becomes $S(e(t)) \to 0$ when $t \to \infty$, then, the attractive control to the invariant manifold is achieved.

The Pseudo-Continuous Exponential Stabilization Control

The pseudo-continuous exponential stabilization control is the method that is realized in one unified manner, instead of conducting the control one by one in the first and second steps described previously. This control excludes the switching, and it can be controlled without a sudden change of the input. The control input of the pseudo-continuous exponential stabilization control is given by adding Eq. (14) and (11), and it is reduced to:

$$u_1 = -2ke_4 - k^2 e_1$$

$$u_2 = \frac{fS(e)}{e_1} - k^2 e_2 - 2ke_5 \tag{16}$$

$\dot{S}(e)$ under the control input of Eq. (16) is written as the following equation:

$$\begin{aligned}\dot{S}(e) &= \dot{e}_3 - \frac{1}{2}\dot{e}_1 e_2 - \frac{1}{2}e_1\dot{e}_2 + \frac{1}{4k}\left(\dot{e}_2 e_4 + e_2\dot{e}_4 - e_1 e_5 - e_1\dot{e}_5\right) \\ &= e_2 e_4 - \frac{1}{2}e_2 e_4 - \frac{1}{2}e_1 e_5 + \frac{1}{4k}\left(e_4 e_5 + e_2 u_1 - e_4 e_5 - e_1 u_2\right) \\ &= \frac{1}{2}e_2 e_4 - \frac{1}{2}e_1 e_5 + \frac{1}{4k}\left[e_2\left(k^2 e_1 - 2ke_4\right) - e_1\left(\frac{fS(e)}{e_1} - k^2 e_2 - 2ke_5\right)\right] \\ &= -\frac{fS(e)}{4k}\end{aligned} \tag{17}$$

Therefore, the time response of $S(e)$ becomes:

$$S(t) = S(e(0))e^{-\frac{f}{4k}t} \tag{18}$$

Thus, $S(t)$ becomes $S(t) \to 0$ when $t \to \infty$. In addition, to keep the input of u_2 finite, the following condition is needed:

$$\left|\frac{S}{e_1}\right| < \infty \tag{19}$$

To confirm the boundedness of the input, the $e_1(t)$ is written as the following equation:

$$e_1(t) = \frac{e_1(0)(1+kt) + e_4(0)t}{e^{kt}} \tag{20}$$

A limited value of $e_1(t)$ can be shown as the following equation using the L'Hopital's rule:

$$\lim_{t\to\infty} e_1(t) = \lim_{t\to\infty} \frac{e_1(0)k + e_4(0)}{ke^{kt}} = 0 \tag{21}$$

Using the same way, the asymptotically stability of $e_4(t)$ is also able to be guaranteed. Therefore, if $\frac{f}{4k} - k > 0$, the following equation is established:

$$\lim_{t\to\infty}\frac{S(t)}{e_1(t)}=\lim_{t\to\infty}\frac{S(e(0))e^{-\frac{f}{4k}t}}{e^{-kt}\left[e_1(0)(1+kt)+e_4(0)t\right]}=0$$
$$=\lim_{t\to\infty}\frac{kS(e(0))e^{-\left(\frac{f}{4k}-k\right)t}}{e_1(0)k+e_4(0)}=0 \tag{22}$$

Here, since:

$$\left|\frac{S}{e_1}\right|<\left|\frac{kS(e(0))}{e_1(0)k+e_4(0)}\right|e^{-\left(\frac{f}{4k}-k\right)t} \tag{23}$$

it follows that $\frac{S(t)}{e_1}$ is attenuated with the exponential coefficient $\left(\frac{f}{4k}-k\right)$.

Additionally, deriving the free response terms in $e_2(t)$ and $e_5(t)$, it can be known that these functions converge to 0 with the exponential coefficient k. Then, if $\frac{f}{4k}-k>k$, namely $f>8k^2$, then $e_2(t)$ and $e_5(t)$ converge to 0. Accordingly, if $f>8k^2$, then $S(e)$, $e_1(t)$, $e_2(t)$, $e_3(t)$, $e_4(t)$, and $e_5(t)$ converge to 0 when $t\to\infty$.

COMBINING OF THE CONTROL USING INVARIANT MANIFOLD WITH THE HPF

The aim of this method is to achieve the kinodynamic motion planning for a two-wheeled drive mobile robot by combining the pseudo-continuous exponential stabilization control with the HPF. Concretely, the kinodynamic motion planning is achieved by adding the directional force in gradient to the control inputs, u_1 and u_2 in the chained form. However, if the control input with the information from the HPF is used in the whole area of the environment, the differential value of the invariant manifold does not become 0, then it is considered that the state does not reach the target point. In other words, if always adding the directional force from the HPF to the controller, the stability of the manifold is broken.

Therefore, in this chapter, basically the pseudo-continuous exponential stabilization control is used, and added the information of the HPF to its control input locally. In this method, the convergence to the target position can be achieved finally, because only the pseudo-continuous exponential stabilization control is used in the area without the obstacles and then the stability of the manifold is not broken. In what follows, it is described a force in the gradient of the HPF and how to design the input torque for the robot by introducing its information as a force, and state the control that combines the pseudo-continuous exponential stabilization control with the HPF.

The Directional Force in Gradient and the Input Torque

From the HPF, the x-directional and y-directional gradient vector of each position of the environment can be calculated. These gradient vectors show the direction that the robot should move. By using the x-directional gradient Δx and the y-directional gradient Δy, the gradient direction θ_k, which is the direction that the robot should move, can be written as follows:

$$\theta_k = \tan^{-1}\frac{\Delta y}{\Delta x} \tag{24}$$

Assuming that the input torque with the gradient directional force is τ', it can be written as the following equation:

$$\tau' = \frac{L}{R}(\tau_1 - \tau_2) + k_1\theta_k \tag{25}$$

with the gain $k_1 > 0$.

Pseudo-Continuous Exponential Stabilization Control Combined With the HPF

In this subsection, the control input combining the pseudo-continuous exponential stabilization with the HPF is derived. As shown in Eqs. (3), (4), and (5), the input torque of the robot is remarkably reflected on u_1. Therefore, the kinodynamic motion planning should be able to be achieved by adding the gradient directional force of the HPF θ_k to u_1. Assuming the gain $k_1 > 0$, the control inputs of the pseudo-continuous exponential stabilization control can be written by using Eqs. (16) and (25) as follows:

$$u_1 = -2ke_4 - k^2e_1 + k_1\theta_k$$

$$u_2 = \frac{fS(e)}{e_1} - k^2e_2 - 2ke_5 \tag{26}$$

SIMULATIONS

In this section, some simulations are conducted to verify the effectiveness of the kinodynamic control for the two-wheeled mobile robot. For the simulations, the MATLAB and PDE Toolbox (one of the toolbox for MATLAB) is used as the calculation tool. The conditions and the results of simulations are described below.

Conditions

In this simulation, it is confirmed that the state of the robot can be converges to the target value $\left[x_d y_d \theta_d v_d \omega_d\right]^T$ by using the pseudo-continuous exponential stabilization control and the HPF. Here, by calculating Eq. (26), the controller converges the initial state $\left[x_0 y_0 \theta_0 v_0 \omega_0\right]^T$ to the target state $\left[x_d y_d \theta_d v_d \omega_d\right]^T$. The physical parameters of the robot are set as $M = 10$ [kg], $I = 2$ [kg/m^2], $R = 0.03$ [m], and $L = 0.06$ [m], whereas the sampling interval and the threshold value are set to $\Delta t = 0.01$ [s] and $\varepsilon = 0.01$, respectively. The feedback gains are set to $k = 1$ and $f = 9$, and Eq. (26) is used as the control input. In the assumed environment, there is a rectangle obstacle, whose apices are located at (15, 15), (35, 15), (15, 35), and (35, 35) [m] respectively. Figure 2 shows the environment and the potential vector when assuming an obstacle exists. The initial and target position are shown as red and blue dots, and the gradient vector in each position is shown as blue arrow. Here, four kinds of the initial state x_0 and the target value x_T are set as follows:

Case A: $x_0 = \left(45, 45, -1.8, 0, 0\right)^T, x_T = \left(5, 10, 0, 0, 0\right)^T$
Case B: $x_0 = \left(45, 45, -2.3, 0, 0\right)^T, x_T = \left(5, 10, 0, 0, 0\right)^T$
Case C: $x_0 = \left(45, 45, -3, 0, 0\right)^T, x_T = \left(5, 10, 0, 0, 0\right)^T$

Figure 2. An assumed environment
(Adapted from [Motonaka, Goto, Watanabe, and Maeyama, (2016)])

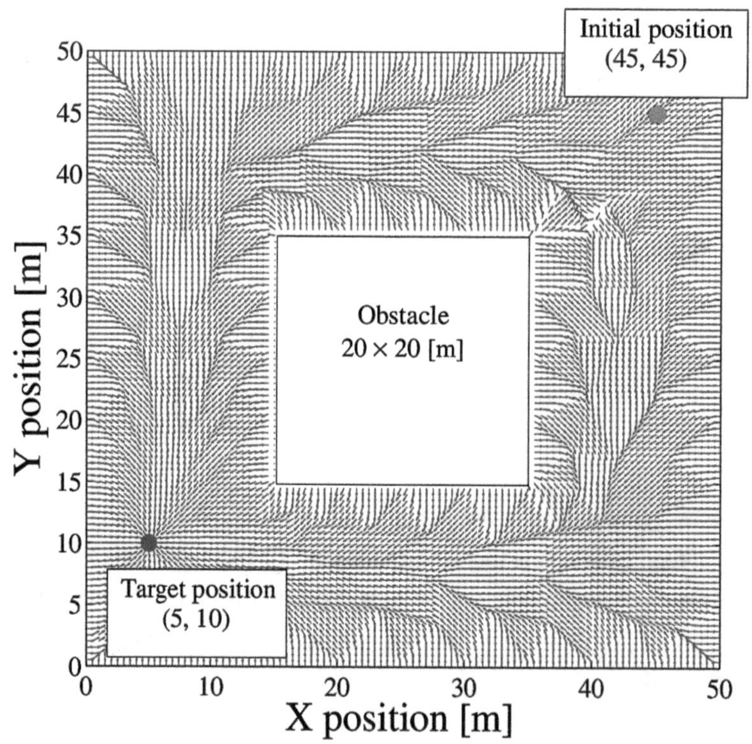

Case D: $x_0 = (45, 45, 1.8, 0, 0)^T, x_T = (5, 10, 0, 0, 0)^T$

Note from Eq. (26) that, the gain k_1 is set to $k_1 = 1$ at the area $0 \le x \le 50$ and $0 \le y \le 5$ in the case A; $k_1 = -1.8$ in the area within 5 [m] from the obstacles in the case B; $k_1 = -1$ at the area $0 \le x \le 50$ and $40 \le x \le 45$ in the case C; and $k_1 = 1$ at the area $35 \le x \le 45$ and $0 \le y \le 50$ in the case D. These gains are decided by trial and error. To design the method to decide this gain is one of the future works of this research.

RESULTS AND DISCUSSIONS

Figures 3 to 6 show the time response of the state $[x, y, \theta]^T$ in (a) and the time response of the state $[v, \omega]^T$ in (b), when using the pseudo-continuous exponential stabilization control with the HPF, respectively. As shown in these graphs, the state of the robot was able to converge to the target value in all the cases A-D. It is confirmed, from these results, that the convergence was achieved when adding the directional force in gradient of the HPF to the pseudo-continuous exponential stabilization control.

In addition, Figure 7 shows the comparison of the trajectories in the X-Y plane when using or not using the HPF. From these figures, it is verified that the obstacle information can be considered by adding the information of the HPF. However, in the case D, the controlled object crashed to the obstacles even though the information of the HPF was used. In the case D, the robot progressed backward toward the target point due to its initial state. Then, it is attributed to the fact that the robot was not able to change the state to avoid the obstacle while keeping its moving direction. Finally, it is confirmed that combining of the pseudo-continuous exponential stabilization control and the HPF is useful, and the autonomous locomotion of the robot was achieved, as long as the robot moves forward.

Figure 3. Simulation results for the case A
(Adapted from [Motonaka, Goto, Watanabe, and Maeyama, (2016)])

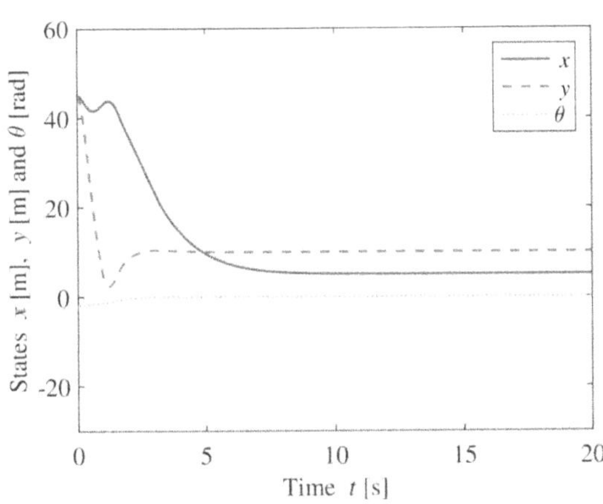

(a) Time response of the states x, y and θ

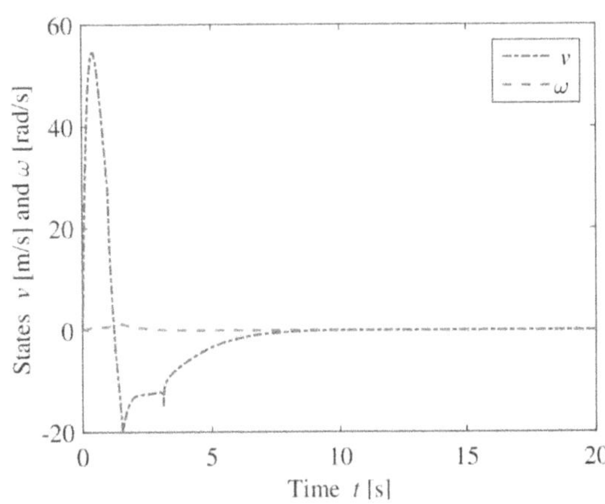

(b) Time response of the states v and ω

Figure 4. Simulation results for the case B
(Adapted from [Motonaka, Goto, Watanabe, and Maeyama, (2016)])

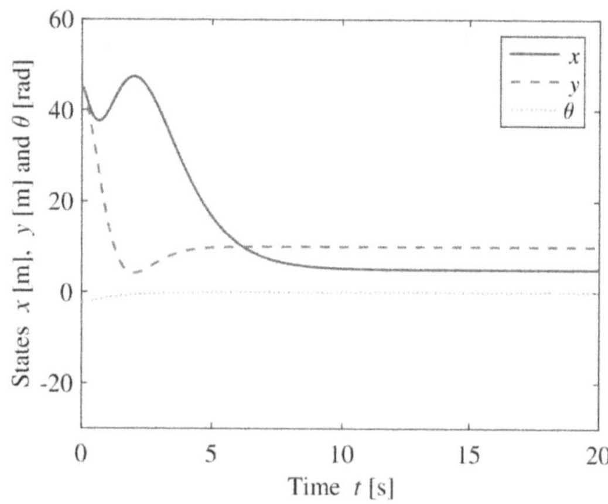

(a) Time response of the states x, y and θ

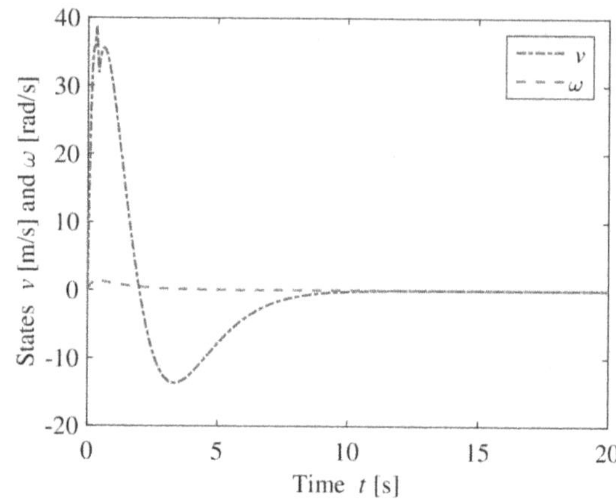

(b) Time response of the states v and ω

Figure 5. Simulation results for the case C
(Adapted from [Motonaka, Goto, Watanabe, and Maeyama, (2016)])

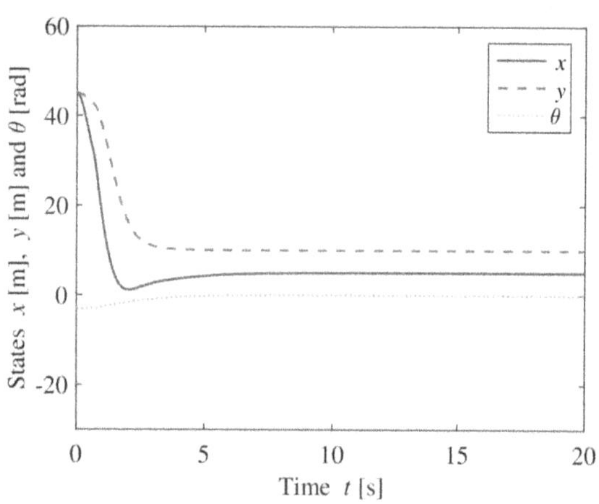

(a) Time response of the states x, y and θ

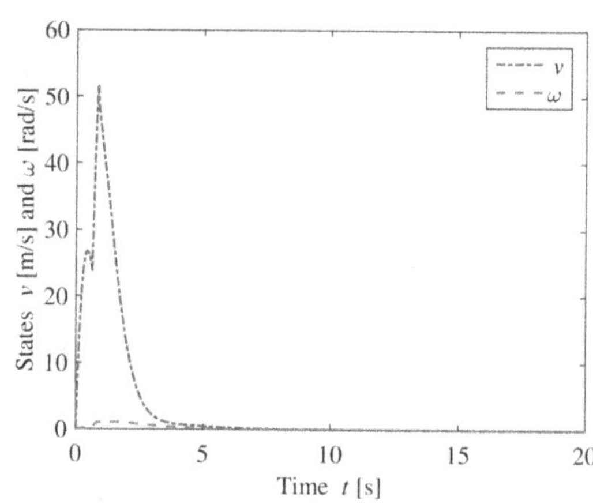

(b) Time response of the states v and ω

CONCLUSION

In this chapter, a kinodynamic control method using the pseudo-continuous exponential stabilization control and the HPF for the two-wheeled drive mobile robot has been described. In the kinodynamic control method, the control input is designed to reach the target point while avoiding obstacles by combining the pseudo-continuous exponential stabilization control with the invariant manifold and the directional force in a gradient of the HPF. To describe the detail of it, we touched the dynamical model of the robot, chained form transformation, the derivation of the invariant manifold, calculation of the gradient of the

Figure 6. Simulation results for the case D
(Adapted from [Motonaka, Goto, Watanabe, and Maeyama, (2016)])

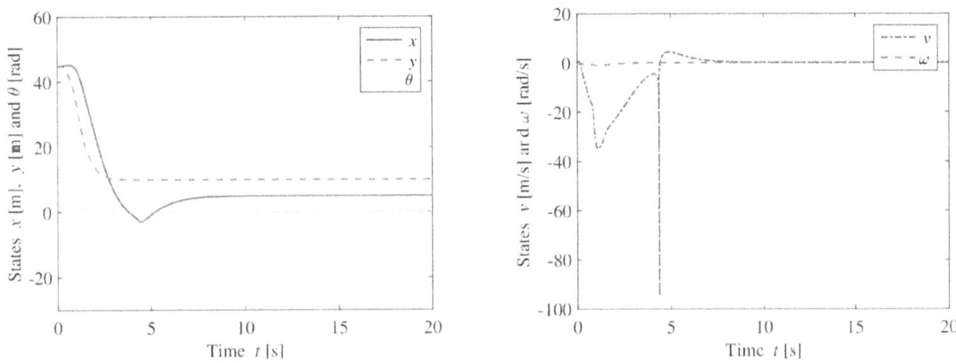

(a) Time response of the states x, y and θ

(b) Time response of the states v and ω

Figure 7. Comparison of the Quasi-continuous exponential stabilization and the proposal method
(Adapted from [Motonaka, Goto, Watanabe, and Maeyama, (2016)])

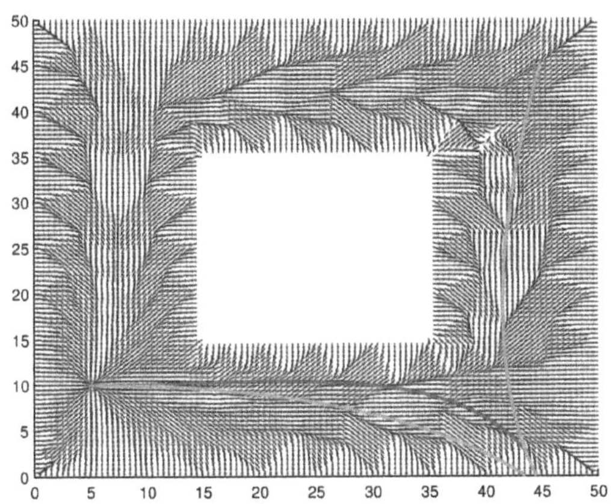

(a) Comparison of the course for the case A

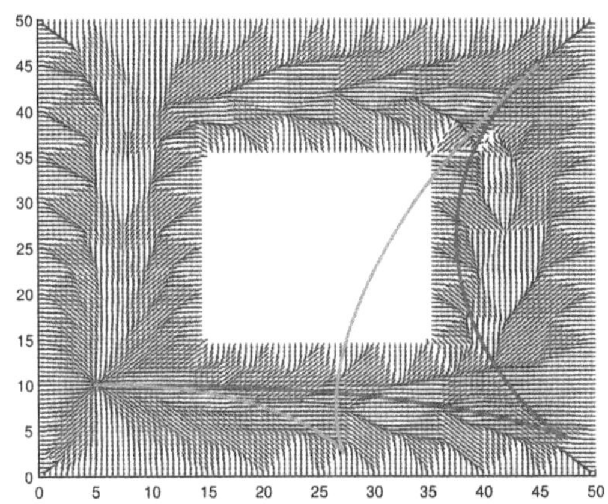

(b) Comparison of the course for the case B

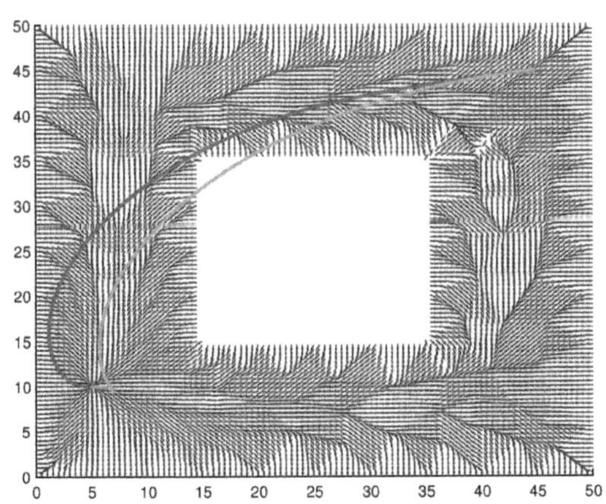

(c) Comparison of the course for the case C

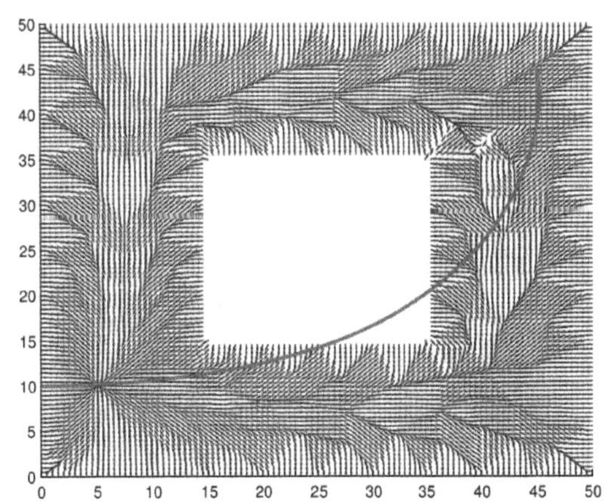

(d) Comparison of the course for the case D

HPF, and how to combine them. Finally, to verify the described control input, some simulations that to guide the robot to the target position were shown. From the simulations, it is confirmed that the robot can reach the target position while avoiding the obstacle in the almost case. However, sometimes the robot was not able to avoid the obstacle even using the kinodynamic control input. The solution of this case is researching still today.

REFERENCES

Donald, B., Xavier, P., Canny, J., & Reif, J. (1993). Kinodynamic motion planning. *Journal of the Association for Computing Machinery*, *40*(5), 1048–1066. doi:10.1145/174147.174150

Goto, T., Watanabe, K., Maeyama, S., & Motonaka, K. (2014). Stabilization control to the arbitrary target point for a two-wheeled drive mobile robot considering the dynamical model. In *Proceedings of SICE System and Information Division Annual Conference* (pp. 599-604). Academic Press. (in Japanese)

Hsu, D., Kindel, R., Latombe, J. C., & Rock, S. (2002). Randomized kinodynamic motion planning with moving obstacles. *The International Journal of Robotics Research*, *21*(3), 233–255. doi:10.1177/027836402320556421

Masoud, A. (2010). Kinodynamic motion planning. *IEEE Robotics & Automation Magazine*, *17*(1), 85–99. doi:10.1109/MRA.2010.935794

Motonaka, K., Goto, T., Watanabe, K., & Maeyama, S. (2016). Kinodynamic motion planning for a two-drive mobile robot using a harmonic potential field. In *Proceedings of Annual Conference of the IEEE Industrial Electronics Society*. IEEE. doi:10.1109/IECON.2016.7793160

Motonaka, K., Watanabe, K., & Maeyama, S. (2013). Kinodynamic motion planning and control using anisotropic damping forces. In *Proceedings of International Symposium on Artificial Life and Robotics*, (pp. 57–60). Oita, Japan: Academic Press. doi:10.1109/IROS.2013.6696989

Peng, J., & Akella, S. (2005). Coordinating multiple robots with kinodynamic constraints along specified paths. *The International Journal of Robotics Research*, *24*(4), 295–310. doi:10.1177/0278364905051974

Rimon, E., & Koditschek, D. (1992, October). Exact robot navigation using artificial potential functions. *IEEE Transactions on Robotics and Automation*, *8*(5), 501–518. doi:10.1109/70.163777

Tsiotras, P., Corless, M., & Longuski, J. M. (1993). Invariant manifold techniques for attitude control of symmetric spacecraft. In *Proceedings of 32th Conference on Decision and Control* (pp. 1470–1475). Academic Press. doi:10.1109/CDC.1993.325431

Watanabe, K., Yamamoto, T., Izumi, K., & Maeyama, S. (2010) Underactuated control for nonholonomic mobile robots by using double integrator model and invariant manifold theory. In *Proceedings of IEEE/RSJ International Conference on Intelligent Robots and Systems (IROS)* (pp. 2862-2867). IEEE. doi:10.1109/IROS.2010.5650951

Chapter 15

Discrimination of Dual-Arm Motions Using a Joint Posterior Probability Neural Network for Human-Robot Interfaces

Taro Shibanoki
Ibaraki University, Japan

Toshio Tsuji
Hiroshima University, Japan

ABSTRACT

This chapter describes a novel dual-arm motion discrimination method that combines posterior probabilities estimated independently for left and right arm movements, and its application to control a robotic manipulator. The proposed method estimates the posterior probability of each single-arm motion through learning using recurrent probabilistic neural networks. The posterior probabilities output from the networks are then combined based on motion dependency between arms, making it possible to calculate a joint posterior probability of dual-arm motions. With this method, all the dual-arm motions consisting of each single-arm motion can be discriminated through leaning of single-arm motions only. In the experiments performed, the proposed method was applied to the discrimination of up to 50 dual-arm motions. The results showed that the method enables relatively high discrimination performance. In addition, the possibility of applying the proposed method for a human-robot interface was confirmed through operation experiments for the robotic manipulator using dual-arm motions.

INTRODUCTION

Gestures include a variety of information no less than voice signals. Estimation of human intentions from them can be applied to the development of communication tools such as gesture translators and human-machine interfaces. However, it is quite difficult to perform accurate gesture discrimination due to the large variety of possible gestures that can be made.

DOI: 10.4018/978-1-5225-2993-4.ch015

This chapter outlines a novel dual-arm motion discrimination method and its application to human-robotic manipulator control using acceleration signals. With the proposed method, all dual-arm motions consisting of individual single-arm movements can be discriminated through just learning of single-arm motions by combining posterior probabilities estimated independently for left and right arm movements. The user can then control the robotic manipulator based on the proposed method using a variety of gestures.

BACKGROUND

Many researchers have tried to accurately measure and discriminate the biological signals generated by gestures using various types of sensors and discriminators. The results clearly indicate that the gestures can be used for purposes such as automatic sign-language recognition, and operation of machine control interfaces.

Gesture discrimination studies investigate either static gestures or dynamic gestures. The former are based on hand/arm shapes, and the latter involve time-series patterns of hand/arm motion. In the area of static gestures, a number of studies using a probabilistic neural network (PNN), hidden Markov model (HMM), etc. have been investigated to discriminate several hand shapes (such as the "V for victory" gesture) (Bowden *et al.*, 2002, Bailador, 2007, Okamoto *et al.*, 2008). These techniques can be used to robotic manipulator operations (Brethes *et al.*, 2004, Raheja *et al.*, 2010, Habib, 2011, Devine *et al.*, 2016, Fall *et al.*, 2017). Additionally, Bu *et al.* proposed a motion discrimination method based on the recurrent probabilistic neural network called a recurrent log-linearized Gaussian mixture network (R-LLGMN) for prosthetic control (Bu *et al.*, 2003, Tsuji *et al.*, 2003, 2006). The generic NN has some drawbacks such as a large-scale network structure and many learning iterations, so that some studies have investigated to integrate domain/task specific knowledge into the architecture of NN (Bridle, 1989, Specht, 1990, Richard and Lippmann, 1991, Caelli *et al.*, 1993). The proposed method integrated hidden Markov model (HMM) and Gaussian mixture model (GMM) into the architecture of NN, and achieves high discrimination performance, even with non-stationary EMG signals during continuous motion. In dynamic gesture discrimination, Liu *et al.* applied a template-matching method to classify gestures based on acceleration signals relating to three dimensions, then discriminated eight dynamic gestures with an accuracy level of 98% and developed a gesture-based interface for a cellular phone. Solís *et al.* also investigated the discrimination of 21 gestures with 93% accuracy using an artificial neural network based on RGB images while gestures were being made. However, these studies focused exclusively on single-arm motion discrimination (Liu *et al.*, 2009, Huang *et al.*, 2015, Solís *et al.*, 2016). As gestures include both left- and right-arm motion among their characteristics, it is necessary to discuss a discrimination method for dual-arm motion in addition to considering single-arm motion.

Sawada *et al.* also applied a dynamic programming (DP) matching method to classify gestures based on the characteristics of hand ACC signals, positions and finger joint angles calculated using ACC sensors and data gloves attached to the subject's left and right arms, and achieved discrimination of nine gestures involving dual-arm motion with 94% accuracy. Kim *et al.* also focused on and investigated the discrimination of 25 dual-arm dynamic gestures using a fuzzy min-max neural network based on hand trajectories and finger angles measured using data gloves. Here, since dual-arm motions have a wide variety of single-arm motion combinations, discriminating them becomes very complex, which in turn makes the learning time potentially excessive for the classifier. Increasing the number of motions also results in similar issues for considerations such as hand shapes and movement directions, which can reduce discrimination rates (Sawada *et al.*, 1998, Kim *et al.*, 1996, Vogler *et al.*, 1997, Zhang *et al.*, 2011, Wang *et al.*, 2016). To solve these problems, a novel method is required for accurate discrimination of a wide variety of dual-arm motions expressed with both left- and right-arm movements.

Against such background, the authors previously proposed a joint posterior probability neural network (J-PNN) that can be used to estimate the joint posterior probability of dual-arm motions based on the learning of individual single-arm motions, supposing that the dual-arm type involves a combination of left- and right-arm movements. However, only the discrimination accuracy for 15 dual-arm motions was verified in the previous study, and the method's validity for the discrimination of more complex motions such as those of sign language has not been investigated (Hiramatsu *et al.*, 2011, Shima *et al.*, 2013). Moreover, applications using a J-PNN for gesture-based human-machine interfaces were performed by relatively dynamic gestures. The proposed method has a great effect under the situation of restriction of the user's motions.

MAIN FOCUS OF THE CHAPTER

This chapter describes a novel approach for the human-machine interfaces based on a J-PNN and a method for dual-arm motion discrimination. The J-PNN consists of two PNNs, and can be used to estimate the joint posterior probability of dual-arm motions by combining the posterior probabilities of individual arm movements estimated independently using each PNN. The posterior probabilities of each arm are combined after modification based on the dependencies between individual arm motions. The network can thus discriminate all dual-arm motions consisting of individual single-arm motions based on learning of single-arm motions only, making it possible to increase the number of motions with no corresponding increase in the learning time or the number of learning samples required.

The rest of this chapter is organized as follows: the proposed neural network and the details of the dual-arm motion discrimination method based on the network are first explained, and an experiment that confirms the effectiveness of the proposed method is then described. Finally, the robot manipulator control experiments is outlined as an example of the application of the proposed method.

A JOINT POSTERIOR PROBABILITY NEURAL NETWORK (J-PNN)

Dual-arm motions are composed of a variety of each arm movement. Figure 1 shows the structure of the proposed recurrent probabilistic NN that estimates the joint posterior probability of dual-arm motions by combining posterior probabilities estimated independently for left and right arm motions (Hiramatsu *et al.*, 2011, Shima *et al.*, 2013). The network structure and the learning algorithm are outlined in the following paragraphs.

The network consists of the two parts of time series pattern discrimination and joint probability estimation. In the time series pattern discrimination stage, the posterior probability $p(c_\alpha \mid x_\alpha)$ for each class $\left(c_\alpha = 1,\ldots,C_\alpha; \alpha \in \{1,2\}\right)$ is estimated from time series vectors $x_\alpha(t)$. These probabilities are then combined to calculate the joint posterior probability after modification based on the dependencies in each input vector. The network therefore allows the estimation of posterior probability for all classes realized by combining the two input vectors. Firstly, the structure of the proposed network is detailed.

Figure 1. Structure of the joint posterior probability neural network (Hiramatsu et al., 2011, Shima et al., 2013). This network consists of time series pattern discrimination and joint probability estimation. Two vectors are input to the network individually, and a joint posterior probability is then calculated based on the motion dependency of each vector using entropy.

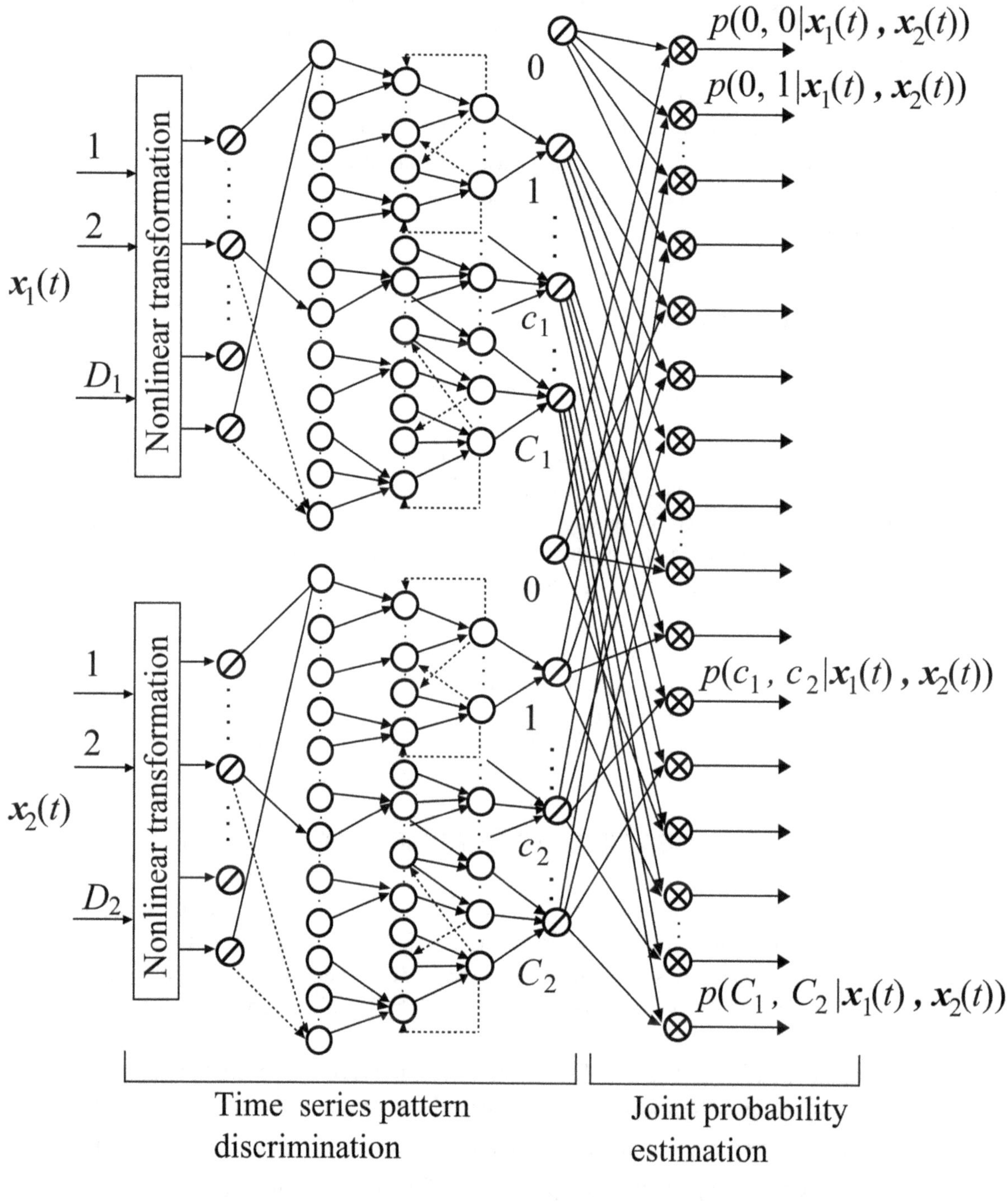

The time series pattern discrimination part consists of dual R-LLGMNs (a type of recurrent probabilistic neural network (Tsuji *et al.*, 2003, 2006)) (Hiramatsu *et al.*, 2011, Shima *et al.*, 2013). It has a Gaussian mixture model (GMM) and an HMM, and copes with the time-varying characteristics of input signals. Two feature vectors $x_\alpha(t) \in \Re^{D_\alpha} \left(\alpha \in \{1,2\}; t = 1,2,\ldots,T\right)$, where T is the time length of the input vector, are input to the network independently for discrimination of individual arm motions.

These input vectors $x_\alpha(t)$ are first preprocessed with a non-linear transformation and converted to form the modified input vector $X_\alpha(t)$ as follows:

$$\begin{aligned} X_\alpha(t) = \Big[& 1, x_\alpha(t)^T, x_{\alpha,1}(t)^2, x_{\alpha,1}(t)x_{\alpha,2}(t), \ldots, \\ & x_{\alpha,1}(t)x_{\alpha,D_\alpha}(t), x_{\alpha,2}(t)^2, x_{\alpha,2}(t)x_{\alpha,3}(t), \ldots, \\ & x_{\alpha,2}(t)x_{\alpha,D_\alpha}(t), \ldots, x_{\alpha,D_\alpha}(t)^2 \Big]^T \end{aligned} \tag{1}$$

This conversion is performed to represent a normalized distribution corresponding to each component of the GMM as weight coefficients of the R-LLGMN (Tsuji *et al.*, 2003, 2006).

The first layer of the network consists of $U_\alpha = 1 + D_\alpha(D_\alpha + 3)/2$ units, which correspond to the dimensions of the input vector $X_\alpha(t) \in \Re^{U_\alpha}$, and the identity function is used for the activation function of each unit.

Units:

$$\{c_\alpha, k_\alpha, k'_\alpha, m_\alpha\}\left(c_\alpha = 1,\ldots,C_\alpha k_\alpha, k'_\alpha = 1,\ldots,K_{\alpha,c}; m_\alpha = 1,\ldots,M_{\alpha,c_\alpha,k_\alpha}\right)$$

in the second layer receives the outputs of the first layer, multiplied by weight ${}^{(2)}w^{c}_{\pm,k'_\alpha,k_\alpha,m_\alpha,u_\alpha}$. The relationship between the input and the output of the second layer is defined as:

$${}^{(2)}I^{c_\alpha}_{\alpha,k'_\alpha,k_\alpha,m_\alpha}(t) = \sum_{h_\alpha=1}^{U_\alpha} {}^{(1)}O_{\alpha,u_\alpha}(l)\, w^{c_\alpha}_{\alpha,k'_\alpha,k_\alpha,m_\alpha,u_\alpha} \tag{2}$$

$${}^{(2)}O^{c_\alpha}_{\alpha,k'_\alpha,k_\alpha,m_\alpha}(t) = \exp\left({}^{(2)}I^{c_\alpha}_{\alpha,k'_\alpha,k_\alpha,m_\alpha}(t)\right) \tag{3}$$

where C_α is the number of classes, $K_{\alpha,c}$ is the number of states in the HMM, and $M_{\alpha,c_\alpha,k_\alpha}$ is the number of components of the Gaussian mixture distribution in class c_α and state k_α. Calculating the probabilities of the input patterns, the exponential function is used (Tsuji *et al.*, 2003, 2006).

The outputs of units $\{c_\alpha, k_\alpha, k'_\alpha, m_\pm\}$ in the second layer are summed and input into a unit $\{c_\alpha, k_\alpha, k'_\alpha\}$ in the third layer. The output of the fourth layer is also fed back to the third layer. The activation function in the third layer is described as:

$$^{(3)}I^{c_\alpha}_{\alpha,k_\alpha^{'},k_\alpha}(t)=\sum_{m_\alpha=1}^{M_{\alpha,c_\alpha,k_\alpha}}{}^{(2)}O^{c_\alpha}_{\alpha,k_\alpha^{'},k_\alpha,m_\alpha}(t) \tag{4}$$

$$^{(3)}O^{c_\alpha}_{\alpha,k_\alpha^{'},k_\alpha}(t)={}^{(4)}O^{c_\alpha}_{\alpha,k_\alpha^{'}}(t-1)\,{}^{(3)}I^{c_\alpha}_{k_\alpha^{'},k_\alpha}(t) \tag{5}$$

where ${}^{(4)}O^{c_\alpha}_{k_\alpha^{'}}(0)=1.0$ is for the initial state.

Unit $\{c_\alpha,k_\alpha\}$ in the fourth layer receives the integrated outputs of unit $\{c_\alpha,k_\alpha,k'_\alpha\}$ in the third layer. The output ${}^{(4)}O^{c_\alpha}_{\alpha,k_\alpha}(t)$ is each input ${}^{(4)}I^{c_\alpha}_{\alpha,k_\alpha}(t)$ in the fourth layer divided by the sum of all inputs in the fourth layer. The relationships in the fourth layer are defined as:

$$^{(4)}I^{c_\alpha}_{\alpha,k_\alpha}(t)=\sum_{k_\alpha^{'}=1}^{K_{\alpha,c}}{}^{(3)}O^{c_\alpha}_{\alpha,k_\alpha^{'},k_\alpha}(t) \tag{6}$$

$$^{(4)}O^{c_\alpha}_{\alpha,k_\alpha}(t)=\frac{{}^{(4)}I^{c_\alpha}_{\alpha,k_\alpha}(t)}{\sum_{c_\alpha^{'}=1}^{C_\alpha}\sum_{k_\alpha^{'}=1}^{K_{\alpha,c_\alpha^{'}}}{}^{(4)}I^{c_\alpha^{'}}_{k_\alpha^{'}}(t)} \tag{7}$$

Finally, unit c in the fifth layer integrates the outputs of $K_{\alpha,c}$ units $\{c_\alpha,k_\alpha\}(k_\alpha=1,\ldots,K_{\alpha,c})$ in the fourth layer. The relationships in this layer are defined as:

$$^{(5)}O^{c_\alpha}_{\alpha}(t)={}^{(5)}I^{c_\alpha}(t)=\sum_{k_\alpha=1}^{K_{\alpha,c}}(4)O^{c_\alpha}_{\alpha,k_\alpha}(t) \tag{8}$$

$$^{(5)}O^{c_\alpha}_{\alpha}(t)={}^{(5)}I^{c_\alpha}(t) \tag{9}$$

Here, ${}^{(5)}O^{c_\alpha}_{\alpha}(t)$ indicates posterior probability $p(c_\alpha\,|\,x_\alpha)(\alpha\in\{1,2\};c_\alpha=1,\ldots,C_\alpha)$ for class c_α. After optimizing the weight coefficients ${}^{(2)}w^{c}_{\pm,k_\alpha^{'},k_\alpha,m_\alpha,u_\alpha}$ between the first layer and the second layer, the R-LLGMN can estimate the posterior probability of each class for each arm.

In the J-PNN, the calculations in the third and fourth layers are associated with feedback connections, so time-varying features in the time series can be used. The R-LLGMN can therefore be adopted to model the posterior probability of each class through GMM-based learning. The joint posterior probability is then calculated using each class' posterior probability (Hiramatsu *et al.*, 2011, Shima *et al.*, 2013).

The joint probability estimation part estimates joint posterior probability $p(c_1,c_2\,|\,x_1,x_2)$ as:

$$p\left(c_1,c_2\,|\,x_1(t),x_2(t)\right)=\prod_{\alpha\in\{1,2\}}\frac{\eta_{\alpha,c_\alpha}p\left(c_\alpha\,|\,x_\alpha(t)\right)}{\sum_{c_\alpha=0}^{C_\alpha}\eta_{\alpha,c_\alpha}p\left(c_\alpha\,|\,x_\alpha(t)\right)} \tag{10}$$

Here, $c_\alpha = 0,1,\ldots,C_\alpha$, where $c_\alpha = 0$ is defined as a complementary event class, and the modifying coefficient of posterior probability η_{α,c_α} calculated based on entropy H_α, which indicates the ambiguity of each posterior probability, is defined as:

$$\eta_{\alpha,c_\alpha} = \begin{cases} 1 - H_\alpha & \left(H_\alpha > H_{th} \mid (c_1,c_2) \in \Xi\right) \\ 1 & (otherwise) \end{cases} \tag{11}$$

$$H_\alpha = -\frac{1}{\log C_\alpha} \sum_{c_\alpha=0}^{C_\alpha} p\left(c_\alpha \mid x_\alpha(t)\right) \log p\left(c_\alpha \mid x_\alpha(t)\right) \tag{12}$$

where ž represents a set consisting of a combination of two class types (c_1, c_2), for which one class may influence the discrimination result of the other, and is defined in advance (these are referred to as *dependent classes*). This means that Eq. (11) modifies the posterior probability of the class belonging to dependent class ž when individual posterior probabilities are ambiguous. Using Eqs. (10) and (11), the J-PNN can be adopted to estimate joint posterior probability based on the dependency between the classes.

Further, $p(0 \mid x_\alpha)$ is a complementary event probability, and is defined as:

$$p\left(0 \mid x_\alpha(t)\right) = \begin{cases} 1 & \left(x_\alpha(t) \in \{\phi\}\right) \\ 0 & \left(x_\alpha(t) \notin \{\phi\}\right) \end{cases} \tag{13}$$

Generally, the PNN cannot calculate posterior probability without input (i.e., a vector with no input: $x_\alpha \in \{\phi\}$). The J-PNN outputs complementary event probability $p(0 \mid x_\alpha) \in \{0.1\}$ according to the state of input vector x_α. If an input vector is empty $(x_\alpha \in \{\phi\})$, the complementary event probability becomes 1. According to Eqs. (10) – (13) then, the J-PNN directly outputs the posterior probability for the other input. On the other hand, if an input vector is not empty $(x_\alpha \notin \{\phi\})$, the complementary event probability becomes zero. The J-PNN can therefore calculate joint posterior probability even if an input vector is empty. The number of discrimination classes J is thus $J = (C_1+1)(C_2+1)-1$. The next paragraph explains the learning procedure of the proposed network.

The PNN modifies the weight to minimize error between teacher vectors and output signals of the network in the supervised learning process. In the case of dual-arm motion discrimination, however, since the relationships between each arm movement and these combinations must be learned in the network, excessive time is required for measuring learning samples and training of the network. Therefore, by learning for left and right arm motions independently, for example, the network can estimate a dual-arm motion's probability. Against such a background, the learning of the J-PNN can be completed based on learning for each network (Hiramatsu *et al.*, 2011, Shima *et al.*, 2013). Here, a set of vectors $x_\alpha(t)^{(n)}\ (n = 1,2,\ldots,N_\alpha; t = 1,2,\ldots,T_\alpha)$ andteachervectors $P_\alpha^{(n)} = \left[P_1^{(n)},\ldots,P_{c_\alpha}^{(n)},\ldots,P_{C_\alpha}^{(n)}\right]^T (\alpha \in \{1,2\})$

are given for the training of the J-PNN. If the vector $x_\alpha(t)^{(n)}$ is set for class c_α, then $P_{c_\alpha}^{(n)} = 1$, and $P_{\hat{c}_\alpha}^{(n)} = 0\left(\hat{c}_\alpha \neq c_\alpha\right)$ for all other classes in this subset. The energy function for the network is defined as:

$$E_\alpha = -\sum_{n=1}^{N_\alpha} \sum_{c_\alpha=1}^{C_\alpha} P_{c_\alpha}^{(n)} \log {}^{(5)}O^{(c_\alpha)}(T_\alpha)^{(n)} \tag{14}$$

and learning is performed to minimize E_α in order to maximize likelihood. Here, ${}^{(5)}O^{(c_\alpha)}(T_\alpha)^{(n)}$ is the posterior probability at given time T_α for time series patterns $x_\alpha(t)^{(n)}$.

Due to the recurrent connection in the J-PNN, the backpropagation through time (BPTT) algorithm is used, and the dynamic of the terminal attractor (TA) is incorporated in the learning rule in order to regulate the convergence time of learning.

DUAL-ARM MOTION DISCRIMINATION METHOD

The structure of the discrimination method for dual-arm motions is shown in Figure 2 (Hiramatsu *et al.*, 2011, Shima *et al.*, 2013). This approach involves first extracting the features of individual arm motions from ACC signals and estimating motion from feature vectors using the J-PNN.

Three-axis ACC sensors are used to measure the acceleration signals of dual-arm motions. The sensors can be attached to any sites in both arms. The signals measured are digitized using an A/D converter and defined as $a_m^v(t)\left(m \in \{r, l\}, v \in \{x, y, z\}\right)$. Here, *r* and *l* represent the indexes of the right and left hands, respectively, and *x*, *y* and *z* are the axis indexes of the acceleration sensors. As the frequency components of motion acceleration signals $g_m^v(t)$ are low, they can be separated in the frequency region using a digital Butterworth low-pass filter (cut-off frequency: f_c (Hz)). The $g_m^v(t)$ values are then normalized to make the maximum value equal to 1 as follows:

$$\hat{g}_m^v(t) = \frac{g_m^v(t)}{\max g_m^v} \tag{15}$$

where $\max g_m^v$ is the maximum value of $g_m^v(t)$ measured previously. The magnitude of vector $\hat{g}_m(t) = \left[g_m^x(t), g_m^y(t), g_m^z(t)\right]^T$ is then calculated as:

$$\hat{g}_m(t) = \sqrt{\hat{g}_m^x(t)^2 + \hat{g}_m^y(t)^2 + \hat{g}_m^z(t)^2} \tag{16}$$

and it is used as a movement detection index.

Figure 2. Overview of the discrimination method for dual-arm motions using the J-PNN (Hiramatsu et al., 2011, Shima et al., 2013). Each arm motion is measured using 3D acceleration sensors, and extracted features are input to J-PNN. J-PNN can discriminate dual-arm motion learning each single-arm motion only.

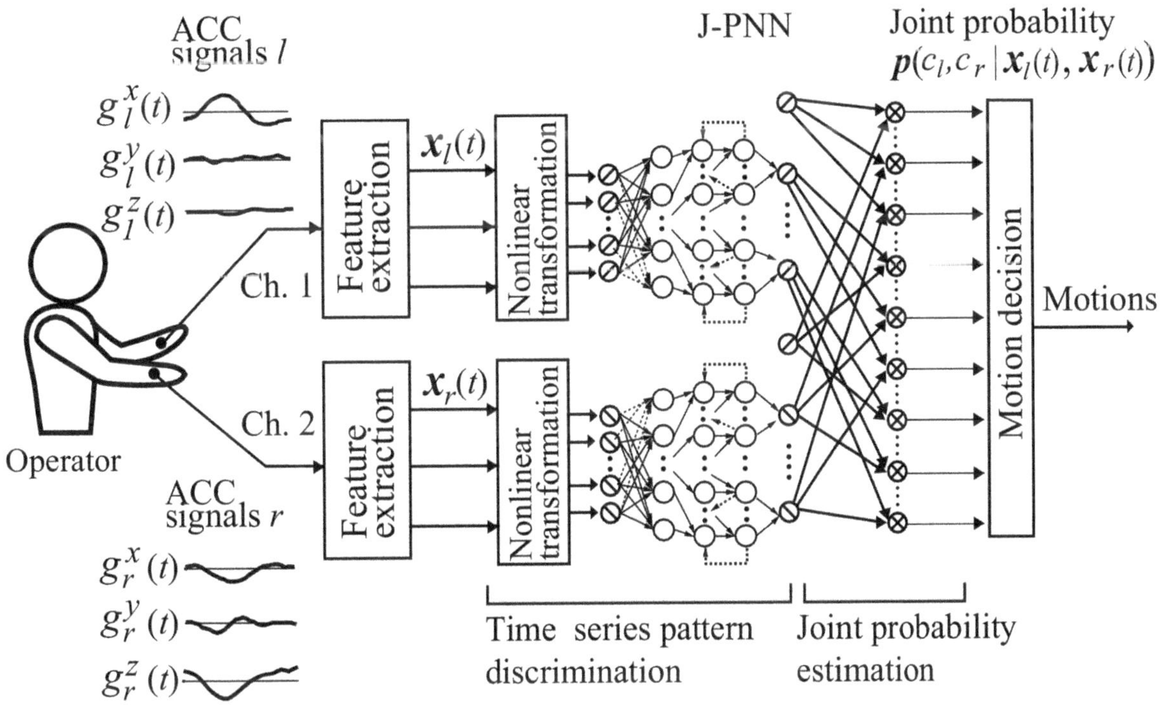

In order to discriminate dual-arm motions, the movement durations are determined. An example of the movement detection duration is shown in Figure 3. The point at which $\hat{g}_m(t)$ first exceeds $\hat{g}_{\text{start}}$ (the motion start threshold) is defined as gesture start time t_{start}, and the initial instance at which it is less than $\hat{g}_{\text{end}}$ (the motion end threshold) during T_{end} (s) is defined as motion end time t_{end}. The time interval from the first to the last local maximum values (first time $\hat{g}_m(t+"t)>\hat{g}_m(t)$ and last time $\hat{g}_m(t-"t)<\hat{g}_m(t)$) in the duration of $(t_{\text{start}}, t_{\text{end}})$ is defined as the movement duration $(t_{\text{start}}, t_{\text{end}})$, and values of $\hat{g}_m(t)\in\Re^3$ in this duration are then extracted as discrimination data.

The data are also resampled so that the number of samples is S since it differs in each trial. The discrimination datum $\hat{g}_m(t)$ whose maximum is normalized to a value of 1 is computed and defined as feature pattern vector $x_m(t)(m\in\{r,l\})$. The feature pattern vectors for each arm motion $x_m(t)$ and $x_{m'}(t)(m'\in\{r,l\}\neq m)$ are input to the J-PNN, and the network then estimates the dual-arm motion posterior probabilities. However, since dual-arm motion involves a time lag between individual arm movements, $x_m(t)$ is set as $x_{m'}(t)=\{\phi\}$ if $x_{m'}(t)$ is not input within the T (s) interval after $x_m(t)$ is input to the J-PNN (Hiramatsu *et al.*, 2011, Shima *et al.*, 2013).

Figure 3. An example of gesture detection results. Motion duration (shaded area) is determined from the root square sum of acceleration signals $\hat{g}_m(t)$ *(Hiramatsu et al., 2011, Shima et al., 2013).*

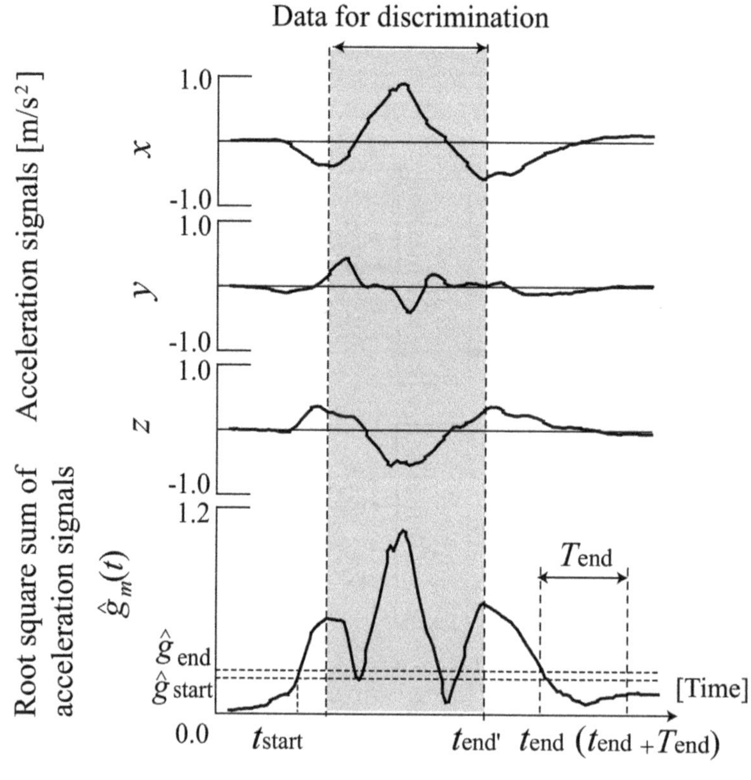

As it is difficult to move each arm at the same time asymmetrically (Guiard 1987), symmetric dual-arm motions are taken as a dependent class of dual-arm motions. In this chapter, therefore, combinations of individual arm motions $(c_r, c_l \mid \arg\max p(c_r \mid x_r)) = \arg\max p(c_l \mid x_l)$ are defined as the factor of the set ž . The class with the maximum value of joint probability $p(c_r, c_l \mid x_r, x_l)$ is thus taken as the discrimination result.

DUAL-ARM MOTION DISCRIMINATION EXPERIMENTS

Gesture discrimination experiments were performed to verify the effectiveness of the J-PNN by dealing with sign language using dual-arm motions. Here, it is known that some Japanese sign languages can be expressed combining ten basic motions as shown in Figure 4 (Sawada *et al.*, 1998). In this chapter, therefore, dual-arm motions combining these basic motions are focused as the discrimination motions.

Figure 5 depicts the hardware components of the proposed discrimination system. Three-axis acceleration sensors (Micro Stone Corporation: MA3-04Ac) and wireless motion recorders (Micro Stone Corporation: MVP-RF) were utilized to measure acceleration signals. Each sensor was attached to both of the subjects' wrists using surgical tapes. The sampling frequency of ACC measurement was 200 (Hz), and the parameters for gesture discrimination were set as f_c = 10.0 (Hz), $\hat{g}_{start}$ = 0.2, $\hat{g}_{end}$ = 0.25,

Figure 4. Dual-arm motions discriminated in the experiments

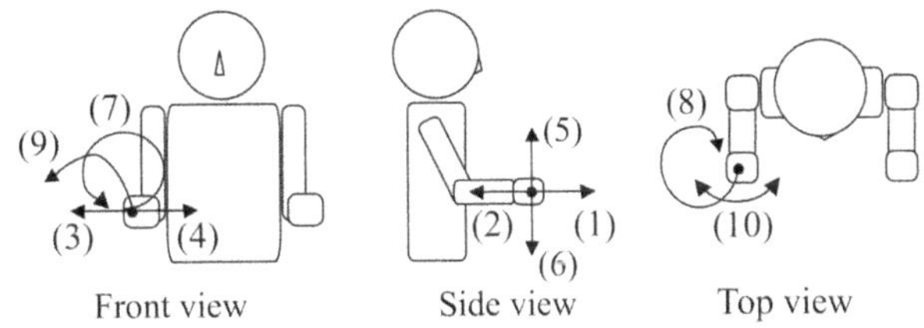

(1): Forward (2): Backward (3): Outward (4): Inward
(5): Upward (6): Downward (7): Circle (8): Horizontal circle
(9): Arch (10): Wave

Figure 5. Hardware components of the proposed dual-arm motion discrimination method. Acceleration signals are measured from both arms and used for classification of dual-arm motions (sample).

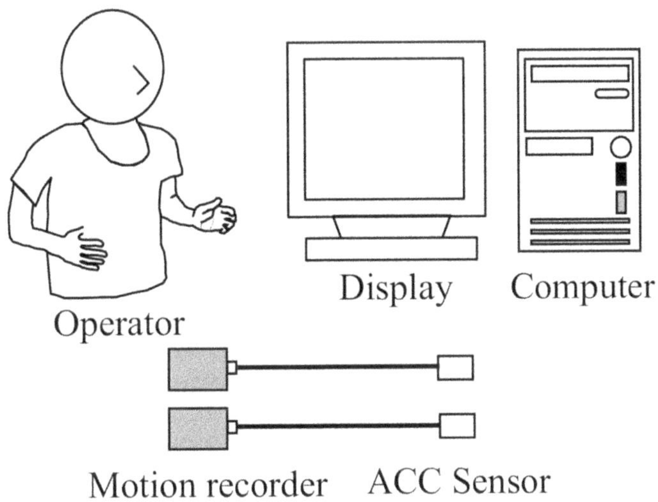

$T_{\text{end}} = 0.15$ (s), $T = 0.4$ (s) and $S = 30$. These parameters were determined in line with the results of preliminary experiments. Informed consent was obtained from all the subjects.

ACC signals measured from seven healthy subjects (males, A-G, average 22 ± 0.0 years old) including three subject's data described in (Hiramatsu *et al.*, 2011, Shima *et al.*, 2013) (A-C, 23 years old). First of all, to verify the basic discrimination performance of the proposed method, the five healthy male subjects performed 15 dual-arm motions ($M_1 - M_{15}$, see Table 1) involving a combination of simple three single-arm motions ((1) forward, (3) outward, (5) upward shown in Figure 4) 50 times. Further, on the basis that sign language recognition is realized by above ten basic motions, 50 dual-arm motions are chosen from 120 motions expressed by combining basic motions (i.e., ten motions for each arm). Subjects F and G were asked to conduct the 50 dual-arm motions 20 times. One data set chosen randomly (defined as the training data) was used for J-PNN training, and the other data are defined as the discrimination data. Average discrimination rates were calculated by changing training and discrimination data. In the 15 dual-arm motion discrimination experiments, 50 sets of each motion data were separated into 10 training data sets and 40 discrimination data sets. It should be noted that the J-PNN

Table 1. Details of motions used in dual-arm motion discrimination experiments. (a) 15 dual-arm motions are consisting of 3 motions; and (b) 50 dual-arm motions are of 10 motions for each arm motion.

(a) 15 Dual-Arm Motions															
Motion	$\mathbf{M}_1$	$\mathbf{M}_2$	$\mathbf{M}_3$	$\mathbf{M}_4$	$\mathbf{M}_5$	$\mathbf{M}_6$	$\mathbf{M}_7$	$\mathbf{M}_8$	$\mathbf{M}_9$	$\mathbf{M}_{10}$	$\mathbf{M}_{11}$	$\mathbf{M}_{12}$	$\mathbf{M}_{13}$	$\mathbf{M}_{14}$	$\mathbf{M}_{15}$
Right	NM	NM	NM	(1)	(3)	(5)	(1)	(3)	(5)	(1)	(3)	(5)	(1)	(3)	(5)
Left	(1)	(3)	(5)	NM	NM	NM	(1)	(1)	(1)	(3)	(3)	(3)	(5)	(5)	(5)
(b) 50 Dual-Arm Motions															
Motion	$\mathbf{M}_1$	$\mathbf{M}_2$	$\mathbf{M}_3$	$\mathbf{M}_4$	$\mathbf{M}_5$	$\mathbf{M}_6$	$\mathbf{M}_7$	$\mathbf{M}_8$	$\mathbf{M}_9$	$\mathbf{M}_{10}$	$\mathbf{M}_{11}$	$\mathbf{M}_{12}$	$\mathbf{M}_{13}$	$\mathbf{M}_{14}$	$\mathbf{M}_{15}$
Right	NM	NM	NM	NM	NM	NM	NM	NM	NM	NM	(1)	(2)	(3)	(4)	(5)
Left	(1)	(2)	(3)	(4)	(5)	(6)	(7)	(8)	(9)	(10)	NM	NM	NM	NM	NM
Motion	$\mathbf{M}_{16}$	$\mathbf{M}_{17}$	$\mathbf{M}_{18}$	$\mathbf{M}_{19}$	$\mathbf{M}_{20}$	$\mathbf{M}_{21}$	$\mathbf{M}_{22}$	$\mathbf{M}_{23}$	$\mathbf{M}_{24}$	$\mathbf{M}_{25}$	$\mathbf{M}_{26}$	$\mathbf{M}_{27}$	$\mathbf{M}_{28}$	$\mathbf{M}_{29}$	$\mathbf{M}_{30}$
Right	(6)	(7)	(8)	(9)	(10)	(1)	(2)	(3)	(4)	(5)	(6)	(7)	(8)	(9)	(10)
Left	NM	NM	NM	NM	NM	(1)	(2)	(3)	(4)	(5)	(6)	(7)	(8)	(9)	(10)
Motion	$\mathbf{M}_{31}$	$\mathbf{M}_{32}$	$\mathbf{M}_{33}$	$\mathbf{M}_{34}$	$\mathbf{M}_{35}$	$\mathbf{M}_{36}$	$\mathbf{M}_{37}$	$\mathbf{M}_{38}$	$\mathbf{M}_{39}$	$\mathbf{M}_{40}$	$\mathbf{M}_{41}$	$\mathbf{M}_{42}$	$\mathbf{M}_{43}$	$\mathbf{M}_{44}$	$\mathbf{M}_{45}$
Right	(5)	(1)	(6)	(1)	(8)	(1)	(4)	(3)	(7)	(3)	(9)	(3)	(5)	(4)	(9)
Left	(1)	(5)	(1)	(6)	(1)	(8)	(3)	(4)	(3)	(7)	(3)	(9)	(4)	(5)	(4)
Motion	$\mathbf{M}_{46}$	$\mathbf{M}_{47}$	$\mathbf{M}_{48}$	$\mathbf{M}_{49}$	$\mathbf{M}_{50}$										
Right	(4)	(6)	(5)	(8)	(7)										
Left	(9)	(5)	(6)	(7)	(8)										

NM: No motion

can discriminate all dual-arm motions by learning just each single-arm motion (three or ten motions for each arm).

Here, since short time learning is required to train the network when applying the proposed method to gesture-based human-machine interface, the learning time (sum of the measurement time of learning samples and the network training time) is measured and discussed. The measurement time of learning samples is defined as the time until measurement of all learning motions with sensors attached.

Relationships between discrimination rates and entropy threshold H_{th} were also investigated. Here, the modifying coefficient of posterior probability w_{α,c_α} is set as $w_{\alpha,c_\alpha} = 1 = 1$ in Eq. (10), which means the motions of each arm are independent, Eq. (10) is rewritten as:

$$p'\left(c_1, c_2 \mid x_1(t), x_2(t)\right) = \prod_{\alpha \in \{1,2\}} p\left(c_\alpha \mid x_\alpha(t)\right) \tag{17}$$

Therefore, ratio of the number of misdiscriminated data › using Eq. (17) and the number of correctly discrimination data Ψ using Eq. (10) was calculated and defined as *the improvement rates* $\frac{\Psi}{\Lambda}$.

Moreover, multi-layered perceptron (MLP) neural network, which is one of the famous NN, and the single R-LLGMN proposed by Tsuji *et al.* (Tsuji *et al.*, 2003) were applied in the comparison experiments in order to demonstrate the discrimination abilities of the J-PNN for gesture recognition. The MLP had 12 layers (10 of them hidden), and the units of the input layer, each hidden layer were 180 (i.e., 3 dimensions × time length S; 3×30×2), and 200, respectively. The number of units of the output layer in the cases of 15 and 50 motions discrimination were 15 and 50, respectively. Sigmoid functions were employed as the output function of each unit. The back propagation algorithm was introduced, its learning rate was set as 0.01, and the error threshold was set as 0.01 for learning. As mentioned above, R-LLGMN is used in time series pattern discrimination of the J-PNN, and has high performances for discrimination using biological signals EMGs and EEGs. The R-LLGMN consists of an input layer including 6 units and an output layer including 15 and 50 units, respectively. The other parameters were the same as those of the J-PNN. It should be noted that 6 dimension vectors consisted by two feature vectors extracted from ACC signals of each arm are input to MLP and R-LLGMN. The computer used in the experiments had an Intel Core (TM) i7 processor, a clock frequency of 3.3 (GHz) and 3.0 (GB) of memory.

Figure 6 shows examples of discrimination scenes of dual-arm gestures (M_7 and M_8) using the J-PNN. Examples of discrimination results when the subject conducted each dual-arm motion in order $M_1 - M_{15}$ are also shown in Figure 7, which plots acceleration signals $g_m^v(t)$ measured, the root square sum of acceleration signals $\hat{g}_m(t)$, and discrimination results. The shaded area describes the no motion duration. It can be seen that dual-arm motions were discriminated accurately using the J-PNN.

Ratio of improvement for the motions, which have dependency between each arm motions (i.e., misdiscriminated data using Eq. (17)), using Eq. (10) is shown in Figure 8. The figure represents the improvement rates of result according to the threshold H_{th} changing from 0.0 to 1.0 with 0.1 increments. From the figure, the discrimination rate is same in $H_{\text{th}} \leq 0.5$, therefore, $H_{\text{th}} = 0.5$ is effective for probability modification based on motion dependency. Experiments in this chapter were performed with $H_{\text{th}} = 0.5$.

The accuracy of discrimination for each dual-arm motion is shown in Figure 9, which plots the average discrimination rates for each motion while the set of training data and discrimination data were changed 200 times. Average discrimination rate was 99.34 ± 0.56% and those of unlearned dual-arm motions $(M_7 - M_{15})$ by J-PNN was 98.96 ± 1.60%.

Here, the motion with the lowest discrimination rate was M_{13} (i.e., moving the left hand in the y-axis direction and the right hand in the x-axis direction). Examples of feature vectors in motion M_{13} and M_{15} (i.e., moving the left hand in the y-axis direction and the right hand in the y-axis direction) are shown in Figure 10.

Details of discrimination results for each subject are shown in Table 2 (a) and (b) plot discrimination rates for 15 and 50 motions discrimination, respectively. In the case of 15 motions discrimination, discrimination rate is high, in decreasing order, that J-PNN, R-LLGMN and MLP (see Table 2 (a)). Average discrimination rates for MLP, R-LLGMN and J-PNN were 90.07 ± 3.80%, 62.03 ± 5.59% and 98.11 ± 1.29%, respectively. On the other hand, in the case of 50 dual-arm motions discrimination, average discrimination rates for each method were 54.04 ± 39.71% (MLP), 87.38 ± 19.73% (R-LLGMN) and 86.95 ± 19.92% (J-PNN) (see Table 2 (b)).

Figure 6. Examples of discrimination scenes using the proposed method. Discrimination result is shown in display.

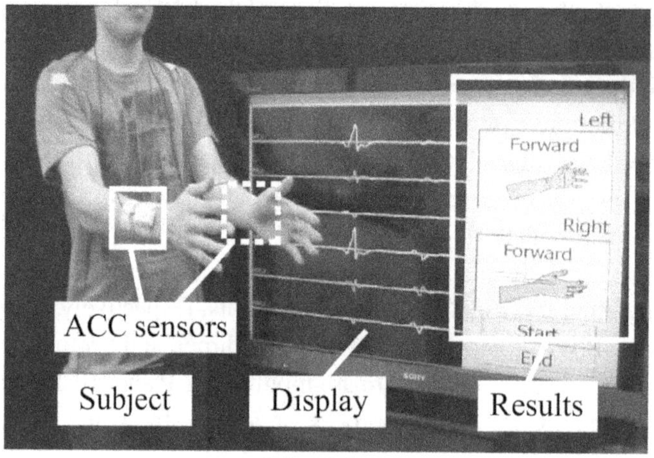

(a) : M 7

(b) : M 8

Figure 11 shows the learning time (sum of the measurement time of learning samples and the network learning time) for all subjects using each method ((a) 15 motions discrimination, (b) 50 motions discrimination). The measurement time of learning samples for both MLP and RLLGMN was 16.60 ± 1.77 s, and J-PNN was 3.88 ± 0.23 s in the 15 motions discrimination. In the case of 50 motions discrimination, it was 93.77 ± 2.25 s and 16.99 ± 1.05 s, respectively. Moreover, network learning time method in 15 motions discrimination for MLP, R-LLGMN and J-PNN was 804.30 ± 6.13, 87.49 ± 0.20 s and 1.18 ± 0.16 s, and in 50 motions discrimination was 16084.24 ± 944.02 s, 3125.06 ± 8.65 s and 21.29 ± 0.04 s, respectively.

From the Figures 6 and 7, it can be seen that the features of acceleration signals are different from those of other motions, and that they were discriminated accurately using the J-PNN. Unlearned dual-arm motions $\left(M_7 - M_{15}\right)$ could also be discriminated accurately, indicating that the network can be

Figure 7. An example of ACC signals and discrimination results (Subject A). The subject conducted each dual-arm motion in order $M_1 - M_{15}$.

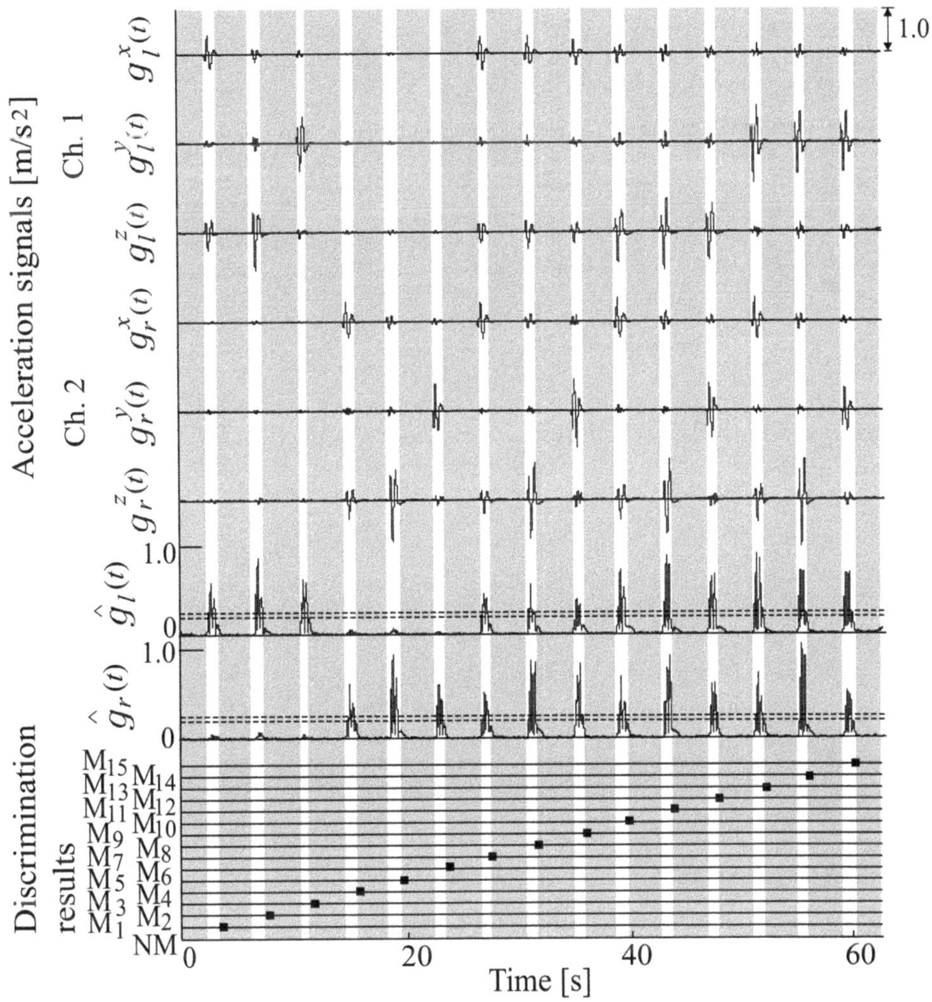

Figure 8. Ratio of improvement in misdiscrimination data for each Hth (Subject A). The graph shows that improvement rates become small at 0.6. Discrimination experiments in this chapter were performed with $H_{\text{th}} = 0.5$.

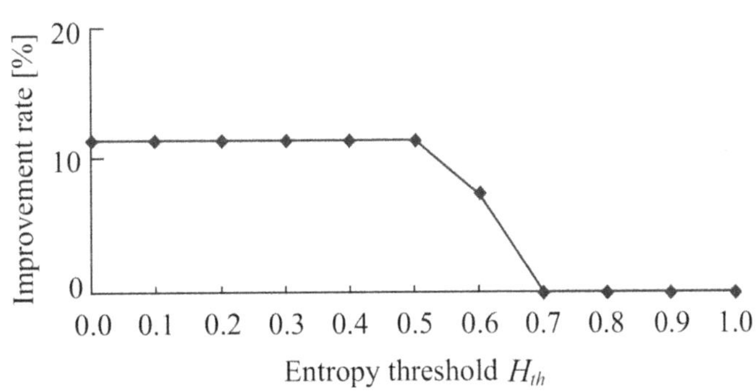

Figure 9. Discrimination rates for each 15 dual-arm motion (Subject A). The average discrimination rate for all dual-arm motion was 99.34 ± 3.56%.

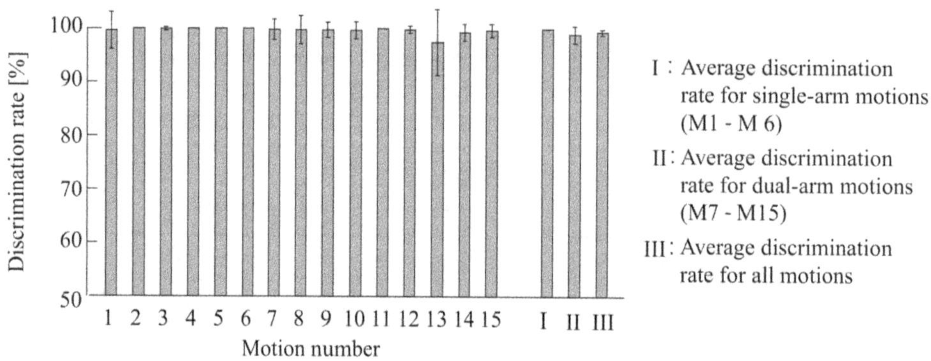

Figure 10. Examples of feature vectors for dual-arm motions (Subject A). M15 is symmetric motion, and M13 is asymmetric motion.

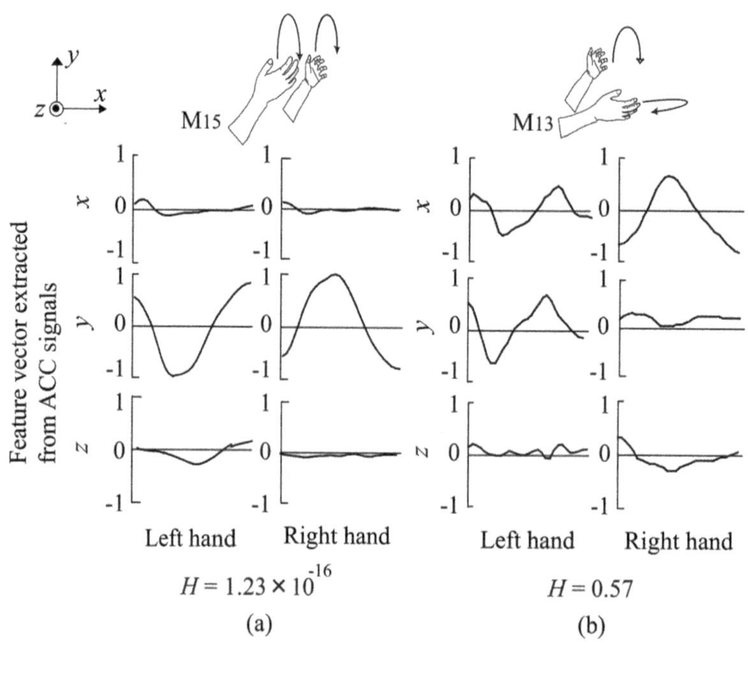

used to estimate the joint posterior probability of dual-arm motions with learning for single-arm motions only.

From Figure 8, improvement rates become small at 0.6, and its indicated that the number of improved motions, which can discriminate by probability modification (i.e., using Eq. (10)) in dependent motions, is decreased since the number of modified motion (i.e., 138 motions) is decreased by increasing the value of H_{th}. From this result, discrimination experiments in this chapter were performed with H_{th} =0.5 for all subjects. However, noted that determination method of optimal entropy threshold H_{th} for individual subjects is required to improve the discrimination rates.

Table 2. Average discrimination rates for 15 and 50 dual-arm motions using MLP, R-LLGMN (Tsuji et al., 2003) and J-PNN for all subjects. The results of each method are (a) 62.03 ± 5.59%, 90.07 ± 3.80% and 98.12 ± 1.29% for 15 dual-arm motions and (b) 53.26 ± 17.96%, 87.54 ± 17.17% and 87.09 ± 18.10% for 50 dual-arm motions.

(a) Results of Discrimination for 15 Dual-Arm Motions			
Subject	**A**	**B**	**C**
MLP	60.68 ± 29.39	56.34 ± 20.39	62.04 ± 15.85
R-LLGMN	93.90 ± 6.61	94.33 ± 5.34	87.97 ± 9.13
J-PNN	99.64 ± 0.56	99.03 ± 2.91	98.03 ± 4.38
Subject	**D**	**E**	
MLP	60.12 ± 26.97	60.41 ± 26.66	
R-LLGMN	85.96 ± 17.68	88.17 ± 13.30	
J-PNN	96.03 ± 4.19	98.15 ± 4.02	
(b) Results of Discrimination for 50 Dual-Arm Motions			
Subject	**F**	**G**	
MLP	53.30 ± 18.03	54.04 ± 39.71	
R-LLGMN	88.13 ± 18.14	86.95 ± 19.92	
J-PNN	86.79 ± 18.39	87.38 ± 19.73	

$Ave. \pm S.D.[\%]$

Figure 9 indicates that the J-PNN has high discrimination performance for the dual-arm motion discrimination, since J-PNN could discriminate dual-arm motions including unlearned motions accurately through learning each arm motion only. Moreover, waveforms of feature vector of M_{15} were large according to the direction in which each arm was moved (shown in Figure 10; left hand: y-axis; right hand: y-axis). On the other hand, in the case of M_{13}, it is also seen that the waveform of the x-axis is large for the left arm, even though the subjects moved their left arm in the y-axis direction. As dual-arm movement in different directions at the same time is difficult for humans, left hand motion may have been influenced by right hand motion (i.e., moving hand in the x-axis direction) because M_{13} was not a symmetric motion. The feature vector shown in Figure 10 (b) was misdiscriminated as M_7 (i.e., moving left hand in the x-axis direction and moving right hand in the x-axis direction) using Eq. (17), since the motion which moving both hands in x-axis direction is the dependent class belonging to the set $\check{z}$. However, the J-PNN was able to discriminate it correctly with Eqs. (10) and (11), indicating its ability to discriminate based on motion dependency for asymmetric dual-arm motions.

Figure 11. Average measurement and learning time for 15 and 50 dual-arm motions using MLP, R-LLGMN and MLP for all subjects. J-PNN can drastically reduce both measurement and learning time.

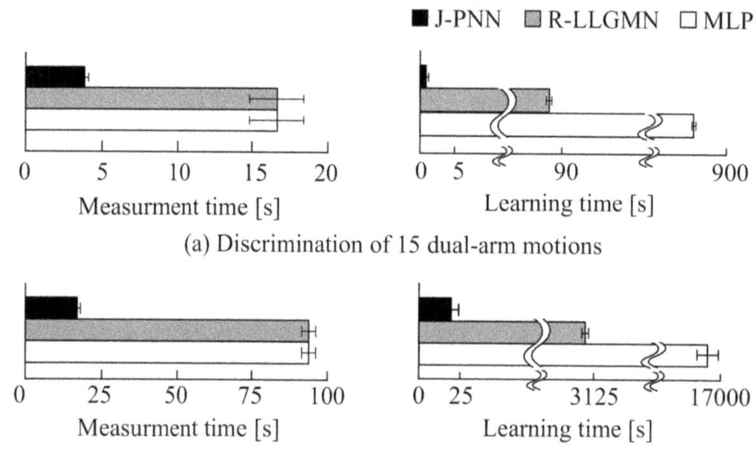

(a) Discrimination of 15 dual-arm motions

(b) Discrimination of 50 dual-arm motions

Table 3 depicts the number of correct classifications achieved using Eq. (10) and incorrect classifications using Eq. (17) for the data belonging to the set of dependent class $\check{z}$ with high entropy (H_{th} = 0.5). The average number of correctly discriminated trials using Eqs. (10) and (11) was 127.6 ± 33.9 for the misdiscriminated 183.6 ± 43.6 trials using Eq. (17). It is thus confirmed that the discrimination performance is increase in all subjects modifying probabilities based on the proposed motion dependencies between left and right arm motion. These results lead us to conclude that discrimination based on the dependency between the arms is effective for dual-arm motion discrimination.

In the discrimination results for all subjects (details are shown in Table 2), it can be seen that the average MLP discrimination rates was low. Conversely, the R-LLGMN and the J-PNN achieved high levels of classification performance because the statistical distribution of time series data could be estimated based on the GMM and the HMM using the R-LLGMN and the J-PNN. It should be note that the J-PNN can also discriminate all dual-arm motions including unlearned motions through learning each single-arm motion only (15 motions discrimination: 98.11 ± 1.29%; 50 motions discrimination: 87.09 ± 0.42%). In addition, J-PNN could discriminate 50 dual-arm motions with about 90% accuracy, although the discrimination rates are lower than that of 15 dual-arm motions. In (Kim *et al.*, 1996), the average discrimination rates for 25 gestures using basic sign language motions have been achieved with 85% accuracy. J-PNN could discriminate 50 dual-arm motions with same accuracy compared with (Kim *et al.*, 1996), indicating that the effectiveness of the J-PNN for many gestures discrimination for such sign language recognition.

Table 3. Relationships between the number of correct and incorrect classification for the data belonging to the set of dependent classes $\check{z}$ with high entropy

Subject	A	B	C	D	E
Number of incorrect classification using Eq. (17)	113	200	227	175	203
Number of correct classification using Eq. (10)	70	149	136	155	128

Finally, from Figure 11, it can be seen that the network learning time using the J-PNN was much shorter than that of the other methods. It can be inferred that a TA is incorporated into the learning rule of the R-LLGMN and the J-PNN in order to regulate the learning convergence time. Moreover, measurement time of learning samples can be reduced because all dual-arm motions can be discriminated using the J-PNN through learning of each single arm motion only, whereas the other method must learn all discrimination motions. As an example, the learning time of MLP and R-LLGMN were increase drastically because of increasing the number of discrimination motions in the case of 50 dual-arm motions discrimination. The results clearly demonstrate that the J-PNN can drastically reduce learning time with high discrimination accuracy.

ROBOTIC MANIPULATOR CONTROL EXPERIMENTS

As an example of the gesture-based human-machine interface using the proposed method, we developed the robot manipulator control system based on gesture recognition. In this chapter, the robot manipulator is controlled with the subject holding bottles in his both hands supposed to work using the glovebox (Miwa MFG Co., Ltd.) equipped robot manipulator. Glovebox is composed of box and groves, specific environment such as vacuum can be created in the box. Users can chemical and biology experiments under such environments using the glove keeping own safety. However, due to the gloves are fixed to the box, it causes significant restriction of the user's motions and works (Tsuji, T. *et al.*, 2014). In this experiment, therefore, we conducted supposed to assist by controlling robot manipulator using both dual-elbow dynamic gestures during working with glovebox using gloves.

Subject was a healthy male (21 years old). The subject attached ACC sensors for his each elbow for controlling of a robot manipulator (MELFA RV-1A, Mitubishi ltd). Eight dual-arm motions involving a combination of two single-arm motions are focused (M_1 ; moving his arm to outward direction, M_2 ; moving his arm to inward direction), and these motions are then allocated to command of moving robot manipulator to each axis direction, hand close and open. Relationships between the gestures and robot motions are shown in Figure 12.

In the experiments, subject was asked to control robot manipulator during holding bottle by following five steps: (i) moving the robot manipulator's endpoint from the initial position to near the subject; (ii)

Figure 12. Discrimination motions used in robot manipulator control experiment. The subject control uses his both dual-elbow motions during holding bottles in his hands.

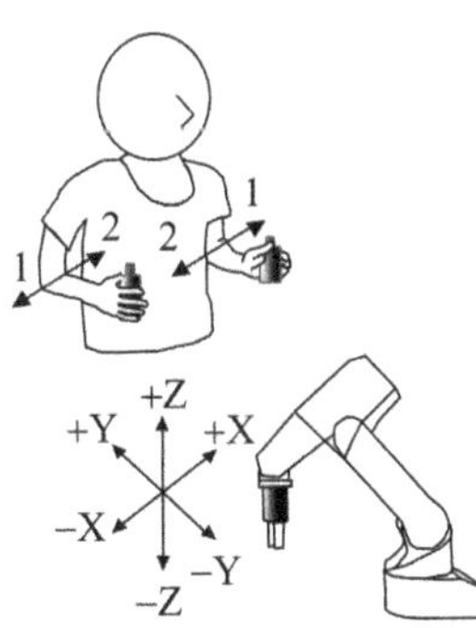

Motion number	Motion		Control command
	Left	Right	
M'_1	1	NM	Move to +Y
M'_2	2	NM	Move to −Y
M'_3	NM	1	Hand open
M'_4	1	1	Move to +Z
M'_5	2	1	No use
M'_6	NM	2	Hand close
M'_7	1	2	No use
M'_8	2	2	Move to −Z

1 : Outward 2 : Inward NM : No motion

holding the bottle from subject hand to the robot hand; (iii) moving the robot manipulator away from the subject; (iv) adjusting the height of the endpoint for putting of the bottle; (v) releasing the bottle holding the robot manipulator on the table; and (vi) adjusting the height of the endpoint once again to move away from the bottle.

Figure 13 shows an example of operation scene of robot manipulator using subject's elbow gestures. In this figure, it is observed that the subject controls the robot manipulator to move away from the subject and to adjust the height of the end effecter by conducting $M_2^{'}$ and $M_8^{'}$. The results clearly indicate that the robot manipulator can be controlled voluntarily according to subject's gestures.

Figure 14 shows an example of experimental results, and it plots measured ACC signals $\hat{g}_m^{v}(t)$, the root square sum of the ACC signals $\hat{g}_m(t)$, discrimination results and coordinates of the end effecter.

Figure 13. Operation scenes of the robot manipulator using dual-arm motions consisted of each elbow motion

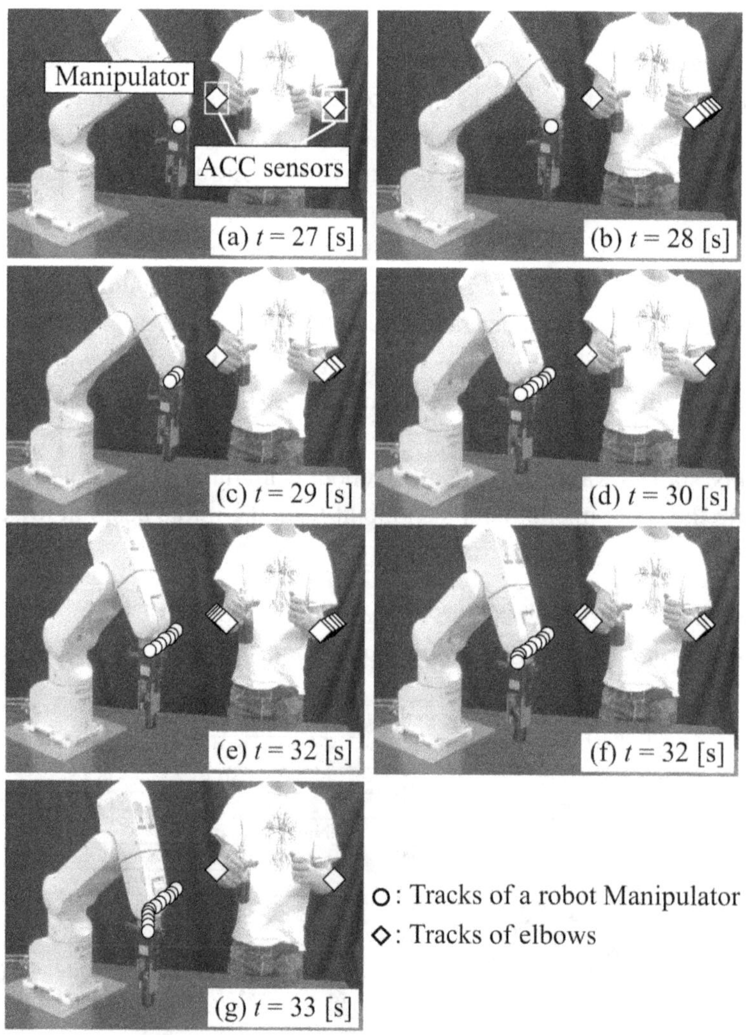

The shaded area describes the movement duration of the robot manipulator. At first, it can be seen that the subject controls the robot manipulator to move from the initial position to near the subject by conducting $M_1^{'}$ (near 14 s), then, the bottle is hold by robot hand from the subject hand conducting $M_6^{'}$ (near 23 s). After that, the robot manipulator moves to $-Y$ direction by the subject conducting $M_2^{'}$ (near 29s). It is confirmed that robot manipulator moves to $-Z$ direction by the subject conducting $M_8^{'}$ near 32 and 35 s, and releases the bottle from the robot hand by conducting $M_3^{'}$ near 41 s. Finally, the subject controls the robot manipulator's endpoint to moves $+Z$ direction by conducting $M_4^{'}$. Dis-

Figure 14. An example of operation results using proposed dual-arm motion discrimination method. The robot manipulator is controlled according to discrimination results.

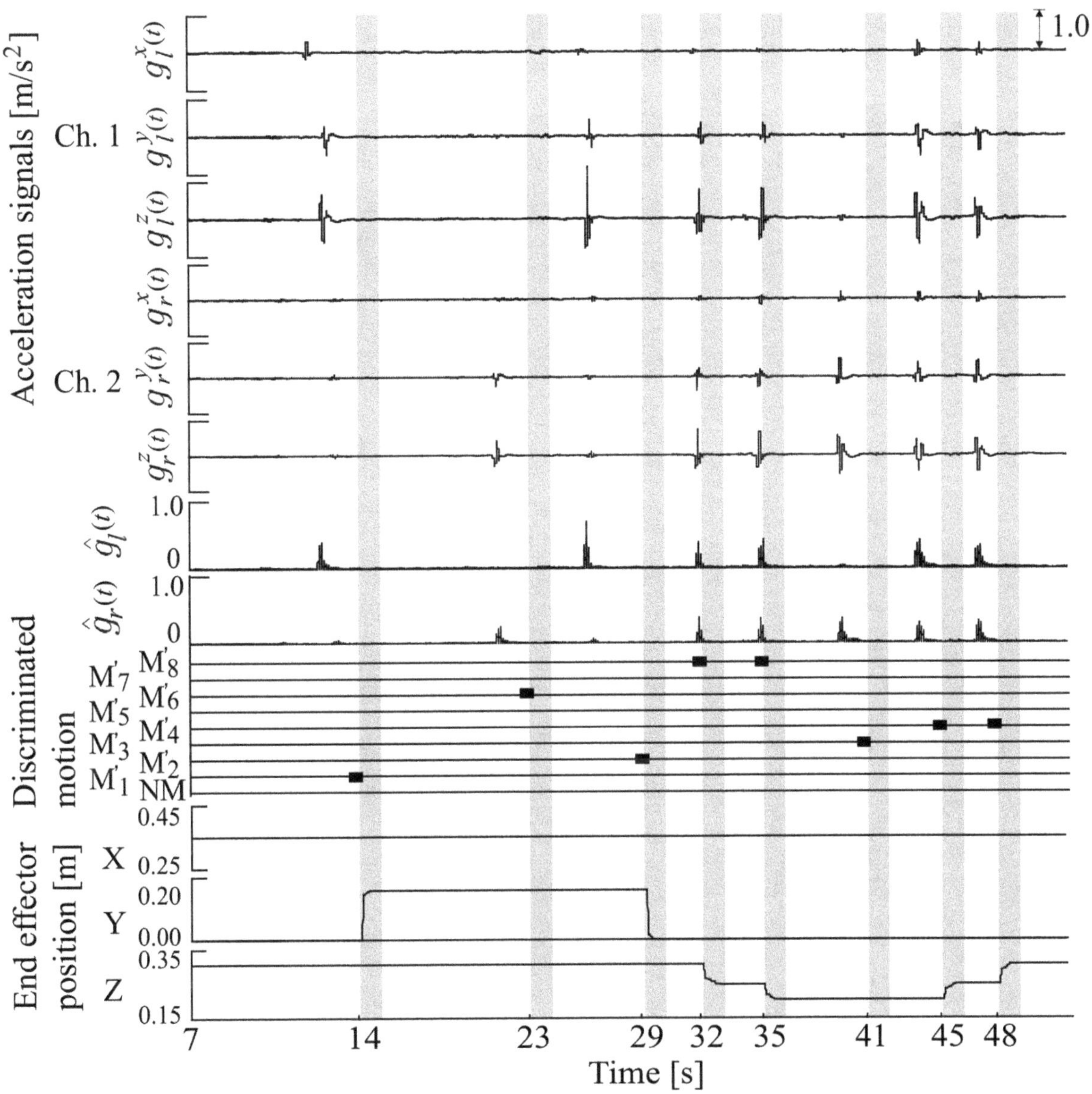

crimination rates of dual-arm motions during operation experiments were 100%, and robot manipulator could thus be controlled without misdiscrimination.

To confirm the effectiveness of the proposed method, glovebox operation experiments using the J-PNN and the R-LLGMN were performed. Relationships between the gestures and operation commands are shown in Figures 15 and 16. Note that the R-LLGMN must learn all discrimination motions. In the experiment, the subject (male, 21 years old) was asked to put on gloves and perform a given task that included four subtasks: (A) grasping and moving the object through gloves, (B) environmental control in the glovebox, (C) moving object to the target position using the task control method of the robotic manipulator, (D) fine-tuning the manipulator position using the direct control method.

Examples of the experimental results are shown in Figures 17 and 18. Figure 17 shows the results of the proposed J-PNN, and Figure 18 shows that of the R-LLGMN. From these figures, it can be seen that the user repeatedly performed each discrimination motion to move and execute operation commands. The robotic manipulator then automatically moved using the task control mode, and after that, the operation was switched to the direct control mode and the position of the robotic manipulator was fine-tuned to the

Figure 15. Discrimination motions used in glovebox with the robotic manipulator control experiment using the J-PNN. Only two single-arm motions are learned to the J-PNN.

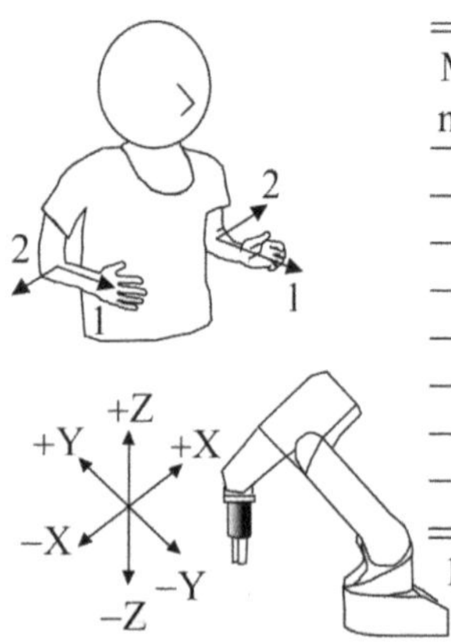

Motion number	Motion Left	Motion Right	Control command	Operation command
M'_1	NM	1	Hand open	Selection
M'_2	NM	2	Move to +X	Previous
M'_3	1	NM	Move to −X	Next
M'_4	1	1	Move to +Y	No use
M'_5	1	2	Move to −Y	No use
M'_6	2	NM	Move to +Z	No use
M'_7	2	1	Move to −Z	No use
M'_8	2	2	Main menu	No use

1 : Forward 2 : Outward NM : No motion

Figure 16. Discrimination motions used in glovebox with the robotic manipulator control experiment using the R-LLGMN. All seven discrimination motions are learned to the R-LLGMN.

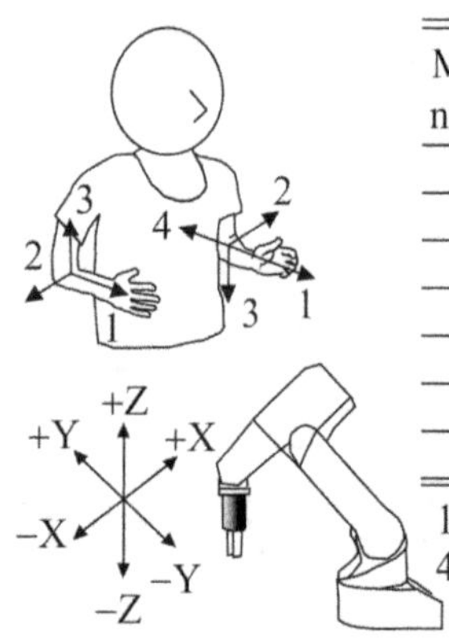

Motion number	Motion Left	Motion Right	Control command	Operation command
M_1	1	1	Main menu	Selection
M_2	2	NM	Move to −X	Previous
M_3	NM	2	Move to +X	Next
M_4	NM	1	Move to +Y	No use
M_5	4	NM	Move to −Y	No use
M_6	NM	3	Move to +Z	No use
M_7	3	NM	Move to −Z	No use

1 : Forward 2 : Outward 3 : Up-/Downward
4 : Backward NM : No motion

Figure 17. An example of operation results using proposed dual-arm motion discrimination method. Note that MMG signals extracted from AAC signals are used to switch to MAC-based control mode of the system.

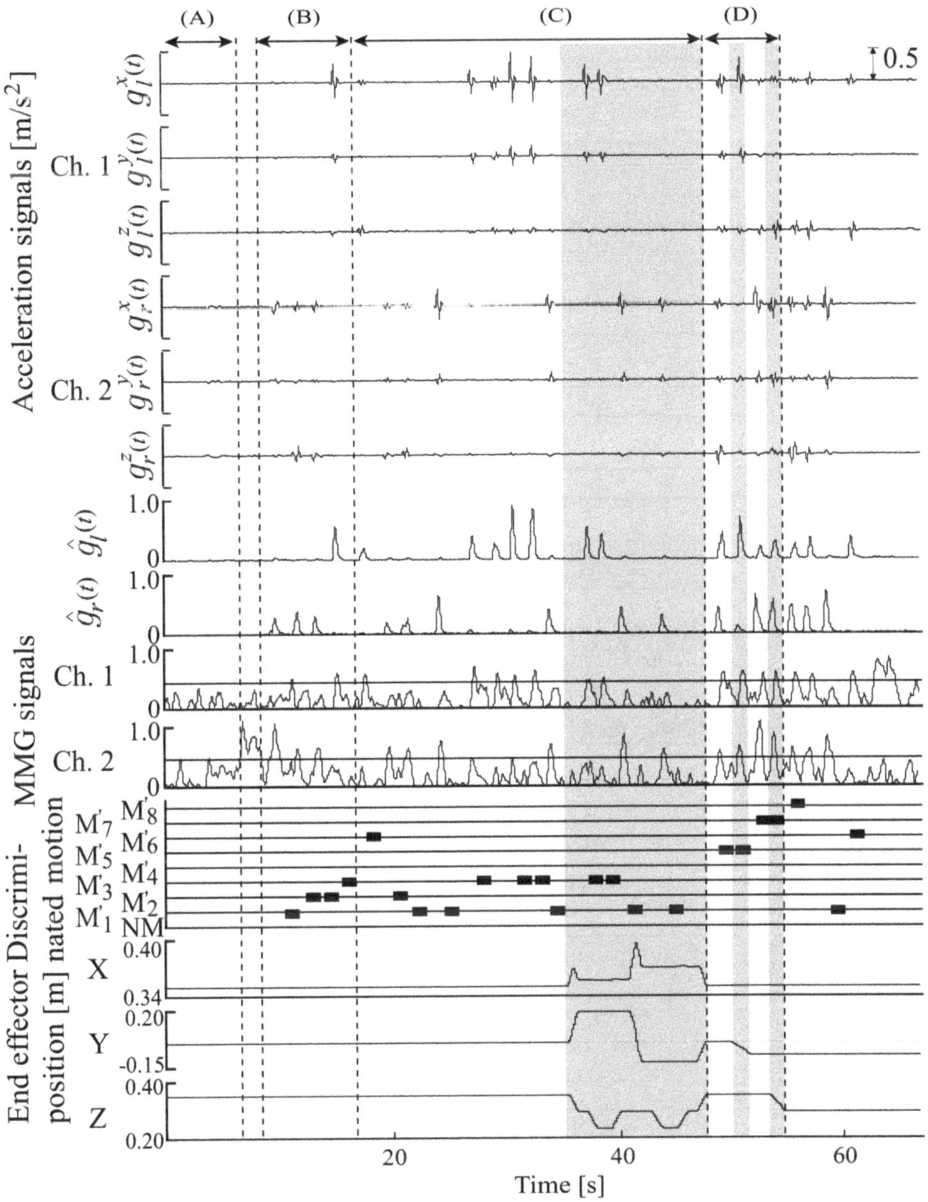

desired position by performing corresponding gestures (see panels C and D in Figures 17 and 18). We conclude that the proposed method can discriminate dual-arm motions and be used to control glovebox and robotic manipulator although the J-PNN only learns each single-arm motion.

These results clearly show that robot manipulator can be controlled according to subject's gestures even if the motions are limited by gloves. Although we applied proposed method to robot manipulator control in this chapter, it should be noted that the other human-machine interfaces could realize based on gesture recognition using proposed method.

Figure 18. An example of operation results using R-LLGMN. Note that MMG signals extracted from AAC signals are used to switch to MAC-based control mode of the system.

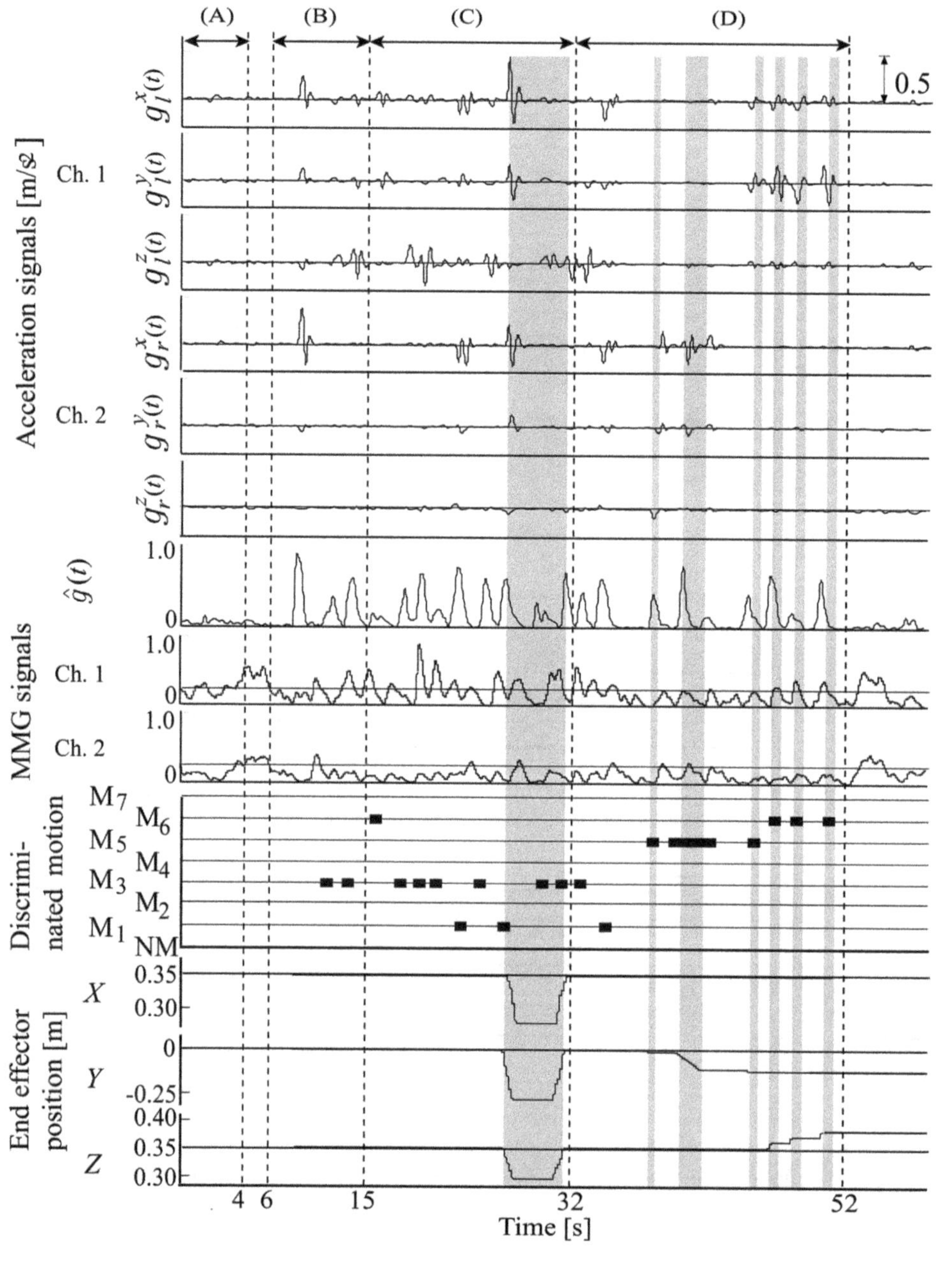

FUTURE RESEARCH DIRECTIONS

In future research, we plan to evaluate the proposed method by increasing the number of subject and discrimination motions. We also would like to investigate the determination method of optimal entropy threshold H_{th} in individually, and discuss the learning algorithm based on the motion dependencies for more accurate discrimination of dual-arm motions for sign language recognition. Since the ACC sensors can measure not only ACC signals but also mechanomyograms (MMGs) that describes the muscle contraction level on the skin surface, we will try the development of discrimination method for static and dynamic gestures such as hand shapes at same process.

CONCLUSION

This chapter outlined the method for dual-arm motion discrimination using a joint probabilistic neural network (J-PNN) involving time series pattern discrimination and joint probability estimation, and applied it to a machine control interface. As the J-PNN can be used to discriminate dual-arm motion based on learning of individual single-arm motions, the learning time for discrimination of a large variety of gestures can be reduced. The method can also be applied to accurate discrimination of all dual-arm motions consisting of single-arm motion combinations based on the dependencies between individual arm movements.

In the experiments, dual-arm motion discrimination and robot manipulator control was performed. Experimental results are summarized below:

- It was confirmed that the J-PNN can discriminate 15 dual-arm motions accurately based on learning of only three motions for each arm (J-PNN:98.11 ± 1.29%, R-LLGMN:90.07 ± 3.80%, MLP:62.03 ± 5.59%). In the case of 50 dual-arm motion discrimination in which combining 10 basic sign language motions, the J-PNN shows the discrimination ability for wide variety of gestures using dual-arm motions compared with the previous method (J–PNN:87.09 ± 0.42%, R-LLGMN:87.38 ± 19.73%, MLP:54.04 ± 39.71%);
- The learning and measurement time of learning samples for dual-arm motion discrimination can be reduced by using the J-PNN (15 motions discrimination;J-PNN:5.06 ± 0.22 s, R-LLGMN:104.09 ± 1.93 s, MLP:820.90 ± 1.93 s, 50 motions discrimination;J-PNN:38.28 ± 1.09 s, R-LLGMN:3218.83 ± 6.40 s, MLP:16178.01 ± 946.27 s);
- Machine control experiment performed indicated that the proposed method is possible to operate the robot manipulator voluntarily using the dual-arm motions.

These results indicate that the proposed method can be used to support robotic manipulator operation based on a variety of sign languages.

ACKNOWLEDGMENT

The authors would like to cordially acknowledge and express appreciation to Mr. Yuki Hiramatsu and Dr. Keisuke Shima for their assistance in the implementation of this research. This work was partially supported by a Grant-in-Aid for Young Scientists B Number 17K12723 and for Scientific Research (C) Number 26462242.

REFERENCES

Bailador, G. (2007). Real time gesture recognition using continuous time recurrent neural networks. *Proceedings International Conference Body Area Networks*, 1-8. doi:10.4108/bodynets.2007.149

Bowden, R., & Sarhadi, M. (2002). A non-linear model of shape and motion for tracking finger spelt American sign language. *Image and Vision Computing*, *20*(9-10), 597–607. doi:10.1016/S0262-8856(02)00049-5

Brethes, L., Menezes, P., Lerasle, F., & Hayet, J. (2004). Face tracking and hand gesture recognition for human-robot interaction. *Proceedings of the International Conference on Robotics and Automation 2004 (ICRA2004)*, 1901-1906. doi:10.1109/ROBOT.2004.1308101

Bridle, J. S. (1989), Neurocomputing: Algorithms, Architectures, and Applications. In Probabilistic interpretation of feedforward classification network outputs, with relationships to statistical pattern recognition (pp. 227-236). New York: Springer-Verlag.

Bu, N., Fukuda, O., & Tsuji, T. (2003). EMG-based motion discrimination using a novel recurrent neural network. *Journal of Intelligent Information Systems*, *21*(2), 113–126. doi:10.1023/A:1024706431807

Caelli, T., Squire, D. M., & Wild, T. P. J. (1993). Model-based neural networks. *Neural Networks*, *6*(5), 613–625. doi:10.1016/S0893-6080(05)80106-6

Devine, S., Rafferty, K., & Ferguson, S. (2016). Real time robotic arm control using hand gestures with multiple end effectors. *2016 UKACC 11th International Conference on Control (CONTROL)*.

Fall, C. L., Gagnon-Turcotte, G., Dubé, J. F., Gagné, J. S., Delisle, Y., Campeau-Lecours, A., ... Gosselin, B. (2017). Wireless sEMG-based body-machine interface for assistive technology devices. *IEEE Journal of Biomedical and Health Informatics*, *21*(4), 967–977. doi:10.1109/JBHI.2016.2642837 PMID:28026793

Guiard, Y. (1987). Asymmetric Division of Laber in Human Skilled Bimanual Action: The Kinematic Chain as a Model. *Journal of Motor Behavior, 19*, 486-517.

Habib, M. K. (2011). Biomimetics: Innovations and Robotics. *International Journal of Mechatronics and Manufacturing Systems*, *4*(2), 113–134. doi:10.1504/IJMMS.2011.039263

Hiramatsu, Y., Shibanoki, T., Shima, K., & Tsuji, T. (2011). A novel dual-arm motion discrimination method using reccurent probability neural networks for automatic gesture recognition. *Proceedings of the 2011 IEEE/SICE International Symposium on System Integration (SII 2011)*, 1346-1351.

Huang, J., Zhou, W., Li, H., & Li, W. (2015). Sign language recognition using 3D convolutional neural networks. *IEEE International Conference on Multimedia and Expo (ICME)*, 1–6. doi:10.1109/ICME.2015.7177428

Kim, J., Jang, W., & Bien, Z. (1996). A Dynamic Gesture Recognition System for the Korean Sign Language (KSL). *IEEE Trans. on Systems, Man and Cybernetics-Part B. Cybernetics*, *26*(2), 354–359. PMID:18263039

Liu, J., Wang, Z., Zhong, L., Wickramasuriya, J., & Vasudevan, V. (2009). uWave Accelerometer-based Personalized Gesture Recognition and Its Applications. *Pervasive and Mobile Computing*, *5*(6), 657–675. doi:10.1016/j.pmcj.2009.07.007

Okamoto, M., Shima, K., Matsubara, Y., & Tsuji, T. (2008). Pattern discrimination method with a boosting approach using hierarchical neural trees. *Proceedings of the Institution of Mechanical Engineers. Part I, Journal of Systems and Control Engineering*, *222*(7), 701–710. doi:10.1243/09596518JSCE563

Raheja, L., Shyam, R., Kumar, U., & Prasad, P. B. (2010). Real-time robotic hand control using hand gestures. *Proceedings of the 2010 Second International Conference on Machine Learning and Computing (ICMLC '10)*, 12-16. doi:10.1109/ICMLC.2010.12

Richard, M. D., & Lippmann, R. P. (1991). Neural network classifiers estimate Bayesian *a posteriori* probabilities. *Neural Computation*, *3*(4), 461–483. doi:10.1162/neco.1991.3.4.461

Sawada, H. (1998). Japanese sign-language recognition based on gesture primitives using acceleration sensors and datagloves. *Proceedings of the 2nd European Conference on Disability, Virtual Reality and Associated Technologies (ECDVRAT)*, 149-157.

Shima, K., Hiramatsu, Y., Shibanoki, T., & Tsuji, T. (2013). Discrimination of Dual-arm Motions Using a Joint Posterior Probability Neural Network [in Japanese]. *Transactions of the Society of Instrument and Control Engineers*, *49*(5), 568–575. doi:10.9746/sicetr.49.568

Solís, F., Martínez, D., & Espinoza, O. (2016). Automatic Mexican Sign Language Recognition Using Normalized Moments and Artificial Neural Networks. *Engineering*, *8*(10), 733–740. doi:10.4236/eng.2016.810066

Specht, D. F. (1990). Probabilistic neural networks. *Neural Networks*, *3*(1), 109–118. doi:10.1016/0893-6080(90)90049-Q PMID:18282828

Tsuji, T., Bu, N., & Fukuda, O. (2006). Neural Networks in Healthcare: Potential and Challenges. In A Recurrent Probabilistic Neural Network for EMG Pattern Recognition (pp.130-153). Idea Group Inc.

Tsuji, T., Bu, N., Fukuda, O., & Kaneko, M. (2003). A recurrent log-linearized Gaussian mixture network. *IEEE Transactions on Neural Networks*, *14*(2), 304–316. doi:10.1109/TNN.2003.809403 PMID:18238014

Tsuji, T., Shibanoki, T., & Shima, K. (2014). EMG-based control of a multi-joint robot for operating a glovebox. In Handbook of Research on Advancements in Robotics and Mechatronics (pp. 36-52). IGI Global.

Vogler, C., & Metaxas, D. (1997). Adapting Hidden Markov Models for American Sign Language Recognition by using Three-dimentional Computer Vision Methods. *IEEE International Conference on Systems, Man and Cybernetics*, 156-161.

Wang, H., Chai, X., Hong, X., Zhao, G., & Chen, X. (2016). Isolated Sign Language Recognition with Grassman Covariance Matrices. *ACM Transactions on Accessible Computing*, *8*(4), 14–21. doi:10.1145/2897735

Zhang, X., Chen, X., Li, Y., Lantz, V., Wang, K., & Yang, J. (2011). A Framework for Hand Gesture Recognition Based on Accelerometer and EMG Sensors. *IEEE Transactions on Systems, Man, and Cybernetics. Part A, Systems and Humans*, *41*(6), 1064–1076. doi:10.1109/TSMCA.2011.2116004

KEY TERMS AND DEFINITIONS

Gesture Recognition: A technique of recognizing bodily movements (such as sign languages) using biological signals.

Human-Robot Interface: A controlling method and controller itself for operation of robots. Biological signals such as electromyograms, electroencephalograms, acceleration signals are often used as a control input.

Joint Posterior Probability: A simultaneous revised probability that can be calculated using J-PNN combining independently estimated posterior probabilities for each arm motion based on entropy.

Joint Probability Neural Network (J-PNN): One of the artificial recurrent probabilistic neural networks (ARPNN) consisting of two ARPNNs incorporated a hidden Markov model and a Gaussian mixture model.

Motion Acceleration Signal (MAC): A record of velocity changes occurred corresponding to bodily movement. Motion acceleration (MAC) and mechanomyogram (MMG) components can be separated from raw acceleration signals using appropriate digital filter because of the difference of frequency band.

Motion Dependency: A property that an arm movement is followed by the other ones performing dual-arm motions (e.g., asymmetric motions).

Recurrent Probabilistic Neural Network: An artificial recurrent neural network that can estimate the probability density function of a given set of time series data. It is used for classification problems because it can compute nonlinear decision boundaries between each discrimination class.

Chapter 16
Off-Line Calibration of Autonomous Wheeled Mobile Robots

Yaser Maddahi
University of Calgary, Canada

ABSTRACT

Wheeled mobile robots (WMRs) are very interesting regarding different applications from in-house activities in assisting elderly people and patients to space exploration. While the design concept and the application of the WMRs determine specifications of the robot, the positional errors occur during the WMR motion. The positional errors are inevitable, as they are caused by imperfections in design to fabrication; therefore, there is a need to rectify them using calibration techniques such as odometry, camera-based error detection, or using gyroscope and compasses. This chapter focuses on the use of odometry as it provides improved short-term accuracy with high sampling rates while it is more economical and requires fewer landmarks to localize the WMR. The context provides an overview of WMRs mechanisms, differential and omnidirectional drive, and then introduces an odometry-based method to correct the motion of both types of WMRs. Experimental results on four robots exhibited that positional error was significantly improved. Using analysis of variance (ANOVA) test, the authors could not detect any change in error improvement when the robot changed.

1. INTRODUCTION

WMRs are employed in many areas such as logistics, inspection and maintenance, security and defense, agriculture, medical care, home care, urban transport, planetary exploration and surveillance operations. Due to the precision required during the performance of a WMR, the robot should properly be calibrated, and the positional error should be improved, before implementation in real field. The accuracy is of importance when complicated mechanisms with tiny structures are proposed. Examples are biologically inspired robots that help scientists to develop intelligent robots to serve in our home, hospital, office, and outdoors (Shi, Habib, Xiao & Hu, 2015; Habib, 2011; Kalvandi & Maddahi, 2012a).

DOI: 10.4018/978-1-5225-2993-4.ch016

Calibration is defined as a set of operations that establishes the relationship between the values of quantities indicated by a measuring instrument and the corresponding values realized by standards (Balazs, 2008). Developing effective calibration techniques have recently been interested in robotic systems (de Wit, 1998). They include odometry (Y. Maddahi & A. Maddahi, 2004b; Y. Maddahi, A. Maddahi, Sepehri, 2013; Y. Maddahi, Sepehri, A. Maddahi & Abdolmohammadi, 2012a), 3D camera error detection (Ko, Hwang, Jung & Kim, 2005), active beacons (Piaggio, Sgorbissa & Zaccaria, 2001), gyroscope (Bury & Hope, 1995) and magnetic compasses (Rahok & Koichi, 2009). Odometry uses the information of positional sensors attached to each actuator to estimate change in position over time (Maddahi et al, 2013). In WMRs, odometry builds an incremental model of the motion using measurements of the wheel angular displacements. The odometry method is applied to correct errors of different types of WMRs including differential drive and omnidirectional robots (Maddahi et al, 2013; Maddahi et al, 2012a).

The calibration of WMRs using odometry approach has been performed by researchers and engineers. Correcting systematic errors with localization based on magnetic fields has been discussed in Rahok and Koichi (2009). A model-based method was developed to formulate odometry of WMRs (Song & Wang, 2008), while the wheel slippage was considered as a significant factor that affects the accuracy of computations. Nistér et al. (2006) presented a platform to estimate the motion of a stereo head based on video input that operated in real time with low delay. University of Michigan Benchmark (UMBmark) method was introduced to measure odometry errors in differential drive mobile robots (Borenstein & Feng, 1994; Borenstein & Feng, 1995; Borenstein & Feng, 1996). The UMBmark was then employed to reduce positional errors of several WMRs including differential drive and omni-mate robots. An integrated system was also proposed for WMRs that rely on a wireless transceiver infrastructure (Marantos, Koveos, Stergiopoulos, Panousopoulou & Tzes, 2008). Maddahi et al. (2004a; 2004b; 2005, 2006, 2012b, 2013) applied the UMBmark benchmark test on different types of wheeled mobile robots. They also proposed a new technique to reduce both systematic (Maddahi et al., 2004a, 2012a, 2013; Maddahi, Maddahi & Monsef, 2012b) and non-systematic (Maddahi et al., 2013; Maddahi et al., 2012a; Maddahi, Monsef & Mastorakis, 2014) positional errors.

This chapter proposes a kinematic-based calibration method, which is simple to implement and applicable to all types of wheeled mobile robots including differential drive, omnidirectional and multi-degree of freedom (DOF) WMRs. The proposed technique and commonly-used method, UMBmark, are employed to correct the motion of a two-wheeled mobile robot, a differential drive robot, a three-wheeled omnidirectional mobile robot, and a four-wheeled omnidirectional mobile robot. The one-way ANOVA) is used to investigate if the error improvements measured are statistically significant, and if the proposed techniques applicable to reduce the positional error regardless of the size, shape, type, and mechanism of the WMR and its wheels.

The remainder of this chapter is organized as follows. Sources of systematic and non-systematic errors are described in Section 2. The characteristics, kinematics modeling and classification of probable errors in WMRs are presented in Section 3. Section 4 documents the special attention of this study which is the development of the proposed kinematic-based calibration method. Section 5 describes the prototyped mobile robots. Experimental results comparing presented in Section 6. Section 7 outlines the conclusions of this research.

2. SOURCES OF ERRORS

The accuracy of robot movement and the amount of positional errors depend on the sources causing both systematic and nonsystematic errors. Systematic errors originate from control and mechanical elements due to imperfections from design to fabrication (Maddahi et al., 2012b). Non-systematic errors are independent of the robot structure and its characteristics, and are defined unwanted errors produced due to surface irregularities. Table 1 shows a classification of positional errors observed in WMRs.

3. STRUCTURE AND KINEMATICS

The characteristics of WMRs are determined by some metrics during the design process. Three types of wheels are used in WMRs: *standard* that are powered to permit the robot to move, *castor* that are used along with the standard wheels to facilitate balancing the robot, or *omnidirectional* that are capable of acting as both castor and standard wheels (Maddahi, 2013). Several omnidirectional mobile robots were prototyped with different design of wheels including universal wheels (West, 1993; Ilon, 1975), ball wheels (West & Asada, 1997) and off-centered wheels (Wada & Mori, 1996). In the following the kinematic formulation of the three typed of WMRs tested in this chapter are explained.

3.1. Two-Wheeled Driven Robot (Differential-Drive Mechanism)

Figure 1 shows the structure of a typical differential drive mobile robot. In this schematic both driving and castor wheels are used to supply required power and stability of the platform, respectively. Figure 1 also illustrates the coordinate system and the essential variables of a two-wheeled driven WMR. Kinematics equations of the robot are obtained using Denavit-Hartenberg notation (Denavit, 1995). Considering the fact that the two driving wheels roll and do not slip, we obtain:

Table 1. Error classifications in mobile robots

Class	Sources of Errors
Systematic errors	**Control factors** - Lack of pulse and/or an error in the pulse - Change or omit data read **Mechanical factors** - Non-circularity of the wheels - Misalignment of the shafts - Unequal wheel diameters - Difference between the averages of actual and nominal wheel diameter - Difference between the actual and nominal wheelbase - Misalignment of wheels and/or rollers in Sweden wheels - Low encoder resolution/accuracy - Gear backlash and belt slippage - Shafts not being properly fixed relative to their bearings
Non-Systematic errors	**Unpredictable phenomena** - Slippery floors that reduces required friction - Over acceleration during movement - Fast turning of WMRs - External forces applied to the robot - Non-point contact between the wheel and ground

Figure 1. Main elements, reference frame, and parameters of a differential drive WMR
Source: Maddahi et al. (2004a)

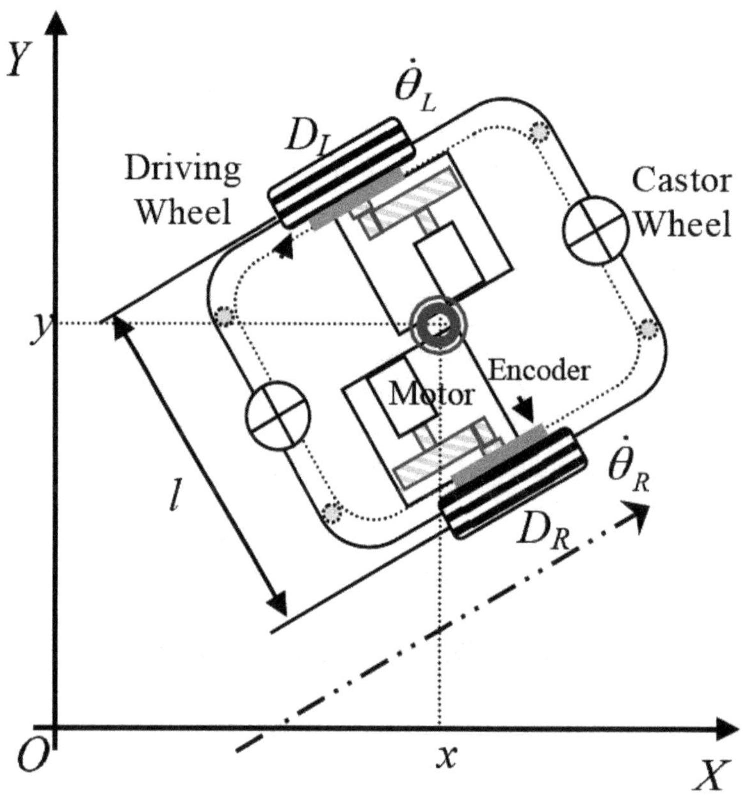

$$\dot{x}\cos\theta + \dot{y}\sin\theta = 0.25\left(D_L\dot{\theta}_L + D_R\dot{\theta}_R\right) \tag{1}$$

$$2l\dot{\theta} = D_R\dot{\theta}_R - D_L\dot{\theta}_L \tag{2}$$

where (x, y) is the coordinate of the robot center of mass, $\dot{x}$ and $\dot{y}$ are the velocities of this center along X and Y axes, respectively. θ denotes the orientation of the robot with respect to the initial position at the start point of the motion, and $\dot{\theta}$ is its derivatives with respect to time. D_R and D_L are the nominal diameters of the right and left wheels, respectively. Angular velocities of the right and left wheels are denoted by $\dot{\theta}_R$ and $\dot{\theta}_L$, respectively. The nominal distance between the two driving wheels is represented by l.

In addition, considering the fact that the mobile robot cannot move in the lateral direction, the third equation is obtained as follows:

$$\dot{x}\sin\theta - \dot{y}\cos\theta = 0 \tag{3}$$

3.2. Three-Wheeled Driven Robot (Omnidirectional Mechanism)

Kinematic diagram and parameters required for modeling a three-wheeled omnidirectional robot are depicted in Figure 2. The robot has three DOFs, two positional (*x* and *y*) and one rotational (θ). The reference coordinate frame is depicted by $\{XY\}$.

We assume that there is a pure rolling at the contact point between each roller of the wheel and the ground that gives us the following equation:

$$\left[\sin(\alpha_i+\beta_i+\gamma_i)-\cos(\alpha_i+\beta_i+\gamma_i)-l_i\cos(\beta_i+\gamma_i)\right]R(\theta)\vec{\mu}-r_i\dot{\varphi}\cos\gamma_i=0 \tag{4}$$

where *i* is the number of wheels (=3 in Figure 2). $R(\theta)\vec{\mu}$ term transforms the motion parameters in the Cartesian coordinate frame, $\{XY\}$, into the base frame attached to the robot center of mass and moves with robot. β_i is the steering angle that is normally a constant number. γ_i denotes the angle between the main wheel plane and the axis of rotation of the small circumferential rollers. l_i is the wheelbase (Maddahi et al., 2012b). α_i is the angle between the wheel shaft and *X* axis when the robot is at home position and $\dot{\varphi}$, *r* are the angular velocity and the radius of the wheel (Maddahi, 2013). Detailed explanations on wheel parameters are discussed by the author in (Maddahi, 2013).

Matrix $R(\theta)$ is expressed as a function of orientation angle (θ), and is expressed as:

$$R(\theta)=\begin{bmatrix}\cos\theta & \sin\theta & 0\\ -\sin\theta & \cos\theta & 0\\ 0 & 0 & 1\end{bmatrix} \tag{5}$$

Figure 2. Coordinate system attached to a three-wheeled omnidirectional WMR
Source: Revised from Maddahi et al. (2013)

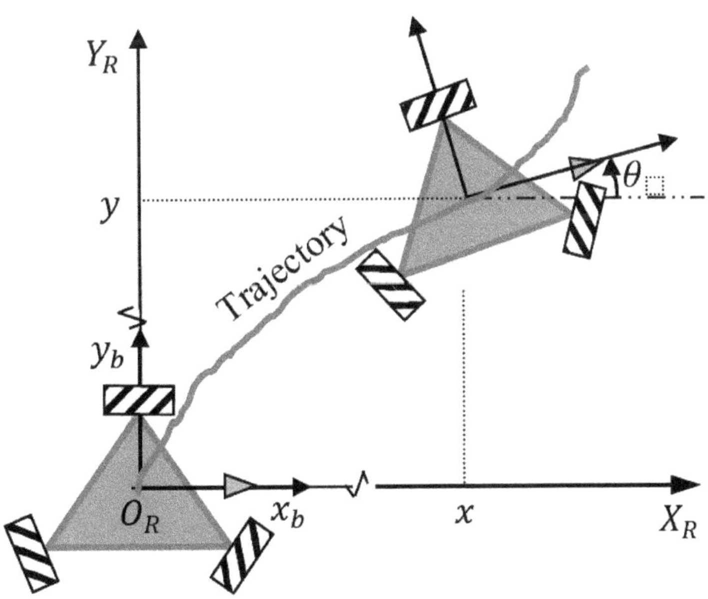

$\vec{\mu}$ represents the robot linear/angular velocity vector with respect to $\{XY\}$, and is defined as:

$$\vec{\mu} = \begin{bmatrix} \dot{x} & \dot{y} & \dot{\theta} \end{bmatrix} \tag{6}$$

Equation (4) can be rewritten as:

$$\vec{\dot{\Phi}} = J^{-1} \vec{\mu} \tag{7}$$

where the wheel velocity vector for a robot having n omnidirectional wheels $\vec{\dot{\Phi}}$, is defined as below:

$$\vec{\dot{\Phi}} = [\dot{\varphi}_1 \quad \dot{\varphi}_2 \cdots \dot{\varphi}_n]^T \tag{8}$$

Using given formulations, the nominal inverse Jacobian matrix, $J_{n\times 3}^{-1}$, is obtained as follows:

$$J_{n\times 3}^{-1} = diag\left[\frac{1}{r_n \cos\gamma_n}\right]\begin{bmatrix} \sin(\alpha_1+\beta_1+\gamma_1) & -\cos(\alpha_1+\beta_1+\gamma_1) & -l_1\cos(\beta_1+\gamma_1) \\ \vdots & \vdots & \vdots \\ \sin(\alpha_n+\beta_n+\gamma_n) & -\cos(\alpha_n+\beta_n+\gamma_n) & -l_n\cos(\beta_n+\gamma_n) \end{bmatrix} R(\theta) \tag{9}$$

Particular application of (9) is presented in Maddahi et al. (2013).

3.3. Four-Wheeled Driven Robot (Omnidirectional Mechanism)

Kinematic diagram of a four-wheeled omnidirectional WMR is illustrated in Figure 3. Using definitions mentioned in Section 3.2, we have (Maddahi, Y., Sepehri, & Maddahi A. 2012c):

$$\begin{bmatrix} \dot{\varphi}_1 \\ \dot{\varphi}_2 \\ \dot{\varphi}_3 \\ \dot{\varphi}_4 \end{bmatrix} = J^{-1} \dot{\mu}_b \tag{10}$$

where $J^{(-1)}$ is the nominal inverse Jacobian matrix and is expressed by (Maddahi et al., 2012c):

Figure 3. Reference coordinate systems of a four-wheeled omnidirectional WMR
Source: Revised from Maddahi et al. (2012c)

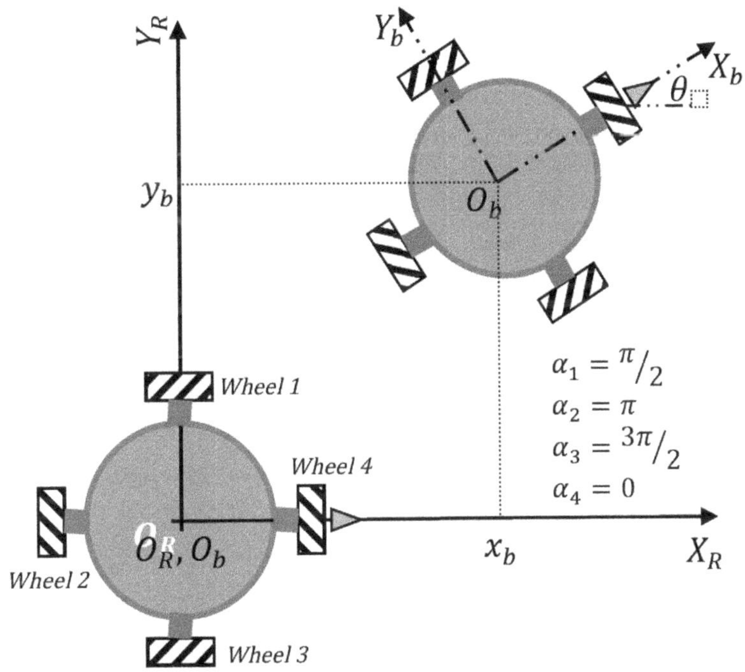

$$J^{-1} = \begin{bmatrix} \frac{1}{r\dot{\varphi}_1} & & \cdots & 0 \\ & \frac{1}{r\dot{\varphi}_2} & & \\ \vdots & & \frac{1}{r\dot{\varphi}_3} & \vdots \\ 0 & & \cdots & \frac{1}{r\dot{\varphi}_4} \end{bmatrix} \begin{bmatrix} \cos(\alpha_1) & \sin(\alpha_1) & -1 \\ \cos(\alpha_2) & \sin(\alpha_2) & -1 \\ \cos(\alpha_3) & \sin(\alpha_3) & -1 \\ \cos(\alpha_4) & \sin(\alpha_4) & -1 \end{bmatrix} R(\theta) \tag{11}$$

where $\dot{\varphi}$ is the vector of the angular velocity the wheels.

4. TESTED ODOMETRY APPROACHES

As mentioned, there exist a number of odometry-based methods used to calibrate the positioning error of mobile robots. The commonly used ones utilize benchmark series test that allow one to draw conclusions about the overall odometric accuracy of the robot. Here, the widely-used benchmark method called 'UMBmark method' is described, followed by describing the proposed alternative method.

4.1. UMBmark Method

The UMBmark technique compensates two types of error sources: the part that is caused by the unequal wheel diameters (D_R and D_L) and the part that relates to the uncertainty on the value of the wheelbase, l (see Figure 4) (Borenstein & Feng, 1996). Using UMBmark method, the robot starts out at a 'start' position and moves along a 4×4 m(L=4 m) uni-directional square path in clockwise (CW) and/or counter-clockwise (CCW) direction (Maddahi et al., 2012b). The robot is programmed to travel along perimeter of the square. However, it will likely not arrive precisely at the start position. Figure 4 illustrates the schematic of expected experimental results in CCW direction. In addition, it depicts the desired (solid line) and actual (dashed line) trajectories obtained during robot motion (Maddahi et al., 2012a). As observed in Figure 4, upon completion of each trial, the position offsets along X and Y axes are measured with respect to the start point. Mean position errors are then calculated. Based on the calculated position offsets, the coefficients α and β are found from simple geometric relations. Note that α and β can be positive or negative depending on the ratio of wheel diameters, actual wheelbase and the actual position of robot. Therefore, the contribution of two types of error labeled with α and β angles can be seen in this figure. Detailed explanations about the equations and formulations of the test is described in Borenstein et al. (1994; 1995; 1996).

4.2. The Proposed Method

This section is a summary of the proposed method presented in Maddahi et al. (2013) and Maddahi et al. (2012a; 2012b). Figure 5 shows the test trajectory and needed parameters. As observed, the robot is programmed to travel along the path A, while regardless of the error sources, the robot follows path B. The positional error has two components: lateral error and longitudinal error. Therefore, two corrective indices are defined to correct these two errors: lateral corrective matrix and longitudinal corrective fac-

Figure 4. Trajectory taken by the differential drive WMR in UMBmark technique in CCW direction

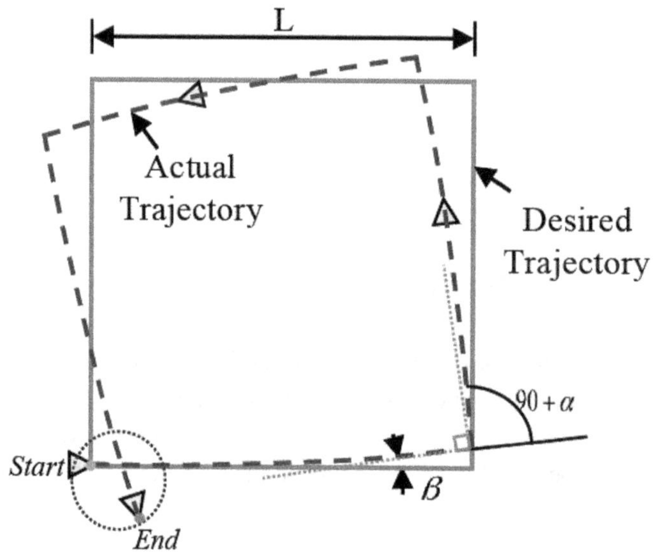

Figure 5. Trajectory taken by the differential drive WMR in the proposed technique

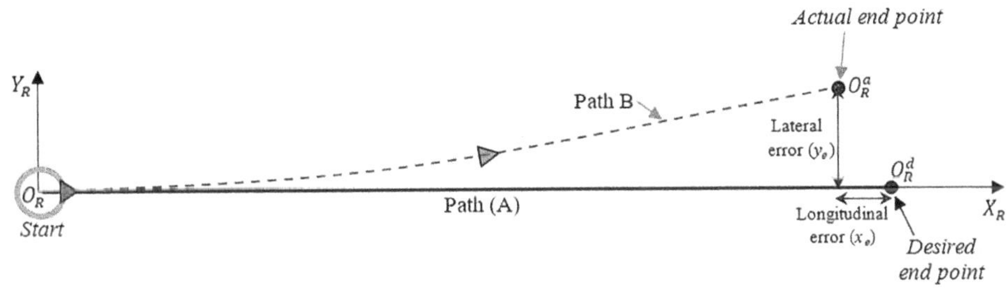

tor. The lateral corrective matrix (F_{lat}) is applied to kinematic equations of robot to ensure that the robot stays along the desired trajectory (Maddahi et al., 2013; Maddahi et al., 2012a). This matrix is calculated using nominal $\left(\vec{\dot{\Phi}}\right)$ and error-driven $\left(\vec{\dot{\Phi}}_e\right)$ wheel angular velocities based on the error angle θ_e, measured during the test (Maddahi et al., 2013). Using Eq. (7), the error-driven wheels angular velocity vector $\left(\vec{\dot{\Phi}}_e\right)$ is calculated as follows:

$$\left(\vec{\dot{\Phi}}_e\right) = J^{-1}\vec{\dot{\mu}}_e \tag{12}$$

or:

$$\begin{bmatrix} \dot{\varphi}_{e,1} \\ \dot{\varphi}_{e,2} \\ \vdots \\ \dot{\varphi}_{e,n} \end{bmatrix} = J^{-1} \begin{bmatrix} 0 \\ 0 \\ -\dot{\theta}_e \end{bmatrix} \tag{13}$$

where, $\dot{\varphi}_{e,i=1,2,\ldots,n}$ represent the angular velocity errors (n is the number of wheels), and $\dot{\theta}_e = \frac{d\theta_e}{dt}$ denotes the derivative of the robot body angular displacement with respect to time (Maddahi et al., 2012a).

The lateral corrective matrix, F_{lat}, is expressed as follow (Maddahi et al., 2013):

$$F_{lat} = diag\left[1 + \frac{\dot{\varphi}_{e,n}}{\dot{\varphi}_n}\right] \tag{14}$$

To compensate the longitudinal position error, the longitudinal corrective factor (F_{lon}) is defined as follows (Maddahi et al., 2012a):

$$F_{lon} = \frac{L}{\sqrt{\left(L + \bar{x}_e\right)^2 + \left(\bar{y}_e\right)^2}} \tag{15}$$

where L is the length of trajectory A. $\overline{x}_e = \sum_{j=1}^{N} x_{e,i}$ and $\overline{y}_e = \sum_{j=1}^{N} x_{e,i}$ are averages of the robot longitudinal and lateral errors over N number of trial runs (Maddahi et al., 2012b).

Implementations of the proposed method for correcting the systematic and non-systematic errors for differential drive WMRs and omnidirectional WMRs are shown in Maddahi et al. (2012b) and Maddahi *et al.* (2013), respectively.

5. PROTOTYPE WMRS

This section describes four prototype mobile robots with a variety of components to examine the performance of the described benchmark techniques. Detailed description of the robots are given in Maddahi et al. (2012a, 2013, 2012b). Specifications of each robot are listed in Table 2. The Jumper (two-wheeled) consists of two driving wheels, two stepping motors with gear boxes, two ultrasonic sensors and a wireless camera to detect the obstacles, as shown in Figure 6a (Maddahi, Seddigh, Mohammadpour & Maleki, 2004b; Maddahi et al., 2012b). Maze (Figure 6b) has two driving wheels that are controlled by a microprocessor. The additional castor wheels are to provide better stability. This robot is designed with high maneuverability in environment. As shown in Figure 6c, the 3W Omni robot comprises three omnidirectional driving wheels with their own transmission systems which are driven by three DC motors. The omnidirectional wheels are centered on the longitudinal axis of the robot. These wheels are driven and unstring wheels powered by three completely independent mechanisms and have single-row roller arrangement. This robot has dimension of 40×40×18 cm^3 and weight of 3.150 kg (Figure 6d) (Maddahi et al., 2012a). The 4W Omni robot has four omnidirectional wheels, and is equipped with some conductor sensors, to detect magnetic objects.

Table 2. Specifications of prototype robots

	Two-Wheeled	Maze	3W Omni	4W Omni
Dimension (L×W×H) (cm)	10×10×22.5	7×7×11	29×29×12	40×40×18
Weight (kg)	0.880	0.105	1.250	3.150
Stall torque of motor (Nm)	0.2	0.1	0.2	0.2
Maximum linear speed (m/min)	0.115	0.820	7.30	4.25
Wheel radius (cm)	5.0	2.5	7	10
Wheelbase (cm)	10.5	3.2	13.8	19
Encoder resolution (pulse/rev)	48	60	360	512

Figure 6. Prototype mobile robots

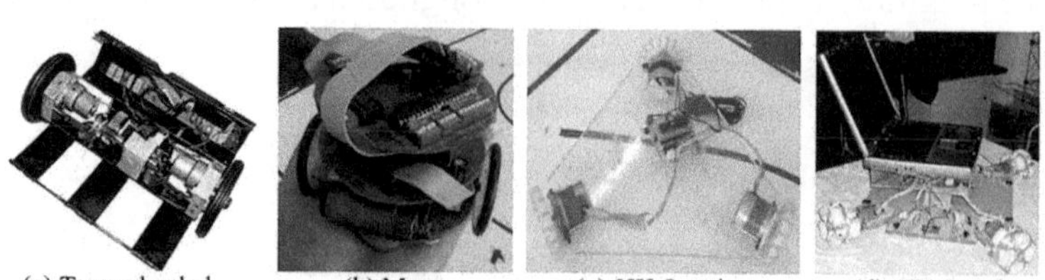

(a) Two-wheeled (b) Maze (c) 3W Omni (d) 4W Omni

6. EXPERIMENTAL RESULTS

This section presents results of experimental tests applied to all prototype robots described in Table 2. The metric used to facilitate the comparison is the radial error (δr) as (Maddahi et al., 2004a):

$$\delta r = \sqrt{(\delta x)^2 + (\delta y)^2} \tag{16}$$

where δx and δy are defined as the positional error along *x*-axis and *y*-axis, respective for UMBmark and as the longitudinal and lateral errors for the proposed method (shown as x_e and y_e in Figure 5). The effectiveness of each method is then measured by comparing the mean error improvement index, δr_m (Maddahi et al., 2012a):

$$\delta r = \frac{\delta r_{m,b} - \delta r_{m,a}}{\delta r_{m,b}} (100\%) \tag{17}$$

where $\delta r_{m,a}$ and $\delta r_{m,b}$ are the mean values of the radial errors before and after calibration, respectively.

Figure 7 shows the position errors before and after calibration using both UMBmark and the proposed methods for the jumper and the maze robots. As seen, the motions of both robots were corrected with either method, i.e., the errors converged to the center of coordinate system (zero positional error) in both approaches. As explained earlier, the UMBmark and the proposed methods are based on robot movement along square and straight trajectories, respectively. Consequently, the final positions of each robot in before calibration are not same for UMBmark and the proposed methods because the robot travels along two different sets of experiment. Results of the omnidirectional robots are illustrated in Figure 8. In Figures 7 and 8, BF and AF show the locations of the robot stop point before and after calibration, respectively. Moreover, the UMBmark and the proposed method are shown by U and P, respectively.

Figure 7. Radial systematic error of (left) jumper robot and (right) maze robot before and after calibration. BF and AF shows the locations of the robot stop point before and after calibration, respectively, in either UMBmark (U) or the proposed (P) method.

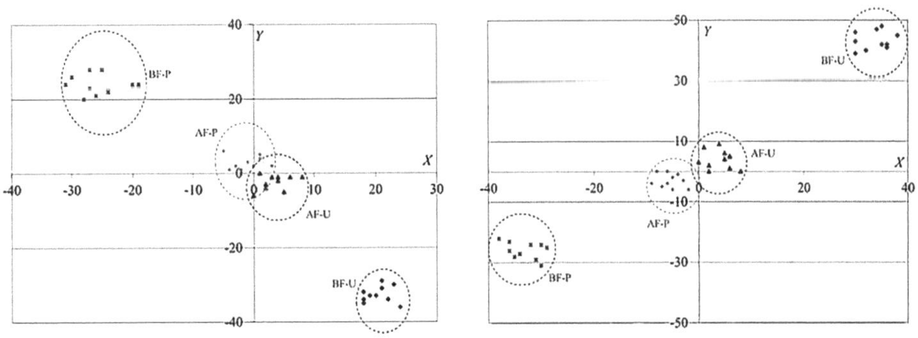

Table 3 shows the calculated parameters required to correct the robot motion for all four robots. These variables were calculated according to the test method described earlier. As shown in this table, using the proposed method, the values F_{lon} vary between 1.024 and 1.104. However, for omnidirectional robots, we have a vector for F_{lat}, and for differential drive robots, we have a number (scalar value). Table 4 lists the values of error improvement for all four robots. As observed, the robots were encountered less positional error at least 83%.

The one-way analysis of variance (ANOVA) was used to analyze the differences between mean values of position errors before and after calibration and their corresponding robot. P-values below 0.05 was considered significant. The p-values for jumper, maze, 3W omnidirectional and 4W omnidirectional were 4.32×10^{-5}, 1.02×10^{-4}, 6.31×10^{-3}, and 1.01×10^{-2}, respectively. Therefore, we detected that type of the robot tested did not change the hypothesis that positional errors reduce using either the proposed or UMBmark methods.

Figure 8. Radial systematic error of (left) three-wheeled and (right) four-wheeled omnidirectional robots before (BF) and after (AF) calibration

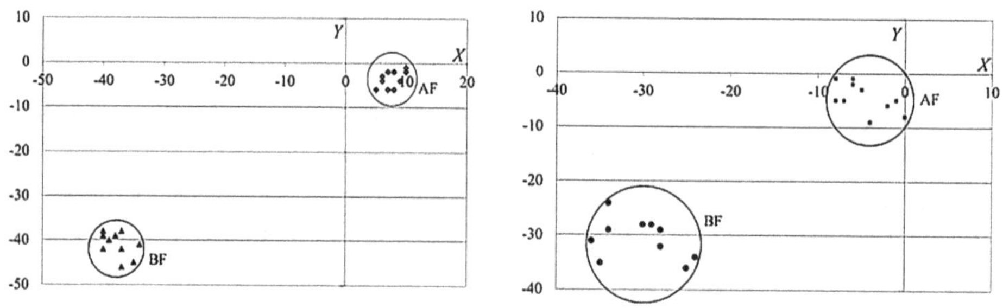

Table 3. Correction factors calculated based on experimental results

	UMBmark Method		Proposed Method	
	α (rad)	β (rad)	F_{Lat}	F_{Lon}
Jumper	-0.19	0.31	1.020	1.024
Maze	-0.32	0.41	1.041	1.028
3W Omni	Does not apply	Does not apply	*diag*[0.991 1.092 1.083]	1.104
4W Omni	Does not apply	Does not apply	*diag*[0.992 0.997 1.006 1.003]	1.075

Table 4. Error improvement (δr_m) for UMBmark and the proposed methods

	UMBmark Method	Proposed Method
Jumper	87.75%	89.24%
Maze	88.71%	83.89%
3W Omni	Does not apply	84.48
4W Omni	Does not apply	83.02

7. CONCLUSION

Odometry errors in mobile robots with differential drives are inevitable as they originate from hard to completely avoid imperfections such as unequal wheels diameters, misalignment at joints, backlash, slippage in encoder pulses and much more. In this study, we proposed a method to reduce the odometry errors, and compared its performance with the UMBmark, a commonly used approach in calibration of both differential drive and omnidirectional mobile robots. The focus was on correcting the systematic errors. Two prototype differential drive mobile robots and two omnidirectional robots were constructed to evaluate the effectiveness of the developed method. Experimental analyses, using the prototype robots, showed that the method can remove the systematic errors by at least 83%., which is comparable with the UMBmark method. The ANOVA test showed that error improvements were significant, therefore, we detected that type of the robot tested did not change the hypothesis that positional errors reduce using wither the proposed or UMBmark methods.

REFERENCES

Balazs, A. (2008). International Vocabulary of Metrology-Basic and General Concepts and Associated Terms. *Chemistry International.*

Borenstein, J., & Feng, L. (1994). *UMBmark: A method for measuring, comparing, and correcting dead-reckoning errors in mobile robots.* Academic Press.

Borenstein, J., & Feng, L. (1995). UMBmark: A benchmark test for measuring odometry errors in mobile robots. Academic Press.

Borenstein, J., & Feng, L. (1996). Measurement and correction of systematic odometry errors in mobile robots. *IEEE Transactions on Robotics and Automation*, *12*(6), 869–880. doi:10.1109/70.544770

Bury, B., & Hope, J. C. (1995). Autonomous mobile robot navigation using a low-cost fibre optic gyroscope. *IFAC Proceedings Volumes, 28*(11), 39-43. doi:10.1016/B978-0-08-042366-1.50011-7

de Wit, C. C. (1998). Trends in mobile robot and vehicle control. In *Control Problems in Robotics and Automation* (pp. 151–175). Berlin: Springer. doi:10.1007/BFb0015082

Denavit, J. (1955). A kinematic notation for lower-pair mechanisms based on matrices. *ASME J. Appl. Mech.*, 215-221.

Habib, M. K. (2011). Biomimetics: Innovations and robotics. *International Journal of Mechatronics and Manufacturing Systems*, *4*(2), 113–134. doi:10.1504/IJMMS.2011.039263

Ilon, B. E. (1975). *U.S. Patent No. 3,876,255.* Washington, DC: U.S. Patent and Trademark Office.

Kalvandi, R., & Maddahi, Y. (2012). On performance evaluation of home health care assistant platforms: Simulation study. *Int J of Biology and Biomedical Engineering*, *2*(6), 135–142.

Ko, J. H., Hwang, D. C., Jung, Y. W., & Kim, E. S. (2005, October). Intelligent mobile robot system for path planning using stereo camera-based geometry information. *Proceedings of the Society for Photo-Instrumentation Engineers*, *6006*, 60060L. doi:10.1117/12.632032

Maddahi, Y. (2005, December). Design and laboratory tests of wheeled mobile robots. In *International Conference on System Science and Simulation in Engineering* (pp. 186-191). Academic Press.

Maddahi, Y., Sepehri, & Maddahi A. (2012c) Calibration of four–wheeled omnidirectional mobile robots. *Int. Conf. of Canadian Society of Mechanical Engineering*.

Maddahi, Y., & Maddahi, A. (2004a). Mobile robot experimental analysis based on kinematics. *WSEAS Transactions on Circuits and Systems*, *3*(8), 1662–1667.

Maddahi, Y., & Maddahi, A. (2006). YMBM: New method for errors measurement in wheeled mobile robots. *WSEAS Transactions on Systems*, *5*(3), 552–557.

Maddahi, Y., Maddahi, A., & Monsef, S. H. (2012b). Design improvement of wheeled mobile robots: Theory and experiment. *J. World Appl. Sci*, *16*(2), 263–274.

Maddahi, Y., Maddahi, A., & Sepehri, N. (2013). Calibration of omnidirectional wheeled mobile robots: Method and experiments. *Robotica*, *31*(6), 969–980. doi:10.1017/S0263574713000210

Maddahi, Y., Monsef, S. H., & Mastorakis, N. E. (2014). How to Improve Positional Accuracy in Redundant Omnidirectional Mobile Robots? In *International Conference on Recent Advances in Electrical Engineering and Educational Technologies* (pp. 22-26). Academic Press.

Maddahi, Y., Seddigh, M., Pour, M. M., & Maleki, M. (2004b). Simulation study and laboratory results of two-wheeledmobile robot. *WSEAS Transactions on Systems*, *3*(9), 2807–2812.

Maddahi, Y., Sepehri, N., Maddahi, A., & Abdolmohammadi, M. (2012a). Calibration of wheeled mobile robots with differential drive mechanisms: An experimental approach. *Robotica*, *30*(6), 1029–1039. doi:10.1017/S0263574711001329

Marantos, P., Koveos, Y., Stergiopoulos, J., Panousopoulou, A., & Tzes, A. (2008, June). Mobile robot odometry relying on data fusion from RF and ultrasound measurements in a wireless sensor framework. In *Control and Automation, 2008 16th Mediterranean Conference on* (pp. 523-528). IEEE. doi:10.1109/MED.2008.4602252

Nistér, D., Naroditsky, O., & Bergen, J. (2006). Visual odometry for ground vehicle applications. *Journal of Field Robotics*, *23*(1), 3–20. doi:10.1002/rob.20103

Piaggio, M., Sgorbissa, A., & Zaccaria, R. (2001). Navigation and localization for service mobile robots based on active beacons. *Systems Science*, *27*(4), 71–83.

Rahok, S. A., & Koichi, O. (2009, February). Odometry correction with localization based on landmarkless magnetic map for navigation system of indoor mobile robot. In *Autonomous Robots and Agents, 2009. ICARA 2009. 4th International Conference on* (pp. 572-577). IEEE.

Shi, L., Habib, M. K., Xiao, N., & Hu, H. (2015). Biologically inspired robotics. *Journal of Robotics*.

Song, X., & Wang, Y. (2008, June). A novel model-based method for odometry calculation of all-terrain mobile robots. In *Intelligent Control and Automation, 2008. WCICA 2008. 7th World Congress on* (pp. 581-586). IEEE.

Wada, M., & Mori, S. (1996, April). Holonomic and omnidirectional vehicle with conventional tires. In *Robotics and Automation, 1996. Proceedings., 1996 IEEE International Conference on* (Vol. 4, pp. 3671-3676). IEEE. doi:10.1109/ROBOT.1996.509272

West, A. M. T. (1993). *U.S. Patent No. 5,186,270*. Washington, DC: U.S. Patent and Trademark Office.

West, M., & Asada, H. (1997). Design of ball wheel mechanisms for omnidirectional vehicles with full mobility and invariant kinematics. *Journal of Mechanical Design*, *119*(2), 153–161. doi:10.1115/1.2826230

Chapter 17
A Real-Time Compressive Tracking System for Amphibious Spherical Robots

Shuxiang Guo
Beijing Institute of Technology, China

Liwei Shi
Beijing Institute of Technology, China

ABSTRACT

Given the special working environments and application functions of the amphibious robot, an improved RGB-D visual tracking algorithm with dual trackers is proposed and implemented in this chapter. Compressive tracking (CT) was selected as the basis of the proposed algorithm to process colour images from a RGB-D camera, and a Kalman filter with a second-order motion model was added to the CT tracker to predict the state of the target, select candidate patches or samples, and reinforce the tracker's robustness to high-speed moving targets. In addition, a variance ratio features shift (VR-V) tracker with a Kalman prediction mechanism was adopted to process depth images from a RGB-D camera. A visible and infrared fusion mechanism or feedback strategy is introduced in the proposed algorithm to enhance its adaptability and robustness. To evaluate the effectiveness of the algorithm, Microsoft Kinect, which is a combination of colour and depth cameras, was adopted for use in a prototype of the robotic tracking system.

ORGANIZATION BACKGROUND

To execute various tasks in ever-changing environments autonomously, it is a critical and important work for a robot to sense and detect the surroundings. As a kind of feasible sensors with advantages of low-cost, low power consumption and strong adaptability (Fabian et al., 2014; Habib, 2011; Shi et al., 2015; Pan et al., 2015), digital cameras have been widely used in robotics to guide electromechanical devices and realize intelligent methods. In a machine vision-based robot, the visual tracking system plays a role of great importance to realize diverse robotic functions such as autonomous navigation (Capi et

DOI: 10.4018/978-1-5225-2993-4.ch017

al., 2010; Wirbel et al., 2013), path planning (Bischoff et al., 2012; Lin et al., 2013), visual servoing (Siradjuddin et al., 2012; Wang, P. et al., 2014), robot-human interaction (Oonishi et al., 2013; Gupta et al., 2014) and et al. The content of this chapter is completely based on the work of Pan et al. (2015).

In general, existing tracking algorithms can be categorized into estimation-based and classification-based (Liu, Q. et al., 2014). Estimation-based or generative algorithms model the target on the basis of appearance features and then search it in each frame of the visual stream (Salti et al., 2012). This kind is mainly motivated by innovations in appearance models and includes MeanShift (Liu, Y. F. et al., 2014), particle filter-based algorithms (Zhang et al., 2013), IVT (Incremental Visual Tracking) (Ross et al., 2008), optical flow-based algorithms (Liu et al., 2013) and et al. Classification-based or discriminative algorithms treat tracking as a binary pattern recognition problem and try to separate the target from the background (Pirzada et al., 2014). This kind is usually built upon pattern recognition algorithms such as SVM (Support Vector Machine) (Avidan, 2004), Bayes Classifier (Chen & Wu, 2013), K-Means (Qi et al., 2011) and et al. Most of studies on visual tracking are performed on high performance computers and evaluated with standard benchmark sequences, each of which contains controlled disturbance from the target or environment.

Compared to normal studies towards tracking algorithms on the basis of commercial computer in a controlled lab environment, designing a reliable visual tracking system for mobile robots is an even more challenging work owing to two aspects. On the one hand, the appearance of targets is variable due to pose or scale changes, random motion and occlusion, which makes it difficult to establish a reliable appearance model (Wang, K. et al., 2014a). On the other hand, the ambient environment of robots is easy to be disturbed because of illumination variation, camera vibration and outside interference, which may cause problems of drift or target missing (Wang, K. et al., 2014b). More than that, most of state-of-the-art studies aimed at improving tracking precision, adaptability and robustness, while the real-time performance and computational consumption of algorithms were often overlooked. That led to a situation in which most existing algorithms were unsuitable to apply to mobile robots equipped with embedded microprocessors and limited power source.

To realize flexible and economical underwater exploration, an amphibious spherical robot inspired by turtles (Guo et al., 2012) was designed by our team in 2012. And the improved version with the conception of "all programmable" was designed in 2014 (Pan et al., 2014). Given the special working environments and application functions of the amphibious robot, an improved visual tracking algorithm with dual trackers was proposed and implemented in this paper. A visible and infrared fusion mechanism was introduced in the proposed algorithm to enhance its adaptability and robustness. To evaluate effectiveness of the algorithm, a Microsoft Kinect, which is a combination of color and depth cameras, was adopted to build a prototype of the robotic tracking system. Compressive tracking (CT) was selected as the basis of the proposed algorism to process color images from the Kinect. And a Kalman filter with a 2-order motion model was added into the CT tracker to predict the state of the target, select candidate patches or samples and reinforce the tracker's robustness to high speed moving target. In the meantime, a CamShift tracker with Kalman prediction mechanism was adopted to process depth images from the Kinect. Through a feedback strategy, the depth tracking results were used to assist the CT tracker in updating classifier parameters in an adaptive rate. Finally, experimental results in various scenarios demonstrated the effectiveness, robustness and real-time performance of the tracking system. And drift, which is an obvious problem of CT system, was partly solved in the proposed algorithm.

The rest of this paper is organized as follows. First, an overview on related works will be introduced, which covers our amphibious spherical robot and CT system. Second, the proposed algorithm will be elaborated in three major areas. Third, the structure of the prototype tracking system for our amphibious spherical robot and experimental results on various benchmark sequences will be provided. Last part will be conclusion and follow-up relevant research work.

RELATED WORKS AND APPLICATION REQUIREMENTS

Amphibious Spherical Robot

As introduced in references (Guo et al., 2012; Pan et al., 2014), an amphibious spherical robot was proposed for covert missions or tasks in narrow room that normal AUV could not complete. As shown in Figure 1, the shell of the robot consisted of a hemisphere upper hull (250 mm in diameter) and two quarter-sphere lower hulls (266 mm in diameter) which can open and close. The hard upper hull was waterproof and can protect electronic devices and batteries inside it from collision. As shown in Figure 2, four actuating units, each of which consisted of two servo motors and a water jet motor, were symmetrically installed under the upper hull. When the robot was in underwater mode, the lower hulls closed and four water jet motors of actuating units provided vectored thrust though strip-type holes of lower hulls to realize 6 DOFs motion. When the robot was in land mode, the lower hulls opened and the actuating units stretched out to quadruped walk under the driven of eight servo motors. To facilitate upgrade, 3-D printing technology, a minimum system of Xilinx Zynq-7000 SoC and an embedded computer equipped with an Intel Atom processor were adopted to fabricate the robot (Pan et al., 2014).

Figure 1. Diagram of our spherical amphibious robot

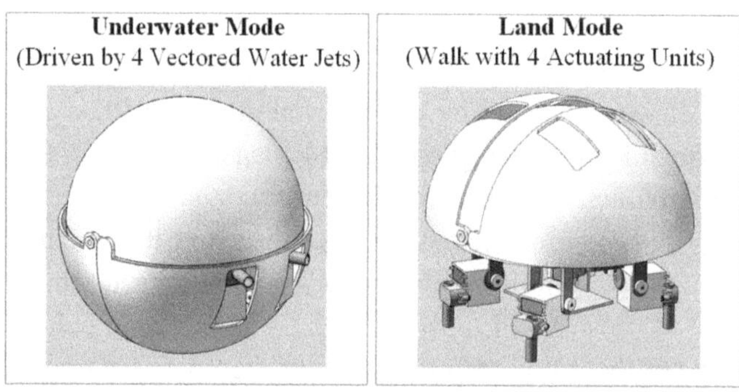

Figure 2. Actuating units and related driving parts

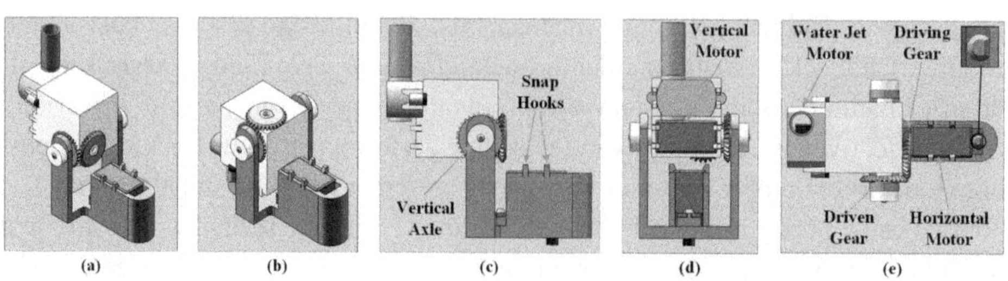

Restricted by the narrow load space and limited power resources of the small-scale robot, the robotic electronic system was fabricated around an embedded computer (Xilinx XC7Z045 SoC, 512 MB DDR3 memory, Linux 3.12.0). The robot was powered by a set of Li-Po batteries, with a total capacity of 24,000 mAh. Sensors including a global positioning system (GPS) module, an inertial measurement unit (IMU), and an industrial camera were used to achieve adaptive motion control of the robot (Li et al., 2017). An acoustic modem and a universal serial bus (USB) radio module were used for communication on land and in water, respectively (Li & Guo, 2016).

Because of the unique mechanical structure and special promising scenarios, it is a challenging work to customize the tracking system of the amphibious spherical robot. Firstly, the upper hull of the robot, in which electronic devices, scientific instruments and batteries were installed, was a very narrow and closed room. Thus high speed computers, which consume a lot of power and generate masses of heat, cannot be adopted in this mobile robot. And real-time performance of the adopting tracking algorithm turned into a critical problem. Secondly, considering that the robot may work in diverse scenarios of land and underwater, robustness to environmental disturbance caused by camera vibration and occlusion was essential to the adopting tracking algorithm. Thirdly, limited by the installation space, the field of view of the camera was very narrow, which made it vital to improve the adaptability of the adopting tracking algorithm to small and high speed moving targets (Wang et al., 2011). Besides, the visual tracking system of the robot was used chiefly in autonomous navigation and guiding electromechanical devices. Consequently, its tracking precision and effectiveness should be acceptable.

Real-Time Compressive Tracking Algorithms

In 2012, Zhang et al. proposed the CT algorithm which provided researchers a novel framework to develop real-time tracking algorithms for mobile robots (Zhang et al., 2012). Figure 3 shows the main components of the CT algorithm. As an effective and efficient discriminative algorithm, CT system consisted 2 stages: tracking and updating. In the tracking stage, candidate image patches of the target of the (n+1)-th frame are sampled around I_n, which is the tracking result at the n-th frame. Then integral vectors of these patches are calculated by accumulation. On the basis of compressive sensing theory, the high-dimensional integral vectors of samples are compressed to exact Haar-like features with a static measurement matrix. The process of compression or measurement can be denoted as $v = Ru$, where $u \in \mathbb{R}^n$ indicates the integral vectors; $v \in \mathbb{R}^m$ indicates the compressed feature vectors and the dimension $m \ll n$. R is a very sparse random matrix, the entries of which were defined as:

$$r_{i,j} = \sqrt{s} \times \begin{cases} +1,\text{with probability} \dfrac{1}{2s} \\ 0,\text{with probability } 1-\dfrac{1}{s} \\ -1,\text{with probability} \dfrac{1}{2s} \end{cases} \tag{1}$$

where s is set to $m/4$. For each row of R, less than 4 entries were non-zero, which makes computational complexity of the compression process very low. Then the low-dimensional feature vectors are entered into an on-line learning naive Bayes classifier. The sample with maximal classifier response will be set to the target for determining I_{n+1}. In the updating stage, training samples of the target and the

Figure 3. Diagram of the original CT algorithm

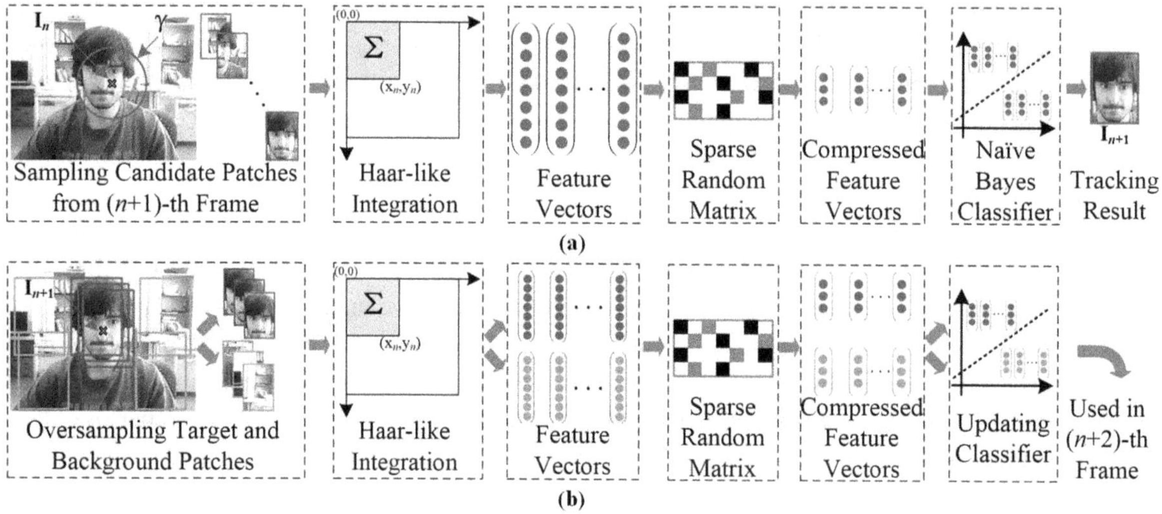

background are oversampled according to the tracking result at the (n+1)-th frame (I_{n+1}). And the compressed feature vectors of training samples are used to update the parameters of a naive Bayes classifier which will be used in the tracking stage of the (n+2)-th frame.

The CT algorithm succeeded in real-time performance and provided a low-cost and lossless way to compress image information and exact appearance features by using compressive sensing theory. However, CT samples candidate patches within a predefined radius around the tracking result of the previous frame. In consequence, the tracking precision relies on the value of the radius heavily and CT is unable to track targets in complex or high speed motion. Moreover, as parameters of the classifier are updated in a constant rate, the tracking results will keep drifting once the deviation occurs. Even worse, the target will be lost if it is fully occluded for a long time. Besides, scale invariance is not taken into account in CT system, which exacerbates the problem of tracking drift.

To make up insufficiencies of CT, Xu et al. proposed an improved CT algorithm containing two weighted classifiers which processed previous and current samples separately (Xu et al., 2013). Chen et al. proposed a scalable CT algorithm by using optical flow to estimate the size of the target (Chen & Wu, 2013). Yan et al combined CT with sparse representation and proposed the kernel sparse tracking algorithm (Yan & Li, 2014). Zhang et al. introduced a coarse-to-fine search strategy into the CT and proposed the fast-compressive tracking (FCT) algorithm (Zhang et al., 2014). But aforementioned improvement studies added complexity of the algorithm significantly or only solved a specialized problem, which limited their applications in our amphibious spherical robot.

Describe the technology utilization, advancements, management practices and philosophies, etc. of the organization prior to initiation of the IT project/experience described in the case, as well as the players involved and their roles within the company.

PROPOSED ALGORITHM

As digital images are projections of three-dimensional objects on two-dimensional plane, visual tracking is an estimation process based on incomplete information. Consequently, stereovision and multi-sensor fusion, which provide additional information of the target (depth, temperature, velocity and et al.), are effective means to improve precision and robustness of tracking algorithms. On the other hand, the drift problem of CT system is associated with tracking error accumulation which is caused by the process of on-line self-updating of the classifier. So adding a semi-supervised or feedback mechanism to the updating process should be a resultful method to handle the drift problem (Liu, Q. et al., 2014).

The Microsoft Kinect, which is a low-cost package of visible and infrared cameras, provides a feasible solution to designing an effective tracking system. Although originally designed for consumer electronics and computer games, the Kinect has attracted numerous interests of robotic researchers recently (Ren et al., 2014). Unlike stereovision system, it acquires 3-D information of environments by capturing color and depth image streams, which reduces the technical threshold and design difficulties.

Given features of the Kinect, an improved visual tracking algorithm for applications of our amphibious spherical robot was presented based on the conception of multi-feature fusion and feedback in this paper. The main components of the proposed algorithm were shown by Figure 4. A Microsoft Kinect was adopted as a compound image sensor for the prototype of tracking system. And dual trackers were implemented to process color and depth images separately. An improved CT tracker with Kalman prediction mechanism was deployed to process color images (C-Frame), while a CamShift tracker was implemented to process depth images (D-Frame). To realize sensor fusion and achieve improved robustness of the proposed algorithm, a feedback strategy inspired by expert control was designed to fuse tracking results of the trackers. The major contribution of the proposed algorithm consisted three areas: motion estimation for CT, CamShift-based depth tracking and the feedback strategy.

Figure 4. Main component of the proposed algorithm

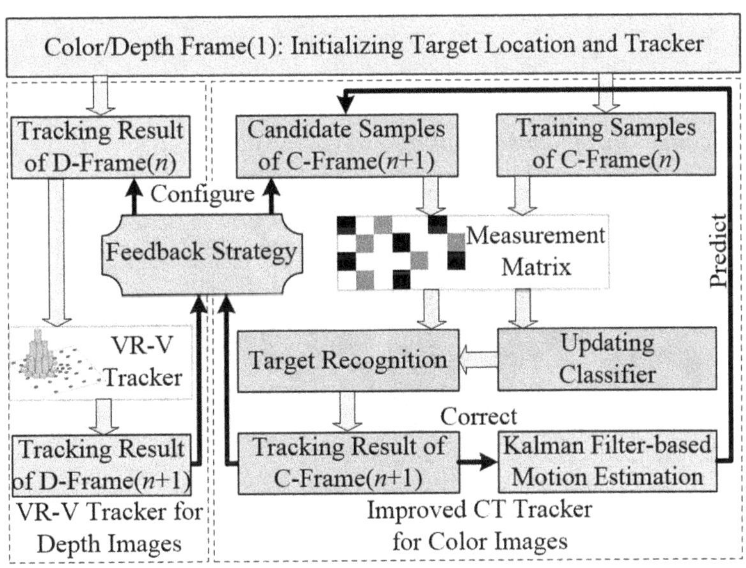

Visual Compressive Tracking With Motion Estimation

As mentioned in Section II, the original CT algorithm samples candidate patches around the previous tracking result in a static radius set manually, which can be denoted as:

$$D^{\gamma} = \left\{ z \middle| \left\| I(z) - I_n \right\| < \gamma \right\} \tag{2}$$

where D^{γ} is a set of candidate samples at the (n+1)-th frame, I_n is the center location of the target at the n-th frame and the sampling radius γ was set to 25 in the sample program of CT (Zhang et al., 2012). That sampling method weakens the robustness of CT to target in complex or high speed motion. As shown in left image of Fig. 5, because the radius is static, the desired tracking result $\tilde{I}_{n+1}$ may be on the edge of the sampling region or even outside the sampling region. That will lead to the visual drift problem or even target missing. On the other side, this problem can be partly solved by setting the sampling radius to a larger value. But the number of candidate patches to be sampled and processed will increase significantly, which decreases the superiority of CT in real-time performance.

As shown in the right image of Figure 5, to balance the real-time performance and robustness of CT, the center of sampling region was modified to the prediction tracking result at the (n+1)-th frame. And a Kalman filter was deployed to estimate the sampling center. Considering the prediction and tracking error of the improved CT tracker, the location and motion tendency of the target can be estimated with the linear dynamic model denoted as:

$$X_{n+1} = \Phi X_n + \beta W_n \tag{3}$$

$$Y_{n+1} = HX_{n+1} + \alpha V_n \tag{4}$$

$$\Phi = \begin{pmatrix} 1 & 0 & \Delta t & 0 \\ 0 & 1 & 0 & \Delta t \\ 0 & 0 & 1 & 0 \\ 0 & 0 & 0 & 1 \end{pmatrix} \tag{5}$$

Figure 5. Diagram of candidate patches sampling

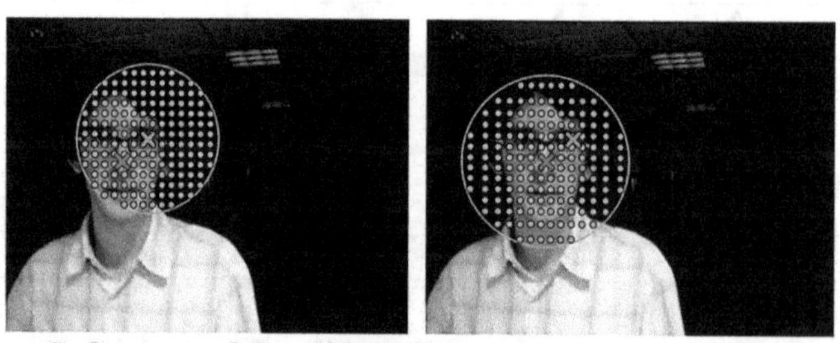

$$H = \begin{pmatrix} 1 & 0 & 0 & 0 \\ 0 & 1 & 0 & 0 \end{pmatrix} \tag{6}$$

$$X_n = \left(X_n, y_n, x_n - x_{n-1}, y_n - y_{n-1}\right)^T \tag{7}$$

$$Y_n = \left(\tilde{x}_n, \tilde{y}_n\right)^T \tag{8}$$

where Φ is the transfer matrix, H is the measurement matrix, α and β are adjustable parameters of the filter, W_n is the noise vector of prediction and V_n is the noise vector of measurement. The real state vector X_n contains two-dimensional position and velocity information of the target. And the measured state vector Y_n contains two-dimensional position information of the target. Considering that most of objects in smooth motion can be described with a three-order motion model and error exists commonly in tracking process, a two-order motion model was adopted in this filter for higher reliability.

The prediction process can be described as:

$$\widehat{X}_{n+1|n} = \Phi \widehat{X}_{n|n} \tag{9}$$

$$P_{n+1|n} = \Phi P_{n|n} \Phi^T + \beta Q_{n+1} \tag{10}$$

where $\mathbf{P}_{n+1|n}$ is the covariance matrix of noise. After predicting the motion state at (n+1)-th frame, the sampling center was set to $\hat{I}_{n+1}$ which was exact from the prediction vector $\widehat{X}_{n+1|n}$. And the candidate samples centered with $\hat{I}_{n+1}$ were collected based on two-dimensional Gaussian distribution. As shown in right image of Fig. 5, the denseness of candidate samples was decreased with the distance to the prediction location, which was closer to motion state of target in reality. Consequently, as the number of samples would not increase exponentially with the sampling radius, it was possible to track high speed moving targets with larger searching regions in real time. Besides, a matrix defining location of Gaussian samples was calculated in advance to accelerate the sampling process.

After the tracking stage of CT, which processed candidate samples and output the tracking result, the correction process can be described as:

$$K_{n+1} = P_{n+1|n} H^T \left(H P_{n+1|n} H^T + \alpha R_{n+1}\right)^{-1} \tag{11}$$

$$\widehat{X}_{n+1|n+1} = \widehat{X}_{n+1|n} + K_{n+1}\left(Y_{n+1} - H\widehat{X}_{n+1|n}\right) \tag{12}$$

$$P_{n+1|n+1} = \left(I - K_{n+1} H\right) P_{n+1|n} \tag{13}$$

where K_{n+1} is the Kalman gain matrix at the (n+1)-th frame, **I** is a unit matrix, $\widehat{X}_{n+1|n}$ is the measured state vector containing tracking result and $\widehat{X}_{n+1|n+1}$ is the filtered state vector. In this way, the correction process corrected parameters of the filter with residual error and tracking result at the (n+1)-th frame.

Visual tracking is an inaccuracy estimation process, the results of which contain error and drift caused by ever-changing environments and target. Therefore, the coefficient vectors (α and β) of noise matrix were adjusted adaptively by the feedback strategy which will be introduced in Section III B.

Depth Tracking on the Basis of Improved CamShift

As mentioned in Section II, the original CT algorithm updates parameters of the naive Bayes classifier in the tracking process for each frame. As a result, if the target is fully occluded for a long time, the classifier will instead learn the features of the covered object and the target will ultimately be lost. As the distance between the camera and most of target would not change toughly, this problem can be alleviated or solved by utilizing the infrared depth information captured by the Kinect. Meanwhile, as the imaging quality of depth images has nothing to do with illumination condition, the adaptability of the entire tracking system was increased by fusing color and depth image features.

As shown in Figure 6, the improved CT tracker cannot be aware of the situation that the target in color images is partly or fully occluded. But the histogram-based tracker for depth images is able to detect the exception and remind the CT tracker of stopping update the classifier temporarily. According to the clue, a CamShift tracker with Kalman prediction mechanism was adopted to process depth image. Prediction and correction processes of the Kalman filter were similar to the improved CT algorithm. And the reference tracking position and target scale provided by the CamShift would be used as reference information for the improved CT tracker.

In actual experiments, the gray pixel value of depth images ranged from 0 to 255. But the effective working distance of the Kinect was only 1.0m to 1.5m, which made the image mainly in dark and most of pixels concentrate upon 200 to 255, just as shown in the left image of Figure 7. To improve the imaging quality and emphasize features of the target, an enhancement process was designed after image acquisition. As show in middle and right images of Figure 7, the value of pixels were flipping and then

Figure 6. Color and depth images captured by a Kinect

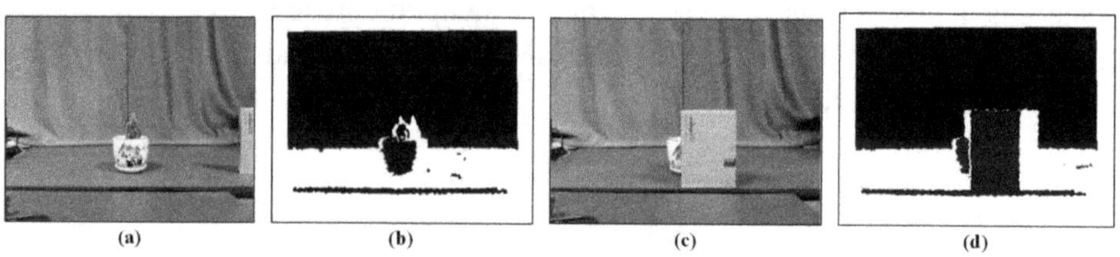

Figure 7. Diagram of depth image enhancement

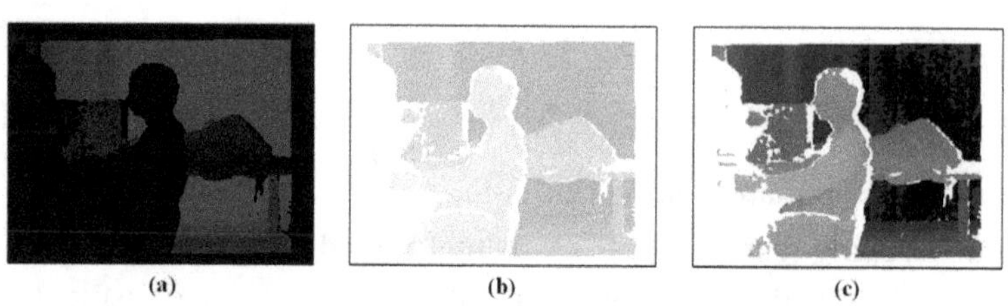

a histogram equalization operation was executed. Besides, depth values of smooth surface and remote objects may be inaccurate, so the range of histogram of the CamShift tracker should be set to a suitable value to avoid instability.

Feedback Strategy for Adaptive On-Line Learning

In practice applications of robotic visual tracking, numerous factors affect the performance of a tracking algorithm. Among these factors, illumination variation and occlusion can appear frequently. To make the proposed tracker more robust against these problems, a feedback strategy inspired by expert control was introduced to realize fusion of tracking results of the CT tracker and the CamShift tracker. By updating parameters of the Bayes Classifier and Kalman filter of CT adaptively, the overall performance of the tracker was improved.

Table 1 shows the status transition of the feedback strategy. The strategy analyzed the quality of tracking results and evaluated the status of ambient environment and trackers. In Table 1, $I_{C(n)}$ ($I_{D(n)}$) represented the tracking results at the n-th color (depth) frame. r_{scale} represented the square residual error of target scale of the CamShift tracker, and $\lambda_0, \varepsilon_0, \delta_0$ were boundary constants which controlled the adjustment of parameters. In condition 1, the CT tracker will update classifier parameters with a constant rate if tracking results of both trackers are consistent. And in condition 2 and 7, the search region of candidate patches will be enlarged if there is acceptable difference between trackers or the scale size of the depth target floats in a small range. While in condition 3 and 8, the CT tracker will stop updating and try to search the target in a much larger region if the tracking results of trackers are inconsistent or the scale size of the depth target changes significantly, in which there is a high likelihood that the environmental condition or target model deteriorates (fully occlusion, illumination variance or pose change).

Table 1. Status transition table of the feedback strategy

No.	$I_{C(n+1)}$-$I_{D(n+1)}$	$I_{D(n+1)}$-$I_{D(n)}$	r_{scale}	Adjust Parameter	Classifier Update
1	$\lvert I_{C(n+1)}-I_{D(n+1)}\rvert$<»$_0$	$\lvert I_{D(n+1)}-I_{D(n)}\rvert<\mu_0$	r_{scale}<'$_0$	$\gamma=\gamma_0$	True
2	»$_0$<$\lvert I_{C(n+1)}-I_{D(n+1)}\rvert$<»$_1$	$\lvert I_{D(n+1)}-I_{D(n)}\rvert<\mu_0$	r_{scale}<'$_0$	$\gamma=1.2\gamma_0$	True
3	$\lvert I_{C(n+1)}-I_{D(n+1)}\rvert$>»$_1$	$\lvert I_{D(n+1)}-I_{D(n)}\rvert<\mu_0$	r_{scale}<'$_0$	$\gamma=1.5\gamma_0$	False
4	$\lvert I_{C(n+1)}-I_{D(n+1)}\rvert$<»$_0$	$\lvert I_{D(n+1)}-I_{D(n)}\rvert>\mu_0$	r_{scale}<'$_0$	$\alpha, \beta, \gamma=1.2\gamma_0$	True
5	»$_0$<$\lvert I_{C(n+1)}-I_{D(n+1)}\rvert$<»$_1$	$\lvert I_{D(n+1)}-I_{D(n)}\rvert>\mu_0$	r_{scale}<'$_0$	$\alpha, \beta, \gamma=1.5\gamma_0$	True
6	$\lvert I_{C(n+1)}-I_{D(n+1)}\rvert$>»$_1$	$\lvert I_{D(n+1)}-I_{D(n)}\rvert>\mu_0$	r_{scale}<'$_0$	Reinitialization	False
7	$\lvert I_{C(n+1)}-I_{D(n+1)}\rvert$<»$_0$	$\lvert I_{D(n+1)}-I_{D(n)}\rvert<\mu_0$	'$_0$<r_{scale}<'$_1$	$\gamma=1.2\gamma_0$	True
8	$\lvert I_{C(n+1)}-I_{D(n+1)}\rvert$<»$_0$	$\lvert I_{D(n+1)}-I_{D(n)}\rvert<\mu_0$	r_{scale}>'$_1$	$\gamma=1.5\gamma_0$	False
9	$\lvert I_{C(n+1)}-I_{D(n+1)}\rvert$>»$_1$	$\lvert I_{D(n+1)}-I_{D(n)}\rvert>\mu_0$	r_{scale}>'$_1$	Reinitialization	False

In conditional 4 and 5, parameters of the CamShift tracker will be adjusted and the search region of the CT tracker will be enlarge, if the tracking results of both trackers are consistent while the depth target moves in a high speed, which may be caused by instability of the infrared camera of the Kinect. In condition 6 and 9, if tracking results of both trackers differ a lot, the target may have run out of the field of view or the target has been lost. Therefore, the classifier updating will terminate for 10 frames and try to reinitialize the valid tracking process after the recovery of the target. In short, this feedback tried to extend the stable margin of the CT tracker by limit the self-updating and enlarge the search region.

EXPERIMENT AND DISCUSSION

Experimental Environment and Tracking System Prototype

As the study aimed at designing a real-time, robust and effective algorithm for our amphibious spherical robot, both robot-based and computer-based experiments were arranged to test its performance in virtual or actual scenarios with occlusion, rough motion, illumination variation and other interferences. Experimental tests on the proposed algorithm consisted of 3 part.

1. In the benchmark test part, the improved CT tracker, which contained motion estimation mechanism and ran on a laptop computer (Intel Core i7-4712 2.30GHz CPU and 8GB RAM), was tested on 7 standard benchmark sequences compared with the original CT algorithm and the online multiple instance learning (MIL) algorithm (Babenko et al., 2011);
2. In the actual test part, the proposed algorithm running on the same computer was deployed to process color and depth image sequences captured by a Kinect;
3. In the robotic test part, the proposed algorithm was implemented to guide the motion of our amphibious spherical robot. As shown in Fig. 8, limited by the size of the current version of our robots, a preliminary prototype of the robotic tracking system for algorithm evaluation was fabricated, in which a Kinect was installed upon the activity region of the robot and monitored the motion region of the robot. And the proposed algorithm running on the same computer tracked a specialized guiding object and controlled the robot follow it through the wireless local network.

Experimental Results

In benchmark and actual test parts, two metrics are used to evaluate the improved CT and proposed algorithm with CT and MIL. The first metric is the success rate (SR) of the benchmark sequences. The success rate of a frame is defined as:

$$score = \frac{area\left(ROI_T \cap ROI_G\right)}{area\left(ROI_T \cup ROI_G\right)} \tag{14}$$

where ROI_T is the tracked bounding box, ROI_G is the ground truth bounding box and $area(\cdot)$ denotes the number of pixels in the region. If the *score* is larger than the given threshold (0.5 in this paper) in a frame, it counts as a success (Zhang et al., 2014). The second metric is the center location error (CLE) which is the Euclidean distance between the central points of the tracked bounding box and the ground truth bounding box. Given that a tracker may be sensitive to the initial conditions or parameters of the target in the first frame, various test results of the same algorithm can be obtained with different initial bounding boxes of the target. For fair performance evaluation, the initial position and scale size of the target were specified according to the data of ground truth bounding boxes provided by (Wu et al., 2013). Therefore, although there was no modification of programs of CT and MIL, the experimental results in this paper may be slightly worse than counterparts reported (Zhang et al., 2014; Babenko et al., 2011).

Table 2 lists the major challenging factors of evaluational image sequences for benchmark and actual test parts. The sequences of, "Coke", "Couple", "Liquid", "Suv", "Tiger1", "Tiger2" and "Fish" were provided by Yi et al. (2013) and used to test the improved CT tracker, which was a single combination of CT and motion estimation based on a Kalman filter (elaborated in Section III A). And the sequences of "Bottle", "Book", "Glove1" and "Glove 2", which consisted of color images and depth images, were recorded with a Kinect by us and used to evaluate the proposed algorithm (elaborated in Section III A, B and C). The total frame number of all image sequences was 6512.

Table 3 shows the quantitative evaluation results of the benchmark test part. In the test towards the improved CT algorithm, the sampling radius was set to γ =50, the dimensionality of compressed feature vectors was set to n=50, and the learning parameter was set to $\lambda = 0.85$. The results indicated that the

Figure 8. Prototype of the robotic tracking system

Table 2. Challenge of the benchmark sequences

Sequences		Major Challenge
Standard	Coke	Fast motion, Occlusion, Background Clutters
	Couple	Fast Motion, Deformation, Background Clutters
	Liquid	Occlusion, Fast Motion, Background Clutters
	Suv	Occlusion, In-Plane Rotation, Out-of-View
	Tiger1	Illumination Variation, Fast Motion, Occlusion
	Tiger2	Illumination Variation, Fast Motion, Occlusion, Out-of-View
	Fish	Illumination variation
Ours	Bottle	Fast Motion, Occlusion
	Book	Illumination variation, Fast Motion, Occlusion
	Glove 1	Random Motion
	Glove 2	Fast Motion, Out-of-View

Figure 9. Tracking result on classical image sequences

robustness to fast motion, occlusion and background clutters of the improved CT algorithm was reinforced with the motion estimation mechanism. And the computational complexity of the improved CT algorithm increased only a little, which did not influence its real-time performance. As shown in Figure 9 (a), the bounding box of CT and MIL drifted obviously when the can of coke moved roughly at #200 and #225, while the improved CT tracker was able to track the target precisely. As shown in Figure 9 (f), the CT tracker thoroughly lost the target when occlusion occurred frequently at #341 and #362, while the im-

Table 3. Success rate (SR) and center location error (CLE) of the improved CT

Sequences		Improved CT		Original CT		MILTrack		Number of Frames
		SR	CLE	SR	CLE	SR	CLE	
Standard	Coke	19.2	31.2	15.8	36.1	9.2	28.8	291
	Couple	83.6	8.6	68.6	35.4	65.7	35.9	140
	Liquor	40.1	141.9	28.1	168.9	21.0	131.8	1741
	Suv	58.5	39.1	17.9	79.2	46.2	59.8	945
	Tiger1	49.3	23.6	46.8	28.0	8.5	104.6	354
	Tiger2	51.7	24.5	45.2	28.6	45.7	28.3	365
	Fish	9.5	37.1	3.8	38.8	81.3	10.2	476
Frames per Second (320×240 pixel)		72.5		73.0		81.5		

proved CT tracker detected the target successful when the target reappeared. As shown in Figure 9 (b), the CT tracker and MIL tracker missed the couple when a car with similar Haar-like features appeared, but the improved CT tracker avoided the failure by predicting the motion trend of the target. Although the improved CT algorithm had higher success rate in the benchmark test, it made little contribution to the performance improvement when illumination variation or fully occlusion happened, as shown in Figure 9 (c) and (g). Even more, as shown in Figure 9 (e), if the target moves random or too fast, the motion estimation based on a Kalman filter may not improve or even exacerbate the effectiveness of the CT algorithm as the Kalman filter only works well for Gaussian linear systems. The fact manifested that an assistant detection mechanism is essential to supervise the improved CT tracker.

Table 4 shows the quantitative evaluation results of the actual test part. In the test towards the proposed algorithm, the sampling radius was set to γ =50, the dimensionality of compressed feature vectors was set to n=50, and the learning parameter was set to $\lambda = 0.85$. The results indicated that the effectiveness and robustness of the proposed algorithm was improved further with the multi-feature fusion mechanism and the feedback strategy. And the real-time performance of the proposed algorithm was still acceptable for robotic applications. As shown in Figure 10 (a), the CT tracker and the MIL tracker missed the target firstly when the target moved in high speed for a long time at #457. And the improved CT tracker failed later after fully occlusion. But the tracker of the proposed algorithm, which stopped updating classifier when fully occlusion occurred, was able to detect the target successfully at #52. As shown in Figure 10 (b) and (c), trackers of the improved CT, CT and MIL failed in tracking successively when the target moved randomly or swerved roughly, but the tracker of the proposed algorithm succeeded by extending the search region and adjusting parameters of the Kalman filter. As shown in Figure 10 (d), the tracker of the proposed algorithm was able to handle the scenario in which the target left the field of view and reappeared later by stopping on-line learning temporarily. Figure 10 (e) shows some examples of depth images which were used by the proposed algorithm. The target can be easily separated from the background and occlusions by comparing the gray value or histogram of regions. The test results verified the effectiveness and robustness of the proposed algorithm and feasibility of designing robust tracking system with compound sensors like the Kinect.

In the robotic test part, a prototype of robotic tracking system was built for algorithm evaluation. As the size of a Kinect is too large to be embedded into our robot, a Kinect was fixed on a frame and

Figure 10. Tracking results of our image sequences in real scenarios

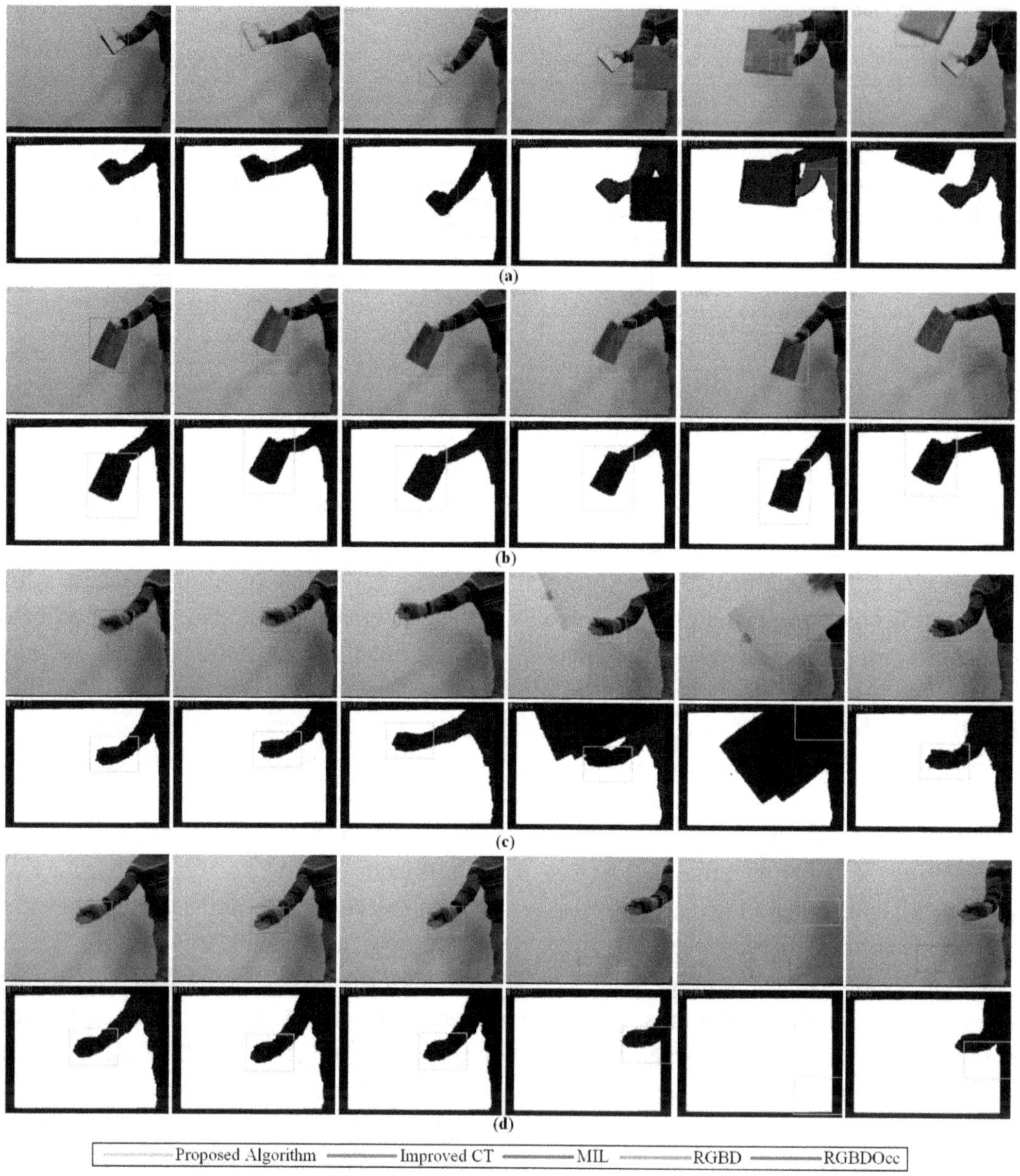

Table 4. Success rate (SR) and center location error (CLE) of proposed algorithm

Sequences		Proposed Algorithm		Improved CT		Original CT		MILTrack		Number of Frames
		SR	CLE	SR	CLE	SR	CLE	SR	CLE	
Customized	Bottle	87.1	10.2	62.1	29.1	56.2	33.2	57.2	63.6	600
	Book	93.3	9.1	36.6	55.3	35.3	57.9	46.3	47.9	600
	Glove1	71.3	8.9	41.5	23.5	34.7	46.7	40.9	42.7	400
	Glove2	82.1	13.3	57.2	22.6	51.7	31.6	55.6	27.2	600
Frames per Second (640×480 pixel)		57.7		69.7		71.1		73.5		

monitored the motion region of our spherical robot from an overhead view, as shown in Figure 11 (a). And a small orange hemisphere which moved along a $0.8\text{m} \times 0.6\text{m}$ black box acted as a guiding object. As shown in Figure 8 and Figure 11 (b), the proposed algorithm running on the laptop tracked the guiding object, calculated its trajectory and controlled the motion of our spherical robot though the wireless local area network. Figure 11 (c) shows the tracking results (red lines) and trajectory of the robot (green lines). There is only slight deviation less than 5cm or 3 pixels between the tracking result and the designated trajectory. And the controller module of the robot was set to omit small deviations or noises. As a result, except drifts occurring in bends of the design trajectory, the robot was able to follow the motion of the guiding object nearly exactly with the guidance of the tracking system.

Figure 11. Robotic tracking experiment of our amphibious spherical robot

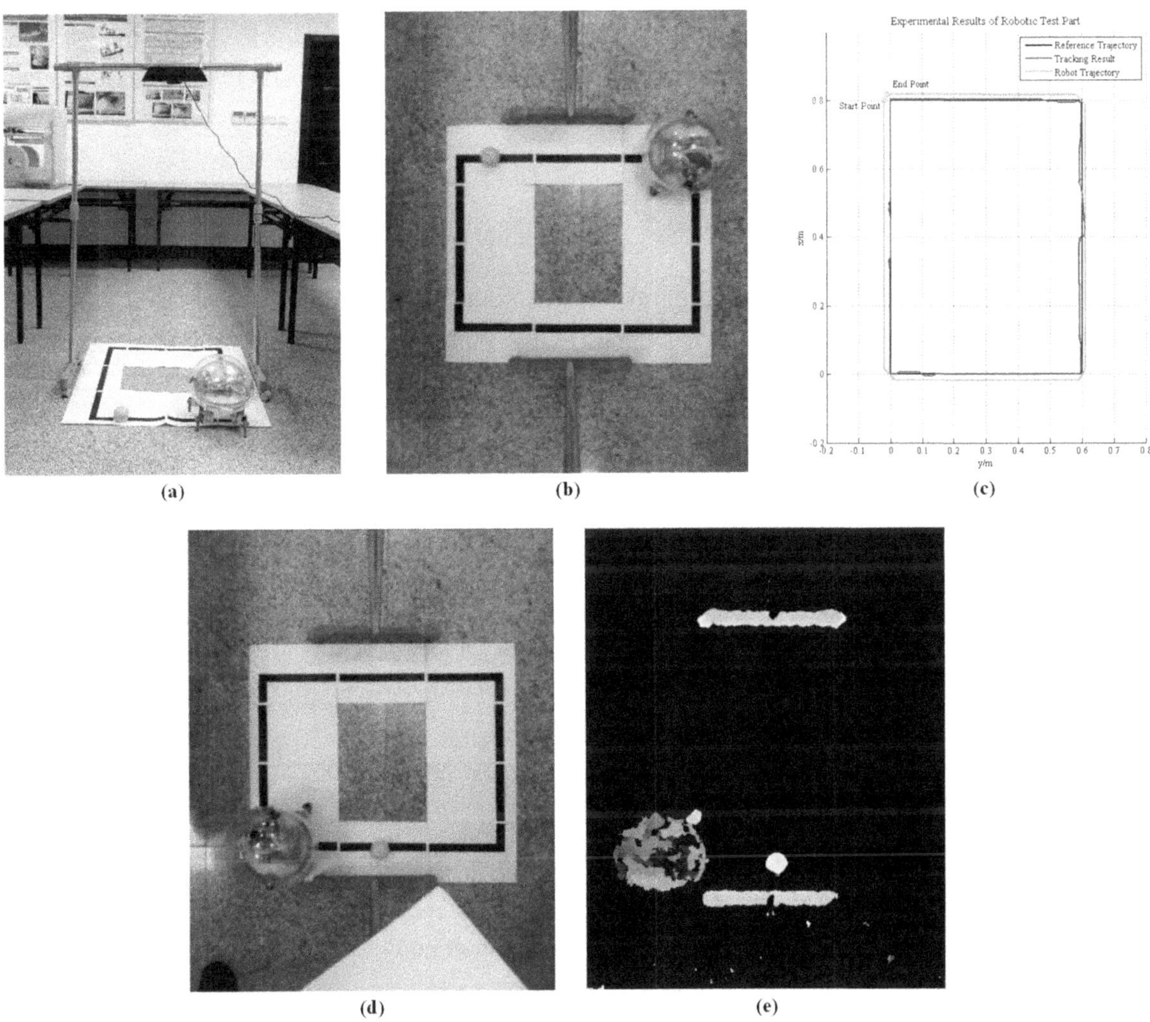

CONCLUSION

In this paper, a real-time visual tracking algorithm was proposed and implemented for our amphibious spherical robot. To meet requirements of robots in robustness, adaptability and effectiveness of tracking systems, a Microsoft Kinect was adopted to capture color and depth images. An improved CT tracker with Kalman perdition mechanism was deployed to process color images from the Kinect. And a Cam-Shift tracker was implemented to process depth images from the Kinect. With a feedback strategy, the updating rate of the CT tracker was adjusted, which partly solved its problems of ineffectiveness in fully occlusion, rough motion target and illumination variation. The experimental results in virtual, actual and robotic scenarios verified the effectiveness, efficiencies and robustness of the proposed algorithm.

This paper only provided a preliminary prototype of robotic tracking system. Limited by the size, a Kinect cannot be installed into our spherical robot. Meanwhile, the imaging quality of depth images from a Kinect is not good enough to realize precise visual servoing in outdoor environments. Besides, the mechanisms of parameter adjustment and adaptive classifier updating in the feedback strategy were experience-based and primitive in some measure. Our future work will focus on realization of a small Kinect-like compound image sensor for robotic applications.

ACKNOWLEDGMENT

This work was supported by National Natural Science Foundation of China (61503028, 61773064, 61375094). This work was supported by National Natural Science Foundation of China and National High Tech. Research and Development Program of China (No.2015AA043202).

REFERENCES

Avidan, S. (2004). Support Vector T racking. *IEEE Transactions on Pattern Analysis and Machine Intelligence*, *26*(8), 1064–1072. doi:10.1109/TPAMI.2004.53 PMID:15641735

Babenko, B., Yang, M. H., & Belongie, S. (2011). Robust Object Tracking with Online Multiple Instance Learning. *IEEE Transactions on Pattern Analysis and Machine Intelligence*, *33*(8), 1619–1632. doi:10.1109/TPAMI.2010.226 PMID:21173445

Bischoff, B., Duy, N. T., Streichert, F., Ewert, M., & Knoll, A. (2012). Fusing vision and odometry for accurate indoor robot localization. In *Proceedings of 2012 12th International Conference on Control Automation Robotics & Vision (ICARCV)*. Guangzhou: IEEE.

Capi, G., Toda, H., & Nagasaki, T. (2010). *A vision based robot navigation and human tracking for social robotics*. Paper presented at 2010 IEEE International Workshop on Robotic and Sensors Environments (ROSE), Phoenix, AZ. doi:10.1109/ROSE.2010.5675255

Chen, X., & Wu, J. (2013). Scalable compressive tracking based on motion. In *Proceedings of 2013 IEEE International Conference on Robotics and Biomimetics (ROBIO)*. Shenzhen: IEEE. doi:10.1109/ROBIO.2013.6739510

Fabian, J., Young, T., Jones, J. C. P., & Clayton, G. M. (2014). Integrating the Microsoft Kinect with Simulink: Real-Time Object Tracking Example. *IEEE/ASME Transactions on Mechatronics*, *19*(1), 249–257. doi:10.1109/TMECH.2012.2228010

Guo, S., Mao, S., Shi, L., & Li, M. (2012). Development of an Amphibious Mother Spherical Robot Used as the Carrier for Underwater Microrobots. In *Proceedings of the 2012 International Conference on Complex Medical Engineering (ICME)*. Kobe: IEEE. doi:10.1109/ICCME.2012.6275640

Gupta, M., Behera, L., Subramanian, V. K., & Jamshidi, M. M. (2014). A Robust Visual Human Detection Approach With UKF-Based Motion Tracking for a Mobile Robot. *IEEE Systems Journal*, *9*(4), 1–13.

Habib, M. K. (2011). Biomimetics: Innovations and Robotics. *International Journal of Mechatronics and Manufacturing Systems*, *4*(2), 113–134. doi:10.1504/IJMMS.2011.039263

Li, M., Guo, S., Hirata, H., & Ishihara, H. (2017). A roller-skating/walking mode-based amphibious robot. *Robotics and Computer-integrated Manufacturing*, *44*, 17–29. doi:10.1016/j.rcim.2016.06.005

Li, Y., & Guo, S. (2016). Communication between spherical underwater robots based on the acoustic communication methods. In *Proceedings of the 2016 IEEE International Conference on Mechatronics and Automation (ICMA)*. Harbin: IEEE. doi:10.1109/ICMA.2016.7558597

Lin, R., Li, M., & Sun, L. (2013). Image features-based mobile robot visual SLAM. In *Proceedings of 013 IEEE International Conference on Robotics and Biomimetics (ROBIO)*. Shenzhen: IEEE.

Liu, J., Subramanian, K. R., & Yoo, T. S. (2013). An optical flow approach to tracking colonoscopy video. *Computerized Medical Imaging and Graphics*, *37*(3), 207–223. doi:10.1016/j.compmedimag.2013.01.010 PMID:23490235

Liu, Q., Zhao, X., & Hou, Z. (2014). Survey of single-target visual tracking methods based on online learning. *IEEE Transactions on Computer Vision*, *8*(5), 419–428. doi:10.1049/iet-cvi.2013.0134

Liu, Y. F., Li, Q., Fang, H., & Xu, H. C. (2014). Research on Embedded System with Implementation of a Moving Object Tracking Algorithm Based on Improved Meanshift on DM6437. *Advanced Materials Research*, *1003*, 207–210. doi:10.4028/www.scientific.net/AMR.1003.207

Oonishi, S., Yanou, A., & Minami, M. (2013). Human-face-tracking using visual servoing by patient robot. In *Proceedings of 2013 SICE Annual Conference (SICE)*. Nagoya: IEEE.

Pan, S., Guo, S., Shi, L., He, Y., Wang, Z., & Huang, Q. (2014). A Spherical Robot based on all Programmable SoC and 3-D Printing. In *Proceedings of 2014 IEEE International Conference on Mechatronics and Automation*. Tianjin: IEEE. doi:10.1109/ICMA.2014.6885687

Pan, S., Shi, L., & Guo, S. (2015). A Kinect-Based Real-Time Compressive Tracking Prototype System for Amphibious Spherical Robots. *Sensors (Basel)*, *15*(4), 8232–8252. doi:10.3390/s150408232 PMID:25856331

Pirzada, N., Nayan, M. Y., Hassan, F. S. M. F., & Khan, M. A. (2014). Device-free Localization Technique for Indoor Detection and Tracking of Human Body: A Survey. *Procedia: Social and Behavioral Sciences*, *129*, 422–429. doi:10.1016/j.sbspro.2014.03.696

Qi, Y., Suzuki, K., Wu, H., & Chen, Q. (2011). EK-means tracker: A pixel-wise tracking algorithm using kinect. In *Proceedings of 2011 Third Chinese Conference on Intelligent Visual Surveillance (IVS)*. Beijing: IEEE.

Ren, H. L., Liu, W., & Lim, A. (2014). Marker-Based Surgical Instrument Tracking Using Dual Kinect Sensors. *IEEE Transactions on Automation Science and Engineering*, *11*(3), 921–924.

Ross, D. A., Lim, J. R., Lin, S., & Yang, M. H. (2008). Incremental learningfor robust visual tracking. *International Journal of Computer Vision*, *77*(1), 125–141. doi:10.1007/s11263-007-0075-7

Salti, S., Cavallaro, A., & Stefano, L. D. (2012). Adaptive Appearance Modeling for Video Tracking: Survey and Evaluation. *IEEE Transactions on Image Processing*, *21*(10), 4334–4348. doi:10.1109/TIP.2012.2206035 PMID:22759454

Shi, L., Habib, M. K., Xiao, N., & Hu, H. (2015). Biologically Inspired Robotics [editorial]. *Journal of Robotics*, *2015*, 1–2. doi:10.1155/2015/894394

Siradjuddin, I., Behera, L., McGinnity, T. M., & Coleman, S. (2012). A position based visual tracking system for a 7 DOF robot manipulator using a Kinect camera. In *Proceedings of the 2012 International Joint Conference on Neural Networks (IJCNN)*. Brisbane: IEEE. doi:10.1109/IJCNN.2012.6252597

Wang, H. S., Chen, W. D., & Wang, Z. L. (2011). Visual tracking of robots in uncalibrated environments. In *Proceedings of 2011 IEEE/RSJ International Conference on Intelligent Robots and Systems (IROS)*. San Francisco: IEEE. doi:10.1109/IROS.2011.6094647

Wang, K., Liu, Y., & Li, L. (2014a). Visual Servoing Trajectory Tracking of Nonholonomic Mobile Robots Without Direct Position Measurement. *IEEE Transactions on Robotics*, *30*(4), 1026–1035. doi:10.1109/TRO.2014.2317891

Wang, K., Liu, Y., & Li, L. (2014b). Visual servoing based trajectory tracking of underactuated water surface robots without direct position measurement. In *Proceedings of 2014 IEEE/RSJ International Conference on Intelligent Robots and Systems (IROS 2014)*. Chicago: IEEE. doi:10.1109/IROS.2014.6942645

Wang, P., Su, J., Li, W., & Qiao, H. (2014). Adaptive visual tracking based on discriminative feature selection for mobile robot. In *Proceedings of 2014 IEEE 4th Annual International Conference on Cyber Technology in Automation, Control, and Intelligent Systems (CYBER)*. Hong Kong: IEEE.

Wirbel, E., Steux, B., Bonnabel, S., & De, L. F. A. (2013). Humanoid robot navigation: From a visual SLAM to a visual compass. In *Proceedings of 2013 10th IEEE International Conference on Networking, Sensing and Control (ICNSC)*. Evry: IEEE.

Wu, Y., Lim, J., & Yang, M. H. (2013). Online Object Tracking: A Benchmark. In *Proceedings of 2013 IEEE Conference on Computer Vision and Pattern Recognition (CVPR)*. Portland: IEEE. doi:10.1109/CVPR.2013.312

Xu, R., Gu, X., & Wei, B. (2013). An Improved Real Time Compressive Tracking. In *Proceedings of 2013 2nd IAPR Asian Conference on Pattern Recognition (ACPR)*. Okinawa: IEEE.

Yan, Q. S., & Li, L. S. (2014). Kernel sparse tracking with compressive sensing. *IET Computer Vision*, *8*(4), 305–315. doi:10.1049/iet-cvi.2013.0095

Zhang, K., Zhang, L., & Yang, M. H. (2012). Real-Time Compressive Tracking. In *Proceedings of European Conference on Computer Vision (ECCV 2012)*. Berlin: IEEE. doi:10.1007/978-3-642-33712-3_62

Zhang, K., Zhang, L., & Yang, M. H. (2014). Fast Compressive Tracking. *IEEE Transactions on Pattern Analysis and Machine Intelligence*, *36*(10), 2002–2015. doi:10.1109/TPAMI.2014.2315808 PMID:26352631

Zhang, M., Xin, M., & Yang, J. (2013). Adaptive multi-cue based particle swarm optimization guided particle filter tracking in infrared videos. *Neurocomputing*, *122*, 163–171. doi:10.1016/j.neucom.2013.05.041

Chapter 18
Self-Powered Height Sensor With ZigBee Networks for Intelligent Systems

Mochen Li
Purdue University, USA

H. Henry Zhang
Purdue University, USA

Li Tan
Purdue University Northwest, USA

Wangling Yu
Purdue University Northwest, USA

ABSTRACT

With the development of micro-electro-mechanical systems (MEMS), wireless communication technology, and embedded systems, wireless sensor network (WSN) has been a focus for research among various fields. Wireless sensor technology inspires many innovations for industrial applications and medical robots, and breaks many limitations and inconvenience for such sensing devices. A self-powered height sensing system with ZigBee technology is presented. It specifically targets to replace an original wired system with an integrated wireless sensor that is comprised of all necessary parts: sensing module, processing module, RF transceiver, and power supply. First, the authors present the system framework design including the layout of a wireless sensor node based on ZigBee. Second, with the vibration environment, a self-powered generator was developed through the comparison between piezoelectric and electromagnetic generation. Then several experiments are conducted to test and analyze the feasibility of the whole system. Finally, a future upgraded design is proposed to improve system performance.

INTRODUCTION

Modern vehicles are intelligentized with various sensors in different systems. These sensing systems will monitor driving conditions, detect the abnormal problems, and alert the driver through the vehicle computer. At the same time, more and more automobile manufactures start to equip their vehicles with

DOI: 10.4018/978-1-5225-2993-4.ch018

automated driving system or develop the autonomous car like the Google Self-Driving Car or Tesla Autopilot. Thanks to utilization of a great variety of sensors in those cars, it can easily obtain real-time surrounding environment information and make proper timely responses. One can assert that more sensors will be applied in vehicles in the near future. In comparison with the current wired sensing system, wireless sensor networks (WSN) can bring more convenience for the layout design of a vehicle. With the development of this burgeoning technology, it shows a great advantage in distributed, automatic detection.

With the development of wireless sensor networks (WSN), this burgeoning technology for distributed, automatic detection can be applied in many parts of a vehicle, and these small computing devices are increasingly embedded in different sensors to work as monitoring devices or for emergency response (Hersent, Boswarthick, & Elloumi, 2011). Thus, it is profitable to research the possibility of replacing a wired sensor with a wireless sensor node in some appropriate parts. Typically, a wireless sensor node has four main parts: a power supply unit, a processing unit, a physical sensor and a RF transceiver (Labiod, Afifi, & Santis, 2007). One of main limitations for a WSN is the power consumption of each node. Normally, one solution is to focus on energy saving techniques by establishing a low-cost, low-power wireless network to extend the functional life with limited battery (Hersent, Boswarthick, & Elloumi, 2011). Another employs the environmental energy harvesting devices as the power supply (Bibo & Daqaq, 2013). The height sensors in truck suspension air spring provide a good way to research these two problems. Originally, the height sensor is powered by an external wire. But considering the inside vibration environment of the truck suspension air spring, an energy scavenger may supply the power for the whole wireless sensor nodes by converting the vibration kinetic energy into available electrical energy.

As a direct method of converting kinetic energy to electrical energy, piezoelectric materials have been the research focus of power supply for small devices (Bibo & Daqaq, 2013). Experiments show that the produced polarized charge is proportional to the applied stress. This feature of the stress-dependent charge output has typically been applied to many different devices, including physical sensors, actuators, microphones, etc. (Galchev & Najafi, 2009). A huge amount of research has been done in the field of power harvesting to enable power up wireless sensor nodes (Bibo & Daqaq, 2013; Galchev & Najafi, 2009; Choi, Jeon, Jeong, Sood, & Kim, 2006; Kim, Kim, & Kim, 2011).

ZigBee is a specification for low-cost, low-power and wireless mesh network based on the IEEE 802.15.4 standard (Hersent, Boswarthick, & Elloumi, 2011). With the development of this specification more and more companies begin to develop their own products (Baronti et al., 2007). ZigBee technology is gradually applied in the field of smart home, industrial control, intelligent transportation, etc. As a basic application of wireless sensor networks, the remote sensor is also one of the research hotspots (Guo, Healy, & Zhou, 2010; Xu et al., 2010).

In fact, the major role of the WSN is to monitor the surrounding physical environment at a low-cost way. Meanwhile, the detected data of surrounding world is precisely processed by biomedical robotics. The current WSN have many types of sensors to apply that can detect a wide range of phenomena including seismic, electromagnetic, temperature, humidity, noise, light intensity, pressure soil composition, velocity etc. On the other hand, the WSN also includes many different communication protocols, such as: WirelessHART, Z-Wave, etc. In this chapter, only one example-the combined application of height sensor and ZigBee-will be described.

SYSTEM FRAMEWORK DESIGN

System Design Principles

The purpose of this research is to develop a self-powered height sensing system with ZigBee technology based on the original wired sensor in a truck suspension air spring. Generally speaking, a wireless sensor node, as shown in Figure 1, commonly consists of four major units: power supply, physical sensor, processing module and RF transceiver. As the design of the generator is determined by the power requirement of the whole system, the other three parts should be designed first. Therefore, it is necessary to specify the design principles for a wireless sensor node:

1. The height sensor should be integrated in the wireless sensor node;
2. The wireless end device should have a small sturdy structure to fulfill the size restrictions in a truck suspension air spring;
3. The RF transceiver should be stable and have a long-life;
4. The wireless end device should have low power consumption;
5. Wireless data transmission should be high reliability and anti-interference;
6. A self-powered generator will replace the original power cable.

The embedded physical sensor senses and collects environmental information from surroundings. Then the produced analog signal is converted into digital signal with an embedded Analog-to-Digital Converter (ADC) and sent to the processing unit to be packed. Last, the wireless end device will transmit the packed data to the coordinator for the further analysis. The self-powered generator will supply the energy for all these units.

Wireless Sensor Node Design

Wireless sensor network (WSN) is a kind of intelligent network in the integration of information collection, processing and transmission (Labiod, Afifi, & Santis, 2007; Baronti et al., 2007). This network has a very broad prospect of application in the field of military and industry with the realization of the

Figure 1. Wireless sensor node block diagram

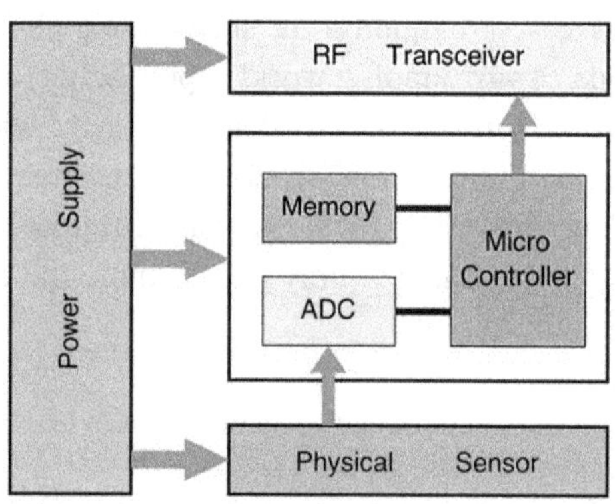

communication between the human world and the physical world. It is one of most important foundations of the Internet of Things (IoT) (Hersent, Boswarthick, & Elloumi, 2011). Through different sensors, the WSN can provide real-time sense and gather all kinds of environmental data and target information. The WSN technology currently is an attractive, emerging and multidisciplinary research focus around the world (see Figure 2).

In fact, more and more low-cost, low-power, multi-function tiny sensors get rapidly developed and applied around various fields nowadays. Integrated with physical sensor, embedded micro-controller and RF transceiver, these tiny devices not only can acquire the ambient information but also have the abilities of data processing and wireless communication. The WSN is such a wireless ad hoc network composing of these intensive deployed nodes to fulfill specific monitoring tasks.

ZigBee Technology

One of the current standards for WSN is IEEE 802.15.4 (Baronti et al., 2007). Compared with the familiar wireless communication technology Wi-Fi (IEEE802.11), IEEE802.15.4 belongs to the category of wireless local area network (WLAN) (Labiod, Aifif, & Santis, 2007). ZigBee standard is just based on IEEE 802.15.4. For the ZigBee protocol stack, there is no definition of the physical layer and MAC layer but directly adopt the IEEE 802.15.4 standard (Su & Han, 2010). From top down, the application layer (APL), the network layer (NWL) and the security part are specified by ZigBee Alliance. The protocol stack is shown in Figure 3. Above the MAC layer is the network layer which is mainly in charge of transmitting and receiving network data and routing. The topmost application layer is divided into several modules: the application support sublayer (APS), which performs application data processing and binding, and the ZigBee device object (ZDO), which provides the equipment management function (including device discovery, service discovery, etc). Apart from these, ZigBee also provides security service named security service provider (SSP) for the network layer and the application layer (Baronti et al., 2007).

The physical layer is the lowest and basic layer of the stack, its main functions including energy detection of current channel, sending the link quality indication, CSMA/CA free channel assessment, the choice of the channel frequency and the sending and receiving of data. Three frequency bands are available for ZigBee protocol: 868 MHz~868.6 MHz with 1 channel, 902~ 928 MHz with 10 channels and 2400~2483.5MHz with 16 channels. Meanwhile, the physical layer can be compatible with other IEEE wireless network standards such as IEEE 802.11 and IEEE 802.15.1 (Hersent, Boswarthick, & Elloumi, 2011) (Labiod, Afifi, & Santis, 2007) (Baronti et al., 2007).

Figure 2. WSN structure block diagram

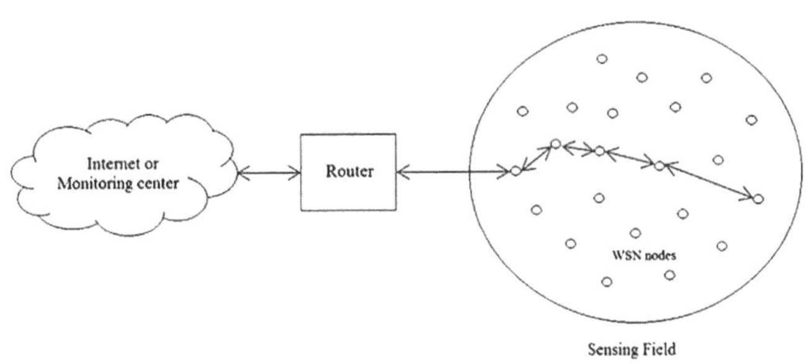

Figure 3. ZigBee protocol stack (Su & Han, 2010)

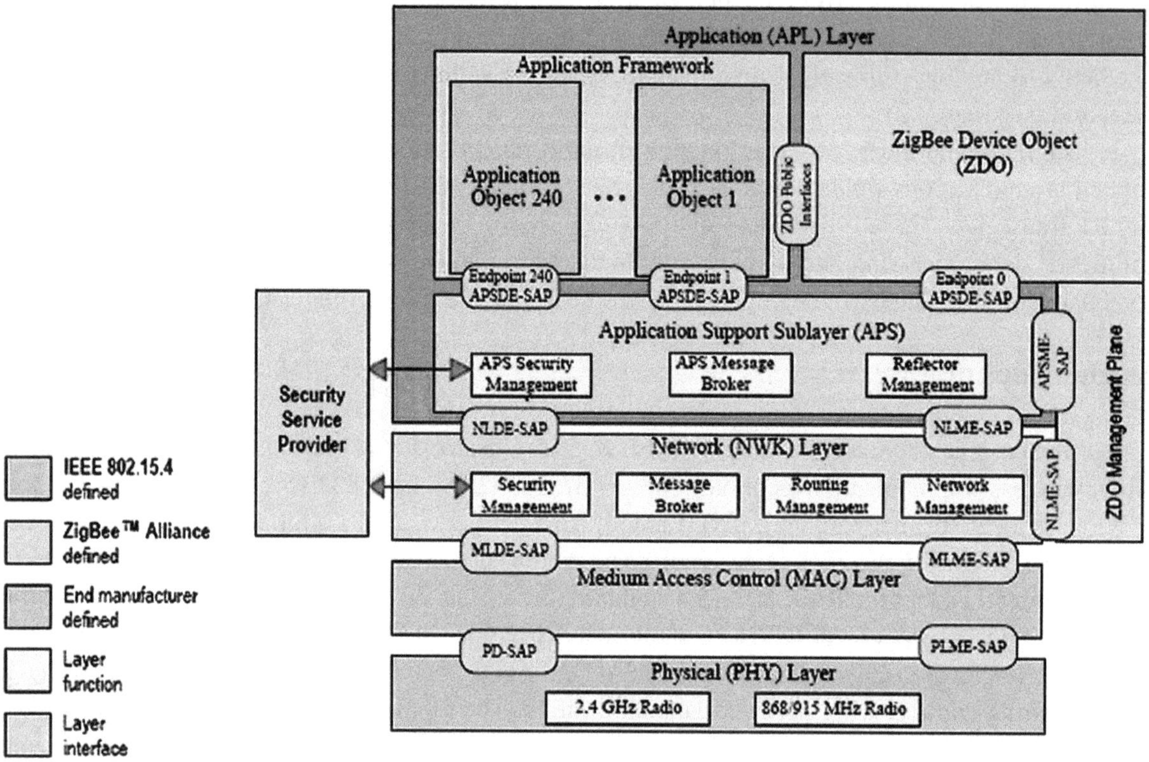

IEEE 802.15.4 standard specifies the following main functions for the MAC layer: producing and synchronization of the beacon, safety specification for the device, processing and maintaining of the guaranteed time slot (GTS) and reliable connection among MAC entity. Given the consideration of the network cost, the MAC layer defines two types of devices: Full Function Devices (FFDs) with full set of functions and Reduced Function Devices (RFDs) with basic function. Besides, IEEE 802.15.4 also defines three device types according to the composition of the network: PAN coordinator, coordinator and end device. FFDs can act as one of the above three devices with implementation of all MAC layer functions but RFDs can only perform the simplest function to work as an end device. The PAN coordinator is responsible for establishing and managing the network, selecting the network identifier, and setting or dismantling the connection with other devices (Hersent, Boswarthcik, & Elloumi, 2011; Baronti et al., 2007; Chaari & Kamoun, 2011).

The ZigBee Alliance defines the other three parts of the protocol stack: the network layer (NWK), the application layer (APL) and the security part. Similar to the general function of other wireless standards, the ZigBee network layer can establish and maintain different network topologies. The application layer defines various types of applications. The application framework (AF) consists of up to 240 application objects (APO). Each one corresponds to a user-defined ZigBee application. ZDO allows multiple APOs to be grouped into a distributed application and meanwhile the APS provides data and management service for APO (Hersent, Boswarthick, & Elloumi, 2011; Labiod, Afifi, & Santis, 2007; Baronti et al, 2007; Chaari & Kamoun, 2011).

The core functions of the network layer are routing and addressing, as well as the establishment and maintenance of the network. ZigBee supports three topologies: star, tree, and mesh topology. After scanning the chosen channel, the ZigBee coordinator will set up corresponding network properties. The connection of each node in this network is achieved by a join procedure. If a separate node wants to join a network, it scans different existing channels to learn about the neighbor routers which broadcast their networks. When the application layer has chosen a network to join, the network layer selects a "father" node and requests to start the association procedure. When successfully receiving the indication of the association request, this node assigns its own network address according to the given address information from the coordinator. Figure 4 shows the three NWK topologies (Hersent, Boswarthick, & Elloumi, 2011; Labiod, Afifi, & Santis, 2007; Baronti et al., 2007; Chaari & Kamoun, 2011).

The application layer consists of a series of different Application Objects (APOs), the ZigBee Device Objects (ZDO) and the Application Support Sublayer (APS). Each APO connects the outside device with a unique application port. The APS provides data transfer, data binding and APL multicast services for the APOs and the ZDO. Through the application layer, users can easily edit and control their commands and requests (Hersent, Boswarthick, & Elloumi, 2011; Labiod, Afifi, & Santis, 2007; Baronti et al., 2007; Chaari & Kamoun, 2011).

Besides, it can sum up the technical characteristics of ZigBee:

1. 2.4 GHz and 868/915 MHz dual PHY modes;
2. Low transmission rate, only 10 kbit/s~250 kbit/s;
3. Short range, from 1 to 75 meters;
4. Low power, with 2AA batteries can continue work for six months;
5. Security, ZigBee provides packet integrity checking function based on cyclic redundancy check(CRC);
6. Low time delay, typically, search time delay is 30ms, activation time delay is 15ms.

Processing Unit, RF Transceiver, and ZigBee Device

The commonly most used processing unit of a wireless sensor node is an integrated micro-controller unit (MCU) including a processor core, memory unit, ADCs, programmable input/output peripherals, counters and timers (Su & Han, 2010; Veerasingam, Karodi, Shukla, & Yeleti, 2009). Therefore, such a processing unit can easily interface with analog sensors and communication devices such as RF transceivers.

Many different types of MCUs are currently integrated in sensor nodes. Generally, those are typically designed to perform simple processing tasks such as reading and processing values from physical

Figure 4. ZigBee NWK topology

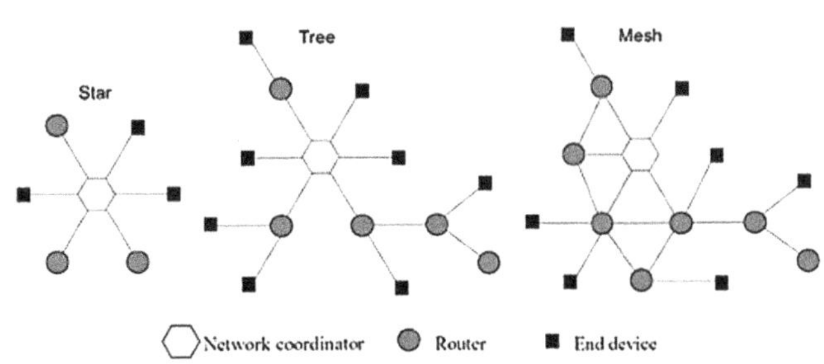

sensors, transmitting data to other nodes, etc. When comparing with general purpose processors, the micro-controller in wireless sensor node does not need high computing capability to processing complicated tasks but should be energy-efficient and have a fast wake-up time (Veerasingam, Karodi, Shukla, & Yeleti, 2009). The embedded MCU should have a programmable system to set and debug the function of the sensor node.

Table 1 shows a comparison of common commercial low-power MCUs. The application of wireless sensor node limits the requirement of a MCU based on its performance and size, the ideal choice is the one that not only realize its function but also has smaller power consumption. The most commonly used MCU in wireless sensor nodes is either ATmega1281 or MSP430. The Arduino board is based on the ATmega128 or ATmega328 with 14 digital input/output (I/O) pins and 6 analog inputs, which is a good test platform for the processing unit of a wireless sensor node. Arduino is an open source micro-controller system which is designed to be easy to learn, easy to test and fast to develop.

The RF transducer just transmits some collected height data from the height sensor instead of a large amount of data like voice or video so that the low-rate ZigBee protocol can easily meet this requirement and create a suitable sensor network. Its devices tend to be used in situations where power must be conserved. A ZigBee network defines three different devices: coordinator, router, and end device (Guo, Healy, & Zhou, 2010). Each network at least contains a coordinator device and one of other two kinds of devices:

- **ZigBee Coordinator (ZC):** This device is responsible for establishing the network, assigning addresses, managing the routing, defining the network, and storing information about the network;
- **ZigBee Router (ZR):** A ZigBee router is a full function node, and it can discover and join an existing ZigBee network. It will transmit data packets from a coordinator to an end device in the network. After a ZR has joined a network, its new devices can join this network by sending a request to it;
- **ZigBee End Device (ZED):** End devices contain just the simple function to only join the network, send and receive data. They do not need to transfer data between other end devices, so they can periodically work in a lower power condition, saving energy by going temporarily into this (sleep) mode.

XBee modules from Digi Inc. are designed for a scalable WSN based on the ZigBee protocols. Its programmable versions make customizing ZigBee applications easy and fast. Arduino and XBees can work extremely well together to form a wireless sensor node. The Arduino board comes with a library for interfacing with XBee module. It can sense the environment by receiving digital input from a physical sensor and transmit the information by sending data with the XBee module:

Table 1. Comparison of commercial micro-controllers

Characteristic	Bits	Flash Memory	Power Consumption	MCU Board
ATmega328	8	32 KB	0.2mA@1MHz@1.8V	Arduino Uno
MSP430	32	64 KB	0.33mA@1MHz@2.2V	MSP432P401R
STM32L0	32	32/64 KB	0.087mA@1MHz@3.3V	STM32VL

- **End Device:** The hardware of the end device consists of an Arduino Uno, XBee Shield and XBee S2 Module, as shown in Figure 5. The height value in Volts is read by the Arduino from the original height sensor and communicated as the digital data to the XBee module;
- **Coordinator:** An XBee module connected to the computer is set to the coordinator, so that it is easy to read the received data from the X-CTU program in the computer to monitor the communication process.

Self-Powered Generator

In this study an energy scavenger converts kinetic energy from the movement of vibration into electrical energy. As one of the most common forms of movement, vibration is around everywhere like a moving train, a rotating windmills or an operating electric drill (Kim, Kim, & Kim, 2011). To best design such a self-powered generator, it is necessary to understand the operation of a specific energy harvester: the characteristics of the energy source, how to transfer the kinetic energy and how to transfer the electrical energy from the energy scavenger to the application (Galchev & Najafi, 2009).

Figure 5. Design of end device

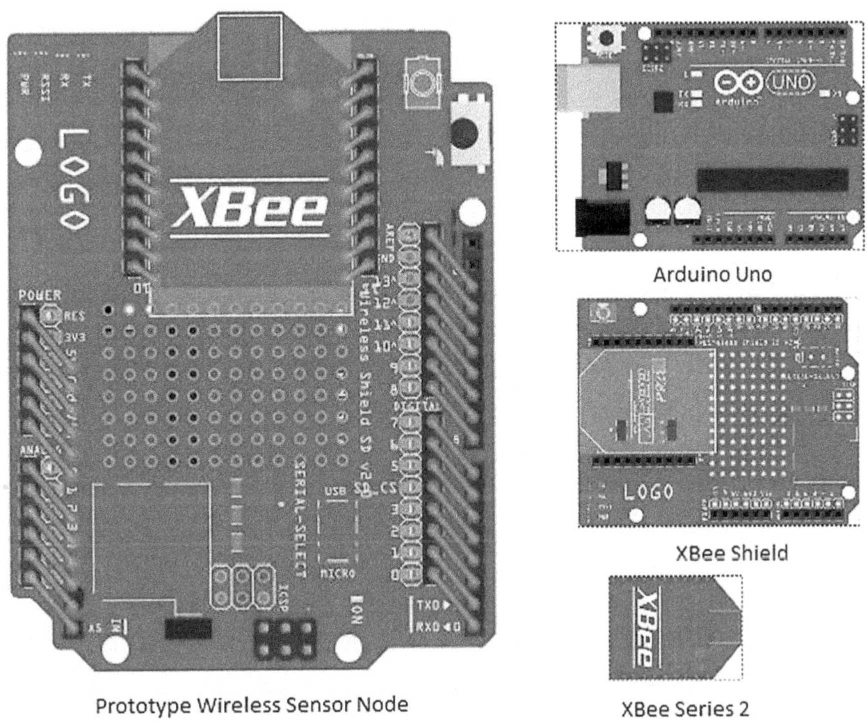

Piezoelectricity and Piezoelectric Materials

In order to use vibration as a power source, three basic methods can be used as the generator: electromagnetic induction, electrostatic generation, and piezoelectric generation (Galchev & Najafi, 2009); Pinna, Dahiya, De Nisi, & Valle, 2010). Although each of these methods has their own advantage, piezoelectric materials have recently drawn more attention with its unique properties (Galchev & Najafi, 2009). Piezoelectricity is the ability of some materials in conversion and reverse conversion between mechanical energy and electrical energy (Bibo & Daqaq, 2013; Kim, Kim, & Kim, 2011; Pinna, Dahiya, De Nishi, & Valle, 2010). When an external force applied on a piezoelectric material, it produces the electric polarization and leads to the bound charge of opposite polarity on antithetical ends. The charge density is proportional to the applied strain. This phenomenon is called positive piezoelectric effect.

The behavior of piezoelectricity can be modeled with the following piezoelectric equations:

Direct Piezoelectric Effect: $D = dT + \varepsilon^T E$ (1)

Converse Piezoelectric Effect: $S = s^E T + d_t E$ (2)

where:

- D is the electric displacement vector;
- d is the piezoelectric strain coefficient;
- $\in^T$ is the dielectric constant of the piezoelectric material;
- E is the electric field;
- T is the mechanical stress;
- S is the mechanical strain;
- s^E is the matrix of compliance coefficient at constant electric field strength;
- d_t is the matrix for the converse piezoelectric effect.

The piezoelectric materials have two modes of operation. For the first mode, named 33 mode, it prefers a polarization direction which is perpendicular to the direction of the external force. The second mode, called 31 mode, is where both the polarization and external force are in the same direction. (Kim, Kim, & Kim, 2011; Pinna, Dahiya, DeNisi, & Valle, 2010; Ramsay & Clark, 2001) The two modes are shown in Figure 6.

Figure 6. Piezoelectric coupling modes

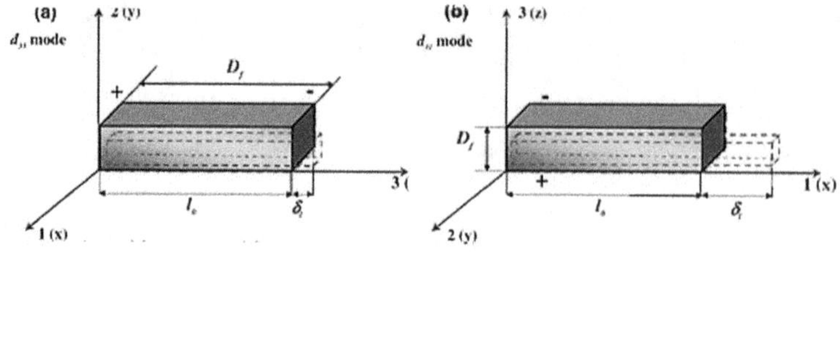

In order to increase the harvested power, multiple pieces of piezoelectric materials are added to the beam. One is a uni-morph cantilever beam, the other is a bimorph structures including two different types. The series type bimorph consists of three layers: one metal layer between two layers of piezoelectric material which are connected in series form. The parallel type has the same structure with the previous type but in a parallel electrical connection. Among these three structures, the series type has a good output power under a high frequency environment compared with the same result of the parallel type in a medium frequency (Ramsay & Clark, 2001).

Figure 7 shows the three cantilevers structures.

Commonly, piezoelectric materials are classified into two groups: inorganic materials and organic materials. Quartz is the most widely used piezoelectric material which will produces an AC signal with the same frequency of vibration. (Bibo & Daqaq, 2013; Choi, Jeon, Jeong, Sood, & Kim, 2006; Saadon & Sidek, 2011) This characteristic makes it the most important piezoelectric material which is applied in the frequency control equipment like oscillator, resonator and narrow-band filter (Choi, Jeon, Jeong, Sood, & Kim, 2006). Organic piezoelectric materials which are also called polymer piezoelectric materials also attracts much attention with special advantages such as flexible, low density, low resistance and high piezoelectric constants. The best polymer materials are polyvinylidene fluoride and its copolymers. This type of copolymer thin film exhibits better piezoelectric properties. Especially its acoustic impedance is very close to the air, water and biological tissue. Table 2 shows the material parameters of these piezoelectric materials (Bibo & Daqaq, 2013; Galchev & Najafi, 2009; Kim, Kim, & Kim, 2011; Ramsay & Clark, 2001).

Figure 7. (a) A series triple layer type cantilever; (b) A parallel triple layer type cantilever; (c) A unimorph cantilever (Ramsay & Clark, 2001)

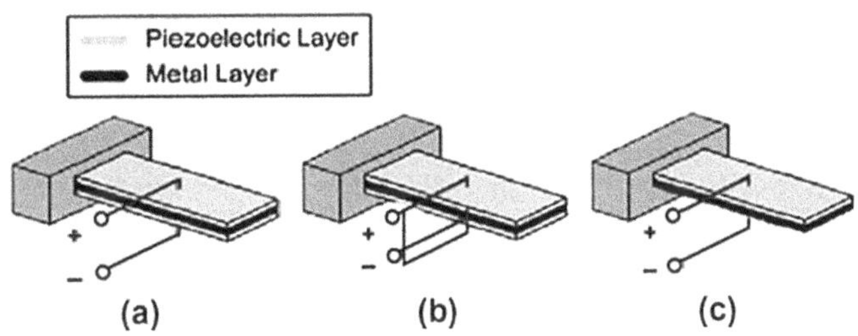

Table 2. Piezoelectric properties of representative piezoelectric materials

Parameter	Quartz	$BaTiO_3$	PZT4	PZT5H	(Pb, Sm) TiO_3	PVDF-TrFE
d_{33}(pC/N)	2.3	190	289	593	65	33
g_{33}(pC/N)	57.8	12.6	26.1	19.7	42	380
k_t	0.09	0.38	0.51	0.5	0.5	0.3
k_p		0.33	0.58	0.65	0.03	
$\varepsilon_3^X / \varepsilon_0$	5	1700	1300	3400	175	6
Q_M	$>10^5$		500	65	900	3-10
T_C(°C)		120	328	193	355	

Because the strain and coupling coefficients in 33 mode are generally much higher than the ones in 31 mode, the commercial piezoelectric material is typically used in 33 mode. In this study, two different kinds of structures, shown in Figure 8, are discussed: one is a single-layer generator; the other is a 2-layer generator. For single-layer generators, when stress is perpendicularly applied to the sheet, voltage is generated attempting to force the ceramic to return to the original status. For 2-layer generators, when force causes bend on the structure, one layer is compressed and the other is stretched so that the produced charges may be collected as the generation.

By now, three kinds of piezoelectric materials are available: PZT, PVDF and PZN-PT. Therefore, the comparison of fundamental properties between different piezoelectric materials is very important. Table 3 shows a number of the commercial piezoelectric materials and their important properties. The strain coefficient (d_{33}) relates the produced mechanical strain to the applied electric field. The coupling coefficient (k) describes the material's ability in conversion of energy from kinetic vibration to electrical or vice versa.

PZT is the most widely used polycrystalline ceramic with good properties; it has several similar versions available with slightly different properties (Saadon & Sidek, 2011) (Ng, 2005). Based on this comparison, two different PZT materials have been chosen as the main material to test. One used here is the PSI-5A4E from Piezo Systems Inc., the other is SM411 from Steniner \& Martins Inc. These two materials are shown in Figure 9.

Figure 8. Two different structures of piezoelectric ceramic

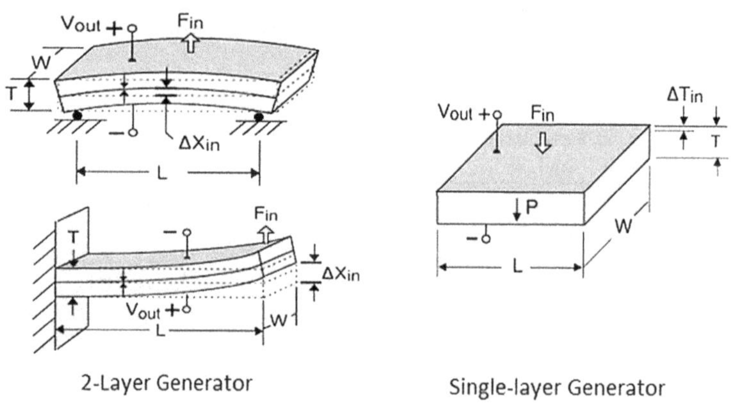

Table 3. Comparison of piezoelectric materials

Property	Units	PZT 4	PZT 5J	PVDF
d_{33}	pC/N	289	500	33
K_t	CV/Nm	0.51	0.45	0.3
k_p	CV/Nm	0.58	0.64	
$\varepsilon_3^X / \varepsilon_0$	N/A	1300	2100	6
Y_{33}	10^{10}N/m^2	7.3	5.4	0.3
g_{33}	10^{-3}Vm/N	26.1	24.2	380

Figure 9. Two set of PZT ceramic

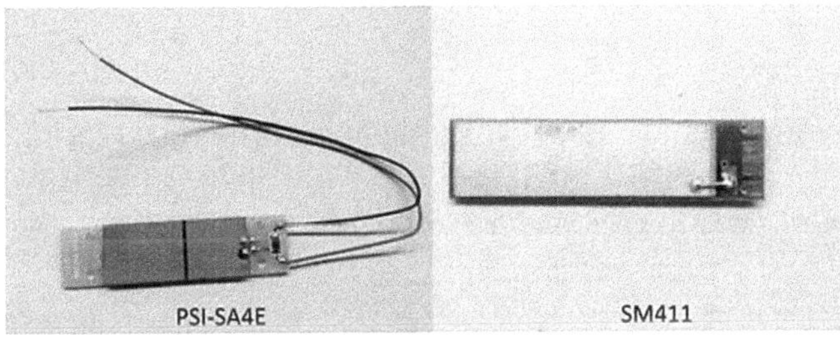

Piezoelectric Beam Modeling

Vibration of a solid body can be easily observed in almost any dynamical system. A single degree of freedom (DOF) mass spring damper model can be used for fundamental analysis of a vibration system which is simply modeled with two parameters: damping constant and natural frequency. The governing equation of motion can be described as:

$$m\ddot{x} + c\dot{x} + kx = 0 \tag{3}$$

where m is the mass, c is the damping coefficient, k is the stiffness of the spring and x is the displacement. Given the real situation, mostly the system response belongs to the underdamping case. Notice that the above equation can be simplified with the two mentioned parameters. In this study, the truck suspension air spring vibration is the addition of a harmonic force for the system which is assumed:

$$F = F_0 \sin(\omega t) \tag{4}$$

Then the new governing equation of motion becomes:

$$m\ddot{x} + c\dot{x} + kx = F_0 \sin(\omega t) \tag{5}$$

From this ordinary differential equation, the steady state solution of this system can be derived as:

$$x(t) = A\sin(\omega t + \Phi_0) \tag{6}$$

$$A = \frac{F_0}{k} \frac{1}{\sqrt{(1-r^2)^2 + (2\zeta r)^2}} \tag{7}$$

where A and Φ_0 are the amplitude and phase shift of the vibration response, respectively, ζ is the damping ratio, and r is the ratio of ω to the natural frequency ω_n. To characterize the amount of damping in a system the damping ratio ζ and natural frequency ω_n are used in analysis of the equation. The formula for the damping ratio (ζ) of this model is:

$$\zeta = \frac{c}{c_c} = \frac{c}{2\sqrt{mk}} \tag{8}$$

where c is the damping coefficient and c_c is the critical damping in this mass spring damper model which is:

$$c_c = 2\sqrt{mk} \tag{9}$$

and the natural frequency is defined by:

$$\omega_n = \sqrt{\frac{k}{m}} \tag{10}$$

For a fixed end beam, the stiffness k is given by $k = 3EI/L^3$, where E is the modulus of elasticity, I is the second moment of area with respect to the x-axis which can be obtained from expression $I = (1/12)bh^3$ for a rectangle cross-section, and L is the length of beam.

The kinetic energy of this system is expressed as:

$$E_k(t) = \frac{mv^2}{2} = \frac{m\dot{x}^2}{2} = \frac{m\omega^2 A^2(\cos(2\omega + 2\Phi_0) + 1)}{4} \tag{11}$$

and the potential energy is given by:

$$E_p(t) = \frac{kx^2}{2} = \frac{m\omega_n^2(1 - \cos(2\omega t + 2\Phi_0))}{4} \tag{12}$$

then the mechanical energy is the sum of $E_k(t)$ and $E_p(t)$ and the average mechanical energy can be expressed as the following equation:

$$\overline{E(t)} = \frac{1}{T}\int_0^T \left(E_k(t) + E_p(t)\right)dt = \frac{m(\omega^2 + \omega_n^2)A^2}{4} \tag{13}$$

because the amplitude of the vibration response is a function of r, when the operating frequency approaches the natural frequency of the system, the above formula can reach the maximum value which is equivalent of the maximum power:

$$P_{\max} = \frac{m\omega_n^2 F_0^2}{2k^2\zeta^2} \tag{14}$$

From this result, it can be noted that with increasing frequency to natural frequency the convertible kinetic energy will get the maximum value.

Rack-and-Pinion Power Generator

As an alternative, a vibration-based electromechanical generator was also considered in Galchev and Najafi (2009). For one of the most common forms of generator, different designs incorporate different sets of constraints. The most important parameters influencing the design of this system are the generator layout and conversion efficiency. Combining that idea with the characteristics of the vibration, a rack-and-pinion generator should be included in this study.

In this design, the pinion would be fixed on the tester and stationary relative to the bottom of the air spring, accordingly the rack will be fixed on the cantilever beam and vibrate with the air spring. There is no difference in terms of the structure between DC motor and DC generator, therefore, a 12 V, 5000 rpm DC motor is selected as the generator and mounted on the pinion. Because the vibration displacement of the air spring is limited, it thus determines that the rack can only move a short distance. For this reason, a 50 rpm planetary gear box is added to improve the rotation speed of the DC motor. The rack-and-pinion generator extracts kinetic energy from the motion aligning with the truck suspension air spring and transforms it to usable electrical energy. The schematic diagram is shown in Figure 10.

Rectifier Circuit

Arduino board can operate on an external supply from 5 to 20 volts and as the core of the wireless sensor node; other two units would connect to it. In other words, it just needs to power the Arduino board which is equivalent for the entire node. Based on this, the produced electricity should be thereafter formatted by an appropriate rectifier circuit in the power unit.

Full wave bridge rectifiers have the benefit that they can transfer both half cycle of AC input into DC output and are also two times better than half wave rectifiers. Following the rectifiers is the regulator. In regulating a low voltage, the most used device is one single IC like LM 7805. LM 7805 is a three-terminal regulator with positive 5V voltage. Figure 11 shows the design of rectifier circuit.

Because the real capacitors are determined by the load resistance and operating frequency, therefore, here just give the reference equation for the selection of capacitor:

$$R_{load} \times C_1 \gg \frac{1}{f} \tag{15}$$

where:

Figure 10. Block diagram of the rack-and-pinion generator

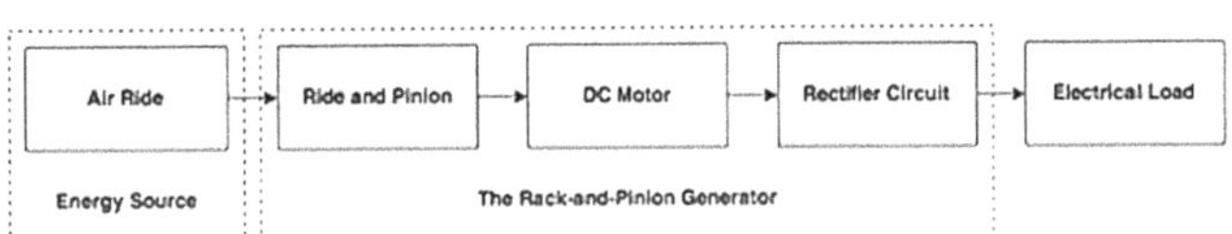

Figure 11. Rectifier circuit

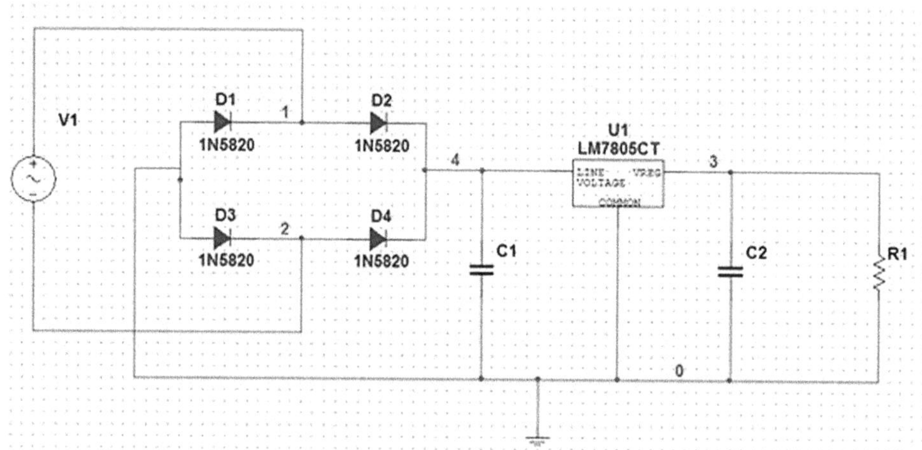

- R_{load} is the resistance load with unit of ohm;
- C_1 is the capacitor in the filter with unit of farad;
- f is the frequency of AC voltage with the unit of hertz.

EXPERIMENT AND RESULTS

Height Sensor Test

Before carrying out the wireless data transmission test, it is necessary to understand the information of the signal from the height sensor. In this study the height sensor in a truck suspension air spring was chosen as the initial system. The output signal of the height sensor is shown in Figure 12. The dynamic voltage readings vary between 0.8 V to 1.6 V when the truck suspension air spring is working.

A static measuring platform is built as shown in Figure 13 to determine the relationship between the output voltage and the vibration height.

The voltage outputs and their corresponding height values are shown in Table 4 and the Figure 14. According to the static test results and the scope screen shots, the relationship between the movement and the output voltage can be achieved and it would be very useful for the wireless data transmission in the following test.

Wireless Data Transmission Test

As explained previously, the wireless sensor node is based on the Arduino-XBee test platform. For this test, in order to verify the accuracy of data transmission, the test platform is set as a continuous work mode. The wireless data transmission test has 4 phases:

- **Preparation and XBee Firmware Configuration:** The construction of the end device which is shown in Figure 15, is rather simple. Plugging the XBee module into a specific shield, the Arduino

Figure 12. Signal of height sensor

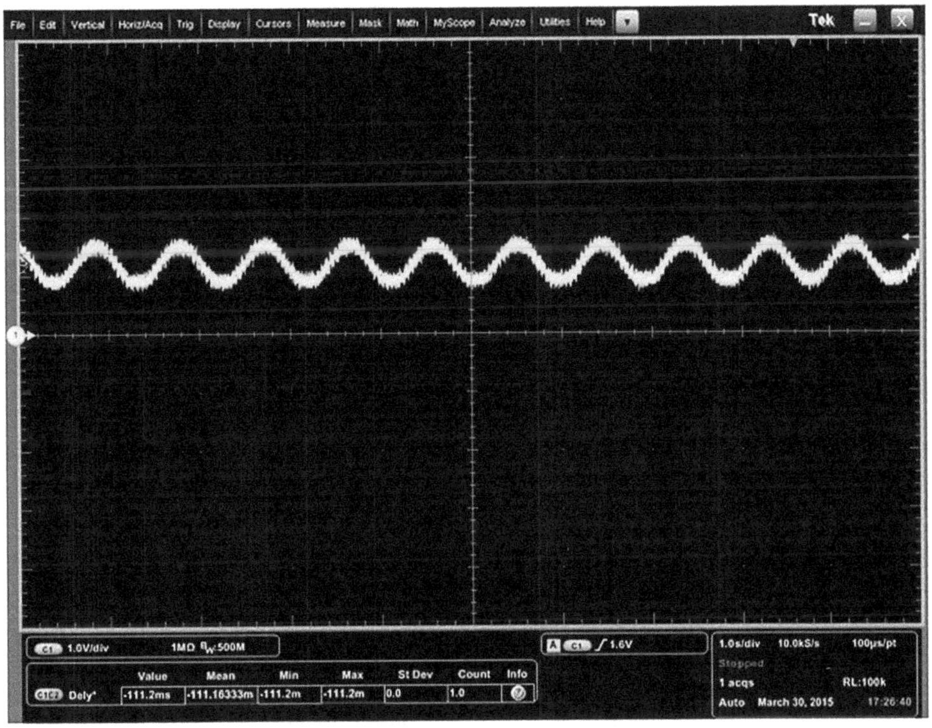

Figure 13. Measuring platform for the height sensor

Table 4. Height sensor output voltage and heights

Test No.	1	2	3	4	5	6	7	8
Vibration Height (mm)	0	8.06	12.79	17.18	24.45	30.85	39.03	50.88
Voltage (V)	0.86	0.96	1.06	1.16	1.26	1.36	1.46	1.56

Figure 14. Voltage vs. vibration height

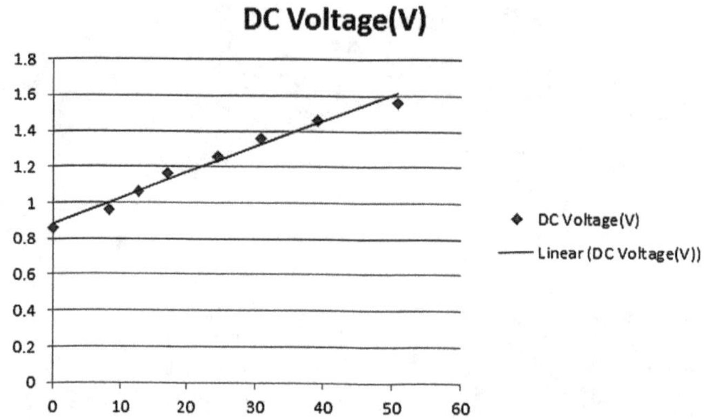

Figure 15. Wireless sensor node end device

board was connected to this shield and the original height sensor. The default working mode in each XBee module is configured as Router AT mode at the factory. Therefore, it is essential to begin with the configuration of each device with the X-CTU program before the test: in this test, one is the router AT mode, the other is the coordinator AT mode. There are three important XBee settings needing to configure in X-CTU: personal area network ID (PAN ID), MY address and destination address. PAN ID is some hexadecimal value between 0 and 0xFFFF, and XBee modules can only communicate with each other when they have the same PAN ID. Each XBee module will be assigned a 16-bit address as their MY address, or the "source" address. For the destination address, one XBee device needs to have the same setting as the other's source. The configuration of each device is shown in Figure 16;

- **Communication and Range Test:** After updating the firmware of XBee modules, two XBee modules need to be configured correctly - on the same network ID and have compatible destination and MY addresses. To test the communication between these two devices, two PCs were connected to each one. In the X-CTU, with the function "discover radio nodes in the same network", any device with the same PAN ID can be found and listed in the program. Additionally, with the serial connection in X-CTU, XBees are configured to talk to each other. In the serial monitor, blue characters indicate the data to be sent and red ones are the received data. The test results are shown in Figure 17. Considering the real work environment in a small truck, the range test was set in 1

Figure 16. XBee modules configuration

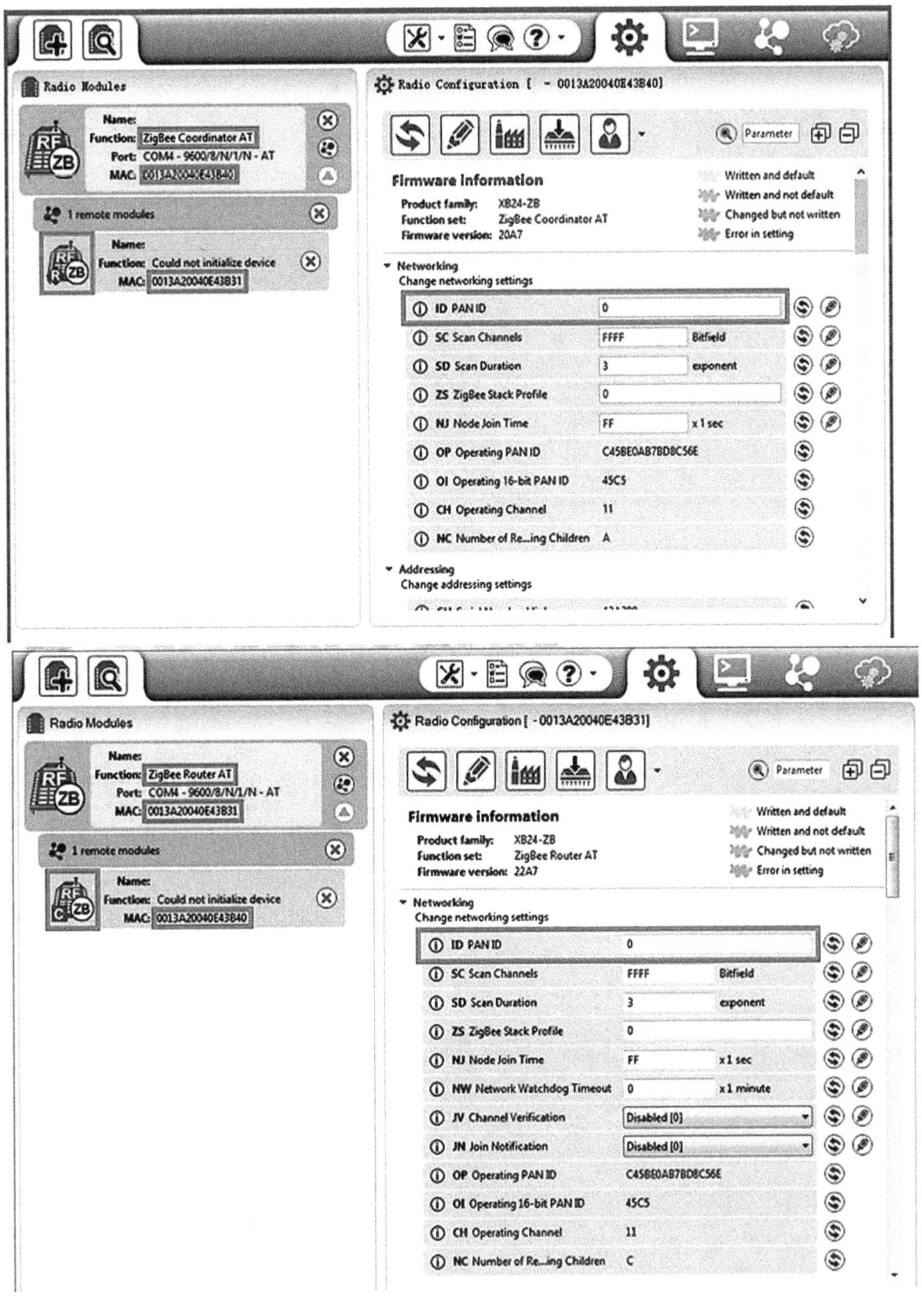

meter, 2 meters and 3 meters. In an open environment, the coordinator device and the end device can quickly discover each other and organize a network for data transmission. However, the complex electromagnetic environment in the vehicle may affect the data transmission in reality test;

- **Data Transmission Test:** The height sensor is connected to the Arduino analog input pin A1 as shown in Figure 15. The output voltage signal from the height sensor will be converted to the digital signal in Arduino with its embedded ADC, and then be sent to the XBee module for further

Figure 17. Communication test between two XBee modules

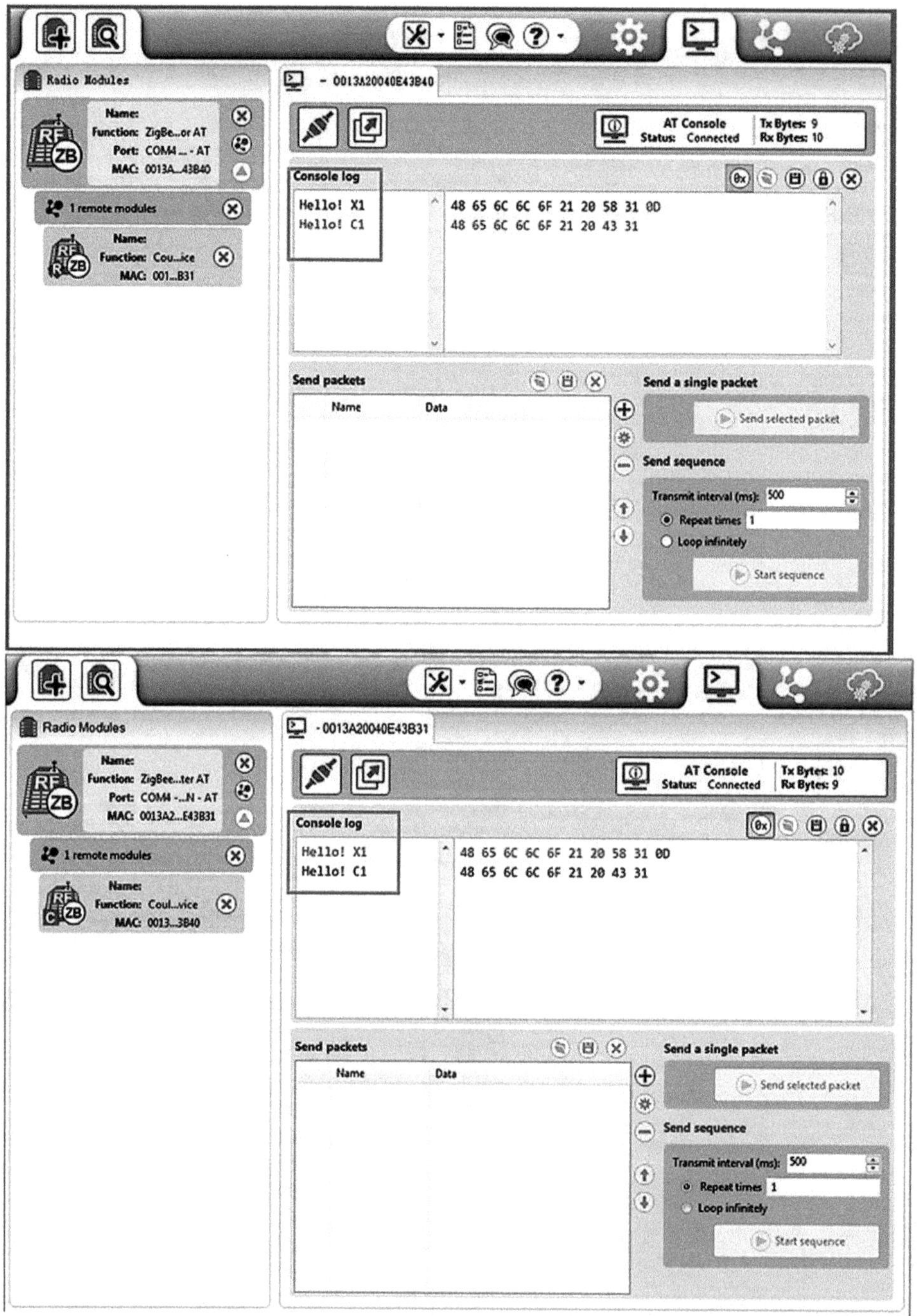

wireless transmission. In the Arduino board, the input analog signal will be read and converted into a digital value between 0 and 1023 which correspond with the voltage from 0 to 5 V. After the transformation by the Arduino program, the produced digital signals are received and transmitted by the XBee module. From the coordinator interface, it can easily analyze the data according to the relationship from the previous height sensor test.

The program used in Arduino board is shown in the following:

```
// SoftwareSerial is used to communicate with the XBee
#include <SoftwareSerial.h>
SoftwareSerial xbee(2,3);// Arduino RX, TX (XBee Dout, Din)
void setup() {
// Initialize XBee Software Serial Port.
// Make sure the baud rate matches XBee setting (Default: 9600)
xbee.begin(9600);
Serial.begin(9600);
}
// The loop routine runs over and over again forever:
void loop() {
// Read the input on analog pin 1,
int sensorValue = analogRead(A1);
// Convert the analog reading (from 0-1023) to a voltage (0-5V),
float voltage = sensorValue*(5.0/1023.0);
// Print the voltage to XBee device
xbee.print(voltage);
// Print out the voltage,
Serial.println(voltage);
// Set the time slot,
delay(1000);
}
```

In order to examine the accuracy of the data transmission, the data from the serial monitor in the Arduino board with XBee end device and the data from the console log in the X-CTU of the coordinator device will be compared and the results are shown in Figure 18. The static output values from the sensor shown in the Arduino serial monitor are 1.24, 1.27, 1.28, etc, which are the input voltages of the Arduino board. Meanwhile, the console log of X-CTU presents the received data as 1.24, 1.27, 1.28, etc. Therefore, it can be concluded that the wireless data transmission is reliable:

- **Vehicle Test:** Finally, the data transmission in the vehicle is tested. Given the truck suspension air spring work environment, a 2006 Jeep Commander was chosen as the test vehicle shown in Figure 19. The coordinator device is located close to the dash board and the end device is placed around the front and rear wheels to test the wireless data transmission. The test results showed that there is a more or less transmission delay in a vehicle. When the end device is located around the spring of the front wheels, data transmission has about a 0.2 seconds delay. And when the end device is placed close to the spring of the rear wheels, this delay will increase to 0.84 seconds in average. According to the location of the end device, it can be inferred that the body of a vehicle can be

Figure 18. Wireless data transmission test

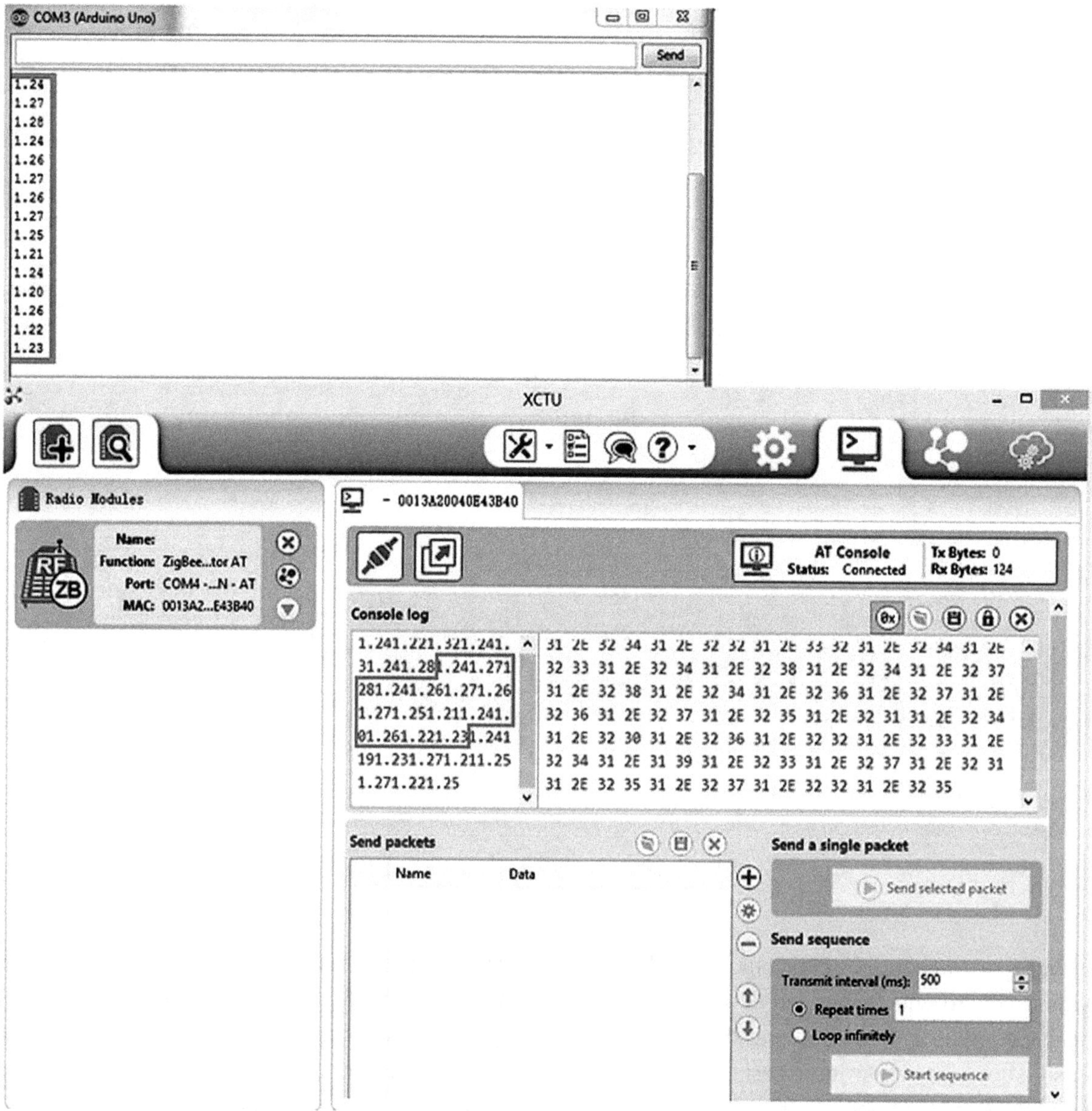

an obstacle for the signal. Through the vehicle test, it is known the wireless transmission will be affected by the structure of a vehicle.

In addition, similar tests with different work modes of the XBee module are also conducted. In these tests, results of data transmission also give support that the wireless data transmission is feasible.

Figure 19. Vehicle test

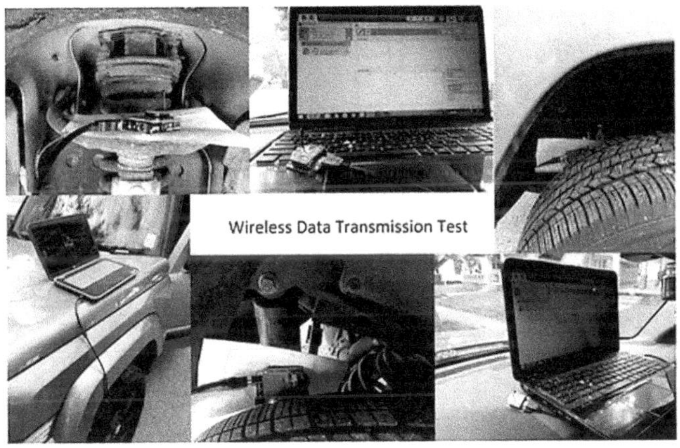

Self-Powered Generator Test

Piezoelectric Generator Test

Before the generator test, the power consumption of the entire system needs to be estimated. Table 5 listed the operate voltage and current of each unit of the wireless sensor node.

Then it gives the estimation of the maximum power consumption which is about 560 mW. But the XBee module can save the power by exchanging different modes. Under the work condition of the sleep mode, the measured operation current of XBee is around 110 μ A at 3 Volts which results in an input power of 330 μ W.

To simulate the real work environment in a vehicle, a vibration tester was designed first. This tester will move the truck suspension air spring as if it is simulating driving down the road on a truck shown in Figure 20. One end of the truck suspension air spring is mounted on the base; the other side was linked to the tester with a cantilever beam. At the same time, the free end of the cantilever was connected to the motor with a mechanical linkage. According to the maximum amplitude of the truck suspension air spring as 1 inch, the free end of the cantilever beam was designed to have a 2 inch vibration amplitude which makes the truck suspension air spring fixed to the midpoint of the beam. The motor speed is 54 rpm and the corresponding frequency of the vibration is about 1.1 Hz.

Because the generator will be located inside the truck suspension air spring, the internal structure of the truck suspension air spring should be taken into account for the design first. Besides the internal environment of the truck suspension air spring, the properties of piezoelectric materials and the size of

Table 5. Power consumption estimation

Unit	Operate Voltage	Operate Current
Arduino Uno	5V	46.5mA
XBee S2 module	3.3V	40mA
Height Sensor	3V	65mA

Figure 20. The truck suspension air spring tester

the ceramic bring about two different piezoelectric generator designs. The primary difference between the two designs is the number of piezoelectric ceramics. The design in Figure 21 is based on the PSI-5H4E piezoelectric ceramic from the Piezo Systems Inc. Three Double Quick-Mount extenders are connected in parallel to achieve a higher current and placed on flat washers with a triangle layout. This design is referred as *Design 1*.

Similarly, four piezoelectric ceramic generators from STEMiNC Inc. are also in a parallel connection and mounted on flat washers but with a rectangle frame. This design is shown in Figure 22 and is referred as *Design 2*.

In the test, these two flat washers are fixed in the center plate through the center axis inside the truck suspension air spring. The piezoelectric ceramics are connected to the top of the truck suspension air spring with a spring aligned with the axis. In such a case, when the truck suspension air spring begins to vibrate, the spring will be compressed and then it will press the piezoelectric ceramic which will generate electricity. The experiment layout is shown in Figure 23.

Figure 24 and Figure 25 show scope screen shots of the output voltage from each generator.

Figure 21. Piezoelectric generator Design 1

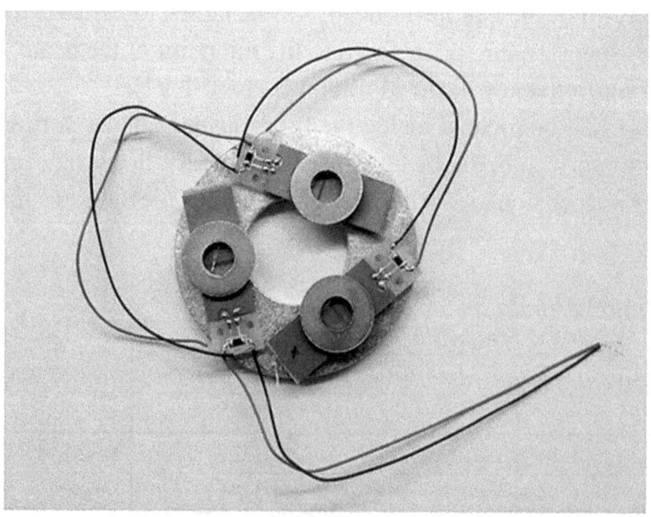

Figure 22. Piezoelectric generator Design 2

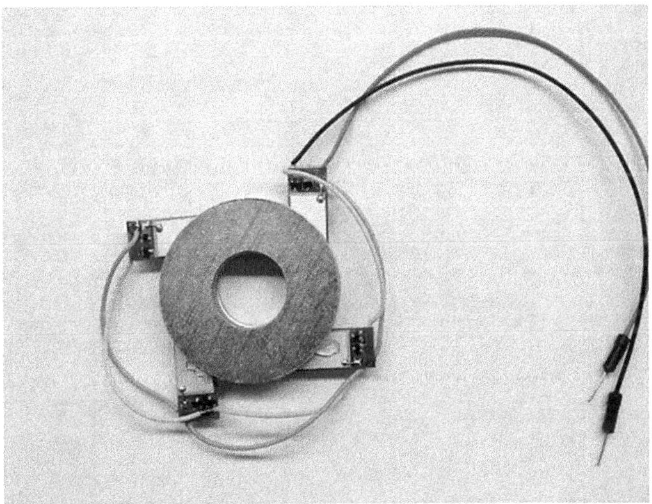

Figure 23. Piezoelectric generator layout

From the test result in *Design 1*, the output voltage is a relatively complete sine wave with the maximum amplitude value of 7 volts and the output current is from 17 μ A to 24 μ A. However, in every relatively stable cycle, there is a sudden voltage drop followed by a zero voltage slot. After a number of repeated experiments, it is found that during the upward movement of the truck suspension air spring, at some point the connected spring would restore to the original position and no longer compress the piezoelectric ceramics. From this time point, the ceramic would not produce any voltage before the next time the spring is compressed. With the readjustment of the layout, the result has been greatly improved. Although the output voltage curve was ideal, the current was quite low - not enough to rectify the signal in a rectifier circuit. Based on the analysis result in the previous section, the power will increase as vibration frequency increases towards the natural frequency. After a series of experiments, the average output current is measured about 21 μ A per 1 Hz. Compared to the power requirement, *Design 1* cannot meet the system requirement even with the maximum generated power. Given the mechanical properties of this material, under the ideal working frequency about of 1500 Hz the output power can reach up to 260 μ W with 3 ceramics.

Figure 24. Design 1 test result

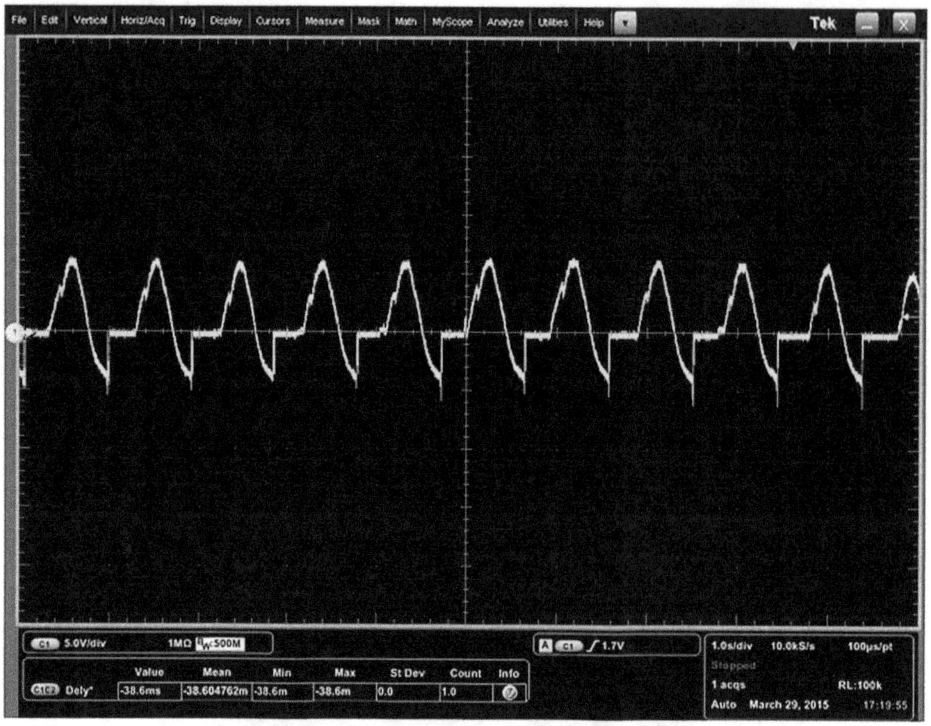

Figure 25. Design 2 test result

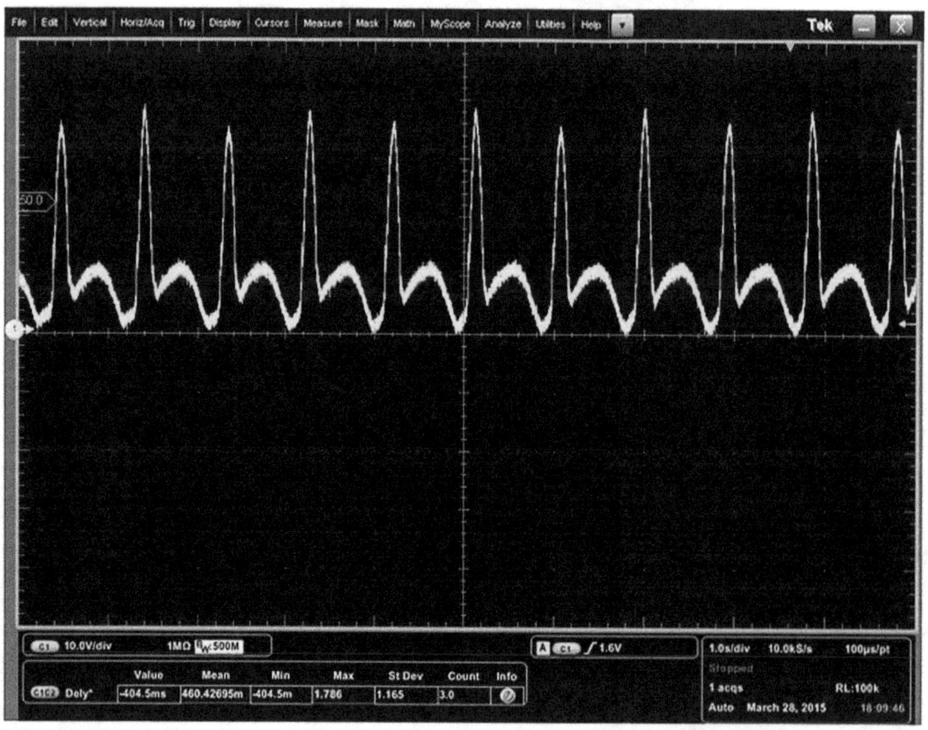

In the test of *Design 2*, in contrast to the result from *Design 1*, due to the 1-layer structure the ceramic will not generate the reverse voltage when the truck suspension air spring is released. Second, in every experiment it is observed that a voltage impulse - about 39 V - appears in each cycle when the truck suspension air spring begins to vibrate from the top point. After analysis of the vibration, it is found that at that time, the ceramic would receive the maximum force from the connected spring. Based on this situation and piezoelectric equations, it can be inferred that this piezoelectric material may produce such a voltage impulse under an instantaneous impact. According to the measurement result, the output current from *Design 2* was also very low and volatile --- from 5 μ A to 18 μ A - even lower than *Design 1*. From this point, *Design 1* is better than *Design 2* as a generator.

Through the above two tests, one can verify that the output power from the piezoelectric generator in a low work frequency cannot supply enough power for such a wireless sensor node. But the result of experiments still give some inspiration that the piezoelectric material contains many properties which can be manipulated to generate power in an appropriate vibration frequency.

Rack-and-Pinion Generator Test

A preliminary rack-and-pinion generator was installed on the tester as shown in Figure 26, and its output voltage wave is shown in Figure 27.

From the scope screen shot, the output voltage with the peak voltage of 22 volts can be approximately regarded as a sine wave. According to the measurement of three devices, the whole load resistance can be estimated to be about 36 Ω. In the case of assumption that the output voltage as an AC power,

Figure 26. Rack-and-pinion generator prototype

Figure 27. Rack-and-pinion generator test result

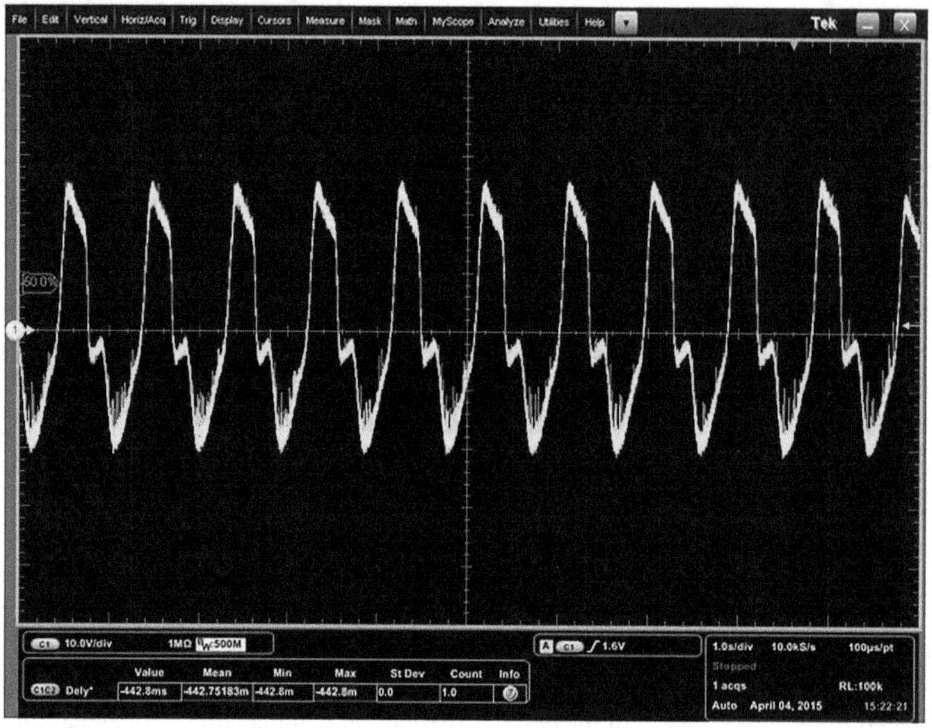

the capacitance in the rectifier circuit can be determined from the Eq.15 which is about 25000 μF and the root mean square of AC voltage is about 16 volts. Utilizing the NI Multisim to simulate the power generation and the result is shown in Figure 28. From the simulation result, the average power generation can be 695.274 mW which is able to meet the demand of power consumption in wireless sensor node. On the other hand, a 25000 μF capacitor is too large for a rectifier circuit. In fact, the smoothed ripple voltage only needs larger than 7 volts to ensure the LM 7805 regulator is in operating conditions. Therefore, with the peak-to-peak ripple voltage equation:

$$\Delta V = \frac{I_{load}}{2fc} \tag{16}$$

where:

- ΔV is the peak-to-peak ripple voltage;
- f is the frequency of the AC power;
- C is the capacitance.

With a help of the voltage regulator, a smaller capacitor of 7000 μF is used. After the simulation with the new capacitance, the new result still meets the requirement for the wireless sensor node.

Figure 28. Simulation result from multisim

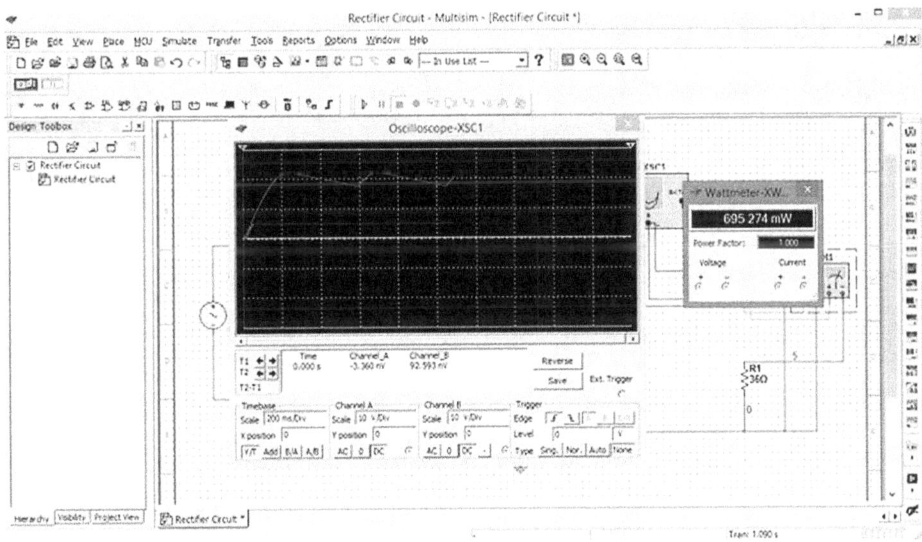

As the input voltage source is only 2.2Hz, it is difficult to obtain such a low frequency voltage power for test. Therefore, a regular lab power supply with 60Hz was applied for experiment. Different values of C_1 from 0 to 1000 μF were tested. Based on Eq.15, the calculation shows that $C_1 \gg 463\mu F$ with 60Hz input voltage. For the particular circuit and load current, when the C_1 is above 40 μF, the output waveform is smooth. When the frequency gets down to 2.2Hz, the capacitor will be much larger. Results were collected with Tektronix MDO 3024 Mixed Domain Oscilloscope. Results were shown in Figures 29 and Figure 30.

Figure 29. Rectifier test result with 470 μF

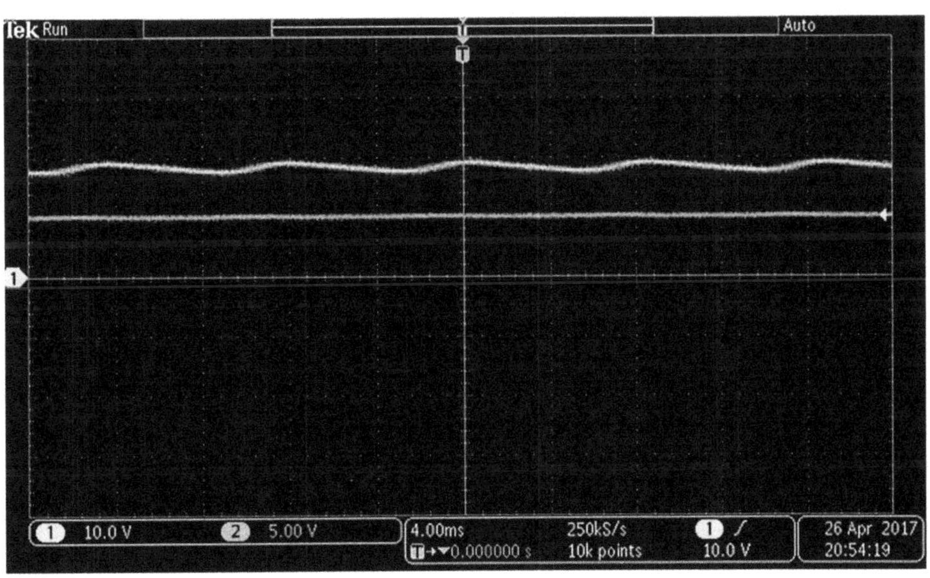

Figure 30. Rectifier test result with 40 μ F

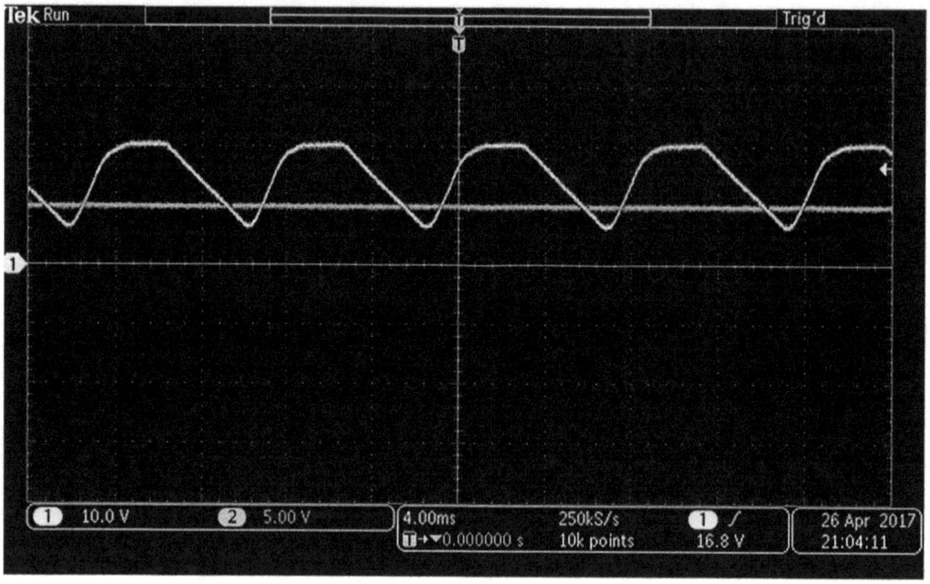

Combining with the real vibration frequency in a truck, the output voltage will have a higher frequency. In this condition, the capacitance will decrease further. Besides, if utilizing a small capacitor in the circuit, a backup battery can be applied to act as voltage compensation or setting the appropriate work cycle for the wireless sensor node to avoid the low voltage time slot.

FUTURE DESIGN

The findings here could make a number of improvements for the future design. First, the previous rack-and-pinion generator is too large for an air ride. Second, the wireless sensor node prototype contains too many unnecessary components resulting larger power consumption for the whole system. Third, a new rectifier circuit and a power management unit need to be applied in the future design. The sketch of the future design is shown in Figure 31.

In this new design, some parts are optimized and redesign:

1. The ZigBee RF antenna will displace the original wire tube;
2. The ZigBee module will be integrated with the processing unit and connected to the height sensor;
3. The new simplified micro-controller with other support devices will be applied to save the size of IC chip and its power consumption;
4. A built-in rack-and-pinion generator will fixed inside and the piston rod will be processed into the rack;
5. A new rectifier circuit should use a low-power regulator to save the power and lower the heat dissipation;
6. A backup rechargeable battery will be installed in the IC chip.

Figure 31. The sketch of the future design

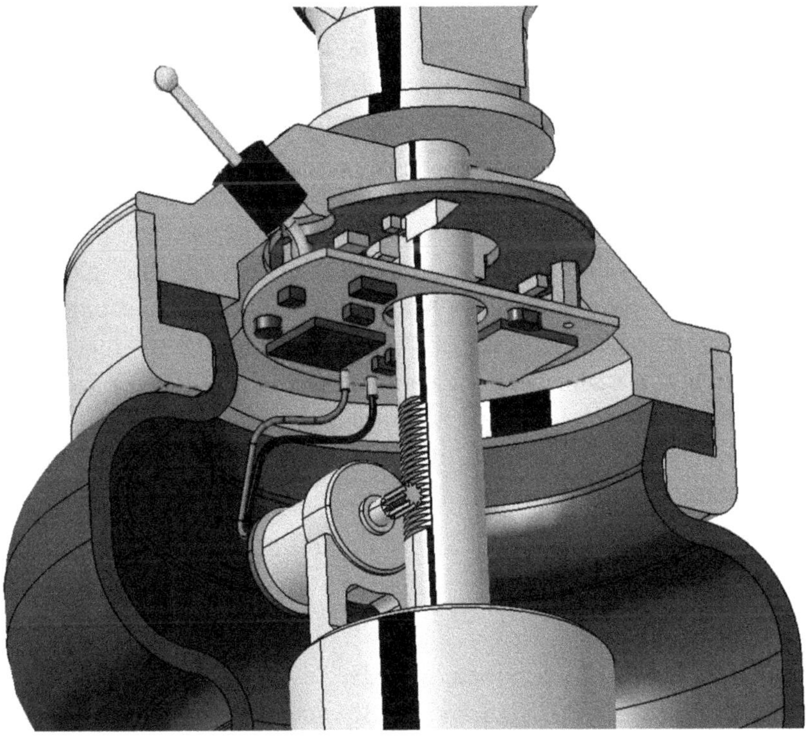

Combined with the data requirement of the electronic control unit (ECU), the appropriate work frequency and the suitable program can guide an optimized processing unit. Based on the power consumption of these units, the power supply circuit and the power management unit can be accurately aligned. Compared with the general circuit, the integrated circuit with micro-controller has a better requirement for the regulated power. An independent low current power supply is also necessary to save the contents of the volatile memory under a sudden power off (which is in order to prolong the dying time). Therefore, in the design of power supply the above characteristics and power consumption need to be considered. In addition to the new design of the hardware, the software of the wireless node will be reconsidered.

Multiple wireless sensor nodes can form a complete WSN including all types of devices. Each air ride will be the base device of the network, and several routers will be fixed in some appropriate places to pass on data to the coordinator which will be integrated in ECU. At the same time, the concept of WPAN also put forward an inspiration that a unique wireless network can be established in a vehicle by replacing the original wired sensor with different types of wireless sensors. Under this assumption, it may save space in a vehicle and simplify the design of vehicle body and the corresponding ECU. In accordance with the assumption of the Internet of things, the WPAN of the vehicle can also be combined with other wireless communication protocols to obtain the information of surrounding environment. In this way, we can make judgment advances and optimize the control of the vehicle.

CONCLUSION

Wireless sensor networks have come out as one of the most important technologies for the future with the availability of small, cheap, smart and multifunctional sensor. The choice of WSN technology should be based on the target of its application. ZigBee, the wireless communication technique used in this study, was tested and observed the acceptable results. Those ZigBee devices can completely satisfy the target of a truck suspension air spring intermittent data transmission. However, as an attempt of self-powered generation the vibration-based piezoelectric generators are still limited by many practical factors and cannot be widely applied in small devices. In fact, under a low frequency situation, although the output voltage is acceptable, the current is too low to be applicable to work as a power supply.

In this chapter, the WSN application was considered with its height sensor's real environment and possible improvement. Different application environments tend to produce different design requirements and energy sources for different sensors. In addition, this chapter didn't fully discuss the comparison of different communication protocols. In reality, it still needs to be mindful of different protocols' best application or suitable places.

REFERENCES

Baronti, P., Pillai, P., Chook, V. W., Chessa, S., Gotta, A., & Hu, Y. F. (2007). Wireless sensor networks: A survey on the state of the art and the 802.15.4 and ZigBee standards. *Computer Communications*, *30*(7), 1655–1695. doi:10.1016/j.comcom.2006.12.020

Bibo, A., & Daqaq, M. F. (2013). Investigation of concurrent energy harvesting from ambient vibrations and wind using a single piezoelectric generator. *Applied Physics Letters*, *102*(24), 1–4. doi:10.1063/1.4811408

Chaari, L., & Kamoun, L. (2011). Performance analysis of IEEE 802.15.4/Zigbee standard under real time constraints. *International Journal of Computer Network & Communications*, *3*(5), 235–251. doi:10.5121/ijcnc.2011.3517

Choi, W. J., Jeon, Y., Jeong, J.-H., Sood, R., & Kim, S. G. (2006). Energy harvesting MEMS device based on thin film piezoelectric cantilevers. *Journal of Electroceramics*, *17*(2-4), 543–548. doi:10.1007/s10832-006-6287-3

Galchev, T., & Najafi, K. (2009). Energy Scavenging From Environmental Vibration. Sandia National Laboratories. doi:10.2172/1001023

Guo, Healy, & Zhou. (2010). ZigBee-wireless mesh networks for building automation and control. *International Conference on Networking, Sensing and Control (ICNSC).*

Hersent, O., Boswarthick, D., & Elloumi, O. (2011). ZigBee. In *The Internet of Things: Key Applications and Protocols* (pp. 93–137). John Wiley & Sons, Ltd. doi:10.1002/9781119958352.ch7

Kim, H. S., Kim, J.-H., & Kim, J. (2011). A review of piezoelectric energy harvesting based on vibration. *International Journal of Precision Engineering and Manufacturing*, *12*(6), 1129–1141. doi:10.1007/s12541-011-0151-3

Labiod, Afifi, & Santis. (2007). *Wi-fi, Bluetooth, ZigBee, and WiMAX*. Dordrecht: Springer.

Ng, T. H., & Liao, W. H. (2005). Sensitivity Analysis and Energy Harvesting for a Self-Powered Piezoelectric Sensor. *Journal of Intelligent Material Systems and Structures*, *16*(10), 785–797. doi:10.1177/1045389X05053151

Pinna, L., Dahiya, R. S., De Nisi, F., & Valle, M. (2010). Analysis of self-powered vibration-based energy scavenging system. *IEEE International Symposium on Industrial Electronics*. doi:10.1109/ISIE.2010.5637866

Ramsay, M. J., & Clark, W. W. (2001). Piezoelectric Energy Harvesting for Bio MEMS Applications. In *Energy Harvesting Technologies* (pp. 429–438). New York: Springer.

Saadon, S., & Sidek, O. (2011). A review of vibration-based MEMS piezoelectric energy harvesters. *Energy Conversion and Management*, *52*(1), 500–504. doi:10.1016/j.enconman.2010.07.024

Su, B., & Han, Y. (2010). Design of Greenhouse Wireless Monitoring System Based on ZigBee. *International Conference on E-Product E-Service and E-Entertainment*. doi:10.1109/ICEEE.2010.5661599

Veerasingam, S., Karodi, S., Shukla, S., & Yeleti, M. (2009). Design of Wireless Sensor Network Node on ZigBee for Temperature Monitoring. *International Conference on Advances in Computing, Control, and Telecommunication Technologies*. doi:10.1109/ACT.2009.14

Xu, M., Ma, L., Xia, F., Yuan, T., Qian, J., & Shao, M. (2010). Design and Implementation of a Wirelss Sensor Network for Smart Homes. *Ubiquitous Intelligence & Computing and 7th International Conference on Autonomic & Trusted Computing (UIC/ATC).*

Chapter 19
Analyzing the Goal-Finding Process of Human Learning With the Reflection Subtask

Tomohiro Yamaguchi
National Institute of Technology, Nara College, Japan

Yuki Tamai
National Institute of Technology, Nara College, Japan

Keiki Takadama
The University of Electro-Communications, Japan

ABSTRACT

This chapter reports the authors' experimental results on analyzing the human goal-finding process in continuous learning. The objective of this research is to make clear the mechanism of continuous learning. To fill in the missing piece of reinforcement learning framework for the learning robot, the authors focus on two human mental learning processes, awareness as pre-learning process and reflection as post-learning process. To observe mental learning processes of a human, the authors propose a new method for visualizing them by the reflection subtask for human to be aware of the goal-finding process in continuous learning with invisible mazes. The two-layered task is introduced. The first layer is the main task of continuous learning designing the environmental mastery task to accomplish the goal for any environment. The second layer is the reflection subtask to make clear the goal-finding process in continuous learning. The reflection cost is evaluated to analyze it.

INTRODUCTION

Recent years in Robotics, there are several researches on biologically inspired approaches (Habib, 2011 & Shi et al., 2015). One of the possible challenges is to make robots more intelligent assistants for a human. The mission of this research is to realize the human-mimetic intelligent system which finds out learning goals from the learning environment. Shi (2015) refers that "Biologically inspired robotics has

DOI: 10.4018/978-1-5225-2993-4.ch019

enabled today's robots to operate in a variety of unstructured and dynamically changing environments" and "learning often plays an important role in achieving adaption". To design the learning robots having human-mimetic learning skill, there are several targets on human's intelligent and mental learning processes such as individual learning or creative thinking (Habib, 2008).

This chapter will contribute the basis of human skill-inspired approach toward designing the human-mimetic intelligent system. Research issues of this chapter are as follows:

1. What is the mechanism of the human's goal finding process in continuous learning?
2. How to design the environmental mastery task to accomplish the goal?
3. How to design the goal finding process in continuous learning?

Our approaches for them are as follows:

1. Designing the reflection subtask for human to be aware of the goal finding process in continuous learning;
2. To design the environmental mastery task, we focus on the complexity of the learning environment which has an affect on the performance of the learning task;
3. We discuss this issue through the analysis of the experiment of human subjects.

Researches on learning process are divided into two fields. As one of them Russell and Norvig (2009) discuss a learning algorithm for robot in robotics and Artificial Intelligence, as the other Marton and Booth (1997) discuss learning of a human in psychology. For the learning robot, reinforcement learning is the major framework since it automatically learns after a learning goal is set in the learning environment. The main feature of reinforcement learning is that the learning goal is given by the human designer. On the other hand, researches on human learning ability have been performed in various research fields such as psychology, education, business, and so on. One of the main features of human learning ability is that it covers a vast territory of learning ability including discovery of learning goals, *awareness*, *reflection*, self-regulated learning (Schunk, & Zimmerman, 2007), or *continuous learning*.

The objective of this research is to bring the learning ability of the learning agent close to that of a human. The authors focus on both reinforcement learning framework for the learning agent and continuous learning model of a human. However, there are two kinds of questions. First question is how to bridge an enormous gap between them. Second question is how to observe mental learning processes of a human. Previous methods of human learning researches mostly depend on observable behaviors or activities. On the other hand, a *learning process* of a human has a major difficulty in observing since it is a mental process. Then a human learning process is yet-to-be-defined. So it is necessary to add a new twist to observe the human learning process.

To solve these problems, the authors propose a new method for *visualizing* mental learning processes with *invisible* mazes. Yamaguchi, Takemori, and Takadama (2013) focus on *continuous learning* and aim for modeling the unified continuous learning process model based on reinforcement learning framework for both a human and the learning agent. Our new approaches are following:

1. To fill in the missing piece of reinforcement learning whose learning process is behavior change, the authors add two mental learning processes, *awareness* as pre-learning process and *reflection* as post-learning process;

2. The learning environment is the *invisible* maze consisting of *invisible* walls which are perceived as a *sign* that suggests the number of walls in the neighborhood;
3. The authors add *the reflection subtask* for expressing and summarizing something to be *aware* of learning from mistake or to be *reflected* on learning from experience. A learner can mark up his/her traces of the actions and signs for future success. It turns out to *visualize* his/her mental *learning processes*;
4. To examine the reason why the non-continuous learner stops learning the task, the authors analyze the learner's performance of both the main learning task by *achievement cost* and the reflection subtask by *reflection cost* in each learning stage.

This paper reports our experimental results on analyzing human's continuous learning ability with the reflection cost. In the previous work of Yamaguchi, Takemori, Tamai, and Takadama (2015a, 2015b), the experimental results showed that there is a strong negative correlation (the correlation factor is -0.78) between the number of discovered solutions and the reflection cost. It is that the learners who found many solutions learn continuously and work smart with a small reflection cost. The main factor whether they learn continuously or not is to find new learning goals. To examine the reason why the non-continuous learners stop learning the task and why the continuous learners keep learning, each learner's performance of the reflection sub task in each learning stage is analyzed. Next section the authors give a theoretical background on *continuous learning* and the continuous learning process model.

BACKGROUND

IEEE Industrial Electronics Society

This section describes a theoretical background of this research, continuous learning, awareness and reflection. The concepts awareness and reflection are viewed differently across the disciplines among Computer-supported cooperative work (CSCW), Computer-supported collaborative learning (CSCL), psychology, business, educational sciences, computer science, and so on. First the authors summarize an overview of continuous learning and its learning process which is compared with that of reinforcement learning, then our usage of awareness and reflection is introduced. For more information, refer (Yamaguchi, Takemori, Tamai, & Takadama, 2015a, 2015b).

An Overview of Continuous Learning

The concept of continuous learning comes from Industrial and organizational psychology. Smita and Trey (2012) review research on continuous learning. There are three levels, individual, group, and organizational level. At the individual level it is self-directed; at the group level it is collaborative. One of conceptual definitions of continuous learning is according to Sessa and London (2006), "Continuous learning at the individual level is regularly changing behavior based on a deepening and broadening of one's skills, knowledge, and worldview."

Learning Process Model to Achieve Continuous Learning

Buckler (1996) has proposed a learning process model to achieve continuous improvement. Learning process is defined as a process that results in changed behavior. Learning process consists of several mental processes. This model assumes a learner and the leader. Dorval et al. (2003) discuss the role of this leader is mentor or coacher rather than teacher. Continuous learning seeks learner-centered rather than teacher-centered.

Comparison of Continuous Learning With Reinforcement Learning

Figure 1 compares learning process between continuous learning and reinforcement learning. Figure 1(a) shows the learning process of continuous learning. The major feature of it is that it circles each process and it has no end. The reason is that continuous learning is designed for an adult learner who is expected to spiral up along the endless learning cycle toward mastery of some professional skill. Thus, it has commitment process to commit a *learning goal*. However, each process is mental process. In contrast with it, Figure 1(b) shows the learning process of reinforcement learning. Its distinction is that its process has a start and an end. Since the objective of existing reinforcement learning is to find an optimal solution for the learning goal given by a human designer. Next subsection reviews *Awareness* and *Reflection* as important mental processes of continuous learning process.

Awareness and Reflection on Learning Processes

The common feature of *Awareness* and *Reflection* is focusing experience of a learner on some information for future improvement. The main differences between them are that *Awareness* relates to the perception, *Reflection* mainly relates to the action or the behavior consisting of perception and action.

Awareness as Pre-Learning Process

As pre-learning process, *awareness* plays an important part to trigger behavior change. Then the authors focus on the meaning of *awareness* as the need for distinction of indistinguishable perceptions or experience between future success and future failure. The authors assume that these indistinguishable perceptions occur by partially observable states (Kaelbling, Littman, & Cassandra, 1998).

Figure 1. A comparison of learning process

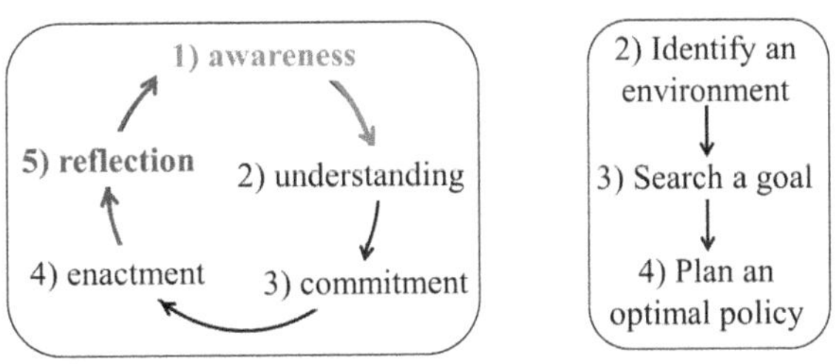

(a) continuous learning process (b) reinforcement learning process

Reflection as Post-Learning Process

There are several meanings of reflection on learning. One of them is as follows; Learning through reflection is defined as "those intellectual and affective activities in which individuals engage to explore their experiences in order to lead to new understandings and appreciations" (Boud, Keogh, & Walker, 1985). The authors take to the concept of Reflective learning (Krogstie, Prilla, & Pammer, 2013) which is "the conscious re-evaluation of experience for the purpose of guiding future behavior".

As post-learning process, *reflection* plays a part to create some meaning out of behavior change or learning result. Then the authors focus on the meaning of *reflection* as creating the becoming explanation or interpretation toward the rule which can lead the learner not to future failure but to future success.

ANALYZING THE GOAL FINDING PROCESS BY THE TWO-LAYERED TASK

This section describes designing the two-layered task to make clear the goal finding process in continuous learning. the goal finding process is a process of meta-learning to discover learning goals while learning the main learning task. First, the environmental mastery task to accomplish the goal for any environment is introduced, then the meta-level learning task is described.

The Environmental Mastery Task

First layer is the main task of continuous learning. The authors define the environmental mastery task which is to accomplish the goal for any environment as one of the continuous learning task. The authors design the goal of the task as the maze sweeping task which is to find the longest paths from the start state to the goal state in the maze as many as possible. As the learning environment for the environmental mastery task, the authors design the sequence of the learning stages in ascending order of difficulty such as from a simple rectangle maze to complex shape one which boundary has concavo-convex shape or which has inner walls. Assisting the exploration to find out sub-goals of a maze in each stage, an action of first contact with an invisible side consisting of one or more walls is rewarded. Note that it is not instructed to a learner. If the learner is aware of this rewarding rule on first contact event, at first he/she is encouraged that the encounter with an invisible wall is positive event, then he/she may become aware of efficient exploration of an invisible wall and an invisible side of the boundary of a maze. The essential of this idea is the potential for the automatic sub-reward generation according to the complexity of a learning environment. In the maze environment of our system, the boundary shape of maze and inner obstacles of maze represents the complexity of the learning environment.

The Meta-Level Learning Task to Find the Learning Goals

Second layer is the meta-level learning task. The authors assign the reflection subtask for human to be aware of the goal finding process in each learning stage to observe the process. The authors design a learning environment with an invisible goal state and invisible walls as the boundary of maze or inner obstacles. To perform the maze sweeping task, it is necessary for a learner to identify them. There are two kinds of objectives of the reflection subtask. First one is that learners are instructed to make out the reflection map of each maze by placing color markers on it what the learner finds on the current maze at the trial of the maze sweeping task in order to give a new learner a clue to perform the stage.

Second objective is to find learning goals which is not instructed directly. It is that the human learners become to see the meaning of the reflection map and the reflection subtask as their learning stages progressed. In the learning experiment, the reflection cost of each learner during the learning stages is analyzed whether stable or not. Note that the reflection cost is the time a learner worked through the reflection subtask per the number of found solutions. If the reflection cost of a learner is stable during the learning stages, it suggests that the learner can perform the reflection subtask in a certain amount of time and can perform a very good job of it, even though it becomes more difficult as the learning stage progressed.

To make clear the reflection mental process of the continuous learners, the authors report the experimental results. The authors have a hypothesis that an intelligent learner utilizes information acquired from the current trial and errors for the forward trials or achievement in the way of the reflection subtask. For example, it is important to be aware of finding the invisible walls through the reflection subtask since the distribution of the markers placed on found invisible walls may suggest the blank area to search as making out the reflection map.

DESIGNING THE CONTINUOUS LEARNING PROCESS WITH THE INVISIBLE MAZE MODEL

This section summarizes the design concepts for the *continuous learning process*, then describes the way to design the *continuous learning support system* based on the reinforcement learning process model to guide a human to achieve *continuous learning*. For modeling the continuous learning process as shown in Figure 1, this paper formalizes them by a *maze model* with *invisible walls* and an invisible goal and designs the *maze sweeping task* which requires discovering and mastering invisible various learning goals of a learner.

The Flow of the Continuous Learning Process

Figure 2 shows the flow of the continuous learning process. This process consists of triple cycle. Innermost cycle is called a *trial*. A *trial* is defined as a transition sequence from start state to encountering either a goal state or a wall. In this cycle, a learner repeats an action and his/her mental process including *awareness* until he/she results in either success or fail of the task. Second cycle is called an *achievement*.

Figure 2. The flow of the continuous learning process

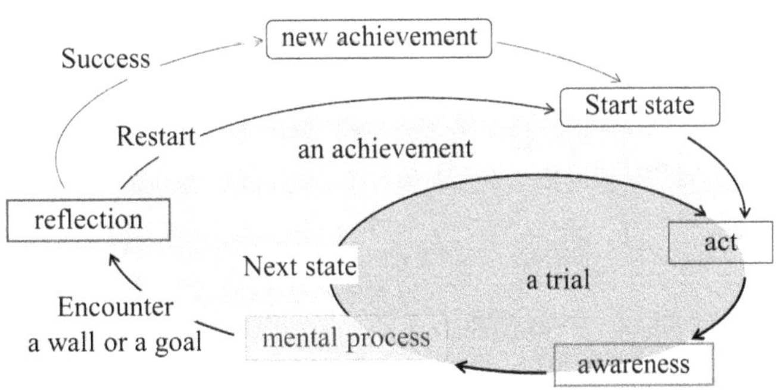

An *achievement* is defined as the learning of a maze sweeping task with the fixed start and the invisible goal. In this cycle, the learner reflects the trial by *the reflection subtask* when the trial terminates by the encounter with a wall or a goal. If current trial is not accomplished, he/she restarts the trial from the start state. Outmost cycle is the *continuous learning* cycle. When the learner accomplished current *achievement*, he/she can challenge next new achievement.

Designing the Learning Environment by an Invisible Maze

Designing a learning environment for a human learner, the authors adopt a grid maze model from start to goal since it is a familiar example to find the path through a trial and error process (Yamaguchi, Takemori, & Takadama, 2013). A *maze model* is defined by five elements, state set, a *sign* of walls, transitions and walls, action set, and meta-actions.

Figure 3 shows the structure of a 2D grid maze. The n x m grid maze with four neighbors consists of the n x m number of 1 x 1 squares. It is called a *simple maze* which is surrounded by walls in a rectangle shape. Figure 3(a) shows a 3 x 2 simple maze with a start and a goal. In a grid maze, every square that touches one of their edges except a wall is connected. Each square in a maze model is called a *state.*

Transitions between states in a maze model is defined whether corresponding square with four neighbors, {up, down, left, right} which is whether connected or not connected by a *wall.* They are represented as the labeled directed graph as shown in figure 3(b). *Action* set is defined as a set of labels to distinguish the possible transitions of a state. In a grid maze model with four neighbors, a learner can execute four kinds of absolute *actions*: {up, right, left, down} or relative *actions*: {forward, turn-right, turn-left, reverse}. The action toward a wall results in the transition to the start state to *restart* the trial. Transition to a goal state results in automatic *restart* with the judgment whether the maze sweeping task is accomplished or not. Next the authors describe a *maze sweeping task* which is performed in an *achievement.* It is defined as to find paths from a fixed start state S to the invisible goal state G which visit all states only at once in the maze.

Visualizing the Awareness Process by a Sign

A *sign* is defined as the number of walls of four neighbors in each state; {0,1,2,3,4}. Figure 4 shows an illustrated example of a non-simple maze with invisible walls. Figure 4(a) shows the 4x4-2 invisible maze and an example of the transition sequence as if wall-following of the learner. Dashed lines are invisible walls which are invisible for a learner during trials. Figure 4(b) shows the distribution of signs among

Figure 3. The structure of a 3 x 2 simple maze

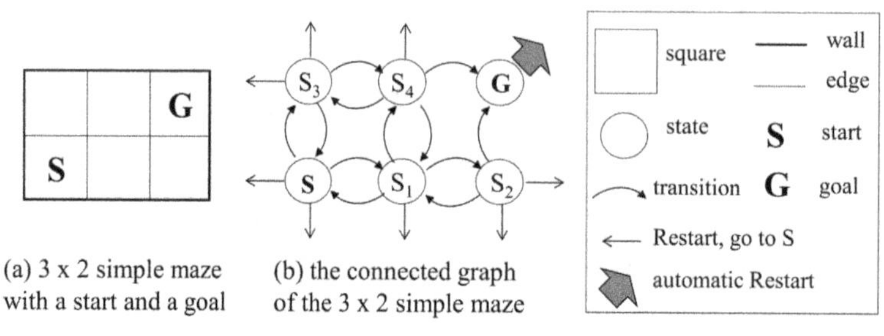

(a) 3 x 2 simple maze with a start and a goal

(b) the connected graph of the 3 x 2 simple maze

Figure 4. An example of the 4x4-2 non-simple maze with invisible walls

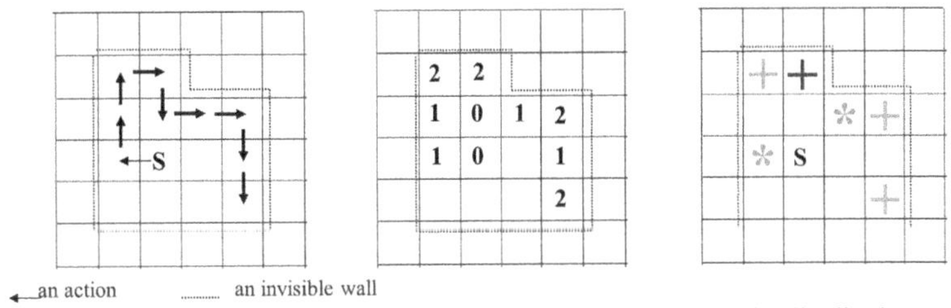

(a) an example of transition sequence (b) the distribution of signs (c) the distribution of marked signs

visited states in the maze. In our previous experiment, a number was displayed as a sign of surrounding walls in each visited state. In this case, the sign '2' means a corner (Yamaguchi, Takemori, Tamai, & Takadama, 2015a, 2015b).

Making the learning tasks more difficult to learn continuously, than previous research the authors introduce a *marked sign* (Yamaguchi, Takemori, Tamai, & Takadama, 2015a, 2015b). Figure 4(c) shows the current version of the distribution of the marked signs which the learner perceives. The marked sign is displayed at the visited state to assist awareness of invisible walls until the end of the trial. A *marked sign* is defined by the sign sequence. A *sign sequence* between two neighbor states are described as follows:

(the sign of previous state, the sign of current state)

In Figure 4(c), the marked sign * with light red which comes from (0, 1) suggests the presence of a wall. A state which sign is 2 suggests whether a corner or a passage, both state is two sides of walls. The marked sign + with light/normal blue which comes from (0/1, 2) suggests the presence of a corner (or a passage). The marked sign + with dark blue which comes from (2, 2) suggests the presence of cascade corners, cascade passages or cascade corner and passage. A state which sign is 3 suggests the dead end. The marked sign with light/normal/dark green which comes from (0/1/2, 3) suggests the presence of a deadend.

Visualizing the Reflection Process by the Reflection Subtask

This subsection describes the way to visualize reflection process. The authors introduce *the reflection subtask* which is used to describe the *awareness* from a sign of the invisible walls to visualize his/her *reflection* process. *The reflection subtask* is defined to attach a marker on a state in the reflection map for expressing and summarizing a learner's mental learning processes. The markers are one yellow marker and many RGB markers. Figure 5 shows the initial reflection map of each stage with ten kinds of markers. RGB markers are red, green, and blue markers, these color density is in three tones, light, normal and deep in each color. Therefore, there are nine kinds of many RGB markers and one yellow marker. Note that right hand of yellow marker is the eraser icon to erase a marker in the reflection map.

At the end of a trial, markers can be attached on any state in the reflection map. Figure 6 shows an example of the reflection subtask of stage 1. Figure 6(a) shows an example of the trial of Stage 1. In each state, whether visited or not is distinguished by its background color of the square either gray (visited)

Figure 5. The initial reflection map of each stage with ten kinds of markers

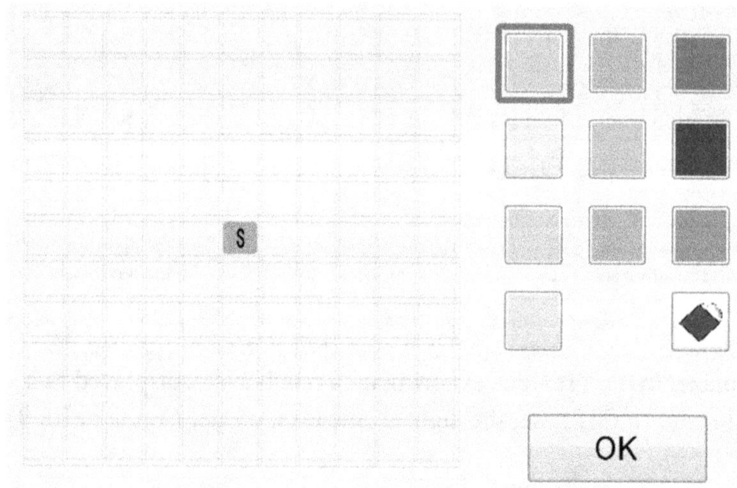

Figure 6. An example of the reflection subtask of stage 1

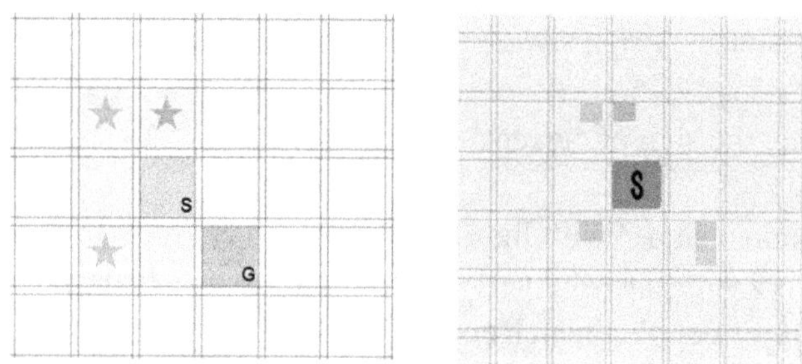

(a) An example of the trial of Stage 1 (b) The reflection map of Stage1

or white (not visited). A star mark is a sign, its color suggests that the wall surrounding situation is different. If the goal state is found, its background color is yellow and small G is displayed. Figure 6(b) shows an example of the reflection map of Stage 1. In each square, maximum four kinds of markers are placeable. The typical usage of them is to memorize the special event such as the encounter with an invisible wall or a goal, or to summarize the sign distribution in the searched maze. The reflection map is displayed during a train of trials.

Designing the Sequence of Learning Goals

Designing the learning goals, even the case of simple mazes may result in the difficulty of a partially observable state, it can be worked out by slightly broadening a learner's perceptual states to distinguish partially observable states. The case of non-simple mazes results in more difficulty of partially observable states since they have various non-distinguishable states. This paper describes learning goals for the simple maze with a partially observable state. Under simple and invisible mazes, the standard steps to master them are follows:

1. Find a goal state if it is invisible;
2. If walls are invisible, estimate the borders of the current maze such like wall-following behaviors;
3. Search all maze sweeping paths from the start state.

The authors focus on learning goals for the second step to estimate the invisible borders of a simple maze. To solve this, the learner should try to consider the sequence of signs toward neighbor state. If the learner encounters either a wall or a sign 0 state, he/she can decide the direction of walls. In this case, an important learning goal to prevision an invisible wall is being aware of sign sequences as (0, 1) or (1, 2). Note that these increasing patterns for provisioning an invisible wall hold good only under simple mazes.

EXPERIMENT

To examine the reason why the non-continuous learner stops learning the task, the authors conducted the more difficult continuous learning experiment than previous research by eight human subjects. In our previous work, the experimental results showed that there is a strong negative correlation (the correlation factor is -0.78) between the number of discovered solutions and the reflection cost (Yamaguchi, Takemori, Tamai, & Takadama, 2015a, 2015b). It is that the learners who found many solutions learn continuously and work smart with a small reflection cost.

The main factor whether they learn continuously or not is to find new learning goals. To examine it, this paper analyzes each learner's reflection cost which is the performance of the reflection sub task in each learning stage. The authors designed the continuous learning task as 30 stages in ascending order of difficulty. Note that there is no limit on the time of the experiment, or the number of encounters with a wall. Main difficulties are as follows:

1. The number of stages is reduced from 54 to 30. So that the degree of increasing difficulty per stage is augmented than our previous experiment;
2. The perceived sign is more difficult to understand as the authors described before. If the learner becomes aware of the meaning of the color of signs, the learner can estimate the presence of walls so that to continue achieving more difficult second-half learning stages.

Experimental Setup

The Learning Environment for the Continuous Learning

The authors prepare the sequence of invisible mazes consists of total thirty stages. In each stage, there is one maze with two or more solutions to be found, excepting 18th, 19th and 26th stages with only one solution. The sequence of stages is arranged in order of increasing difficulty. First three stages are practice stages, and first nine stages are simple mazes, 10th stage or later are all non-simple mazes. Note that a non-simple maze is surrounded by outside walls in concavo-convex shape, or it has an internal dividing wall.

Visualizing Reflection Process by the Reflection Subtask

The authors prepared the reflection subtask with ten kinds of color markers as the authors described. The method of operation is instructed, but the role of each color markers is not assigned.

The First Instruction for Subjects

It is instructed at the beginning of the experiment:

- A solution is a path from the start state to the goal state;
- The number of discovered solutions is limited by a ceiling in each stage;
- The objective of the task is to find more sweeping solutions in each maze;
- A sweeping solution is the path to visit all states from the start state to the goal state;
- First three stages are practice stages.

The Second Instruction for Subjects

It is instructed after first three stages are finished:

- The real purpose of the task is to learn the way to get better by challenging the maze task one after another;
- In the reflection subtask, place the minimum markers in the reflection map to hint the way to solve the maze required by a new challenger of this maze task;
- You may decide freely what to get better;
- If you are hard to get better, you may terminate the maze task.

The Measurement Item

The measurements of the experiment are as follows:

a. The total time of the experiment;
b. The number of challenged stages;
c. The number of discovered sweeping solutions;
d. The number of encounters with a wall;
e. The total time of the achievements;
f. The total time of the reflection subtask;
g. The reflection cost;
h. The achievement cost.

Note that the total time of the experiment (a) is defined by Eq. (1). The reflection cost (g) is defined by Eq. (2), it evaluates the quality of the reflection process. The achievement cost (h) is defined by Eq. (3), it evaluates the quality of the achievement process:

$$(a) = (e) + (f) \quad (1)$$

$$(g) = (f) / (c) \quad (2)$$

$$(h) = (e) / (c) \quad (3)$$

Experimental Results

Summary of the Experimental Results

Figure 7 shows summary of the experimental results compared to our previous experiment. This shows that the learning task becomes more difficult to continue learning than previous one. The experimental results are as follows:

1. 1. The total time of the experiment (minutes):

Average time is 42, maximum is 91, minimum is 20

2. 2. The number of challenged stages:

Average is 17.4, maximum is 30, minimum is 9

3. 3. The number of discovered sweeping solutions:

Average is 29.3, maximum is 64, minimum is 17

The reason that the standard deviation in (C) is bigger than that of (b) is that there are about two sweeping solutions in each stage.

Figure 7. Summary of the experimental results compared to previous experiment

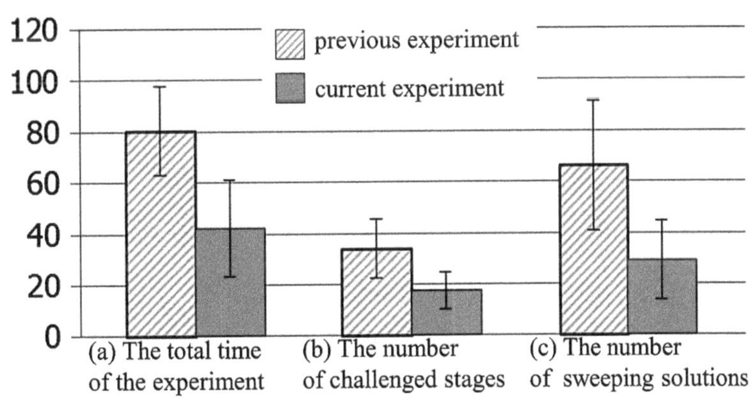

Two subjects within eight continued to perform the experiment more than 90 percent stages (more than 27 stages within total 30 stages). Considering the number of stages as the continuous learning ability, the authors call these subjects as *continuous learners.* Subjects who perform the experiment less than 90 percent stages are called *non-continuous learners.*

Analyzing the Experimental Results

To examine the reason why the *non-continuous learners* stop learning the task, the authors analyze the learner's performance of both the main learning task and the reflection subtask in each learning stage. As the experimental results, there are three cases whether the human learning is continuous or not. Figure 8, 9, 10, and 11 shows the relationship between the reflection cost and the progression of stages. Horizontal axis is the progression of stages, vertical axis is the reflection cost in each stage (second/solution). Figure 8 shows an example of the continuous learning case in which the reflection cost of the continuous learner is stable during the learning stages. It suggests that the continuous learner can perform the reflection subtask in a certain amount of time, even though it becomes more difficult as the learning stage progressed.

On the other hand, Figure 9 shows two kinds of non-continuous learners. Figure 9(a) shows an example of increasing the reflection cost case that it becomes too large to continue the task. In contrast

Figure 9. The reflection cost in each stage of non-continuous learners

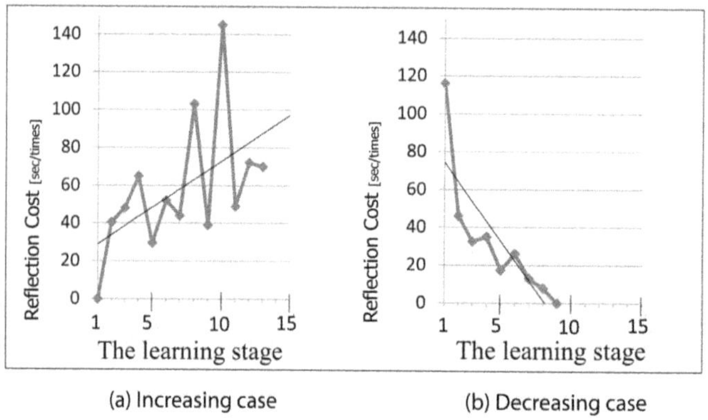

Figure 8. The reflection cost in each stage of the continuous learner

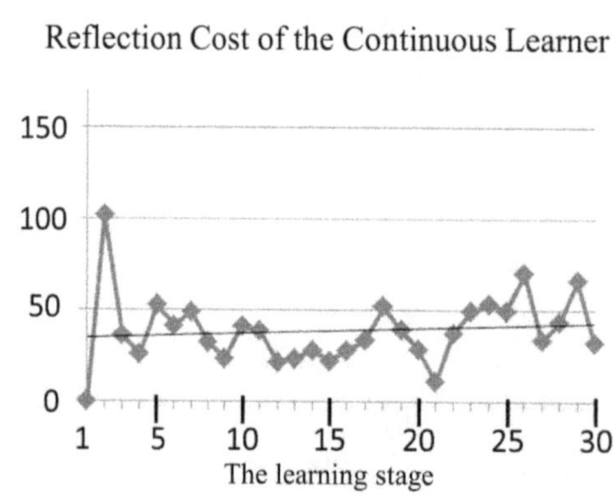

to this, Figure 9(b) shows an example of decreasing case, it suggests that the motivation of the learner becomes too small to continue the task.

Interestingly, one non-continuous learner is compound case. Figure 10 shows the reflection cost and the achievement cost of the compound non-continuous learner, which increases in first half stages, then it decreases in second half stages. The reason why the reflection cost of this learner decreases after the 10th stage is that it becomes difficult from 10th stage on since each maze is the non-simple maze from 10th stage on. At the peak of both the reflection cost and the achievement cost is 11th stage as shown in Figure 10 since the learner struggles against the non-simple mazes. After 11th stage, these costs decrease then fade out.

Figure 11 shows the reflection cost in each stage among eight subjects. Each line segment shows the linear approximation of this relationship of each subject. Right side's end point of a line segment is the last stage that the subject achieved. Figure 12 shows the achievement cost in each stage among

Figure 10. an example of compound non-continuous learning case

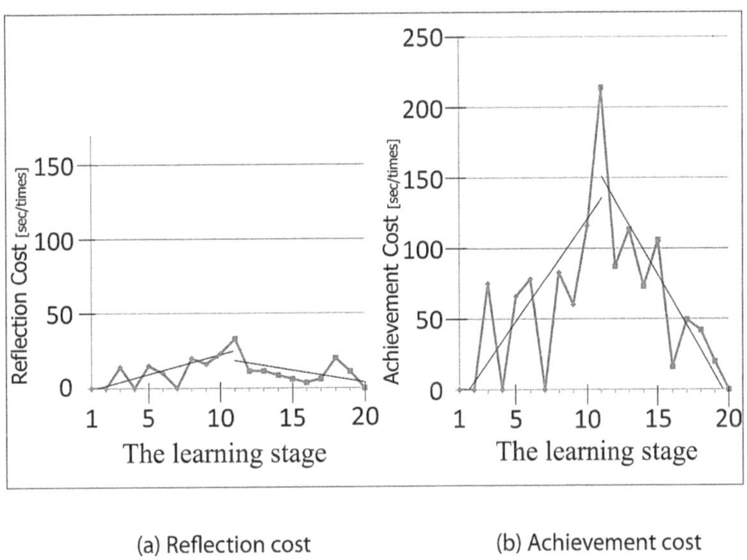

Figure 11. The reflection cost in each stage among eight subjects

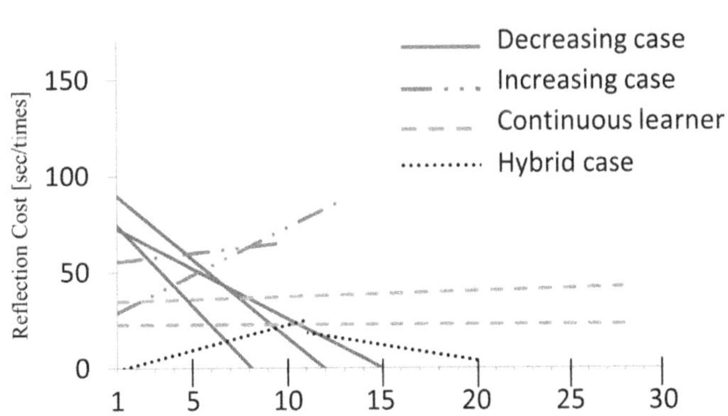

Figure 12. The achievement cost in each stage among eight subjects

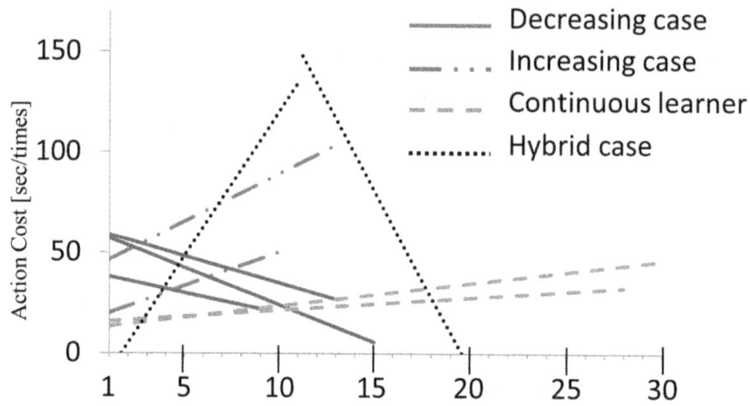

them. As the experimental results, both the reflection cost and the achievement cost of the continuous learner is stable during the learning stages as compared to non-continuous learners. It suggests that the continuous learner can perform continuously and can work smart with small and stable reflection costs.

FUTURE RESEARCH DIRECTIONS

First, the meaning of the reflection cost and its usage are discussed. Regarding the reflection cost as the level of proficiency on the reflection skill, as a learner gets more proficient at reflection, the reflection cost comes down then the learner becomes to solve invisible mazes more easily. Therefore, the learner can achieve more difficult learning stages within the range of achievement cost.

Second, the authors discuss the way to generalize our proposed methods. Since the objective of the continuous learning task is to learn the way to learn continuously. One of good ways is to find a new interpretation of the task. One of the applications of the continuous learning task is human robot interaction task. A learning stage is a context of interactions between a human and the robot. The maze in the stage is a situation of the interaction, surrounding walls are the frame to set the interaction in it. Sign from the human in the interactions suggests that the action of the robot will be out of the situation.

The interaction starts from the human's request such like a question. In each interaction context, the learning robot achieves to find the goal which is the intention of the human. When the robot encounters the failure such as wall, it is marked in the context map to avoid it after the trial. Then the robot achieves to find the way to the human goal in the interaction context. If the reflection cost on the human robot interaction task is stable, it may suggest that their interaction grows well. If the interaction cost increases, the interaction context is too hard for the learning robot. Then the robot can express the hard situation to the human.

CONCLUSION

The authors described the way to design the continuous learning support system and reported our experimental results on analyzing human's goal finding process for continuous learning with the reflection cost. As the experimental results, both the reflection cost and the achievement cost of the continuous learner is stable during the learning stages as compared to non-continuous learners. It suggests that the continuous learner can perform continuously and can work the reflection subtask smart with small and stable reflection costs in a certain amount of time, even though it becomes more difficult as the learning stage progressed. Future work is to design the collaborative continuous learning system between a human learner and the learning robot.

ACKNOWLEDGMENT

The authors would like to thank Prof. Habib and Prof. Shimohara for offering a good opportunity to present this research. This work was supported by JSPS KAKENHI (Grant-in-Aid for Scientific Research ©) Grant Number 16K00317.

REFERENCES

Boud, D., Keogh, R., & Walker, D. (1985). Reflection: Turning experience into learning. Abingdon-on-Thames, UK: Routledge.

Buckler, B. A. (1996). Learning process model to achieve continuous improvement. *The Learning Organization*, *3*(3), 31–39. doi:10.1108/09696479610119660

Dorval, K. B., Isaksen, S. G., & Noller, R. B. (2003). Leadership for learning: Tips for effective mentoring and coaching. In *Mentoring for Talent Development*. Albion, MI: Reclaiming Youth International. Retrieved August 1, 2017, from http://www.cpsb.com/resources/downloads/public/Mentoring_Coaching.pdf

Habib, M. K. (2008). Interdisciplinary mechatronics: Problem solving, creative thinking and concurrent design synergy. *International Journal of Mechatronics and Manufacturing Systems*, *1*(1), 264–269. doi:10.1504/IJMMS.2008.018272

Habib, M. K. (2011). Biomimetics: Innovations and robotics. *International Journal of Mechatronics and Manufacturing Systems*, *4*(2), 113–134. doi:10.1504/IJMMS.2011.039263

Kaelbling, L. P., Littman, M. L., & Cassandra, A. R. (1998). Planning and acting in partially observable stochastic domains. *Artificial Intelligence*, *101*(1-2), 99–134. doi:10.1016/S0004-3702(98)00023-X

Krogstie, B., Prilla, M., & Pammer, V. (2013). Understanding and supporting reflective learning processes in the workplace. In *Proceedings of the Eighth European Conference on Technology Enhanced Learning* (EC-TEL 2013). Berlin, Germany: Springer-Verlag.

Marton, F., & Booth, S. (1997). Learning and awareness. Abingdon-on-Thames, UK: Routledge.

Russell, S., & Norvig, P. (2009). *Artificial intelligence: A modern approach* (3rd ed.). Upper Saddle River, NJ: Prentice Hall.

Schunk, D. H., & Zimmerman, B. J. (2007). Motivation and self-regulated learning: Theory, research, and applications. Abingdon-on-Thames, UK: Routledge.

Sessa, V. I., & London, M. (2006). *Continuous learning in organizations: Individual, group, and organizational perspectives*. Hove, UK: Psychology Press.

Shi, L., Habib, M. K., Xiao, N., & Hu, H. (2015). Biologically inspired robotics. *Journal of Robotics, 2015*, 1–2. doi:10.1155/2015/894394

Smita, J., & Trey, M. (2012). *Facilitating continuous learning: A review of research and practice on individual learning capabilities and organizational learning environments*. Paper presented at the Annual Meeting of the AECT International Convention, Louisville, KY. Retrieved August 1, 2017, from http://www.aect.org/pdf/proceedings12/2012i/12_09.pdf

Yamaguchi, T., Takemori, K., & Takadama, K. (2013). Modeling a human's learning processes toward continuous learning support system. In M. K. Habib & J. P. Davim (Eds.), *Mechatronics Engineering* (pp. 69–94). Hoboken, NJ: Wiley-ISTE. doi:10.1002/9781118577516.ch4

Yamaguchi, T., Takemori, K., Tamai, Y., & Takadama, K. (2015a). Analyzing human's continuous learning ability toward the intelligent robotics. In *Proceedings of the Twentieth International Symposium on Artificial Life and Robotics 2015 (AROB 20th 2015)*. Chiyoda-ku, Japan: Springer Japan.

Yamaguchi, T., Takemori, K., Tamai, Y., & Takadama, K. (2015b). Analyzing human's continuous learning processes with the reflection sub task. *Journal of Communication and Computer, 12*(1), 20–27.

KEY TERMS AND DEFINITIONS

Awareness: Increasingly related to finding appropriate learning objects, peers and experts, or the "right" learning path.

Continuous Learning: Continuous learning at the individual level is regularly changing behavior based on a deepening and broadening of one's skills, knowledge, and worldview.

Environmental Mastery Task: To accomplish the goal for any environment as one of the continuous learning tasks.

Goal-Finding Process in Continuous Learning: A process of meta-learning to discover learning goals while learning the main learning task.

Invisible Maze: A learning environment with an invisible goal state and invisible walls as the boundary of maze or inner obstacles.

Learning Process: A process that consists of several mental processes. It results in changed behavior.

Maze-Sweeping Task: A maze task to find paths from a fixed-start state S to the invisible goal state G which visit all states only once in the maze.

Reflection: Those intellectual and affective activities in which individuals engage to explore their experiences in order to lead to new understandings and appreciations.

Reflection Cost: The time a learner worked through the reflection subtask per the number of found solutions. It suggests that the continuous learner can perform the reflection subtask in a certain amount of time and can perform a very good job of it.

Reflection Subtask: To describe the awareness from a sign of the invisible walls to visualize his/her reflection process. It is defined to attach a marker on a state in the reflection map for expressing and summarizing a learner's mental learning processes.

Reinforcement Learning: A learning algorithm for a robot or a software agent to take actions in an environment so as to maximize the sum of rewards through trial and error.

Chapter 20
Designing the Learning Goal Space for Human Toward Acquiring a Creative Learning Skill

Takato Okudo
National Institute of Technology, Nara College, Japan

Tomohiro Yamaguchi
National Institute of Technology, Nara College, Japan

Keiki Takadama
The University of Electro-Communications, Japan

ABSTRACT

This chapter presents the way to design a learning support system toward acquiring a creative skill on learning. There are two research goals. One is to establish designing the creative learning task. The other is to make clear the human sense of creativity. As the background of this research, the jobs with high creativity or social skills will remain in the future. However, acquiring human's creativity is too difficult for computers. To solve this problem, the authors focus on the way to utilize higher creativity of human than that of computers. The main method is the visualization of learning traces to support awareness for creativity on the learning. The authors conducted the preliminary learning experiment with three human subjects. After that, the questionnaire and the hearing investigation were conducted. As the future work, the authors are planning to conduct an updated version of the experiment.

INTRODUCTION

Recent years in Robotics, there are several researches on biologically inspired approaches (Habib, 2011 & Shi et al., 2015). One of the challenges is to make robots more intelligent learner as a human. This paper describes the learning ability for both human and robot. The missions of this research are to clarify the

DOI: 10.4018/978-1-5225-2993-4.ch020

mechanism of human's creativity on the learning and to improve the learning abilities both for a human and for intelligent robot. Shi (2015) refers that "Biologically inspired robotics has enabled today's robots to operate in a variety of unstructured and dynamically changing environments" and "learning often plays an important role in achieving adaption". To design the learning robots having human-equivalent skill level, there are several targets on human's intelligent and mental processes such as individual learning, team-based learning or creative thinking (Habib, 2008). This research will contribute the basis of human skill-inspired approach toward designing the creative learning algorithm for intelligent robot to the field of biomimetics.

The progress of information technology will make that about half of human jobs are replaced by computers in near future. The remained jobs having technical difficulty for both AI and computers need high creativity or social skills. According to Boden (2003), "creativity is the ability to come up with ideas or artifacts that are novel and valuable". Previous research on human creativity suggests that "one process of creating ideas involves making unfamiliar combinations of familiar ideas, requiring a rich store of knowledge" (Frey, C. B., 2015, p.26). However, it is too difficult for robot or computer to acquire human's creativity. The reason is that creativity is not clear, and the way to learn human's creativity is not clear. To solve this problem, the authors focus on the interactive way to utilize higher creativity of human than that of robots or computers, and analyze the human's behaviors with the creativity in the task.

This paper describes how the interactive system inspires the learner to find more learning goals, and this is instrumental in considering the effect of finding of learning goals toward creativity. Research issues of this research are as follows:

1. What is the mechanism of human's creative learning skill?
2. How is it possible to realize difficult creative learning skill for previous machine learning methods?
3. Why is it important that intelligent systems have human-like creative learning ability?

Our approaches for them are as follows:

1. Designing the creative learning task, then analyzing the human's learning behaviors to clarify the mechanism of human's creative learning skill;
2. To make it a goal to realize the human-mimetic learning mechanism;
3. It is important that a learning system is easy to understand the learning property to realize the human friendly intelligent robots or systems.

The authors proposed the continuous learning support system for a human learner (Yamaguchi, 2015). The mechanism proposed in this paper is based on its framework and extends it to suggest the creativity on the learner's own learning. In the proposed mechanism, the support system generates a derived learning achievement by combining the learner's taken achievement and the learner's found solution. Then the learner can reflect the learner's own learning trace on the learning goal space. Owing to them, the support system suggests the awareness toward both unclear learning results and the sense of values for the learner. There are three kinds of the proposed support methods. First, it is the visualization of learning traces to support awareness of creativity on the learning. The authors design the learning goal space to visualize learning traces. They are the distribution of the learning goals found by the learner who learns the original achievement and the derived one generated by the support system. It makes easier to reflect the learning orientation by means of showing the relative positioning of the learning goal to the learning trace. Second is the discovery support for unknown solutions by generating the derived achievement based on negation of the learner's found shortest solution. It encourages the learner to perceive the learner's unclear solutions. Third is the generating the derived achievement by the justification of the found redundant solution. It encourages the learner to perceive his/her unclear sense of values.

BACKGROUND

This section describes the theoretical background of this research. After the research on the creativity is described, the authors summarize an overview of continuous learning because it is the basic framework of the creative learning process, and then the authors describe the creative learning skill.

Continuous Learning

The Definition of Continuous Learning

The concept of continuous learning comes from industrial and organizational psychology. Smita and Trey (2012) reviews research on continuous learning. There are three levels, individual, group and organizational level. At the individual level it is self-directed; at the group level it is collaborative. This paper discusses the continuous learning at the individual level. The authors adopt Sessa's definition as follows; "Continuous learning at the individual level is regularly changing behavior based on a deepening and broadening of one's skills, knowledge, and worldview" (Sessa & London, 2006).

Learning Process Model to Achieve Continuous Learning

A learning process model to achieve continuous improvement has been proposed (Buckler, B., 1996). Learning process is defined as a process that results in changed behavior. Learning process consists of several mental processes. This model assumes a learner and the leader. The role of this leader is mentor or coacher rather than teacher (Dorval, Isaksen, & Noller, 2003). There are two kinds of major learning methods, whether teacher-centered (Taught model) or learner-centered (Discovery model). The features of Taught model are extrinsic motivation, controlled and consistency. Its theoretical basis is Gestalt or cognitive psychology. Continuous learning seeks learner-centered rather than teacher-centered.

Comparison of Continuous Learning With Reinforcement Learning

The learning process between continuous learning and reinforcement learning are described. *Figure 1(a)* shows the learning process of continuous learning. The major feature of it is that it circles each process and it has no end. The reason is that continuous learning is designed for an adult learner who is expected to spiral up along the endless learning cycle toward mastery of some professional skill. Thus, it has commitment process to commit a learning goal. However, each process is mental process. *Figure 1 (b)* shows the learning process of reinforcement leaning. Its distinction is that its process has a start and an end. Since the objective of existing reinforcement learning is to find an optimal solution for the learning goal given by a human designer, the learning does not continue.

Figure 1. A comparison of learning process
(Yamaguchi, Tamai, & Takadama, 2015)

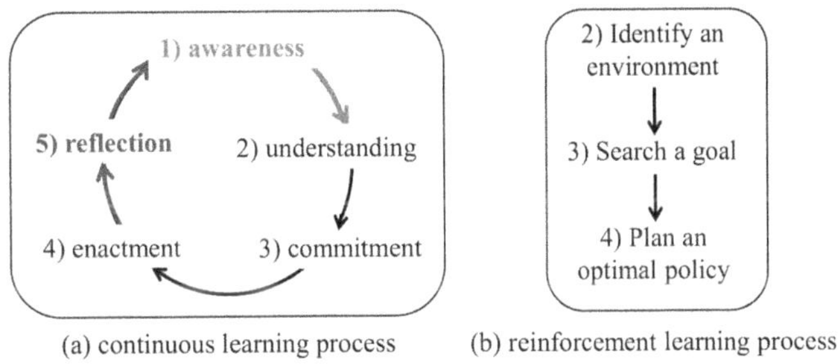

Creative Learning

The Research on Creativity

The authors consider creativity at the base of J.P. Guilford's approach. Guilford (1987) says creativity has primary characteristics, *sensitivity to problems*, *fluency in generating ideas*, *flexibility* and *novelty of ideas*, and *the ability to synthesize and reorganize information* (Finke, Ward, & Smith, 1992). *Sensibility to problems* is the skill to find the problem. The authors consider that it is the skill to comprehend the learning task. *Fluency in generating ideas* shows how many ideas a human create. *Flexibility* is the skill to create various ideas. *Novelty of ideas* is the skill to create unusual ideas. *The ability to synthesize and reorganize information* is the skill to utilize a thing for the divergent purposes. The authors consider it needs to focus on the interpretation of the meta-learning process.

Then the authors describe several characteristics which are concerned in their research. As *sensibility to problems* in the task, the learner can comprehend the structure of the learning environment through the trial and errors. As *fluency in generating ideas* in the task, the learner can find many solutions since the task gives the achievement to him/her. As *flexibility* in the task, the learner can find the various solutions by seeing his/her learning trace in the learning goal space.

The Definition of Creative Learning

Creative learning is defined as the continuous learning with discoveries of unusual solutions from achievements in the learner's own. In their previous research (Yamaguchi, Tamai, & Takadama, 2015), the human designed the achievements for a learner as the sequence of mazes. However, creative learner needs a new achievement continuously. In other words, it is necessary for the creative learner to discover the new achievements by himself/herself, but it is not easy. So the authors propose the interactive mechanism between a human learner and the learning support system in which the system derives the achievements from his/her found solutions with two kinds of heuristics. Once the learner found an unusual solution, the system can derive the new achievements from the unusual one.

The Flow of the Creative Learning Process

Figure 2 shows the flow of the creative learning process based on the continuous learning process. This process consists of triple cycle. Innermost cycle is called *a trial*. *A trial* is defined as a transition sequence from start state to encountering either a goal state or a wall. In this cycle, a learner repeats an action and his/her mental process including awareness until he/she results in either success or fail of the task. Second cycle is called *an achievement*. *An achievement* is defined as a unit of the main task which is the learning of a maze with the start and the invisible goal. In this cycle, when the trial terminates by the encounter with a goal, the learner finds the solution of the achievement. Then, the learner reflects the trial by the reflection of viewing his/her learning traces on the learning goal space. This process is described at Section Designing Automatically Generating the Derived Achievement. If current trial is not accomplished, he/she restarts the trial from the start state. Outmost cycle is the creative learning cycle. When the learner accomplished current achievement, the system generates a derived achievement according to the learner's solution, and then, He/she can challenge next new achievement. Section Designing Automatically Generating the Derived Achievement describes this process.

The Creative Learning Skill

The creative learning skill is defined as the learning skill to try to find more creative solutions on the given tasks or problems having optimal or entrenched solution. The authors propose the interactive mechanism consisting of two parts. The human part is to find a new solution from the achievement. The support system's part is to generate a new achievement derived from the human learner's solution by adding the sub-reward on it randomly to support the learner to find more creative solutions.

Figure 2. The flow of the creative learning process

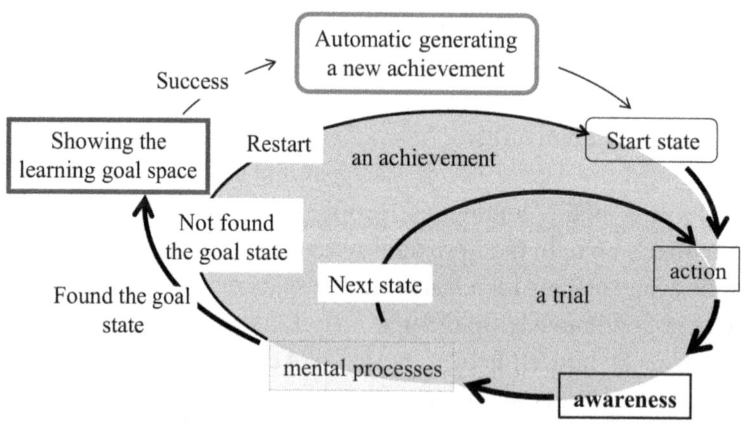

DESIGNING THE CREATIVE LEARNING SUPPORT SYSTEM

The Learning Environment by an Maze Model

As the learning environment for a human learner, the authors adopt a grid maze from start to goal since it is a familiar example to find the path through a trial and error process. First, the authors define *a maze*, *a path* in the maze, and *a solution* in the maze. *A maze* is the shape of two-dimensional maze defined by three kinds of states (start, goal and normal state) and the walls surrounding the states. In detail, it is described later as the maze model. A *path* consists of states and action transition sequence from the start state to the goal state. *A solution* is a path of the achievement of the maze.

A maze model for creative learning version consists of five elements, state set, transitions and walls, action set, and rewards. *Figure 3* shows the structure of a two-dimensional grid maze. The n x m grid maze with four neighbors consists of the n x m number of 1 x 1 squares. It is called a *simple maze* which is surrounded by walls in a rectangle shape. *Figure 3* (a) shows a 3 x 2 simple maze with a start and a goal. In a grid maze, every square touches one of their edges except for a wall. Each square in a maze model is called *a state*. A state can be visited at once.

Transitions between states in a maze model is defined whether corresponding square with four neighbors, {up, down, left, right} is connected or not connected by a wall. They are represented as the labeled directed graph as shown in *Figure 3* (b). Action set is defined as a set of labels to distinguish the possible transitions of a state. In a grid maze, the learner can take four kinds of actions: up, right, down, left.

Note that a trial is a transition sequence from the start state to encountering either a goal state or a wall, and the action toward a wall results in the transition to the start state to restart the trial. Transition to the goal state results in the success of the achievement, then the learner finds a solution and obtains a main reward (+1).

Designing the Creative Learning Support System

This section describes the way to automatically generate a new achievement as shown in *Figure 2*. First the authors describe a stage and an achievement in the creative learning task. A stage is a set of achievements of the same maze shape. An achievement of the creative learning task is defined as the learning of a maze to find a path from the start state to the goal state. An achievement consists of a maze shape and generated sub-rewards if any. It is a unit of the learning which is either an original achievement which consists of only maze shape or a derived achievement which contains generated sub-rewards.

Figure 3. The structure of a 3 x 2 simple maze
(Yamaguchi, 2015)

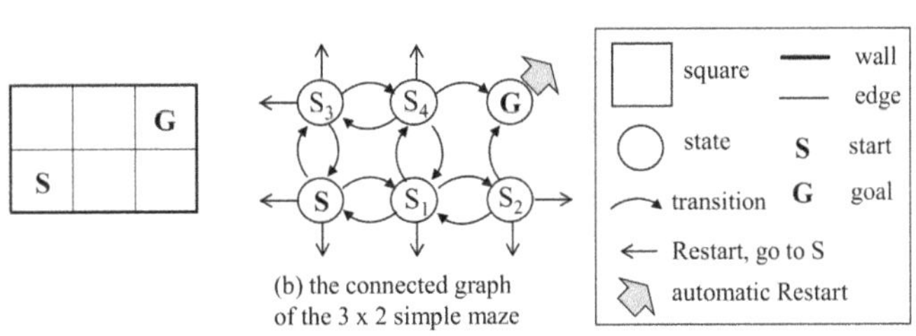

Figure 4 shows the flow of generating a new achievement by the system. The inner loop in *Figure 4* shows the interactive process of generating a new achievement by the system from the solution the learner searched. After a solution is found by the learner, if it is a new one, it is displayed on the learning goal space as the found learning goal for the reflection of the learner, and then the system derives the achievement by adding a reward according to the type of the solution. The system resets the achievement and the learner tries it. The outer loop in *Figure 4* shows the progress of the learning stage. It is two ways. One is when the learner decides to leave a current stage. Another is the decision of the system when the learner finds same solution in series. There are six stages. Each stage is the rectangle-shaped maze labeled as weight x height. First stage is 2 x 3 maze, second stage is 3 x 3 maze, third stage is 3 x 4 maze, fourth stage is 4 x 4 maze, fifth stage is 4 x 5 maze, and sixth stage is 5 x 5 maze.

Figure 4. The flow of generating a new achievement by the system

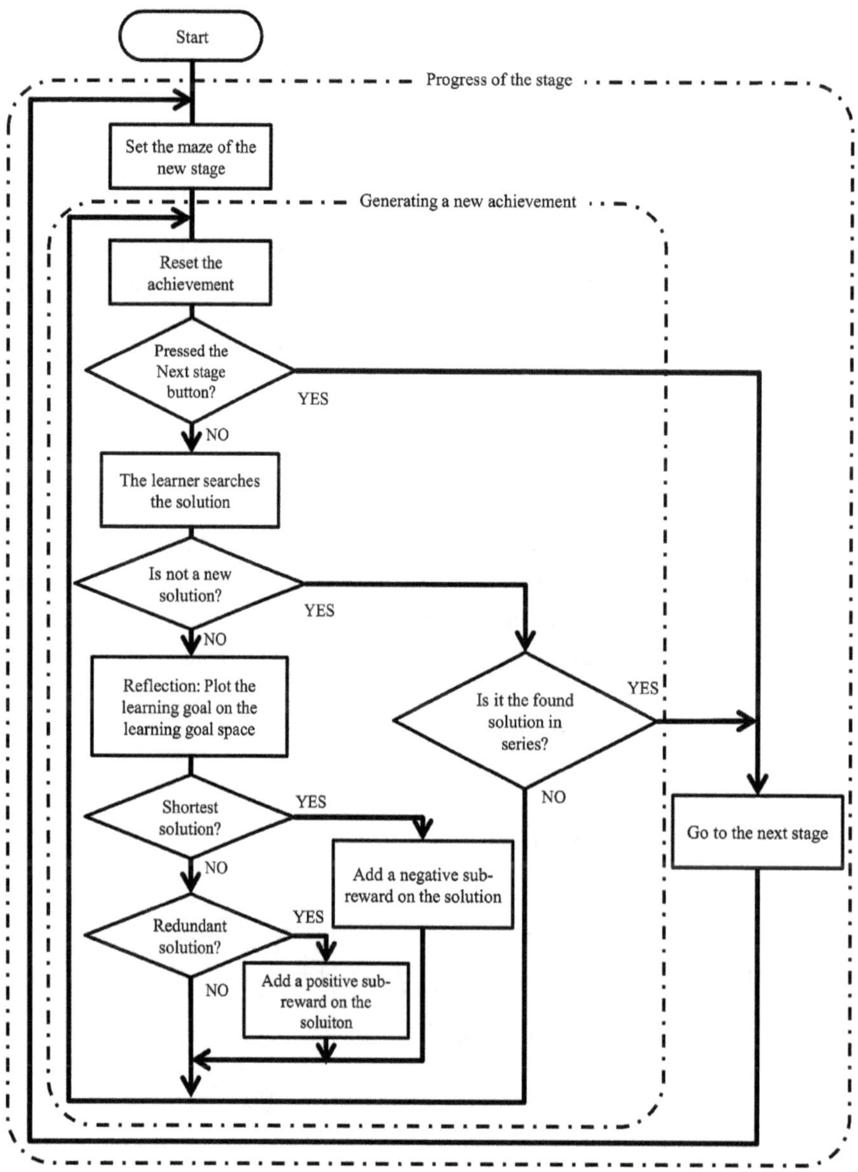

Designing Automatically Generating the Derived Achievement

Classification of the Solutions for Creative Learning

This section describes the classification of the solutions for creative learning. First, the authors mention the size of solution. *Table 1* shows the classifying solutions based on the length of them. The authors classify them whether it is shortest or not. Note that to make a redundant solution into the learning goal, it is necessary to introduce some optimality.

Second, the authors introduce the optimality of a solution to define the quality of solution. This paper adopts average reward reinforcement learning framework. In it, optimal solution is defined as a solution with the maximum average reward. Note that average reward of it is the sum of the acquired rewards divided by the solution length. Therefore, the shortest solution with acquiring rewards has a tendency to be optimal.

In the field of reinforcement learning, the way to find an optimal solution has been investigated in recent years. However, it is not the end of learning on creative learning. So the authors focus on the learning after an optimal solution found, and also focus on redundant, i.e. non-optimal solutions to utilize them since they have not been drawn attention as learning goals. Next subsection describes how to derive the learning goal from a shortest solution and from a redundant solution.

Deriving a New Achievement From Negating an Shortest Solution

This subsection describes the method to generate a derived achievement. When a learner finds a shortest solution, the system adds a negative sub-reward on one of transitions in the found path to try to negate it. This roughly negation of the optimal solution derives a new achievement creatively since the rest of redundant solutions encourage him/her to be creative for avoiding the negative sub-reward to find new solutions. Note that the position of negative sub-reward is random on the path, and its value is -1.

Deriving a New Achievement From Justifying a Redundant Solution

This subsection describes the method to generate a derived achievement from the justification of a non-optimal solution. When a learner finds a redundant solution, the system adds a positive sub-reward on one of transitions in the found path to try to justify it. This roughly justification of the redundant solution derives a new achievement creatively since there may be better solutions with the positive sub-reward than this redundant solution. Note that the position if positive sub-reward is random on the path and its value is +1.

Table 1. Classifying solutions based on the length of them

The Types of a Solution	Description
The shortest solutions	The shortest paths from a start state to an encountering goal state
The redundant solutions	The paths besides shortest solutions

Designing the Learning Goal Space

An Overview of the Learning Goal Space

The learning goal space is the space in which learning goals are positioned to display. The learning goal space has the vertical axis as the solution quality and the horizontal axis as the solution size. *A learning goal* is a set of solutions which both the solution quality and the solution size are the same in solutions found at achievements by the learner. It is positioned as a point on the learning goal space. In this paper, the vertical axis is defined as the total self-entropy based on the acquired rewards, and the horizontal axis is defined as the length of the solution. Note that these definitions are not presented to a human learner. *Figure 5* shows the illustrated example of the learning goal space. LG*i* is the *i*th learning goal found by the learner. The number of LG expresses the order in which the learner finds them. The transition from LG1 to LG2 shows that the direction of learning is right, and it means the learning only increases the solution size towards the horizontal learning goal. LG3 is transited from LG2 towards the vertical direction in which the solution quality only grows. LG4 is transited from LG3 towards the direction simultaneously to increase both the solution size and the solution quality.

The Total Self-Entropy Based on the Acquired Reward

In this research, the authors employ the difficulty to obtain positive sub-rewards in solutions as the index expressing the solution quality. Based on the self-entropy in the field of information theory, the self-entropy is proportional to the reciprocal of the occurrence probability of an acquired reward. In this paper, the self-entropy based on the acquired reward equation is as follows:

$$I(r) = -\log_{10} P(r)$$

Figure 5. The illustrated example of the learning goal space

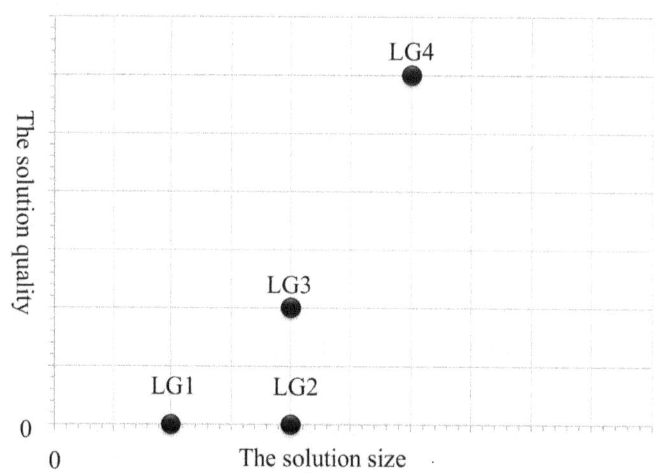

where r is acquiring the positive sub-reward as the event and $P(r)$ is the occurrence probability of the event per a step. Here, the occurrence probability of the positive sub-rewards is as follows:

$$P(r) = 1 / L$$

The total self-entropy based on the acquired reward is as follows:

$$\mathrm{S} = \sum_{r \in \mathbb{R}} I(r)$$

where S is the total self-entropy based on the acquired reward and $\mathbb{R} = \{r_1, r_2, \ldots\}$ is the set of the positive sub-reward.

PRELIMINARY EXPERIMENT

To examine the effects of this creative learning support system, the authors conducted the preliminary experiment by three subjects.

Experimental Setup

The Creative Learning Task

The creative learning task is the maze learning task which is the sequence of stages which consist of different size of maze. The objective is both to find various solutions and to get as many scores implemented as rewards as possible. A stage is a set of achievements of the same maze. The authors employ six kinds of simple mazes with visible walls. The number of stages is six, 2 x 3, 3 x 3, 3 x 4, 4 x 4, 4 x 5, and 5 x 5. The learning goal space is displayed to the learner through the task. It is updated when a subject find a new learning goal.

The GUI of Creative Learning Support System

Creative learning support system consists of two windows, Maze window and Learning goal space window. *Figure 6* shows Maze window. As shown in *Figure 6*, Maze window says Stage number, score, and the number of the found solutions, and has the grid area and two buttons, Derive button and Restart button. The grid area in Maze window shows the maze surrounded by black rectangles. Start state labeled as S in the square is visible in the maze, while goal state labeled as G is invisible. The green rectangle expresses a positive sub-reward, and the purple rectangle expresses a negative sub-reward. The gray square is the visited state. *Figure 7* shows Learning goal space window. As shown in *Figure 7*, Learning goal space window shows the found solution in the grid area and the learning goal space. The learning goal space has two kinds of points, red and blue. A red point is the learner's recent found solution. Blue points are the learner's found solutions except for the learner's recent found solution.

Figure 6. Maze window

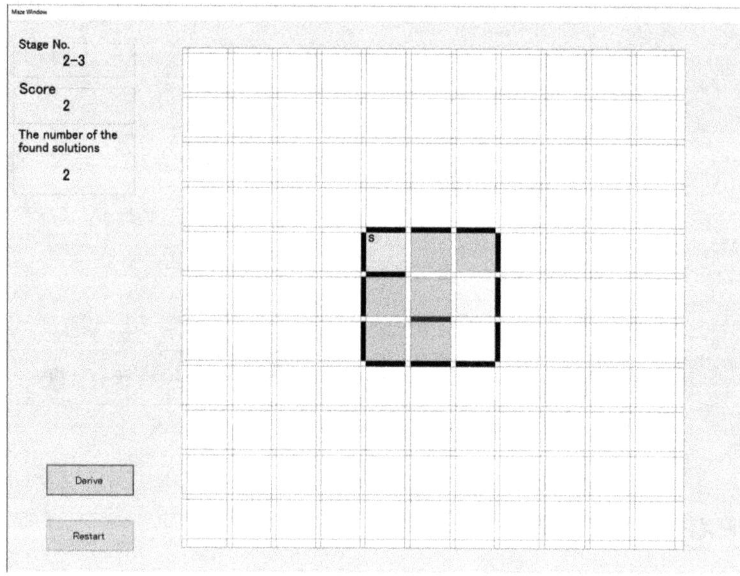

Figure 7. Learning goal space window

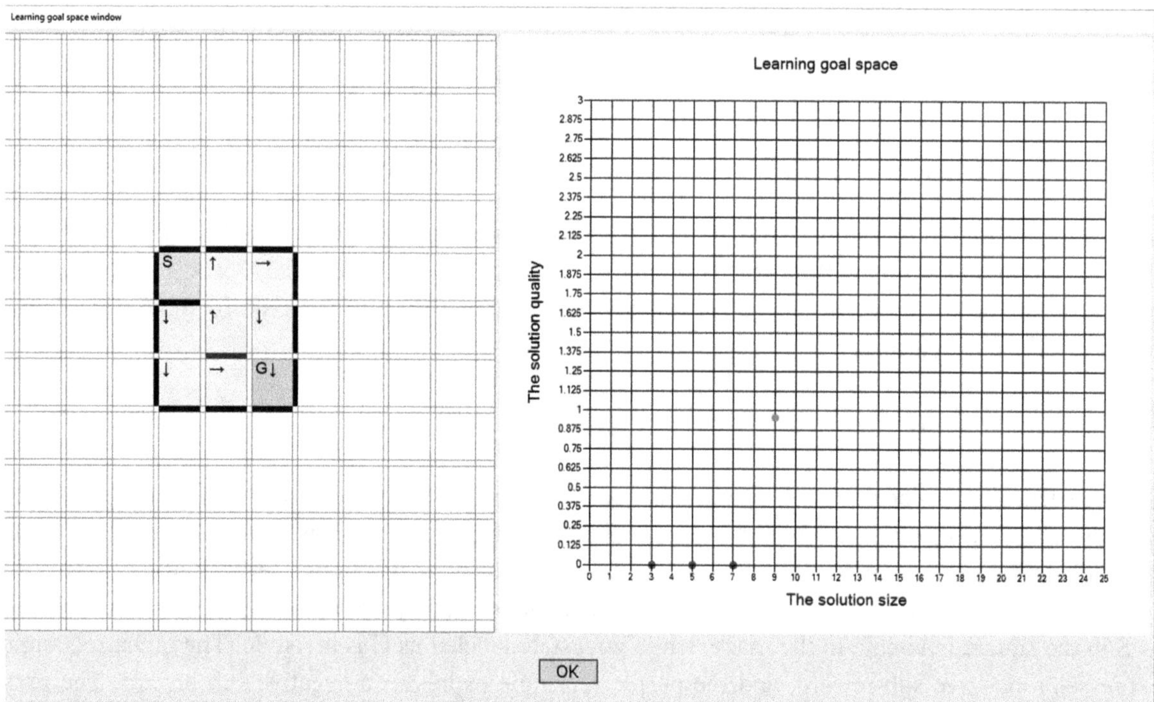

The Instruction for Subjects

A brief summary of the instruction for subjects includes the following points:

- The objective of the creative learning task is both to find more solutions in each maze and as many scores as possible;
- If you obtain a positive score on a transition, the color changes green at the corresponding rectangle grid between the neighboring square grids;
- If you obtain a negative score on a transition, the color changes purple at the corresponding rectangle grid between the neighboring square grids;
- If you find a same solution in series, the stage progresses.

The Measurement Items

The major measurement items of the experiment are as follows:

1. The number of the trials;
2. The number of the found solutions;
3. The number of the learning goals;
4. The total number of the acquired positive sub-rewards;
5. The total number of the acquired negative sub-rewards;
6. Score;
7. The experiment duration;
8. The rate of learning goals.

Note that (h) = (c) / (b) * 100. The learning goals rate among found solutions means that how many the solutions discover the new learning goals. Closer this is to one hundred, more efficient finding the learning goals is.

Experimental Results

The quantity of the continuous learning is analyzed. *Table 2* shows the experimental results.

As shown in *Table 2*, the average experiment duration (g) of three subjects is twenty five minutes (one thousand five hundred seconds), and the average number of the found solutions of the subject is

Table 2. Experimental results

Measurement Items / Subjects	(a)	(b)	(c)	(d)	(e)	(f)	(g)	(h)
Subject 1	202	169	79	1061	183	1000	1628	46.7
Subject 2	140	96	67	395	78	415	1425	69.8
Subject 3	122	102	51	328	91	370	1447	50.0
Average	154.7	122.3	65.7	594.7	117.3	595.0	1500.0	55.5
Standard deviation	42.0	40.5	14.0	405.2	57.2	351.5	111.4	12.5

one hundred twenty two. It follows from this that three subjects found many solutions on the continuous learning.

Next, the quantity of the creative learning is analyzed. As shown in *Table 2*, the mean rate of learning goals (h) is fifty five point five. The each rate of three subjects is approximately fifty percent. Since Subject2 has the highest rate of three subjects, it is likely that subject2 aims to find the different solution. In other words, the learning goal of subject2 is to find the different solution. On the other hand, subject1 acquires more than twice as many the total number of the acquired positive sub rewards (e) as the other two subjects. It follows from this that it is likely that the learning goal of subject1 is to make the total number of the acquired positive sub-rewards more. As stated above, it is suggested that the difference between learning results of the learners depends on which of unclear learning goals the learner focuses on in the creative learning experiment.

FUTURE RESEARCH DIRECTIONS

First, the authors analyze how effect this proposed creative learning support methods gave to the subject's creative learning from the questionnaires. *Table 3* shows the motivating factors and the number of the motivated subjects in each question form. In the questionnaire, all subjects answered maze window motivated them to continue the task. In the hearing investigation, two subjects answered that learning goal space window motivated them to continue the task and they do not pay attention the sub-rewards. By contrast, the other subject answered that he don't see the learning goal space and he find out how to add a sub-reward displayed in maze window.

Second, the authors discuss the effect of the reflection of learning by presenting the learning goal space. *Figure 8(a)*, *(b)*, and *(c)* show the final distributions of the learning trace on the learning goal space

Table 3. The motivating factors and the number of the motivated subjects

Factor / Form	Maze Window	Learning Goal Space Window
Questionnaire	3subjects of 3	1 subjects of 3
Hearing investigation	1subject of 3	2 subjects of 3

Figure 8. The final distributions on the learning goal space

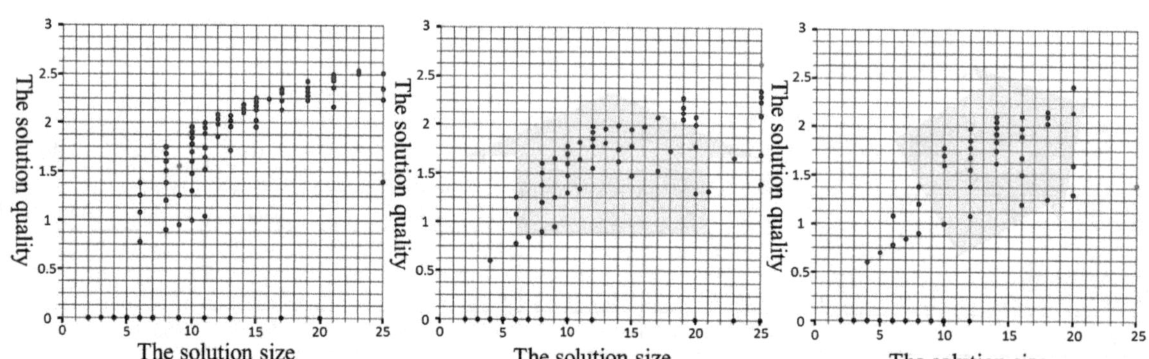

(a) The learning goal space of subject1 (b) The learning goal space of subject2 (c) The learning goal space of subject3

of each subject. The arrows in *Figure 8* (b) and (c) shows the learning direction of each subject from the hearing investigation. In *Figure 8* (a), there is no arrow since this subject did not watch the learning goal space during the experiment. The distribution of his learning trace seems to become convergent toward upper right side of the learning goal space. In contrast to it, *Figure 8* (b) and (c) shows the learning traces of subject2 and subject3 who watched the learning goal space and were motivated their learning from their learning traces on the learning goal space. The common feature on their learning is that they were aware of the open space in the learning goal space as the learning directions.

Third, the authors discuss the meaning of this creative learning support which is directing the creative learning by adding two kinds of sub-rewards. Automatically generating the derived achievement supports both convergent thinking and divergent thinking on creative thinking. Adding a positive sub-reward to the found path supports convergent thinking because the learner tries to approach the similar solution to the found one to obtain the positive sub-reward. In contrast, adding a negative sub reward to the found path supports divergent thinking because the learner tries to find different solution against the found one to avoid the negative sub-reward. So the learning diverges.

Fourth, the authors discuss the way to apply our proposed method to another domain. First point is how to design the learning goal space in the target domain. To design this, there are two kinds of axes, one is the solution quantity such as size, and the other is the solution quality such as complexity. Second point is how to cluster similar solutions in the target domain. In that, each cluster represents a sub-goal on the learning goal space.

CONCLUSION

The authors presented the interactive learning support system for the learner to learn creatively and proposed the learning goal space to the learner as the reflection of learning. The learning experiment is conducted to examine the effect of the learning goal space. In the experiment, the learning goal space shows a subject the learning trace of the subject. As the result of the experiment, two subjects of three answered that the learning goal space motivated them to continue the learning task.

As the future work, the authors are planning to investigate the condition for inspiration in the process of deciding the learning goals.

ACKNOWLEDGMENT

The authors would like to thank Prof. Habib and Prof. Shimohara for offering a good opportunity to present this research. This work was supported by JSPS KAKENHI (Grant-in-Aid for Scientific Research ©) Grant Number 16K00317.

REFERENCES

Boden, M. A. (2003). *The creative mind: Myths and mechanisms.* Abingdon, UK: Routledge.

Buckler, B. (1996). *A learning process model to achieve continuous improvement. In The Learning Organization* (pp. 31–39). MCB University Press.

Dorval, K. B., Isaksen, S. G., & Noller, R. B. (2003). Leadership for Learning: Tips for Effective Mentoring and Coaching, in Mentoring for talent development Unknown Binding. *Reclaiming Youth International.* Retrieved from http://www.cpsb.com/resources/downloads/public/Mentoring_Coaching.pdf

Finke, R. A., Ward, T. B., & Smith, S. M. (1992). *Creative Cognition: Theory, Research, and Applications.* Cambridge, MA: The MIT Press.

Frey, C. B., & Osborne, M. A. (2013). *The future of employment: how susceptible are jobs to computerization?* Oxford Martin School Working Paper, No.7, 1-72.

Guilford, J. P. (1987). Creativity Research: Past, Present and Future. In *Frontiers of creativity research: Beyond the basics* (pp. 34-64). Buffalo, NY: Bearly Ltd. Retrieved from http://www.cpsb.com/research/articles/creativity-research/Creativity-Research-Guilford.pdf

Habib, M. K. (2008). Interdisciplinary mechatronics: Problem solving, creative thinking and concurrent design synergy. *International Journal of Mechatronics and Manufacturing Systems*, *1*(1), 264–269. doi:10.1504/IJMMS.2008.018272

Habib, M. K. (2011). Biomimetics: Innovations and Robotics. *International Journal of Mechatronics and Manufacturing Systems*, *4*(2), 113–134. doi:10.1504/IJMMS.2011.039263

Sessa, V. I., & London, M. (2006). *Continuous Learning in Organizations: Individual, Group, and Organizational Perspectives.* Psychology Press.

Shi, Habib, & Xiao. (2015). Biologically Inspired Robotics. *Journal of Robotics*. doi:10.1155/2015/894394

Smita, J., & Trey, M. (2012). *Facilitating continuous learning: review of research on individual learning capabilities and organizational learning environments.* In The annual meeting of the AECT International Convention, Louisville, KY. Retrieved from http://www.memphis.edu/icl/idt/clrc/clrc-smita-research.pdf

Yamaguchi, T., Takemori, K., Tamai, Y., & Takadama, K. (2015). Analyzing human's continuous learning processes with the reflection sub task. *Journal of Communication and Computer*, *12*(1), 20–27.

Yamaguchi, T., Tamai, Y., & Takadama, K. (2015). Analyzing human's continuous learning ability with the reflection cost. *Proceedings of 41st Annual Conference of the IEEE Industrial Electronics Society (IECON2015)*, 2920-2925. doi:10.1109/IECON.2015.7392546

KEY TERMS AND DEFINITIONS

Creative Learning: Continuous learning with discoveries of unusual solutions from achievements of the learner's own.

Deriving a New Achievement from Justifying a Redundant Solution: When a learner finds a shortest solution, a negative sub-reward is added on one of transitions in the found path to try to negate it.

Deriving a New Achievement from Negating and Shortest Solution: When a learner finds a shortest solution, the system adds a negative sub-reward on one of transitions in the found path to try to negate it.

Learning Goal: A set of solutions in which both the solution quality and the solution size are the same in solutions found at achievements by the learner.

Learning Goal Space: The space in which learning goals are positioned to display.

Redundant Solution: The paths besides shortest solutions and longest solutions.

Self-Entropy Based on the Acquired Reward: The difficulty to obtain positive sub-rewards in solutions as the index expressing the solution quality.

Shortest Solution: The shortest paths from a start state to an encountering goal state.

Compilation of References

Abdel-Aal, H. A., & El Mansori, M. (2011). A study on the frictional response of reptilian skin for green tribology. *Proceedings of 3rd European Conference on Tribology.*

Abdel-Aal, H. A., & El Mansori, M. (2011). *Metrological Characterization of Reptilian skin for Green Tribo Surfaces.* ASPE Spring Topical Meeting, Freeform Surfaces, Charlotte, NC.

Abdel-Aal, H. A., & El Mansori, M. (2011). Python Regius (Ball Python) shed skin: Biomimetic analogue for function-targeted design of tribo-surfaces In Biomimetics - Materials Structures and Processes Examples Ideas and Case Studies. Springer.

Abdel-Aal, H. A., & El Mansori, M. (2015). The Fractal Structure of the Ventral Scales in Legless Reptiles. In *15th International Conference on Metrology and Properties of Engineering Surfaces*. UNC Charlotte.

Abdel-Aal, H. A. (2013). On Surface Structure and Friction Regulation in Reptilian Locomotion. *Journal of the Mechanical Behavior of Biomedical Materials, 22*, 115–135. doi:10.1016/j.jmbbm.2012.09.014 PMID:23582565

Abdel-aal, H. A. (2015). The structure of Ventral Scale Textures in Snakes in comparison to Texturing of Deterministic Tribological Surfaces. In J. P. Davim (Ed.), *Processing Techniques and Tribological Behavior of Composite Materials, Rajnesh Tyagi* (pp. 268–316). IGI Global Publication. doi:10.4018/978-1-4666-7530-8.ch010

Abdel-Aal, H. A. (under consideration). Comparison of friction profiles of two snakes. *Journal of the Royal Society, Interface.*

Abdel-Aal, H. A., & El Mansori, M. (2013). Tribological analysis of ventral scale structure in a python regius in relation to laser textured surfaces IOP-J. *Surface Topography: Metrology and Properties.*, *1*(1), 015001. doi:10.1088/2051-672X/1/1/015001

Abdel-Aal, H. A., & El Mansori, M. (2014). Characterization of Load Bearing Metrological Parameters in Reptilian Exuviae In Comparison To Plateau Honed Surfaces. *IOP-J. Surf. Top. Metrol. Prop.*, *2*, 045002. doi:10.1088/2051-672X/2/4/045002

Abdel-Aal, H. A., Vargiolu, R., Zahouani, H., & El Mansori, M. (2011). A study on the frictional response of reptilian shed skin. *Journal of Physics: Conference Series*, *311*, 012016. doi:10.1088/1742-6596/311/1/012016

Abdel-Aal, H. A., Vargiolu, R., Zahouani, H., & El Mansori, M. (2012). Preliminary investigation of the frictional response of reptilian shed skin. *Wear*, *290–291*, 51–60. doi:10.1016/j.wear.2012.05.015

Abdollahi, F., Case Lazarro, E. D., Listenberger, M., Kenyon, R. V., Kovic, M., Bogey, R. A., ... Patton, J. L. (2014). Error augmentation enhancing arm recovery in individuals with chronic stroke: A randomized crossover design. *Neurorehabilitation and Neural Repair*, *28*(2), 120–128. doi:10.1177/1545968313498649 PMID:23929692

Abernethy, B., Kippers, V., Hanrahan, S., Pandy, M., McManus, A., & Mackinnon, L. (2013). *The Biophysical foundations of Human Movement* (3rd ed.). Champaign: Human Kinetics.

Aggarwal, L., Aggarwal, K., & Urbanic, R. J. (2014). Use of artificial neural networks for the development of an inverse kinematic solution and visual identification of singularity zone(s). *Procedia CIRP*, *17*, 812–817. doi:10.1016/j.procir.2014.01.107

Akdogan, E., & Mehmet, A. A. (2011). The design and control of a therapeutic exercise for lower limb rehabilitation: Physiotherabot. *Mechatronics*, *21*(3), 509–522. doi:10.1016/j.mechatronics.2011.01.005

Alessandro Scano, Andrea Chiavenna, Matteo Malosio, and Lorenzo Molinari Tosatti, "Kinect V2 Performance Assessment in Daily-Life Gestures: Cohort Study on Healthy Subjects for a Reference Database for Automated Instrumental Evaluations on Neurological Patients," Applied Bionics and Biomechanics, vol. 2017, Article ID 8567084, 16 pages, 2017. doi:10.1155/2017/8567084

Alexander, A. A., & Gans, C. (1966). The pattern of dermal-vertebral correlation in snakes and amphisbaenians. *Zoölogische Mededeelingen*, *41*(11), 171–190.

Alibardi, L. (1996). Scale morphogenesis during embryonic development in the lizard Anolis lineatopus. *Journal of Anatomy*, *188*, 713–725. PMID:8763488

Alibardi, L. (1998). Differentiation of the epidermis during scale formation in embryos of lizard. *Journal of Anatomy*, *192*(2), 173–186. doi:10.1046/j.1469-7580.1998.19220173.x PMID:9643418

Alibardi, L., & Toni, M. (2007). Alpha and beta keratins of snake epidermis. *Zoology (Jena)*, *110*(1), 41–47. doi:10.1016/j.zool.2006.07.001 PMID:17169542

Allington, S., Klein, J., Buell, M., Reinkensmeyer, D. J., & Bobrow, J. (2011). Supinator extender (SUE): A pneumatically actuated robot for forearm/wrist rehabilitation after stroke. *Proceedings of Annual International Conference of the IEEE Engineering in Medicine and Biology Society*. doi:10.1109/IEMBS.2011.6090459

Amarello, M., Novak, E. M., Taylor, E. N., Schuett, G. W., Repp, R. A., Rosen, P. C., & Hardy, D. L. (2010). Potential environmental influence on variation in body size and sexual size dimorphism among Arizona populations of the western diamond-backed rattlesnake (Crotalus atrox). *Journal of Arid Environments*, *74*(11), 1443–1449. doi:10.1016/j.jaridenv.2010.05.019

Ambrosini, E., Ferrante, S., Gfoeler, M., Reichenfelser, W., Karner, J., Schauer, T., . . . Pedrocchi, A. (2009). *A hybrid assistive system to support daily upper limb activities*. Academic Press.

Anderson, A. M., Streitlien, K., Barrett, D. S., & Triantafyllou, M. S. (1998). Oscillating Foils of High Propulsive Efficiency. *Journal of Fluid Mechanics*, *360*, 41–72. doi:10.1017/S0022112097008392

Ariizumi, R., & Matsuno, F. (2013). Dynamical analysis of sidewinding locomotion by a snake-like robot. *Proceedings - IEEE International Conference on Robotics and Automation*, 5149–5154. doi:10.1109/ICRA.2013.6631313

Armstrong, W. P. (2007). Do Pearls Come From Coconuts? *The Drifting Seed*, *13*(1), 13–17.

Army Technology. (2015). *Raytheon XOS 2 Exoskeleton, Second-Generation Robotics Suit - Army Technology*. Retrieved from http://www.army-technology.com/projects/raytheon-xos-2-exoskeleton-us/

Arnold, S. (1988). Quantitative genetics and selection in natural populations: Microevolution of vertebral numbers in the garter snake Thamnophis eleguns. *Proc. 2nd Intl. Conf. Quant. Genet.*, *2*, 619-636.

Arnold, S. J. (1983). Morphology, performance and fitness. *American Zoologist*, *23*(2), 347–361. doi:10.1093/icb/23.2.347

Arnold, S., & Bennett, A. (1988). Behavioral variation in natural populations. V. Morphological correlations of locomotion in the garter snake. Thamnophis radix. *Biological Journal of the Linnean Society. Linnean Society of London*, *34*(2), 175–190. doi:10.1111/j.1095-8312.1988.tb01955.x

Asada, N., Matsuki, H., Minami, K., & Esashi, M. (1994). Silicone micromachined two-dimensional galvano optical scanner. *IEEE Transactions on Magnetics*, *30*(6), 4647–4649. doi:10.1109/20.334177

Aschenbeck, K. S., Kern, N. I., Bachmann, R. J., & Quinn, R. D. (2006). Design of a Quadruped Robot Driven by Air Muscles. *The IEEE/RAS-EMBS International Conference on Biomedical Robotics and Biomechatronics (BioRob'06)*, 875-880. doi:10.1109/BIOROB.2006.1639201

Ashton, K. G., & Feldman, C. R. (2003). Bergmann's rule in nonavian reptiles: Turtles follow it, lizards and snakes reverse it. *Evolution; International Journal of Organic Evolution*, *57*(5), 1151–1163. doi:10.1111/j.0014-3820.2003.tb00324.x PMID:12836831

Ates, S., Haarman, C. J. W., & Stienen, A. H. A. (2016). SCRIPT passive orthosis: Design of interactive hand and wrist exoskeleton for rehabilitation at home after stroke. *Autonomous Robots*, *41*(3), 711–723. doi:10.1007/s10514-016-9589-6

Aubret, F., & Shine, R. (2009). The origin of evolutionary innovations: Locomotor consequences of tail shape in aquatic snakes. *Functional Ecology*, *22*(2), 317–322. doi:10.1111/j.1365-2435.2007.01359.x

Avidan, S. (2004). Support Vector T racking. *IEEE Transactions on Pattern Analysis and Machine Intelligence*, *26*(8), 1064–1072. doi:10.1109/TPAMI.2004.53 PMID:15641735

Ayers, J., Wilbur, C., & Olcott, C. (2000b). Lamprey Robots. In *The International Symposium on Aqua Biomechanisms*. Tokai University.

Ayers, J., Witting, J., McGruer, N., Olcott, C., & Massa, D. (2000a). Lobster Robots. In *The International Symposium on Aqua Biomechanisms*. Tokai University.

Babenko, B., Yang, M. H., & Belongie, S. (2011). Robust Object Tracking with Online Multiple Instance Learning. *IEEE Transactions on Pattern Analysis and Machine Intelligence*, *33*(8), 1619–1632. doi:10.1109/TPAMI.2010.226 PMID:21173445

Bagley, D. H., & Cubler-Goodman, A. (1990). Radiation exposure during ureteroscopy. *The Journal of Urology*, *144*(6), 1356–1358. doi:10.1016/S0022-5347(17)39739-2 PMID:2231925

Bailador, G. (2007). Real time gesture recognition using continuous time recurrent neural networks. *Proceedings International Conference Body Area Networks*, 1-8. doi:10.4108/bodynets.2007.149

Balazs, A. (2008). International Vocabulary of Metrology-Basic and General Concepts and Associated Terms. *Chemistry International*.

Banala, S. K., Agrawal, S. K., Fattah, A., Rudolph, K., & Scholz, J. P. (2004). A gravity balancing leg orthosis for robotic rehabilitation. In *Robotics and Automation, 2004. Proceedings. ICRA'04. 2004 IEEE International Conference on* (Vol. 3, pp. 2474-2479). IEEE. doi:10.1109/ROBOT.2004.1307432

Barbareschi, G., Richards, R., Thornton, M., Carlson, T., & Holloway, C. (2015). Statically vs dynamically balanced gait: Analysis of a robotic exoskeleton compared with a human. In *Engineering in Medicine and Biology Society (EMBC), 2015 37th Annual International Conference of the IEEE* (pp. 6728-6731). IEEE.

Bar-Cohen, Y. (2011). *Biomimetics: Nature-Based Innovation*. CRC Press.

Bar-Cohen, Y. (Ed.). (2001). Electroactive Polymer (EAP) Actuators as Artificial Muscles: Reality, Potential and Challenges. SPIE Press.

Bar-Cohen, Y. (2002, January). Biologically Inspired Robots as Artificial Inspectors. *e-Journal of NDT*, *7*(1).

Bar-Cohen, Y. (2003). Biologically Inspired Intelligent Robotics. *The SPIE Smart Structures Conference*.

Baronti, P., Pillai, P., Chook, V. W., Chessa, S., Gotta, A., & Hu, Y. F. (2007). Wireless sensor networks: A survey on the state of the art and the 802.15.4 and ZigBee standards. *Computer Communications*, *30*(7), 1655–1695. doi:10.1016/j.comcom.2006.12.020

Basteris, A., Nijenhuis, S. M., Stienen, A. H., Buurke, J. H., Prange, G. B., & Amirabdollahian, F. (2014). Training modalities in robot-mediated upper limb rehabilitation in stroke: A framework for classification based on a systematic review. *Journal of Neuroengineering and Rehabilitation*, *11*(111), 15. PMID:25012864

Bauchot, R. (n.d.). *Snakes: A Natural History*. New York, NY: Sterling Publishing Co., Inc.

Baum, M. J., Heepe, L., Fadeeva, E., & Gorb, S. N. (2014). Dry friction of microstructured polymer surfaces inspired by snake skin. Beilstein Journal of Nanotechnology, 5, 1091-1103. doi:10.3762/bjnano.5.122

Baumann, R., & Clavel, R. (1998). Haptic interface for virtual reality based minimally invasive surgery simulation. *Proc. IEEE Int. Conf. on Robotics and Automation*, 381-386. doi:10.1109/ROBOT.1998.676435

Baum, M. J., Kovalev, A. E., Michels, J., & Gorb, S. N. (2014). Anisotropic Friction of the Ventral Scales in the Snake Lampropeltis getula californiae. *Tribology Letters*, *54*(2), 139–150. doi:10.1007/s11249-014-0319-y

Bayraktaroglu, Z. Y., Butel, F., Pasqui, V., & Blazevic, P. (2000). Snake-like locomotion: Integration of geometry and kineto-statics. *Advanced Robotics*, *14*(6), 447–458. doi:10.1163/156855300741942

Beiko, D. T., & Denstedt, J. D. (2007). Advances in ureterorenoscopy. *The Urologic Clinics of North America*, *34*(3), 397–408. doi:10.1016/j.ucl.2007.05.003 PMID:17678989

Bell, D. J., Lu, T. J., Fleck, N. A., & Spearing, S. M. (2005). MEMS actuators and sensors: Observations on their performance and selection for purpose. *Journal of Micromechanics and Microengineering*, *15*(7), S153–S164. doi:10.1088/0960-1317/15/7/022

Benjamin, E. J., Blaha, M. J., Chiuve, S. E., Cushman, M., Das, S. R., Deo, R., & Muntner, P. (2017). Hearth Disease Statistics – 2017 Update: A Report From the American Heart Association. In Heart Disease and Stroke Statistics—2017 Update. American Health Association.

Bennet, S., McConnell, T., & Trubatch, S. L. (1974). Quantitative analysis of the speed of snakes as a function of peg spacing. *The Journal of Experimental Biology*, *60*(1), 161–165.

Benyus, J. (1997). *Biomimicry: Innovation Inspired by Nature*. New York: William Morrow & Company, Inc.

Benz, M. J., Kovalev, A. E., & Gorb, S. N. (2012). Anisotropic frictional properties in snakes. *SPIE Smart Structures and Materials+ Nondestructive Evaluation and Health Monitoring*.

Benz, M. J., Kovalev, A. E., & Gorb, S. N. (2012). Anisotropic frictional properties in snakes. *Proc. of SPIE*, 8339. doi:10.1117/12.916972

Bergamasco, M., Allotta, B., Bosio, L., Ferretti, L., Perrini, G., Prisco, G. M., . . . And, G. Sartini, (1994). An Arm Exoskeleton System for Teleoperation and Virtual Environment Applications. *IEEE Int'l Conf. Robot. Automat.*, *2*, 1449–1454.

Berthé, R. A., Westhoff, G., Bleckmann, H., & Gorb, S. N. (2009). Surface structure and frictional properties of the skin of the Amazon tree boa Corallus hortulanus (Squamata Boidae). *J. Comp. Phys. A, 195*(3), 311–318. doi:10.1007/s00359-008-0408-1 PMID:19137315

Bhardwaj, J. K., & Ashraf, H. (1995). Advanced silicon etching using high-density plasmas. *Proc. of SPIE Micromachining and Micro fabrication Process Technology, 2639*, 224-233. doi:10.1117/12.221279

Bhattacharya, S., & Agrawal, S. K. (2000). Spherical rolling robot: A design and motion planning studies. *IEEE Transactions on Robotics and Automation, 16*(16), 835–839. doi:10.1109/70.897794

Bibo, A., & Daqaq, M. F. (2013). Investigation of concurrent energy harvesting from ambient vibrations and wind using a single piezoelectric generator. *Applied Physics Letters, 102*(24), 1–4. doi:10.1063/1.4811408

Biddiss, E., & Chau, T. (2008). Dielectric elastomers as actuators for upper limb prosthetics: Challenges and opportunities. *Medical Engineering & Physics, 30*(4), 403–418. doi:10.1016/j.medengphy.2007.05.011 PMID:17632030

Binns, C. (2006). New Material Collects Water From Thin Air. *LiveScience*. Retrieved from http://www.livescience.com/animals/060616_beetle_material.html

Bioness Inc. (2015). *H200 Wireless Hand.* Retrieved from http://www.bioness.com/Products/H200_for_Hand_Paralysis.php

Bischoff, B., Duy, N. T., Streichert, F., Ewert, M., & Knoll, A. (2012). Fusing vision and odometry for accurate indoor robot localization. In *Proceedings of 2012 12th International Conference on Control Automation Robotics & Vision (ICARCV)*. Guangzhou: IEEE.

Blackwell, M., Nikou, C., DiGioia, A. M., & Kanade, T. (2000). An image overlay system for medical data visualization. *Medical Image Analysis, 4*(1), 67–72. doi:10.1016/S1361-8415(00)00007-4 PMID:10972322

Boback, S. M. (2003). Body size evolution in snakes: Evidence from island population. *Copeia, 2003*(1), 81–94. doi:10.1643/0045-8511(2003)003[0081:BSEISE]2.0.CO;2

Boback, S. M., & Guyer, C. (2003). Empirical evidence for an optimal body size in snakes. *Evolution; International Journal of Organic Evolution, 3*, 195–211. PMID:12683530

Boden, M. A. (2003). *The creative mind: Myths and mechanisms*. Abingdon, UK: Routledge.

Bogert, C. M. (1947). Rectilinear locomotion in snakes. *Copeia, 1947*(4), 253–254. doi:10.2307/1438921

Bogue, R. (2009). Exoskeletons and robotic prosthetics: A review of recent developments. *Industrial Robot: An International Journal, 36*(5), 421–427. doi:10.1108/01439910910980141

Bogue, R. (2014). Snake robots: A review of research, products and applications. Industrial Robot. *International Journal (Toronto, Ont.), 41*(3), 253–258.

Bogue, R. (2015). Robotic exoskeletons: A review of recent progress. *Industrial Robot: An International Journal, 42*(1), 5–10. doi:10.1108/IR-08-2014-0379

Borenstein, J., & Feng, L. (1994). *UMBmark: A method for measuring, comparing, and correcting dead-reckoning errors in mobile robots*. Academic Press.

Borenstein, J., & Feng, L. (1995). UMBmark: A benchmark test for measuring odometry errors in mobile robots. Academic Press.

Borenstein, J., & Feng, L. (1996). Measurement and correction of systematic odometry errors in mobile robots. *IEEE Transactions on Robotics and Automation, 12*(6), 869–880. doi:10.1109/70.544770

Boud, D., Keogh, R., & Walker, D. (1985). Reflection: Turning experience into learning. Abingdon-on-Thames, UK: Routledge.

Bowden, F. P., & Tabor, D. (1964). *The friction and lubrication of solids. Part II.* Oxford, UK: Clarendon Press.

Bowden, R., & Sarhadi, M. (2002). A non-linear model of shape and motion for tracking finger spelt American sign language. *Image and Vision Computing*, *20*(9-10), 597–607. doi:10.1016/S0262-8856(02)00049-5

Breda, A., Ogunyemi, O., Leppert, J. T., & Schulam, P. G. (2009). Flexible Ureteroscopy and Laser Lithotripsy for Multiple Unilateral Intrarenal Stones. *European Urology*, *55*(5), 1190–1197. doi:10.1016/j.eururo.2008.06.019 PMID:18571315

Brethes, L., Menezes, P., Lerasle, F., & Hayet, J. (2004). Face tracking and hand gesture recognition for human-robot interaction. *Proceedings of the International Conference on Robotics and Automation 2004 (ICRA2004)*, 1901-1906. doi:10.1109/ROBOT.2004.1308101

Bretl, T. (2006). Motion planning of multi-limbed robots subject to equilibrium constraints: The free-climbing robot problem. *The International Journal of Robotics Research*, *25*(4), 317–342. doi:10.1177/0278364906063979

Bridle, J. S. (1989), Neurocomputing: Algorithms, Architectures, and Applications. In Probabilistic interpretation of feedforward classification network outputs, with relationships to statistical pattern recognition (pp. 227-236). New York: Springer-Verlag.

Brown, J. H., Gillooly, J. F., Allen, A. P., Savage, V. M., & West, G. B. (2004). Toward a metabolic theory of ecology. *Ecology*, *85*(7), 1771–1789. doi:10.1890/03-9000

Buckler, B. (1996). *A learning process model to achieve continuous improvement. In The Learning Organization* (pp. 31–39). MCB University Press.

Buckler, B. A. (1996). Learning process model to achieve continuous improvement. *The Learning Organization*, *3*(3), 31–39. doi:10.1108/09696479610119660

Buehler, M. (2000). Dynamic Locomotion with One, Four and Six-Legged Robots. *Journal of the Robotics Society of Japan*, *20*(3), 15–20.

Bu, N., Fukuda, O., & Tsuji, T. (2003). EMG-based motion discrimination using a novel recurrent neural network. *Journal of Intelligent Information Systems*, *21*(2), 113–126. doi:10.1023/A:1024706431807

Bury, B., & Hope, J. C. (1995). Autonomous mobile robot navigation using a low-cost fibre optic gyroscope. *IFAC Proceedings Volumes, 28*(11), 39-43. doi:10.1016/B978-0-08-042366-1.50011-7

Caelli, T., Squire, D. M., & Wild, T. P. J. (1993). Model-based neural networks. *Neural Networks*, *6*(5), 613–625. doi:10.1016/S0893-6080(05)80106-6

Cafolla, D., & Carbone, G. (2014). A study of feasibility of a human finger exoskeleton. *Service Orientation in Holonic and Multi-Agent Manufacturing and Robotics, Studies in Computational Intelligence, Springer*, *544*, 355–364. doi:10.1007/978-3-319-04735-5_24

Caimmi, M., Guanziroli, E., Malosio, M., Pedrocchi, N., Vicentini, F., Tosatti, L. M., & Molteni, F. (2015). Normative Data for an Instrumental Assessment of the Upper-Limb Functionality. *BioMed Research International*, *2015*, 1–23. doi:10.1155/2015/484131 PMID:26539500

Cameirão, M. S., Smailagic, A., Miao, G., Siewiorek, D. P., Sander, M., Oxlund, B., ... Brooke, J. (2016). Coaching or gaming? Implications of strategy choice for home based stroke rehabilitation. *Journal of Neuroengineering and Rehabilitation*, *13*(1), 18. doi:10.1186/s12984-016-0127-8 PMID:26921185

Capi, G., Toda, H., & Nagasaki, T. (2010). *A vision based robot navigation and human tracking for social robotics.* Paper presented at 2010 IEEE International Workshop on Robotic and Sensors Environments (ROSE), Phoenix, AZ. doi:10.1109/ROSE.2010.5675255

Cappozzo, A., Catani, F., Leardini, A., Benedetti, M. G., & Della Croce, U. (1996). Position and orientation in space of bones during movement: Experimental artefacts. *Clinical Biomechanics (Bristol, Avon)*, *11*(2), 90–100. doi:10.1016/0268-0033(95)00046-1 PMID:11415604

Carbone, G., Cavero, C. A., Ceccarelli, M., & Altuzarra, O. (2017). A Study of Feasibility for a Limb Exercising Device. Advances in Italian Mechanism Science, 47, 11-21.

Carbone, G., Gherman, B., Ulinici, I., Vaida, C., & Pisla, D. (2017). Design Issues for an Inherently Safe Robotic Rehabilitation Device. *Proceedings of the 26th International Conference on Robotics in Alpe-Adria-Danube Region RAAD 2017*, 967-974.

Carignan, C. R., & Cleary, K. R. (2000, February 23). Closed- Loop Force Control for Haptic Simulation of Virtual Environments. *Haptics-e*, *1*(2), 1–14. http://www.haptics-e.org

Carpi, F., Mannini, A., & De-Rossi, D. (2008). *Elastomeric contractile actuators for hand rehabilitation splints.* Paper presented at the The 15th International Symposium on: Smart Structures and Materials & Nondestructive Evaluation and Health Monitoring. doi:10.1117/12.774644

Carpino, G. I., Accoto, D. I., Tagliamonte, N. L., Ghilardi, G. I., & Guglielmelli, E. U. (2013). Lower limb wearable robots for physiological gait restoration: State of the art and motivations. *Medic*, *21*(2), 72–80.

Carrera, J., Moreno, H. A., Saltaren, R., Pérez, C., Puglisi, L., & Garcia, C. (2011, October). ROAD: Domestic assistant and rehabilitation robot. *Medical & Biological Engineering & Computing*, *49*(10), 1201–1211. doi:10.1007/s11517-011-0805-4 PMID:21789672

Casadio, M., Sanguineti, V., Morasso, P. G., & Arrichello, V. (2006). Bracio di Ferro: A new haptic workstation for neuromotor rehabilitation. *Technology and Health Care*, *14*(3), 123–142. PMID:16971753

Cassanova, R. A. (2002). *Planetary Exploration using Biomimetics an Entomopter for Flight on Mars.* Phase II Final Report, NASA Institute for Advanced Concepts Project NAS5-98051, October 2002.

Cassemiro, E. R., Rosario, J. M., & Dumur, D. (2005). Robot axis dynamics control using a virtual robotics environment. *2005 IEEE Conference on Emerging Technologies and Factory Automation.* doi:10.1109/ETFA.2005.1612539

Cauraugh, J. H., & Summers, J. J. (2005). Neural plasticity and bilateral movements: A rehabilitation approach for chronic stroke. *Progress in Neurobiology*, *75*(5), 309–320. doi:10.1016/j.pneurobio.2005.04.001 PMID:15885874

Cauraugh, J., & Summers, J. (2004). Natural plasticity and bilateral movements: A rehabilitation approach for chronic stroke. *Progress in Neurobiology*, 5(75), 309–320.

Ceccarelli, M., & Romdhane, L. (2010). 2010 Design issues for human-machine platform interface in cable-based parallel manipulators for physiotherapy applications. *Journal of Zhejiang University. Science A*, *11*(4), 231–239. doi:10.1631/jzus.A1000027

Chaari, L., & Kamoun, L. (2011). Performance analysis of IEEE 802.15.4/Zigbee standard under real time constraints. *International Journal of Computer Network & Communications*, *3*(5), 235–251. doi:10.5121/ijcnc.2011.3517

Chae, J., Sheffler, L., & Knutson, J. (2008). Neuromuscular electrical stimulation for motor restoration in hemiplegia. *Topics in Stroke Rehabilitation*, *15*(5), 412–426. doi:10.1310/tsr1505-412 PMID:19008202

Chakarov. (2015). *Torque measuring device*. Patent Application (BG, N°111934/20.02.2015).

Cham, J. G., Karpick, J. K., & Cutkosky, M. R. (2004). Stride period adaptation for a biomimetic running hexapod. *The International Journal of Robotics Research*, *23*(2), 141–153. doi:10.1177/0278364904041323

Chandrapal, M. (2012). *Intelligent assistive knee orthotic device utilizing pneumatic artificial muscles*. Academic Press.

Chandrapa, M., Xiao, Ch., & Wenhui, W. (2013). Preliminary Evaluation of Intelligent Intention Estimation Algorithms for an Actuated Lower-Limb Exoskeleton. *International Journal of Advanced Robotic Systems*, *10*(2), 147. doi:10.5772/56063

Chen, B., Ma, H., Qin, L.-Y., Gao, F., Chan, K.-M., Law, S.-W., ... Liao, W.-H. (2006). Recent developments and challenges of lower extremity exoskeletons. *Journal Orthopaedic Translation*, *5*, 26–37. doi:10.1016/j.jot.2015.09.007

Cheng, G., Qi, P., Guo, Z., & Yu, H. (2016). Mechanical design and evaluation of a compact portable knee-ankle -foot robot for gait rehabilitation. *Mechanism and Machine Theory*, *103*, 51–64. doi:10.1016/j.mechmachtheory.2016.04.012

Chen, X., & Wu, J. (2013). Scalable compressive tracking based on motion. In *Proceedings of 2013 IEEE International Conference on Robotics and Biomimetics (ROBIO)*. Shenzhen: IEEE. doi:10.1109/ROBIO.2013.6739510

Cherouvim, N., & Papadopoulos, E. (2006). Energy Saving Passive-Dynamic Gait for a One-Legged Hopping Robot. *Robotica*, *24*(04), 491–498. doi:10.1017/S026357470500250X

Chin, J. H., & Vora, N. (2014). The global burden of neurologic diseases. *Neurology*, *83*(4), 349–351. doi:10.1212/WNL.0000000000000610 PMID:25049303

Chirikjian, G. S. (1992). *Theory and applications of hyper-redundant robotic manipulators*. California Institute of Technology.

Chirikjian, G. S. (1994). Hyper-redundant manipulator dynamics: A continuum approximation. *Advanced Robotics*, *9*(3), 217–243. doi:10.1163/156855395X00175

Chirikjian, G. S., & Burdick, J. W. (1995). The Kinematics of Hyper-Redundant Robot Locomotion. *IEEE Transactions on Robotics and Automation*, *11*(6), 781–793. doi:10.1109/70.478426

Choi, W. J., Jeon, Y., Jeong, J.-H., Sood, R., & Kim, S. G. (2006). Energy harvesting MEMS device based on thin film piezoelectric cantilevers. *Journal of Electroceramics*, *17*(2-4), 543–548. doi:10.1007/s10832-006-6287-3

Cicconofri, G., & DeSimone, A. (2015). Limbless Locomotion on Solid Surfaces: A Case Study in Soft Bio-Inspired Robotics. *Mathematical Modelling*, *8*(1), 827–828.

Clarke, D. J., & Forster, A. (2015). An Improving post-stroke recovery: The role of the multidisciplinary health care team. *Journal of Multidisciplinary Healthcare*, *8*, 433–442. doi:10.2147/JMDH.S68764 PMID:26445548

Clark, J. E., Cham, J. G., Bailey, S. A., Froehlich, E. M., Nahata, P. K., Full, R. J., & Cutkosky, M. R. (2001). Biomimetic Design and Fabrication of a Hexapedal Running Robot. *The IEEE International Conference on Robotics and Automation (ICRA'2001)*, 4, 3643–3649. doi:10.1109/ROBOT.2001.933183

Cohen, R. (2008). *Beetle-Based Water Harvesting*. Retrieved from http://www.rexresearch.com/qinetiq/qinetiq.htm

Colomer, C., Noe, E., & Llorens, R. (2016). Mirror therapy in chronic stroke survivors with severely impaired upper limb function: A randomized controlled trial. *European Journal of Physical and Rehabilitation Medicine*, *52*(3), 271–278. PMID:26923644

Contreras-Vidal, J. L., & Grossman, R. G. (2013). NeuroRex: A clinical neural interface roadmap for EEG-based brain machine interfaces to a lower body robotic exoskeleton. In *Engineering in medicine and biology society (EMBC), 2013 35th annual international conference of the IEEE* (pp. 1579-1582). IEEE.

Corke, P. (1996). A robotics toolbox for MATLAB. *IEEE Robotics & Automation Magazine*, *3*(1), 24–32. doi:10.1109/100.486658

Cowles, R. B. (1956a). Some of the activities of the side winder. *Copeia, 4*, 220-224.

Cowles, R. B. (1956b). Sidewinding locomotion in snakes. *Copeia*, *4*(4), 211–214. doi:10.2307/1440272

Cox, C. L., Boback, S. M., & Guyer, C. (2011). Spatial dynamics of body size frequency distributions for North American squamates. *Evolutionary Biology*, *38*(4), 453–464. doi:10.1007/s11692-011-9132-9

Crema, A., Mancuso, M., Frisoli, A., Salsedo, F., Raschella, F., & Micera, S. (2015). *A hybrid NMES-exoskeleton for real objects interaction.* Paper presented at the Neural Engineering (NER), 2015 7th International IEEE/EMBS Conference on. doi:10.1109/NER.2015.7146710

Crosta, E. (2012). *An assistive device based on the detection of the user's intention from residual EMG to drive an upper limb neuroprosthesis*. Academic Press.

Cuervo, P., López, D., Cano, J., Sanchez, J., Rudas, S., Estupiñan, H., . . . Abdel-Aal, H. A. (2015). Topographic characterization of snake skin ventral scales for tribological purposes. *VIII Congress of Materials.*

Cuervo, P., López, D. A., Cano, J. P., Sánchez, J. C., Rudas, S., Estupiñán, H., ... Abdel-Aal, H. A. (2016). Development of low friction snake-inspired deterministic textured surfaces. *Surface Topography: Metrology and Properties*, *4*(2), 24013. doi:10.1088/2051-672X/4/2/024013

Cyber Glove Systems Web Page. (n.d.). Retrieved from http://www.cyberglovesystems.com/cybergrasp/#specs

Dal Corso, F., Misseroni, D., Pugno, N. M., Movchan, A. B., Movchan, N. V., & Bigoni, D. (2017). Serpentine locomotion through elastic energy release. *Journal of the Royal Society, Interface*, *14*(130), 20170055. doi:10.1098/rsif.2017.0055 PMID:28566512

Daltorio, K. A. (2006). A robot that climbs walls using micro-structured polymer feet. Proc. of Climbing and Walking Robots, 131–138. doi:10.1007/3-540-26415-9_15

Date, H., & Takita, Y. (2007). Adaptive locomotion of a snake like robot based on curvature derivatives. *IEEE International Conference on Intelligent Robots and Systems*. doi:10.1109/IROS.2007.4399635

de Castro, M. P., Abreu, S. C., Sousa, H., Machado, L., Santos, R., & Vilas-Boas, J. P. (2014). In-shoe plantar pressures and ground reaction forces during overweight adults' overground walking. *Research Quarterly for Exercise and Sport*, *85*(2), 188–197. doi:10.1080/02701367.2014.893055 PMID:25098014

De Luca, C. J. (2002). Surface electromyography: Detection and recording. *DelSys Incorporated, 10*, 2011.

de Melo, L. F., Rodrigues, E. J., & Rosário, J. M. (2015). Wheelchair Secure Navigation with RF Signal Triangulation and Genetic Algorithm Optimization. In *Handbook of Research on Advancements in Robotics and Mechatronics* (pp. 750–792). IGI Global. doi:10.4018/978-1-4666-7387-8.ch024

de Wit, C. C. (1998). Trends in mobile robot and vehicle control. In *Control Problems in Robotics and Automation* (pp. 151–175). Berlin: Springer. doi:10.1007/BFb0015082

Del-Ama, A. J., Koutsou, A. D., Moreno, J. C., De-Los-Reyes, A., Gil-Agudo, Á., & Pons, J. L. (2012). Review of hybrid exoskeletons to restore gait following spinal cord injury. *Journal of Rehabilitation Research and Development*, *49*(4), 497–514. doi:10.1682/JRRD.2011.03.0043 PMID:22773254

Delcomyn, F. (2007). Biologically inspired robots. In *Bioinspiration and Robotics Walking and Climbing Robots*. InTech. doi:10.5772/5506

Denavit, J. (1955). A kinematic notation for lower-pair mechanisms based on matrices. *ASME J. Appl. Mech.*, 215-221.

Denavit, J. (1955). A kinematic notation for lower-pair mechanisms based on matrices. *Trans. of the ASME. Journal of Applied Mechanics*, *22*, 215–221.

Desai, M. M., Aron, M., Gill, I. S., Pascal-Haber, G., Ukimura, O., Kaouk, J. H., ... Moll, F. (2008). Flexible Robotic Retrograde Renoscopy: Description of Novel Robotic Device and Preliminary Laboratory Experience. *Urology*, *72*(1), 42–46. doi:10.1016/j.urology.2008.01.076 PMID:18372029

Desai, M. M., Grover, R., Aron, M., Ganpule, A., Joshi, S. S., Desai, M. R., & Gill, I. S. (2011). Robotic Flexible Ureteroscopy for Renal Calculi: Initial Clinical Experience. *The Journal of Urology*, *186*(2), 563–568. doi:10.1016/j.juro.2011.03.128 PMID:21683380

Devine, S., Rafferty, K., & Ferguson, S. (2016). Real time robotic arm control using hand gestures with multiple end effectors. *2016 UKACC 11th International Conference on Control (CONTROL)*.

Di, S., Industriale, I., & Informazione, E. D. (2015). *Design and development of a planar cable-driven VSA based robot for upper-limb*. Academic Press.

Do Carmo, M. P., & Do Carmo, M. P. (1976). *Differential geometry of curves and surfaces* (Vol. 2). Prentice-hall Englewood Cliffs.

Dobrolyubov, A. I. (1986). The mechanism of locomotion of some terrestrial animals by travelling waves of deformation. In *Journal Theor* (pp. 457–466). Oxford, UK: Biology. doi:10.1016/S0022-5193(86)80195-3

Donald, B. R., Levey, C. G., McGray, C. D., Paprotny, I., & Rus, D. (2006). An untethered, electrostatic, globally controllable MEMS micro-robot. *Journal of Microelectromechanical Systems*, *15*(1), 1–15. doi:10.1109/JMEMS.2005.863697

Donald, B., Xavier, P., Canny, J., & Reif, J. (1993). Kinodynamic motion planning. *Journal of the Association for Computing Machinery*, *40*(5), 1048–1066. doi:10.1145/174147.174150

Dorval, K. B., Isaksen, S. G., & Noller, R. B. (2003). Leadership for Learning: Tips for Effective Mentoring and Coaching, in Mentoring for talent development Unknown Binding. *Reclaiming Youth International*. Retrieved from http://www.cpsb.com/resources/downloads/public/Mentoring_Coaching.pdf

Dorval, K. B., Isaksen, S. G., & Noller, R. B. (2003). Leadership for learning: Tips for effective mentoring and coaching. In *Mentoring for Talent Development*. Albion, MI: Reclaiming Youth International. Retrieved August 1, 2017, from http://www.cpsb.com/resources/downloads/public/Mentoring_Coaching.pdf

Doucet, B. M., Lam, A., & Griffin, L. (2012). Neuromuscular electrical stimulation for skeletal muscle function. *The Yale Journal of Biology and Medicine*, *85*(2), 201. PMID:22737049

Drucker, E. G., & Lauder, G. V. (1999). Locomotor Forces on a Swimming Fish: Three-Dimensional Vortex Wake Dynamics Quantified using DPIV. *The Journal of Experimental Biology*, *202*, 2393–2412. PMID:10460729

Dubrovkiy, V., & Fedorova, V. (2008). *Biomechanics*. Moscow: Vlados-press.

Duka, A. V. (2014). Neural network based inverse kinematics solution for trajectory tracking of a robotic arm. *Procedia Technology, 12*, 20–27. doi:10.1016/j.protcy.2013.12.451

Duka, A. V. (2015). ANFIS Based Solution to the Inverse Kinematics of a 3DOF Planar Manipulator. *Procedia Technology, 19*, 526–533. doi:10.1016/j.protcy.2015.02.075

Duret, C., Courtial, O., Grosmaire, A. G., & Hutin, E. (2015). Use of a Robotic Device for the Rehabilitation Severe Upper Limb Paresis in Subacute Stroke: Exploration of Patient/Robot Interaction and the Motor Recovery Process. *BioMed Research International, 2015*, 482389. doi:10.1155/2015/482389 PMID:25821804

Edqvist, E., Snis, N., Mohr, R. C., Scholz, O., Corradi, P., Gao, J., ... Johansson, S. (2009). Evaluation of building technology for mass producible millimeter-sized robots using flexible printed circuit boards. *Journal of Micromechanics and Microengineering, 19*(7), 1–11. doi:10.1088/0960-1317/19/7/075011 PMID:20161118

Elkoushy, M. A., & Andonian, S. (2011). Prevalence of Orthopedic Complaints Among Endourologists and Their Compliance with Radiation Safety Measures. *Journal of Endourology, 25*(10), 1609–1613. doi:10.1089/end.2011.0109 PMID:21819221

Ellis, L., Jackson, S., Liu, C., Molloy, P., & Paterson, P. (2013). *Lower Limb Exoskeleton Final Report.* Academic Press.

Elnady, A. M., Zhang, X., Xiao, Z. G., Yong, X., Randhawa, B. K., Boyd, L., & Menon, C. (2015). A single-session preliminary evaluation of an affordable BCI-controlled arm exoskeleton and motor-proprioception platform. *Frontiers in Human Neuroscience*, 9. PMID:25870554

Ergina, P. L., Barkun, J. S., Mcculloch, P., Cook, J. A., & Altman, D. G. (2013). IDEAL framework for surgical innovation 2: Observational studies in the exploration and assessment stages. *BMJ (Clinical Research Ed.), 346*. doi:10.1136/bmj.f3011 PMID:23778426

Esquenazi, A., Talaty, M., Packel, A., & Saulino, M. (2012). The ReWalk powered exoskeleton to restore ambulatory function to individuals with thoracic-level motor-complete spinal cord injury. *American Journal of Physical Medicine & Rehabilitation, 91*(11), 911–921. doi:10.1097/PHM.0b013e318269d9a3 PMID:23085703

Fabian, J., Young, T., Jones, J. C. P., & Clayton, G. M. (2014). Integrating the Microsoft Kinect with Simulink: Real-Time Object Tracking Example. *IEEE/ASME Transactions on Mechatronics, 19*(1), 249–257. doi:10.1109/TMECH.2012.2228010

Fall, C. L., Gagnon-Turcotte, G., Dubé, J. F., Gagné, J. S., Delisle, Y., Campeau-Lecours, A., ... Gosselin, B. (2017). Wireless sEMG-based body-machine interface for assistive technology devices. *IEEE Journal of Biomedical and Health Informatics, 21*(4), 967–977. doi:10.1109/JBHI.2016.2642837 PMID:28026793

Farris, R. J., Quintero, H. A., & Goldfarb, M. (2011). Preliminary evaluation of a powered lower limb orthosis to aid walking in paraplegic individuals. *IEEE Journal on Neural Systems and Rehabilitation Engineering, 19*(6), 652–659. doi:10.1109/TNSRE.2011.2163083 PMID:21968791

Fearing, R.S. (1998). Powering 3 Dimensional Microrobots: Power Density Limitations. *IEEE Int. Conf. Robotics and Automation.*

Feldman, A., & Meiri, S. (2013). Length-mass allometry in snakes. *Biological Journal of the Linnean Society. Linnean Society of London, 108*(1), 161–172. doi:10.1111/j.1095-8312.2012.02001.x

FernandezR.TanY. K.KaberleW.BestS. L.OlwenyE. O.PearleM. S.CadedduJ. A. (2012).

Ferrati, F., Bortoletto, R., Menegatti, E., & Pagello, E. (2013), November. Socio-economic impact of medical lower-limb exoskeletons. In *Advanced Robotics and its Social Impacts (ARSO), 2013 IEEE Workshop on* (pp. 19-26). IEEE.

Fichtinger, Deguet, Masamune, Balogh, Fischer, Mathieu, … Fayad. (2005, August). Image overlay guidance for needle insertion in CT scanner. *Biomedical Engineering. IEEE Transactions on*, *52*(8), 1415–1424.

Finke, R. A., Ward, T. B., & Smith, S. M. (1992). *Creative Cognition: Theory, Research, and Applications*. Cambridge, MA: The MIT Press.

Fischer. (2007). MRI image overlay: Application to arthrography needle insertion. *Computer Aided Surgery*, *12*(1), 2–14. PMID:17364654

Fish, F. E. (2006). Limits of Nature and Advances of Technology: What does Biomimetics have to Offer to Aquatic Robots? *ABBI*, *3*(1), 49–60.

Fish, F. E., & Hui, C. A. (1991). Dolphin Swimming: A Review. *Mammal Review*, *21*(4), 181–195. doi:10.1111/j.1365-2907.1991.tb00292.x

Floriano-Batista, R., & Rosario, J. M. (2013). *Control Architecture Proposal for Robotic Exoskeleton to Human Gait in Assistive Technology*. Campinas, São Paulo: Unicamp.

Fokker, A. D. (1927). De Voortbeweging Der Slangen. *Phisica, 7*(65).

Forsich, Gorb, Einwögerer, Trappl, Heim, Schift, & Bergmair. (2015). UV-NIL fabricated bio-inspired inlays for injection molding to influence the friction behavior of ceramic surfaces. *Microelectronic Engineering, 141*, 140-144. doi:10.1016/j.mee.2015.02.051

Frey, C. B., & Osborne, M. A. (2013). *The future of employment: how susceptible are jobs to computerization?* Oxford Martin School Working Paper, No.7, 1-72.

Frisoli, A., Rocchi, F., Marcheschi, S., Dettori, A., Salsedo, F., & Bergamasco, M. (2005). A new force-feedback arm exoskeleton for haptic interaction in virtual environments. *WHC 2005. First Joint Eurohaptics Conf. and Symposium on Haptic Interfaces for Virtual Environment and Teleoperator Systems*, 195–201. doi:10.1109/WHC.2005.15

Frisoli, Borelli, Montagner, Marcheschi, Procopio, Salsedo, … Rossit. (2007). Arm rehabilitation with a robotic exoskeleleton in Virtual Reality. *Proc. of the 2007 IEEE 10th International Conference on Rehabilitation Robotics*.

Frisoli, A., Salsedo, F., Bergamasco, M., Rossi, B., & Carboncini, M. C. (2009, June). Fabio Salsedo, Massimo Bergamasco, Bruno Rossi And Maria C. Carboncini, (2009). A force-feedback exoskeleton for upper-limb rehabilitation in virtual reality. *Applied Bionics and Biomechanics*, *6*(2), 115–126. doi:10.1155/2009/378254

Galchev, T., & Najafi, K. (2009). Energy Scavenging From Environmental Vibration. Sandia National Laboratories. doi:10.2172/1001023

Galvez, J. A., Gonzalez de Santos, P., & Pfeiffer, F. (2001). Intrinsic Tactile Sensing for the Optimization of Force Distribution in a Pipe Crawling Robot. *IEEE/ASME Transactions on Mechatronics*, *6*(1), 26–35. doi:10.1109/3516.914388

Gans, C., & Mendelssohn, H. (1972). Sidewinding and jumping progression of vipers. In Toxins of Animal and Plant Origin (pp. 17–38). London: Gordon & Breach Science Publications.

Gans, B. C., & Gasc, J. P. (1990). Kinematics and performance capacity for the concertina locomotion of a Snake (Coluber Constrictor). *The Journal of Experimental Biology*, *1991*(156), 539–556.

Gans, C. (1962). Terrestrial locomotion without limbs. *American Zoologist*, *2*(2), 167–182. doi:10.1093/icb/2.2.167

Gans, C. (1994). Approaches to the evolution of limbless locomotion. *Cusdemos de Herpetologia*, *8*, 12–17.

Gans, C., & Gasc, J. P. (1990). Tests on the locomotion of the elongate and limbless reptile Ophisaurus apodus (Sauria: Anguidae). *Journal of Zoology (London, England)*, *220*(4), 517–536. doi:10.1111/j.1469-7998.1990.tb04731.x

Gan, Y., Dai, X., & Li, D. (2013). Off-line programming techniques for multi-robot Cooperation System. *International Journal of Advanced Robotic Systems*, *10*(7), 1–17. doi:10.5772/56506

Geim, A. K., Dubonos, S. V., Grigorieva, I. V., Novoselov, K. S., Zhukov, A. A., & Shapoval, S. Y. (2003). Microfabricated Adhesive Mimicking Gecko Foot-Hair. *Nature Materials*, *2*(7), 461–463. doi:10.1038/nmat917 PMID:12776092

Gijbels, D., Lamers, I., Kerkhofs, L., Alders, G., Knippenberg, E., & Feys, P. (2011). The Armeo Spring as training tool to improve upper limb functionality in multiple sclerosis: A pilot study. *Journal of Neuroengineering and Rehabilitation*, *8*(5), 5. doi:10.1186/1743-0003-8-5 PMID:21261965

Gong, C., Tesch, M., Rollinson, D., & Choset, H. (2014). Snakes on an inclined plane: Learning an adaptive sidewinding motion for changing slopes. *IEEE International Conference on Intelligent Robots and Systems*. doi:10.1109/IROS.2014.6942697

Goodsell, D. S. (1996). *Our molecular Nature: The body's Motors*. Copernicus, NY: Machines and Messages. doi:10.1007/978-1-4612-2336-8

Gorb, S. N., Beutel, R. G., Gorb, E. V., Jiao, Y., Kastner, V., Niederegger, S., ... Vötsch, W. (2002). Structural design and biomechanics of friction-based releasable attachment devices in insects. *Integrative and Comparative Biology*, *42*(6), 1127–1139. doi:10.1093/icb/42.6.1127 PMID:21680397

Goto, T., Watanabe, K., Maeyama, S., & Motonaka, K. (2014). Stabilization control to the arbitrary target point for a two-wheeled drive mobile robot considering the dynamical model. In *Proceedings of SICE System and Information Division Annual Conference* (pp. 599-604). Academic Press. (in Japanese)

Graham-Rowe, D. (2003, May). Fancy A Walk on the Ceiling? *New Scientist*, *178*(2395), 15.

Graighead, H. G. (2000). Nanoelectromechanical Systems. *Science*, *290*(5496), 1532–1535. doi:10.1126/science.290.5496.1532 PMID:11090343

Gray, J. (1946). The mechanism of locomotion in snakes. *The Journal of Experimental Biology*, *23*(2). PMID:20281580

Gray, J., & Lissmann, H. W. (1950). The kinetics of locomotion of the grass-snake. *The Journal of Experimental Biology*, *26*, 354–367.

Greenwood, J. A., & Tripp, J. H. (1971). The Contact of Two Nominally Flat Rough Surfaces. *Proceedings - Institution of Mechanical Engineers*, *185*(1), 625–633. doi:10.1243/PIME_PROC_1970_185_069_02

Greiner, C., & Schäfer, M. (2015). Bio-inspired scale-like surface textures and their tribological properties. *Bioinspiration & Biomimetics*, *10*(4), 044001. doi:10.1088/1748-3190/10/4/044001 PMID:26125522

Guiard, Y. (1987). Asymmetric Division of Laber in Human Skilled Bimanual Action: The Kinematic Chain as a Model. *Journal of Motor Behavior, 19*, 486-517.

Guidance for the Content of Premarket Submissions for Software Contained in Medical Devices. (2015, May 11). U.S. Department of Health and Human Services, Food and Drug Administration.

Guilford, J. P. (1987). Creativity Research: Past, Present and Future. In *Frontiers of creativity research: Beyond the basics* (pp. 34-64). Buffalo, NY: Bearly Ltd. Retrieved from http://www.cpsb.com/research/articles/creativity-research/Creativity-Research-Guilford.pdf

Guo, Healy, & Zhou. (2010). ZigBee-wireless mesh networks for building automation and control. *International Conference on Networking, Sensing and Control (ICNSC).*

Guo, S., Mao, S., Shi, L., & Li, M. (2012). Development of an Amphibious Mother Spherical Robot Used as the Carrier for Underwater Microrobots. In *Proceedings of the 2012 International Conference on Complex Medical Engineering (ICME).* Kobe: IEEE. doi:10.1109/ICCME.2012.6275640

Guo, Z. V., & Mahadevan, L. (2008). Limbless undulatory propulsion on land. *Proceedings of the National Academy of Sciences of the United States of America, 105*(9), 3179–3184. doi:10.1073/pnas.0705442105 PMID:18308928

Gupta, A., & O'Malley, M. (2006, June). Design of a haptic arm exoskeleton for training and rehabilitation. *IEEE/ASME Transactions on Mechatronics, 11*(3), 280–289. doi:10.1109/TMECH.2006.875558

Gupta, M., Behera, L., Subramanian, V. K., & Jamshidi, M. M. (2014). A Robust Visual Human Detection Approach With UKF-Based Motion Tracking for a Mobile Robot. *IEEE Systems Journal, 9*(4), 1–13.

Gutiérrez-Carvajal, R. E., de Melo, L. F., Rosário, J. M., & Machado, J. T. (2016). Condition-based diagnosis of mechatronic systems using a fractional calculus approach. *International Journal of Systems Science, 47*(9), 2169–2177. do i:10.1080/00207721.2014.978833

Guyer, C., & Donnelly, M. A. (1990). Length–mass relationships among an assemblage of tropical snakes in Costa Rica. *Journal of Tropical Ecology, 6*(1), 65–76. doi:10.1017/S0266467400004041

Gwynne, P. (2013). Technology: Mobility machines. *Nature, 503*(7475), 16–17. doi:10.1038/503S16a PMID:24227002

Habib, M. K. (2011). Article. *Biomimetics: Innovations and Robotics, 4*(2), 113–134.

Habib, M. K., Watanabe, K., & Izumi, K. (2007b). Biomimetics Robots: From Bio-inspiration to Implementation. *The 33rd Annual Conference of the IEEE Industrial Electronics Society IECON'2007*, 143-148.

Habib, M. K. (2008). Interdisciplinary mechatronics: Problem solving, creative thinking and concurrent design synergy. *International Journal of Mechatronics and Manufacturing Systems, 1*(1), 264–269. doi:10.1504/IJMMS.2008.018272

Habib, M. K. (2011). Biomimetics: Innovations and Robotics. *Int. J. Mechatronics and Manufacturing Systems, 4*(2), 113–134. doi:10.1504/IJMMS.2011.039263

Habib, M. K., & Watanabe, K., & Izumi, K. (2007a). Biomimetics Intelligent Robots and Biological Inspiration. *International Conference on Control, Instrumentation and Mechatronics 'CIM'2007'*, 824-832.

Hara, Y. (2013). Rehabilitation with Functional Electrical Stimulation in Stroke Patients. *International Journal of Physical Medicine & Rehabilitation, 1*(147), 2.

Hartopanu, S., Poboroniuc, M., Serea, F., & Livint, G. (2014). *Towards human arm rehabilitation in stroke patients by means of a hybrid FES&robotic glove.* Paper presented at the Electrical and Power Engineering (EPE), 2014 International Conference and Exposition on. doi:10.1109/ICEPE.2014.6969886

Hartopanu, S., Poboroniuc, M., Serea, F., Irimia, D., & Livint, G. (2013). *Design of a hybrid FES-mechanical intelligent haptic robotic glove.* Paper presented at the System Theory, Control and Computing (ICSTCC), 2013 17th International Conference. doi:10.1109/ICSTCC.2013.6689040

Hartopanu, S., Poboroniuc, M., Serea, F., Irimia, D., & Livint, G. (2015). *New Issues on FES and Robotic Glove Device to Improve the Hand Rehabilitation in Stroke Patients.* Paper presented at the 6th International Conference on Modern Power System.

Hasanzadeh, S., & Tootoonchi, A. A. (2010). Ground adaptive and optimized locomotion of snake robot moving with a novel gait. *Autonomous Robots*, *28*(4), 457–470. doi:10.1007/s10514-010-9179-y

Hatton, R. L., & Choset, H. (2010). Generating gaits for snake robots: Annealed chain fitting and keyframe wave extraction. *Autonomous Robots*, *28*(3), 271–281. doi:10.1007/s10514-009-9175-2

Hazel, J., Stone, M., Grace, M. S., & Tsukruk, V. V. (1999). Nanoscale design of snake skin for reptation locomotions via friction anisotropy. *Journal of Biomechanics*, *32*(5), 477–484. doi:10.1016/S0021-9290(99)00013-5 PMID:10327001

Hellawell, G., Mutch, S., Thevendran, G., Wells, E., & Morgan, R. (2005). Radiation Exposure and The Urologist: What Are The Risks? *The Journal of Urology*, *174*(3), 948–952. doi:10.1097/01.ju.0000170232.58930.8f PMID:16094003

Hersent, O., Boswarthick, D., & Elloumi, O. (2011). ZigBee. In *The Internet of Things: Key Applications and Protocols* (pp. 93–137). John Wiley & Sons, Ltd. doi:10.1002/9781119958352.ch7

Herzog, W. A., Nigg, B. M., Read, L. J., & Olsson, E. W. (1989, February 1). Asymmetries in ground reaction force patterns in normal human gait. *Medicine and Science in Sports and Exercise*, *21*(1), 110–114. doi:10.1249/00005768-198902000-00020 PMID:2927295

Hesse, S., Schulte-Tigges, G., Konrad, M., Bardeleben, A., & Werner, C. (2003). Robot-assisted arm trainer for the passive and active practice of bilateral forearm and wrist movements in hemiparetic subjects. Archives of Physical Medicine and Rehabilitation, 84(6), 915-920.

Hesse, S., Waldner, A., & Tomeller, C. (2010). Innovative gait robot for the repetitive practice of floor walking and stair climbing up and down in stroke patients. *Journal of Neuroengineering and Rehabilitation*, *7*(1), 30. doi:10.1186/1743-0003-7-30 PMID:20584307

Hiramatsu, Y., Shibanoki, T., Shima, K., & Tsuji, T. (2011). A novel dual-arm motion discrimination method using reccurent probability neural networks for automatic gesture recognition. *Proceedings of the 2011 IEEE/SICE International Symposium on System Integration (SII 2011)*, 1346-1351.

Hirose, S., & Fukushima, E. F. (2002). Snakes and Strings: New Robotic Components for Rescue Operations. *The 41st Annual SICE Conference*, 338-343.

Hirose, S., Nagakubo, A., & Toyama, R. (1991). Machine that can Walk and Climb on Floors, Walls and Ceilings. *The 5th International Conference on Advanced Robotics*, 753-758.

Hirose, S. (1993). *Biologically inspired robot*. Oxford University Press.

Hirose, S. (1993). *Biologically inspired robots (snake-like locomotor and manipulator)*. Oxford University Press.

Hiroshi, K., Hasegawa, S., & Nozaki, H. (2008). Development and Application of a Muscle Force Enhancement Wear: Muscle Suit. *Proceedings of the 11th Symposium on Construction Robotics in Japan*, 93-100.

Hoang, G., Min, J. H., Lee, G. M., Jun, B. H., Kim, H. K., & Kim, S. B. (2014, October), Omni-directional walking control for a six-legged robot using differential kinematics algorithm, In *Proceedings of 2014 14th International Conference on Control, Automation and Systems (ICCAS 2014)* (pp. 1163–1168). Academic Press. doi:10.1109/ICCAS.2014.6987735

HOCOMA Webpage. (n.d.). Retrieved from http://www.hocoma.com

Hogan, N., Krebs, H. I., Charnnarong, J., Srikrishna, P., & Sharon, A. (1992). MIT-MANUS:a workstation for manual therapy and training. Robot and Human Communication. *Proceedings., IEE International Workshop*. DOI: .1992.253895 doi:10.1109/ROMAN

Holden, T., Pedro, R. N., Hendlin, K., Durfee, W., & Monga, M. (2008). Evidence-Based Instrumentation for Flexible Ureteroscopy: A Review. *Journal of Endourology*, *22*(7), 1423–1426. doi:10.1089/end.2007.0327 PMID:18613778

Hoover, M. A., Steltz, E., & Fearing, S. R. (2008). RoACH: An autonomous 2.4 g crawling hexapod robot. *Proceedings of the 2008 IEEE/RSJ International Conference on Intelligent Robots and Systems*, 26–33. doi:10.1109/IROS.2008.4651149

Hopkins, J. K., Spranklin, B. W., & Gupta, S. K. (2009). A survey of snake-inspired robot designs. *Bioinspiration & Biomimetics*, *4*(2), 21001. doi:10.1088/1748-3182/4/2/021001 PMID:19158415

Housman, S., Scott, K., & Reinkensmeyer, D. (2009). Randomized Controlled Trial of Gravity-Supported, Computer-Enhanced Arm Exercise for Individuals With Severe Hemiparesis. *Neurorehabil Neural Repair.* doi:10.1177/1545968308331148

Hsieh, C.-Y., Chen, C.-H., Li, C.-Y., & Lai, M.-L. (2015). Validating the diagnosis of acute ischemic stroke in a National Health Insurance claims database. *Journal of the Formosan Medical Association, 114*(3), 254–9. 10.1016/j.jfma.2013.09.009

Hsu, D., Kindel, R., Latombe, J. C., & Rock, S. (2002). Randomized kinodynamic motion planning with moving obstacles. *The International Journal of Robotics Research*, *21*(3), 233–255. doi:10.1177/027836402320556421

Huang, J., Zhou, W., Li, H., & Li, W. (2015). Sign language recognition using 3D convolutional neural networks. *IEEE International Conference on Multimedia and Expo (ICME)*, 1–6. doi:10.1109/ICME.2015.7177428

Huang, T. J., Liu, Y., Brough, B., Flood, A. H., Bonvallet, P., Tseng, H.-R., ... Ho, C.-M. (2005). *A Nano-Chemo-Mechanical Actuator Based on Artificial Molecular Machines*. Miami, FL: MEMS.

Hu, D. L., Nirody, J., Scott, T., & Shelley, M. J. (2009). The mechanics of slithering locomotion. *Proceedings of the National Academy of Sciences of the United States of America*, *106*(25), 10081–10085. doi:10.1073/pnas.0812533106 PMID:19506255

Hutchinson, H. (1879). About Snakes. *Nature*. Retrieved from http://adsabs.harvard.edu/abs/1879Natur.20.528H

Hutchinson, H. F. (1879). About Snakes. *Nature*, *20*(518), 528–530. doi:10.1038/020528c0

Iida, F. (2003). Biologically Inspired Visual Odometer for Navigation of a Flying Robot. *Journal of Robotics and Autonomous Systems*, *44*(3-4), 201–208. doi:10.1016/S0921-8890(03)00070-8

Ijspeert, A. J. (2008). Central pattern generators for locomotion control in animals and robots: A review. *Neural Networks*, *21*(4), 642–653. doi:10.1016/j.neunet.2008.03.014 PMID:18555958

Ilon, B. E. (1975). *U.S. Patent No. 3,876,255*. Washington, DC: U.S. Patent and Trademark Office.

Inman, V. T., Ralston, H. J., & Todd, F. (1981). *Human walking*. Williams & Wilkins.

Irani, R., Bauer, R., North, L., Nicholson, M., Nolan, D., & West, B. (2015). Analysis of joint failures on the lateral undulation gait of a robotic snake. *Transactions of the Canadian Society for Mechanical Engineering*, *39*(2).

Ishii, M., Yamamoto, K., & Hyodo, K. (2005). Stand-Alone Wearable Power Assist Suit –Development and Availability-. *Journal of Robotics and Mechatronics*, *17*(5), 575–583. doi:10.20965/jrm.2005.p0575

Ito, S., Kawasaki, H., Ishigure, Y., Natsume, M., Mouri, T., & Nishimoto, Y. (2011). A design of fine motion assist equipment for disabled hand in robotic rehabilitation system. *Journal of the Franklin Institute*, *348*(1), 79–89. doi:10.1016/j.jfranklin.2009.02.009

Jack, A. J. (2001). Inflatable Robotics for Planetary Applications. *The 6th International Symposium on Artificial Intelligence, Robotics, and Automation in Space*. I-SAIRAS, Canadian Space Agency.

Jackson, A., Holt, R., Culmer, P., Makower, S., Levesley, M., Richardson, R., & Bhakta, B. B. (2007). Dual robot system for upper limb rehabilitation after stroke: The design process. *Journal of Mechanical Engineering Science. Proceedings of the Institution of Mechanical Engineers, Part C., 221*(7), 845–857.

Jafari, A. (2014). Coupling between the Output Force and Stiffness in Different Variable Stiffness Actuators. *Actuators, 3*(3), 270–284. doi:10.3390/act3030270

Jarrassé, N., Proietti, T., Crocher, V., Robertson, J., Sahbani, A., Morel, G., & Roby-Brami, A. (2014). Robotic exoskeletons: A perspective for the rehabilitation of arm coordination in stroke patients. *Frontiers in Human Neuroscience*, 8. PMID:25520638

Jatsun, S., Savin, S., & Yatsun, A. (2016). Comparative analysis of global optimization-based controller tuning methods for an exoskeleton performing push recovery. In *System Theory, Control and Computing (ICSTCC), 2016 20th International Conference on* (pp. 107-112). IEEE.

Jatsun, S., Savin, S., & Yatsun, A. (2016). Improvement of energy consumption for a lower limb exoskeleton through verticalization time optimization. In *Control and Automation (MED), 2016 24th Mediterranean Conference on* (pp. 322-326). IEEE.

Jatsun, S., Savin, S., & Yatsun, A. (2016). Study of controlled motion of an exoskeleton performing obstacle avoidance during a single support walking phase. In *System Theory, Control and Computing (ICSTCC), 2016 20th International Conference on* (pp. 113-118). IEEE.

Jatsun, S., Savin, S., Lushnikov, B., & Yatsun, A. (2016). Algorithm for motion control of an exoskeleton during verticalization. In *ITM Web of Conferences* (Vol. 6, p. 1-6). EDP Sciences.

Jatsun, S., Savin, S., Lushnikov, B., & Yatsun, A. (2016). System analysis of sagittal plane human motion wearing an exoskeleton using marker technology. In *ITM Web of Conferences* (Vol. 6). EDP Sciences.

Jatsun, S., Savin, S., Yatsun, A., & Turlapov, R. (2015). Adaptive control system for exoskeleton performing sit-to-stand motion. In *Mechatronics and its Applications (ISMA), 2015 10th International Symposium on* (pp. 1-6). IEEE.

Jatsun, S., Yatsun, A., Savin, S., & Postolnyi, A. (2016). Approach to motion control of an exoskeleton in "verticalization-to-walking" regime utilizing pressure sensors. In *Cyber Technology in Automation, Control, and Intelligent Systems (CYBER), 2016 IEEE International Conference on* (pp. 452-456). IEEE.

Jatsun, S., Savin, S., & Yatsun, A. (2016). Motion Control Algorithm for a Lower Limb Exoskeleton Based on Iterative LQR and ZMP method for trajectory generation. In *International Workshop on Medical and Service Robots* (pp. 305-317). Springer.

Jatsun, S., Savin, S., Yatsun, A., & Malchikov, A. (2016). Study of controlled motion of exoskeleton moving from sitting to standing position. In *Advances in Robot Design and Intelligent Control* (pp. 165–172). Springer International Publishing. doi:10.1007/978-3-319-21290-6_17

Jayne, B. C. (1986). Kinematics of terrestrial snake locomotion. *Copeia, 1986*(4), 915–927. doi:10.2307/1445288

Jayne, B. C. (1988). Muscular mechanisms of snake locomotion: An electromyographic study of lateral undulation of the Florida banded water snake (Nerodia fasciata) and the yellow rat snake (Elaphe obsoleta). *Journal of Morphology, 197*(2), 159–181. doi:10.1002/jmor.1051970204 PMID:3184194

Jayne, B. C., & Herrmann, M. P. (2011). Perch size and structure have species-dependent effects on the arboreal locomotion of rat snakes and boa constrictors. *The Journal of Experimental Biology, 214*(13), 2189–2201. doi:10.1242/jeb.055293 PMID:21653813

Jeong, Y., Lee, Y., Kim, K., Hong, Y., & And, J. Park, (2001). A 7 DOF Wearable Robotic Arm using Pneumatic Actuators. *Proceedings of the 32nd ISR International Symposium on Robotics*, 388-393.

Jin, H., Zhu, L., & Chen, Z. (2016). Trajectory Control of Snake-Like Robots in Natural Oscillation. *Asian Journal of Control*. 10.1002/asjc.1271

Johnson, M., Van der Loos, H., Burgar, C., & Leifer, L. (1999). Driver's SEAT: Simulation Environment for Arm Therapy. *Proceedings of 6th International Conference on Rehabilitation Robotics (ICORR'99)*, 227-234.

Johnson, W., Onuma, O., Owolabi, M., & Sachdev, S. (2016). Stroke: A global response is needed. *Bulletin of the World Health Organization*, *94*(9), 634–634A. doi:10.2471/BLT.16.181636 PMID:27708464

Jones, B. A., McMahan, W., & Walker, I. D. (2006). Practical Kinematics for Real-Time Implementation of Continuum Robots. *The 2006 IEEE International Conference on Robotics and Automation (ICRA'2006)*, 1840-1847.

Jones, A. T. (1933). The gliding of a snake. *Physics*, *4*(4), 164–165. doi:10.1063/1.1745173

Jung, Y., Jung, M., Lee, K., & Koo, S. (2014). Ground reaction force estimation using an insole-type pressure mat and joint kinematics during walking. *Journal of Biomechanics*, *47*(11), 2693–2699. doi:10.1016/j.jbiomech.2014.05.007 PMID:24917473

Kaelbling, L. P., Littman, M. L., & Cassandra, A. R. (1998). Planning and acting in partially observable stochastic domains. *Artificial Intelligence*, *101*(1-2), 99–134. doi:10.1016/S0004-3702(98)00023-X

Kalauni, K. (2014). *K. M. Gupta, and Isha Bharti*. Reptiles Inspired Biomimetic Materials and Their Novel Applications.

Kalvandi, R., & Maddahi, Y. (2012). On performance evaluation of home health care assistant platforms: Simulation study. *Int J of Biology and Biomedical Engineering*, *2*(6), 135–142.

Kano, T., Sato, T., Kobayashi, R., & Ishiguro, A. (2012). Local reflexive mechanisms essential for snakes' scaffold-based locomotion. *Bioinspiration & Biomimetics*, *7*(4), 46008. doi:10.1088/1748-3182/7/4/046008 PMID:22918023

Kawamoto, , & Lee, , Kanbe, & Sankai. (2003). Power assist method for hal-3 using emg-based feedback controller. *IEEE International Conference, Man and Cybernetics*, 2, 1648–1653.

Kazerooni, H. (2008). Exoskeletons for human performance augmentation. In *Springer handbook of robotics* (pp. 773–793). Springer Berlin Heidelberg. doi:10.1007/978-3-540-30301-5_34

Kelley, K. C., Arnold, S. J., & Gladstone, J. (1997). The effects of substrate and vertebral number on locomotion in the grater snake Thamnophis elegans. *Functional Ecology*, *11*(2), 189–198. doi:10.1046/j.1365-2435.1997.00077.x

Kiguchi, K., & Fukuda, T. (2004). A 3 DOF Exoskeleton for Upper Limb Motion Assist: Consideration of the Effect of Bi-Articular Muscles. *IEEE Int'l Conf. Robot. Automat.*, *3*, 2424–2429. doi:10.1109/ROBOT.2004.1307424

Kim, S., Asbeck, A., Cutkosky, M., & Provancher, W. (2005). SpinybotII: Climbing hard walls with compliant microspines. *Proc. of the IEEE Int. Conf. on Advanced Robotics (ICAR)*, 601–606.

Kim, S., Clark, J. E., & Cutkosky, M. R. (2004). iSprawl: Autonomy, and the Effects of Power Transmission. *The 7th International Conference on Climbing and Walking Robots CLAWAR'2004*.

Kim, H. S., Kim, J.-H., & Kim, J. (2011). A review of piezoelectric energy harvesting based on vibration. *International Journal of Precision Engineering and Manufacturing*, *12*(6), 1129–1141. doi:10.1007/s12541-011-0151-3

Kim, J., Jang, W., & Bien, Z. (1996). A Dynamic Gesture Recognition System for the Korean Sign Language (KSL). *IEEE Trans. on Systems, Man and Cybernetics-Part B. Cybernetics*, *26*(2), 354–359. PMID:18263039

Kim, K. P., Miller, D. L., Gonzalez, A. B., Balter, S., Kleinerman, R. A., Ostroumova, E., ... Linet, M. S. (2012). Occupational Radiation Doses to Operators Performing Fluoroscopically-Guided Procedures. *Health Physics*, *103*(1), 80–99. doi:10.1097/HP.0b013e31824dae76 PMID:22647920

Kim, S. (2008). Smooth vertical surface climbing with directional adhesion. *IEEE Transactions on Robotics*, *24*(1), 65–74. doi:10.1109/TRO.2007.909786

Kim, S., Spenko, M., Trujillo, S., Heyneman, B., Mattoli, V., & Cutkosky, M. R. (2007). Whole body adhesion: Hierarchical, Directional and Distributed control of Adhesive forces for a climbing robot. *The IEEE International Conference on Robotics and Automation (ICRA'2007)*, 1268-1273. doi:10.1109/ROBOT.2007.363159

Kinsman Enterprises Product Catalog Web Page. (n.d.). Retrieved from http://www.kinsmanenterprises.com/

Klaassen, B., Linnemann, R., Spenneberg, D., & Kirchner, F. (2003). Biologically Inspired Robot Design and Modeling. *The International Conference on Advanced Robotics*, 576-581.

Klamroth-Marganska, V., Blanco, J., Campen, K., Curt, A., Dietz, V., Ettlin, T., ... Riener, R. (2014). Three-dimensional, task-specific robot therapy of the arm after stroke: A multicentre, parallel-group randomised trial. *Lancet Neurology*, *13*(2), 159–166. doi:10.1016/S1474-4422(13)70305-3 PMID:24382580

Klein & Gorb. (2014). Ultrastructure and wear patterns of the ventral epidermis of four snake species (Squamata, Serpentes). *Zoology, 117*(5), 295-314. doi:10.1016/j.zool.2014.01.004

Klein, M-C. G., & Gorb, S. N. (2012). Epidermis architecture and material properties of the skin of four snake species. *J. R. Soc. Interface*. doi:1742-566210.1098/rsif.2012.0479

Klein, J., Spencer, S. J., Allington, J., Minakata, K., Wolbrecht, E. T., Smith, R., & Bobrow, J. E. (2008). Biomimetic orthosis for the neurorehabilitation of the elbow and shoulder (BONES) 2008. *Proceedings of the 2nd Biennial IEEE/RAS-EMBS International Conference on Biomedical Robotics and Biomechatronics.*

Klein, M.-C. G., Deuschle, J., & Gorb, S. N. (2010). Material properties of the skin of the Kenyan sand boa Gongylophis colubrinus (Squamata, Boidae). *J. of Comp. Phys. A: Neuroethology, Sensory, Neural, and Behav. Phys.*, *196*, 659–668.

Knight, A. (2009). *Hidden Histories: The Story of Sustainable Design*. Retrieved from http://www.csa.com/discoveryguides/design/review.pdf

Knoll, T., Jessen, J. P., Honeck, P., & Wendt-Nordahl, G. (2011). Flexible ureterorenoscopy versus miniaturized PNL for solitary renal calculi of 10–30 mm size. *World Journal of Urology*, *29*(6), 755–759. doi:10.1007/s00345-011-0784-y PMID:22037633

Ko, J. H., Hwang, D. C., Jung, Y. W., & Kim, E. S. (2005, October). Intelligent mobile robot system for path planning using stereo camera-based geometry information. *Proceedings of the Society for Photo-Instrumentation Engineers*, *6006*, 60060L. doi:10.1117/12.632032

Koker, R. (2013, February). A genetic algorithm approach to a neuralnetwork-based inverse kinematics solution of robotic manipulators based on error minimization. *Information Sciences*, *222*, 528–543. doi:10.1016/j.ins.2012.07.051

Koker, R., Oz, C., Çakar, T., & Ekiz, H. (2004, December). A study of neural network based inverse kinematics solution for a three-joint robot. *Robotics and Autonomous Systems*, *49*(3), 227–234. doi:10.1016/j.robot.2004.09.010

Krabben, T., Prange, G. B., Molier, B. I., Stienen, A. H., Jannink, M. J., Buurke, J. H., & Rietman, J. S. (2012). Influence of gravity compensation training on synergistic movement patterns of the upper extremity after stroke, a pilot study. *Journal of Neuroengineering and Rehabilitation*, *9*(1), 44. doi:10.1186/1743-0003-9-44 PMID:22824488

Krebs, H., Hogan, N., Volpe, B. T., Aisen, M. L., Edelstein, L., & Diels, C. (1999). Overview of clinical trials with MIT-MANUS: A robot-aided neuro- rehabilitation facility. *Technology and Health Care*, *7*, 419–423. Retrieved from http://www.scopus.com/inward/record.url?eid=2-s2.0-0033400738&partnerID=40&md5=1cfad3a95596bfa282229f338a7fb624 PMID:10665675

Krishnaprasad, P. S., & Tsakiris, D. P. (1994). G-snakes: Nonholonomic kinematic chains on Lie groups. In *Decision and Control, 1994., Proceedings of the 33rd IEEE Conference on* (*Vol. 3*, pp. 2955–2960). IEEE.

Krogstie, B., Prilla, M., & Pammer, V. (2013). Understanding and supporting reflective learning processes in the workplace. In *Proceedings of the Eighth European Conference on Technology Enhanced Learning* (EC-TEL 2013). Berlin, Germany: Springer-Verlag.

Kubow, T. M., & Full, R. J. (1999). The Role of the Mechanical System in Control: A Hypothesis of Self-Stabilization in Hexapedal Runners. *Philosophical Transactions of the Royal Society of London. Series B, Biological Sciences*, *354*(1385), 849–861. doi:10.1098/rstb.1999.0437

Kumazawa, Y. (2007). Mitochondrial genomes from major lizard families suggest their phylogenetic relationships and ancient radiations. *Gene*, *388*(1-2), 1–2, 19–26. doi:10.1016/j.gene.2006.09.026 PMID:17118581

Kuznetsov, V. M., Lugovtsov, B. A., & Sher, Y. N. (1967). On the motive mechanism of snakes and fish. *Archive for Rational Mechanics and Analysis*, *25*(5), 367–387. doi:10.1007/BF00291937

Labiod, Afifi, & Santis. (2007). *Wi-fi, Bluetooth, ZigBee, and WiMAX*. Dordrecht: Springer.

Lancaster, J. K. (1968). The effect of carbon fibre reinforcement on the friction and wear of polymers. *Journal of Physics. D, Applied Physics*, *1*(5), 549–560. doi:10.1088/0022-3727/1/5/303

Landmann, L. (1979). Keratin formation and barrier mechanisms in the epidermis of Natrix natrix (Reptilia: Serpentes): An ultrastructural study. *J. Morphol.*, *162*, 93-126.

Landmann, L. (1986). The skin of Reptiles: epidermis and dermis. In Biology of the integument: Vertebrate. New York: Springer Verlag.

Laver, K. E., George, S., Thomas, S., Deutsch, J. E., & Crotty, M. (2015). Virtual reality for stroke rehabilitation. *Cochrane Database of Systematic Reviews*, (2): CD008349. doi:10.1002/14651858.CD008349. PMID:25927099

Laver, K., George, S., Thomas, S., Deutsch, J., & Crotty, M. (2012). Virtual reality for stroke rehabilitation. *Stroke*, *91*, 1–16. doi:10.1097/PHM.0b013e31826bce79.Comparison PMID:22713539

Lavrentiev, M.A., & M. M. L. (1962). On a principle for creating a tractive force of motion. *Journal of Applied Mechanics and Technical Physics*, 4.

Lawrence, D. (1992). Stability and transparency in bilateral teleoperation. *Proc. IEEE Conf. Decision and Control.*

Lee, S., & Sankai, Y. (2003). The natural frequency-based power assist control for lower body with hal-3. *IEEE International Conference on Systems, Man and Cybernetics,* 2, 1642–1647. doi:10.1109/ICSMC.2003.1244648

Legnani, G., Casalo, F., Righettini, P., & Zappa, B. (1996). A homogeneous matrix approach to 3D kinematics and dynamics - II. Applications to chains of rigid bodies and serial manipulators. *Mechanism and Machine Theory*, *31*(5), 589–605. doi:10.1016/0094-114X(95)00101-4

Legnani, G., Tosi, D., Fassi, I., Giberti, H., & Cinquemani, S. (2010). The "point of isotropy" and other properties of serial and parallel manipulators. *Mechanism and Machine Theory*, *45*(10), 1407–1423. doi:10.1016/j.mechmachtheory.2010.05.007

Lewinger, W. A., Harley, C. M., Ritzmann, R. E., Branicky, M. S., & Quinn, R. D., R. D (2005). Insect-like Antennal Sensing for Climbing and Tunneling Behavior in a Biologically-Inspired Mobile Robot. *The IEEE International Conference on Robotics and Automation (ICRA'05)*, 4176-4181. doi:10.1109/ROBOT.2005.1570761

Lewinger, W. A., Watson, M. S., & Quinn, R. D. (2006). Obstacle Avoidance Behavior for a Biologically-Inspired Mobile Robot using Binaural Ultrasonic Sensors. *The IEEE/RSJ International Conference on Intelligent Robots and Systems (IROS'06)*, 6. doi:10.1109/IROS.2006.282461

Li, L., Ding, L., Chen, N., Mao, Y., Huang, D., & Li, Le. (2015). Improved walking ability with wearable robot-assisted training in patients suffering chronic stroke. *Proceeding of the 4th International Conference on Biomedical Engineering and Biotehnology*. doi:10.3233/BME-151320

Liljeback, Stavdahl, Pettersen, & Gravdahl. (2014). Mamba-A waterproof snake robot with tactile sensing. *Intelligent Robots and Systems (IROS 2014), 2014 IEEE/RSJ International Conference on* (pp. 294-301). IEEE.

Liljebäck, P., Pettersen, K. Y., Stavdahl, O., & Gravdahl, J. T. (2010). Hybrid modelling and control of obstacle-aided snake robot locomotion. *IEEE Transactions on Robotics*, *26*(5), 781–799. doi:10.1109/TRO.2010.2056211

Liljebäck, P., Pettersen, K. Y., Stavdahl, O., & Gravdahl, J. T. (2011). Experimental investigation of obstacle-aided locomotion with a snake robot. *IEEE Transactions on Robotics*, *27*(4), 792–800. doi:10.1109/TRO.2011.2134150

Liljebäck, P., Pettersen, K. Y., Stavdahl, O., & Gravdahl, J. T. (2012). *Snake robots: modelling, mechatronics, and control*. Springer Science & Business Media.

Lillywhite, H. B., & Maderson, P. F. A. (1982). Skin structure and permeability. In Gans, Lillywhite HB, Henderson RW. 1993. Behavioral and functional ecology of arboreal snakes. In R. A. Seigel & J. T. Collins (Eds.), *Snakes: Ecology and behavior* (pp. 1–48). New York: McGraw-Hill.

Li, M., Guo, S., Hirata, H., & Ishihara, H. (2017). A roller-skating/walking mode-based amphibious robot. *Robotics and Computer-integrated Manufacturing*, *44*, 17–29. doi:10.1016/j.rcim.2016.06.005

Lin, R., Li, M., & Sun, L. (2013). Image features-based mobile robot visual SLAM. In *Proceedings of 013 IEEE International Conference on Robotics and Biomimetics (ROBIO)*. Shenzhen: IEEE.

Li, P. Y. (2004). Design and Control of a Hydraulic Human Power Amplifier. *ASME International Mechanical Engineering Conference and Exposition*. doi:10.1115/IMECE2004-60868

Lissmann, H. W. (1950). Rectilinear locomotion in a snake (Boa occidentalis). *The Journal of Experimental Biology*, *26*(4), 368–379. Retrieved from http://jeb.biologists.org/content/26/4/368.short

Liu, K., Liu, M., Zhang, X., & Li, H. (2016). A bio-inspired geomagnetic navigation model based on course constraint strategy under anomalies field disturbing for AUV. OCEANS 2016 - Shanghai. doi:10.1109/OCEANSAP.2016.7485469

Liu, T., Li, C., Inoue, Y., & Shibata, K. (2010). A Self-controlled Master-Slave Robot and Its Application for Upper Limb Rehabilitation. Lecture Notes in Electrical Engineering, 75, 219-235. doi:10.1007/978-3-642-15687-8_12

Liu, J., Subramanian, K. R., & Yoo, T. S. (2013). An optical flow approach to tracking colonoscopy video. *Computerized Medical Imaging and Graphics*, *37*(3), 207–223. doi:10.1016/j.compmedimag.2013.01.010 PMID:23490235

Liu, J., Wang, Z., Zhong, L., Wickramasuriya, J., & Vasudevan, V. (2009). uWave Accelerometer-based Personalized Gesture Recognition and Its Applications. *Pervasive and Mobile Computing*, *5*(6), 657–675. doi:10.1016/j.pmcj.2009.07.007

Liu, Q., Zhao, X., & Hou, Z. (2014). Survey of single-target visual tracking methods based on online learning. *IEEE Transactions on Computer Vision*, *8*(5), 419–428. doi:10.1049/iet-cvi.2013.0134

Liu, Y. F., Li, Q., Fang, H., & Xu, H. C. (2014). Research on Embedded System with Implementation of a Moving Object Tracking Algorithm Based on Improved Meanshift on DM6437. *Advanced Materials Research*, *1003*, 207–210. doi:10.4028/www.scientific.net/AMR.1003.207

Y. Liu, & D. Sun (Eds.). (2011). *Biologically inspired robotics*. CRC Press. doi:10.1201/b11365

Li, Y., & Guo, S. (2016). Communication between spherical underwater robots based on the acoustic communication methods. In *Proceedings of the 2016 IEEE International Conference on Mechatronics and Automation (ICMA)*. Harbin: IEEE. doi:10.1109/ICMA.2016.7558597

Lo, A. C., Guarino, P. D., Richards, L. G., Haselkorn, J. K., Wittenberg, G. F., Federman, D. G., ... Peduzzi, P. (2010). Robot-Assisted Therapy for Long-Term Upper-Limb Impairment after Stroke. *The New England Journal of Medicine*, *362*(19), 1772–1783. doi:10.1056/NEJMoa0911341 PMID:20400552

Lo, H. S., & Xie, S. Q. (2012). Exoskeleton robots for upper-limb rehabilitation: State of the art and future prospects. *Medical Engineering & Physics*, *34*(3), 261–268. doi:10.1016/j.medengphy.2011.10.004 PMID:22051085

Looned, R., Webb, J., Xiao, Z. G., & Menon, C. (2014). Assisting drinking with an affordable BCI-controlled wearable robot and electrical stimulation: A preliminary investigation. *Journal of Neuroengineering and Rehabilitation*, *11*(51), 1–13. PMID:24708603

Lorenz, R. D., Jones, J. A., & Wu, J. J. (2003). Mars Magnetometry from a Tumbleweed Rover. *IEEE Aerospace Conference*, 597-602.

Loureiro, R. C., Harwin, W. S., Nagai, K., & Johnson, M. (2011). Advances in upper limb stroke rehabilitation: A technology push. *Medical & Biological Engineering & Computing*, *49*(10), 1103–1118. doi:10.1007/s11517-011-0797-0 PMID:21773806

Low, K. H. (2011). Robot-assisted gait rehabilitation: From exoskeletons to gait systems. *Defense Science Research Conference and Expo (DSR)*, 1–10. doi:10.1109/DSR.2011.6026886

Lu, Q., Ortega, C., & Ma, O. (2011). Passive Gravity Compensation Mechanisms: Technologies and Applications. *Recent Patents on Engineering*, *5*, 32–44. Retrieved from http://cat.inist.fr/?aModele=afficheN&cpsidt=24159482

Lu, E. C., Wang, R. H., Hebert, D., Boger, J., Galea, M. P., & Mihailidis, A. (2011). The development of an upper limb stroke rehabilitation robot: Identification of clinical practices and design requirements through a survey of therapists. *Disability and Rehabilitation. Assistive Technology*, *6*(5), 420–431. doi:10.3109/17483107.2010.544370 PMID:21184626

Luk, B. L., Collie, A. A., Piefort, V., & Virk, G. S. (1996). Robug III: A Tele-Operated Climbing and Walking Robot. *UKACC International Conference on Control*, 1, 347-352. doi:10.1049/cp:19960577

Lum, P., Reinkensmeyer, D., & Lehman, S. (1993). Robotic Assist Devices for Bimanual Physical Therapy: Preliminary Experiments. *IEEE Transactions on Rehabilitation Engineering, 1*(3), 185-191. DOI: 10.1109/86.279267

Lum, P., Lehman, S., & Reinkensmeyer, D. (1995). The bimanual Lifting Rehabilitator: An Adaptive Machine for Therapy of Stroke Patients. *IEEE Transactions on Rehabilitation Engineering*, *3*(2), 166–174. doi:10.1109/86.392371

Lum, P., Van der Loos, M., Shor, P., & Burgar, C. (1999). A robotic system for upper-limb exercises to promote recovery of motor function following stroke. *Proceedings of Sixth International Conference on Rehabalition Robotics*, 235–9.

Luo, M., Pan, Y., Skorina, E. H., Tao, W., Chen, F., Ozel, S., & Onal, C. D. (2015). Slithering towards autonomy: A self-contained soft robotic snake platform with integrated curvature sensing. *Bioinspiration & Biomimetics*, *10*(5), 55001. doi:10.1088/1748-3190/10/5/055001 PMID:26335857

MacDonald, H. N. (1983). *Trees and networks in biological models*. Wiley.

Maciejasz, P., Eschweiler, J., Gerlach-Hahn, K., Jansen-Troy, A., & Leonhardt, S. (2014). A survey on robotic devices for upper limb rehabilitation. *J. Neuroeng. Rehabil, 11*(3).

MacQuitty, M. (2011). *Shark*. Dorling Kindersley Ltd.

Madangopal, R., Khan, Z. A., & Agrawal, S. K. (2005). Biologically Inspired Design of Small Flapping Wing Air Vehicles using Four-Bar Mechanisms and Quasi-Steady Aerodynamics. *Journal of Mechanical Design, 127*, 809–816.

Maddahi, Y. (2005, December). Design and laboratory tests of wheeled mobile robots. In *International Conference on System Science and Simulation in Engineering* (pp. 186-191). Academic Press.

Maddahi, Y., Monsef, S. H., & Mastorakis, N. E. (2014). How to Improve Positional Accuracy in Redundant Omnidirectional Mobile Robots? In *International Conference on Recent Advances in Electrical Engineering and Educational Technologies* (pp. 22-26). Academic Press.

Maddahi, Y., Sepehri, & Maddahi A. (2012c) Calibration of four–wheeled omnidirectional mobile robots. *Int. Conf. of Canadian Society of Mechanical Engineering*.

Maddahi, Y., & Maddahi, A. (2004a). Mobile robot experimental analysis based on kinematics. *WSEAS Transactions on Circuits and Systems, 3*(8), 1662–1667.

Maddahi, Y., & Maddahi, A. (2006). YMBM: New method for errors measurement in wheeled mobile robots. *WSEAS Transactions on Systems, 5*(3), 552–557.

Maddahi, Y., Maddahi, A., & Monsef, S. H. (2012b). Design improvement of wheeled mobile robots: Theory and experiment. *J. World Appl. Sci, 16*(2), 263–274.

Maddahi, Y., Maddahi, A., & Sepehri, N. (2013). Calibration of omnidirectional wheeled mobile robots: Method and experiments. *Robotica, 31*(6), 969–980. doi:10.1017/S0263574713000210

Maddahi, Y., Seddigh, M., Pour, M. M., & Maleki, M. (2004b). Simulation study and laboratory results of two-wheeled-mobile robot. *WSEAS Transactions on Systems, 3*(9), 2807–2812.

Maddahi, Y., Sepehri, N., Maddahi, A., & Abdolmohammadi, M. (2012a). Calibration of wheeled mobile robots with differential drive mechanisms: An experimental approach. *Robotica, 30*(6), 1029–1039. doi:10.1017/S0263574711001329

Maderson, P. F. A. (1965). Histological changes in the epidermis of snakes during the sloughing cycle. *Journal of Zoology, 146*, 98–113.

Maderson, P. F. A. (1966). Histological changes in the epidermis of the Tokay (Gekko gecko) during the sloughing cycle. *Journal of Morphology, 119*(1), 39–50. doi:10.1002/jmor.1051190105

Maderson, P. F. A. (1970). Lizard hands and lizard glands: Models for evolutionary study. *Forma. Et. Functio., 3*, 179–204.

Maderson, P. F. A. (1972). When? Why? and How? Some speculations on the evolution of the vertebrate integument. *American Zoologist, 12*(1), 159–171. doi:10.1093/icb/12.1.159

Maderson, P. F. A., & Licht, P. (1967). The epidermal morphology and sloughing frequency in normal and prolactin injected Anolis carolizensis (Iguanidae, Lacertilia). *Journal of Morphology, 123*(2), 157–172. doi:10.1002/jmor.1051230205 PMID:6073684

Maderson, P. F. A., Rabinowitz, T., Tandler, B., & Alibardi, L. (1998). Ultrastructural contributions to an understanding of the cellular mechanisms involved in lizard skin shedding with comments on the function and evolution of a unique Lepidosaurian phenomenon. *Journal of Morphology*, *236*(1), 1–24. doi:10.1002/(SICI)1097-4687(199804)236:1<1::AID-JMOR1>3.0.CO;2-B

Maeda, Y., Fujiwara, T., & Ito, H. (2014, September). Robot control using high dimensional neural networks. In *Proceedings of SICE Annual Conference 2014 -International Conference on Instrumentation, Control, Information Technology and System Integration-* (pp. 738–743). Academic Press. doi:10.1109/SICE.2014.6935220

Maffiuletti, N. A., Minetto, M. A., Farina, D., & Bottinelli, R. (2011). Electrical stimulation for neuromuscular testing and training: State-of-the art and unresolved issues. *European Journal of Applied Physiology*, *111*(10), 2391–2397. doi:10.1007/s00421-011-2133-7 PMID:21866361

Magee, D., Zhu, Y., Ratnalingam, R., Gardner, P., & Kessel, D. (2007). An augmented reality simulator for ultrasound guided needle placement training. *Medical & Biological Engineering & Computing*, *45*(10), 957–967. doi:10.1007/s11517-007-0231-9 PMID:17653784

Maggiore, M., & Consolini, L. (2013). Virtual Holonomic Constraints for Euler–Lagrange Systems. *IEEE Transactions on Automatic Control*, *58*(4), 1001–1008. doi:10.1109/TAC.2012.2215538

Major, K. A., Major, Z. Z., Carbone, G., Pîslă, A., Vaida, C., Gherman, B., & Pîslă, D. (2016). Ranges of motion as basis for robot-assisted post-stroke rehabilitation. *In. J Bioflux Society, Human & Veterinary Medicine*, *8*(4), 192–196.

Malešević, N. M., Maneski, L. Z. P., Ilić, V., Jorgovanović, N., Bijelić, G., Keller, T., & Popović, D. B. (2012). A multi-pad electrode based functional electrical stimulation system for restoration of grasp. *Journal of Neuroengineering and Rehabilitation*, *9*(1), 1. doi:10.1186/1743-0003-9-66 PMID:23009589

Malosio, M., Corbetta, F., & Ram, F. (n.d.). *The kinematic architecture of a two-dof parallel and orthogonal variable stiffness actuator.* Academic Press.

Malosio, M., Caimmi, M., Cotti Cottini, M., Crema, A., Dinon, T., Mihelj, M., ... Spagnuolo, G. (2016). An affordable, adaptable, and hybrid assistive device for upper-limb neurorehabilitation. *Journal of Rehabilitation and Assistive Technologies Engineering*, *3*, 205566831668098. doi:10.1177/2055668316680980

Malosio, M., Pedrocchi, N., Vicentini, F., & Tosatti, L. M. (2011). Analysis of elbow-joints misalignment in upper-limb exoskeleton. *IEEE International Conference on Rehabilitation Robotics*, 1–6. doi:10.1109/ICORR.2011.5975393

Manzoor, S., & Choi, Y. (2016). A unified neural oscillator model for various rhythmic locomotions of snake-like robot. *Neurocomputing*, *173*, 1112–1123. doi:10.1016/j.neucom.2015.08.069

Mao, C., Solis, D. J., Reiss, B. D., Kottmann, S. T., Sweeney, R. Y., Hayhurst, A., ... Belcher, A. M. (2004). Virus-Based Toolkit for the Directed Synthesis of Magnetic and Semiconducting Nanowires. *Science*, *303*(5655), 213–217. doi:10.1126/science.1092740 PMID:14716009

Mao, Z., & Hsia, T. C. (1997, January). Obstacle avoidance inverse kinematics solution of redundant robots by neural networks. *Robotica*, *15*(1), 3–10. doi:10.1017/S0263574797000027

Marantos, P., Koveos, Y., Stergiopoulos, J., Panousopoulou, A., & Tzes, A. (2008, June). Mobile robot odometry relying on data fusion from RF and ultrasound measurements in a wireless sensor framework. In *Control and Automation, 2008 16th Mediterranean Conference on* (pp. 523-528). IEEE. doi:10.1109/MED.2008.4602252

Marchal-Crespo, L., & Reinkensmeyer, D. J. (2009). Review of control strategies for robotic movement training after neurologic injury. *Journal of Neuroengineering and Rehabilitation*, *6*(1), 20. doi:10.1186/1743-0003-6-20 PMID:19531254

Martins, M., Araujo, M. S., Sawaya, R. J., & Nunes, R. (2001). Diversity and evolution of macrohabitat use, body size and morphology in a monophyletic group of Neotropical pitvipers (Bothrops). *Journal of Zoology*, *254*(4), 529–538. doi:10.1017/S0952836901001030

Marton, F., & Booth, S. (1997). Learning and awareness. Abingdon-on-Thames, UK: Routledge.

Marvi, H., & Hu, D. L. (2012). Friction enhancement in concertina locomotion of snakes. *Journal of The Royal Society Interface*.

Marvi, H., Bridges, J., Meyers, G., Russell, G., & Hu, D. L. (2011). Scalybot: A snake-inspired robot with active control of friction. Proc. Of ASME Dynamic Systems and Control Conf., 443–450. doi:10.1115/DSCC2011-6174

Marvi, H., Bridges, J., & Hu, D. L. (2013). Snakes mimic earthworms: Propulsion using rectilinear travelling waves. *Journal of the Royal Society, Interface*, *10*(84), 20130188. doi:10.1098/rsif.2013.0188 PMID:23635494

Marvi, H., Gong, C., Gravish, N., Astley, H., Travers, M., Hatton, R. L., ... Goldman, D. I. (2014). Sidewinding with minimal slip: Snake and robot ascent of sandy slopes. *Science*, *346*(6206), 224–229. doi:10.1126/science.1255718 PMID:25301625

Ma, S., Araya, H., & Li, L. (2001). Development of a creeping snake-robot. *Proceedings of IEEE International Symposium on Computational Intelligence in Robotics and Automation, CIRA*. doi:10.1109/CIRA.2001.1013176

Masiero, S., Celia, A., Rosati, G., & Armani, M. (2005). Robotic-Assisted Rehabilitation of the Upper Limb After Acute Stroke. Archives of Physical Medicine and Rehabilitation, 88(2), 142-149.

Masoud, A. (2010). Kinodynamic motion planning. *IEEE Robotics & Automation Magazine*, *17*(1), 85–99. doi:10.1109/MRA.2010.935794

Massie, T., & Salisbury, J. K. (1994). The PHANTOM haptic interface: A device for probing virtual objects. *Proc. ASME Winter Annual Meeting: Symposium on Haptic Interfaces for Virtual Environment and Teleoperator Systems*.

Mattison, C. (2006). *The Encyclopedia of Snakes*. HarperCollins Publishers Limited.

Mcculloch, P., Cook, J. A., Altman, D. G., Heneghan, C., & Diener, M. K. (2013). IDEAL framework for surgical innovation 1: The idea and development stages. *BMJ (Clinical Research Ed.)*, *346*. doi:10.1136/bmj.f3012 PMID:23778427

Meister, E., Stepanenko, S., & Kernbach, S. (2011). *Adaptive locomotion of multibody snake-like robot.* arXiv Preprint arXiv:1108.6175

Meyer, J. A., & Guillot, A. (2008). Biologically inspired robots. In *Springer Handbook of Robotics* (pp. 1395–1422). Springer Berlin Heidelberg. doi:10.1007/978-3-540-30301-5_61

Michaud, F., & Caron, S. (2002). Roball, the Rolling Robot. *Journal of Autonomous Robots*, *12*(2), 211–222.

Michelson, R. C., & Naqvi, M. A. (2003). Beyond Biologically-Inspired Insect Flight. von Karman Institute for Fluid Dynamics RTO/AVT Lecture Series on Low Reynolds Number Aerodynamics on Aircraft Including Applications in Emerging UAV Technology.

Michelson, R. C., & Reece, S. (1998). Update on Flapping Wing Micro Air Vehicle Research-Ongoing Work to Develop a Flapping Wing, Crawling Entomopter. *13th Bristol International RPV/UAV Systems Conference*, 30.1-30.12.

Mikołajewska, E., & Mikołajewski, D. (2011). Exoskeletons in neurological diseases-current and potential future applications. *Advances in Clinical and Experimental Medicine*, *20*(2), 227–233.

Mohammadi, A., Rezapour, E., Maggiore, M., & Pettersen, K. Y. (2016). Maneuvering Control of Planar Snake Robots Using Virtual Holonomic Constraints. *IEEE Transactions on Control Systems Technology*, *24*(3), 884–899. doi:10.1109/TCST.2015.2467208

Molinari-Tosatti, L., Fassi, I., & Legnani, G. (2003). Kineto-static optimisation of PKMs. *CIRP Annals - Manufacturing Technology, 52*(1), 337–341. 10.1016/S0007-8506(07)60596-4

Monga, M., Dretler, S. P., Landman, J., Slaton, J. W., Conradie, M. C., & Clayman, R. V. (2002). Maximizing uretero-scope deflection: "play it straight. *Urology*, *60*(5), 902–905. doi:10.1016/s0090-4295(02)01891-5

Moon, B. R., & Gans, C. (1998). Kinematics, muscular activity and propulsion in gopher snakes. *The Journal of Experimental Biology*, *201*(19), 2669–2684. PMID:9732322

Morasso, P., Casadio, M., Giannoni, P., Masia, L., Sanguineti, V., Squeri, V., & Vergaro, E. (2009). Desirable features of a "Humanoid" robot-therapist. *Proceedings of the 31st Annual International Conference of the IEEE Engineering in Medicine and Biology Society: Engineering the Future of Biomedicine, EMBC 2009*, 2418–2421. doi:10.1109/IEMBS.2009.5334954

Mosauer, W. (1932). On the locomotion of snakes. *Science, 76*(1982), 583–585.

Mosauer, W. (1930). A note on the sidewinding locomotion of snakes. *American Naturalist*, *64*(691), 179–183. doi:10.1086/280308

Motonaka, K., Goto, T., Watanabe, K., & Maeyama, S. (2016). Kinodynamic motion planning for a two-drive mobile robot using a harmonic potential field. In *Proceedings of Annual Conference of the IEEE Industrial Electronics Society*. IEEE. doi:10.1109/IECON.2016.7793160

Motonaka, K., Watanabe, K., & Maeyama, S. (2013). Kinodynamic motion planning and control using anisotropic damping forces. In *Proceedings of International Symposium on Artificial Life and Robotics*, (pp. 57–60). Oita, Japan: Academic Press. doi:10.1109/IROS.2013.6696989

Muratori, L. M., Lamberg, E. M., Quinn, L., & Duff, S. V. (2014). Article. *Extremity Rehabilitation*, *26*(2), 94–103. doi:10.1016/j.jht.2012.12.007.Applying

Mutlu, M., Melo, K., Vespignani, M., Bernardino, A., & Ijspeert, A. J. (2016). Where to place cameras on a snake robot: Focus on camera trajectory and motion blur. *IEEE International Symposium on Safety, Security, and Rescue Robotics*. doi:10.1109/SSRR.2015.7442948

Nagata, F., Inoue, S., Fujii, S., Otsuka, A., Watanabe, K., & Habib, M. K. (2015, August). *Learning of inverse kinematics using a neural network and its application to kinematic control of position-based servo motor*. Paper presented at The 2015 World Congress on Advances in Aeronautics, Nano, Bio, Robotics, and Energy (ANBRE15), Incheon, South Korea.

Nagata, F., Otsuka, A., Sakakibara, K., Watanabe, K., & Habib, M. K. (2013, September), Experiment Systems Using Three Types of Motors for Biomimetic Machine Research, In *Proceedings of SICE Annual Conference 2013 -International Conference on Instrumentation, Control, Information Technology and System Integration* (pp. 2711–2717). Academic Press.

Nagata, F., & Watanabe, K. (2016). Neural network-based inverse kinematics for Motoman HS20 and its efficient learning method. *Journal of the Institute of Industrial Applications Engineers*, *4*(4), 166–171. doi:10.12792/JIIAE.4.166

Nakai, A., Ohashi, T., & Hashimoto, H. (1998). 7 dof arm type haptic interface for teleoperation and virtual reality systems. *Proc. IEEE/RSJ International Conference on Intelligent Robots and Systems*, 2, 1266–1271. doi:10.1109/IROS.1998.727473

NASA-JPL. (2000). *Leaping into the Future: One Hop at a Time*. Retrieved from http://www.jpl.nasa.gov/releases/2000/frog.html

NASA-JPL. (2006). *Limber Robot Might Hitchhike to Space*. Retrieved from http://www.jpl.nasa.gov/news/features.cfm?feature=1101

Nef, Mihelji, & Riener. (2007). Armin: A robot for patient-cooperative arm therapy. *Medical & Biological Engineering & Computing, 45*, 887–900.

Neuhaus, P. D., O'Sullivan, M., Eaton, D., Carff, J., & Pratt, J. E. (2004). Concept Designs for Underwater Swimming Exoskeletons. *The IEEE International Conference on Robotics and Automation (ICRA'2004)*, 4893-4898. doi:10.1109/ROBOT.2004.1302493

Neves, J. C. L., & Francke, I. C. M. (2012). Creative Product Design Using Biomimetics. *WIT Transactions on Ecology and the Environment, 160*, 149–155. doi:10.2495/DN120141

Ng, T. H., & Liao, W. H. (2005). Sensitivity Analysis and Energy Harvesting for a Self-Powered Piezoelectric Sensor. *Journal of Intelligent Material Systems and Structures, 16*(10), 785–797. doi:10.1177/1045389X05053151

Nicolau. (2005). A complete augmented reality guidance system for liver punctures: First clinical evaluation. *Medical Image Computing and Computer-Assisted Intervention*, 539–547.

Nicolau, S. A., Pennec, X., Soler, L., Buy, X., Gangi, A., Ayache, N., & Marescaux, J. (2009). An augmented reality system for liver thermal ablation: Design and evaluation on clinical cases. *Medical Image Analysis, 13*(3), 494–506. doi:10.1016/j.media.2009.02.003 PMID:19282234

Nicolau, S., Soler, L., Mutter, D., & Marescaux, J. (2011). Augmented reality in laparoscopic surgical oncology. *Surgical Oncology, 20*(3), 189–201. doi:10.1016/j.suronc.2011.07.002 PMID:21802281

Nikou, C., Digioia, A. M. III, Blackwell, M., Jaramaz, B., & Kanade, T. (2000). Augmented reality imaging technology for orthopaedic surgery. *Operative Techniques in Orthopaedics, 10*(1), 82–86. doi:10.1016/S1048-6666(00)80047-6

Nistér, D., Naroditsky, O., & Bergen, J. (2006). Visual odometry for ground vehicle applications. *Journal of Field Robotics, 23*(1), 3–20. doi:10.1002/rob.20103

Nor, N. M., & Ma, S. (2014). Smooth transition for CPG-based body shape control of a snake-like robot. *Bioinspiration & Biomimetics, 9*(1), 16003. doi:10.1088/1748-3182/9/1/016003 PMID:24343201

Ohno, H., & Hirose, S. (2000). Study on Slim Slime Robot. *Proceedings of RSJ/IEEE International Conference on Intelligent Robots and Systems (IROS'2000)*. 2218-2223.

Okamoto, M., Shima, K., Matsubara, Y., & Tsuji, T. (2008). Pattern discrimination method with a boosting approach using hierarchical neural trees. *Proceedings of the Institution of Mechanical Engineers. Part I, Journal of Systems and Control Engineering, 222*(7), 701–710. doi:10.1243/09596518JSCE563

Okazaki, K., Ogiwara, T., Yang, D., Sakata, K., Saito, K., Sekine, Y., & Uchikoba, F. (2011). Development of pulse control type MEMS micro robot with hardware neural network. *Artificial Life and Robotics, 16*(2), 229–233. doi:10.1007/s10015-011-0925-9

Oonishi, S., Yanou, A., & Minami, M. (2013). Human-face-tracking using visual servoing by patient robot. In *Proceedings of 2013 SICE Annual Conference (SICE)*. Nagoya: IEEE.

Ottaviano, E., Ceccarelli, M., Paone, A., & Carbone, G. (2005). A Low-Cost Easy-Operation 4-Cable Driven Parallel Manipulator. *IEEE International Conference on Robotics and Automation ICRA'05*, 4019-4024. doi:10.1109/ROBOT.2005.1570734

Palmer, L. R., Diller, E. D., & Quinn, R. D. (2010). Toward a rapid and robust attachment strategy for vertical climbing. *Proc. of the IEEE Int. Conf. on Robotics and Automation*, 2810–2815. doi:10.1109/ROBOT.2010.5509680

Panovko, G. Y., Savin, S. I., Yatsun, S. F., & Yatsun, A. S. (2016). Simulation of exoskeleton sit-to-stand movement. *Journal of Machinery Manufacture and Reliability*, *45*(3), 206–210. doi:10.3103/S1052618816030110

Pan, S., Guo, S., Shi, L., He, Y., Wang, Z., & Huang, Q. (2014). A Spherical Robot based on all Programmable SoC and 3-D Printing. In *Proceedings of 2014 IEEE International Conference on Mechatronics and Automation*. Tianjin: IEEE. doi:10.1109/ICMA.2014.6885687

Pan, S., Shi, L., & Guo, S. (2015). A Kinect-Based Real-Time Compressive Tracking Prototype System for Amphibious Spherical Robots. *Sensors (Basel)*, *15*(4), 8232–8252. doi:10.3390/s150408232 PMID:25856331

Papanek, V. (1971). *Design for the real world: human ecology and social change*. London: Thames and Hudson.

Park, J. H., & Kim, K. D. (1998). Biped Robot Walking Using Gravity-Compensated Inverted Pendulum Mode and Computed Torque Control. *The IEEE International Conference on Robotics and Automation (ICRA'1998)*, 3528–3533. doi:10.1109/ROBOT.1998.680985

Park, W., Jeong, W., Kwon, G. H., Kim, Y. H., & Kim, L. (2013) A Rehabilitation Device to Improve the Hand Grasp Function of Stroke using a Patient-Driven Approach. *Proceeding of International Conference on Rehabilitation Robotics*. Doi:10.1109/ICORR.2013.6650482

Patel, A., & Fuchs, G. J. (1997). Expanding the Horizons of SWL Through Adjunctive Use of Retrograde Intrarenal Surgery: New Techniques and Indications. *Journal of Endourology*, *11*(1), 33–36. doi:10.1089/end.1997.11.33

Patton, J. L., Kovic, M., & Mussa-Ivaldi, F. A. (2006). Custom-designed haptic training for restoring reaching ability. *Journal of Rehabilitation Research and Development*, *43*(5), 643–656. doi:10.1682/JRRD.2005.05.0088 PMID:17123205

Patton, J., Small, S. L., & Zev Rymer, W. (2008). Functional restoration for the stroke survivor: Informing the efforts of engineers. *Topics in Stroke Rehabilitation*, *15*(6), 521–541. doi:10.1310/tsr1506-521 PMID:19158061

Paulson, L. D. (2004). *Biomimetic Robots*. IEEE Computer.

Pedrocchi, A., Ferrante, S., Ambrosini, E., Gandolla, M., Casellato, C., Schauer, T., ... Gföhler, M. (2013). MUNDUS project: MUltimodal Neuroprosthesis for daily Upper limb Support. *Journal of Neuroengineering and Rehabilitation*, *10*(1), 66. doi:10.1186/1743-0003-10-66 PMID:23822118

Peng, J., & Akella, S. (2005). Coordinating multiple robots with kinodynamic constraints along specified paths. *The International Journal of Robotics Research*, *24*(4), 295–310. doi:10.1177/0278364905051974

Perry, J., Rosen, J., & Burns, S. (2007, August). Upper-limb powered exoskeleton design. *IEEE/ASME Transactions on Mechatronics*, *12*(4), 408–417. doi:10.1109/TMECH.2007.901934

Petuskey, K., Bagley, A., Abdala, E., James, M. A., & Rab, G. (2007). Upper extremity kinematics during functional activities: Three-dimensional studies in a normal pediatric population. *Gait & Posture*, *25*(4), 573–579. doi:10.1016/j.gaitpost.2006.06.006 PMID:16875821

Pfeifer, R., Iida, F., & Gomez, G. (2006). Morphological computation for adaptive behavior and cognition. *International Congress Series*, *1291*, 22–29. doi:10.1016/j.ics.2005.12.080

Pfeifer, R., Lungarella, M., & Iida, F. (2007). Self-organization, embodiment, and biologically inspired robotics. *Science*, *318*(5853), 1088–1093. doi:10.1126/science.1145803 PMID:18006736

Piaggio, M., Sgorbissa, A., & Zaccaria, R. (2001). Navigation and localization for service mobile robots based on active beacons. *Systems Science*, *27*(4), 71–83.

Pinna, L., Dahiya, R. S., De Nisi, F., & Valle, M. (2010). Analysis of self-powered vibration-based energy scavenging system. *IEEE International Symposium on Industrial Electronics*. doi:10.1109/ISIE.2010.5637866

Pirzada, N., Nayan, M. Y., Hassan, F. S. M. F., & Khan, M. A. (2014). Device-free Localization Technique for Indoor Detection and Tracking of Human Body: A Survey. *Procedia: Social and Behavioral Sciences*, *129*, 422–429. doi:10.1016/j.sbspro.2014.03.696

Plitea, N., Gherman, B., Carbone, G., Ceccarelli, M., Vaida, C., Banica, A., & Pisla, D. (2017). Kynematic analysis of an exoskeleton-based robot for elbow and wrist rehabilitation. *6th International Symposium on Multibody Systems and Mechatronics-MuSMe*.

Plitea, N., Pisla, D., Carbone, G., Vaida, C., Gherman, B., & Banica, A. (2017). *Robotic system for elbow and wrist rehabilitation*. Patent pending.

Plitea, N., Pisla, D., Carbone, G., Vaida, C., Gherman, B., & Ulinici, I. (2017). *Spherical mechanism for shoulder rehabilitation*. Patent pending

Plitea, N., Vaida, C., Carbone, G., Pisla, A., Ulinici, I., & Pisla, D. (2017). On the Kinematics of an Innovative Spherical Parallel Robot for the Shoulder Rehabilitation. *6th International Symposium on Multibody Systems and Mechatronics-MuSMe*.

Pons, J. L. (2008). *Wearable robots: biomechatronic exoskeletons*. John Wiley & Sons. doi:10.1002/9780470987667

Pons, J. L., Ceres, R., & Calderón, L. (2008). Introduction to Wearable Robotics. In J. L. Pons (Ed.), *Wearable Robots: Biomechatronic Exoskeletons*. Chichester, UK: John Wiley & Sons, Ltd; doi:10.1002/9780470987667.ch1

Pough, F. (1980). The advantage of ectothermy for tetrapods. *American Naturalist*, *115*(1), 92–112. doi:10.1086/283547

Pough, F. H., & Groves, J. D. (1983). Specialization of the body form and food habits of snakes. *American Zoologist*, *23*(2), 443–454. doi:10.1093/icb/23.2.443

Preminger G. M., Tiselius H. G., Assimos D. G., Alken P., Buck C., . . . Wolf J. S. (2007) Guideline for the Management of Ureteral Calculi. *American Urological Association, 178*(6), 2418-2434.doi:10.1016/j.juro.2007.09.107

Proietti, T., Crocher, V., Roby-Brami, A., & Jarrasse, N. (2016). Upper-limb robotic exoskeletons for neurorehabilitation: A review on control strategies. *IEEE Reviews in Biomedical Engineering*, *9*, 4–14. doi:10.1109/RBME.2016.2552201 PMID:27071194

Pugin, F., Bucher, P., & Morel, P. (2011). History of robotic surgery: From AESOP® and ZEUS® to da Vinci®. *Journal of Visceral Surgery*, *148*(5), e3–e8. doi:10.1016/j.jviscsurg.2011.04.007 PMID:21974854

Pylatiuk, C., Kargov, A., Gaiser, I., Werner, T., Schulz, S., & Bretthauer, G. (2009). *Design of a flexible fluidic actuation system for a hybrid elbow orthosis*. Paper presented at the Rehabilitation Robotics, 2009. ICORR 2009. IEEE International Conference on. doi:10.1109/ICORR.2009.5209540

Qi, Y., Suzuki, K., Wu, H., & Chen, Q. (2011). EK-means tracker: A pixel-wise tracking algorithm using kinect. In *Proceedings of 2011 Third Chinese Conference on Intelligent Visual Surveillance (IVS)*. Beijing: IEEE.

Quigley, M., Conley, K., Gerkey, B., Faust, J., Foote, T., & Leibs, J. … Ng, A. Y. (2009). ROS: an open-source Robot Operating System. In ICRA workshop on open source software (Vol. 3, p. 5). Kobe, Japan: ICRA.

Raheja, L., Shyam, R., Kumar, U., & Prasad, P. B. (2010). Real-time robotic hand control using hand gestures. *Proceedings of the 2010 Second International Conference on Machine Learning and Computing (ICMLC '10)*, 12-16. doi:10.1109/ICMLC.2010.12

Rahman, Ouimet, Saad, & Kenne. (2010). *Development and Control of a Wearable Robot for Rehabilitation of Elbow and Shoulder Joint Movements*. IEEE.

Rahmani, M., Ghanbari, A., & Ettefagh, M. M. (2016). Robust adaptive control of a bio-inspired robot manipulator using bat algorithm. *Expert Systems with Applications*, *56*, 164–176. doi:10.1016/j.eswa.2016.03.006

Rahok, S. A., & Koichi, O. (2009, February). Odometry correction with localization based on landmarkless magnetic map for navigation system of indoor mobile robot. In *Autonomous Robots and Agents, 2009. ICARA 2009. 4th International Conference on* (pp. 572-577). IEEE.

Raibert, M. H. (1986). *Legged Robots that Balance*. Cambridge, MA: MIT Press.

Rajamahanty, S., & Grasso, M. (2008). Flexible ureteroscopy update: Indications, instrumentation and technical advances. *Indian Journal of Urology*, *24*(4), 532. doi:10.4103/0970-1591.44263 PMID:19468513

Ramsay, M. J., & Clark, W. W. (2001). Piezoelectric Energy Harvesting for Bio MEMS Applications. In *Energy Harvesting Technologies* (pp. 429–438). New York: Springer.

Ranchev. (2015). *A fluid drive device*. Patent Application (BG, N°112191/28.12.2015).

Rashedi, E., Mirbagheri, B., Taheri, B., Farahmand, F., Vossoughi, G. R., & Parnianpour, M. (2009) Design and Development of a Hand Robotic Rehabilitation Device for Post Stroke Patients. In *Proceeding of the Annual International Conference of the IEEE Engineering in Medicine and Biology Society*. IEEE Engineering in Medicine and Biology Society. Doi:10.1109/IEMBS.2009.5333827

Rassweiler, J. J., Goezen, A. S., Jalal, A. A., Schulze, M., Pansadoro, V., Pini, G., ... Turner, C. (2012). A New Platform Improving the Ergonomics of Laparoscopic Surgery: Initial Clinical Evaluation of the Prototype. *European Urology*, *61*(1), 226–229. doi:10.1016/j.eururo.2011.09.018 PMID:21978421

Rassweiler, J., Safi, K. C., Subotic, S., Teber, D., & Frede, T. (2005). Robotics and telesurgery – an update on their position in laparoscopic radical prostatectomy. *Minimally Invasive Therapy & Allied Technologies*, *14*(2), 109–122. doi:10.1080/13645700510010908 PMID:16754625

Rau, G., Disselhorst-Klug, C., & Schmidt, R. (2000). Movement biomechanics goes upwards: From the leg to the arm. *Journal of Biomechanics*, *33*(10), 1207–1216. doi:10.1016/S0021-9290(00)00062-2 PMID:10899329

Reinkensmeyer, D. J., Dewald, J. P. A., & Rymer, W. Z. (1999). Guidance-based quantification of arm impairment following brain injury: A pilot study. *IEEE Transactions on Rehabilitation Engineering*, *7*(1), 1–11. doi:10.1109/86.750543 PMID:10188602

Reinkensmeyer, D., Lehman, S., & Lum, P. (1993) A bimanual therapy robot: controller design and prototype experiments. In *Proceedings of the 15th Annual International Conference of the IEEE*. Engineering in Medicine and Biology Society. Doi:10.1109/IEMBS.1993.978943

Ren, H. L., Liu, W., & Lim, A. (2014). Marker-Based Surgical Instrument Tracking Using Dual Kinect Sensors. *IEEE Transactions on Automation Science and Engineering*, *11*(3), 921–924.

Resquín, F., Gómez, A. C., Gonzalez-Vargas, J., Brunetti, F., Torricelli, D., Rueda, F. M., ... Pons, J. L. (2016). Hybrid robotic systems for upper limb rehabilitation after stroke: A review. *Medical Engineering & Physics*, *38*(11), 1279–1288. doi:10.1016/j.medengphy.2016.09.001 PMID:27692878

Rezapour, E., Pettersen, K. Y., Gravdahl, J. T., & Liljeback, P. (2014). Body shape and orientation control for locomotion of biologically-inspired snake robots. In *Biomedical Robotics and Biomechatronics 2014 5th IEEE RAS & EMBS International Conference on* (pp. 993-1000). IEEE. doi:10.1109/BIOROB.2014.6913910

Richard, M. D., & Lippmann, R. P. (1991). Neural network classifiers estimate Bayesian *a posteriori* probabilities. *Neural Computation*, *3*(4), 461–483. doi:10.1162/neco.1991.3.4.461

Rimon, E., & Koditschek, D. (1992, October). Exact robot navigation using artificial potential functions. *IEEE Transactions on Robotics and Automation*, *8*(5), 501–518. doi:10.1109/70.163777

Rivera, G., Savitzky, A. H., & Hinkley, J. A. (2005). Mechanical properties of the integument of the common gartersnake, Thamnophis sirtalis (Serpentes: Colubridae). *The Journal of Experimental Biology*, *208*(15), 2913–2922. doi:10.1242/jeb.01715 PMID:16043596

Rocon, E., Belda-Lois, J., Ruiz, A., Manto, M., Moreno, J. C., & Pons, J. (2007). Design and validation of a rehabilitation robotic exoskeleton for tremor assessment and suppression. *Neural Systems and Rehabilitation Engineering. IEEE Transactions on*, *15*(3), 367–378. PMID:17894269

Roderick, H. (1988). Myths and Symbols in Pagan Europe: Early Scandinavian and Celtic Religions. Manchester University Press.

Rohmer, E., Singh, S. P. N., & Freese, M. (2013). V-REP: A versatile and scalable robot simulation framework. In *Intelligent Robots and Systems (IROS), 2013 IEEE/RSJ International Conference on* (pp. 1321–1326). IEEE. doi:10.1109/IROS.2013.6696520

Rohm, M., Schneiders, M., Kreilinger, A., Müller-Putz, G., & Rupp, R. (2012). First evaluation results of a BCI-controlled hybrid neuroprosthesis for restoration of grasping in a high spinal cord injured individual. *Proceedings of TOBI Workshop LII.*

Rohm, M., Schneiders, M., Müller, C., Kreilinger, A., Kaiser, V., Müller-Putz, G., & Rupp, R. (2013). Hybrid brain–computer interfaces and hybrid neuroprostheses for restoration of upper limb functions in individuals with high-level spinal cord injury. *Artificial Intelligence in Medicine*, *59*(2), 133–142. doi:10.1016/j.artmed.2013.07.004 PMID:24064256

Rollinson, D., Alwala, K. V., Zevallos, N., & Choset, H. (2014). Torque control strategies for snake robots. *IEEE International Conference on Intelligent Robots and Systems*. doi:10.1109/IROS.2014.6942694

Rollinson, D., Bilgen, Y., Brown, B., Enner, F., Ford, S., & Layton, C., … Choset, H. (2014). Design and architecture of a series elastic snake robot. *IEEE International Conference on Intelligent Robots and Systems*. doi:10.1109/IROS.2014.6943219

Rollinson, D., Buchan, A., & Choset, H. (2012). Virtual Chassis for Snake Robots: Definition and Applications. *Advanced Robotics*, *26*(17), 2043–2064. doi:10.1080/01691864.2012.728695

Rollinson, D., & Choset, H. (2016). Pipe Network Locomotion with a Snake Robot. *Journal of Field Robotics*, *33*(3), 322–336. doi:10.1002/rob.21549

Rong, W., Tong, K., Hu, X., & Ho, N. (2012). *Combined Electromyography (EMG)-driven robotic system with Functional Electrical Stimulation (FES) for rehabilitation.* Paper presented at the Bioengineering Conference (NEBEC), 2012 38th Annual Northeast.

Rong, W., Tong, K. Y., Hu, X. L., & Ho, S. K. (2013). Effects of electromyography-driven robot-aided hand training with neuromuscular electrical stimulation on hand control performance after chronic stroke. *Disability and Rehabilitation. Assistive Technology*, *10*(2), 149–159. doi:10.3109/17483107.2013.873491 PMID:24377757

Rose, J., & Gamble, J. G. (1994). *Human Walking. 2nd*. Baltimore, MD: Williams and Williams.

Rossa. (2016). Adaptive quasi-static modelling of needle deflection during steering in soft tissue. *IEEE Robotics and Automation Letters*.

Rossa, C., & Tavakoli, M. (2017). Issues in closed-loop needle steering. *Control Engineering Practice*, *62*, 55–69. doi:10.1016/j.conengprac.2017.03.004

Ross, D. A., Lim, J. R., Lin, S., & Yang, M. H. (2008). Incremental learningfor robust visual tracking. *International Journal of Computer Vision*, *77*(1), 125–141. doi:10.1007/s11263-007-0075-7

Roux, E., Bouilland, S., Godillon-Maquinghen, A.-P., & Bouttens, D. (1999).. . *Evaluation of the GO Method Within the Upper Limb Kinematics Analysis*, *35*(1), 1–14.

Royal College of Physicians. (2010). *Medical rehabilitation in 2011 and beyond. Report of a working party*. London: RCP.

Rupp, R., Rohm, M., Schneiders, M., Weidner, N., Kaiser, V., Kreilinger, A., & Müller-Putz, G. (2013). *Think2grasp-BCI-controlled neuroprosthesis for the upper extremity*. Biomedical Engineering/Biomedizinische Technik.

Russell, S., & Norvig, P. (2009). *Artificial intelligence: A modern approach* (3rd ed.). Upper Saddle River, NJ: Prentice Hall.

Ryu, J., Jeong, Y., Tap, Y., Kim, B., Kim, B., & Park, J. (2002). A Ciliary Motion Based 8-legged Walking Micro Robot Using Cast IPMC Actuators. *Proc. of 2002 International Symposium on Micromechatronics and Human Science*.

Ryu, J.-K., Chong, N. Y., You, B. J., & Christensen, H. I. (2009). Locomotion of snake-like robots using adaptive neural oscillators. *Intelligent Service Robotics*, *3*(1). doi:10.1007/s11370-009-0049-4

Saadon, S., & Sidek, O. (2011). A review of vibration-based MEMS piezoelectric energy harvesters. *Energy Conversion and Management*, *52*(1), 500–504. doi:10.1016/j.enconman.2010.07.024

Saglam, R., Muslumanoglu, A. Y., Tokatlı, Z., Çaşkurlu, T., Sarica, K., Taşçi, A. İ., ... Traxer, O. (2014). A new robot for flexible ureteroscopy: Development and early clinical results (IDEAL stage 1–2b). *European Urology*, *66*(6), 1092–1100. doi:10.1016/j.eururo.2014.06.047 PMID:25059998

Saito, K., Maezumi, K., Naito, Y., Hidaka, T., Iwata, K., Okane, Y., ... Uchikoba, F. (2014). Neural Networks Integrated Circuit for Biomimetics MEMS Microrobot. *Robotics*, *3*(3), 235–246. doi:10.3390/robotics3030235

Saito, K., Sugita, K., Ishihara, Y., Iwata, K., Asano, Y., Okane, Y., ... Uchikoba, F. (2017). Insect-Type MEMS Microrobot with Mountable Bare Chip IC of Artificial Neural Networks. *Artificial Life and Robotics, Springer*, *22*(1), 118–124. doi:10.1007/s10015-016-0324-3

Saito, K., Takato, M., Sekine, Y., & Uchikoba, F. (2012). Biomimetics Micro Robot with Active Hardware Neural Networks Locomotion Control and Insect-Like Switching Behaviour. *International Journal of Advanced Robotic Systems, In Tech*, *9*(226), 1–6.

Sale, P., Lombardi, V., & Franceschini, M. (2012). Hand Robotics Rehabilitation: Fesability and Preliminary Results of a Robotic Treatment in Patients with Hemiparesis. *Stroke Research and Treatment*. 10.1155/2012/820931

Salti, S., Cavallaro, A., & Stefano, L. D. (2012). Adaptive Appearance Modeling for Video Tracking: Survey and Evaluation. *IEEE Transactions on Image Processing*, *21*(10), 4334–4348. doi:10.1109/TIP.2012.2206035 PMID:22759454

Sankay, Y. (2007). HAL: Hybrid Assistive Limb based on Cybernics. New York, NY: Springer

Sarfati, J. (2003). Gecko Foot Design—Could it Lead to a Real Spiderman'? *Creation*, *26*(1), 54–55.

Sauer. (2002). An augmented reality navigation system with a single-camera tracker: System design and needle biopsy phantom trial. *Medical Image Computing and Computer-Assisted intervention*, 116–124.

Sauvage, J.-P., & Dietrich-Buchecker, C. (1999). Molecular Catenanes Rotaxanes and Knots. Wiely-VCH. doi:10.1002/9783527613724

Sawada, H. (1998). Japanese sign-language recognition based on gesture primitives using acceleration sensors and datagloves. *Proceedings of the 2nd European Conference on Disability, Virtual Reality and Associated Technologies (ECDVRAT)*, 149-157.

Scano, A., Caimmi, M., Chiavenna, A., Malosio, M., & Tosatti, L. M. (2017). A Kinect-Based Biomechanical Assessment of Neurological Patients' Motor Performances for Domestic Rehabilitation. *Virtual Reality Enhanced Robotic Systems for Disability Rehabilitation*, 252–279. 10.4018/978-1-4666-9740-9.ch013

Scano, A., Caimmi, M., Chiavenna, A., Malosio, M., & Tosatti, L. M. (2015). Kinect One-based biomechanical assessment of upper-limb performance compared to clinical scales in post-stroke patients. *Proceedings of the Annual International Conference of the IEEE Engineering in Medicine and Biology Society, EMBS, 2015*, 5720–5723. doi:10.1109/EMBC.2015.7319691

Scano, A., Caimmi, M., Malosio, M., & Tosatti, L. M. (2014). Using Kinect for upper-limb functional evaluation in home rehabilitation: A comparison with a 3D stereoscopic passive marker system. *5th IEEE RAS/EMBS International Conference on Biomedical Robotics and Biomechatronics*, 561–566. doi:10.1109/BIOROB.2014.6913837

Scano, A., Spagnuolo, G., Caimmi, M., Chiavenna, A., Malosio, M., Legnani, G., & Tosatti, L. M. (2015). Static and dynamic characterization of the LIGHTarm exoskeleton for rehabilitation. *IEEE International Conference on Rehabilitation Robotics,* (1), 428–433. doi:10.1109/ICORR.2015.7281237

Schaefer, S. Y., & Hengge, C. R. (2016). Testing the concurrent validity of a naturalistic upper extremity reaching task. *Experimental Brain Research*, *234*(1), 229–240. doi:10.1007/s00221-015-4454-y PMID:26438508

Schill, O., Rupp, R., & Reischl, M. (2009). *Signal processing concepts for optimal myoelectric sensor placement in a modular hybrid FES orthosis.* Paper presented at the 4th European Conference of the International Federation for Medical and Biological Engineering. doi:10.1007/978-3-540-89208-3_433

Schill, O., Wiegand, R., Schmitz, B., Matthies, R., Eck, U., Pylatiuk, C., ... Rupp, R. (2011). OrthoJacket: An active FES-hybrid orthosis for the paralysed upper extremity. *Biomedizinische Technik. Biomedical Engineering*, *56*(1), 35–44. doi:10.1515/bmt.2010.056 PMID:21210758

Schmidt, C. V., & Gorb, S. N. (2012). Snake Scale Microstructure: Phylogenetic Significance and Functional Adaptations. *Zoologica*, *187*, 1–106.

Schmidt, R. (1975). A Schema Theory of Discrete Motor Skill Learning. *Psychological Review*, *82*(4), 225–260. doi:10.1037/h0076770

Schulz, S., Pylatiuk, C., Kargov, A., Gaiser, I., Schill, O., Reischl, M., . . . Rupp, R. (2009). *Design of a hybrid powered upper limb orthosis.* Paper presented at the World Congress on Medical Physics and Biomedical Engineering, Munich, Germany. doi:10.1007/978-3-642-03889-1_125

Schulz, S., Schmitz, B., Wiegand, R., Pylatiuk, C., & Reischl, M. (2011). *The Hybrid Fluidic Driven Upper Limb Orthosis-Orthojacket*. Academic Press.

Schunk, D. H., & Zimmerman, B. J. (2007). Motivation and self-regulated learning: Theory, research, and applications. Abingdon-on-Thames, UK: Routledge.

Schwind, W. J., & Koditchek, D. E. (1995). Control of Forward Velocity for a Simplified Planar Hopping Robot. *The IEEE International Conference on Robotics and Automation (ICRA'1995)*, 691–696. doi:10.1109/ROBOT.1995.525364

Secor, M. S., & Bruce, C. (1992). Jayne Albert F. Bennett, Locomotor performance and energetic cost of sidewinding by the snake crotalus cerastes. *The Journal of Experimental Biology*, *163*, 1–14.

Secor, S. M., Jayne, B. C., & Bennett, A. F. (1992). Locomotor performance and energetic cost of sidewinding by the snake Crotalus cerastes. *The Journal of Experimental Biology*, *163*(1), 1–14.

Senelick, R. C. (2010). Technological Advances in Stroke Rehabilitation—High Tech Marries High Touch. *US Neurology*, *6*(2), 102–104. doi:10.17925/USN.2010.06.02.102

Seok, S., Wang, A., Chuah, M. Y., Hyun, D. J., Lee, J., Otten, D. M., ... Kim, S. (2015). Design Principles for Energy-Efficient Legged Locomotion and Implementation on the MIT Cheetah Robot. *IEEE/ASME Transactions on Mechatronics*, *20*(3), 1117–1129. doi:10.1109/TMECH.2014.2339013

Sepkoski, J. (2002). A compendium of fossil marine animal genera (Reptilia entry). *Bulletins of American Paleontology*, *364*, 560.

Serea, F., Poboroniuc, M., Hartopanu, S., & Irimia, D. (2015). *Towards Clinical Implementation of an FES&Exoskeleton to Rehabilitate the Upper Limb in Disabled Patients.* Paper presented at the Control Systems and Computer Science (CSCS), 2015 20th International Conference on. doi:10.1109/CSCS.2015.114

Serea, F., Poboroniuc, M., Hartopanu, S., & Olaru, R. (2014). *Preliminary tests on a hybrid upper arm exoskeleton for upper arm rehabilitation for disabled patients.* Paper presented at the Electrical and Power Engineering (EPE), 2014 International Conference and Exposition on. doi:10.1109/ICEPE.2014.6969887

Serea, F., Poboroniuc, M., Irimia, D., Hartopanu, S., & Olaru, R. (2013). *Preliminary results on a hybrid FES-exoskeleton system aiming to rehabilitate upper limb in disabled people.* Paper presented at the System Theory, Control and Computing (ICSTCC), 2013 17th International Conference. doi:10.1109/ICSTCC.2013.6689046

Sessa, V. I., & London, M. (2006). *Continuous learning in organizations: Individual, group, and organizational perspectives*. Hove, UK: Psychology Press.

Sessa, V. I., & London, M. (2006). *Continuous Learning in Organizations: Individual, Group, and Organizational Perspectives*. Psychology Press.

Sharpe, S. S., Koehler, S. A., Kuckuk, R. M., Serrano, M., Vela, P. A., Mendelson, J. III, & Goldman, D. I. (2015, February 1). Locomotor benefits of being a slender and slick sand swimmer. *The Journal of Experimental Biology*, *218*(Pt 3), 440–450. doi:10.1242/jeb.108357 PMID:25524983

Sheng, B., Zhang, Y., Meng, W., Deng, C., & Shengquan, X. (2016). Bilateral robots for upper-limb stroke rehabilitation: State of the art and future prospects. *Medical Engineering & Physics*, *38*(7), 587–606. doi:10.1016/j.medengphy.2016.04.004 PMID:27117423

Shen, W. M., Will, P., Galstyan, A., & Chuong, C. M. (2004). Hormone-Inspired Self-Organization and Distributed Control of Robotic Swarms. *Autonomous Robots*, *17*(1), 93–105. doi:10.1023/B:AURO.0000032940.08116.f1

Sherwood, D. E., & Lee, T. D. (2003). Schema theory: Critical review and implications for the role of cognition in a new theory of motor learning. *Research Quarterly for Exercise and Sport*, *74*(4), 376–382. doi:10.1080/02701367.2003.10609107 PMID:14768838

Shigeki, T., & Yonetake, J. (2006). UltraSonicMotor Powered Assisted Suit System. *Society of Biomechanism Japan*, *30*(4), 189–193. doi:10.3951/sobim.30.189

Shi, L., Habib, M. K., Xiao, N., & Hu, H. (2015). Biologically Inspired Robotics [editorial]. *Journal of Robotics*, *2015*, 1–2. doi:10.1155/2015/894394

Shi, L., Habib, M. K., Xiao, N., & Hu, H. (2015). Biologically inspired robotics. *Journal of Robotics*.

Shima, K., Hiramatsu, Y., Shibanoki, T., & Tsuji, T. (2013). Discrimination of Dual-arm Motions Using a Joint Posterior Probability Neural Network [in Japanese]. *Transactions of the Society of Instrument and Control Engineers*, *49*(5), 568–575. doi:10.9746/sicetr.49.568

Shine, R. (1994). Allometric patterns in the ecology of Australian snakes. *Copeia*, *1994*(4), 851–867. doi:10.2307/1446709

Shuster-Amft, C., Eng, K., Lehmann, I., Schmid, L., Kobashi, N., Thaler, I., … Kiper, D. (2014). Using mixed methods to evaluate efficacy and user expectations of a virtual reality–based training system for upper-limb recovery in patients after stroke: a study protocol for a randomized controlled. *Trial*, *15*(3). DOI 1745-6215-15-350

Simmons, R., & Krotkov, E. (1991). An Integrated Walking System for the Ambler Planetary Rover. *The IEEE International Conference on Robotics & Automation (ICRA'1991)*, 2086–2091. doi:10.1109/ROBOT.1991.131935

Siradjuddin, I., Behera, L., McGinnity, T. M., & Coleman, S. (2012). A position based visual tracking system for a 7 DOF robot manipulator using a Kinect camera. In *Proceedings of the 2012 International Joint Conference on Neural Networks (IJCNN)*. Brisbane: IEEE. doi:10.1109/IJCNN.2012.6252597

Sivan, M., O'Connor, R. J., Makower, S., Levesley, M., & Bhakta, B. (2011). Systematic review of outcome measures used in the evaluation of robot-assisted upper limb exercise in stroke. *Journal of Rehabilitation Medicine*, *43*(3), 181–189. doi:10.2340/16501977-0674 PMID:21305232

Sledd, A., & O'Malley, M. K. (2006). Performance Enhancement of a Haptic Arm Exoskeleton. *Symposium on Haptic Interfaces for Virtual Environment and Teleoperator Systems*.

Smita, J., & Trey, M. (2012). *Facilitating continuous learning: A review of research and practice on individual learning capabilities and organizational learning environments*. Paper presented at the Annual Meeting of the AECT International Convention, Louisville, KY. Retrieved August 1, 2017, from http://www.aect.org/pdf/proceedings12/2012i/12_09.pdf

Smita, J., & Trey, M. (2012). *Facilitating continuous learning: review of research on individual learning capabilities and organizational learning environments*. In The annual meeting of the AECT International Convention, Louisville, KY. Retrieved from http://www.memphis.edu/icl/idt/clrc/clrc-smita-research.pdf

Sniegowski, J. J., & Garcia, E. J. (1996). Surface-micromachined gear trains driven by an on-chip electrostatic micro-engine. *IEEE Electron Device Letters*, *17*(7), 366–368. doi:10.1109/55.506369

Solís, F., Martínez, D., & Espinoza, O. (2016). Automatic Mexican Sign Language Recognition Using Normalized Moments and Artificial Neural Networks. *Engineering*, *8*(10), 733–740. doi:10.4236/eng.2016.810066

Somani, B. K., Srivastava, A., Traxer, O., & Aboumarzouk, O. (2013). Flexible ureterorenoscopy: Tips and tricks. *Urology Annals*, *5*(1). doi:10.4103/0974-7796.106869

Song, X., & Wang, Y. (2008, June). A novel model-based method for odometry calculation of all-terrain mobile robots. In *Intelligent Control and Automation, 2008. WCICA 2008. 7th World Congress on* (pp. 581-586). IEEE.

Spagnuolo, G., Malosio, M., Dinon, T., Tosatti, L. M., & Legnani, G. (2017). Analysis and synthesis of LinWWC-VSA, a Variable Stiffness Actuator for linear motion. *Mechanism and Machine Theory, 110*, 85–99. 10.1016/j.mechmachtheory.2016.11.002

Spagnuolo, G., Malosio, M., Scano, A., Caimmi, M., Legnani, G., & Tosatti, L. M. (2015). Passive and active gravity-compensation of LIGHTarm, an exoskeleton for the upper-limb rehabilitation. *IEEE International Conference on Rehabilitation Robotics, 2015*, 440–445. doi:10.1109/ICORR.2015.7281239

Specht, D. F. (1990). Probabilistic neural networks. *Neural Networks*, *3*(1), 109–118. doi:10.1016/0893-6080(90)90049-Q PMID:18282828

Spenko, M., Haynes, G. C., Saunders, J. A., Cutkosky, M. R., Rizzi, A. A., Full, R. J., & Koditschek, D. E. (2008). Biologically inspired climbing with a hexapedal robot. *Journal of Field Robotics*, *25*(4-5), 223–242. doi:10.1002/rob.20238

Stetten. (2001b). Real time tomographic reflection: phantoms for calibration and biopsy. *Augmented Reality, 2001. Proceedings. IEEE and ACM International Symposium on*, 11–19.

Stetten, G. D., & Chib, V. S. (2001). Overlaying ultrasonographic images on direct vision. *Journal of Ultrasound in Medicine*, *20*(3), 235–240. doi:10.7863/jum.2001.20.3.235 PMID:11270527

Stewart, G. R., & Daniel, R. S. (1973). Scanning electron microscopy of scales from different body regions of three lizard species. *Journal of Morphology*, *139*(4), 377–388. doi:10.1002/jmor.1051390402

Stinear, J., & Byblow, W. (2002). Disinhibition in the human motor cortex is enhanced by synchronous upper limb movements. *The Journal of Physiology*, *543*(Pt 1), 307–316. doi:10.1113/jphysiol.2002.023986 PMID:12181301

Stinear, J., & Byblow, W. (2004). Rhythmic Bilateral Movement Training Modulates Corticomotor Excitability and Enhances Upper Limb Motricity Postroke: A Pilot Study. *Journal of Clinical Neurophysiology*, *21*(2), 124–131. doi:10.1097/00004691-200403000-00008 PMID:15284604

Su, B., & Han, Y. (2010). Design of Greenhouse Wireless Monitoring System Based on ZigBee. *International Conference on E-Product E-Service and E-Entertainment*. doi:10.1109/ICEEE.2010.5661599

Sullivan, S. B., Schmitz, T. J., & Fulk, G. D. (2014). *Physical rehabilitation* (6th ed.). Philadelphia, PA: F.A. Davis Co.

Surbled, P., Clerc, C., Pioufle, B. L., Ataka, M., & Fujita, H. (2001). Effect of the composition and thermal annealing on the transformation temperature sputtered TiNi shape memory alloy thin films. *Thin Solid Films*, *401*(1-2), 52–59. doi:10.1016/S0040-6090(01)01634-0

Suzuki, Y., Tani, K., & Sakuhara, T. (1999). Development of a new type piezo electric Micromotor. *Proc. of Transduceres*, *99*, 1748–1751.

Swinnen, S. P. (2002). Intermanual Coordination: From Behavioural Principles To Neural-Network Interactions. *Nature Reviews. Neuroscience*, *3*(5), 348–359. doi:10.1038/nrn807 PMID:11988774

Takato, M., Naito, Y., Maezumi, K., Ishihara, Y., Okane, Y., Oku, H., ... Uchikoba, F. (2015). Development of impact-type rotary actuator and application for MEMS microrobot by bare chip IC of hardware neuron model. *Artificial Life and Robotics*, *20*(4), 359–365. doi:10.1007/s10015-015-0239-4

Talso, M., Emiliani, E., Haddad, M., Berthe, L., Baghdadi, M., Montanari, E., & Traxer, O. (2016). Laser Fiber and Flexible Ureterorenoscopy: The Safety Distance Concept. *Journal of Endourology*, *30*(12), 1269–1274. doi:10.1089/end.2016.0209

Tanaka, M., Kon, K., & Tanaka, K. (2015). Range-Sensor-Based Semiautonomous Whole-Body Collision Avoidance of a Snake Robot. *IEEE Transactions on Control Systems Technology*, *23*(5), 1927–1934. doi:10.1109/TCST.2014.2382578

Tanaka, M., Nakajima, M., & Tanaka, K. (2016). Smooth control of an articulated mobile robot with switching constraints. *Advanced Robotics*, *30*(1), 29–40. doi:10.1080/01691864.2015.1102646

Tang, W. C., Nguyen, T. H., & Howe, R. T. (1989). Laterally driven poly silicon resonant microstructure. *Proc. of IEEE Micro Electro Mechanical Systems. An Investigation of Micro Structures, Sensors, Actuators, Machines and Robots*, 53-59.

Teber, D., Guven, S., Simpfendörfer, T., Baumhauer, M., Güven, E. O., Yencilek, F., ... Rassweiler, J. (2009). Augmented reality: A new tool to improve surgical accuracy during laparoscopic partial nephrectomy? Preliminary in vitro and in vivo results. *European Urology*, *56*(2), 332–338. doi:10.1016/j.eururo.2009.05.017 PMID:19477580

Tejomurtula, S., & Kak, S. (1999). Inverse kinematics in robotics using neural networks. *Information Sciences*, *116*(2-4), 147–164. doi:10.1016/S0020-0255(98)10098-1

Terrible, L. C., Diniz-Filho, J. A. F., Rodríguez, M. A., & Rangel, T. F. L. V. B. (2009). Ecological and evolutionary components of body size: Geographic variation of venomous snakes at the global scale. *Biological Journal of the Linnean Society. Linnean Society of London*, *98*(1), 94–109. doi:10.1111/j.1095-8312.2009.01262.x

Toni, M., Dalla, L. V., & Alibardi, L. (2007). Hard (Beta-) Keratins in the Epidermis of Reptiles: Composition Sequence and Molecular Organization. *Journal of Proteome Research*, *6*(9), 3377–3392. doi:10.1021/pr0702619 PMID:17705524

Tóth, A., Arz, G., Fazekas, G., Bratanov, D., & Zlatov, N. (2004). Post stroke shoulder-elbow physiotherapy with industrial robots. Advances in Rehabilitation Robotics, 391–411.

Toyoshima, S., Tanaka, M., & Matsuno, F. (2014). A study on sinus-lifting motion of a snake robot with sequential optimization of a hybrid system. Automation Science and Engineering. *IEEE Transactions on*, *11*(1), 139–144.

Transeth, A. A., Leine, R. I., Glocker, C., Pettersen, K. Y., & Liljebäck, P. (2008). Snake robot obstacle-aided locomotion: Modeling, simulations, and experiments. *IEEE Transactions on Robotics*, *24*(1), 88–104. doi:10.1109/TRO.2007.914849

Transeth, A. A., Pettersen, K. Y., & Liljebäck, P. (2009). A survey on snake robot modeling and locomotion. *Robotica*, *27*(7), 999–1015. doi:10.1017/S0263574709005414

Triantafyllou, M. S., Techet, A. H., & Hover, F. S. (2004). Review of Experimental Work in Biomimetic Foils. *IEEE Journal of Oceanic Engineering*, *29*(3), 585–594. doi:10.1109/JOE.2004.833216

Tsagarakis, N. G., & Caldwell, D. (2003). Development and control of a "soft-actuated" exoskeleton for use in physiotherapy and training. *Autonomous Robots*, *15*(1), 21–33. doi:10.1023/A:1024484615192

Tsagarakis, N. G., Tsachouridis, V., & Caldwell, D. G. (2003). Modeling and control of a pneumatic muscle actuated joint using on/off solenoid valves. *Proceedings of the IEEE International Conference on Advanced Robotics*, 929–934.

Tsiotras, P., Corless, M., & Longuski, J. M. (1993). Invariant manifold techniques for attitude control of symmetric spacecraft. In *Proceedings of 32th Conference on Decision and Control* (pp. 1470–1475). Academic Press. doi:10.1109/CDC.1993.325431

Tsuji, T., Bu, N., & Fukuda, O. (2006). Neural Networks in Healthcare: Potential and Challenges. In A Recurrent Probabilistic Neural Network for EMG Pattern Recognition (pp.130-153). Idea Group Inc.

Tsuji, T., Shibanoki, T., & Shima, K. (2014). EMG-based control of a multi-joint robot for operating a glovebox. In Handbook of Research on Advancements in Robotics and Mechatronics (pp. 36-52). IGI Global.

Tsuji, T., Bu, N., Fukuda, O., & Kaneko, M. (2003). A recurrent log-linearized Gaussian mixture network. *IEEE Transactions on Neural Networks*, *14*(2), 304–316. doi:10.1109/TNN.2003.809403 PMID:18238014

Tu, X., He, J., Wen, Y., Huang, J., Huang, X., Huang, H., . . . Yuan, Y. (2014). *Cooperation of electrically stimulated muscle and pneumatic muscle to realize RUPERT bi-directional motion for grasping.* Paper presented at the Engineering in Medicine and Biology Society (EMBC), 2014 36th Annual International Conference of the IEEE.

Tu, X., Huang, J., Yu, L., Xu, Q., & He, J. (2012). *Design of a wearable rehabilitation robot integrated with functional electrical stimulation.* Paper presented at the Biomedical Robotics and Biomechatronics (BioRob), 2012 4th IEEE RAS & EMBS International Conference on. doi:10.1109/BioRob.2012.6290720

Tully, S., & Choset, H. (2016). A filtering approach for image-guided surgery with a highly articulated surgical Snake Robot. *IEEE Transactions on Biomedical Engineering*, *63*(2), 392–402. doi:10.1109/TBME.2015.2461531 PMID:26241966

Tung, W. Y. (2013). *Design and Operation of Minimally Actuated Medical Exoskeletons for Individuals with Paralysis* (Doctoral dissertation). University of California, Berkeley, CA.

Unver, O., Uneri, A., Aydemir, A., & Sitti, M. (2006). Geckobot: A Gecko Inspired Climbing Robot using Dry Adhesives. *The IEEE Robotics and Automation Conference (ICRA'2006)*, 2329-2335.

Vaca Benitez, L. M., Tabie, M., Will, N., Schmidt, S., Jordan, M., & Kirchner, E. A. (2013). Exoskeleton technology in rehabilitation: Towards an EMG-based orthosis system for upper limb neuromotor rehabilitation. *Journal of Robotics.*

Vaida, C., Carbone, G., Major, K., Major, Z., Plitea, N., & Pisla, D. (2017). On Human Robot Interaction Modalities in the Upper Limb Rehabilitation After Stroke. *Acta Tehnica Napocensis, Applied Mathematics, Mechanics and Engineering*, *60*(1), 91–102.

van Delden, L., Peper, C. E., Kwakkel, G., & Beek, P. J. (2012, November). A Systematic Review of Bilateral Upper Limb Training Devices for Poststroke Rehabilitation. Stroke Research Treatment.

Van Delden, A. E. Q., Peper, C. E., Kwakkel, G., & Beek, P. J. (2012). A systematic review of bilateral upper limb training devices for poststroke rehabilitation. *Stroke Research and Treatment*, *2012*, 1–17. doi:10.1155/2012/972069 PMID:23251833

Vanderborght, B., Albu-Schaeffer, A., Bicchi, A., Burdet, E., Caldwell, D. G., Carloni, R., ... Wolf, S. (2013). Variable impedance actuators: A review. *Robotics and Autonomous Systems*, *61*(12), 1601–1614. doi:10.1016/j.robot.2013.06.009

Vannuccini, S. (1999). Shark utilization, marketing, and trade vol. 389. Food and Agriculture Organization of the United Nations (FAO) Fisheries Technical Papers.

Varoto, R., Barbarini, E. S., & Cliquet, A. Jr. (2008). A hybrid system for upper limb movement restoration in quadriplegics. *Artificial Organs*, *32*(9), 725–729. doi:10.1111/j.1525-1594.2008.00597.x PMID:18684204

Veerasingam, S., Karodi, S., Shukla, S., & Yeleti, M. (2009). Design of Wireless Sensor Network Node on ZigBee for Temperature Monitoring. *International Conference on Advances in Computing, Control, and Telecommunication Technologies*. doi:10.1109/ACT.2009.14

Veneman, J. F., Ekkelenkamp, R., Kruidhof, R., van der Helm, F. C., & van der Kooij, H. (2006). A series elastic-and bowden-cable-based actuation system for use as torque actuator in exoskeleton-type robots. *The International Journal of Robotics Research, 25*(3), 261–281. doi:10.1177/0278364906063829

Veneva. (2015). *Lower limb with active joints*. Patent Application (BG, N°112139/06.11.2015).

Vidal N., S. (2005). The phylogeny of squamate reptiles (lizards, snakes, and amphisbaenians) inferred from nine nuclear protein-coding genes. *Comptes Rendus Biologies, 328*(10-11), 1000-1008. DOI: 10.1016/j.crvi.2005.10.001

Videler, J. (1993). *Fish Swimming*. London, UK: Chapman and Hall. doi:10.1007/978-94-011-1580-3

Viollet, S., & Franceschini, N. (2001). *Aerial Minirobot that Stabilizes and Tracks with a Bio-Inspired Visual Scanning Sensor. In* B. Webb & T. Consi (Eds.), *Biorobotics* (pp. 67–73). Cambridge, MA: MIT Press.

Virgala, I., Dovica, M., Kelemen, M., Prada, E., & Bobovský, Z. (2014). Snake Robot Movement in the Pipe Using Concertina Locomotion. *Applied Mechanics and Materials, 611*, 121–129. doi:10.4028/www.scientific.net/AMM.611.121

Viteckova, S., Kutilek, P., & Jirina, M. (2013). Wearable lower limb robotics: A review. *Biocybernetics and Biomedical Engineering, 33*(2), 96–105. doi:10.1016/j.bbe.2013.03.005

Vitiello, N., Lenzi, T., Roccella, S., De Rossi, S. M. M., Cattin, E., Giovacchini, F., ... Carrozza, M. C. (2013). NEUROExos: A powered elbow exoskeleton for physical rehabilitation. *IEEE Transactions on Robotics, 29*(1), 220–235. doi:10.1109/TRO.2012.2211492

Vitt, L. J., Pianka, E. R., Cooper, W. E. Jr, & Schwenk, K. (2003). History and the global ecology of squamate reptiles. *American Naturalist, 162*(1), 44–60. doi:10.1086/375172 PMID:12856236

Vogler, C., & Metaxas, D. (1997). Adapting Hidden Markov Models for American Sign Language Recognition by using Three-dimentional Computer Vision Methods. *IEEE International Conference on Systems, Man and Cybernetics*, 156-161.

Voigt, D., Karguth, A., & Gorb, S. (2012). Shoe soles for the gripping robot: Searching for polymer-based materials maximising friction. *Robotics and Autonomous Systems, 60*(8), 1046–1055. doi:10.1016/j.robot.2012.05.012

Vucobratovic, M., & Borovac, B. (2004). Zero-moment point—thirty five years of its life. *International Journal of Humanoid Robotics, 1*(01), 157–173. doi:10.1142/S0219843604000083

Vukobratovic, M., Borovac, B., Surla, D., & Stokic, D. (2012). *Biped locomotion: dynamics, stability, control and application* (Vol. 7). Springer Science & Business Media.

Wacker, F. K., Vogt, S., Khamene, A., Jesberger, J. A., Nour, S. G., Elgort, D. R., ... Lewin, J. S. (2006). An augmented reality system for MR image–guided needle biopsy: Initial results in a swine model 1. *Radiology, 238*(2), 497–504. doi:10.1148/radiol.2382041441 PMID:16436814

Wada, M., & Mori, S. (1996, April). Holonomic and omnidirectional vehicle with conventional tires. In *Robotics and Automation, 1996. Proceedings., 1996 IEEE International Conference on* (Vol. 4, pp. 3671-3676). IEEE. doi:10.1109/ROBOT.1996.509272

Walker, I. D. (2013). Continuous backbone "continuum" robot manipulators. *ISRN Robotics, 2013.*

Walker, I. D., Choset, H., & Chirikjian, G. S. (2016). Snake-Like and Continuum Robots. In *Springer Handbook of Robotics* (pp. 481–498). Springer. doi:10.1007/978-3-319-32552-1_20

Walton, M., Jayne, B. C., & Bennett, A. F. (1990). The energetic cost of limbless locomotion. *Science*, *249*(4968), 524–527. doi:10.1126/science.249.4968.524 PMID:17735283

Wang, L., Wang, S., van Asseldonk, E. H., & van der Kooij, H. (2013). Actively controlled lateral gait assistance in a lower limb exoskeleton. In *Intelligent Robots and Systems (IROS), 2013 IEEE/RSJ International Conference on* (pp. 965-970). IEEE.

Wang, P., Su, J., Li, W., & Qiao, H. (2014). Adaptive visual tracking based on discriminative feature selection for mobile robot. In *Proceedings of 2014 IEEE 4th Annual International Conference on Cyber Technology in Automation, Control, and Intelligent Systems (CYBER)*. Hong Kong: IEEE.

Wang, H. S., Chen, W. D., & Wang, Z. L. (2011). Visual tracking of robots in uncalibrated environments. In *Proceedings of 2011 IEEE/RSJ International Conference on Intelligent Robots and Systems (IROS)*. San Francisco: IEEE. doi:10.1109/IROS.2011.6094647

Wang, H., Chai, X., Hong, X., Zhao, G., & Chen, X. (2016). Isolated Sign Language Recognition with Grassman Covariance Matrices. *ACM Transactions on Accessible Computing*, *8*(4), 14–21. doi:10.1145/2897735

Wang, K., Liu, Y., & Li, L. (2014a). Visual Servoing Trajectory Tracking of Nonholonomic Mobile Robots Without Direct Position Measurement. *IEEE Transactions on Robotics*, *30*(4), 1026–1035. doi:10.1109/TRO.2014.2317891

Wang, K., Liu, Y., & Li, L. (2014b). Visual servoing based trajectory tracking of underactuated water surface robots without direct position measurement. In *Proceedings of 2014 IEEE/RSJ International Conference on Intelligent Robots and Systems (IROS 2014)*. Chicago: IEEE. doi:10.1109/IROS.2014.6942645

Watanabe, K., Yamamoto, T., Izumi, K., & Maeyama, S. (2010) Underactuated control for nonholonomic mobile robots by using double integrator model and invariant manifold theory. In *Proceedings of IEEE/RSJ International Conference on Intelligent Robots and Systems (IROS)* (pp. 2862-2867). IEEE. doi:10.1109/IROS.2010.5650951

Weiss, C. R., Marker, D. R., Fischer, G. S., Fichtinger, G., Machado, A. J., & Carrino, J. A. (2011). Augmented reality visualization using image overlay for MR-guided interventions: System description, feasibility, and initial evaluation in a spine phantom. *AJR. American Journal of Roentgenology*, *196*(3), W305–W307. doi:10.2214/AJR.10.5038 PMID:21343479

West, A. M. T. (1993). *U.S. Patent No. 5,186,270*. Washington, DC: U.S. Patent and Trademark Office.

West, M., & Asada, H. (1997). Design of ball wheel mechanisms for omnidirectional vehicles with full mobility and invariant kinematics. *Journal of Mechanical Design*, *119*(2), 153–161. doi:10.1115/1.2826230

Westwood. (2006). MRI image overlay: applications to arthrography needle insertion. In *Medicine Meets Virtual Reality 14: Accelerating Change in Healthcare: Next Medical Toolkit*, (vol. 119, p. 150). Academic Press.

Whittle, M. W. (2014). *Gait analysis: An introduction*. Butterworth-Heinemann.

Wiegand, R., Schmitz, B., Pylatiuk, C., & Schulz, S. (2011). Mechanical Performance of Actuators in an Active Orthosis for the Upper Extremities. *Journal of Robotics*.

Williams, R., Murphy, M., North, D., Berlin, J., & Krier, M. (1998). Kinesthetic force/moment feedback via active exoskeleton. *Proc. 1998 Image Conference*.

Winter Green Research. (2014). *Rehabilitation robots, active prostheses, and exoskeletons: Market shares, strategies, and forecasts, worldwide, 2014 to 2020*. Retrieved from http://wintergreenresearch.com/reports/RehabilitationRobots.html

Winter, D. A. (1984). Kinematic and kinetic patterns in human gait: Variability and compensating effects. *Human Movement Science*, *3*(1), 51–76. doi:10.1016/0167-9457(84)90005-8

Winter, D. A. (2009). *Biomechanics and motor control of human movement*. John Wiley & Sons. doi:10.1002/9780470549148

Wirbel, E., Steux, B., Bonnabel, S., & De, L. F. A. (2013). Humanoid robot navigation: From a visual SLAM to a visual compass. In *Proceedings of 2013 10th IEEE International Conference on Networking, Sensing and Control (ICNSC)*. Evry: IEEE.

Wood, R. J., Finio, B., Karpelson, M., Ma, K., Pérez-Arancibia, N. O., Sreetharan, P. S., ... Whitney, J. P. (2016). Robotics Research. In H. I. Christensen & O. Khatib (Eds.), *Progress on "Pico" Air Vehicles* (pp. 3–19). Springer.

World Health Organization. (2002). Towards a Common Language for Functioning, Disability and Health ICF. *International Classification, 1149*, 1–22.

World Heart Federation. (2015). *Stroke*. Retrieved from http://www.world-heart-federation.org/cardiovascular-health/stroke/

World Stroke Campaign. (2015). *Facts and Figures*. Retrieved from http://www.worldstrokecampaign.org/learn/facts-and-figures.html

Wright, C., Buchan, A., Brown, B., Geist, J., Schwerin, M., & Rollinson, D. … Choset, H. (2012). Design and architecture of the unified modular snake robot. *Proceedings - IEEE International Conference on Robotics and Automation*. doi:10.1109/ICRA.2012.6225255

Wright, A. E., Rukin, N. J., & Somani, B. K. (2014). Ureteroscopy and stones: Current status and future expectations. *World Journal of Nephrology, 3*(4), 243. doi:10.5527/wjn.v3.i4.243 PMID:25374818

Wu, X. (2011). *Sensor-driven Neural Controller for Self-adaptive Collision-free Behavior of a Snake-like Robot*. Academic Press.

Wu, J., Gao, J., Song, R., Li, R., Li, Y., & Jiang, L. (2016). The design and control of a 3DOF lower limb rehabilitation robot. *Mechatronics, 33*, 13–22. doi:10.1016/j.mechatronics.2015.11.010

Wu, Y., Lim, J., & Yang, M. H. (2013). Online Object Tracking: A Benchmark. In *Proceedings of 2013 IEEE Conference on Computer Vision and Pattern Recognition (CVPR)*. Portland: IEEE. doi:10.1109/CVPR.2013.312

Xia, Y., & Wang, J. (2001, February). A dual neural network for kinematic control of redundant robot manipulators. *IEEE Transactions on Systems, Man, and Cybernetics. Part B, Cybernetics, 31*(1), 147–154. doi:10.1109/3477.907574 PMID:18244777

Xie, H. (2009, September). MEMS-based 3D optical microendoscopy. In *Engineering in Medicine and Biology Society, 2009. EMBC 2009. Annual International Conference of the IEEE* (pp. 6703-6705). IEEE.

Xu, M., Ma, L., Xia, F., Yuan, T., Qian, J., & Shao, M. (2010). Design and Implementation of a Wirelss Sensor Network for Smart Homes. *Ubiquitous Intelligence & Computing and 7th International Conference on Autonomic & Trusted Computing (UIC/ATC)*.

Xu, R., Gu, X., & Wei, B. (2013). An Improved Real Time Compressive Tracking. In *Proceedings of 2013 2nd IAPR Asian Conference on Pattern Recognition (ACPR)*. Okinawa: IEEE.

Yadav, R. K., Sharma, R., Borah, D., & Kothari, S. Y. (2016). Efficacy of Modified Constraint Induced Movement Therapy in the Treatment of Hemiparetic Upper Limb in Stroke Patients: A Randomized Controlled Trial. *Journal of Clinical and Diagnostic Research: JCDR, 10*(11), YC01–YC05. PMID:28050492

Yamada, H., & Hirose, S. (2006). Study on the 3D shape of active cord mechanism. In *Proceedings - IEEE International Conference on Robotics and Automation* (Vol. 2006). IEEE. doi:10.1109/ROBOT.2006.1642140

Yamada, H., Chigisaki, S., Mori, M., Takita, K., Ogami, K., & Hirose, S. (2005). Development of Amphibious Snake-like Robot ACM-R5. In *Proceedings of 36th International Symposium on Robotics (ISR'2005).* TH3C4.

Yamaguchi, T., Takemori, K., Tamai, Y., & Takadama, K. (2015a). Analyzing human's continuous learning ability toward the intelligent robotics. In *Proceedings of the Twentieth International Symposium on Artificial Life and Robotics 2015 (AROB 20th 2015).* Chiyoda-ku, Japan: Springer Japan.

Yamaguchi, T., Takemori, K., & Takadama, K. (2013). Modeling a human's learning processes toward continuous learning support system. In M. K. Habib & J. P. Davim (Eds.), *Mechatronics Engineering* (pp. 69–94). Hoboken, NJ: Wiley-ISTE. doi:10.1002/9781118577516.ch4

Yamaguchi, T., Takemori, K., Tamai, Y., & Takadama, K. (2015). Analyzing human's continuous learning processes with the reflection sub task. *Journal of Communication and Computer*, *12*(1), 20–27.

Yamaguchi, T., Takemori, K., Tamai, Y., & Takadama, K. (2015b). Analyzing human's continuous learning processes with the reflection sub task. *Journal of Communication and Computer*, *12*(1), 20–27.

Yamaguchi, T., Tamai, Y., & Takadama, K. (2015). Analyzing human's continuous learning ability with the reflection cost. *Proceedings of 41st Annual Conference of the IEEE Industrial Electronics Society (IECON2015)*, 2920-2925. doi:10.1109/IECON.2015.7392546

Yang, Q., Cao, D., & Zhao, J. (2013). Analysis on State of the Art of Upper Limb Rehabilitation Robots. *Journal of Robotics*, *5*, 630–640.

Yan, J., Wood, R. J., Avadhanula, S., Sitti, M., & Fearing, R. S. (2001). Towards Flapping Wing Control of a Micromechanical Flying Insect. *The IEEE Robotics and Automation Conference (ICRA'2001)*, 3901-3908. doi:10.1109/ROBOT.2001.933225

Yan, Q. S., & Li, L. S. (2014). Kernel sparse tracking with compressive sensing. *IET Computer Vision*, *8*(4), 305–315. doi:10.1049/iet-cvi.2013.0095

Yap, H. K., Lim, J. H., & Nasrallah, F. (2015) A Soft Exoskeleton for Hand Assistive and Rehabilitation Application using Pneumatic Actuators with Variable Stiffness. In *Proceeding of International Conference on Robotics and Automation.* Seattle, WA: IEEE. Doi:10.1109/ICRA.2015.7139889

Yeh, R., Hollar, S., & Pister, K. S. J. (2002). Single Mask, Large Force, and Large Displacement Electrostatic Linear Inchworm Motors. *Journal of Microelectromechanical Systems, 11*(4), 330–336. doi:10.1109/JMEMS.2002.800937

Yokoi, H., Yamashita, J., Fukui, Y., & Shimojo, M. (1994). Development of the virtual shape manipulating system. *Proc. of the 4th Int. Conf. on Artifcial Reality and Tele-Existence (ICAT'94)*, 43-48.

Yoo, D. H., & Kim, S. K. (2015). Effects of upper limb robot-assisted therapy in the rehabilitation of stroke patients. *Journal of Physical Therapy Science*, *27*(3), 677–679. doi:10.1589/jpts.27.677 PMID:25931706

Yurtkuran, S., Kirh, G., & Taneli, T. (2013). Learning from Nature: Biomimetic Design in Architectural Education. *Procedia: Social and Behavioral Sciences*, *89*, 633–639. doi:10.1016/j.sbspro.2013.08.907

Zhang, S., Guo, S., Pang, M., Gao, B., & Guo, P. (2014). *Mechanical Design and Control Method for SEA and VSA-based Exoskeleton Devices for Elbow Joint Rehabilitation.* Academic Press.

Zhang, K., Zhang, L., & Yang, M. H. (2012). Real-Time Compressive Tracking. In *Proceedings of European Conference on Computer Vision (ECCV 2012).* Berlin: IEEE. doi:10.1007/978-3-642-33712-3_62

Zhang, K., Zhang, L., & Yang, M. H. (2014). Fast Compressive Tracking. *IEEE Transactions on Pattern Analysis and Machine Intelligence*, *36*(10), 2002–2015. doi:10.1109/TPAMI.2014.2315808 PMID:26352631

Zhang, M., Xin, M., & Yang, J. (2013). Adaptive multi-cue based particle swarm optimization guided particle filter tracking in infrared videos. *Neurocomputing*, *122*, 163–171. doi:10.1016/j.neucom.2013.05.041

Zhang, X., Chen, X., Li, Y., Lantz, V., Wang, K., & Yang, J. (2011). A Framework for Hand Gesture Recognition Based on Accelerometer and EMG Sensors. *IEEE Transactions on Systems, Man, and Cybernetics. Part A, Systems and Humans*, *41*(6), 1064–1076. doi:10.1109/TSMCA.2011.2116004

Zhao. (2012). Tracking micro tool in a dynamic 3D ultrasound situation using kalman filter and ransac algorithm. *Biomedical Imaging (ISBI), IEEE International Symposium on. IEEE*, 1076–1079.

Zhou, X., Jun, S. K., & Krovi, V. (2014). Stiffness modulation exploiting configuration redundancy in mobile cable robots. *Proceedings - IEEE International Conference on Robotics and Automation*, 5934–5939. doi:10.1109/ICRA.2014.6907733

Zhu. (2006). A virtual ultrasound imaging system for the simulation of ultrasound-guided needle insertion procedures. *Proceedings of Medical Image Understanding and Analysis*, 61–65.

About the Contributors

Maki Habib obtained his Doctor of Engineering Sciences in Intelligent and Autonomous Robot from the University of Tsukuba, Japan. He was a senior research scientist with GMD, Japan, leading the Telecooperation group, Associate Professor with Monash University and lead the Mechatronics Engineering Programme. He was appointed as a full Professor of Robotics and Mechatronics at Swinburne University. Then, he was an invited Professor at KAIST, Korea, Visiting Professor at Saga University, Japan, and now a full Professor at AUC. His main area of research is focusing on human adaptive and friendly Mechatronics, autonomous navigation, service robots and humanitarian demining, telecooperation, distributed teleoperation and collaborative control, wireless sensor networks and ambient intelligence, biomemtic, bioinspiration and biomedical robots.

* * *

Hisham A Abdel-Aal, PhD, obtained his first degree in Mechanical Engineering from the Alexandria University, Aleandria Egypt. He received his Doctoral degree in Precision Engineering from the University of North Carolina in Charlotte in 1998. Since his graduation he has been active in research and teaching in the US and Europe. He occupied several Professorship positions in Europe, South America and Asia. Including an invited Research Professorship at the Laboratoire de Materiux et Procedes at Arts et Métiers ParisTech in France where he founded the first research group for Bio-inspired surfaces for manufacturing. Professor Abdel-Aal, has lectured extensively on the subjects of tribology, Bio-inspired surfaces, and thermodynamics of wear: He is the author or co-author of numerous research papers on the subjects.

Mark Adams is a Physiotherapist registered with the Physiotherapy Board of New Zealand. He has obtained a BSc (Hons) Physiotherapy from the University of Limerick, Limerick, Ireland in 2006, and a Postgraduate Diploma in Rehabilitation from the Research, Training and Rehabilitation Unit (RTRU), University of Otago, Wellington in 2014. He is currently completing his Master's thesis through RTRU. He works as Physiotherapist in the Brain Injury Rehabilitation and Neurological outpatient services at Burwood Hospital, Christchurch, New Zealand. His research interests are technology in neurological rehabilitation, and personal factors in relation to pain in neurological populations.

Gabriel Campos is a student of Doctoral Degree in Electrical Engineering from the State University of Londrina (2017). Master in Computer Science from the State University of Londrina (2016). Graduated in Computer Science from the State University of Londrina (2014). He works in the areas of Computational Vision and Pattern Recognition.

Giuseppe Carbone is Assistant Professor at University of Cassino and South Latium, Italy. He has carried out several periods of research abroad such as in Germany, Japan, Spain, China. His research interests include stiffness of multibody robotic systems, robotic hands and grippers, mechatronic designs, design of experimental test-beds. He has published more than 250 peer reviewed papers on the above-mentioned topics in Proceedings of well-known peer-reviewed International Journals or Conferences. Among others, he is currently Associate Editior of the International Journals Robotica (Cambridge), Journal of Mechanics and Control (Levrotto & Bella), Advanced Robotic Systems (Sage). He is and has been Principal investigator and/or member of several National, European, or International research projects. He has received several awards including three IFToMM "Young Delegate" Awards, two "JSPS Short-Term Invitation Fellowship", several Conference Best Paper Awards.

Mehmet Çetinkaya received his B.S. degree in electrical engineering from Middle East Technical University (Ankara, Turkey) in 2014. After his graduation, he worked at Transportation, Security, Energy and Automation Systems Business Sector of ASELSAN (Ankara, Turkey) for two years, where his tasks included multi-body force distribution analysis and control of lightweight railway vehicles as well as software implementation of these on the target hardware. He is currently a research assistant and working towards his M.S. degree at the same department. His current research is focused on the prediction and gating of renal respiratory motion of patients under general anesthesia using both external sensors and internal imaging tools.

Dimitar Chakarov was born in Bulgaria on August 1, 1956. He received the B.S. degree in mechanical engineering and cutting machines in 1981, the M.S. degree in mechanical engineering and robotics in 1983, both from Technical University of Sofia. He received the Ph.D. degree in applied mechanics from Institute of Mechanics and Biomechanics, Bulgarian Academy of Sciences (BAS) in 1992. From 1992 - 2001 he was research associate in the Mechatronic Systems Lab - Institute of Mechanics (BAS). Since 2001 he has been an associate professor on Mechatronics & Robotics in Mechatronic Systems Department - Institute of Mechanics (BAS). His research areas include mechanics and control of robots, mechatronic devices and mechanisms. His research interests are parallel manipulators, linear actuators, redundant actuation, antagonistic stiffness, passive compliance, stiffness control, contact tasks, micro robots, "human-robot" interaction.

XiaoQi Chen is a professor in the Department of Mechanical Engineering at the University of Canterbury, Christchurch, New Zealand. After obtaining his BE in 1984 from South China University of Technology, Guangzhou, he received the China-UK Technical Co-Operation Award for his MS study in the Department of Materials Technology at Brunel University, London, UK (1985–1986) and his Ph.D study in the Department of Electrical Engineering and Electronics at the University of Liverpool, UK (1986–1989). He has been a senior scientist at the Singapore Institute of Manufacturing Technology (1992–2006). His research interests spans over mechatronics, robotics, materials processing, and manufacturing automation. He has been elected to Fellow of IPENZ, and Fellow of SME.

Andrea Chiavenna is a Physiotherapist graduated at Università Vita e Salute San Raffaele and affiliated as a research fellow to the National Research Council of Italy. Principal research topics: neuro-rehabilitation, robotic therapy, stroke rehabilitation, human motion analysis, ICF evaluation.

Tito Dinon is an Industrial designer and Research Fellow at the Institute of Industrial Technologies and Automation (ITIA) in Milan in the area of Intelligent Robot and Autonomous Systems (IRAS). Research field: surface and solid 3D CAD modelling of simple to complex parts and assemblies; design and development of rehabilitation devices with extensive use of additive manufacturing technologies and FDM 3D printing technologies; design and manufacturing of mechatronic prototypes with use of conventional and non-conventional manufacturing technologies (ranging from CNC manufacturing to advanced composites lamination).

Didier Dumur. CentraleSupélec, Automatic Control Department, Control and Systems Pole-Laboratory of Signals and Systems (L2S, UMR 8506)-CentraleSupélec, CNRS, Université Paris-Sud, Université Paris Saclay, Professor, HDR, Head of Automatic Control Department, Supervisor of 3rd year Major "Control and Systems."

Aydan M. Erkmen received her B.Sc. from Bogazici University, Turkey, in Electrical Engineering double majoring in Mathematics, her M.Sc. in ECE from Drexel University, USA, Ph.D. in Info. Tech. & Eng. from George Mason University, USA, in 1989. She is full professor in Electrical Eng. Dept. Middle East Technical University, Turkey. Her research interests include mechatronic design, search & rescue robotics, robot hands, sensor fusion, bioinspired control, biorobotics, multi-robots networks, human robot interactions. She served 2 terms in IEEE Robotics and Automation Soc. (RAS) Admin Committee; was IEEE RAS vice president 2000 to 2003; was till 2010 associate editor Int. Journal of Robotics & Automation. She is since 2009 Editor in Springer Verlag-Paladyn. Journal of Behavioral Robotics. She served 3 years on the editoral board of IEEE Robotics and Automation Magazine. She was 4 years at TUBITAK Scientific&Tech Research Council of Turkey as vice president for industrial funds (TIDEB), the FP6 national coordinator till 2002, IST NCP till 2003. Between November 2007-Dec2009, she was the Vice Dean of the Faculty of Eng. in Middle East Technical University. Since 2009, she serves in EURON as an Award Jury member for George Giralt PhD Theses Awards.

Ismet Erkmen received his B.Sc. and M.Sc. degrees in EE from the Middle East Technical University, (METU), Turkey, in 1970 and 1972 respectively, and Ph.D. degree from Drexel University, in 1984. His employment experience includes the Turkish Electricity Authority (Ankara, Turkey); Brown Boveri (Baden, Switzerland); Drexel University (Philadelphia, USA); Electric Power Research Institute - Carlsen and Fink Associates, (Virginia, USA); and Turkish Scientific and Technological Research Council (Ankara, Turkey). He served as the chairman of the Electrical Engineering Department of METU for 9 years, and is the founder of the EE Mechatronics Lab. He has served as an executive board member of METU-BILTER CAD/CAM, Robotics and Automation Research Center, and has been awarded Excellency in Education in 93-94 at METU. He was a member of IEEE and Counsellor of IEEE student branch. Ismet has been Professor Emeritus at METU since 2013.

Bogdan Gherman received his PhD from the Technical University of Cluj-Napoca in 2011. In his thesis he developed a parallel robotic system for minimally invasive surgical interventions. Further on he continued his research in the medical robotics area, especially robotic systems for cancer diagnosis and treatment, for surgery, urology and post-stroke upper and lower limb rehabilitation. the first Romanian surgical robot for minimally invasive interventions. He currently works at the Technical University of Cluj-Napoca as Lecturer and Senior Researcher within the Research Center for Industrial Robots Simulation and Testing (CESTER). He has published over 60 scientific publications and has applied for 11 patents.

Cid Gilani was a mechatronics graduate in the year 2014 at the University of Canterbury. He is currently completing Master's Thesis on the Design and Control of a Snake Robot.

Ozlem Gumuskanat graduated from Middle East Technical University obtaining her Bachelor's Degree in Electrical and Electronics Engineering in 2016. She is currently a master's student at the same department. Her current research interests include intelligent biomedical pattern recognition and robot-arm control systems.

Shuxiang Guo received the B.S. and the M.S. degrees in mechanical engineering from the Changchun Institute of Optics and Fine Mechanics, Changchun, China, in 1983 and in 1986, respectively, and the Ph.D. degree in mechano-informatics and systems from Nagoya University, Nagoya, Japan, in 1995. In 1995, he was a Faculty Member at Mie University, Mie, Japan and in 1998 at Kagawa University, Kagawa, Japan. Currently, he is a Professor at Beijing Institute of Technology, China. He has published about 270 refereed journal and conference papers. His current research interests include microrobotics and mechatronics, microrobotics system for minimal invasive surgery, micro catheter system, micro pump, and smart material (SMA, ICPF) based on actuators.

Sergey Jatsun has a Scientific degree: Doctor of technical sciences, academic title: Professor In 1979 Sergey Jatsun has received PhD degree in Mishkole technical University in Hungary. In 1990 received degree Doctor science in a field of Dynamics and durability of machines devices and apparatus in Riga, Latvia. In 1990 he defended his doctoral dissertation: "Theory and Synthesis of Vibratory, Technological Processes and Vibromachines for the Processing of Dry Materials". He is author of more than 250 publications and has more than 100 patents on inventions. He is a member of the International Academy of Ecology and Safety and was awarded the token "High Achiever of Invention and Rationalization" and "USSR's Inventor". He has the silver and golden medal of exhibition of achievements of USSR and Russia national economy. He awarded the title of Honored scientist of the Russian Federation. Sergey Jatsun is a member of the editorial council of Izvestiya SWSU journal. He has prepared more than 40 PhD students.

Mohammadali Javaheri Koopaee received his BS in electrical engineering from Amirkabir University of Technology and his MS from Tarbiat Modares University in 2013 and 2015 respectively. He joined Mechanical engineering department of University of Canterbury, where he is perusing his PhD. His research interest includes Control Systems, Intelligent Control and Bio inspired robotics.

A. Sinan Kabakci received his BS degree from Middle East Technical University, Ankara, Turkey in Electrical and Electronics Engineering (1989). In 1991, he became the co-founder of ELMED Electronics and Medical Inc., working on the production of ESWL (Extracorpereal Shock Wave Lithotripsy) devices for the first time in Turkey. He obtained his masters degree from Middle East Technical University and Gazi University, Ankara, Turkey. He is currently working on the completion of PhD in Hacettepe University, Department of Bioengineering, Ankara, Turkey.

Minami Kaneko received the bachelor, master, and the Ph.D. degree from Nihon University in 2010 2012 and 2015, respectively. Her research is the ceramic material and process applying for micro energy systems and microrobot systems. She is now a Research Assistant at Department of Precision Machinery Engineering, College of Science and Technology, Nihon University.

McNiel Keri received his B.Sc degree in the Biomedical stream in Electrical and Computer Engineering from the University of Alberta in 2015. He is currently pursuing a master degree in Mechanical Engineering at the University of Alberta, Edmonton, Canada.

Mochen Li received the B.S. degree from China University of Mining & Technology, Beijing, China, in 2012 and the M.S. degree in Engineering Technology from Purdue University, West Lafayette in 2015. Mochen Li is currently a Ph.D. student in Purdue Polytechnic Institute. His research interests are wireless sensor networks (WSN), and data analysis.

Yaser Maddahi received his BSc and MSc in Mechanical Engineering and Biomedical Engineering in 2001 and 2003, respectively. He obtained his PhD degree in Mechanical Engineering from the University of Manitoba, Canada in 2013. He was an Eyes High postdoctoral fellow in Department of Clinical Neurosciences at University of Calgary and a visiting academic scholar in Department of Mechanical Engineering at the University of Manitoba. He is currently a Killam postdoctoral laureate at University of Calgary. The research interests of Dr. Maddahi include surgical robotics, MR-compatible haptic devices, networked robotics, and human-machine interaction.

Matteo Malosio is a Mechanical Engineer. Ph.D. in Applied Mechanics. Researcher in the Intelligent and Autonomous Robot Systems Division at CNR-ITIA. Research field: rehabilitation robotics; variable stiffness actuators; complex mechanical architectures; force-based human-robot interaction. Scientific coordinator of the Robotics and Physically Interactive Systems for Rehabilitation Lab.

João Maurício Rosário is a Full Professor at UNICAMP, Mechanical Design Department, responsible for the Laboratory of Integrated Automation and Robotics, working in the area of Control and Automation Engineering, with emphasis on the following topics: mechatronics, robotics, control, modeling and industrial automation. He holds a degree in Mechanical Engineering from the State University of Campinas (1981), Specialization Diploma in Industrial Automation: Automated Production - University of Nancy I (1986), Master of Mechanical Engineering from the State University of Campinas (1984), PhD in Automation Robotics Specialty - Ecole Centrale des Arts Et Manufactures Paris (1990), postdoctoral degree in institutions: Geestach Technological Institute - GKSS, Germany (1992), SUPELEC, France (2005), and ISEP, Portugal (2006). CENTRALE-SUPELEC (2015-2016).

Leonimer Flávio de Melo is graduated in Electrical Engineering from the State University of Campinas (1985), Masters in Electrical Engineering from State University of Campinas (2002) and Ph.D. in Mechanical Engineering from the State University of Campinas (2007). He is currently a professor at State University of Londrina (UEL). He has experience in the Electrical Engineering and Mechatronic Engineering, working mainly in the following areas: industrial automation, manufacturing engineering, embedded control systems, robotics, industrial mechatronic systems, reconfigurable architecture, distance education and traffic engineering.

Kimiko Motonaka received the B.E. degree from the Department of System Engineering at Okayama University in 2010, the M. E. and D.E. degree from the Department of Natural Science and Technology at Okayama University in 2012 and 2015. She is currently an Assistant Professor of Electrical and Electronic Engineering with Kansai University, Osaka, Japan. Her research interest includes kinodynamics and motion planning of robots.

Iuliu Nadas is a Technological Development Engineer at National Institute for Research and Development of Isotopic and Molecular Technologies Cluj-Napoca, Romania. Since 2016 he is PhD student at the Technical University of Cluj-Napoca, having a collaboration with CESTER (Research Center for Industrial Robots Simulation and Testing).

Fusaomi. Nagata received the B.E. degree from the Department of Electronic Engineering at Kyushu Institute of Technology in 1985, and the D.E. degree from the Faculty of Engineering Systems and Technology at Saga University in 1999. He was a research engineer with Kyushu Matsushita Electric Co. from 1985 to 1988, and a special researcher with Fukuoka Industrial Technology Centre from 1988 to 2006. He is currently a Professor at the Department of Mechanical Engineering, Faculty of Engineering, Tokyo University of Science, Yamaguchi, Japan and a Dean of the Faculty of Engineering. His research interests include intelligent control of industrial robots and its application, 3D printer-like data interface for machining robots, ORiN middleware for robotic standardization, deep neural networks for industrial inspection systems.

Takato Okudo is currently a undergraduate student of National Institute of Technology, Nara College. His research interests include learning support system, mastery process, human-agent interaction, autonomous learning agent and human learning process.

Anup Patel is a Consultant Urological Surgeon in London with FRCS(Urol), MS in Surgery (U. London), FRCS (Royal College of Surgeons England), MBBS and BSc (1st Class Hons) at St. Bartholomew's Hospital, London. Double certified supra-specialty fellowships in Complex Endourology-Laparoscopic Urology and Uro-Oncology at UCLA Department of Urology. 1st Chair of Clinical Studies Committee EAU Research Foundation (8 years), Consultant to EAU International Relations Office (4 years), EUSP Board Member (3 years), Faculty ESU (18 years) & EUREP (6 years); Last Scientific Chair + Ex-Executive Board Member (4 years), Past Board of Directors Member (3 years) + Member at Large (1 year), and Fellowship Committee member (3 years) Endourology Society; Examiner Intercollegiate Specialty Examinations Board in Urology (6 years), Member UK - NCRI Prostate Cancer CSG, European Urology Editorial Board Member, Peer reviewer multiple international urology journals. Winner, 1st Challenge the Experts, EAU Winter Forum, Davos.

Doina Pisla is Full Professor of robotics and PhD supervisor within the Technical University of Cluj-Napoca. She is the Director of the Research Center for Industrial Robots Simulation and Testing - CESTER, where in the last 12 years she coordinated the activity of a team of researchers involved in medical, space exploration and industrial robotics. Under her coordination multiple national premieres were achieved in Surgical Robotics, Oncology, Urology and Post-Stroke Rehabilitation, encompassing 6 innovative parallel robotic structures and 5 specialized medical instruments. She successfully coordinated 7 PhD theses. She has over 15 patents or patent applications and over 140 scientific publications with 700 citations. She coordinated over 10 national and international research projects and published as Editor 6 books in Springer.

Christopher G. Pretty is currently a senior lecturer in the Department of Mechanical Engineering at the University of Canterbury, NZ. He received his Masters Degree in Mechanical Engineering at the University of Canterbury in 2002. Between 2002 and 2008, he worked as a consulting engineer in NZ and the UK. In 2008, He returned to the University of Canterbury and completed his PhD in Bioengineering in 2012. Following his PhD, he moved to Belgium to undertake a postdoctoral fellowship at the University of Liege in the GIGA Cardiovascular Sciences research group. Dr Pretty's research interests predominantly lie in modelling, sensing, and control of physiological systems, with a particular emphasis on clinical applications. He is currently involved in research on cardiovascular, respiratory, and metabolic systems, as well as rehabilitative robotics.

Alessio Prini is a research fellow in ITIA-CNR since 2014, Mechanical Engineer. Research field: development and modelling of mechatronics systems, design and development of embedded software on open-source and PC-based platform, particularly focused on neurorehabilitation robotics.

Carlos Rossa is an assistant professor in the Faculty of Engineering and Applied Science at the University of Ontario Institute of Technology, Oshawa, ON, Canada. He received his Engineering and the M.Sc degrees in Mechatronics from the Ecole Nationale d'Ingénieurs de Metz, Metz, France, both in 2010, and earned his Ph.D. degree in Mechatronics and Robotics from the University of Paris VI, France in 2013. From 2014 to 2017, he was a postdoctoral research fellow at the University of Alberta, Edmonton, Canada. Dr. Rossa's current research interests include the design and control of haptic interfaces, actuators and sensors technologies, mechatronics, and medical robotics.

Ken Saito was born in Japan in 1978. He received the B.S. degree in electronic engineering, the M.S. degree in electronic engineering, and the Ph.D. degree in engineering from Nihon University in 2001, 2004 and 2010, respectively. He was a Research Assistant at Department of Physics, College of Humanities & Sciences, Nihon University from 2007 to 2010, Research Assistant at Department of Precision Machinery Engineering, College of Science and Technology, Nihon University from 2010 to 2011, and Assistant Professor from 2011 to 2017. He was also as visiting researcher at University of California, Berkeley from 2016 to 2017. He is now an Associate Professor at Department of Precision Machinery Engineering, College of Science and Technology, Nihon University. He is currently also a researcher at University of California, Berkeley. His current research interests include hardware neural network, and microrobot. Dr. Saito is a member of IEICE, IEEJ, INNS and so on.

Sergei Savin graduated from Southwest State University (SWSU) in 2011, and received PhD in 2014. Since 2013 works at the department of Mechanics, Mechatronics and Robotics of SWSU. His research interests include application of Optimal Control Theory and convex optimization to mobile robotics, especially to legged locomotion, exoskeletons control and in-pipe mobile robots. He is author of 3 books, 40 scientific papers and over 20 patents.

Alessandro Scano is a Biomedical engineer, PhD, works as researcher at the Institute of Industrial Technologies and Automation (ITIA) in Milan in the area of Intelligent Robot and Autonomous Systems (IRAS). His main research topics involve human motor control, stroke rehabilitation, robotic and low-cost devices for rehabilitation, biological signal analysis.

Callum Scott completed a Bachelor of Engineering in Mechatronics in 2015. Currently he is working on a Masters of Engineering in Mechanical Engineering. His research interest is Pose Estimation for Feedback Control in Snake Robots.

Liwei Shi received the B.S. degree in mechanical engineering from Harbin Engineering University, China, in 2006, the M.S. and the Ph.D. degrees in intelligent machine system from Kagawa University, Japan, in 2009 and in 2012, respectively. Currently, he is an Associate Professor at Beijing Institute of Technology, China. He researches on underwater microrobot and spherical underwater robots.

Taro Shibanoki received the B.E. degree from the University of Tokushima, Tokushima, Japan, in 2008, and the M.E. and D.Eng. degrees from Hiroshima University, Hiroshima, Japan, in 2010 and 2012, respectively. He was a Research Fellow of the Japan Society for the Promotion of Science from 2011 to 2012 (DC2) and from 2012 to 2013 (PD), and an Assistant Professor (Special Appointment) in Hiroshima University from 2013 to 2014 and an Assistant Professor in Ibaraki University since 2014. He is currently a Lecturer in College of Engineering, Ibaraki University. His current research interests focus on dimension elimination, biological signal processing, and human–machine interface. Dr. Shibanoki won the Best Paper Award from the Society of Instrument and Control Engineers in 2014. Dr. Shibanoki is a member of the Institute of Electrical and Electronics Engineers (IEEE), the Society of Instrument and Control Engineers in Japan, and the Robotics Society of Japan.

Giulio Spagnuolo is a Research fellow in Robotics, Institute of Industrial Technologies and Automation - National Council of Research, Milan, Italy. Member of the Rehabilitation section of the Intelligent and Autonomous Robot Systems group (Rehab - IRAS). Research sectors: design and development of automated biomedical devices aimed at neurorehabilitation and domestic aid. Mechatronic designer of rehabilitation robotic devices. Ph.D. Candidate in Applied Mechanics, Mechanical and Industrial Engineering, University of Brescia, 15, Piazza del Mercato, Brescia (Italy)

Ashley M. Stewart received a bachelor's degree in engineering in 2012 from the University of Canterbury, New Zealand. She is currently working toward the Ph.D. degree at the University of Canterbury, New Zealand. Her current research interests include functional electrical stimulation and exoskeletons.

Keiki Takadama received his M.E. degree from Kyoto University, Japan, in 1995 and got Doctor of Engineering Degree from the University of Tokyo, Japan, in 1998, respectively. He joined Advanced Telecommunications Research Institute (ATR) International from 1998 to 2002 as the a visiting researcher and worked at Tokyo Institute of Technology from 2002 to 2006 as a lecturer. He moved to The University of Electro-Communications as associate professor in 2006 and is currently a professor from 2011. His research interests include multiagent system, autonomous system, human-agent interaction, reinforcement learning, learning classifier system, and emergent computation. He is a member of IEEE and a member of major AI- and informatics-related academic societies in Japan.

Yuki Tamai graduated the Faculty of Advanced Engineering, National Institute of Technorogy, Nara College, Japan and received his B.A. degree from The National Institution for Academic Degrees and University Evaluation, Japan in 2017. Currently he joined NTT-Neomeit from 2017. His research interests include learning support system, motivating human, learning process, human-agent interaction and human learning process.

Li Tan received the B.S. degree from Southeast University, Nanjing, China, in 1984 and the M.S. degree in Engineering Mechanics and the M.S. and Ph.D. degrees in Electrical Engineering from the University of New Mexico, Albuquerque, in 1987, 1989 and 1992, respectively. Dr. Tan is currently a Professor with the Department of Electrical and Computer Engineering, Purdue University Northwest, Indiana. He is a senior member of the IEEE. His research interests include the areas of digital signal processing, adaptive signal processing, active noise control, control systems and digital communications. He authored and co-authored two textbooks: Digital Signal Processing: Fundamentals and Applications, Second Edition (Elsevier, 2013) and Analog Signal Processing and Filter Design, Second Edition (Linus Publications, 2016). He is the holder of a granted US patent. He has served as an associate editor for the International Journal of Engineering Research and Innovation.

Mahdi Tavakoli received the B.Sc. degree from Ferdowsi University of Mashhad, Mashhad, Iran, and the M.Sc. degree from the K.N. Toosi University of Technology, Tehran, Iran, in 1996 and 1999, respectively, both in electrical engineering, and the Ph.D. degree in electrical and computer engineering from the University of Western Ontario, London, ON, Canada, in 2005. He is an Associate Professor in the Department of Electrical and Computer Engineering, University of Alberta, Edmonton, AB, Canada. In 2006, he was a Postdoctoral Researcher at Canadian Surgical Technologies and Advanced Robotics, Canada. In 20072008, he was an NSERC Postdoctoral Fellow at Harvard University, Cambridge, MA, USA. His research interests include the areas of robotics and systems control. Specifically, his research focuses on haptics and teleoperation control, medical robotics, and image-guided surgery.

Dimitar Trifonov was born on November 30, 1948 in Sofia. Eng. Dimitar Trifonov graduated from the Technical University-Sofia with a degree in Mechanical Engineering and Sofia University "St. Kl. Ohridski "with diploma mathematician. Dimitar Trifonov was assistant professor at the Institute of Information and Communication Technologies-BAS from 1987 to 2009 and from 2010 to 2015 was assistant professor at the Institute of Mechanics-BAS as 1 year is an assistant professor of applied mechanics at the same Institute. He has 43 years experience in the research and development of automated robotic systems, manipulators, mechatronic devices and mechanisms, including robotic systems for in vitro fertilization and cell microinjection.

Toshio Tsuji received the B.E. degree in industrial engineering, and the M.E. and D.Eng. degrees in systems engineering from Hiroshima University, Hiroshima, Japan, in 1982, 1985, and 1989, respectively. He was a Research Associate from 1985 to 1994, and an Associate Professor from 1994 to 2002, in the Faculty of Engineering, Hiroshima University. He was a Visiting Professor at the University of Genova, Genova, Italy, from 1992 to 1993. He is currently a Professor in the Department of System Cybernetics, Hiroshima University. His current research interests focus on human–machine interface and computational neural sciences, in particular, biological motor control. Dr. Tsuji won the Best Paper Award from the

Society of Instrumentation and Control Engineers in 2002, 2008 and 2014, and the K. S. Fu Memorial Best Transactions Paper Award of the IEEE Robotics and Automation Society in 2003. He is a member of the Institute of Electrical and Electronics Engineers (IEEE), the Japan Society of Mechanical Engineers, the Robotics Society of Japan, and the Society of Instrument and Control Engineers in Japan.

Fumio Uchikoba received the bachelor degree from Waseda University in 1983. He received the master degree from University of Electro-Communications in 1985. He entered the TDK Corporation in 1985. He was a visiting scientist at the Massachusetts Institute of Technology form 1988 to 1990. He joined department of precision machinery engineering, College of Science and Technology, Nihon University in 2003 as an associate professor. He is currently a professor in same university. He is the doctor of engineering.

Calin Vaida received his PhD from the Technical University of Cluj-Napoca in 2009, developing the first Romanian surgical robot for minimally invasive interventions. In the following period and up to the current date he continued his research activity in medical robotics covering innovative robotic solutions and specialized instruments for minimally invasive surgery, oncology, urology and rehabilitation. He currently works at the Technical University of Cluj-Napoca as Associate Professor and Senior Researcher in CESTER (Research Center for Industrial Robots Simulation and Testing). He has over 15 patents applications in medical devices and over 100 scientific publications.

Pavel Venev is an Electronics technician in Mechatronic Department at Institute of Mechanics, Bulgarian Academy of Sciences and a student at The Technical University Sofia. Currently he is developing new skills for 3D visualization and control for exoskeleton systems, using motion capturing and virtual reality devices. His current research interests cover computer programing, electronic design, 3D CAD modelling and prototyping of robotic systems for Rehabilitation. He has been a team member of the international FP7 project VERE - 25769 (FP7), whose main task is the design and development of an active exoskeleton for human body interacting with the virtual reality.

Ivanka Veneva is an experienced engineer and scientist specialized in electronics, automation and human-machine-interaction with comprehensive expertise in rehabilitation robotics and exoskeleton systems interacting with virtual reality. Ivanka Veneva's research interests are in the field on design and control of robotic systems, and advanced human computer interfaces, Brain Computer Interfaces. Currently she is studying new exoskeleton systems with pneumatic actuation, haptics, motion capturing and virtual reality communication devices and interfaces. She has been a team leader in VERE FP7 project and leader of the work package for design and development of an active exoskeleton for human body interacting with the virtual reality.

Tomohiro Yamaguchi received his M.E. degree from Osaka University, Japan, in 1987. He joined Mitsubishi Electric Corporation in 1987 and moved to Matsushita Electric Industrial in 1988. He worked at Osaka University from 1991 to 1998 as a research associate and got Doctor of Engineering Degree from Osaka University in 1996. He moved to Nara National College of Technology as associate professor in 1998 and is currently a professor from 2007. His research interests include interactive recommender system, music information retrieval, multiagent reinforcement learning, autonomous learning agent, human-agent interaction, learning support system, human learning process and mastery process. He is a member of The Japanese Society for Artificial Intelligence and The Society of Instrument and Control Engineers, Japan.

Andrey Yatsun graduated from Southwest State University (SWSU, Kursk, Russia) in 2009, and received PhD in 2012. He is the head of the laboratory of Mechatronics and Robotics and a senior lecturer at the department of Mechanics, Mechatronics and Robotics of SWSU. The scientific interests of Andrey Yatsun are mobile robots, exoskeletons design and control, parallel mechanisms, Mechatronics. He is author of 3 books, 80 scientific papers and 8 patents.

Wangling Yu is an assistant professor in the Electrical & Computer Engineering Technology Department of the Purdue University Northwest. He was a test engineer over 15 years, providing technical leadership in the certification, testing and evaluation of custom integrated security systems. He received his PhD degree in Electrical Engineering from the City University of New York in 1992, specializing in control theory and electronic technology.

Jessica Zamaia graduated in electrical engineering from the State University of Londrina, attending the master's degree strictu sensu in electrical engineering by the State University of Londrina.

H. Henry Zhang is an associate professor in Purdue Polytechnic Institute at Purdue University. He is also director of Center for Technology Development. He received his PhD degree in Mechanical Engineering at University of Michigan-Ann Arbor in 1996. He also received degrees in Hydraulic Control, Neural Networking, and Electrical & Computer Engineering. Before he joined Purdue in 2007, he was a senior specialist of automatic transmission engineering in Chrysler Technology Center with 13 years of automotive industry experience. He established Multidisciplinary Design Laboratory to support his research interests in analytical Mechatronics and vehicle powertrain/automatic transmission design and manufacturing.

Index

R

S

T

U

V

W

Z

www.ingramcontent.com/pod-product-compliance
Lightning Source LLC
LaVergne TN
LVHW080407011026
849161LV00009B/2360